SOME USEFUL PHYSICAL PROPERTIES
(USCS UNITS AND SI UNITS)

Property[1]	USCS	SI
Aluminum		
specific weight	169 lb/ft^3	26.6 kN/m^3
mass density	5.26 slugs/ft^3	2710 kg/m^3
Concrete[2]		
specific weight	150 lb/ft^3	23.6 kN/m^3
mass density	4.66 slugs/ft^3	2400 kg/m^3
Steel		
specific weight	490 lb/ft^3	77.0 kN/m^3
mass density	15.2 slugs/ft^3	7850 kg/m^3
Water (fresh)		
specific weight	62.4 lb/ft^3	9.81 kN/m^3
mass density	1.94 slugs/ft^3	1000 kg/m^3
Acceleration of gravity (g)		
(at earth's surface)		
recommended value	32.2 ft/sec^2	9.81 m/sec^2
standard international value	32.1740 ft/sec^2	9.80665 m/sec^2
Atmospheric pressure		
(at sea level)		
recommended value	14.7 psi	101 kPa
standard international value	14.6959 psi	101.325 kPa

[1]Except as noted, the tabulated values are typical, or average, values.

[2]The specific weight of concrete made of cement, sand, and stone or gravel ranges from 140 to 150 lb/ft^3. The values tabulated here are average values for steel-reinforced concrete. The specific weight of concrete made with lightweight aggregate ranges from 90 to 115 lb/ft^3.

MECHANICS OF MATERIALS

MECHANICS OF MATERIALS

ROY R. CRAIG, JR.

JOHN WILEY & SONS

New York ■ Chichester ■ Brisbane ■ Toronto ■ Singapore

ACQUISITIONS EDITOR   Charity Robey
MARKETING MANAGER   Debra Riegert
SENIOR PRODUCTION MANAGER   Charlotte Hyland
DESIGNER   Lee Goldstein
COLOR INSERT   Pedro Noa
MANUFACTURING MANAGER   Mark Cirillo
PHOTO EDITOR   Mary Ann Price
ILLUSTRATION EDITOR   Sigmund Malinowski
PRODUCTION MANAGEMENT   Ingrao Associates
ELECTRONIC ILLUSTRATIONS   Precision Graphics
COVER PHOTO   David Madison/Tony Stone Image

This book was set in 10/12 Times Roman by CRWaldman Graphic Communications and
printed and bound by R. R. Donnelley/Willard. The cover was printed by Lehigh Press.

Recognizing the importance of preserving what has been written, it is a
policy of John Wiley & Sons, Inc. to have books of enduring value published
in the United States printed on acid-free paper, and we exert our best
efforts to that end.

The paper in this book was manufactured by a mill whose forest management programs include
sustained yield harvesting of its timberlands. Sustained yield harvesting principles ensure that
the number of trees cut each year does not exceed the amount of new growth.

Library of Congress Cataloging in Publication Data:
Craig, Roy R., Jr. 1934–
    Mechanics of materials / Roy R. Craig, Jr.
      p.   cm.
    Includes bibliographical references (p.   ).
    ISBN 0-471-50284-7 (cloth : alk. paper)
    1. Strength of materials.   2. Mechanics of solids.
  TA405.C89   1996
  620.1′1292—dc20   95-34726
                                CIP

Printed in the United States of America

10 9 8 7 6 5 4 3 2

PREFACE

This textbook is an introduction to the topic of *mechanics of materials*, an engineering subject that also goes by the names *mechanics of solids*, *mechanics of deformable bodies*, and *strength of materials*. You should already have a thorough background in statics, that is, you should know how to analyze the equilibrium of ''rigid'' bodies by constructing free-body diagrams and formulating and solving the appropriate equilibrium equations. In this course you will learn how to extend equilibrium analysis to *deformable bodies*, specifically to various members that make up structures and machines. This requires consideration of the behavior of the material (e.g., aluminum, steel, or wood) of which the member is made, and also consideration of the geometry of deformation. Therefore, you will learn to apply the three fundamental concepts of solid mechanics—*Equilibrium*, *Force-Temperature-Deformation Behavior*, and *Geometry of Deformation*.

With the aid of this textbook you will learn *systematic problem-solving methods*, and you should also learn ways to assess the probable accuracy of your solutions to homework problems, since these are two skills that are essential, especially so as you begin to use the computer to solve solid-mechanics problems. You will have the opportunity to see how readily a computer program can be developed when you know how to set up the solution in a very systematic fashion, and you should enjoy using the computer program **MechSOLID** that is available for use with this textbook. Nevertheless, the emphasis in this textbook remains on your developing an understanding of the fundamentals of elementary solid mechanics, not just learning to use an existing computer program.

Please take time now to look at the color photographs that are included as an insert in this book. They illustrate how the *finite element method*, a direct extension of this introduction to solid mechanics, is used to design and analyze everything from airplanes to cars, from skyscrapers and bridges to bicycle frames and tennis racquets, from offshore oil rigs to computer chips. The goal of this book is to prepare you to study further courses in solids and structures that will enable you to carry out such complex analyses, for example, courses in matrix structural analysis or finite element analysis. By learning the concepts and procedures that are presented in this textbook, you will be opening the door to the exciting world of *Computer-Aided Engineering*, whether your application is to aerospace engineering, architectural engineering, civil engineering, mechanical engineering, petroleum engineering, or even to electrical engineering.

The overriding philosophy guiding the development of this introductory solid-mechanics textbook has been that students learn engineering topics best when (1) they are made aware of the *principal concepts* involved in the subject, (2) they are taught *systematic problem-solving procedures*, and (3) they are given *real engineering examples* and shown the relevance of what they are studying. Items (1) and (2) are all the more important as students move from using ''hand-crank'' problem-solving methods to computer-based problem-solving approaches. Instead of learning problem-solving approaches that can be used to solve only a limited group of problems, students are best served if they are exposed to the powerful and systematic methods that form the basis of computer algorithms for solving complex problems in the mechanics of deformable bodies. To implement this philosophy, the following features have been incorporated in this textbook.

A Strong Emphasis on the Three Basic Concepts of Deformable-Body Mechanics.

Throughout this book students are reminded that solid mechanics problems invariably involve three basic concepts: *Equilibrium*, *Material Behavior*, and *Geometry of Deformation*. It is clearly indicated that these are three <u>distinct</u> concepts, and that each must be properly accounted for in the solution of problems involving deformable solids. In the Example Problems, the equations that correspond to each of these three concepts are highlighted and identified by name. The student should not leave this course without knowing the importance of understanding the separate role played by each one of the concepts in the ''BIG 3'':

- Equilibrium,
- Force-Temperature-Deformation Behavior of Materials, and
- Geometry of Deformation.

Systematic Problem-Solving Procedures.

The use of systematic problem-solving procedures is not only of great benefit to the student carrying out the solution of a relatively simple engineering problem ''by hand,'' but it becomes absolutely essential when even moderately complex problems are solved by hand or when a computer program is developed for use in solving the problems. In this textbook, over 130 Example Problems provide the student with detailed illustrations of systematic procedures for solving solid-mechanics problems.

As noted above, the distinct contribution of each of the three basic concepts is highlighted and identified by name. Once the basic equations have been written down, solutions are completed by combining them in either a *force-method* sequence or in a *displacement-method* sequence.[1] Although the equations of equilibrium, material behavior, and geometry of deformation could be solved without following one of these two special sequences, there are strong reasons to acquaint the student with these procedures even in this introductory course in solid mechanics. Those reasons are: (1) that systematic problem-solving is easier than less-systematic problem-solving; (2) that more advanced courses in solid mechanics (e.g., energy methods or theory of elasticity) employ problem-solving strategies that fall into one or the other of these two categories; and (3) that all finite element computer programs for solving structures problems and other solid-mechanics problems are based on a specific one of these two solution procedures, namely, they are based on the displacement method. For example, whereas most textbooks on mechanics of materials employ only force-method superposition to solve statically indeterminate beam problems (Section 7.6 of this book), here an equal or even greater emphasis is placed on displacement-method

---

[1]These sequences are sometimes referred to as the *flexibility method* and the *stiffness method* respectively.

solutions (Section 7.7) because of the usefulness of this approach in the development of computer programs to solve beam-deflection problems (Appendix G.6) and other solid-mechanics problems.

Rigorous *sign conventions* for forces, displacements, etc. are established and followed. For example, only in the very special case of buckling of columns (Chapter 10) is an axial force considered to be positive in compression; otherwise, axial forces are always taken to be positive in tension, so that the sign of the answer will immediately identify whether the force is tension, or compression. Wherever equilibrium equations are required, a complete *free-body diagram* is drawn. (It is never assumed that the naming of unknowns or that the sign convention employed is "readily apparent" from the equilibrium equations; these are always shown on the relevant free-body diagram(s).) Because equilibrium analysis plays such a central role in mechanics of solids, a special section in Chapter 1 is devoted to reviewing statics of rigid bodies and to introducing the concept of internal resultants for deformable bodies.

A Four-Step Problem-Solving Procedure.

Two of the most important steps in solving engineering problems are seldom mentioned and never, or almost never, explicitly included in Example Problems in a textbook; they are: planning the solution process at the outset, and assessing whether the answers obtained in the solution have a high probability of being correct (without using an answer book). Therefore, the following four steps are included in the solution of most of the Example Problems in this book.[2]

- State the Problem
- Plan the Solution
- Solve the Problem
- Review the Solution

Once an engineering student leaves the university environment, where teachers play a major role in planning how problems are solved and answers in the back of the book and graded exams indicate whether problem solutions are correct, and once the practicing engineer has powerful computer programs to carry out the detailed problem solution, the importance of being able to *Plan the Solution* and *Review the Solution* for probable accuracy will be apparent.

Computer Exercises and Computer Software.

As noted above, the systematic problem-solving methods employed in this book serve as the basis for straightforward development of computer algorithms for solving solid-mechanics problems of the type covered here. Computer exercises are included as Homework Problems with Chapters 3–5 and 7–9. The computer is especially useful for solving design problems, where it is helpful to plot the effect of changes in design parameters (e.g., area of member cross sections) on performance measures (e.g., maximum stress in a member or maximum deflection). There are several such parameter studies for computer solution, and a "capstone" *computer-aided-design* project is posed in Problem C9.4-3.

The discontinuity-function method presented in Section 5.5 forms a natural basis for the development of a computer program to graph shear and moment diagrams. The displacement method, discussed in Sections 3.5–3.7, Section 4.4, and Section 7.7 serves as the basis for computer algorithms for solving load-deflection problems for axial deformation, torsion, and bending, respectively. Students are presented the opportunity to develop their own computer programs, and the *MechSOLID* software, described in Appendix G, is available in Macintosh format or Windows format. It includes computer programs for solving the following types of problems: calculation of geometric properties of a cross

---

[2]Length considerations precluded use of the four-step solution procedure for all Example Problems.

section; solution of simultaneous, linear, algebraic equations; plotting of shear and moment diagrams for beams; solution of axial-deformation, torsion, and bending problems to determine member deformation and internal-force distribution; and plotting of Mohr's circles for stress and strain.

Thorough Coverage of "Traditional" Topics in Mechanics of Materials. Although the book includes special features, such as the incorporation of Plan the Solution and Review the Solution steps in Example Problems and the introduction of computer methods, you will also find that "traditional" topics, like energy methods, buckling, and stress concentrations are thoroughly covered. Thus, there are ample topics for a three-hour, semester-long course that does not include discussion of computer implementation. For example, the methods of *virtual work* and *complementary virtual work* presented in the extensive chapter on Energy Methods, Chapter 11, could serve as a valuable link between the present course and succeeding courses in matrix structural analysis or finite element analysis that the student might take.

High-Quality Color and Black/White Photographs. A special effort has been made to include high-quality photographs in this textbook, for example, photographs of testing machines, test specimens before and after testing, examples of buckling of axially-compressed members, etc. The insert of color photos is included to give the student a preview of the power of the methods introduced in this textbook, especially when those methods are implemented in a finite element computer program with color pre-processing and post-processing capability.

Accuracy. Special efforts have been made to provide as error-free a book as possible. Of particular importance is the detailed technical and grammatical editing of every chapter of the textbook by Mr. Art Hale of Bell Laboratories. At least two independent solutions have been obtained for every homework problem, and the Solutions Manual was prepared directly by the author in order to insure consistency and accuracy.

All inquiries and comments regarding this book, the Solutions Manual, or the *MechSOLID* software should be addressed to the following: Dr. Roy R. Craig, Jr., ASE-EM Department, Mail Code C0600, The University of Texas at Austin, Austin, TX 78712; or by e-mail ⟨Roy_Craig@mail.utexas.edu⟩.

■■■■■■■■■■■ SUPPLEMENTS

Solutions Manual. The Solutions Manual contains a complete solution for every problem in the book. Each solution appears with the original problem statement and, where appropriate, the problem figure. Thus, the instructor will more readily be able to use the Solutions Manual in making up homework assignments and, if desired, will be able to post solutions or make overhead transparencies for use in class. The Solutions Manual is also available in CD-ROM format from the publisher.

Computer Software. The *MechSOLID* software and User's Manual, an expanded version of Appendix G, are available from the publisher. The software is available in both Macintosh and Windows formats. This is not interactive tutorial software; it is user-friendly, menu-driven software that enables the student to solve problems like many of the Example Problems and Homework Problems in this book, not just specific pre-programmed problems. The software should be especially useful for carrying out iterative design solutions. The *MechSOLID* software is available from the publisher via ftp. To download the files, go to the Wiley homepage at the following address: http://www.wiley.com. Then go to the "ftpArchive" listed on the main menu.

ACKNOWLEDGMENTS

I express my sincerest gratitude to Mr. Albert Hale of Bell Laboratories for his meticulous checking of practically every detail of the manuscript for the entire book, from sentence construction to example-problem solutions to numerical entries in tables. The following colleagues in the teaching profession reviewed parts or all of the manuscript, and I am deeply indebted for their constructive criticism and words of encouragement.

Joel I. Abrams, *University of Pittsburgh*; Sunder H. Advani, *Lehigh University*; William L. Bingham, *North Carolina State University*; Robert D. Cook, *University of Wisconsin*; Donald A. Grant, *University of Maine*; Klod Kokini, *Purdue University*; David J. McGill, *Georgia Institute of Technology*; Don H. Morris, *Virginia Polytechnic Institute and State University*; Al P. Moser, *Utah State University*; Robert G. Oakberg, *Montana State University*; David B. Oglesby, *University of Missouri—Rolla*; Jerome Sackman, *University of California—Berkeley*; Kaspar Willam, *University of Colorado—Boulder*; Alvin J. Yorra, *Northeastern University*; Sam Y. Zamrik, *Penn State University*

Professors Bingham, Kokini, Oakberg, and Zamrik reviewed computer exercises and Appendix G, which describes the *MechSOLID* software, and gave many helpful suggestions.

I am indebted to my teachers at the University of Illinois and to my colleagues at The University of Texas at Austin for the knowledge of Mechanics they have shared with me and for the exceptionally high standards they have set as teachers of Mechanics. Since the early days of my teaching career, Dean John J. "Johnny" McKetta, an inspired teacher and prolific writer, has been a constant source of inspiration and support. I am grateful to those students and practicing engineers who have expressed their appreciation for my earlier book *Structural Dynamics—An Introduction to Computer Methods* (Wiley, 1981), and also to former students who have encouraged me to write this book emphasizing systematic procedures for solving Mechanics of Materials problems.

Charity Robey insisted that I choose John Wiley & Sons as the publisher, and she has provided an excellent staff to assist in the editing and production of the book: Charlotte Hyland, Sigmund Malinowski, Pedro Noa, and Mary Ann Price. It has been a real pleasure to work with Susanne Ingrao, who provided copyediting services and managed the various stages of production, and Lee Goldstein, who skillfully handled the design and layout of the book. Finally, to Tom Brucker of Precision Graphics, who guided the process of converting my rough sketches into attractive two-color artwork; and to Karen Lemens and Tzu-Jeng Su, who eased my task of creating the original LaTeX© manuscript, I owe a sincere debt of gratitude.

Ms. Robey graciously provided for the inclusion of black-and-white photographs and a color-photo insert. For the color photographs I am indebted to Bell Helicopter—Textron (James Kronkite), Motorola (Ashok Shrikantapa), NASA-Johnson Space Center (Erik Evenson), the Structural Dynamics Corporation (Chris Flanigan), and Greg Swadener; and for the black-and-white photographs to Dr. David Bushnell, Prof. Kenneth Liechti, Prof. Oleg Sherby, the Boeing Company (Charles R. Pickrel), Lockheed—Fort Worth Division (Billy J. Pendley), Measurements Group, Inc., MTS Systems Corp., and Tinius Olsen. The skill and persistence of photographer Timothy Valdez is evident in the quality of the black-and-white photos of materials-test specimens and the other original photographs.

I appreciate the suggestions of Professors John Breen and Neils Lind regarding Load and Resistance Factor design (Section 2.12), and Coach Rock Gullickson regarding the Computer-Aded Design exercise in Chapter 9. Greg Cabe, Steve Golab, Aly Khawaja, Brett Pickin, Clint Slatton, and Shawn Van DeWiele have been diligent in assisting me with the solution of the homework problems. Without the consummate programming skill and the dedication of Eric Blades, creation of the Macintosh and Windows versions of the *MechSOLID* computer software would never have been possible.

Finally, for her constant love and support, I sincerely thank my wife, Jane.

ABOUT THE AUTHOR

Roy R. Craig, Jr. is the John J. McKetta Energy Professor in Engineering in the Department of Aerospace Engineering and Engineering Mechanics at The University of Texas at Austin. He received his B.S. degree in Civil Engineering from the University of Oklahoma, and M.S. and Ph.D. degrees in Theoretical and Applied Mechanics from the University of Illinois at Urbana-Champaign. Since 1961 he has been on the faculty of The University of Texas at Austin. His industrial experience has been with the U.S. Naval Civil Engineering Laboratory, the Boeing Company, Lockheed Palo Alto Research Laboratory, Exxon Production Research Corporation, NASA, and IBM.

Dr. Craig's research and publications have been principally in the areas of structural dynamics analysis and testing, structural optimization, control of flexible structures, and the use of computers in engineering education. He is the developer of the Craig-Bampton Method of component mode synthesis, which is extensively used throughout the world for analyzing the dynamic response of complex structures. Dr. Craig has received several citations for his contributions to aerospace structures technology, including a NASA citation for contributions to the U.S. manned space flight program, and he has presented invited lectures in England, France, Taiwan, The Netherlands, and The People's Republic of China.

Dr. Craig has received numerous teaching awards and faculty leadership awards, including the General Dynamics Teaching Excellence Award in the College of Engineering, a University of Texas Students' Association Teaching Excellence Award, and the Halliburton Award of Excellence. Dr. Craig is the author of many technical papers and reports and of one previous textbook, *Structural Dynamics—An Introduction to Computer Methods* (Wiley, 1981). He is a Fellow of the American Institute of Aeronautics and Astronautics and is a licensed Professional Engineer.

Roy R. Craig, Jr.
Austin, Texas
October 1995

TABLE OF CONTENTS

1 INTRODUCTION TO MECHANICS OF MATERIALS 1

1.1 What is Mechanics of Materials? 1
1.2 The Fundamental Concepts of Deformable-Body Mechanics, 4
1.3 Problem-Solving Procedures, 5
1.4 Review of Static Equilibrium; Equilibrium of Deformable Bodies, 7
1.5 Problems, 16

2 STRESS AND STRAIN; DESIGN 19

2.1 Introduction, 19
2.2 Normal Stress, 19
2.3 Extensional Strain; Thermal Strain, 25
2.4 Stress-Strain Diagrams; Mechanical Properties of Materials, 29
2.5 Elasticity and Plasticity; Temperature Effects, 36
2.6 Linear Elasticity; Hooke's Law and Poisson's Ratio, 39
2.7 Shear Stress and Shear Strain; Shear Modulus, 42
2.8 Stresses on an Inclined Plane in an Axially loaded Member, 47
2.9 General Definitions of Stress and Strain, 49
2.10 Cartesian Components of Stress; Generalized Hooke's Law for Isotropic Materials, 57
2.11 The Relationship Between E and G, 62
2.12 Introduction to Design of Deformable Bodies, 64
2.13 Problems, 70

3 AXIAL DEFORMATION 94

3.1 Introduction, 94
3.2 Basic Theory of Axial Deformation, 95
3.3 Saint-Venant's Principle, 101
3.4 Elastic Behavior of a Uniform Axially loaded Member, 102
3.5 Structures with Uniform Axial-Deformation Members, 103
3.6 Thermal Stresses in Axial Deformation, 118
3.7 Geometric "Misfits," 129
*3.8 Introduction to the Analysis of Planar Trusses, 134
*3.9 Inelastic Axial Deformation, 140
3.10 Problems, 152

4 TORSION 173

4.1 Introduction, 173
4.2 Basic Theory of Torsion of Circular Bars, 174
4.3 Linearly Elastic Behavior of a Uniform Torsion Member, 184
4.4 Assemblages of Uniform Torsion Members, 185
4.5 Stress Distribution in Circular Torsion Bars; Torsion Testing, 194
4.6 Power-Transmission Shafts, 198

*4.7 Thin-wall Torsion Members, 201
*4.8 Torsion of Noncircular Prismatic Bars, 206
*4.9 Inelastic Torsion of Circular Rods, 208
4.10 Problems, 214

5 EQUILIBRIUM OF BEAMS   229

5.1 Introduction, 229
5.2 Equilibrium of Beams Using Finite Free-body Diagrams, 231
5.3 Equilibrium Relationships Among Loads, Shear Force, and Bending Moment, 235
5.4 Shear Force and Bending Moment Diagrams, 238
*5.5 Discontinuity Functions to Represent Loads, Shear, and Moment, 243
5.6 Problems, 250

6 STRESSES IN BEAMS   258

6.1 Introduction, 258
6.2 Strain-Displacement Analysis, 259
6.3 Flexural Stress in Linearly Elastic Beams, 266
6.4 Design of Beams for Strength, 274
6.5 Flexural Stress in Nonhomogeneous Beams, 280
*6.6 Unsymmmetric Bending, 287
*6.7 Inelastic Bending of Beams, 295
6.8 Shear Stress and Shear Flow in Beams, 304
6.9 Limitations on the Shear-Stress Formula, 310
6.10 Stresses in Thin-wall Beams, 312
6.11 Shear in Built-up Beams, 323
*6.12 Shear Center, 326
6.13 Problems, 332

7 DEFLECTION OF BEAMS   358

7.1 Introduction, 358
7.2 Differential Equations of the Deflection Curve, 359
7.3 Slope and Deflection by Integration—Statically Determinate Beams, 364
7.4 Slope and Deflection by Integration—Statically Indeterminate Beams, 375

*7.5 Use of Discontinuity Functions to Determine Beam Deflections, 385
7.6 Slope and Deflection of Beams by Superposition—Force Method, 391
7.7 Slope and Deflection of Beams by Superposition—Displacement Method, 406
7.8 Problems, 413

8 TRANSFORMATION OF STRESS AND STRAIN; MOHR'S CIRCLE   429

8.1 Introduction, 429
8.2 Plane Stress, 430
8.3 Stress Transformation for Plane Stress, 432
8.4 Principal Stresses and Maximum Shear Stress, 438
8.5 Mohr's Circle for Plane Stress, 444
*8.6 Three-Dimensional Stress States—Principal Stresses, 450
8.7 Absolute Maximum Shear Stress, 453
8.8 Plane Strain, 459
8.9 Transformation of Strains in a Plane, 460
8.10 Mohr's Circle for Two-Dimensional Strain, 463
*8.11 Analysis of Three-Dimensional Strain, 469
8.12 Measurement of Strain; Strain Rosettes, 471
8.13 Problems, 476

9 STRESSES DUE TO COMBINED LOADING; PRESSURE VESSELS   487

9.1 Introduction, 487
9.2 Thin-Wall Pressure Vessels, 488
9.3 Stress Distribution in Beams, 493
9.4 Stresses Due to Combined Loads, 498
9.5 Problems, 505

10 BUCKLING OF COLUMNS   511

10.1 Introduction, 511
10.2 The Ideal Pin-Ended Column, 514
10.3 The Effect of End Conditions on Column Buckling, 519

*10.4 Eccentric Loading; The Secant Formula, 526

*10.5 Imperfections in Columns, 532

*10.6 Inelastic Buckling of Ideal Columns, 533

10.7 Design of Centrally Loaded Columns, 536

10.8 Problems, 542

11 ENERGY METHODS 554

11.1 Introduction, 554

11.2 Work and Complementary Work; Strain Energy and Complementary Energy, 555

11.3 Elastic Strain Energy for Various Types of Loading, 561

11.4 Work-Energy Principle for Calculating Deflections, 567

*11.5 Virtual Work, 571

*11.6 Strain Energy Methods, 575

*11.7 Complementary Energy Methods, 580

11.8 Castigliano's Second Theorem; The Unit-Load Method, 588

*11.9 Dynamic Loading; Impact, 598

11.10 Problems, 603

12 SPECIAL TOPICS RELATED TO DESIGN 617

12.1 Introduction, 617

12.2 Stress Concentrations, 617

*12.3 Failure Theories, 623

*12.4 Fatigue and Fracture, 631

12.5 Problems, 635

APPENDICES

APPENDIX A NUMERICAL ACCURACY; APPROXIMATIONS A-2

A.1 Numerical Accuracy; Significant Digits, A-2

A.2 Approximations, A-3

APPENDIX B SYSTEMS OF UNITS A-4

B.1 Introduction, A-4

B.2 SI Units, A-4

B.3 U.S. Customary Units: Conversion of Units, A-6

APPENDIX C GEOMETRIC PROPERTIES OF PLANE AREAS A-7

C.1 First Moments of Area; Centroid, A-7

C.2 Moments of Inertia of an Area, A-10

C.3 Product of Inertia of an Area, A-14

C.4 Area Moments of Inertia About Inclined Axes; Principal Moments of Inertia, A-16

APPENDIX D SECTION PROPERTIES OF SELECTED STRUCTURAL SHAPES A-23

D.1 Properties of Steel W Shapes (U.S. Cust.), A-24

D.2 Properties of Steel W Shapes (SI), A-25

D.3 Properties of Steel S Shapes, A-26

D.4 Properties of Steel C Shapes, A-27

D.5 Properties of Steel Angle Sections— Equal Legs, A-28

D.6 Properties of Steel Angle Sections— Unequal Legs, A-29

D.7 Properties of Standard Weight Steel Pipe, A-30

D.8 Properties of Structural Lumber, A-31

D.9 Properties of Aluminum Association Standard I Beams, A-32

D.10 Properties of Aluminum Association Standard Channels, A-33

APPENDIX E DEFLECTIONS AND SLOPES OF BEAMS; FIXED-END ACTIONS A-34

E.1 Deflections and Slopes of Cantilever Uniform Beams, A-34

E.2 Deflections and Slopes of Simply-Supported Uniform Beams, A-36

E.3 Fixed-End Actions for Uniform Beams, A-38

xiii

APPENDIX F   MECHANICAL PROPERTIES OF SELECTED ENGINEERING MATERIALS   A-39

Table F.1   Specific Weight and Mass Density, A-40
Table F.2   Modulus of Elasticity, Poisson's Ratio, and Coefficient of Thermal Expansion, A-41
Table F.3   Yield Strength, Ultimate Strength, and Percent Elongation in 2 Inches, A-42

APPENDIX G   COMPUTATIONAL MECHANICS—THE *MechSOLID* COMPUTER SOFTWARE   A-43

G.1   AREAPROP—Computation of Geometric Properties of Plane Areas, A-44
G.2   LINSOLVE—Solution of Simultaneous Algebraic Equations Expressed in Matrix Notation, A-45
G.3   AXIALDEF—Displacement-Method Solution of Axial-Deformation Problems, A-47
G.4   TORSDEF—Displacement-Method Solution of Torsional-Deformation Problems, A-54
G.5   V/M DIAG—Shear and Bending-Moment Diagrams, A-56
G.6   BEAMDEF—Displacement-Method Solution of Beam-Deflection Problems, A-59
G.7   MOHR—Mohr's Circle for Stress; Mohr's Circle for Strain; Strain Rosettes, A-63

ANSWER SECTION   AN-1

REFERENCES   R-1

INDEX   I-1

INTRODUCTION TO MECHANICS OF MATERIALS

1.1 WHAT IS MECHANICS OF MATERIALS?

Mechanics is the physical science that is concerned with the conditions of rest or motion of bodies acted on by forces or by thermal disturbances. The study of bodies at rest is called *statics,* whereas *dynamics* is the study of bodies in motion. You have been introduced to the fundamental principles of statics and dynamics and have applied these principles to *particles* and to *rigid bodies,* which are both simplified idealizations of real physical systems. The principles of statics and dynamics are also fundamental to the *mechanics of solids* and to the *mechanics of fluids,* two major branches of applied mechanics that deal, respectively, with the behavior of solids and with the behavior of fluids. This book is an introduction to *mechanics of materials,* a topic that is also known by several other names, including "strength of materials," "mechanics of solids," and "mechanics of deformable bodies."

We can begin to answer the question, What is "mechanics of materials?" by considering Fig. 1.1. First, a *deformable body* is a solid that changes size and/or shape as a result of loads that are applied to it or as a result of temperature changes. The diving board in Fig. 1.1 visibly changes shape due to the weight of the diver standing on it. Changes of size and/or shape are referred to as *deformation*. The deformation may even be so small that it is invisible to the naked eye, but it is still very important. And to relate the deformation to the applied loading, it is necessary to understand how materials (i.e., solids) behave under loading.

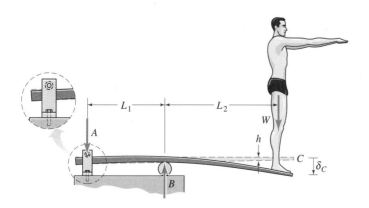

FIGURE 1.1. A diving board as an example of a deformable body.

Whereas it would be possible from rigid-body equilibrium alone, given the weight of the diver and the lengths L_1 and L_2, to determine the diving-board support reactions at A and B in Fig. 1.1, questions of the following type can only be answered by employing the principles and procedures of mechanics of materials:

1. What weight W would cause the given diving board to break, and where would the break occur?
2. For a given diving board and given position of roller B, what is the relationship between the tip deflection at C, δ_C, and the weight, W, of the diver standing on the board at C?
3. Would a tapered diving board be ''better'' than one of constant thickness? If so, how should the thickness, h, vary from A to C?
4. Would a diving board made of fiberglass be ''preferable'' to one made of aluminum?

All of the preceding questions require consideration of the diving board as a *deformable body;* that is, they require consideration not only of the external forces applied to the diving board by the diver and by the diving-board supports at A and B, but they also involve the localized effects of these forces within the diving board (i.e., the *stress distribution* and the *strain distribution*) and the behavior of the material from which the diving board is constructed (i.e., the *stress-strain behavior* of the material).[1]

All deformable-body mechanics problems fall into one of two categories—*strength* problems, and *stiffness* problems. A structure or machine must be ''strong enough''; that is, it must satisfy prescribed strength criteria. It must also be ''stiff enough''; that is, its deformation must be within acceptable limits. The first diving-board question is a strength question; the second one addresses stiffness.

The first two diving-board questions above fall under the category of *analysis*. That is, given the system (in this case the diving board) and the loads applied to it, your task is to analyze the behavior of this particular system subjected to this particular loading condition. Questions 3 and 4, on the other hand, are *design* questions. Given certain information about the loading and the performance criteria (e.g., the range of diver weight to be accommodated, and what constitutes a better diving board), your task is to select the configuration of the diving board and the material to be used in its construction.

The *design process* usually involves an iterative procedure whereby a *design* (a specific configuration made of specific materials) is proposed, the response of the designed system to given loads is analyzed, and the response is compared with the response of other designs and with the stated design criteria. The ''best'' of several candidate designs is then selected. Finally, there may be a requirement that a prototype based on the selected design be manufactured and tested to verify that the system meets all the design requirements in an acceptable manner. Figure 1.2 shows an airplane wing undergoing an ultimate-load test, that is, a test to determine the maximum wing loading that can be applied without causing the wing to break. The designers' expectations were exceeded when, during the test, the wings were pulled to approximately 24 ft above their normal position. (Of course, in normal service the wing will only undergo deflections that are much smaller than those experienced in an ultimate-load test like this.)

The applications of deformable-body mechanics are practically endless and can be found in every engineering discipline. The impressive Brooklyn Bridge, shown in Fig. 1.3,

---

[1]*Stress* and *strain* are formally defined in Chapter 2.

Finite Element Analyses

The solid mechanics principles and the computational procedures presented in this *Mechanics of Materials* textbook form the basis for the finite element method, which was used to create these color photos.

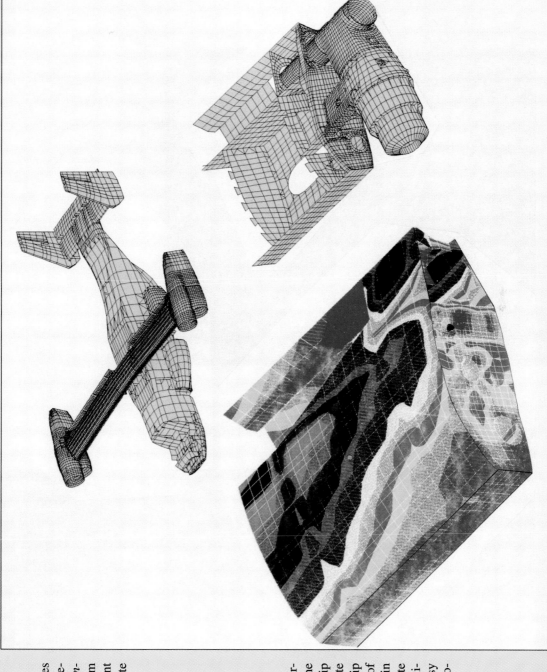

Right. V-22 Osprey tilt-rotor aircraft structure. Shown are the complete aircraft and a wing-tip engine-mount section with finite element grids, and a wing-tip section that illustrates the use of color graphics to interpret strain contours. (*MSC/NASTRAN* finite element analysis by Bell Helicopter Textron. Photo courtesy of American Institute of Aeronautics and Astronautics.)

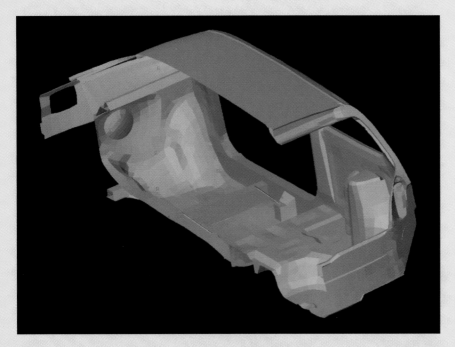

Above and below. Examples of the use of finite element analysis to support the design of automobile components. Flat-plate and thin-shell elements were used in modeling the car body and the wheel. (*SDRC I-DEAS* finite element analysis photos by Renault (above) and AirBoss (below). Courtesy of Structural Dynamics Research Corporation.)

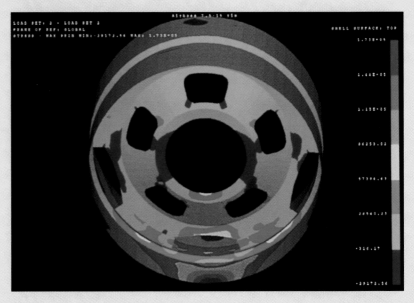

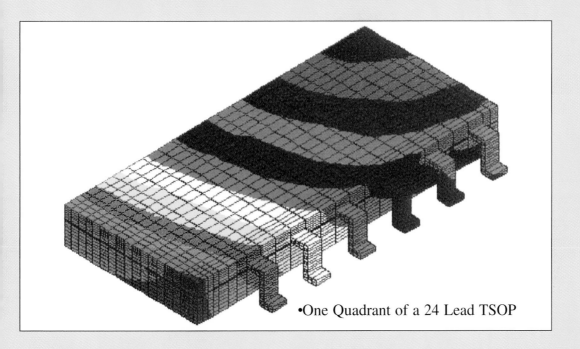

•One Quadrant of a 24 Lead TSOP

Above and below. Examples of the use of finite element analysis to support the design of computer chips. The photo above depicts thermally induced deformation of the chip; the photo below is a shear stress plot showing areas of stress concentration. (*ANSYS* finite element analysis photos courtesy of Motorola.)

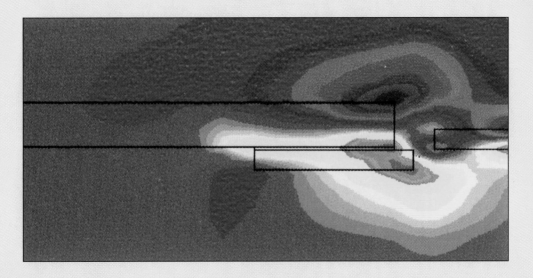

▲
Centerline   •Cross-section of a TSOP

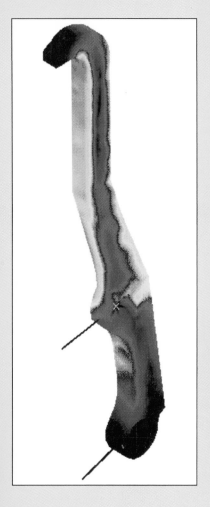

Left. An example of the use of finite element analysis in support of the design of a mechanical latch member from a Space Shuttle payload. (*MSC/NASTRAN* finite element analysis with *SDRC I-DEAS* postprocessing. Photo courtesy of NASA-Johnson Space Center.)

Below. Finite element analysis of the stress concentration due to a hole in an axially loaded flat bar, as described in Section 12.2. The adjacent color bar indicates that the stress depicted is the von Mises stress, which is discussed in Section 12.3. (*ABAQUS* finite element analysis by Greg Swadener.)

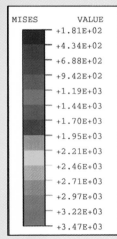

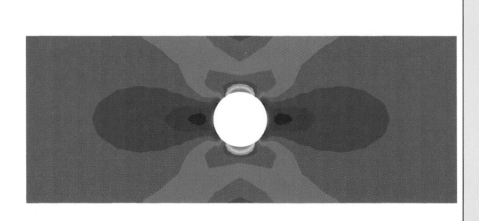

FIGURE 1.2. An airplane wing undergoing an ultimate-load test. (Used with permission, The Boeing Company, 1995.)

is truly an engineering marvel. It was designed by Johann Roebling in 1867–1869 and built under the supervision of his son Washington Roebling in 1870–1883 at a cost of $25 million. Its span of 1595 ft is suspended from towers that are 271 ft tall.

Since the 1960s, computer programs (e.g., NASTRAN, ANSYS, ABAQUS, GT-STRUDL) have been employed extensively for carrying out the numerical computations required in the design of mechanical systems and structures. Very detailed finite element

FIGURE 1.3. The Brooklyn Bridge, New York. (Andreas Feininger/*Life Magazine*)

models were used to create the color stress plots shown in the color-photo insert. In designing modern aeroplanes, automobiles, and other mechanical systems, Computer-Aided-Design (CAD) plays an essential role in defining the geometry of components, creating mathematical models of these components, and then performing the deformable-body analysis of these components. Color-insert Fig. 2 illustrates how CAD and finite element analysis are combined to determine the stress distribution and deformation of automobile components.

Not only are the principles and procedures of deformable-body mechanics used to analyze and design large objects like bridges and space vehicles, but they also find application to very small objects as well. Color-insert Fig. 3 shows a computer chip and the color-coded thermal deformation of the chip at one stage in the heating–cooling cycle that occurs when the computer is turned on and off. Such analyses are critical to the design of computer chips for high reliability and low fabrication cost.

Although many of the tasks involved in the design and analysis of the systems illustrated in the color-photo insert require a knowledge of mechanics of materials that is beyond the scope of this introductory textbook, the principles and procedures introduced in this book form a foundation on which more advanced topics build, and on which the design of complex applications, like those illustrated in the color-photo insert, depends.

1.2 THE FUNDAMENTAL CONCEPTS OF DEFORMABLE-BODY MECHANICS

The *three fundamental concepts* that are used in solving strength and stiffness problems of deformable-body mechanics are:

1. The *equilibrium* conditions must be satisfied.
2. The *geometry of deformation* must be described.
3. The *material behavior* (i.e., the force-temperature-deformation relationships of the materials) must be characterized.

In this textbook, these concepts are applied to fairly simple deformable bodies, but the same fundamentals apply, in more advanced mathematical form in many cases, to all studies of deformable solids.

Equilibrium. We have already noted that the principles of statics, that is, the *equations of equilibrium,* are fundamental to the study of deformable-body mechanics. Section 1.4 gives a brief review of static equilibrium and introduces the equilibrium concepts that are particularly important in the study of mechanics of solids. It also stresses the importance of drawing complete, accurate free-body diagrams. An entire chapter, Chapter 5, is devoted to the topic of equilibrium of beams.

Geometry of Deformation. There are several ways in which the *geometry of deformation* enters the solution of deformable-body mechanics problems, including:

1. Definitions of extensional strain and shear strain (Chapter 2).
2. Simplifications and idealizations (e.g., ''rigid'' member, ''fixed'' support, plane sections remain plane, displacements are small).
3. Connectivity of members, or geometric compatibility.
4. Boundary conditions and other constraints.

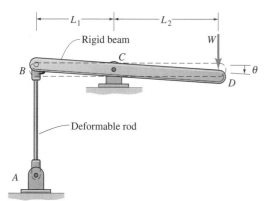

FIGURE 1.4. A system that illustrates several deformation assumptions.

Several of these may be illustrated by a comparison of Fig. 1.1 with Fig. 1.4. In Fig. 1.1, the diving board itself was considered to be deformable, but the supports at *A* and *B* were assumed to be rigid. Therefore, the *idealized model* in Fig. 1.1 is a *deformable beam* with *rigid constraints* at *A* and *B*. By contrast, the beam *BD* in Fig. 1.4 is assumed to be "rigid" under the loading and support conditions shown. Although *BD* does actually deform, that is, change shape, its deformation is assumed to be small in comparison to the rotation, θ, that it undergoes if the rod *AB* stretches significantly when load *W* is applied to the beam at *D*. Hence, the *idealized model* depicted in Fig. 1.4 is a *rigid beam, BD*, connected by a frictionless pin at end *B* to a *deformable rod AB*. As rod *AB* stretches, beam *BD* rotates through a small angle about a fixed, frictionless pin at *C*.

Material Behavior. The third principal ingredient in deformable-body mechanics is *material behavior*. Unlike equilibrium and geometry of deformation, which are purely analytical in nature, the constitutive behavior of materials, that is, the *force-temperature-deformation relationships* that describe the materials, can only be established by conducting experiments. These are discussed in Chapter 2.

It will be of great help to you in solving problems in the mechanics of deformable bodies if you will keep in mind these three distinct ingredients: **equilibrium, geometry of deformation,** and **material behavior**.

1.3 PROBLEM-SOLVING PROCEDURES

A consistent, systematic procedure is required for solving most problems in engineering practice, and this certainly applies to solving problems involving the mechanics of deformable bodies. The five steps in such a problem-solving procedure are:

1. Select the *system* of interest. This may be based on an existing physical system, or it may be defined by a set of design drawings and specifications.
2. Make simplifying assumptions that reduce the real system to an *idealized model*, or idealization of the system. For example, Figs. 1.1 and 1.4 illustrate two different idealized models of a diving board.
3. Apply the principles of deformable-body mechanics to the idealized model to create a *mathematical model* of the system, and solve the resulting equations to predict the *response* of the system to the applied disturbances (applied forces and/ or temperature changes).

4. Perform a *test* to compare the predicted responses to the behavior of the actual system. (A full-scale prototype or a scale model may have to be constructed if the physical system does not already exist.)

5. If the response predicted in step 3 does not agree with the response of the tested system, repeat steps 1–4, making changes as necessary until agreement is achieved.

In the future, as a practicing engineer you will find steps 1, 2, and 4 to be very important and very challenging. However, the main purpose of this textbook is to introduce you to the fundamental concepts of mechanics of solids and to enable you to solve deformable-body mechanics problems. Therefore, attention here will be devoted primarily to carrying out step 3, in which a mathematical model is formulated and its behavior analyzed.

So, how do you apply the principles of deformable-body mechanics to create mathematical equations, and how do you solve these equations to obtain the response of the system? A glance at the Example Problems in this textbook will indicate that the following four steps are clearly identified:

1. State the Problem.
2. Plan the Solution.
3. Carry out the Solution.
4. Review the Solution.

State the Problem— This step involves listing the given data, drawing any figures needed to describe the problem data, and listing the results that are to be obtained.

Plan the Solution— While you probably have not seen this step treated in a formal manner in previous textbooks, your success in solving problems quickly and accurately depends on how carefully you plan your solution strategy in advance. You should think about the given data and the results desired, identify the basic principles involved and recall the applicable equations, and plan the steps that will be needed to carry out the solution.

Carry Out the Solution— As noted in Section 1.2, your solution will involve three basic ingredients: equilibrium, geometry of deformation, and material behavior. Example problems and homework problems in this textbook appear in one of two forms—problems where numerical values are employed directly in the solution, and ones where algebraic symbols are used to represent the quantities involved and where the final answer is essentially a formula. One advantage of the symbolic form of solution is that a check of dimensional homogeneity may be easily made at each step of the solution. A second advantage is that the symbols serve to focus attention on the physical quantities involved. For example, the effect of a force designated by the symbol P can be traced through the various steps of the solution. Finally, since the end result is an equation in symbolic form, different numerical values can be substituted into the equation if desired.

The importance of checking for *dimensional homogeneity* at each step of a solution cannot be overemphasized! The principal quantities involved in static deformable-body mechanics problems are force (dimension F) and length (dimension L). If, at some step in the solution of a problem, you check an equation for dimensional homogeneity and obtain the result that

$$F \cdot L = F/L^2$$

you know that some error has been made. It is a waste of time and effort to proceed further without rectifying the error and establishing a dimensionally homogeneous equation! Other checks for accuracy should also be made frequently as the solution progresses.

Appendix B provides a discussion of the *units* used in solving deformable-body mechanics problems, and Appendix A discusses the number of *significant digits* required.

Review the Solution— This step, like the Plan the Solution step, may be one that you have not previously encountered in a formal manner. However, it is important for you to get into the habit of always checking your results by asking yourself the following types of questions:

- Is the answer dimensionally correct?
- Do the quantities involved appear in the final solution in a reasonable manner?
- Is the sign of the final answer reasonable, and is the numerical magnitude reasonable?
- Is the final result consistent with the assumptions that were made in order to achieve the solution? (For example, an assumption that "the slope is small" is violated if the final slope turns out to be 45°.)

You may be tempted to think that Review the Solution means for you to compare your answer with one in the back of the book. However, in the "real world" there are no "answer books." Hence, you should begin now, if you have not already done so, to make a habit of testing by any means possible the reasonableness, dimensional homogeneity, etc. of your own answers.

Although the Carry out the Solution step will undoubtedly occupy a major portion of your problem-solving time as you work on problems from this textbook, you will probably find that, as a practicing engineer, most of your time is involved in the other three steps, since a computer may quickly and obediently carry out the actual numerical solutions of your engineering problems. It then becomes your task to set up the problem correctly (State the Problem and Plan the Solution) and to evaluate carefully the results of the computer solution (Review the Solution). Finally, you should heed the advice of the authors of a popular statics text:[2]

> It is also important that all work be neat and orderly. Careless solutions that cannot be easily read by others are of little or no value. The discipline involved in adherence to good form will in itself be an invaluable aid to the development of the abilities for formulation and analysis. Many problems that at first may seem difficult and complicated become clear and straightforward when begun with a logical and disciplined method of attack.

Communication is a vital part of any engineering project, and only work that is neat and orderly can serve to communicate technical information, like the solution of a deformable-body mechanics problem.

1.4 REVIEW OF STATIC EQUILIBRIUM; EQUILIBRIUM OF DEFORMABLE BODIES

In this textbook we consider deformable bodies at rest, that is, bodies whose acceleration and velocity are both zero. In your previous study of statics, you learned the equations of equilibrium and you learned how to apply these equations to particles and to rigid bodies

---

[2]See [Ref. 1-1].

through the use of free-body diagrams. In this section we will review the fundamental equations and problem-solving procedures of statics and will begin to indicate how they apply to the study of deformable bodies.

Equations of Equilibrium. Recall that the *necessary conditions* for equilibrium of a body (rigid or deformable) are:[3]

$$\sum F = 0, \qquad \left(\sum M\right)_O = 0 \qquad (1.1)$$

That is, if a body is in equilibrium,

- the sum of the external forces acting on the body is zero, and
- the sum of the moments, about any arbitrary point O, of all the external forces acting on the body is zero.

These equations are usually expressed in component form with the components referred to a set of rectangular Cartesian axes x, y, z. Then, the resulting scalar equations are:

$$\sum F_x = 0, \qquad \left(\sum M_x\right)_O = 0$$

$$\sum F_y = 0, \qquad \left(\sum M_y\right)_O = 0 \qquad (1.2)$$

$$\sum F_z = 0, \qquad \left(\sum M_z\right)_O = 0$$

When the number of *independent* equilibrium equations available is equal to the number of unknowns, the problem is said to be *statically determinate.* When there are more unknowns than available independent equations of equilibrium, the problem is said to be *statically indeterminate.* For example, if a body has more external supports or constraints than are required to maintain the body in a stable equilibrium state, the body is statically indeterminate. Supports that could be removed without destroying the equilibrium of the body are called *redundant supports,* and their reactions are called, simply, *redundants.*

In order to apply equilibrium equations to a body, it is <u>always</u> wise to draw a *free-body diagram* (FBD) of the body. However, before reviewing the procedure for drawing a free-body diagram, let us consider the types of external loads that may act on a body and several ways in which the body may be supported or connected to other bodies.

External Loads. The *external loads* acting on a deformable body may be classified in four categories, or types. These types, together with their appropriate dimensions, are:

- *Concentrated loads,* including *point forces* (F) and *couples* ($F \cdot L$),
- *Line loads* (F/L),
- *Surface loads* (F/L^2), and
- *Body forces* (F/L^3).

---

[3]For a rigid body, Eqs. 1.1 also constitute *sufficient conditions* for equilibrium of the body. That is, if Eqs. 1.1 are satisfied, then the rigid body is in equilibrium. However, the *necessary and sufficient condition* for equilibrium of a deformable body is that the sets of external forces that act on the body and on every possible subsystem isolated out of the original body all be sets of forces that satisfy Eqs. 1.1. (See, for example, [Ref. 1-2] p. 16.)

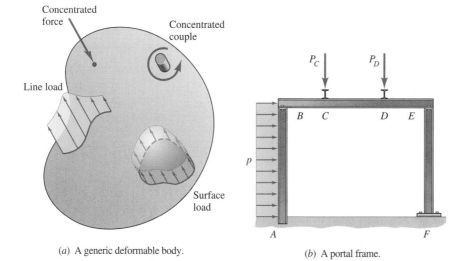

(a) A generic deformable body.

(b) A portal frame.

FIGURE 1.5. External loads acting on deformable bodies.

The first three types of external loads are illustrated on the generic deformable body in Fig. 1.5a. Body forces are produced by action-at-a-distance. Like the force of gravity (weight), they are proportional to volume, and they act on particles throughout the body.

Although, in reality, all external loads that act on the surface of a deformable body must act on a finite area of that surface, line loads and concentrated forces are considered to act along a ''line'' or to act at a single ''point,'' respectively, as indicated in Fig. 1.5a. Concentrated loads and line loads are, therefore, idealizations. Nevertheless, they permit accurate analysis of the behavior of the deformable body, except in the immediate vicinity of the loads.[4] In Fig. 1.5b a cross-beam at C (shown in end view as an I) exerts a downward concentrated force P_C on the horizontal frame member BE, and a horizontal line load of uniform intensity p acts on the vertical frame member AB.

Concentrated forces have the units of force [e.g., Newtons (N) or pounds (lb)], and line loads have the units of force per unit length (e.g., N/m or lb/ft). Other external loads are expressed in units appropriate to their dimensionality.

Support Reactions and Member Connections. The external loads that are applied to a member must generally be transmitted to adjacent members that are connected to the given member, or carried directly to some form of support.[5] For example, the vertical loads P_C and P_D that act on the horizontal beam BE in Fig. 1.5b are eventually ''transmitted'' to the ground at supports A and F. Where there is a *support,* as at points A and F in Fig. 1.5b, the displacement (i.e., the change of position) is specified to be zero, but the force is unknown. Therefore, forces (including couples) at supports are called *reactions,* since they react to the loads that are applied elsewhere. We say that the support enforces a *constraint,* that is, the support constrains (i.e., makes) the displacement to be zero.

Table 1.1 gives the symbols that are used to represent idealized supports and member connections. Also shown are the force components and the couples that correspond to these. For the most part, in this text we will consider loading and supports that lie in a

---

[4]See the discussion of St. Venant's Principle in Section 3.3.

[5]The exception is a self-equilibrated system, like an airplane, whose (upward) distributed lift force equals its (downward) weight.

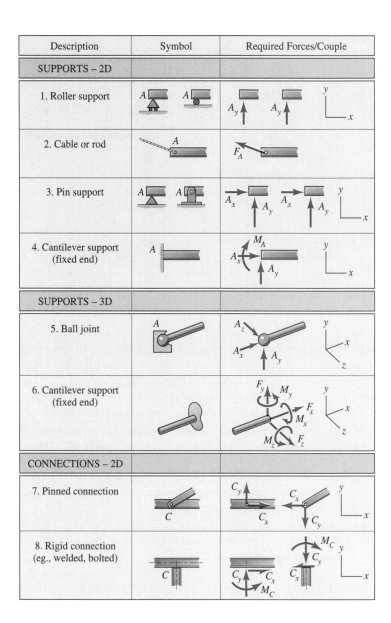

Description	Symbol	Required Forces/Couple
SUPPORTS – 2D		
1. Roller support		
2. Cable or rod		
3. Pin support		
4. Cantilever support (fixed end)		
SUPPORTS – 3D		
5. Ball joint		
6. Cantilever support (fixed end)		
CONNECTIONS – 2D		
7. Pinned connection		
8. Rigid connection (eg., welded, bolted)		

single plane, that is coplanar loading and support. Occasionally, however, we will consider a three-dimensional situation.

Internal Resultants. In the study of mechanics of deformable bodies, we must consider not only external forces and couples, that is, the applied loads and reactions, but we also must consider *internal resultants,* that is, forces and couples that are internal to the original body. For example, to analyze the L-shaped two-force linkage in Fig. 1.6*a*,[6] it is necessary to imagine a cutting plane, like the one indicated in Fig. 1.6*a*, and to show the (unknown) internal resultants acting on this plane, as has been done in Fig. 1.6*b*.

---

[6]You may recall from statics that, if a body in equilibrium is subjected only to concentrated forces acting at two points in the body, the forces must be equal and opposite and must be directed along the line joining the points of application of the forces, as illustrated in Fig. 1.6*a*. Such a body is referred to as a *two-force member.*

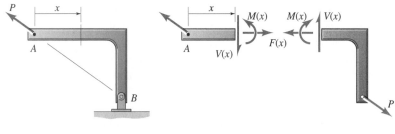

(*a*) A two-force member.   (*b*) Internal resultants.

FIGURE 1.6. An illustration of internal resultants.

The engineering theories that are developed in this textbook apply to deformable bodies for which one dimension is significantly greater than the other two dimensions; that is, we will consider long, thin members.[7] The six internal resultants that result from general loading of such a member are indicated on the sketch in Fig. 1.7, where the x axis is taken to lie along the longitudinal direction of the member, and a cutting plane normal to the x axis, called a *cross section,* is passed through the member at coordinate x.

On an arbitrary cutting plane through a body subject to general three-dimensional loading there will be three components of the *resultant force* and three components of the *resultant moment.* When the body is slender, as in Fig. 1.7, these resultants are given special names. The force normal to the cross section, labeled $F(x)$, is called the *normal force,* or *axial force.* The two components of the resultant force that are tangent to the cutting plane, $V_y(x)$ and $V_z(x)$, are called transverse shear forces, or just *shear forces.* The component of moment about the axis of the member is called the *torque,* or *twisting moment,* and is labeled $T(x)$. Finally, the other two components of moment, $M_y(x)$ and $M_z(x)$, are called *bending moments.* Much of the remainder of this book is devoted to the determination of how these six resultants are distributed over the cross section.

Free-body Diagram (FBD).

Let us now review the steps that are involved in drawing a complete free-body diagram. They may be summarized as follows:

* Determine the extent of the body to be included in the FBD. Completely *isolate* this body from its supports and from any other bodies attached to it. When internal

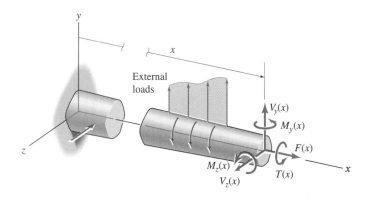

FIGURE 1.7. The six internal resultants on an arbitrary cross section of a slender member.

[7]The only exception is the thin-wall pressure vessels discussed in Section 9.2.

resultants are to be determined, pass a sectioning plane through the member at the desired location. Sketch the contour of the resulting free body.

- Indicate on the sketch all of the *applied loads,* that is, all known external forces and couples, acting <u>on</u> the body. These include distributed and concentrated forces applied to the body and also, when it is not negligible, the distributed weight of the body itself. The location, magnitude, and direction of each applied load should be clearly indicated on the sketch.

- Where the body is supported or is connected to other bodies, or where it has been sectioned, show the *unknown forces and couples* that are exerted on this body by the adjacent bodies. Assign a symbol to each such force (or force component or couple) and, where the direction of an unknown force or couple is known, use this information. Often there is a *sign convention* that establishes the proper sense to be assumed as positive. This is particularly true for the internal resultants. However, in some cases the sense of an unknown can be assumed arbitrarily.

- Label significant points and include significant *dimensions.* Also, if reference axes are needed, show these on the sketch.

- Finally, keep the FBD as simple as possible so that it conveys the essential equilibrium information quickly and clearly.

Except in Chapter 10, where we will examine *stability of equilibrium* and where it will be necessary to draw a free-body diagram of the deformed system, we will assume that deformations are small enough that the *free-body diagram can be drawn showing the body in its undeformed configuration,* even though the forces acting on it are those associated with the deformed configuration.

The following two example problems will serve as a review of the way that free-body diagrams are chosen, and will illustrate how equilibrium equations are used to determine internal resultants.

■■■■■■■■■■■■□□□ E X A M P L E 1 . 1 □□□□□■■■■■■■■■■■

The simple planar truss in Fig. 1 consists of two straight, two-force members, *AB* and *BC*, that are pinned together at *B*. The truss is loaded by a downward

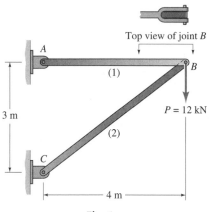

Fig. 1

force of $P = 12$ kN acting on the pin at B. Determine the internal axial forces F_1 and F_2 in members AB and BC, respectively. Neglect the weight of the truss members.

Plan the Solution This problem asks for the internal axial forces, so we need to consider equilibrium. The external force P and the truss members AB and BC all act on the pin at B. Therefore, a free-body diagram that contains the pin at B plus an arbitrary length of member AB and an arbitrary length of member BC will permit us to write equilibrium equations that relate F_1 and F_2 to P.

   Since P acts downward, we can expect the force in member BC to be compressive (i.e., member BC pushes upward on the pin at B).

Solution In Fig. 2 we draw a free-body diagram of the joint, taking F_1 and F_2 to be <u>positive in tension</u>. Also, we select reference axes x and y as shown. Equilibrium must be satisfied in both the x and y directions.

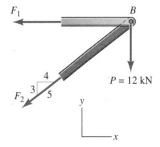

Fig. 2. Free-body diagram.

$$\xrightarrow{+} \sum F_x = 0: \qquad -F_1 - (4/5)F_2 = 0$$

$$+\uparrow \sum F_y = 0: \qquad -(3/5)F_2 - P = 0$$

$$F_2 = -(5/3)P = -(5/3)12 \text{ kN} = -20 \text{ kN}, \quad F_2 = 20 \text{ kN (C)} \qquad \text{Ans.}$$

$$F_1 = -(4/5)F_2 = -(4/5)(-20 \text{ kN}) = 16 \text{ kN}, \quad F_1 = 16 \text{ kN (T)} \qquad \text{Ans.}$$

Review the Solution The force F_2 has turned out to be in compression, as we expected, which means that it is pushing upward to the right. Then, F_1 must be in tension to maintain horizontal equilibrium of the pin at B. Finally, the magnitudes are reasonable, so our solution appears to be correct.[8]

Based on our experience in solving equilibrium problems, we could have assumed at the outset that the force in member BC acts upward to the right (i.e., in compression). Had we done so (by reversing the sense of the arrow representing force F_2), we would have gotten the answer $F_2 = 20$ kN, without the minus sign. Instead, we chose to show F_1 and F_2 on the free-body diagram assuming tension to be positive and, as a consequence, the answer for F_2 turned out to be $F_2 = -20$ kN. That is, the minus sign indicates that the force F_2 is a compressive force rather than a tensile force.

   As problems get more complex (e.g., several interconnected bodies) it will become impossible to mentally solve all of the resulting equilibrium equations to the extent that the "correct" sense of every force can be established at the outset when the free-body diagram is drawn. **The procedure of assuming internal axial forces to be positive in tension makes it both easy to draw the free-body diagram and easy to interpret the meaning of the answers (positive forces are tension; negative forces are compression).** Hence, this sign convention will be followed throughout this textbook.

---

[8]Note that the symbol P has been used up to the last steps in the solution above, and that the units (kN) are stated when numerical values are inserted. It is good practice to show the proper force units (F) and length units (L) in the solution of numerical problems. Also, note that the answers are marked, and that tension (T) and compression (C) are identified in the answers.

An electrical worker stands in the bucket that hangs from a pin at end D of the boom of the cherry picker in Fig. 1. The worker and bucket together weigh a total of 200 lb. Between A and C the boom weighs 1.1 lb/in., and between C and D it weighs 0.8 lb/in. Assume that AC and CD are uniform beams.

Determine the normal force, the transverse shear force, and the bending moment that act at cross section E, midway between A and B.

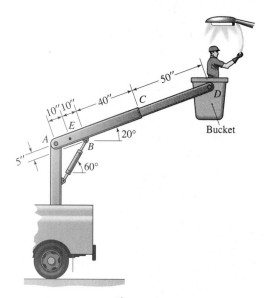

Fig. 1

Plan the Solution   Since there will be three unknowns on the cross section at E and, in addition, an unknown force at the pin B, we cannot solve for all four unknowns using a single free-body diagram. Hence, we will have to determine the pin force first using a separate free-body diagram; then we can determine the two components of the internal force at E and the moment at E.

Solution

Pin Reaction at B: First, we use the free-body diagram in Fig. 2 to determine the pin reaction at B.

$$W_{AC} = (1.1 \text{ lb/in.})(60 \text{ in.}) = 66 \text{ lb}$$

$$W_{CD} = (0.8 \text{ lb/in.})(50 \text{ in.}) = 40 \text{ lb}$$

$$\left(\sum M \right)_A = 0:$$

$$(66 \text{ lb})(30 \text{ in.})(\cos 20°) + (40 \text{ lb})(85 \text{ in.})(\cos 20°)$$

$$+ (200 \text{ lb})(110 \text{ in.})(\cos 20°) - (B \cos 50°)(20 \text{ in.})$$

$$- (B \cos 40°)(5 \text{ in.}) = 0$$

$$B = 1542 \text{ lb}$$

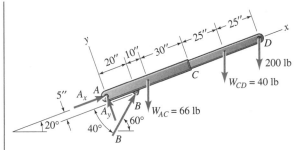

Fig. 2. A free-body diagram of boom *AD*.

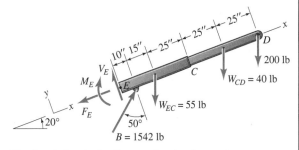

Fig. 3. A free-body diagram showing the resultants at section *E*.

Internal Resultants: Next, we pass a section through the beam at *E* and determine the forces and moment on this cross section (Fig. 3). (Let point *E* be on the centerline of the beam *AD*.) On the section at *E* we show the unknown normal force F_E, the unknown transverse shear force V_E, and the unknown bending moment M_E.

$$W_{EC} = (1.1\text{ lb/in.})(50\text{ in.}) = 55\text{ lb}$$

$+\nearrow \sum F_x = 0:$

$$-F_E + (1542\text{ lb})(\sin 50°) - (55\text{ lb} + 40\text{ lb} + 200\text{ lb})(\sin 20°) = 0$$

$$F_E = 1080\text{ lb}$$

$+\nwarrow \sum F_y = 0:$

$$V_E + (1542\text{ lb})(\cos 50°) - (55\text{ lb} + 40\text{ lb} + 200\text{ lb})(\cos 20°) = 0$$

$$V_E = -714\text{ lb}$$

(i.e., 714 lb acting opposite to the direction shown on the free-body diagram)

$\left(\sum M\right)_E = 0:$

$$M_E + (55\text{ lb})(\cos 20°)(25\text{ in.}) + (40\text{ lb})(\cos 20°)(75\text{ in.})$$
$$+ (200\text{ lb})(\cos 20°)(100\text{ in.}) - (1542\text{ lb})(\cos 50°)(10\text{ in.})$$
$$- (1542\text{ lb})(\sin 50°)(5\text{ in.}) = 0$$
$$M_E = -7088\text{ lb·in.}$$

The answers, rounded to the proper number of significant digits are:

$$F_E = 1080\text{ lb}, \quad V_E = -710\text{ lb}, \quad M_E = -7090\text{ lb·in.} \qquad \text{Ans.}$$

Review the Solution Because of the long moment arm of the 200-lb load compared with the moment arm of the force at *B*, it is reasonable for the magnitude of *B* to be much larger than the magnitude of the total load. The magnitude and sense of F_E, V_E, and M_E also seem to be reasonable in view of the magnitude of *B* and the magnitude and location of the other loads.

15

*In the truss equilibrium problems, **Probs. 1.4-1 through 1.4-6,** adopt the sign conversion for axial force that a <u>tensile force is positive</u>. That is, on your free-body diagrams, show all unknown forces F_i as tensile forces. Then, a compressive force will be negative.*

Prob. 1.4-1. For the pin-jointed truss in Fig. P1.4-1, (a) determine the reactions at the pin supports at C and E, and (b) determine the axial force in each of the following members: F_3 (in member BC), F_4 (in member CD), and F_5 (in member DE).

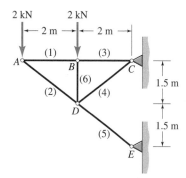

P1.4-1

Prob. 1.4-2. A pin-jointed truss ACE is part of a cable-hoist system that is used to lift cargo boxes, as shown in Fig. P1.4-2. The cable from the lift motor to the cargo sling passes over a 6-in. pulley that is supported by a frictionless pin at C. The weight of the cargo box being lifted is 1500 lb. Neglecting the weight of the truss members and the cables, (a) determine the reactions at the pin supports at A and E, and (b) determine the axial force in each of the following members: F_1 (in member AB), F_2 (in member AD), and F_3 (in member DE). (c) Explain the answer you got for the value of F_2.

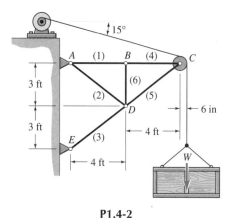

P1.4-2

Prob. 1.4-3. Each member in the pin-jointed planar truss in Fig. P1.4-3 is 5 ft long. The truss is attached to a firm base by a frictionless pin at A, and it rests on a roller support at B. For the loading shown, (a) determine the reactions at A and B, and

(b) determine the force in each of the three members, labeled (1) through (3).

Prob. 1.4-4. Solve Prob. 1.4-3 with the position of the pin support and the roller support interchanged. That is, support the truss by a roller at A and by a pin at B.

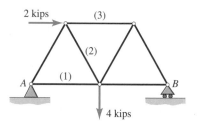

P1.4-3 and P1.4-4

Prob. 1.4-5. The inclined, pin-jointed truss in Fig. P1.4-5 is made of seven identical members of length L. The truss has a pin support at A and a roller support at C. The truss is inclined at an angle of 20° and is loaded by a single vertical force P at pin B. (a) Determine the reactions at A and C, and (b) determine the axial force in each of the three members, labeled (1) through (3) on the figure.

Prob. 1.4-6. Solve Prob. 1.4-5 with the position of the pin support and the roller support interchanged. That is, support the truss by a roller at A and by a pin at C.

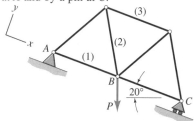

P1.4-5 and P1.4-6

Prob. 1.4-7. Beam AD has a frictionless-pin support at A and a roller support at D, as shown in Fig. P1.4-7. The beam supports a uniformly distributed load of intensity w_0 (force per unit length) over two-thirds of its length. (a) Determine the reactions at A and D, and (b) determine the internal resultants (axial force, shear force, and bending moment) on the cross section at point B.

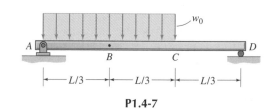

P1.4-7

Prob. 1.4-8. A weight W is supported by a cable attached to beam BE at E, as shown in Fig. P1.4-8. The beam is supported by a fixed, frictionless pin at B and by an inclined cable from A to D. (a) Determine the tension in cable AD, and determine the reaction force at pin B. (b) Determine the internal resultants (axial force, shear force, and bending moment) on the cross section at point C.

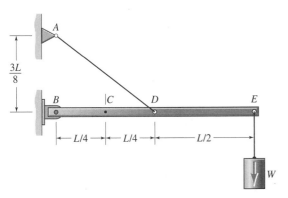

P1.4-8

Prob. 1.4-9. For the beam shown in Fig. P1.4-9, determine: (a) the reactions at supports B and C, and (b) the internal resultants—F_D, V_D, and M_D—at section D, midway between the supports.

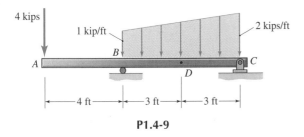

P1.4-9

Prob. 1.4-10. For the beam shown in Fig. P1.4-10, determine: (a) the reactions at supports B and C, and (b) the internal resultants at section E, midway between the supports.

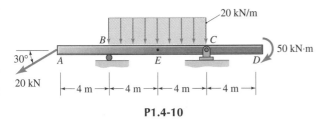

P1.4-10

Prob. 1.4-11. A log that weighs 3600 lb and is 30 ft long is being hoisted by a cable that passes over two pulleys, as shown in Fig. P1.4-11. In the position shown, the cable from the log to the first pulley is vertical, and the log is inclined at an angle of 30° to the ground. Assume that neither the ground nor the block at A provides any moment restraint, and neglect the radial distance from the center of the log to the two points of contact at A.

(a) Determine the tension in the cable and the reaction force at end A, and (b) determine the internal resultants at cross section C at the midpoint of the log.

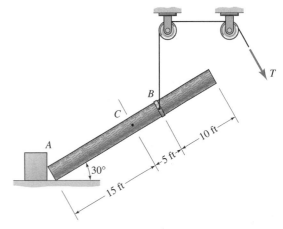

P1.4-11

Prob. 1.4-12. Beam AB is supported by a frictionless pin at B and by a cable AC attached at end A, as shown in Fig. P1.4-12. The beam supports a uniformly distributed downward load of intensity w_0 (force per unit length). Determine expressions for the internal resultants $F(x)$, $V(x)$, and $M(x)$ as functions of distance x measured from A.

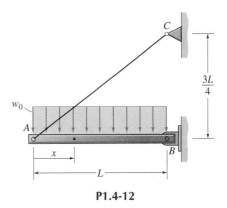

P1.4-12

Prob. 1.4-13. For the overhanging beam shown in Fig. P1.4-13: (a) determine expressions for the internal resultants $V_1(x)$ and $M_1(x)$ in interval AB ($0 \leq x < 6$ in.), and (b) determine the internal resultants V_D and M_D at section D, midway between the two supports.

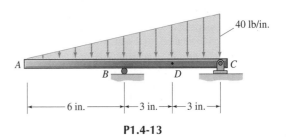

P1.4-13

17

Prob. 1.4-14. The right-angle frame in Fig. P1.4-14 has equal legs of length L and is supported by a pin and roller as shown. If the total weight of the frame is W, (a) determine the reactions at supports A and B, and (b) determine the internal resultants (axial force, shear force, and bending moment) on the cross section at point C.

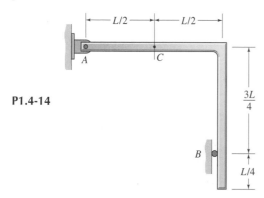

P1.4-14

*Prob. 1.4-15. A force $P = 200$ lb is applied to a frame ABE as shown in Fig. P1.4-15. The frame is supported, through friction-less pins, by a wheel at B and a fixed-pin support block at E. (a) Determine the reaction at E, and (b) determine the internal resultants (axial force, shear force, and bending moment) on the cross section at point C. Use the given dimensions to eliminate the angle ϕ from your answers.

P1.4-15

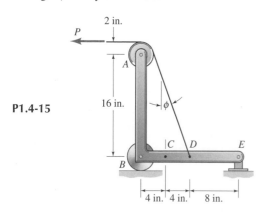

Prob. 1.4-16. A shop crane lifts a weight $W = 2$ kN, as shown in Fig. P1.4-16. With the crane in the configuration shown, determine the internal resultants $F(x)$, $V(x)$, and $M(x)$ at an arbitrary cross section x between A and B.

Prob. 1.4-17. One of the lift arms of a fork-lift truck has the loading and support shown in Fig. P1.4-17. The load consists of three identical crates, each weighing $W = 400$ lb. (Each of the two lift arms supports half of the total load.) The "support" consists of a hoist cable attached to the lift arm at B, and frictionless rollers that react against the truck frame at A and C. Neglect the weight of the lift arm. (a) Determine the roller reactions R_A and R_C at A and C, respectively. (b) Determine the internal resultants (axial force, shear force, and bending moment) on a horizontal cross section at D. (c) Determine the internal resultants on a vertical cross section at E.

18

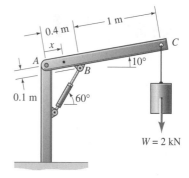

P1.4-16

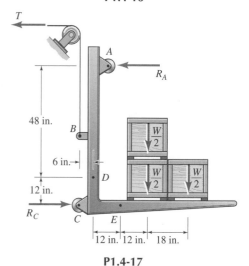

P1.4-17

Prob. 1.4-18. Determine the internal resultants F_G, V_G, and M_G on the cross section at G of the horizontal frame member in Fig. P1.4-18. The uniformly distributed load has a magnitude $w_0 = 200$ lb/ft. (See the inset for a definition of the resultants.)

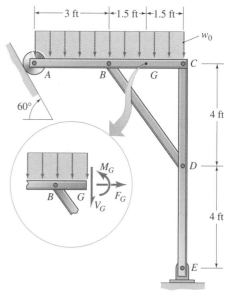

P1.4-18

STRESS AND STRAIN; DESIGN $\quad$ 2

2.1 INTRODUCTION

The photographs in the color photo insert give a visual indication of the complexity of the internal behavior of the members pictured in response to the given external loads.[1] Figure 1.6*a* shows an L-shaped bracket loaded as a two-force member, and Fig. 1.6*b* shows the internal resultants, $F(x)$, $V(x)$, and $M(x)$, that are required to maintain the equilibrium of the two sectioned parts of the bracket. Although we could compute the internal resultants shown on Fig. 1.6*b* by using static equilibrium procedures (free-body diagrams and equations of equilibrium), those procedures are clearly insufficient for determining the complex *internal* force distribution making up those resultants. The concept of *stress* is introduced in this chapter to enable us to quantify internal force distributions.

The shape of the bracket also changes due to the applied loads; that is, the member deforms. The concept of *strain* is introduced to permit us to give a detailed analytical description of such deformation. Finally, stress and strain are related to each other. This relationship, which depends on the material(s) used in the fabrication of the member, must be determined by performing certain *stress-strain tests* that are described in this chapter.

2.2 NORMAL STRESS

To introduce the concepts of stress and strain, we begin with the relatively simple case of a straight bar undergoing *axial loading,* as shown in Fig. 2.1.[2] In this section we consider the stress in the bar, and in Section 2.3 we treat the corresponding strain.

Equal and opposite forces of magnitude P acting on a straight bar cause it to *elongate,* and also to get narrower, as can be seen by comparing Figs. 2.1*a* and 2.1*b*. The bar is said to be in *tension.* If the external forces had been applied in the opposite sense, that is, pointing toward each other, the bar would have shortened and would then be said to be in *compression.*

---

[1] See color-photo insert.

[2] *Axial loading* is discussed here in order to introduce the concepts of stress and strain and the relationship of stress to strain. Chapter 3 treats axial deformation in greater detail.

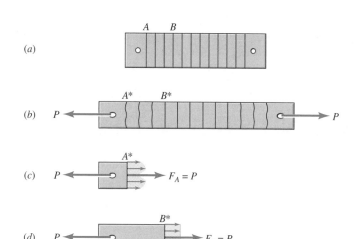

FIGURE 2.1. A straight bar undergoing axial loading. (*a*) The undeformed bar, with vertical lines indicating cross sections. (*b*) The deformed bar. (*c*) The distribution of internal force at section *A*. (*d*) The distribution of internal force at section *B*.

Definition of Normal Stress. The dashed arrows in Figs. 2.1*c* and 2.1*d* represent the distribution of force on cross sections at *A* and *B*, respectively. (A *cross section* is a plane that is perpendicular to the axis of the bar.) Near the ends of the bar, for example at section *A*, the normal force F_A is not uniformly distributed over the cross section; but at section *B*, farther from the point of application of force *P*, the force distribution is uniform. In mechanics, the term *stress* is used to describe the distribution of a force over the area on which it acts and is expressed as force intensity, that is, as force per unit area.

$$\text{Stress} = \frac{\text{Force}}{\text{Area}}$$

The units of stress are units of force divided by units of area. In the U.S. Customary System of units (USCS), stress is normally expressed in pounds per square inch (psi) or in kips per square inch, that is, kilopounds per square inch (ksi). In the International System of units (SI), stress is specified using the basic units of force (newton) and length (meter) as newtons per meter squared (N/m^2). This unit, called the *pascal* ($1\ Pa = 1\ N/m^2$), is quite small, so in engineering work stress is normally expressed in kilopascals ($1\ kPa = 10^3\ N/m^2$), megapascals ($1\ MPa = 10^6\ N/m^2$), or gigapascals ($1\ GPa = 10^9\ N/m^2$).

There are two types of stress, called *normal stress* and *shear stress*. In this section we will consider only normal stress; shear stress is introduced in Section 2.7. In words, normal stress is defined by

$$\textbf{Normal stress} = \frac{\text{Force } \textbf{normal} \text{ to an area}}{\text{Area on which the force acts}}$$

The symbol used for normal stress is the lowercase Greek letter sigma (σ). The *normal stress at a point* is defined by the equation.

$$\sigma(x, y, z) = \lim_{\Delta A \to 0} \left(\frac{\Delta F}{\Delta A} \right) \tag{2.1}$$

where, as shown in Fig. 2.2*a*, ΔF is the normal force (assumed positive in tension) acting on an elemental area ΔA containing the point (x, y, z) where the stress is to be determined.

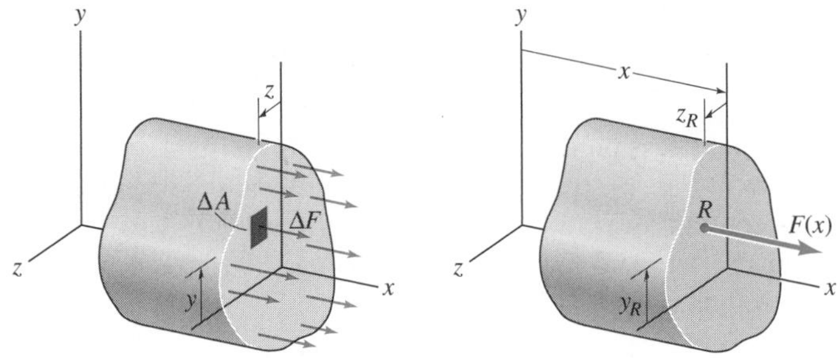

(a) Distribution of normal
force on a cross section.

(b) Stress resultant on the
cross section.

FIGURE 2.2. Normal force on a cross section.

The *sign convention for normal stress* is as follows:

- A positive value for σ indicates *tensile stress,* that is, the stress due to a force ΔF that <u>pulls</u> on the area on which it acts.
- A negative value for σ indicates *compressive stress.*

Thus, the equation $\sigma = 6.50$ MPa signifies that σ is a *tensile stress* of magnitude 6.50 MPa, or 6.50 MN/m^2, and the equation $\sigma = -32.6$ ksi indicates a *compressive stress* of magnitude 32.6 kips/in^2.

Stress Resultant. In Fig. 2.2*b*, the resultant force on the cross section at x is labeled $F(x)$, and it acts at point (y_R, z_R) in the cross section. Given the distribution of normal stress on a cross section, $\sigma \equiv \sigma(x, y, z)$, we can integrate over the cross section to determine the magnitude and point of application of the *resultant normal force:*[3]

$$\sum F_x: \qquad F(x) = \int_A \sigma \, dA$$

$$\sum M_y: \qquad z_R F(x) = \int_A z\sigma \, dA \qquad (2.2)$$

$$\sum M_z: \qquad -y_R F(x) = -\int_A y\sigma \, dA$$

(Note that the sign convention for σ implies that the force F in Eq. 2.2 is to be taken positive in tension. This is the reason that we will consistently, as we did in Chapter 1, take normal-force resultants to be positive in tension.)

---

[3]For generality, the normal force has been permitted to be a function of x in Fig. 2.2*b* and in Eqs. 2.2. Of course, $F(x) = P = $ const in the axial-loading case illustrated in Fig. 2.1.

EXAMPLE 2.1

The normal stress on the cross section at x is constant over the cross section, as shown in Fig. 1a. (In Section 3.2 you will learn the conditions under which $\sigma(x)$ is constant over the cross section.) Prove that this stress distribution corresponds to an axial force $F(x) = A\sigma(x)$ acting through the centroid of the cross section at x.

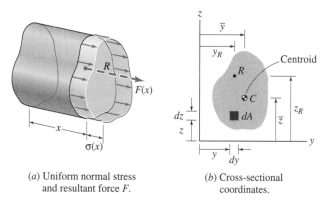

Fig. 1. Uniform normal stress.

(a) Uniform normal stress
and resultant force F.

(b) Cross-sectional
coordinates.

Plan the Solution Let the resultant be assumed to be a force $F(x)$ acting parallel to the x axis and passing through point (y_R, z_R), as in Fig. 2.2b. We must show that

$$F(x) = A\sigma(x), \quad y_R = \bar{y}, \quad z_R = \bar{z}$$

For this we can use Eqs. 2.2.

Solution Substituting the condition $\sigma(x, y, z) = \sigma(x)$ into Eqs. 2.2, we get

$$\sum F_x: \qquad F(x) = \sigma(x) \int_A dA = \sigma(x) A$$

$$\sum M_y: \qquad z_R F(x) = \sigma(x) \int_A z\, dA = \sigma \bar{z} A$$

$$\sum M_z: \qquad -y_R F(x) = -\sigma(x) \int_A y\, dA = -\sigma(x)\bar{y} A$$

Therefore, $F(x) = A\sigma(x)$, $z_R = \bar{z}$, and $y_R = \bar{y}$, which corresponds to a tensile force $F(x)$ acting at the centroid of the cross section, as shown in Fig. 2. Q.E.D.

Review the Solution We would certainly expect uniform stress on a circular rod to correspond to a force acting along the axis of the rod, and similarly for a square or rectangular bar. Hence, it is "reasonable" that a uniform normal stress distribution acting over a cross section of general shape produces a resultant force acting through the centroid of the cross section.

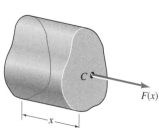

Fig. 2. The resultant of uniform normal stress $\sigma(x)$.

If the normal stress is uniform on a cross section, as in the preceding example, the normal stress on the cross section is given by

$$\sigma(x) = \frac{F(x)}{A(x)} \tag{2.3}$$

(In most cases, the cross-sectional area is constant throughout the length of the member, but Eq. 2.3 may also be used if the cross-sectional area varies slowly with x. See Example 3.2.)

Average Normal Stress. Even when the normal stress varies over a cross section, as it does in Fig. 2.1c, we can compute the *average normal stress* on the cross section by letting

$$\sigma_{\text{avg}} = \frac{F}{A} \tag{2.4}$$

Thus, for Figs. 2.1c and 2.1d we get

$$(\sigma_{\text{avg}})_A = \frac{F_A}{A} = \frac{P}{A}, \quad (\sigma_{\text{avg}})_B = \frac{F_B}{A} = \frac{P}{A}$$

However, we are usually more interested in knowing the maximum value of stress on the cross section, rather than just the average stress. Therefore, much of the rest of this textbook is devoted to determining how stress is distributed on cross sections of structural members under various loading conditions.

Uniform Normal Stress in an Axially Loaded Bar. Under certain assumptions, an axially loaded bar will have the same uniform normal stress on every cross section; that is, $\sigma(x, y, z) = \sigma = $ constant. These assumptions are:

- The bar is *prismatic;* that is, the bar is straight, and it has the same cross section throughout its length.
- The bar is *homogeneous;* that is, the bar is made of the same material throughout.
- The load is applied as equal and opposite uniform stress distributions over the two end cross sections of the bar. See Fig. 2.3a.

So long as the axial loads are applied at the centroids of the ends of the bar, the last assumption—that the loads are applied as uniform normal stress distributions on the end cross sections—can be relaxed. As illustrated in Figs. 2.1, the stress is uniform on cross sections except very near the points of application of load.[4]

The normal stress on cross sections of an axially loaded member, like the one in Fig. 2.3, is called the *axial stress.* Since the resultant force, $F(x)$, on every cross section of the bar is equal to the applied load P, and since the cross-sectional area is constant, from Eq. 2.3 we get

$$\sigma = \frac{F(x)}{A(x)} = \frac{P}{A} = \text{const} \tag{2.5}$$

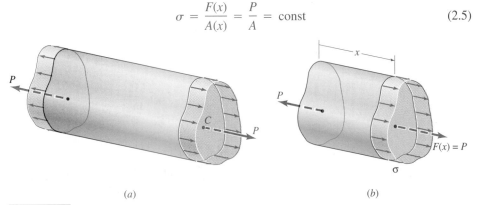

FIGURE 2.3. Uniform stress in an axially loaded bar.

(a) $\qquad\qquad\qquad (b)$

---

[4]This is an application of St. Venant's Principle, which is discussed further in Section 3.3.

EXAMPLE 2.2

Two solid circular rods are welded to a plate at B to form a single rod, as shown in Fig. 1. Consider the 30-kN force at B to be uniformly distributed around the circumference of the collar at B and the 10 kN load at C to be applied at the centroid of the end cross section. Determine the axial stress in each portion of the rod.

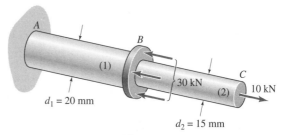

$d_1 = 20$ mm

$d_2 = 15$ mm

Fig. 1

Plan the Solution Since each segment of the rod satisfies the conditions for uniform axial stress, we can use Eq. 2.3 (or Eq. 2.5) to calculate the two required axial stresses. First, however, we need to compute the force in each rod by using an appropriate free-body diagram and equation of equilibrium.

Solution

Free-body Diagrams: First we draw free-body diagrams that expose the rod forces F_1 (or F_{AB}) and F_2 (or F_{BC}). We show F_1 and F_2 positive in tension.

Equations of Equilibrium: From free-body diagram 1 (Fig. 2a),

$$\xrightarrow{+} \left(\sum F \right)_1 = 0: \quad -F_1 - 30 \text{ kN} + 10 \text{ kN} = 0, \quad F_1 = -20 \text{ kN}$$

and, from free-body diagram 2 (Fig. 2b),

$$\xrightarrow{+} \left(\sum F \right)_2 = 0: \quad -F_2 + 10 \text{ kN} = 0, \quad F_2 = 10 \text{ kN}$$

$$A_1 = \frac{\pi}{4} d_1^2 = \frac{\pi}{4} (20 \text{ mm})^2 = 314.2 \text{ mm}^2$$

$$A_2 = \frac{\pi}{4} d_2^2 = \frac{\pi}{4} (15 \text{ mm})^2 = 176.7 \text{ mm}^2$$

(a) Free-body diagram 1.

F_2 (2) 10 kN

C

(b) Free-body diagram 2.

Fig. 2

Axial Stresses: Using Eq. 2.3, we obtain the axial stresses

$$\sigma_1 = \frac{F_1}{A_1} = \frac{-20 \text{ kN}}{314.2 \text{ mm}^2} = -63.7 \frac{\text{MN}}{\text{m}^2}$$

$$\sigma_2 = \frac{F_2}{A_2} = \frac{10 \text{ kN}}{176.7 \text{ mm}^2} = 56.6 \frac{\text{MN}}{\text{m}^2}$$

$$\left. \begin{array}{l} \sigma_1 = -63.7 \text{ MPa (63.7 MPa C)} \\ \sigma_2 = 56.6 \text{ MPa (56.6 MPa T)} \end{array} \right\} \quad \text{Ans.}$$

Review the Solution In this problem we could "mentally" solve the equilibrium problems and "see" that AB is in compression and that BC is in tension.

When a solid body is subjected to external loading and/or temperature changes, it deforms; that is, changes occur in the size and/or the shape of the body. The general term *deformation* includes both changes of lengths and changes of angles. For example, consider the axial deformation of a bar as shown in Fig. 2.4. To illustrate how the bar deforms locally when it is stretched, two squares are drawn on the surface of the undeformed bar (Fig. 2.4*a*); one square is aligned with the axis of the bar, and one square is oriented at 45° to the axis of the bar. Local changes in length, such as that of line segment *BC* when it elongates to become segment *B*C**, are described by *extensional strain*.[5] Local angle changes, like the change of right angle *DEF* to become acute angle *D*E*F** are described by *shear strain*. Both types of strain are important in describing the *geometry of deformation* of deformable bodies. (Shear strain is discussed in Section 2.7.)

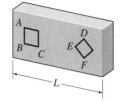

(*a*) The undeformed bar

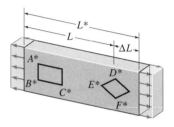

(*b*) The deformed bar

FIGURE 2.4. The deformation of a bar under axial loading.

Definition of Extensional Strain. To define extensional strain, let us consider again the case of axial deformation, as illustrated by Fig. 2.4. The *total elongation* of the bar is designated by ΔL,[6] and the *extensional strain,* or *normal strain,* is designated by the lowercase Greek letter epsilon (ϵ).[7] The *average extensional strain* is defined as the ratio of the *total elongation* ΔL to the original length L, that is,

$$\epsilon_{avg} = \frac{\Delta L}{L} = \frac{L^* - L}{L} \tag{2.6}$$

If the bar stretches (i.e., $L^* > L$), the strain ϵ is positive and is called *tensile strain.* A shortening of the bar results in a negative value for ϵ and is referred to as *compressive strain.* Although strain is a dimensionless quantity, it is common practice to report strain values in units of in./in., or μin./in. (1 microinch per inch = 10^{-6} in./in.), or μm/m. The magnitude of extensional strain is generally quite small, say < 0.001, so the latter two microstrain units are appropriate. Frequently just the symbol μ is used, like 100μ.

For the remainder of this chapter we will consider only the case where the extensional strain is uniform along the length of a member. Thus, the extensional strain is given by

$$\epsilon = \epsilon_{avg} = \frac{\Delta L}{L} \tag{2.7}$$

(Nonuniform extensional strain is discussed in Sections 2.9 and 3.2.)

Strain-Displacement Analysis. Equations 2.6 and 2.7 are definitions that involve only the *geometry of deformation.* They enable us to relate strain quantities to displacement quantities, as will be illustrated in the following example of *strain-displacement analysis.*

---

[5]Asterisk superscripts denote "after deformation has occurred."

[6]The Greek capital letter delta (Δ) is frequently used to designate the change in a quantity, so ΔL is a change in the length L.

[7]In Section 2.4 it is shown that the extensional strain ϵ is related to the normal stress σ. Therefore, *extensional strain* is sometimes called *normal strain.*

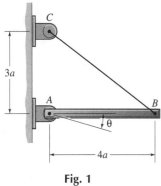

Fig. 1

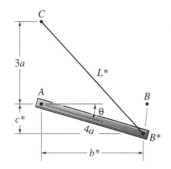

Fig. 2. Deformation diagram.

When the "rigid" beam AB in Fig. 1 is horizontal, the rod BC is strain free. (a) Determine an expression for the average extensional strain in rod BC as a function of the angle θ of clockwise rotation of AB in the range $0 \leq \theta \leq \pi/2$. (b) Determine an approximation for $\epsilon(\theta)$ that gives acceptable accuracy for values of ϵ when $\theta \ll 1$ rad.

Plan the Solution   The defining equation for extensional strain, Eq. 2.6, can be used to determine the required expression for the average extensional strain of rod BC. To determine the geometrical relationship between the extended length L^* and the angle θ, we can draw a sketch of the deformed rod-beam system (i.e., a *deformation diagram*).[8] Since beam AB is assumed to be rigid, end B moves in a circle about end A.

Solution
(a) Obtain an expression for $\epsilon(\theta)$. From Eq. 2.6,

$$\epsilon = \frac{L^* - L}{L}$$

From the original figure,

$$L \equiv \overline{BC} = 5a$$

Deformation Diagram: Rigid beam AB rotates about A, while rod BC stretches and rotates about C, as indicated in Fig. 2.

Geometry of Deformation: Using the Pythagorean theorem and dimensions from the sketch of the deformed configuration in Fig. 2, we get

$$L^* = \sqrt{(3a + c^*)^2 + (b^*)^2}$$

where

$$b^* = 4a \cos \theta, \qquad c^* = 4a \sin \theta$$

Then, from Eq. 2.6,

$$\epsilon = \frac{L^* - L}{L} = \frac{a\sqrt{(3 + 4 \sin \theta)^2 + (4 \cos \theta)^2} - 5a}{5a}$$

This can be simplified to

$$\epsilon(\theta) = \sqrt{1 + \left(\frac{24}{25}\right) \sin \theta} - 1 \qquad \textbf{Ans.} \quad (1)$$

(b) Approximate $\epsilon(\theta)$ for $\theta \ll 1$ rad.[9] If $\theta \ll 1$, then $\sin \theta \doteq \theta$. Furthermore, the term under the radical has the form $1 + \beta$, with $\beta \ll 1$. According to the

---

[8]A deformation diagram is useful in the analysis of the geometry of deformation of a system, analogous to the way a free-body diagram is useful in an equilibrium analysis.

[9]See Appendix A.2 for information regarding approximations of this nature.

binomial expansion theorem, for small values of β

$$\sqrt{1 + \beta} \doteq 1 + \frac{\beta}{2}$$

Thus, for $\theta \ll 1$, $\epsilon(\theta)$ takes the form

$$\epsilon(\theta) = \left(\frac{12}{25}\right) \theta, \quad \theta \ll 1 \text{ rad} \qquad \text{Ans.} \quad (2)$$

Review the Solution In both answers, $\epsilon(\theta)$ is dimensionless, as it should be. If Eq. (1) is evaluated at $\theta = \pi/2$, we get $\epsilon(\pi/2) = 2/5$. Since $L^* = 3a + 4a = 7a$ when B^* is directly below A, this value of $\epsilon(\pi/2)$ is correct.

Note that nothing was said in the preceding example about what caused the deformation illustrated in Fig. 2. In fact, a downward force at B would cause beam AB to rotate clockwise; but heating of rod BC could also be the cause. Thus, Example 2.3 illustrates that we do not need to know what caused the deformation in order to relate the extensional strain ϵ to the rotation angle θ. That is purely a geometry problem!

Small-Displacement/Small-Strain Behavior. In Part (b) of Example 2.3, an expression for the strain $\epsilon(\theta)$ was obtained that is valid for very small values of the angular displacement θ, namely

$$\epsilon(\theta) = \left(\frac{12}{25}\right) \theta, \quad \theta \ll 1 \text{ rad}$$

Two things may be noted about this approximation: (1) the strain is a linear function of the displacement variable θ, and (2) since the displacement angle θ is assumed to be very small, the strain ϵ will also be very small. This small-displacement/small-strain behavior is typical of the normal behavior of engineering structures (e.g., building structures and machines) for which deformations are usually too small to be seen with the naked eye, and for which strains are usually of the order of 0.001 in./in. (i.e., 1000μ) or less.[10]

Thermal Strain. We turn now to the relationship between extensional strain and the change of temperature of a body, that is, to *thermal strain.*[11] Although there are some exceptions, most engineering materials respond to a uniform increase in temperature, ΔT, by *expanding* in all directions by a uniform amount

$$\boxed{\epsilon_t = \alpha \, \Delta T} \qquad (2.8)$$

where ϵ_t is the *thermal strain,* α is the *coefficient of thermal expansion,* and ΔT is the change in temperature. (A positive ΔT corresponds to an *increase* in temperature above the reference temperature.) When the temperature of a body decreases, ΔT is negative, and

---

[10]There are situations where large displacements occur, but the strain remains very small, like the deflection of the tip of a very flexible diving board under the weight of a heavy diver. However, throughout this textbook we will assume that displacements are small.

[11]Although some texts refer to *thermal stress,* a uniform temperature change only results in stresses in a body if the body is restrained, that is, if it is not completely free to expand. Here we will assume that the solids that are heated or cooled are completely free to expand or contract. We will solve for thermally induced stresses in Section 3.6.

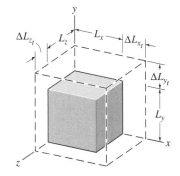

FIGURE 2.5. The free expansion of a uniform block caused by a uniform increase in the temperature of the block.

the body shrinks a corresponding amount. The units of α and ΔT must be consistent. For example, the U.S. Customary units of α are $1/°F$ (the reciprocal of degrees Fahrenheit). The SI units are $1/K$ (the reciprocal of kelvins) or $1/°C$ (the reciprocal of degrees Celcius), depending on the units of ΔT.

Let the block of material in Fig. 2.5 be subjected to a uniform temperature increase ΔT. Since it is free to expand in all directions,

$$\epsilon_{x_t} = \epsilon_{y_t} = \epsilon_{z_t} = \alpha \, \Delta T$$

Since this strain is uniform throughout the block, Eqs. 2.7 and 2.8 can be combined to give the elongations of the block:

$$\Delta L_{x_t} = (\alpha \, \Delta T)L_x, \qquad \Delta L_{y_t} = (\alpha \, \Delta T)L_y, \qquad \Delta L_{z_t} = (\alpha \, \Delta T)L_z \qquad (2.9)$$

The coefficient of thermal expansion is a property of the material; it is determined experimentally by applying a change in temperature and measuring the change in dimensions of the specimen, as in Fig. 2.5. Values for α are given in tables of Mechanical Properties of Engineering Materials, such as those in Appendix F.2.

■■■■■■■■■■■■■■ E X A M P L E   2 . 4 ■■■■■■■■■■■■■■

If the rod BC in Example 2.3 is made of high-strength steel ($\alpha = 8.0 \times 10^{-6}/°F$), and if BC is uniformly heated by $100°F$, through what angle θ will beam AB rotate? Assume that AB and BC are weightless, and that AB is horizontal before the heating of BC occurs.

Plan the Solution   Equation 2.8 relates the temperature change ΔT to the thermal strain, and the answers in Example 2.3 relate strain to θ. Therefore, we can simply combine these.

Solution

$$\epsilon_t = \alpha \, \Delta T = (8.0 \times 10^{-6}/°F)(100°F) = 800 \times 10^{-6} \text{ in./in.}$$

Since the expression in Eq. (2) of Example 2.3 is simpler than the expression in Eq. (1), let us try Eq. (2) first.

$$\theta = \frac{25}{12}\,\epsilon = \frac{25}{12}\,(8.0 \times 10^{-4}) = 1.67 \times 10^{-3}\ \text{rad}$$

or

$$\theta = 0.0955\ \text{deg} \qquad\qquad \textbf{Ans.}$$

Since this answer satisfies the requirement that $\theta \ll 1$ rad, we do not need to resort to Eq. (1).

Review the Solution   Although a tenth of a degree rotation for a temperature increase of $100°F$ may seem to be very small, the formulas are so simple that all we can do to check our result is just to double-check the calculations and the conversion from radians to degrees.

2.4 STRESS-STRAIN DIAGRAMS; MECHANICAL PROPERTIES OF MATERIALS

In order to relate the loads on engineering structures to the deformation produced by the loads, experiments must be performed to determine the *load-deformation behavior* of the materials (e.g., aluminum, steel, and concrete) used in fabricating the structures. Many useful mechanical properties of materials are obtained from tension tests or from compression tests, and these properties are listed in tables like those in Appendix F. This section describes how a tension test is performed and discusses the material properties that are obtained from this type of test.

Tension Tests and Compression Tests. Figure 2.6 shows a computer-controlled, hydraulically actuated testing machine that may be used to apply a tensile load or a compressive load to a test specimen, like the steel tension specimen in Fig. 2.7a or the concrete compression specimen in Fig. 2.7c.[12] Figure 2.7b shows a close-up view of a ceramic tension specimen mounted in special testing-machine grips. Electromechanical *extensometers* are mounted on the specimens in Figs. 2.7a and 2.7b to measure the extension (i.e., the elongation) that occurs over the *gage length* of the test section.[13]

Figure 2.8a illustrates an undeformed tension specimen with two points on the specimen marking the *original gage length, L_0.*[14] An axial load P causes the portion of the specimen between the gage marks to elongate, as indicated in Fig. 2.8b. As the specimen is pulled, the load P is measured by the testing machine and recorded. The extensometer provides a simultaneous measurement of the corresponding length, $L^* \equiv L^*(P)$, of the test section, or else it directly measures the elongation

$$\Delta L = L^* - L_0 \qquad\qquad (2.10)$$

---

[12]Specimen dimensions and procedures for preparing and testing specimens are prescribed by various standards organizations, like the *American Society for Testing Materials* (ASTM) and the *American Concrete Institute* (ACI).

[13]Instead of the extensometer, an electrical resistance strain gage, described in Section 8.12, may be used.

[14]The notation L_0 is used here to emphasize that this is the original gage length, not the total length of the specimen.

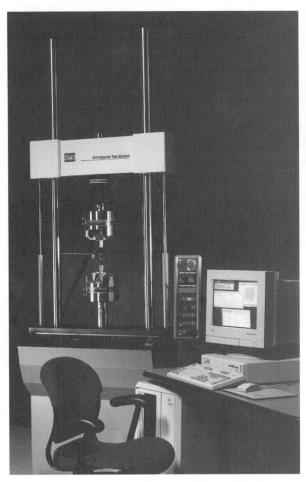

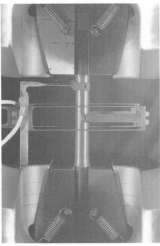

(*a*) A metal tension specimen with extensometer attached.

(*b*) A ceramic tension specimen with extensometer attached. (MTS Systems Corp. photo.)

FIGURE 2.6. A computer-controlled hydraulically actuated testing machine. (MTS Systems Corp. photo.)

(*c*) A concrete cylinder before and after compression testing.

FIGURE 2.7. Tension and compression test specimens.

In a *static tension test* the length of the specimen is increased very slowly, in which case the loading rate need not be measured. In some situations, however, a dynamic test must be performed. Then, the rate of loading must be measured and recorded, since material properties are affected by high rates of loading.

Stress-Strain Diagrams. A plot of stress versus strain is called a *stress-strain diagram,* and from such stress-strain diagrams we can deduce a number of significant *mechanical properties of materials.*[15] The values of normal stress and extensional strain that

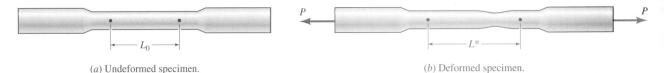

(*a*) Undeformed specimen.

(*b*) Deformed specimen.

FIGURE 2.8. A typical tension-test specimen.

---

[15]These material properties are also called *constitutive properties of materials.*

30

are used in plotting a *conventional stress-strain diagram* are the *engineering stress* (load divided by <u>original</u> cross-sectional area of the test section) and *engineering strain* (elongation divided by <u>original</u> gage length), that is,

$$\sigma = \frac{P}{A_0} \ , \quad \epsilon = \frac{L^* - L_0}{L_0}$$

(2.11)

Mechanical Properties of Materials. Figures 2.9*a* and 2.9*b* are stress-strain diagrams for *structural steel* (also called *mild steel,* or *low-carbon steel*), which is the metal commonly used in fabricating bridges, buildings, automotive and construction vehicles, and many other machines and structures. A number of important mechanical properties of materials that can be deduced from stress-strain diagrams are illustrated in Fig. 2.9. In Fig. 2.9*a* the stress is plotted accurately, but the strain is plotted to a variable scale so

FIGURE 2.9. Stress-strain diagrams for structural steel in tension. (*a*) Strain not plotted to scale. (*b*) Strain plotted to two different scales.

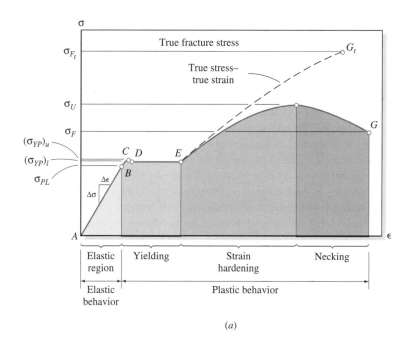

(*a*)

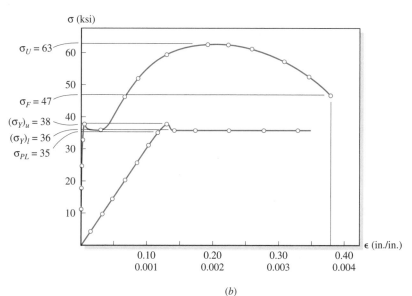

(*b*)

that all important features can be shown and discussed. In Fig. 2.9b, which gives typical numerical values of stress and strain for structural steel, one stress-strain curve, the lower one, is plotted against a strain scale that emphasizes the low-strain region; the upper curve is plotted against a strain scale that emphasizes the high-strain region and puts the entire stress-strain history into perspective.

Starting at the origin A in Fig. 2.9a and continuing to point B, there is a linear relationship between stress and strain. The stress at point B is called the *proportional limit,* σ_{PL}. The ratio of stress to strain in this linear region of the stress-strain diagram is called *Young's modulus,*[16] or the *modulus of elasticity,* and is given by

$$E = \frac{\Delta\sigma}{\Delta\epsilon} \quad , \quad \sigma < \sigma_{PL} \tag{2.12}$$

Typical units for E are ksi or GPa.

At B the specimen begins *yielding,* that is, smaller and smaller increments of load are required to produce a given increment of elongation. The stress at C is called the *upper yield point,* $(\sigma_{YP})_u$, while the stress at D is called the *lower yield point,* $(\sigma_{YP})_l$. The upper yield point has little practical importance, so the lower yield point is usually referred to simply as the *yield point,* σ_{YP}. From D to E the specimen continues to elongate without any increase in stress. The region DE is referred to as the *perfectly plastic* zone. The stress begins to increase at E, and the region from E to F is referred to as the zone of *strain hardening.* The stress at F is called the *ultimate stress,* or *ultimate strength,* σ_U. At F the load begins to drop, and the specimen begins to "neck down." This neck-down behavior continues until, at G, fracture occurs at the *fracture stress,* σ_F. Figure 2.10a, shows a hot-rolled steel specimen at three stages of tensile testing: (1) before testing, (2) as removed from the testing machine at a point between F and G with pronounced reduction in area referred to as *necking* or *neck-down,* and (3) after fracture. Figure 2.10b shows the typical *cup and cone fracture* of a hot-rolled steel tensile specimen.

The *true stress,* σ_{true}, is the load at some instant during the test divided by the actual minimum cross-sectional area of the specimen at that instant. Thus, when a specimen starts to neck down, the true stress is taken as the load divided by the minimum cross-sectional area in the neck-down region. The *true strain,* ϵ_{true}, is the instantaneous change in length of a test section divided by the instantaneous length of that test section. True stress and true strain are given by the formulas[17]

$$\sigma_{\text{true}} = \frac{P}{A_{\min}} \quad , \quad \epsilon_{\text{true}} = \ln\left(\frac{A_0}{A_{\min}}\right) \tag{2.13}$$

The solid-line curve in Fig. 2.9a is a *conventional stress-strain diagram* of engineering stress versus engineering strain, while the dashed curve is a sketch of true stress versus true strain. The curves differ only when strain is large and when the cross-sectional area is decreasing significantly.

Figures 2.9a and 2.9b both illustrate the stress-strain behavior of structural steel. Materials scientists and metallurgists have developed a number of processes for altering the

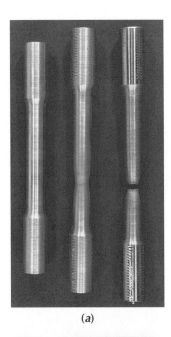

(*a*)

(*b*)

FIGURE 2.10. A hot-rolled steel tensile specimen.

[16]In two volumes entitled *A Course of Lectures on Natural Philosophy and Mechanical Arts,* London, 1807, Thomas Young (1773–1829) introduced the modulus of elasticity and discussed many other interesting topics in mechanics of deformable bodies [Ref. 2-1].

[17]True strain can also be expressed in terms of area change. The formula for true strain in Eq. 2.13 is given by D. C. Drucker in *Introduction to Mechanics of Deformable Solids,* [Ref. 2-2], p. 12.

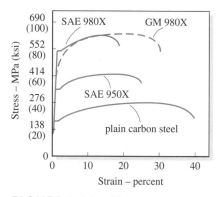

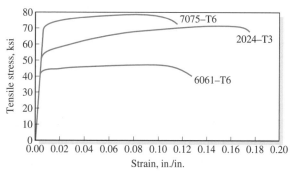

FIGURE 2.11. The stress-strain curves of plain carbon steel and three high-strength/low-alloy steels. (Used with the permission of ASM International.)

FIGURE 2.12. The stress-strain curve of three typical aluminum alloys. (Used with the permission of McGraw-Hill, Inc.)

mechanical properties of metals, including alloying, work-hardening, and tempering. Figure 2.11 contrasts the tensile stress-strain behavior of several ferrous metals. Stress-strain curves for several aluminum alloys are shown in Fig. 2.12.[18]

It is apparent from a comparison of Figs. 2.9 and 2.12 that several properties exhibited by structural steel, for example, a definite yield point followed by a significant zone of yielding at constant stress, are not characteristic of all other materials. For materials like aluminum that have no clearly defined yield point, a stress value called the *offset yield stress,* σ_{YS}, is used in lieu of a yield-point stress. As illustrated in Fig. 2.13, the offset yield stress is determined by first drawing a straight line that best fits the data in the initial (linear) portion of the stress-strain diagram. A second line is then drawn parallel to the original line but offset by a specified amount of strain. The intersection of this second line with the stress-strain curve determines the offset yield stress. In Fig. 2.13 the commonly used offset (strain) value of 0.002 (or 0.2%) is illustrated.

Design Properties. Now that you have some idea of the stress-strain behavior of several common metallic materials, let us note the material properties that are of primary interest to an engineer designing some structure or machine. From the design standpoint the most significant stress-strain properties can be categorized under the three headings—*strength, stiffness, and ductility.*

- *Strength*—There are three strength values of interest.[19] (1) The *yield strength,* σ_Y, is the highest stress that the material can withstand without undergoing significant yielding. The yield-point stress or the offset yield stress, whichever is appropriate for the particular material, is taken as the yield strength (i.e., $\sigma_Y = \sigma_{YP}$ or $\sigma_Y = \sigma_{YS}$ as appropriate). (2) The *ultimate strength,* σ_U, is the maximum value of stress (i.e., the maximum value of engineering stress) that the material can withstand. Finally, (3) the *fracture stress,* σ_F, if different from the ultimate stress, may be of interest. It is the value of the stress at fracture.
- *Stiffness*—The *stiffness* of a material is basically the ratio of stress to strain.

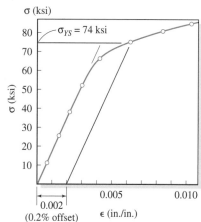

FIGURE 2.13. The procedure for determining the offset yield stress.

---

[18]ASM International, Materials Park, OH, 44073-0002, is an excellent sources for information on the mechanical properties of materials. ASM International publishes the *Metals Handbook* and many other reference works.

[19]The word *strength* is used to designate various critical stress quantities exhibited by materials.

33

Stiffness is of interest primarily in the linearly elastic region; therefore, Young's modulus, E, is the value used to represent the stiffness of a material.[20]

• *Ductility*—Materials that can undergo large strain before fracture are classified as *ductile materials*; those that fail at small values of strain are classified as *brittle materials*.[21]

Ductility. The two commonly used measures of ductility are the *percent elongation* (the final elongation expressed as a percentage of the original gage length) and the *percent reduction in area* at the section where fracture occurs (the area reduction expressed as a percentage of the original area). The *percent elongation* is given by the formula

$$\text{Percent elongation} = \left(\frac{L_F - L_0}{L_0} \right) (100\%)$$

where L_0 is the original gage length and L_F is the length of the gage section at fracture. When percent elongation is stated, the gage length should also be stated, since the elongation at fracture is not uniform over the entire gage length but is concentrated in the necked-down region. The *percent reduction in area* is given by the formula

$$\text{Percent reduction in area} = \left(\frac{A_0 - A_F}{A_0} \right) (100\%)$$

where A_0 is the original cross-sectional area of the test section, and A_F is the final cross-sectional area at the fracture location.

Structural steel is a highly ductile material with a percent elongation of about 30% and a percent reduction in area of about 50%. High ductility permits this type of steel to be pressed to form automobile sheet-metal parts and bent to form concrete-reinforcing bars. Ductile materials permit large local deformation to occur near cracks, rivet holes, and other stress concentrations, thereby preventing the occurrence of sudden, catastrophic failures. Other materials that may be classified as ductile include pure aluminum and some of its alloys, brass, copper, nickel, nylon, and teflon. Figure 2.14 shows the 1100% elongation of a *superplastically deformed* alloy-steel tensile specimen.

As can be seen in Fig. 2.11, high-strength steels generally tend to be far less ductile than mild steel; the dashed-line stress-strain curve labeled ''GM 980X'' is for a special high-strength, high-ductility steel. Although there is no sharp dividing line between ductile materials and brittle materials, glass, other ceramics, gray cast iron, and concrete are among the materials that are classified as brittle materials. Figure 2.15a shows a typical stress-strain curve for a brittle material, with the curve for a ductile material shown for comparison. The fracture surface of a brittle tensile-test specimen is illustrated in Fig. 2.15b. Note that brittle fracture is a direct result of the tensile stress on the cross section, and contrast this to the fracture of a ductile specimen (Fig. 2.10b). Ordinary glass is a nearly ideal brittle material, exhibiting linear stress-strain behavior up to a fracture stress in the neighborhood of 10^4 psi. It has been found that the strength of glass is highly dependent

---

[20]In Chapter 3 we will define *stiffness of a member*, in contrast to the *stiffness of a material*.

[21]Strictly speaking, the terms ductile and brittle refer to *modes of fracture*, and a material like structural steel, which behaves in a ductile manner at room temperature, may exhibit brittle behavior at very low temperatures. Therefore, when we speak of a ''brittle material'' or a ''ductile material,'' we are referring to the normal (room temperature) behavior of the material.

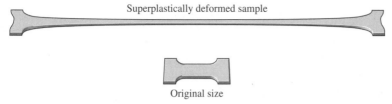

Superplastically deformed sample

Original size

A steel-alloy specimen tested to over 1100% elongation.

FIGURE 2.14. A superplastically deformed steel-alloy specimen tested to over 1100% elongation. (Courtesy of Prof. Oleg D. Sherby, Stanford University.)

on the type of glass, the size of the specimen, and surface defects. *Glass fibers* with diameters on the order of 10^{-4} in. may have an ultimate strength as high as 10^6 psi or more.

Compression Tests.

In *compression,* a ductile metal like steel or aluminum yields at a stress magnitude approximately equal to its tensile yield stress, and Young's modulus in compression is equal to the tensile modulus. Therefore, compression tests are seldom performed on these materials. By contrast, periodic casting and testing of concrete compression specimens, like the one illustrated in Fig. 2.7c, is required for quality control on reinforced-concrete construction projects. The compressive strength of concrete increases with age, as illustrated by the compression stress-strain diagrams for Portland-cement concrete specimens, as shown in Figure 2.16. Concrete is a brittle material with very little tensile strength, so it is usually just assumed that the tensile strength of plain (i.e., unreinforced) concrete is zero.

Plastics and Composites.

Since the introduction of Bakelite in 1906, hundreds of polymeric materials, called *plastics,* have been developed, and many of these, like nylon, have found structural application. Their advantages, which include their light weight, re-

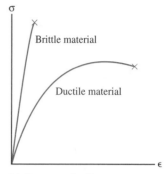

(a) Stress-strain diagrams.

(b) Brittle-fracture surface.
(MTS Systems Corp. photo.)

FIGURE 2.15. Tensile-test behavior of a brittle material.

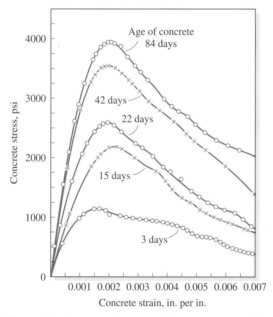

FIGURE 2.16. Stress-strain diagrams for concrete in compression. (From [Ref. 2-3], Bureau of Reclamation.)

FIGURE 2.17. A graphite-epoxy tensile test specimen; $\pm 45°$ graphite-fiber lamina.

sistance to corrosion, ease of molding, and good electrical insulation properties, have made them increasingly popular. On the other hand, their low stiffness, tendency to creep (i.e., to continue to deform under constant load) and to absorb moisture, and the strong dependence of their strength and stiffness properties on temperature, are disadvantages that must be carefully weighed against their advantages.

Composite materials are materials that combine two constituent materials in a manner that leads to improved mechanical properties. Reinforced concrete can be considered to be a composite material, in which the steel reinforcement is used to overcome the inherent weakness of concrete in tension. It was previously noted that glass can be made extremely strong when drawn into thin fibers and properly protected against surface damage. It was also noted that some plastics have several undesirable properties, including low stiffness. A material that is stronger and stiffer than the plastic matrix material can be produced by embedding glass fibers or other reinforcing fibers in the matrix. Furthermore, by placing the fibers in specific orientations in the matrix, a material with direction-dependent properties can be fabricated. Figure 2.17 shows the $\pm 45°$ lay-up of graphite-epoxy layers to form a tensile test specimen and the resulting tensile-test failure. In sports, composite materials are being used in many applications, including tennis racquets, bicycle frames, skis, and boat hulls.[22]

Tables of *Mechanical Properties of Selected Engineering Materials* are provided in Appendix F. Because of the extremely large number of commercially available plastics, these tables include only a small sample of the properties of engineering plastics.

2.5 ELASTICITY AND PLASTICITY; TEMPERATURE EFFECTS ■■■■■■■■■■■■

In the previous section we considered the stress-strain behavior of tension specimens under *loading*. Specifically, the tensile tests were assumed to be performed in a relatively short time with monotonically increasing tensile strain. We consider now what happens when the loading is reversed, that is, when the strain is allowed to decrease.

Elastic Behavior and Plastic Behavior. Consider the loading and unloading behavior of a material, as illustrated in Fig. 2.18. The heavier curves in Fig. 2.18*a* and Fig. 2.18*b* are the loading curves, that is, the curve that would be followed for monotonically increasing initial loading. The stress-strain behavior of a material is said to be *elastic* if the unloading path retraces the loading path. The stress at C is called the *elastic limit,* σ_{EL}. Unloading from a point below point C (e.g., point B) retraces the loading path indicated in Fig. 2.18*a*, but unloading from a point beyond C (e.g., point D) follows a path that is different than the loading path. Unloading from a point on the stress-strain curve beyond the elastic limit typically follows a straight-line path whose slope is parallel to the tangent to the stress-strain curve at the origin. The strain that remains at point E when the stress returns to zero is called the *permanent set,* or *residual strain,* as illustrated in Fig. 2.18*b*.

For structural steel the elastic limit is very close to the proportional limit and also to the yield point. Therefore, for structural steel the proportional limit stress and the elastic limit stress are usually assumed to be equal to the yield-point stress. It is sometimes convenient to approximate the stress-strain behavior of mild steel and similar materials by

---

[22]An excellent source for information on engineering composites is the *International Encyclopedia of Composites*. [Ref. 2-4]

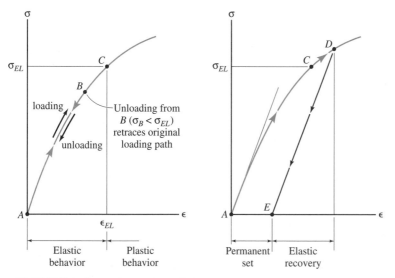

FIGURE 2.18. Illustrations of elastic and plastic stress-strain behavior.

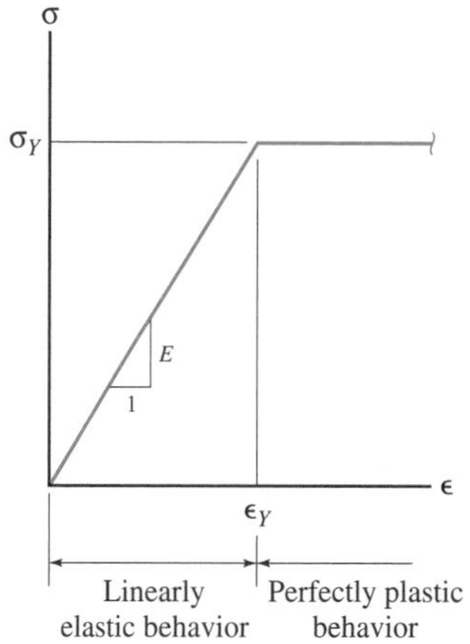

FIGURE 2.19. Linearly elastic, perfectly plastic material model.

the *linearly elastic, perfectly plastic* representation of Fig. 2.19. In Sections 3.9, 4.9, and 6.7, we will use this model of stress-strain behavior in describing the inelastic behavior of members undergoing axial deformation, torsion, and bending, respectively.

Time-dependent Stress-Strain Behavior. Stress-dependent plastic deformation such as the yielding of structural steel described above, is referred to as *slip*. It is essentially an instantaneous process resulting from the slip that occurs within crystals and along the boundaries between crystals that make up the solid.[23] Depending on the material and its temperature, a time-dependent plastic deformation may occur. One form of time-dependent deformation is called *creep*. Creep behavior may be demonstrated by the *constant-stress experiment* illustrated in Fig. 2.20a. A constant load is placed on a specimen, and the

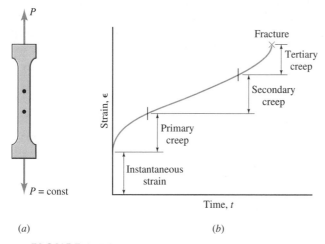

(a)   (b)

FIGURE 2.20. A constant-stress creep experiment.

---

[23]The slip that occurs during yielding will be discussed further in Section 2.8.

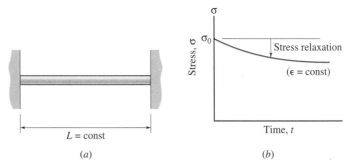

FIGURE 2.21. A constant-strain stress-relaxation experiment.

elongation (or strain) of the specimen is then plotted versus time, as illustrated in Fig. 2.20b. Creep elongation is negligible in many materials at room temperature. However, when heated sufficiently, most materials will exhibit creep behavior. Hence, creep is a significant design consideration in high-temperature applications like turbine engines, boilers, and so on. Since plastics exhibit creep behavior at much lower temperatures, they may be unsuitable for certain applications even though they possess adequate static strength and stiffness at room temperature.

A related form of time-dependent stress-strain behavior, called *stress relaxation,* can be demonstrated by a *constant-strain experiment.* Figure 2.21a shows a specimen that has been stretched to give it an initial tensile stress σ_0 and then attached to immovable supports, and Fig. 2.21b is a graph of the stress in the specimen versus time. This *stress relaxation* at constant strain is a second manifestation of time-dependent stress-strain behavior.

Temperature Effects on Material Properties. The coefficient of thermal expansion, discussed in Section 2.3, is a measure of the direct effect that temperature change has on the deformation of a body. However, temperature also has a significant effect on material properties like yield strength and modulus of elasticity. Materials developed specifically to perform at very high temperatures are required for applications like turbine blades in aircraft engines or the structure of a hypersonic aircraft. Figure 2.22 illustrates the profound influence that temperature has on the strength and ductility properties of a particular stainless steel. Note that strength decreases with increasing temperature and vice versa. As indicated by the strain at fracture, ductility decreases with decreasing temperature

FIGURE 2.22. The influence of temperature on the stress-strain behavior of type 304 stainless steel. (Used with the permission of ASM International.)

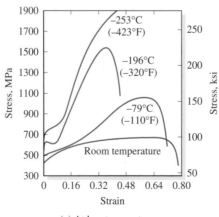

(*a*) At low temperatures.

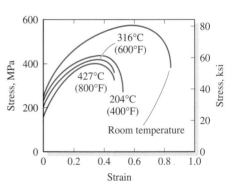

(*b*) At elevated temperatures.

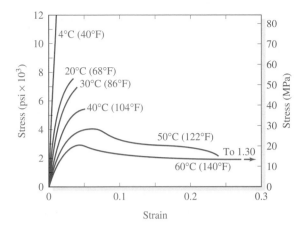

FIGURE 2.23. The influence of temperature on the stress-strain behavior of polymethyl methacrylate. (From [Ref. 2-5]. Reprinted with the permission of the American Society for Testing Materials.)

and increases with increasing temperature. Temperature also affects the stiffness of a material (i.e., the slope of the initial portion of the stress-strain curve), as can be clearly seen in Fig. 2.23.

2.6 LINEAR ELASTICITY; HOOKE'S LAW AND POISSON'S RATIO

In order to keep deformations small and stresses at safe levels, most structures and machine parts are designed so that stresses remain well below the yield stress. Fortunately, most engineering materials exhibit a linear stress-strain behavior at these lower stress levels. Let us, for the present, restrict our discussion to the case of uniaxial stress applied to a homogeneous (same properties throughout), isotropic (same properties in every direction) member oriented along the x axis (Fig. 2.24). The linear relationship between stress and strain, given by Eq. 2.12, applies for $0 \le \sigma \le \sigma_Y$. Therefore,[24]

$$\boxed{\sigma_x = E\epsilon_x} \tag{2.14}$$

This equation is called *Hooke's Law*.[25]

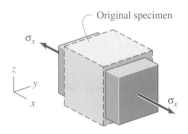

FIGURE 2.24. The deformation (much exaggerated) of a homogeneous, isotropic specimen under uniaxial stress.

---

[24]Subscripts on σ and ϵ identify the axis of the particular stress and strain.

[25]Robert Hooke (1635–1703) was, for several years, curator of experiments of the Royal Society (London). In a paper published in 1678, Hooke discussed his experiments with elastic bodies, describing the relationship between the force applied to wires of various lengths and the elongation of the wires. This is the first published paper in which the elastic properties of materials are discussed. The linear relationship between force and deformation, called *Hooke's Law*, became the foundation upon which further development of the mechanics of elastic bodies was built. [Ref. 2-1]

Hooke's Law is valid for uniaxial tension or compression within the linear portion of the stress-strain diagram. As noted earlier, E is called the *modulus of elasticity,* or *Young's modulus.* It has the units of stress, typically ksi (or psi) in U.S. Customary units, and GPa (or MPa) in SI units. Representative approximate values of E are: 30×10^3 ksi (200 GPa) for steel, 10×10^3 ksi (70 GPa) for aluminum, and 300 ksi (2 GPa) for nylon.[26]

Associated with the elongation of a member in axial tension, there is a transverse contraction, which is illustrated in Fig. 2.24. The transverse contraction during a tensile test is related to the longitudinal elongation by

$$\epsilon_{\text{transv.}} = -\nu\epsilon_{\text{longit.}} \tag{2.15}$$

where ν (Greek symbol nu) is *Poisson's ratio.*[27] This expression also holds when the longitudinal strain is compressive; then, the lateral strain results in an expansion of the transverse dimensions. Poisson's ratio is dimensionless, with typical values in the 0.25–0.35 range. For the orientation of axes in Fig. 2.24 the transverse strains are related to the longitudinal strain by

$$\epsilon_y = \epsilon_z = -\nu\epsilon_x \tag{2.16}$$

Equations 2.14 and 2.16 apply to the simple case of uniaxial stress. More general cases of linearly elastic behavior are treated in Section 2.10.

---

[26]Values of the modulus of elasticity for selected materials are listed in Appendix F.2.

[27]In 1829 and in 1831 S. D. Poisson (1781–1840), a French mathematician and physicist, wrote important memoirs on the mechanics of solids. He noted that, for simple tension of a prismatic bar, the axial elongation ϵ is accompanied by a lateral contraction of magnitude $\nu\epsilon$. [Ref. 2-1]

■■■■■■■■■■■■■■□ E X A M P L E   2 . 5 ■□□□□□□□□□□□□□□■■

The cylindrical rod in Fig. 1 is made of steel with $E = 30 \times 10^3$ ksi, $\nu = 0.3$, and $\sigma_Y = 50$ ksi. If the initial length of the rod is $L = 4$ ft and its original diameter is $d = 1$ in., what is the change in length, ΔL, and the change in diameter, Δd, due to the application of an axial load $P = 10$ kips?

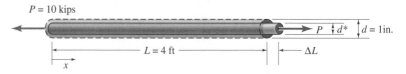

Fig. 1

Plan the Solution From the load P and the cross-sectional area A we can determine the stress σ. Then, if $\sigma \leq \sigma_Y$, we can use Hooke's Law to relate the stress σ to the strain ϵ. Then we can relate the uniform strain, ϵ, to the elongation ΔL. The change in diameter is due to the Poisson's-ratio effect.

Solution

Equilibrium: From the free-body diagram in Fig. 2,

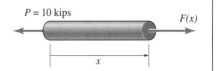

$$\sum F_x = 0: \qquad\qquad F(x) = P = \text{const}$$

Fig. 2. Free-body diagram.

From Eq. 2.5,

$$\sigma_x = \frac{P}{A} = \frac{10 \text{ kips}}{\pi(0.5 \text{ in.})^2} = 12.73 \text{ ksi} < 50 \text{ ksi} \qquad (1)$$

Therefore, linearly elastic behavior occurs when the load $P = 10$ kips is applied to the bar.

Material Behavior: From Eq. 2.14,

$$\epsilon_x = \frac{\sigma_x}{E} = \frac{P}{AE} \qquad (2a)$$

and, from Eq. 2.15,

$$\epsilon_{\text{radial}} = -\nu\epsilon_x = -\frac{\nu P}{AE} \qquad (2b)$$

Strain-Displacement: From Eq. 2.7,

$$\Delta L = L\epsilon_x = \frac{PL}{AE} \qquad (3a)$$

and

$$\Delta d = d\epsilon_{\text{radial}} = \frac{-\nu P d}{AE} \qquad (3b)$$

Substituting numerical values into Eqs. (3), we get

$$\Delta L = L\epsilon_x = \frac{(10 \text{ kips})(48 \text{ in.})}{\pi(0.5 \text{ in.})^2(30 \times 10^3 \text{ ksi})} = 20.4(10^{-3}) \text{ in.} \quad \textbf{Ans.} \quad (4a)$$

and

$$\Delta d = d\epsilon_{\text{radial}} = -\frac{0.3(10 \text{ kips})(1 \text{ in.})}{\pi(0.5 \text{ in.})^2(30 \times 10^3 \text{ ksi})} \quad \textbf{Ans.} \quad (4b)$$

$$= -127(10^{-6}) \text{ in.}$$

Review the Solution The changes in length and diameter are quite small in comparison with the original length and the original diameter, respectively, as they should be. Δd should definitely be smaller than ΔL, and the signs should be different, which is the case. The extensometer in Fig. 2.7a measures both ΔL and Δd.

41

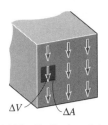

(a) The distribution of shear force on a sectioning plane.

(b) The resultant shear force on the sectioning plane.

FIGURE 2.25. Shear force on a sectioning plane.

In the preceding sections of Chapter 2, you were introduced to stress and strain through a discussion of normal stress and extensional strain. We turn now to a discussion of *shear stress* and *shear strain,* which are used, respectively, to quantify the distribution of force acting tangent to a surface and the angle change produced by tangential forces.

Definition of Shear Stress. Referring to Fig. 2.25, we define the *shear stress at a point* by the equation[28]

$$\tau = \lim_{\Delta A \to 0} \left(\frac{\Delta V}{\Delta A} \right) \tag{2.17}$$

where ΔV is the tangential (shear) force acting on an infinitesimal area ΔA at the point where the shear stress is to be determined. As in the case of normal stress, σ, the units of shear stress are force/area; hence, usually psi or ksi in the USCS units, and kPa or MPa in the SI system.

The *resultant shear force,* shown in Fig. 2.25b, is obtained by summing the ΔV's over the cross section, giving

$$\sum F: \qquad V = \int_A \tau \, dA \tag{2.18}$$

Equations 2.17 and 2.18 will be used in Chapters 4 and 6, where we will determine the shear stress distribution for torsion of circular rods and bending of beams, respectively.

Average Shear Stress; Direct Shear. Even when the exact shear stress distribution on a surface cannot be readily determined, it is sometimes useful to calculate the *average shear stress* on the surface. This is given by

$$\tau_{\text{avg}} = \frac{V}{A_s} \tag{2.19}$$

where V is the total shear force on area A_s.

The average shear stress can be readily calculated in the case of *direct shear,* examples of which are shear in lap splices and shear in bolts, pins, and rivets.[29] Consider the *lap joint,* or lap splice, in Fig. 2.26a, where two rectangular bars are glued together to form a tension member. (Assume that, because the bars are very thin, the moment Pt caused by misalignment of the tensile forces, P, may be neglected.) Applying $\Sigma F = 0$ to the free-body diagram in Fig. 2.26b, we get $V = P$. The area on which the shear V acts is $A_s = L_s w$. Therefore, from Eq. 2.21, the average shear stress on the splice area is

$$\tau_{\text{avg}} = \frac{V}{A_s} = \frac{P}{L_s w}$$

---

[28]Here we assume that all ΔV's act in the same direction on the sectioning plane. More general cases are treated in Section 2.9.

[29]The actual shear stress distribution in these situations is quite nonuniform and would be very difficult to determine. Therefore, the average shear stress value is calculated, and an allowance is made, through factors of safety (see Section 2.12), for the approximate nature of the calculated average stress.

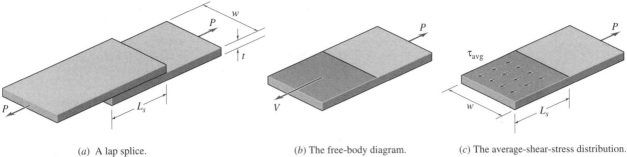

(a) A lap splice.   (b) The free-body diagram.   (c) The average-shear-stress distribution.

FIGURE 2.26. An illustration of direct shear—a lap splice.

In the following two examples of direct shear, note that in order to determine τ_{avg} we must first determine what area has shear stress acting on it, and then, using a free-body diagram, we must determine the value of the shear force acting on this area.

■■■■■■■■■■■■■■■■□□□ **EXAMPLE 2.6** □□□□□□□□□□□□□□□■■■

As illustrated in Fig. 1, a metal punch (similar in principle to a paper punch) is used to punch holes in thin sheet steel that will be used to make a metal cabinet. To punch a 0.25-in.-diameter disk, or "slug," out of the sheet metal that is 0.040 in. thick requires a punch force of $P = 1000$ lb. Determine the average shear stress in the sheet metal resulting from the punching operation.

Plan the Solution The punch rod exerts a compressive normal force on the top of the metal disk, and this force is resisted by a shear force on the cylindrical (vertical) surface of the disk, since there is no force on the bottom.

Solution

Equilibrium: A free-body diagram of the disk during the punching operation is needed to determine the shear force V. This is shown in Fig. 2.

$$\sum F = 0: \qquad V = P = 1000 \text{ lb}$$

Average Shear Stress: The equation for the average shear stress is (Eq. 2.19)

$$\tau_{avg} = \frac{V}{A_s}$$

The cylindrical area on which V acts is $A_s = \pi t d$. Therefore,

$$\tau_{avg} = \frac{V}{A_s} = \frac{1000 \text{ lb}}{\pi(0.040 \text{ in.})(0.25 \text{ in.})} = 31.8 \text{ ksi} \qquad \textbf{Ans.}$$

Review the Solution Since the calculations are so simple, the answer should be checked by carefully checking all calculations. The answer, 31.8 ksi, is a reasonable value of shear stress at "failure," that is, when the slug is punched out from the sheet metal.

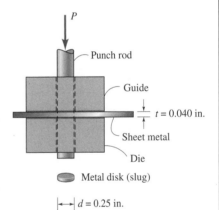

Fig. 1. A sheet-metal punch.

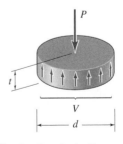

Fig. 2. Free-body diagram.

43

Although the transfer of load from one member to another through a riveted or bolted joint is not easily analyzed when there are several bolts or rivets at the joint, the case of load transfer through a single bolt or a single pin can be treated as direct shear. The following example illustrates the calculation of shear stress in *single-shear* and *double-shear* joints.[30]

---

[30]The book *Structures, or Why Things Don't Fall Down,* by J. E. Gordon [Ref. 2-6] contains an interesting discussion of joints.

■■■■■■■■■■■■■■■□□ E X A M P L E 2 . 7 □□■■■■■■■■■■■■■■■

The special transition link shown in Fig. 1 is used to connect a rectangular bar on the left to a circular rod on the right. At A, a bolted lap joint connects the rectangular bar to the link. At B, a pin passes through the yoke end of the link and through an eye at the end of the rod. The cross-sectional area of the bolt is A_b and of the pin is A_p. If the axial load is P, determine the average shear stress in the bolt on surface S_1, and the average shear stress in the pin on surfaces S_2 and S_3.

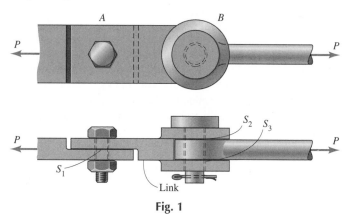

Fig. 1

Plan the Solution   By drawing appropriate free-body diagrams, we can determine the shear force transmitted across S_1 from the rectangular bar on the left to the bolt, and the shear forces transmitted across surfaces S_2 and S_3 from the rod on the right to the pin. Then we can apply Eq. (2.19), the formula for average shear stress.

Solution

Free-body Diagrams and Equilibrium Equations: We pass cutting planes forming S_1, S_2, and S_3 as shown on the free-body diagrams in Fig. 2, and write the equation of equilibrium for each diagram.

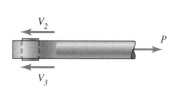

Fig. 2. Free-body diagrams.

$$\left(\sum F \right)_{\text{bar}} = 0: \qquad V_1 = P$$

$$\left(\sum F \right)_{\text{rod}} = 0: \qquad V_2 + V_3 = P$$

Because of symmetry,

$$V_2 = V_3 = \frac{P}{2}$$

Therefore, for the bolt, the average shear stress is

$$(\tau_b)_{\text{avg}} = \frac{V_1}{A_b} = \frac{P}{A_b} \qquad \text{Ans.}$$

and, for the pin the average shear stress is

$$(\tau_p)_{\text{avg}} = \frac{V_2}{A_p} = \frac{V_3}{A_p} = \frac{P}{2A_p} \qquad \text{Ans.}$$

Review the Solution   The above answers account for the fact that a single-shear surface, S_1, transmits the force P into the transition link, but that there are two shear surfaces to transmit the force from the transition link to the rod. The bolt at A is said to be in *single shear,* and the pin at B is said to be undergoing *double shear.*

Equilibrium Requirements; Pure Shear. In our study of shear stress, we have, so far, only considered the average shear stress on a particular surface. Before proceeding further to define shear strain and to discuss the distribution of shear stress under various loading conditions, we must examine the equilibrium requirements that must be satisfied by shear stresses.

Suppose that we have a rectangular parallelepiped of deformable material bonded to two ''rigid'' plates, as shown in Fig. 2.27a and Fig. 2.27b. The horizontal load P in Fig. 2.27b definitely produces shear stress on horizontal planes, as indicated at one point on section D—D in Fig. 2.27c. Let us examine the stresses that act on the six faces of the elemental volume highlighted in Fig. 2.27c and isolated in Fig. 2.27d.[31] First, there is no stress, normal stress or shear stress, on the front $(+z)$ face or the back $(-z)$ face. The shear force acting to the right on the top face has been labeled ΔV_P. Let the average shear stress on the top face be τ_P. Then,

$$\Delta V_P = \tau_P \, \Delta A_y = \tau_P \, (t \, \Delta x)$$

To satisfy the force-equilibrium equation, $\Sigma F_x = 0$, an equal force, ΔV_P, must act to the left on the bottom $(-y)$ face. Since the top face and bottom face have the same area, the average shear stress on the bottom face is also τ_P.

FIGURE 2.27.
Equilibrium requirements for shear stresses.

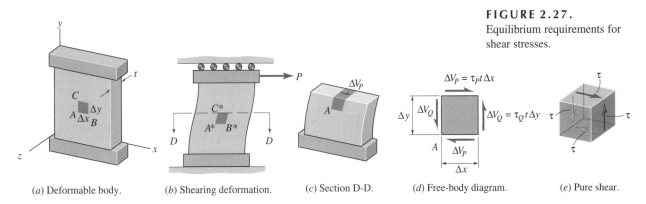

(a) Deformable body.   (b) Shearing deformation.   (c) Section D-D.   (d) Free-body diagram.   (e) Pure shear.

---

[31]It is reasonable to assume that, because the element in Fig. 2.27d is from near the center of the body, the normal stresses on its x and y faces are zero.

If the only forces on the free-body diagram in Fig. 2.27*d* were the ΔV_P forces on the top and bottom faces, we would be unable to satisfy moment equilibrium about an axis in the z direction, for example, about corner A $((\Sigma M_z)_A = 0)$. Therefore, in addition to the horizontal forces ΔV_P, there must also be equal and opposite vertical forces, labeled ΔV_Q, on the free body. Let

$$\Delta V_Q = \tau_Q \, \Delta A_x = \tau_Q \, (t \, \Delta y)$$

So, to satisfy moment equilibrium, we get

$$\left(\sum M_z \right)_A = 0: \qquad (\tau_P \, t \, \Delta x) \, \Delta y - (\tau_Q \, t \, \Delta y) \, \Delta x = 0$$

or

$$\tau_P = \tau_Q \equiv \tau$$

The results of the above equilibrium analysis are summarized in Fig. 2.27*e*, which illustrates an elemental volume in *pure shear*. We can conclude, therefore, that the shear stresses on an element in pure shear satisfy the following statements:

- The shear stresses on parallel faces are equal in magnitude and opposite in sense.
- On adjacent faces at right angles to each other, the shear stresses are equal in magnitude, and they must both point either toward the intersection of the faces or away from it (i.e., the shear-stress arrows on an element must be "head-to-head" and "tail-to-tail").

These conclusions about shear stresses are valid even if normal stresses as well as shear stresses act on the faces of the element (see Section 2.9).

Shear Strain. Referring to Fig. 2.28 let us now consider the shear strain that is associated with shear stress. As a result of the shear stress τ, the original right angle at A becomes an acute angle θ^*. The *shear strain* γ (lowercase Greek letter gamma) at A is defined as the *change in angle* between two originally perpendicular line segments that intersect at A. Thus,

$$\gamma = \frac{\pi}{2} - \theta^* \qquad (2.20)$$

where $\pi/2$ is the angle at A before deformation, and θ^* is the angle at A after deformation. The corresponding shear stresses point toward the two corners where the original right angle is *decreased* by γ, and they point away from the two corners where the angle is *increased* by γ. Although γ is dimensionless, it is frequently stated in the same "dimensionless units" as extensional strain, that is, in./in., and so on, or, since shear strain is an angle, it may be stated in radians. Since shear strains, like extensional strains, are usually very small in magnitude, we can use the small-angle approximations $\tan(\gamma) \doteq \gamma$ and $\sin(\gamma) \doteq \gamma$. Then, γ can be computed using the formula

$$\gamma = \frac{\pi}{2} - \theta^* \doteq \tan\left(\frac{\pi}{2} - \theta^* \right) = \frac{\delta_s}{L_s} \qquad (2.21)$$

where δ_s and L_s are defined in Fig. 2.28*b*.

(*a*) Original
(undeformed) element.

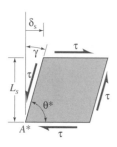

(*b*) Pure shear
deformation.

FIGURE 2.28. Illustrations for a definition of shear strain.

Material Properties in Shear. Material properties relating to shear, like those for normal-stress-extensional-strain behavior, must be determined experimentally. The material properties in shear, such as yield stress in shear, shear modulus of elasticity, and so on, may be obtained from a torsion test, which will be discussed later in Section 4.5. For example, linearly elastic behavior in shear is described by *Hooke's Law for shear,* which is

$$\tau = G\gamma \tag{2.22}$$

The constant of proportionality, G, is called the *shear modulus of elasticity,* or, simply, the *shear modulus.* Like E, the shear modulus G is usually expressed in units of ksi or GPa. The shear properties are closely related to the extensional properties through equations of equilibrium and geometry of deformation. For example, it is shown in Section 2.11 that G, E, and ν are related through the equation

$$G = \frac{E}{2(1 + \nu)} \tag{2.23}$$

2.8 STRESSES ON AN INCLINED PLANE IN AN AXIALLY LOADED MEMBER

Figure 2.4, illustrating the deformation of an axially loaded bar, clearly shows that there are shear strains in the bar. For example, the right angle *DEF* in Fig. 2.4*a* becomes the acute angle *D*E*F** in Fig. 2.4*b*. Let us now consider how the normal stress and the shear stress on an oblique plane, such as the plane whose normal direction is labeled n in Fig. 2.29, are related to the axial stress, $\sigma_x = P/A$.

To satisfy equilibrium of the free body in Fig. 2.29*b*, we get

$$+\nearrow \sum F_n = 0: \qquad\qquad N = P \cos \theta$$

$$\searrow \sum F_t = 0: \qquad\qquad V = P \sin \theta$$

where θ is the angle measured *positive counterclockwise* from the x axis (normal to the cross section) to the n axis (normal to the oblique plane). The area of the oblique plane is related to the cross-sectional area of the bar, A, by

$$A_n = \frac{A}{\cos \theta}$$

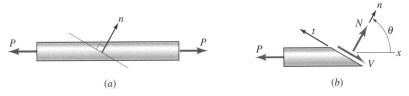

FIGURE 2.29. The force resultants on an oblique section through an axial tension member.

The average normal stress, σ_n, and the average shear stress, τ_{nt}, on the oblique plane are obtained by dividing N and $-V$, respectively, by the area A_n on which they act:

$$\sigma_n = \frac{N}{A_n} = \sigma_x \cos^2 \theta$$

$$\tau_{nt} = \frac{-V}{A_n} = -\sigma_x \cos \theta \sin \theta$$

(2.24)

where the subscript n in σ_n and τ_{nt} designates the outward normal to the face on which these stresses act. The subscript t identifies τ_{nt} as ''the shear stress acting in the $+t$ direction on the n face.'' (Shear stress τ_{nt} is negative in Eq. 2.24b, since shear force V acts in the $-t$ direction on the n face.)

Using trigonometric identities, we can write σ_n and τ_{nt} as functions of the double-angle 2θ. The resulting equations:

$$\sigma_n = (\sigma_x/2)(1 + \cos 2\theta)$$
$$\tau_{nt} = -(\sigma_x/2) \sin 2\theta$$

(2.25)

show that σ_n and τ_{nt} are periodic in θ with a period of 180°. These expressions are plotted in Fig. 2.30. For example, consider an element rotated at $\theta = 45°$ with respect to the x axis, as shown in Fig. 2.31. The subscripts 1 and 2 refer to the points designated 1 and 2 on Fig. 2.30 and to faces designated n_1 and n_2 in Fig. 2.31. The shear stresses on the 45°-rotated element in Fig. 2.31 account for the shear deformation of the 45°-rotated element at point E in Fig. 2.4.

Figure 2.31 illustrates the fact that shear stresses of magnitude $\sigma_x/2$ occur on the planes oriented at 45° to the axis of a member undergoing axial deformation with axial stress σ_x. One evidence of this shear can be obtained by performing a tensile test of a low-carbon steel bar with polished surfaces. When the bar is loaded to the yield point, *slip bands* can be observed to form at approximately 45° to the axis of the bar. These slip bands are called *Lüders' bands* or *Piobert's bands*.[32] Mild steel and other highly ductile materials exhibit this type of failure in shear, rather than a direct tensile failure.

Chapter 8 treats more general cases in which equilibrium is again used to relate the stresses on various planes to each other.

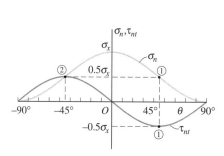

FIGURE 2.30. The normal stress and shear stress on arbitrary oblique planes.

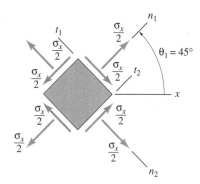

FIGURE 2.31. The stresses on a 45°-rotated element.

---

[32]Such slip bands were first observed by G. Piobert in 1842 and then by W. Lüders in 1860.

Several short pieces of timber are to be glued together end-to-end to form a single longer piece of timber. Figure 1 shows a simple diagonal splice joint. The glue that is to be used in the splice joint is 50% stronger in shear than in tension. Is it possible to take advantage of this higher shear strength by selecting a splice angle θ such that the magnitude of the average shear stress on the joint is 50% higher than the average normal stress? If so, what is the appropriate angle?

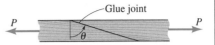

Fig. 1. A diagonal splice joint.

Plan the Solution The stresses on an oblique plane are given by Eqs. 2.24 and plotted in Fig. 2.30. From Fig. 2.30 it appears that the answer is, Yes, the cut should be somewhere between 50° and 60° (or between −50° and −60°).

Solution We want to determine a splice angle θ_s such that $|\tau_{nt}(\theta_s)| = 1.5\sigma_n(\theta_s)$. Therefore, from Eqs. 2.24 we have

$$1.5(\sigma_x \cos^2 \theta_s) = \pm \sigma_x \cos \theta_s \sin \theta_s$$

or

$$\tan \theta_s = \pm 1.5$$

Therefore, the shear stress exceeds the normal stress by 50% on planes oriented at

$$\theta_s = \pm 56.3° \qquad \text{Ans.}$$

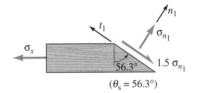

$(\theta_s = 56.3°)$

Review the Solution When $\theta = 0$ we get no shear stress on the splice joint. Therefore, a "long" splice, like one at ±56.3°, makes sense as a splice on which shear stress predominates over normal stress. The stresses for these two cases are illustrated in Fig. 2.

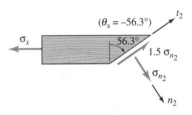

Fig. 2

2.9 GENERAL DEFINITIONS OF STRESS AND STRAIN   ■■■■■■■■■■■

As has been illustrated in Section 2.8, the values of normal stress and shear stress depend on the orientation of the plane on which the stresses act; they may also depend on the point in the plane where the stresses are to be obtained. Therefore, we extend our previous definitions of normal stress (Eq. 2.3) and shear stress (Eq. 2.17) and give definitions for *normal stress and shear stress at an arbitrary point in an arbitrarily oriented plane* passing through an arbitrarily loaded three-dimensional body. The stresses on a plane are then related to the stress resultants on the plane.

Definitions of Normal Stress and Shear Stress.
To define the normal stress and shear stress at an arbitrary point P in a body, on the plane that has an outward normal vector $\boldsymbol{n}$ and that passes through point P, we start with the force vector $\Delta\boldsymbol{R}(P)$ at point P acting on an infinitesimal area ΔA of the $\boldsymbol{n}$ plane, as shown in Fig. 2.32. In Fig. 2.33a the force vector $\Delta\boldsymbol{R}(P)$, whose magnitude is denoted by $\Delta R(P)$, is resolved into normal and tangential components. In Fig. 2.33b the tangential component is further resolved into components along two orthogonal directions, s and t, in the $\boldsymbol{n}$ plane. Then, the normal

49

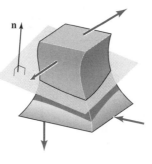

(a) A 3-D body with cutting plane.

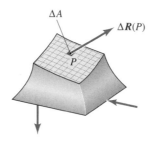

(b) The force on an infinitesimal area ΔA at point P in plane n.

FIGURE 2.32. The cutting plane whose normal vector is $\boldsymbol{n}$ and which passes through a given point P in a 3-D body.

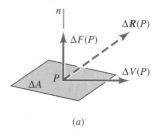

(a)

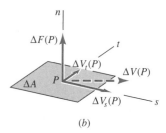

(b)

FIGURE 2.33. The components of force on infinitesimal area ΔA.

stress σ_n and the shear stresses τ_{ns} and τ_{nt} at point P on the plane whose outward normal is n are defined by the expressions

$$
\sigma_n(P) = \lim_{\Delta A \to 0} \left(\frac{\Delta F(P)}{\Delta A} \right) , \; P \text{ always in } \Delta A
$$

$$
\tau_{ns}(P) = \lim_{\Delta A \to 0} \left(\frac{\Delta V_s(P)}{\Delta A} \right) , \; P \text{ always in } \Delta A \qquad (2.26)
$$

$$
\tau_{nt}(P) = \lim_{\Delta A \to 0} \left(\frac{\Delta V_t(P)}{\Delta A} \right) , \; P \text{ always in } \Delta A
$$

Sign Convention: The subscript n in σ_n, τ_{ns} and τ_{nt} designates the direction of the outward normal to the face on which these stresses act. The shear stress τ_{ns} is the shear stress acting in the $+s$ direction on the $+n$ face and also the shear stress acting in the $-s$ direction on the $-n$ face. (If τ_{ns} is negative, it is a shear stress that acts in the $-s$ direction on the $+n$ face, etc.) The same shear stress sign convention applies to τ_{nt}, with, of course, t substituted for s.

Stress Resultants. Equations 2.26 are the basis for very important equations that relate the stresses on an area to the stress resultants on that area. Consider the stresses on a cross

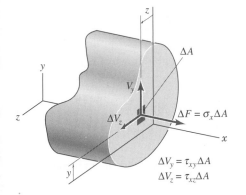

(a) The force components at point (x, y, z).

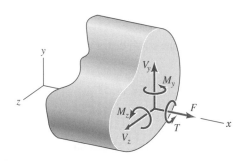

(b) The stress resultants on the cross section at x.

FIGURE 2.34. The stresses and stress resultants on a cross section.

section of a slender member, as shown in Fig. 2.34. Here the subscripts xyz correspond to nst in Eqs. 2.26. The stresses σ_x, τ_{xy}, and τ_{xz} on a small element of area ΔA at point (x, y, z) give rise to forces $\Delta F = \sigma_x \, \Delta A$, $\Delta V_y = \tau_{xy} \, \Delta A$, and $\Delta V_z = \tau_{xz} \, \Delta A$. These forces, in turn, contribute to the force and moment resultants shown in Fig. 2.34b, which are related to the stresses by the following integrals over the cross section.

Force Resultants:

$\sum F_x$:

$\sum F_y$:

$\sum F_z$:

$$F(x) = \int_A \sigma_x \, dA$$
$$V_y(x) = \int_A \tau_{xy} \, dA$$
$$V_z(x) = \int_A \tau_{xz} \, dA$$

(2.27)

Moment Resultants:

$\sum M_x$:

$\sum M_y$:

$\sum M_z$:

$$T(x) = \int_A y\tau_{xz} \, dA - \int_A z\tau_{xy} \, dA$$
$$M_y(x) = \int_A z\sigma_x \, dA$$
$$M_z(x) = -\int_A y\sigma_x \, dA$$

(2.28)

In Eqs. 2.27, F is the *normal force* on the x face, tension positive, while V_y and V_z are components of *shear force* that act on the x face in the y direction and the z direction, respectively. In Eqs. 2.28, T is the *torque*, or twisting moment, while M_y and M_z are *bending moments* about y and z, respectively. The *right-hand rule* is used to establish the sign convention for the torque and the two bending moments. The dimensions of torque and bending moment are $F \cdot L$. Typical units used are lb · ft, kN · m, and so on.[33]

The relationship of stresses to stress resultants is illustrated by the following example.

---

[33]The units may be stated in the reverse order (e.g., in. · kip rather than kip · in.).

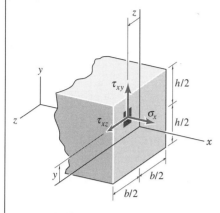

(a) The stresses at a point.

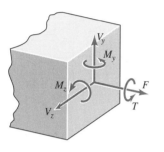

(b) The stress resultants.

Fig. 1

The stress distribution on the rectangular cross section shown in Fig. 1 is given by

$$\sigma_x = (800y - 400z + 1200) \text{ psi}$$

$$\tau_{xy} = 0, \ \tau_{xz} = 300(9 - z^2) \text{ psi}$$

Determine the net internal force system (i.e., the resultant forces and moments) on this cross section. Let $b = 6$ in. and $h = 8$ in.

Plan the Solution   At the cross section there can, in general, be three components of force—F, V_y, and V_z—and three moments—T, M_y, and M_z. Using the given distribution of stresses and using Eqs. 2.27 and 2.28, we can calculate these stress resultants.

Solution

$$F = \int_A \sigma_x \, dA = 800 \int_A y \, dA - 400 \int_A z \, dA + 1200 \int_A dA$$

$$= 800\bar{y}A - 400\bar{z}A + 1200A$$

$$= (1200 \text{ psi})(8 \text{ in.})(6 \text{ in.}) = 57{,}600 \text{ lb}$$

(Since the origin of y and z is at the centroid of the cross section, $\bar{y} = \bar{z} = 0$.) Continuing,

$$V_y = \int_A \tau_{xy} \, dA = 0$$

$$V_z = \int_A \tau_{xz} \, dA = 2700 \int_A dA - 300 \int_A z^2 \, dA$$

From Appendix C.2, for the rectangular cross section of "base" b and "height" h, $\int_A z^2 \, dA = \frac{1}{12}hb^3$ and $\int_A y^2 \, dA = \frac{1}{12}bh^3$. Therefore,

$$V_z = 2700A - \frac{300}{12}(hb^3)$$

$$= (2700 \text{ psi})(8 \text{ in.})(6 \text{ in.}) - (25 \text{ lb/in.}^4)(8 \text{ in.})(6 \text{ in.})^3$$

$$= 86{,}400 \text{ lb}$$

$$T = \int_A y\tau_{xz} \, dA - \int_A z\tau_{xy} \, dA$$

$$= 2700 \int_A y \, dA - 300 \int_A yz^2 \, dA$$

$$= 2700\bar{y}A - 300 \int_{-3}^{3} z^2 \int_{-4}^{4} y \, dy \, dz = 0$$

52

$$M_y = \int_A z\sigma_x \, dA$$

$$= 800 \int_A yz \, dA - 400 \int_A z^2 \, dA + 1200 \int_A z \, dA$$

$$= -\frac{400}{12}(hb^3) = -\left(\frac{400}{12} \text{ lb/in.}^3\right)(8 \text{ in.})(6 \text{ in.})^3$$

$$= -57{,}600 \text{ lb} \cdot \text{in.}$$

$$M_z = -\int_A y\sigma_x \, dA = -800 \int_A y^2 \, dA$$

$$+ 400 \int_A yz \, dA - 1200 \int_A y \, dA$$

$$= -\frac{800}{12}(bh^3) = -\left(\frac{800}{12} \text{ lb/in.}^3\right)(6 \text{ in.})(8 \text{ in.})^3$$

$$= -204{,}800 \text{ lb} \cdot \text{in.}$$

In summary,

$$\left. \begin{array}{llll} F = 57.6 \text{ kips}, & V_y = 0, & & V_z = 86.4 \text{ kips} \\ T = 0, & M_y = -57.6 \text{ kip} \cdot \text{in.}, & M_z = -205 \text{ kip} \cdot \text{in.} \end{array} \right\} \text{ Ans.}$$

Review the Solution   The only way to check the above answers is to go back over the calculations, using information from Appendix C to check all of the integrals. We can also spot-check some of the magnitudes. For example, the 1200-psi term in σ_x represents constant normal stress on the cross section, which would produce an axial force $F = \sigma_x A = (1200 \text{ psi})(8 \text{ in.})(6 \text{ in.}) = 57.6$ kips. We can also see that $0 \le \tau_{xz} \le 2700$ psi everywhere on the cross section. Hence, V_z must be less than $(2700 \text{ psi})(8 \text{ in.})(6 \text{ in.}) = 129.6$ kips. So the value of $V_z = 86.4$ kips seems reasonable.

(Stress distributions like the one in this example will arise in Chapter 6 in the study of bending of beams.)

Let us turn our attention now to general definitions of extensional strain and shear strain to complement the definitions of normal stress and shear stress in Eqs. 2.26.

General Definition of Extensional Strain.

Recall that extensional strain is the change in length of a line segment divided by the original length of the line segment. To define the *extensional strain* in a direction n at a point P in a body, we take an infinitesimal line segment of length Δs, in direction n, starting at P as shown in Fig. 2.35a.[34] That is, we take the infinitesimal line segment PQ of length Δs as the original line segment. After deformation, the line segment PQ becomes the infinitesimal arc P^*Q^* with arclength Δs^*,

---

[34]In the present discussion the term *line segment* refers to the collection of material particles that lie along a straight line connecting specified points, say P and Q, in the undeformed body. Such a line segment is sometimes referred to as a *fiber*.

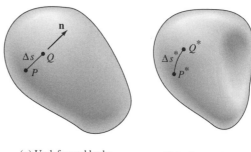

(*a*) Undeformed body.   (*b*) Deformed body.

FIGURE 2.35.   The infinitesimal line segment used to define extensional strain.

as shown in Fig. 2.35*b*. To determine the extensional strain right at point *P*, we need, of course, to start with a very short length Δs; that is, we must pick *Q* very close to point *P*. By picking *Q* closer and closer to *P*, we get, in the limit as $\Delta s \to 0$, the extensional strain right at point *P*. Then, the *extensional strain at point P in direction n,* denoted by $\epsilon_n(P)$, is defined by

$$\epsilon_n(P) = \lim_{Q \to P \text{ along } n} \left(\frac{\Delta s^* - \Delta s}{\Delta s} \right) \tag{2.29}$$

EXAMPLE 2.10

The thin, square plate *ABCD* in Fig. 1a undergoes deformation in which no point in the plate displaces in the *y* direction. Every horizontal line in the plate, except line *AD*, is uniformly shortened as the edge *AB* remains straight and rotates clockwise. Determine an expression for $\epsilon_x(x, y)$.

Plan the Solution   We can use the definition of extensional strain $\epsilon_n(P)$, which, in this case, is $\epsilon_x(x, y)$. We will have to use the geometry of deformation to determine an expression for Δs^* in Eq. 2.29.

Solution   To determine $\epsilon_x(x, y)$, let Eq. 2.29 be written as

$$\epsilon_x(x, y) = \lim_{\Delta x \to 0} \left(\frac{\Delta x^* - \Delta x}{\Delta x} \right)$$

where Δx and Δx^* are defined in Fig. 2.

To get an expression for $\epsilon_x(x, y)$ we need an expression for Δx^*. We are told that every horizontal line is uniformly shortened, and we see that *AD* remains its original length while B^*C^* is shorter than *BC* by 20%. Furthermore, the shortening of horizontal lines is linearly related to *y*. Therefore,

$$\Delta x^* = \Delta x - \left(\frac{y}{a} \right)(0.2 \, \Delta x)$$

So,

$$\epsilon_x(x, y) = \lim_{\Delta x \to 0} \left[\frac{-(y/a)(0.2 \, \Delta x)}{\Delta x} \right]$$

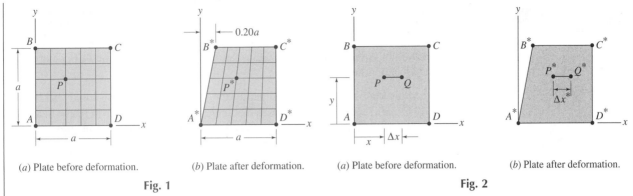

(a) Plate before deformation.	(b) Plate after deformation.	(a) Plate before deformation.	(b) Plate after deformation.

Fig. 1 **Fig. 2**

or

$$\epsilon_x(x,\ y) = -\left(\frac{y}{a}\right)(0.2) \qquad\qquad \textbf{Ans.}$$

Review the Solution The answer is negative, which indicates a shortening, and the answer is dimensionless, as it should be for strain. Also, according to the answer, there is no strain at $y = 0$ (along AD), and there is a 20% shortening at $y = a$. These results agree with the stated geometry of deformation.

Definition of Shear Strain. When a body deforms, the change in angle that occurs between two line segments that were originally perpendicular to each other is called *shear strain*. To define the shear strain, let us consider the undeformed body in Fig. 2.36*a* and the deformed body in Fig. 2.36*b*. Let PQ and PR be infinitesimal line segments in the n direction and t direction, respectively, in the undeformed body. After deformation, line segments PQ and PR become arcs P^*Q^* and P^*R^*. Secant lines P^*Q^* and P^*R^* define an angle θ^* in the deformed body. In the limit, as we pick Q and R closer and closer to P, the angle θ^* approaches the angle between tangents to the arcs at P^*, shown as dashed lines in Fig. 2.36*b*. The *shear strain between line segments extending from P in directions n and t* is defined by the expression

$$\gamma_{nt}(P) = \lim_{\substack{Q\to P \text{ along } n \\ R\to P \text{ along } t}} \left(\frac{\pi}{2} - \theta^*\right) \tag{2.30}$$

This definition is illustrated in the following example.

FIGURE 2.36. The angles used to define shear strain.

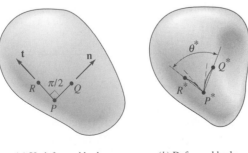

(a) Undeformed body.	(b) Deformed body.

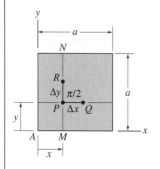

(a) Plate before deformation.

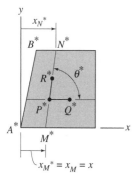

(b) Plate after deformation.

Fig. 1

For the square plate and the deformation described in Example 2.10, determine an expression for $\gamma_{xy}(x, y)$.

Plan the Solution By comparing the "before deformation" and "after deformation" figures, we can see that there is a definite change in the right angle between x lines and y lines. We need to determine this change in angle so that we can evaluate the shear strain from Eq. 2.30.

Solution Equation 2.30 can be written as

$$\gamma_{xy}(P) \equiv \gamma_{xy}(x, y) = \lim_{\substack{\Delta x \to 0 \\ \Delta y \to 0}} \left(\frac{\pi}{2} - \theta^* \right)$$

where Δx, Δy, and θ^* are indicated in Fig. 1.

$$\theta^* = \frac{\pi}{2} - \tan^{-1} \left(\frac{x_{N*} - x_{M*}}{a} \right)$$

$$x_{N*} = x + \left(1 - \frac{x}{a} \right)(0.2a)$$

$$x_{N*} - x_{M*} = \left(1 - \frac{x}{a} \right)(0.2a)$$

Therefore,

$$\theta^* = \frac{\pi}{2} - \tan^{-1} \left[0.2 \left(1 - \frac{x}{a} \right) \right]$$

Since the x lines and y lines remain straight, θ^* doesn't depend on the lengths of Δx and Δy, and we don't need the limit operation. So,

$$\gamma_{xy}(x, y) = \frac{\pi}{2} - \theta^* = \tan^{-1} \left[0.2 \left(1 - \frac{x}{a} \right) \right] \qquad \textbf{Ans.}$$

Review the Solution From Fig. 1b, we see that γ_{xy} should be independent of y, as our final result indicates. Also, from Fig. 1b we see that γ_{xy} should be greatest when $x = 0$ and should be zero at $x = a$. These observations are consistent with our answer. Finally, the largest shear strain is at $x = 0$, where $\gamma_{xy}(x, y) = \tan^{-1}(0.2) = 0.1974 \doteq 0.2$. Therefore, γ_{xy} could be approximated by $\gamma_{xy}(x, y) = 0.2 \left(1 - \frac{x}{a} \right)$.

In Chapters 3, 4, and 6 we will make extensive use of the definitions of extensional strain and shear strain, Eqs. 2.29 and 2.30, respectively, to determine key strain-displacement equations for theories of axial deformation, torsion, and bending.

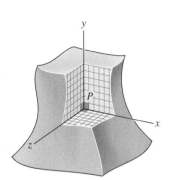

FIGURE 2.37. A set of three mutually orthogonal planes through an arbitrary point P.

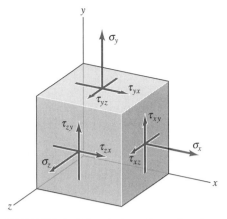

FIGURE 2.38. A three-dimensional state of stress referred to rectangular cartesian axes.

2.10 CARTESIAN COMPONENTS OF STRESS; GENERALIZED HOOKE'S LAW FOR ISOTROPIC MATERIALS

Definitions of normal stress and shear stress were given in Eqs. 2.26 of the previous section. These equations define the components of stress on a particular plane, the n plane, at the given point. However, we can pass three mutually orthogonal planes through any point in a deformable body, so we need to consider the stresses on all three mutually orthogonal planes. Consider the x, y, and z planes passing through point P in Fig. 2.37. It is customary to sketch the components of stress, as defined by Eqs. 2.26, on a cube having x, y, and z faces, as shown in Fig. 2.38.

The sign convention for the stresses shown on Fig. 2.38 is the sign convention associated with Eqs. 2.26. The normal stresses are always taken positive in tension. Therefore, σ_x, σ_y, and σ_z are all shown in tension on the respective x, y, and z faces. The first subscript on a shear stress refers to the plane on which the shear stress acts, while the second subscript indicates the direction in which the shear stress acts. Hence, on the "$+x$ face," that is, the face whose outward normal is the $+x$ axis, the shear stress τ_{xy} acts in the $+y$ direction, and the stress τ_{xz} acts in the $+z$ direction. Conversely, on the "$-x$ face," τ_{xy} acts in the $-y$ direction, and τ_{xz} acts in the $-z$ direction. (To avoid "cluttering," the stresses acting on the hidden ($-$) faces are not shown in Fig. 2.38.)

Shear Stress Equilibrium Requirements. In the discussion of pure shear in Section 2.7, you learned that, in order to satisfy moment equilibrium, the shear stresses on faces that intersect at right angles must be equal (e.g., see Fig. 2.27e). Let us now examine the relationship of shear stresses on perpendicular faces on which normal stresses also act. To simplify the free-body diagram, only those stresses that contribute to moment about the z axis are shown in Fig. 2.39. Stresses on exposed faces are indicated by a $+$ superscript; those on the three hidden faces have no subscript.[35] Consider first the moment about the z axis due to σ_x and σ_x^+. We get

$$(\Delta M_z)_{\sigma_x} = (\sigma_x \, \Delta y \, \Delta z)\left(\frac{\Delta y}{2}\right) - (\sigma_x^+ \, \Delta y \, \Delta z)\left(\frac{\Delta y}{2}\right)$$

---

[35]Since the $+x$ face and the $-x$ face are Δx distance apart, the stress on the $+x$ face may be slightly different than the stress on the $-x$ face; this distinction is indicated by the $+$ superscript notation.

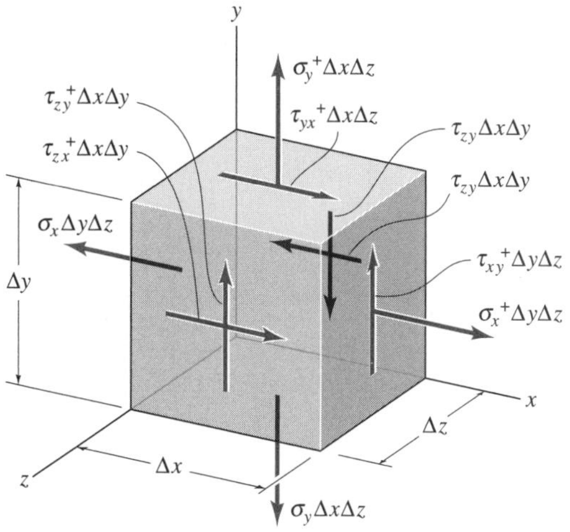

FIGURE 2.39. A three-dimensional free-body diagram.

Since Δx is small, we can write

$$\sigma_x^+ = \sigma_x + \Delta\sigma_x$$

where $\Delta\sigma_x$ is a small quantity of the same order as Δx. Then,

$$(\Delta M_z)_{\sigma_x} = -\frac{1}{2}\Delta\sigma_x\,\Delta y^2\,\Delta z$$

Note that the right-hand side is a product of four Δ-terms. The stresses σ_y, τ_{zx}, and τ_{zy} contribute similar "four-Δ" amounts to ΔM_z. On the other hand, consider the moment due to τ_{xy}^+ and τ_{yx}^+.

$$(\Delta M_z)_{\tau_{xy}} = (\tau_{xy}^+\,\Delta y\,\Delta z)\,\Delta x - (\tau_{yx}^+\,\Delta x\,\Delta z)\,\Delta y$$
$$= (\tau_{xy} - \tau_{yx})\,\Delta x\,\Delta y\,\Delta z + (\Delta\tau_{xy} - \Delta\tau_{yx})\,\Delta x\,\Delta y\,\Delta z$$

Collecting all contributions to moment about the z axis, we get

$$\sum M_z = (\tau_{xy} - \tau_{yx})\,\Delta x\,\Delta y\,\Delta z + \text{higher-order terms}$$

As Δx, Δy, and Δz approach zero, the higher-order terms become negligible, leaving only the first term. However, to satisfy moment equilibrium, we must have $\Sigma M_z = 0$, so it is necessary that

$$\tau_{xy} = \tau_{yx}$$

From the above analysis of moment equilibrium, we can conlude that, **even if there are normal stresses acting on an element, the shear stresses must satisfy the following equations**:

$$\boxed{\tau_{yx} = \tau_{xy}, \quad \tau_{yz} = \tau_{zy}, \quad \tau_{zx} = \tau_{xz}} \tag{2.31}$$

Generalized Hooke's Law for Isotropic Materials. A solid whose material properties, for example E and ν, are independent of orientation in the body, is said to be *isotropic*. We now consider the stress-strain-temperature relationships for a linearly elastic,

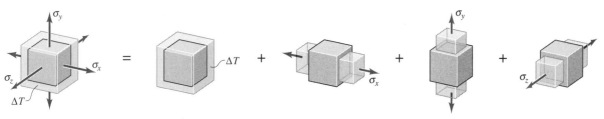

(a) The total extensional strain.   (b) Uniform thermal strain.   (c) Strains due to σ_x.   (d) Strains due to σ_y.   (e) Strains due to σ_z.

FIGURE 2.40.
Superposition of extensional strains.

isotropic body. Let the body be subjected to stresses σ_x, σ_y, σ_z, τ_{xy}, τ_{xz}, and τ_{yz}, and to a temperature change ΔT. By the *principle of linear superposition*[36] we can get the combined strain response by adding together the separate responses due to σ_x, σ_y, and so on. For example, the strains produced by σ_x are

$$\epsilon_x = \frac{\sigma_x}{E}, \quad \epsilon_y = \epsilon_z = \frac{-\nu\sigma_x}{E}$$

Figure 2.40*a* illustrates the combined effect of the σ's and ΔT. Figures 2.40*b* through 2.40*e* illustrate the strains produced separately by the three normal stresses, σ_x, σ_y, and σ_z, and the strain produced by a temperature increase ΔT. Let

$$\epsilon_{x_t} = \epsilon_{y_t} = \epsilon_{z_t} = \alpha\,\Delta T \qquad \text{(Fig. 2.38}b)$$

$$\epsilon'_x = \frac{\sigma_x}{E}, \quad \epsilon'_y = \epsilon'_z = \frac{-\nu\sigma_x}{E} \qquad \text{(Fig. 2.38}c)$$

$$\epsilon''_y = \frac{\sigma_y}{E}, \quad \epsilon''_x = \epsilon''_z = \frac{-\nu\sigma_y}{E} \qquad \text{(Fig. 2.38}d)$$

$$\epsilon'''_z = \frac{\sigma_z}{E}, \quad \epsilon'''_x = \epsilon'''_y = \frac{-\nu\sigma_z}{E} \qquad \text{(Fig. 2.38}e)$$

By the superposition principle, the total strains are given by

$$\epsilon_x = \frac{1}{E}\left[\sigma_x - \nu(\sigma_y + \sigma_z)\right] + \alpha\,\Delta T$$

$$\epsilon_y = \frac{1}{E}\left[\sigma_y - \nu(\sigma_x + \sigma_z)\right] + \alpha\,\Delta T \qquad (2.32)$$

$$\epsilon_z = \frac{1}{E}\left[\sigma_z - \nu(\sigma_x + \sigma_y)\right] + \alpha\,\Delta T$$

For an isotropic material, the shear stresses are related to the shear strains by the following equations (Hooke's Law):

$$\gamma_{xy} = \frac{1}{G}\tau_{xy}, \quad \gamma_{xz} = \frac{1}{G}\tau_{xz}, \quad \gamma_{yz} = \frac{1}{G}\tau_{yz} \qquad (2.33)$$

These shear strains are illustrated in Fig. 2.41.

(a) τ_{xy} produces γ_{xy} only.

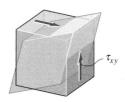

(b) τ_{xz} produces γ_{xz} only.

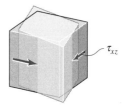

(c) τ_{yz} produces γ_{yz} only.

FIGURE 2.41. Illustration of shear strains.

---

[36]The strains may be added linearly if the deformation is small and the material remains linearly elastic.

59

Note that, in an isotropic material, shear stresses do not enter into the expressions for extensional strains, and, likewise, normal stresses do not enter into the expressions for shear strains. In addition, the three components of shear are uncoupled.

Equations 2.32 (without the ΔT terms) and Eqs. 2.33 are referred to as the *generalized Hooke's Law* for isotropic materials. Solving Eqs. 2.32 and 2.33 for the stresses in terms of the strains and ΔT, we get

$$\sigma_x = \frac{E}{(1 + \nu)(1 - 2\nu)} [(1 - \nu)\epsilon_x + \nu(\epsilon_y + \epsilon_z) - (1 + \nu)(\alpha \Delta T)]$$

$$\sigma_y = \frac{E}{(1 + \nu)(1 - 2\nu)} [(1 - \nu)\epsilon_y + \nu(\epsilon_z + \epsilon_x) - (1 + \nu)(\alpha \Delta T)]$$

$$\sigma_z = \frac{E}{(1 + \nu)(1 - 2\nu)} [(1 - \nu)\epsilon_z + \nu(\epsilon_x + \epsilon_y) - (1 + \nu)(\alpha \Delta T)]$$

$$\tau_{xy} = G\gamma_{xy}, \quad \tau_{yz} = G\gamma_{yz}, \quad \tau_{zx} = G\gamma_{zx}$$

(2.34)

As noted in Section 2.7, there is an equation that relates G to E and ν, namely

$$G = \frac{E}{2(1 + \nu)}$$

This relationship is derived in Section 2.11.

The following example illustrates the relationship of Poisson's ratio to the change in volume of a body subjected to triaxial stress. The change in volume per unit volume is called the *volumetric strain,* or *dilatation,* ϵ_V.

■■■■■■■■■■■■■■■■■■ E X A M P L E 2 . 1 2 ■■■■■■■■■■■■■■■■■■

Determine the volumetric strain, ϵ_V, of a rectangular parallelepiped of linearly elastic, isotropic material subjected to general triaxial stress σ_x, σ_y, σ_z. Assume that $\epsilon \ll 1$ for all three coordinate directions.

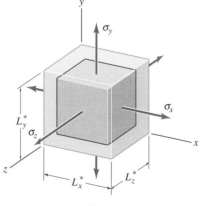

Fig. 1

Plan the Solution   We can write the change in volume in terms of the change in length of each of the three sides of the body, which can, in turn, be expressed in terms of the extensional strains ϵ_x, ϵ_y, and ϵ_z. We can use Hooke's Law, Eqs. 2.32, to relate these three strains to the three stresses.

Solution   The basic equation for volumetric strain (dilatation) is

$$\epsilon_V = \frac{\Delta V}{V} = \frac{L_x^* L_y^* L_z^* - L_x L_y L_z}{L_x L_y L_z}$$

where the L^* length are the after-deformation dimensions of the body, and the L's are the corresponding pre-deformation dimensions of the body. For uniform strain in each coordinate direction,

$$L^* = (1 + \epsilon)L$$

so

$$\epsilon_V = (1 + \epsilon_x)(1 + \epsilon_y)(1 + \epsilon_z) - 1$$
$$= \epsilon_x + \epsilon_y + \epsilon_z + \epsilon_x \epsilon_y + \epsilon_y \epsilon_z + \epsilon_z \epsilon_x + \epsilon_x \epsilon_y \epsilon_z$$

Since all three strains satisfy $\epsilon \ll 1$, the squared and cubed terms can be dropped in the preceding equation, leaving the approximation

$$\epsilon_V = \epsilon_x + \epsilon_y + \epsilon_z \qquad (1)$$

From Eqs. 2.32,

$$\epsilon_x = \frac{1}{E}[\sigma_x - \nu(\sigma_y + \sigma_z)]$$

$$\epsilon_y = \frac{1}{E}[\sigma_y - \nu(\sigma_x + \sigma_z)] \qquad (2)$$

$$\epsilon_z = \frac{1}{E}[\sigma_z - \nu(\sigma_x + \sigma_y)]$$

Finally, Eqs. (1) and (2) may be combined to give following expression for the dilatation ϵ_V:

$$\epsilon_V = \frac{1 - 2\nu}{E}(\sigma_x + \sigma_y + \sigma_z) \qquad \text{Ans.}$$

Review the Solution   From the above answer we see that the change in volume per unit volume is of the order of magnitude of the strain, and is therefore small, as we would expect. It is interesting to note that, because of the $(1 - 2\nu)$ factor, a tensile stress causes an increase in volume, which is what we would expect to happen, but only if $0 \leq \nu \leq 0.5$. As noted earlier, most materials have a value of ν that falls within the range $\nu = 0.25$ to $\nu = 0.35$.

It was noted in Section 2.7 and again in Section 2.10 that Young's modulus E and the shear modulus G for linearly elastic behavior of isotropic materials are related by the equation

$$G = \frac{E}{2(1 + \nu)} \tag{2.35}$$

By considering the case of *pure shear* and employing equations of equilibrium, geometry of deformation, and isotropic material behavior, we will now derive this relationship.[37] Consider a square plate of unit thickness subjected to pure shear relative to the x, y axes, as shown in Fig. 2.42a, and let the n and t axes be the diagonals of the square.

By considering the shear stresses on the faces of an n, t element and relating them to the elongation of the diagonal AC (i.e., the elongation in the n direction), we can derive Eq. 2.35.

Equilibrium: From the free-body diagram in Fig. 2.42d,

$$+\nearrow \sum F_n = 0: \qquad \sigma_n a\sqrt{2} - 2\tau a(\sqrt{2}/2) = 0, \quad \sigma_n = \tau$$

$$\nwarrow \sum F_t = 0: \quad \tau_{nt} a\sqrt{2} - \tau a(\sqrt{2}/2) + \tau a(\sqrt{2}/2) = 0, \quad \tau_{nt} = 0$$

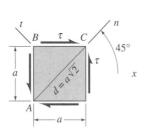

(a) Stresses on an element in pure shear.

(b) Deformation of the element in pure shear.

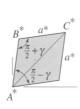

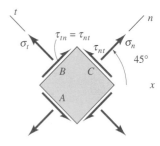

(c) Stresses on n and t faces.

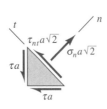

(d) Free-body diagram - n face.

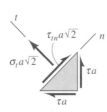

(e) Free-body diagram - t face.

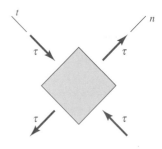

(f) Stresses on n and t faces.

FIGURE 2.42. Illustrations for relating E and G.

---

[37]The relationship in Eq. 2.35 is general and applies to any stress state in an isotropic, linearly elastic body. It is most easily demonstrated by use of the case of *pure shear,* as is done here.

and from the free-body diagram in Fig. 2.42e,

$$+\nearrow \sum F_n = 0: \quad \tau_{tn}a\sqrt{2} - \tau a(\sqrt{2}/2) + \tau a(\sqrt{2}/2) = 0, \quad \tau_{tn} = 0$$

$$+\nwarrow \sum F_t = 0: \qquad\qquad \sigma_t a\sqrt{2} + 2\tau a(\sqrt{2}/2) = 0, \quad \sigma_t = -\tau$$

Material Behavior: From the generalized Hooke's Law in Eqs. 2.32 and 2.33 we have the following:

1. Relating the shear stresses shown in Fig. 2.42a to the shear strain shown in Fig. 2.42b is the equation

$$\tau_{xy} = G\gamma_{xy}, \quad \text{or} \quad \tau = G\gamma \qquad (a)$$

2. Relating the normal stresses $\sigma_x = \sigma_y = 0$ in Fig. 2.42a to the extensional strains in Fig. 2.42b, we get

$$\epsilon_x = \epsilon_y = 0, \quad \text{so} \quad a^* = a$$

3. Relating the normal stresses, $\sigma_n = \tau$ and $\sigma_t = -\tau$, in Fig. 2.42f to the extensional strain ϵ_n along diagonal A^*C^* in Fig. 2.42b, we have

$$\epsilon_n = \frac{1}{E}(\sigma_n - \nu\sigma_t) = \frac{\tau}{E}(1 + \nu) \qquad (b)$$

Geometry of Deformation: The law of cosines can be applied to the triangle $A^*B^*C^*$ in Fig. 2.42b to give

$$\overline{A^*C^*}^2 = \overline{A^*B^*}^2 + \overline{B^*C^*}^2 - 2\overline{A^*B^*}\,\overline{B^*C^*}\cos(\angle B^*)$$

or

$$[a\sqrt{2}(1 + \epsilon_n)]^2 = a^2 + a^2 - 2a^2\cos\left(\frac{\pi}{2} + \gamma\right)$$

$$= 2a^2(1 + \sin\gamma)$$

$$1 + 2\epsilon_n + \epsilon_n^2 = 1 + \sin\gamma$$

For small ϵ_n and γ, the ϵ_n^2 term can be dropped, and $\sin\gamma$ can be approximated by γ. Then,

$$\epsilon_n = \frac{\gamma}{2} \qquad (c)$$

Combining Eqs. (a), (b), and (c), we get the desired result,

$$G = \frac{\tau}{\gamma} = \frac{E}{2(1 + \nu)}$$

The design of a structure[38] involves selecting a promising configuration for the structure and applying the principles and equations of deformable-body mechanics to select materials and dimensions of individual members, or components (beams and columns of a building, ribs and spars of an airplane wing, etc.) so that no failure occurs under the prescribed loading conditions. Several possible *modes of failure* typically need to be considered. These can be grouped under the three headings—*design for strength, design for stiffness,* and *design for ductility*—that were mentioned earlier in Section 2.4.

- Design for Strength—the actual stresses in a member must not exceed certain values based on one or more of the following: yield strength, ultimate strength, creep strength, fatigue strength.
- Design for Stiffness—the actual deformations must not exceed certain values, and/ or the actual load must not exceed the specified buckling load.
- Design for Ductility—the actual energy that is to be absorbed must not exceed a certain value, and there must be no sudden, catastrophic failure.

The design might be based on only one of the preceding criteria; however, in some cases several criteria may have to be considered.

Factor of Safety; Allowable Stress. In selecting member sizes and materials, the designer must ensure that the *failure load* (the minimum value of the load that, according to specified criteria, would cause failure of the member) is safely above the *allowable load* (the maximum load that the member is allowed to experience during its service lifetime). The *factor of safety, FS,* is defined as

$$FS = \frac{\text{failure load}}{\text{allowable load}} \tag{2.36}$$

Of course, $FS > 1$. If there is a linear relationship between the loads on a structure and the stresses caused by the loads, it is permissible to define the factor of safety as the ratio of two stresses, the *failure stress* and the *allowable stress*. For axial deformation, the tensile (or compressive) yield strength σ_Y is taken as the stress corresponding to failure by yielding; in direct shear, the shear yield strength τ_Y is used.[39]

$$FS = \frac{\text{failure stress}}{\text{allowable stress}} \tag{2.37}$$

For simple loading situations like axial loading and direct shear, Eq. 2.37 can be used to define a factor of safety with respect to *ultimate failure* by using σ_U (or τ_U) as the failure stress. Equations 2.37 is the design equation that will be used most frequently in this text, but Eq. 2.36 is required for the design of columns (Section 10.7).

---

[38]The term *structure* will be used as a generic term covering the wide range of deformable bodies including bridges, buildings, machines, land vehicles, and airplanes, as well as the many individual components that make up each of these.

[39]Section 12.3 discusses *Failure Theories* for more complex loadings.

Design based on Eq. 2.37 is frequently referred to as *allowable-stress design (ASD).* Design based on either Eq. 2.36 or Eq. 2.37 may be referred to as *factor-of-safety design (FSD).* In either case, a single factor of safety is used to incorporate all of the uncertainty related to the loads and all of the uncertainty associated with the structure's ability to resist the loads. The *margin of safety, MS,* defined by

$$MS = FS - 1.0 \qquad (2.38)$$

is sometimes used rather than the factor of safety.

The normal range of values for the factor of safety is from about 1.3 to 3.0, although a value as high as 10 might occasionally be applied. The use of a small value of factor of safety (e.g., $FS = 1.1$) is justified only when it is possible, by analysis and testing, to minimize uncertainties sufficiently, and when there is no likelihood that failure will result in unacceptable circumstances such as serious personal injury or death. On the other hand, it is undesirable to use a factor of safety that is unnecessarily large (e.g., $FS > 3$), since that would lead to excess structural weight, which, in turn, entails excess initial and operating costs. Since the choice of a value of factor of safety has such important economic and legal implications, design specifications, including the relevant factor(s) of safety to be used, conform to design codes or other standards developed by groups of experienced engineers in engineering societies or in various government agencies. Two examples of such codes and specifications are: (1) for steel: *Code of Standard Practice for Steel Buildings and Bridges,* by the American Institute of Steel Construction, and (2) for concrete: *Building Code Requirement for Reinforced Concrete (ACI 318–89),* by the American Concrete Institute.

There are two ways in which the preceding information is used in practice:

- *Design of a New Structure:* The configuration of a structure is given, but the sizes of its components (tension rods, pins, beams, etc.) are to be chosen so that maximum stresses in the components do not exceed the allowable stress (Eq. 2.37), or, if loads are to be considered, the maximum applied load does not exceed the allowable load (Eq. 2.36).
- *Evaluation of an Existing Structure:* The configuration of a structure is known, together with the sizes of all of its components, the materials used, and so on.
  (a) If the present loading is known, it may be required that an effective factor of safety be computed to see if additional load could be applied without exceeding a given code-specified *FS.*
  (b) If the code-specified *FS* is known, then the allowable load can be calculated.

Example 2.13 illustrates the sizing of components; Example 2.14 illustrates the calculation of allowable load.

■■■■■■■■■■■■■■■■□ E X A M P L E 2 . 1 3 □■■■■■■■■■■■■■■■

The shop crane in Fig. 1 consists of a boom *AC* that is supported by a pin at *A* and a rectangular tension bar *BD*. Details of the pin joints at *A* and *B* are shown in Views $a - a$ and $b - b$, respectively. The tension bar is to be made of structural steel with $\sigma_Y = 36$ ksi, while the pins at *A* and *B* are to be of high-strength steel with $\tau_Y = 48$ ksi.

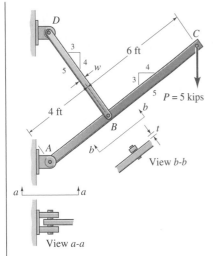

Fig. 1. A shop crane.

The design (i.e., allowable) load is $P = 5$ kips, and there is to be a factor of safety with respect to yielding of $FS = 3.0$. (a) If the width of the bar BD is $w = 2$ in., determine the required thickness, t, to the nearest $1/8$ in. (b) Determine the required pin diameters at A and B to the nearest $1/8$ in.

Plan the Solution Equilibrium of boom AC can be used to determine the tension F_B in two-force member BD and the resultant force F_A on the pin at A. Then we can use allowable stress design (Eq. 2.37) to determine the cross-sectional area of bar BD and the required pin diameters, noting that the pin at A is in *double shear* and the pin at B is in *single shear*.

Solution

Equilibrium: Equilibrium equations were used to solve for the forces F_A and F_B that are shown on the free-body diagram in Fig. 2.

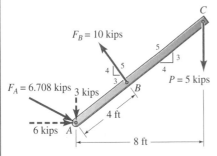

Fig. 2. Free-body diagram of boom AC.

Design of Bar BD: From Eq. 2.37,

$$FS = \frac{\sigma_Y}{\sigma_{allow.}} \rightarrow \sigma_{allow.} = \frac{36 \text{ ksi}}{3.0} = 12 \text{ ksi}$$

$$F_{BD} = \sigma_{allow.} A_{BD} \rightarrow A_{BD} = (2 \text{ in.}) t_{BD} = \frac{10 \text{ kips}}{12 \text{ ksi}}$$

$$t_{BD} = 0.417 \text{ in.} \qquad \text{Select } t_{BD} = 0.5 \text{ in.} \qquad \textbf{Ans.}$$

Design of Pins at A and B:

$$FS = \frac{\tau_Y}{\tau_{allow.}} \rightarrow \tau_{allow.} = \frac{48 \text{ ksi}}{3.0} = 16 \text{ ksi}$$

$$\frac{1}{2} F_A = \tau_{allow.} A_A \rightarrow A_A = \frac{6.708 \text{ kips}}{2(16 \text{ ksi})} = 0.210 \text{ in}^2$$

$$A_A = \frac{\pi d_A^2}{4} \rightarrow d_A = 0.517 \text{ in.} \qquad \text{Select } d_A = 0.625 \text{ in.} \qquad \textbf{Ans.}$$

$$F_B = \tau_{allow.} A_B \rightarrow A_B = \frac{10 \text{ kips}}{16 \text{ ksi}} = 0.625 \text{ in}^2$$

$$A_B = \frac{\pi d_B^2}{4} \rightarrow d_B = 0.892 \text{ in.} \qquad \text{Select } d_B = 1.0 \text{ in.} \qquad \textbf{Ans.}$$

Review the Solution These calculations are very straightforward, but should be double-checked, especially to make sure that the FS has been properly applied. The answers seem to be "reasonable" numbers.

The pin-jointed planar truss in Fig. 1 is subjected to a single downward force P at joint A. All members have a cross-sectional area of 500 mm². The allowable stress in tension is $(\sigma_T)_{allow.} = 300$ MPa, while the allowable stress (magnitude) in compression is $(\sigma_C)_{allow.} = 200$ MPa.[40] Determine the *allowable load*, $P_{allow.}$.

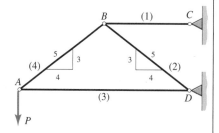

Fig. 1. A pin-jointed truss.

Plan the Solution   Equilibrium (e.g., the "method of sections") can be used to determine all member forces in terms of the load P. The stress in the member with the largest tensile axial force and the stress in the member with the largest compressive axial force should be set equal to $(\sigma_T)_{allow.}$ and $(\sigma_C)_{allow.}$, respectively, to determine values of $P_{allow.}$ based on each. The <u>smaller</u> of the two values governs.

Solution

Equilibrium: The results of an equilibrium analysis are shown in Fig. 2. (The reader should verify the values shown in Fig. 2.)

Allowable Force Based on Member BC: Members AB and BC are in tension; member BC has the larger tensile force. Therefore,

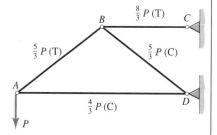

$$\frac{8}{3}\frac{P_1}{A_1} = (\sigma_T)_{allow.} = 300 \text{ MPa}$$

$$P_1 = \frac{3}{8}\left(300 \times 10^6 \frac{\text{N}}{\text{m}^2}\right)(500 \times 10^{-6} \text{ m}^2) = 56.25 \text{ kN}$$

Fig. 2. Member forces in a pin-jointed truss.

Allowable Force Based on Member BD:

$$\frac{5}{3}\frac{P_2}{A_2} = (\sigma_C)_{allow.} = 200 \text{ MPa}$$

$$P_2 = \frac{3}{5}\left(200 \times 10^6 \frac{\text{N}}{\text{m}^2}\right)(500 \times 10^{-6} \text{ m}^2) = 60.00 \text{ kN}$$

Since P_1, the force based on the tension allowable, is smaller than P_2, the allowable force is

$$P_{allow.} = 56 \text{ kN}$$ **Ans.**

Design Philosophies.   Examples 2.13 and 2.14 illustrate one approach to design, the approach based on a single factor of safety. Since the 1970s there has been an increasing awareness of some of the limitations of this type of factor-of-safety design. It has been realized that the design of complex systems, from airplanes to buildings to chemical plants,

---

[40]The magnitude of the compressive allowable stress is given. The allowable stress in compression is frequently governed by elastic buckling or inelastic buckling, topics that are discussed in Chapter 10.

and even to computers, requires more sophisticated *probability-based* (or, *reliability-based*) *design methods.*[41]

Let us consider briefly some of the key concepts in probability-based design. Suppose that an engineer's task is to design the legs of the tension-leg offshore oil platform depicted in Fig. 2.43. The platform is basically a floating barge that is held in place by a number of large cables that are anchored to the ocean floor. In this design, as is typical of all designs, there are two factors to consider, which we will refer to as the *load L* and the *resistance R.*[42] The structural member, in this case the tension leg, must have the *resistance* capacity R to resist the *load* demand L, that is, the design must satisfy the inequality

$$L < R \tag{2.39}$$

For the tension leg under consideration, the "load" L might be the maximum tensile force that could occur in the leg during the service life of the platform. Correspondingly, the "resistance" R would be the resisting force that a cable of a certain diameter made of a certain type of steel would be able to provide without exceeding its yield strength, that is, the strength of the cable. However, the actual *service load, L_s*, and the actual *service resistance, R_s*, can never be known exactly. There is only the engineer's *best estimate* of L based on assumed wind forces, wave forces, and so on, and based on other assumptions that are made and calculations that are performed to obtain this estimate. Likewise, there is only the engineer's *best estimate* of the resisting force R that the cable would be able to supply, taking into account manufacturing tolerances, variations in material properties, quality of fabrication, and other factors. Thus, there are many factors that enter into the determination of both L and R, and these factors are all subject to variability and uncertainty. Some of these factors are listed in Table 2.1.

To illustrate the key concepts of probability-based engineering design, suppose that an engineering firm has obtained 15 independent calculations of the load L and that these 15 estimates are recorded on the bar chart in Fig. 2.44a. For example, four of the estimates indicate that the maximum cable load that might ever occur in service is the load designated as L_3 in Fig. 2.44a. Likewise, fifteen independent estimates of the strength of the cable

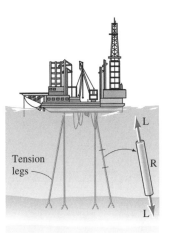

FIGURE 2.43. A tension-leg offshore oil platform.

TABLE 2.1. Uncertainties and Judgments Affecting Load (L) and Resistance (R).

Uncertainties Affecting L	Uncertainties and Judgments Affecting R
• Assumed loads on the platform (e.g., location, magnitude, direction) • Assumptions and approximations made in computing estimates of L, the force in the cable (e.g., boundary conditions, joints, linear behavior)	• Manufacturing uncertainties (e.g., material properties, cable dimensions, tolerances) • Assumptions made in calculating the strength of the cable • Construction uncertainties (e.g., workmanship) • Importance of failure of a single member to overall platform failure • Nature of the failure (e.g., "slow" ductile failure, or "fast" brittle failure) • "Cost," economic and/or social, of failure

[41]Typical references are: *Probabilistic Methods in Structural Engineering,* [Ref. 2-7]; *Methods of Structural Safety,* [Ref. 2-8]; *Introduction to Reliability Engineering,* [Ref. 2-9]; and *Reliability-Based Design,* [Ref. 2-10].

[42]Other pairs of terms used in the literature are: *load* and *capacity, demand* and *resistance, load* and *strength,* and *required strength* and *design strength.*

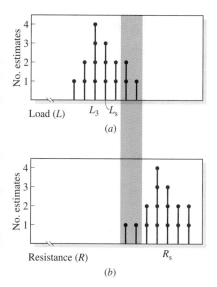

FIGURE 2.44. Probability densities of load and resistance.

are plotted on the bar chart in Fig. 2.44b. These bar charts are simple examples of *probability distributions* of load and resistance. The professional design engineer's task is to ensure that there is a very high probability that the resistance R_s of the member that actually gets placed in the structure will be greater than the maximum load L_s that will actually occur in the member in service, that is, that $R_s > L_s$. There is always some chance—a very slight chance, we hope—that the load and resistance will fall in the shaded region, with $L > R$; in which case the member will fail in service.

In recent years, versions of probability-based design have been adopted by major standards organizations, for example, the *Load and Resistance Factor Design (LRFD)* method adopted in 1986 by the American Institute of Steel Construction [Ref. 2-11]. This *probability-based design method* accounts separately for uncertainties in load and resistance and attempts to ensure that resistance exceeds load. Each individual "load" (e.g., the tension-leg force due to wind forces, to wave forces, to the dead weight of the platform, etc.) is multiplied by a *load factor,* λ_i, that reflects the uncertainty associated with that particular load. The *factored load,* $L_{\text{fact.}}$ is then given by

$$L_{\text{fact.}} \equiv \sum \lambda_i L_i \qquad (2.40)$$

Values of λ_i typically range from 1.2 to 1.6. To reflect uncertainties in the resistance of the member, the *nominal resistance,* $R_{\text{nom.}}$, is reduced by a single *strength-reduction factor* (or, *resistance factor*) $\phi < 1.$[43] Values of ϕ from 0.75 to 0.90 are commonly used. Therefore, in this method a "safe" design is obtained if

$$\sum \lambda_i L_i < \phi R_{\text{nom.}} \qquad (2.41)$$

Probabilistic methods are applied in only one example problem in this textbook (Example Problem 3.16). However, this brief introduction to design philosophies has been included here in order to caution you about some limitations of factor-of-safety design and to indicate that other design methods are also important and will, more than likely, be covered in your future courses in structural design or mechanical component design.

---

[43]The symbol λ is the lowercase Greek letter *lambda;* ϕ is the lowercase Greek letter *phi.*

For all problems in this chapter, assume unknown axial forces to be positive in tension. As in Example 2.2, label tensile-stress answers with (T) and compressive-stress answers with (C).

Prob. 2.2-1. A 1-in.-diameter solid bar (1), a square solid bar (2), and a circular tubular member with 0.2-in. wall thickness (3), each supports an axial tensile load of 5 kips. (a) Determine the axial stress in bar (1). (b) If the axial stress in each of the other bars is 6 ksi, what is the dimension, b, of the square bar, and what is the outer diameter, c, of the tubular member?

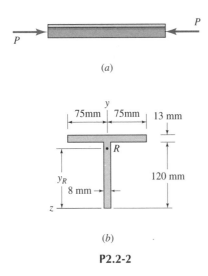

P2.2-1

Prob. 2.2-2. The structural tee shown in Fig. P2.2-2 supports a compressive load $P = 200$ kN. (a) Determine the coordinate y_R of the point R in the cross section where the load must act in order to produce uniform compressive axial stress in the member, and (b) determine the magnitude of that compressive stress.

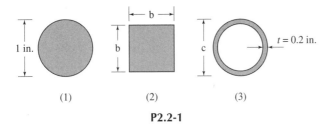

P2.2-2

Prob. 2.2-3. At a particular cross section of a rectangular member, the normal stress on the cross section varies linearly from $\sigma_b = 3$ ksi at the bottom edge to $\sigma_t = 2$ ksi at the top edge, as shown in Fig. P2.2-3b. (a) Determine the magnitude of the resultant of this stress distribution, and (b) determine the coordinate y_R of the point R in the cross section through which the resultant force acts. [Due to symmetry of the stress distribution about the y axis, $z_R = 0$, as shown in Fig. P2.2-3a.]

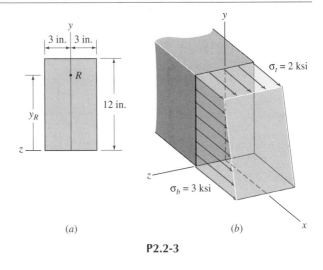

P2.2-3

Prob. 2.2-4. A uniform tensile stress $\sigma_t = 1$ MPa acts over the top half of a rectangular cross section, and a uniform tensile stress of $\sigma_b = 3$ MPa acts over the bottom half of the cross section, as shown in Fig. P2.2-4b. (a) Determine the magnitude of the resultant force on the cross section, and (b) determine the coordinate y_R of the point R in the cross section through which the resultant force acts.

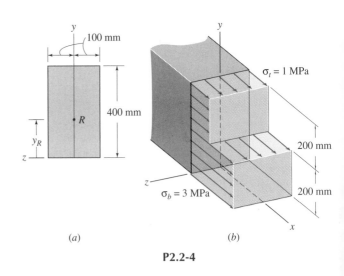

P2.2-4

Prob. 2.2-5. On a particular cross section of a rectangular bar there is a constant normal stress of magnitude σ_0 over the bottom half of the bar, and the stress tapers linearly to zero at the top edge of the cross section, as shown in Fig. P2.2-5b. (a) Determine an expression for the magnitude of the resultant force on the cross section, and (b) determine the coordinate y_R of the point R in the cross section through which the resultant force acts.

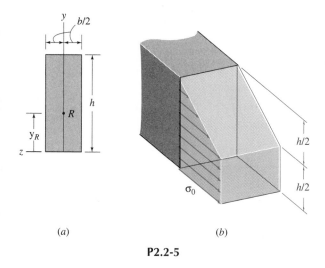

(a)   (b)

P2.2-5

Prob. 2.2-6. A 12-ft beam AB that weighs $W_b = 180$ lb supports an air conditioner that weighs $W_a = 1000$ lb. The beam, in turn, is supported by hanger rods (1) and (2), as shown in Fig. P2.2-6. (a) If the diameter of rod (1) is $\frac{3}{8}$ in., what is the stress, σ_1, in the rod? (b) If the stress in rod (2) is to be the same as the stress in rod (1), what should the diameter of rod (2) be (to the nearest $\frac{1}{32}$ in.)?

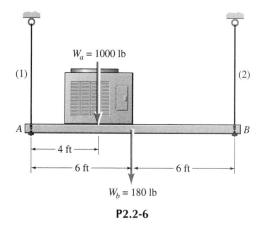

P2.2-6

Prob. 2.2-7. The three-part axially loaded member in Fig. P2.2-7 consists of a tubular segment (1) with outer diameter $(d_o)_1 = 1.25$ in. and inner diameter $(d_i)_1 = 0.875$ in., a solid circular rod segment (2) with diameter $d_2 = 1.25$ in., and another solid circular rod segment (3) with diameter $d_3 = 0.875$ in. The line of

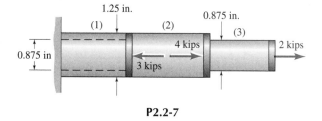

P2.2-7

action of each of the three applied loads is along the centroidal axis of the member. Determine the axial stresses σ_1, σ_2, and σ_3 in each of the three respective segments.

Prob. 2.2-8. A column in a two-story building is fabricated from square structural tubing having the cross-sectional dimensions shown in Fig. P2.2-8b. Axial loads $P_A = 200$ kN and $P_B = 300$ kN are applied to the column at levels A and B, as shown in Fig. P2.2-8a. Determine the axial stress σ_1 in segment AB of the column and the axial stress σ_2 in segment BC of the column. Neglect the weight of the column itself.

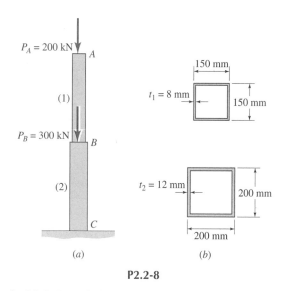

(a)   (b)

P2.2-8

Prob. 2.2-9. In static-load tests of airplane wings, like the one depicted in Fig. 1.2, the loading is applied to the wing in a distributed manner through a "whiffle tree" system that consists of horizontal load-distribution beams, vertical tension bars, and loading "pads," as shown in Fig. P2.2-9a. Determine the stresses

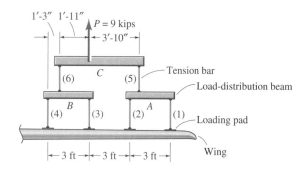

(a) Whiffle-tree loading system.

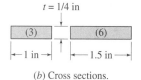

(b) Cross sections.

P2.2-9

σ_3 and σ_6 in tension rods (3) and (6), respectively, if they have the cross-sectional dimensions shown in Fig. P2.2-9b, and if the total load applied to the system is $P = 9$ kips.

*For **Problems 2.2-10 and 2.2-11,** assume that the pin joints apply **axial loading** to each truss member. That is, assume that the force in each truss member acts along the centroidal axis of the member.*

Prob. 2.2-10. Each member of the truss in Fig. P1.4-1 is a solid circular rod with diameter $d = 25$ mm. Determine the axial stress σ_1 in the truss member (1) and the axial stress σ_6 in the truss member (6). (See Prob. 1.4-1.)

Prob. 2.2-11. Each member of the truss in Fig. P1.4-2 is a solid circular rod with diameter $d = 0.75$ in. Determine the axial stresses σ_1, σ_2, and σ_3 in members (1), (2), and (3), respectively. (See Prob. 1.4-2.)

Prob. 2.2-12. The pins at B and D in Fig. P1.4-18 apply an axial load to diagonal bracing member BD. If BD has a rectangular cross section measuring 0.50 in. $\times$ 2.00 in., what is the axial stress in member BD when the load is $w_0 = 200$ lb/ft?

Prob. 2.2-13. The three-member frame in Fig. P2.2-13 is subjected to a horizontal load P at pin E. The pins at C and D apply an axial load to cross-brace member CD, which has a rectangular cross section measuring 30 mm $\times$ 50 mm. If $P = 210$ kN, what is the axial stress in member CD?

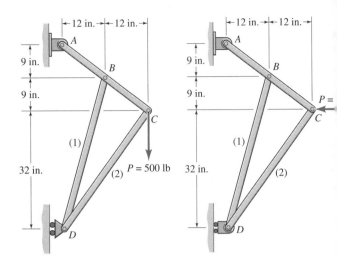

P2.2-14 and P2.2-15

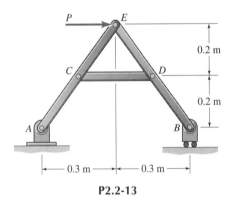

P2.2-13

Prob. 2.2-14. The three-member structure in Fig. P2.2-14 is subjected to a vertical load $P = 500$ lb at pin C. The pins at B, C, and D apply axial loads to members BD and CD. Determine the normal stresses σ_1 and σ_2 in these two members if $A_1 = 0.5$ in² and $A_2 = 1.0$ in².

Prob. 2.2-15. Solve for the normal stresses σ_1 and σ_2 in members BD and CD, respectively, if the load $P = 500$ lb acts horizontally to the left at joint C, rather than vertically downward as in Prob. 2.2-14.

Where both undeformed and deformed configurations are shown, the undeformed configuration is shown with dashed lines and the deformed configuration is shown with solid lines. Points on the deformed body are indicated by an asterisk (*).

Prob. 2.3-1. A wire is used to hang a lantern over a pool. Neglect the weight of the wire, and assume that it is taut, but strain free, before the lantern is hung. When the lantern is hung, it causes a 6-in. sag in the wire. Determine the extensional strain in the wire with the lantern hanging as shown.

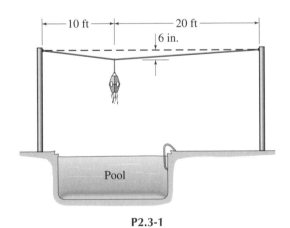

P2.3-1

Prob. 2.3-2. Wire AB of length $L_1 = 3$ m and wire BC of length $L_2 = 4$ m are attached to a ring at B. Upon loading, point B moves vertically downward by an amount $\delta_B = 100$ mm. Determine the extensional strains ϵ_1 and ϵ_2 in wires (1) and (2), respectively.

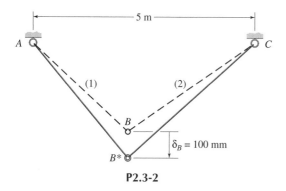

P2.3-2

Prob. 2.3-3. A rigid beam BC of length L is supported by a fixed pin at C and by an extensible rod AB, whose original length is also L. When $\theta = 45°$, rod AB is vertical and strain free, that is, $\epsilon(\theta = 45°) = 0$. Determine an expression for $\epsilon(\theta)$, the strain in rod AB as a function of the angle θ shown in Fig. P2.3-3, valid for $45° \leq \theta \leq 90°$.

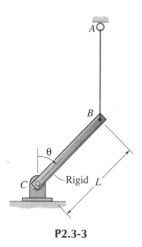

P2.3-3

Prob. 2.3-4. A rod AB, whose unstretched length is L, is originally oriented an angle θ counterclockwise from the $+x$ axis. (a) If the rod is free to rotate about a fixed pin at A, what is the extensional strain in the rod when end B moves a distance u in the $+x$ direction to point B^*? Express your answer in terms of the displacement u, the original length L, and the original angle θ. (b) Simplify the answer you obtained for Part (a), obtaining a small-displacement approximation that is valid if $u \ll L$. (Note: This result will be used in Section 3.8.)

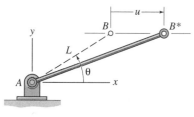

P2.3-4

Prob. 2.3-5. A rod AB, whose unstretched length is L, is originally oriented at angle θ counterclockwise from the $+x$ axis. The rod is free to rotate about a fixed pin at A. (a) What is the extensional strain in the rod when end B moves a distance v in the $+y$ direction to point B^*? Express your answer in terms of v, L, and θ. (b) Simplify the answer you obtained for Part (a), obtaining a small-displacement approximation that is valid if $v \ll L$. (Note: This result will be used in Section 3.8.)

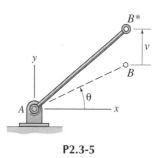

P2.3-5

Prob. 2.3-6. When $P = 0$, the "rigid" beam AD is horizontal, and support rods AB and DE are strain free. With a downward load P on the beam, it is found that the extensional strain in rod DE is $\epsilon_2 = 0.05 \frac{\text{in.}}{\text{in.}}$. Assume (and later show that this is a valid assumption) that the angle θ through which beam AD rotates is small enough that points A and D essentially move vertically, even though both points actually move in circular paths about the fixed pin at C. (a) Determine the value of the (counterclockwise) beam angle θ that corresponds to the measured strain $\epsilon_2 = 0.05 \frac{\text{in.}}{\text{in.}}$. (b) Determine the corresponding extensional strain ϵ_1 in rod AB.

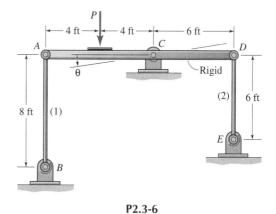

P2.3-6

Prob. 2.3-7. For small loads, P, the rotation of "rigid" beam AF in Fig. P2.3-7 is controlled by the stretching of rod AB. For larger loads, the beam comes into contact with the top of column DE, and further resistance to rotation is shared by the rod and the column. Assume (and later show that this is a valid assumption)

73

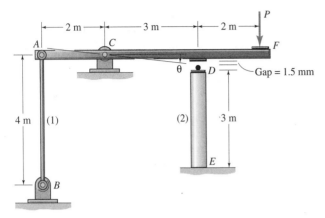

P2.3-7

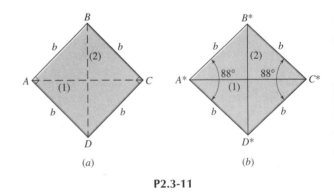

that the angle θ through which beam AF rotates is small enough that points on the beam essentially move vertically, even though they actually move on circular paths about the fixed pin at C. (a) A load P is applied at end F that is just sufficient to close the 1.5-mm gap between the beam and the top of the column at D. What is the strain, ϵ_1, in rod AB for this value of load P? (b) If load P is increased further until $\epsilon_1 = 0.001 \frac{mm}{mm}$, what is the corresponding strain, ϵ_2, in column DE?

Prob. 2.3-8. Vertical rods (1), (2), and (3) are all strain free when they are initially pinned to a straight, rigid, horizontal beam BF. Subsequentially, heating of the rods causes them to elongate and leaves the beam in the position denoted by $B*D*F*$. Point D moves vertically downward by a distance $\delta_D = 0.24$ in., and the inclination angle of the beam is $\theta = 0.5°$ in the counterclockwise sense, as indicated on Fig. P2.3-8. Determine the strains ϵ_1, ϵ_2, and ϵ_3 in the three rods.

Prob. 2.3-9. The "relaxed" rubber band in Fig. P2.3-9a measures $L = 3.5$ in. from end to end. In Fig. P2.3-9b the rubber band is uniformly stretched around a solid circular cylinder of diameter $d = 2.5$ in. What is the extensional strain experienced by the rubber band in the second figure? (Neglect the thickness of the rubber band in determining its stretched diameter.)

Prob. 2.3-10. When a rubber band is uniformly stretched around the solid circular cylinder in Fig. P2.3-10b, its extensional strain is $\epsilon = 0.20$. If the diameter of the cylinder is $d = 4$ in., what is the unstretched length of the rubber band [i.e., length L in Fig. P2.3-10a]? (Neglect the thickness of the rubber band.)

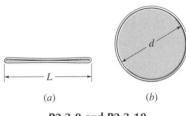

(a)   (b)

P2.3-9 and P2.3-10

Prob. 2.3-11. A thin sheet of rubber in the form of a square (Fig. P2.3-11a) is uniformly deformed into the parallelogram shape shown in Fig. P2.3-11b. All edges remains the same length, b, as the sheet deforms. (a) Compute the extensional strain ϵ_1 of diagonal AC. (b) Compute the extensional strain ϵ_2 of diagonal BD.

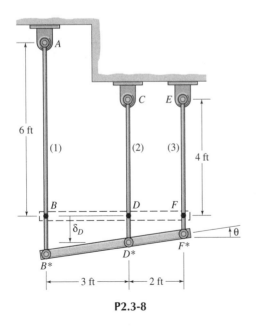

P2.3-8

P2.3-11

Prob. 2.3-12. A thin sheet of rubber in the form of a square (Fig. P2.3-12a) is uniformly deformed into the parallelogram shape shown in Fig. P2.3-12b. All edges remain the same length, b, as the sheet deforms. (a) Compute the extensional strain ϵ_1 of diagonal AC. (b) Compute the extensional strain ϵ_2 of diagonal BD.

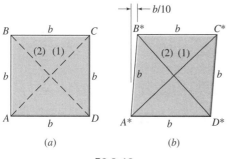

(a)

(b)

P2.3-12

Prob. 2.3-13. A titanium-alloy boom AB in Fig. P2.3-13 supports a very flexible solar array on a communications satellite. At 20°C, the nominal length of the boom is $L_{20} = 10.000$ m. As the satellite passes into and out of the sunlight, the temperature of the boom can vary between $+40$°C and -60°C. What is the length of the boom at each of these temperatures if the coefficient of thermal expansion is $\alpha = 8 \times 10^{-6}$/°C? Record your answers to the nearest millimeter.

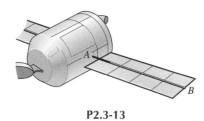

P2.3-13

Prob. 2.3-14. Gage blocks are used as calibrated references for measuring length. The rectangular gage block shown in Fig. P2.3-14 has a nominal thickness of $t_{70} = 1.000$ in. at 70°F. (a) What is the error in thickness per °F if the block is made of tungsten with $\alpha = 2.4 \times 10^{-6}$/°F? (b) What is the error in thickness per °F if the block is made of bronze with $\alpha = 10 \times 10^{-6}$/°F?

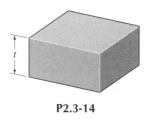

P2.3-14

Prob. 2.3-15. A steel pipe ($\alpha = 8 \times 10^{-6}$/°F) has a nominal inside diameter $d_i = 4.06$ in. at 70°F. (a) What is the inside diameter if the pipe carries steam that raises its temperature to 212°F? (Assume that the outside of the pipe is insulated so that the pipe reaches a uniform temperature of 212°F.) (b) How much would the steam-carrying pipe increase in length if it is originally 40 ft long, if its ends are not restrained against axial motion, and if the temperature increases from 70°F to 212°F?

Prob. 2.3-16. A steel restraining ring in the form of a toroid with rectangular cross section is to be heated, slipped over the end of a solid steel shaft, and cooled to shrink-fit onto the shaft. At 20°C the shaft has an outside diameter of $d_{os} = 50.000 \pm 0.005$ mm, and the ring has an inside diameter of $d_{ir} = 49.950 \pm 0.005$ mm. What is the maximum amount that the ring would have to be heated in order to slip it over the end of the shaft if the shaft temperature is held at 20°C? $\alpha = 14 \times 10^{-6}$/°C.

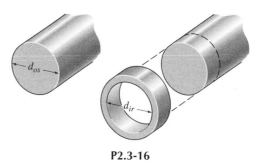

P2.3-16

Prob. 2.3-17. A rigid beam BC of length L is supported by a pin at C and by a rod AB, as illustrated in Fig. P2.3-17. When rod AB is at the reference temperature, its length is L and the angle θ is 45°. Determine an expression for the angle θ as a function of the temperature of the rod, that is, determine an expression for $\theta(\Delta T)$. The properties of rod AB are: A, E, and α. (Hint: Use the trigonometric law of cosines to obtain an expression for L_1^*.)

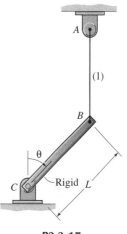

P2.3-17

75

Prob. 2.3-18. At the reference temperature, three identical rods of length L form a pin-jointed truss in the shape of an equilateral triangle, ABC, as shown in Fig. P2.3-18. Determine the angle θ as a function of ΔT, the temperature increase of tie rod AB. Rods AC and BC remain at the reference temperature. The properties of the rods are: area $= A$, modulus of elasticity $= E$, and coefficient of thermal expansion $= \alpha$.

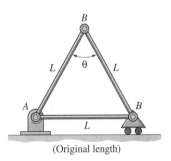

(Original length)

P2.3-18

Prob. 2.3-19. The angular orientation θ of a mirror that is used to reflect a beam of laser light is controlled by the lengths of rods (1) and (2), as shown in Fig. P2.3-19. At the reference temperature, the rods are the same length: $L_1 = L_2 = L$. If the properties of the rods are A_1, E_1, α_1, and A_2, E_2, α_2, respectively, determine an expression for θ as a function of ΔT_1 and ΔT_2. (Assume that the rods are uniformly heated along their lengths.)

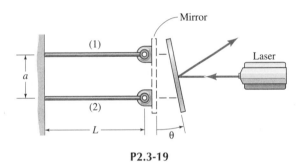

P2.3-19

Prob. 2.3-20. As illustrated in Fig. P2.3-20, a rigid beam BC of length $L_{BC} = 30$ in. is supported by a cable AB that is 40 in. long

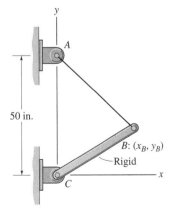

P2.3-20

at the reference temperature of $T_0 = 70°F$. Determine an expression that relates the horizontal coordinate of point B to the temperature T of the rod AB in °F. Let $\alpha = 8 \times 10^{-6}/°F$. (Hint: Use the trigonometric law of cosines.)

Prob. 2.4-1. A *mechanical extensometer* uses the lever principle to magnify the elongation of a test specimen enough to make the elongation (or contraction) readable. The extensometer shown in Fig. P2.4-1 is held against the test specimen by a spring that forces two sharp points against the specimen at A and B. The pointer AD pivots about a pin at C, so that the distance between the contact points at A and B is exactly $L_0 = 6$ in. (the *gage length*, or gauge length, of this extensometer) when the pointer points to the origin, O, on the scale. In a particular test, the extensometer arm points "precisely" at point O when the load P is zero. Later in the test, the 10-in.-long pointer points a distance $d = 0.12$ in. below point O. What is the current extensional strain in the test specimen at this reading?

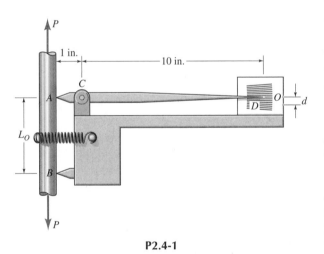

P2.4-1

Prob. 2.4-2. A "pencil" *laser extensometer,* like the mechanical lever extensometer in Prob. 2.4-1, measures elongation (from which extensional strain can be computed) by multiplying the elongation. In Fig. P2.4-2 the laser extensometer is being used to measure strain in a reinforced concrete column. The target is set up across the room from the test specimen so that the distance from the fulcrum, C, of the laser to the reference point O on the target is $d_{OC} = 200$ in. Also, the target is set so that the laser beam points directly at point O on the target when the extensometer points are exactly $L_0 = 6.00$ in. apart on the specimen, and the cross section at B does not move vertically. At a particular value of (compressive) load P, the laser points upward by an angle that is indicated on the target to be $\phi = 0.0030$ rad. Determine the extensional strain in the concrete column at this load value.

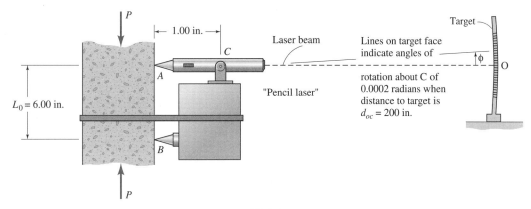

P2.4-2

Problems 2.4-3 through 2.4-6. You are strongly urged to use a computer program (e.g., a spreadsheet program) to plot the stress-strain diagrams for these problems. In some cases it will be advantageous to make two plots, one covering the initial few points and one covering the entire dataset.

Prob. 2.4-3. The data in Table P2.4-3 was obtained in a tensile test of a flat-bar steel specimen having the dimensions shown in Fig. P2.4-3. (a) Plot a curve of engineering stress, σ, versus engineering strain, ϵ, using the given data. (b) Determine the modulus of elasticity of this material. (c) Use the 0.2%-offset method to determine the yield strength of this material, σ_{YS}.

TABLE P2.4-3.   Tension-test Data; Flat Steel Bar

P(kips)	ΔL (in.)	P(kips)	ΔL (in.)
1.2	0.0008	6.25	0.0060
2.4	0.0016	6.50	0.0075
3.6	0.0024	6.65	0.0100
4.8	0.0032	6.85	0.0125
5.7	0.0040	6.90	0.0150
5.95	0.0050	—	—

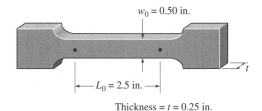

$w_0 = 0.50$ in.

$L_0 = 2.5$ in.

Thickness = $t = 0.25$ in.

P2.4-3

Prob. 2.4-4. A standard ASTM tension specimen (diameter = $d_0 = 0.505$ in., gage length = $L_0 = 2.0$ in.) was used to obtain the load-elongation data given in Table P2.4-4. (a) Plot a curve of engineering stress, σ, versus engineering strain, ϵ, using the given data. (b) Determine the modulus of elasticity of this material. (c) Use the 0.2%-offset method to determine the yield strength of this material, σ_{YS}.

TABLE P2.4-4.   Tension-test Data; ASTM Tension Specimen

P(kips)	ΔL (in.)	P(kips)	ΔL (in.)
1.9	0.0020	10.0	0.0145
3.8	0.0040	10.4	0.0180
5.7	0.0060	10.65	0.0240
7.6	0.0080	11.00	0.0300
9.0	0.0100	11.05	0.0360
9.5	0.0120	—	—

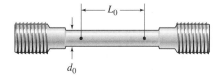

L_0

d_0

P2.4-4

Prob. 2.4-5. A tension specimen (diameter = $d_0 = 0.500$ in., gage length = $L_0 = 2.0$ in.) was used to obtain the load-elongation data given in Table P2.4-5. (a) Plot a curve of engineering stress, σ, versus engineering strain, ϵ, using the given data. (b) Determine the modulus of elasticity of this material. (c) Use the 0.2%-offset method to determine the yield strength of this material, σ_{YS}. (d) Determine the tensile ultimate stress, σ_{TU}.

TABLE P2.4-5.   Tension-test Data

P(kips)	ΔL (in.)	P(kips)	ΔL (in.)
0.0	0.000	5.9	0.067
2.0	0.002	6.1	0.080
3.2	0.008	6.1	0.093
3.8	0.013	6.2	0.107
4.8	0.027	6.1	0.120
5.4	0.040	5.9	0.133
5.7	0.053	5.6	0.147

Prob. 2.4-6. A tension specimen (diameter $= d_0 = 0.500$ in., gage length $= L_0 = 2.0$ in.) was used to obtain the load-elongation data given in Table P2.4-6. (a) Plot a curve of engineering stress, σ, versus engineering strain, ϵ, using the given data. (b) Determine the modulus of elasticity of this material. (c) Use the 0.2%-offset method to determine the yield strength of this material, σ_{YS}. (d) Determine the tensile ultimate stress, σ_{TU}.

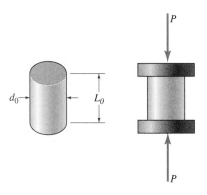

P2.6-2

TABLE P2.4-6. Tension-test Data

P(kips)	ΔL (in.)	P(kips)	ΔL (in.)
0.0	0.000	12.5	0.060
5.2	0.005	12.7	0.070
9.4	0.009	12.9	0.080
9.7	0.010	13.0	0.090
10.0	0.013	13.1	0.100
10.6	0.020	13.2	0.110
11.3	0.030	13.2	0.120
11.8	0.040	13.0	0.130
12.2	0.050	12.6	0.138

elasticity, E, for this brass specimen? (b) If the diameter increases due to the load P by 0.0058 mm, what is the value of Poisson's ratio, ν?

Prob. 2.6-3. A rectangular aluminum bar ($w_0 = 2.0$ in., $t_0 = 0.5$ in.) is subjected to a tensile load P by pins at A and B (Fig. P2.6-3). Strain gages (which will be described in Section 8.12) measure the following strains in the longitudinal (x) and transverse (y) directions: $\epsilon_x = 566\mu$, and $\epsilon_y = -187\mu$. (a) What is the value of Poisson's ratio for this specimen? (b) If the load P that produces these values of ϵ_x and ϵ_y is $P = 6$ kips, what is the modulus of elasticity, E, for this specimen? (c) What is the change in volume, ΔV, of a segment of bar that is initially 2 in. long? (Hint: $\epsilon_z = \epsilon_y$.)

*In **Problems 2.6-1 through 2.6-10**, dimensions that are shown on the figures, or dimensions that are labeled with subscript **0** (e.g., $\mathbf{d_0}$, $\mathbf{L_0}$) are dimensions of the specimen without any load applied.*

Prob. 2.6-1. A tensile test is performed on an aluminum specimen that is 0.505 in. in diameter using a gage length of 2 in., as shown in Fig. P2.6-1. (a) When the load is increased by an amount $P = 2$ kips, the distance between gage marks increases by an amount $\Delta L = 0.00196$ in. Calculate the value of the modulus of elasticity, E, for this specimen. (b) If the proportional limit stress for this specimen is $\sigma_{PL} = 45$ ksi, what is the distance between gage marks at this value of stress?

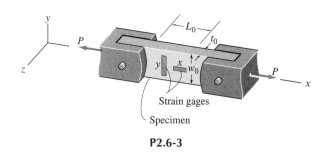

P2.6-3

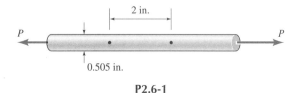

P2.6-1

Prob. 2.6-2. A short brass cylinder ($d_0 = 15$ mm, $L_0 = 25$ mm) is compressed between two perfectly smooth, rigid plates by an axial force $P = 20$ kN. (a) If the measured shortening of the cylinder due to this force is 0.0283 mm, what is the modulus of

Prob. 2.6-4. A tensile force of 500 kN is applied to a uniform segment of a brass bar. The cross section is a 50 mm × 50 mm square, and the length of the segment being tested is 200 mm. Upon application of the load, the cross-sectional dimension changes to 49.967 mm and the length changes to 200.381 mm. Calculate the values of the modulus of elasticity, E, and Poisson's ratio, ν.

Prob. 2.6-5. Under a compressive load of $P = 24$ kips, the length of the concrete cylinder in Fig. P2.6-5 is reduced from 12 in. to 11.9970 in., and the diameter is increased from 6 in., to 6.0003 in. Determine the value of the modulus of elasticity, E, and the value of Poisson's ratio, ν. Assume linearly elastic deformation.

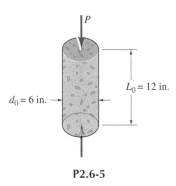

P2.6-5

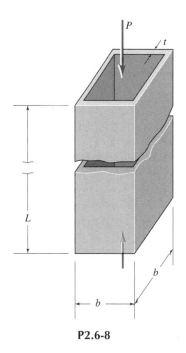

P2.6-8

Prob. 2.6-6. A tensile specimen of a certain alloy has a diameter of 12.7 mm and a gage length of 200 mm. Under a load $P = 20$ kN, the specimen reaches its proportional limit and is elongated by 0.30 mm. At this load the diameter is reduced by 0.0064 mm. Determine the following material properties: (a) the modulus of elasticity, E, and (b) Poisson's ratio, ν, and (c) the proportional limit, σ_{PL}.

Prob. 2.6-7. The cylindrical rod in Fig. P2.6-7 is made of annealed (soft) copper with modulus of elasticity $E = 17 \times 10^3$ ksi and Poisson's ratio $\nu = 0.33$, and it has an initial diameter $d_o = 1.9998$ in. For compressive loads less than a "critical load" P_{cr}, a ring with inside diameter $d_r = 2.0000$ in. is free to slide along the cylindrical rod. What is the value of the critical load P_{cr}?

40 mm, $t_1 = 20$ mm) and ($w_2 = 50$ mm, $t_2 = 25$ mm), respectively. (a) At what location, b, must the load P act if the rigid bar BD is to remain horizontal when the load is applied? (b) If the longitudinal strain in the hanger rods is $\epsilon_1 = \epsilon_2 = 500\mu$ when the load is $P = 205$ kN, what is the value of the modulus of elasticity, E? (c) If application of load P in the manner described in Parts (a) and (b) above causes the dimension w_2 of hanger rod (2) to be reduced from 50 mm to 49.9918 mm, what is the value of Poisson's ratio, ν, for this material?

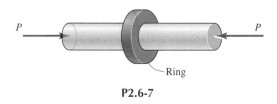

P2.6-7

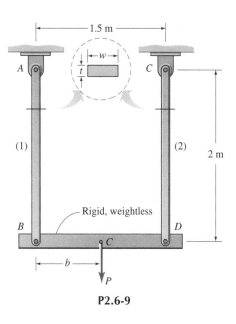

P2.6-9

Prob. 2.6-8. A square tubular steel column of length $L_0 = 12$ ft, outer dimension $b_0 = 4$ in., and wall thickness $t_0 = 0.5$ in. is subjected to an axial compressive load $P = 125$ kips as shown in Fig. P2.6-8. If the steel has a modulus of elasticity $E = 29 \times 10^3$ ksi and Poisson's ratio $\nu = 0.30$, determine: (a) the change, ΔL, in the length of the column, and (b) the change, Δt, in the wall thickness.

Prob. 2.6-9. A rigid, weightless beam BD supports a load P and is, in turn, supported by two hanger rods, (1) and (2), as shown in Fig. P2.6-9. The rods are initially the same length $L = 2$ m and are made of the same material and $L_{BD} = 1.5$ m. Their rectangular cross sections have original dimensions ($w_1 = $

Prob. 2.6-10. In the evening, a contractor attaches a steel wire of initial length $L_0 = 8$ ft and diameter $d_0 = \frac{3}{16}$ in. at point A on an air compressor that weighs 600 lb, as illustrated in Fig. P2.6-10. The other end of the wire is attached to a bolt B on the end of the boom of a construction crane. The boom is then raised until the air compressor is suspended about 10 ft above ground, a safe distance to prevent mischief from being done to the compressor overnight. (a) If the modulus of elasticity of the steel wire is $E = 29 \times 10^3$ ksi, how much does wire AB elongate when the compressor is lifted off the ground? (b) If $\nu = 0.30$, how much change is there in the diameter of the wire?

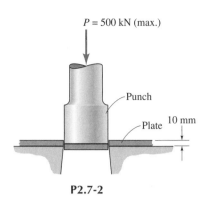

P2.7-2

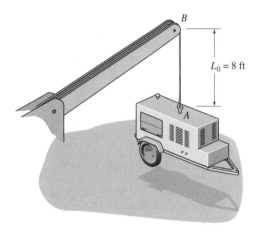

P2.6-10

Prob. 2.7-3. The hole in a hasp plate is punched out by a hydraulic punch press similar to the one in Prob. 2.7-2, with a punch in the shape of the "rectangular" hole as illustrated in Fig. P2.7-3. If the hasp plate is $\frac{1}{16}$-in.-thick steel with an average punching shear resistance of 40 ksi, what is the required punch force P?

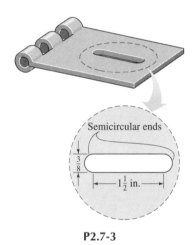

P2.7-3

Prob. 2.7-1. Two bolts are used to form a joint connecting rectangular bars in tension, as shown in Fig. P2.7-1. Determine the required diameter of the bolts if the average shear stress for the bolts is not to exceed 20 ksi for the given loading.

Prob. 2.7-4. A contractor, who finds it necessary to splice together two 2×4 boards (see Appendix D.8) to form a long tension member, does so by using four $\frac{3}{8}$-in.-diameter bolts to attach two steel splice plates, as shown in Fig. P2.7-4. If the 2×4s carry a tensile load of 2 kips, what is the average shear stress experienced by the bolts?

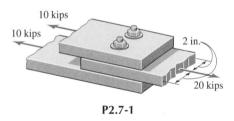

P2.7-1

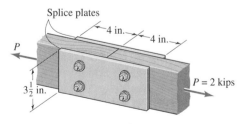

P2.7-4 and P2.7-5

Prob. 2.7-2. A 500-kN-capacity hydraulic punch press is used to punch circular holes in a 10-mm-thick aluminum plate, as illustrated in Fig. P2.7-2. If the average punching shear resistance of this plate is 200 MPa, what is the maximum diameter of hole that can be punched?

Prob. 2.7-5. If, instead of using bolts, as described in Prob. 2.7-4, the contractor had used glue to attach the two $3\frac{1}{2} \times 8 \times \frac{1}{4}$-in. steel splice plates to the two 2×4s, what would be the value of the average shear stress in the glue joints for the loading and the positioning of splice plates shown in Fig. P2.7-5?

Prob. 2.7-6. The high-wire for a circus act is attached to a vertical beam AC and is kept taut by a tensioner cable BD as illustrated in Fig. P2.7-6. At C, the beam AC is attached by a 10-mm-diameter bolt to the bracket shown in View a-a. Determine the average shear stress in the bolt at C if the tension in the high-wire is 5 kN. (Assume that the high-wire is horizontal, and neglect the weight of AC.)

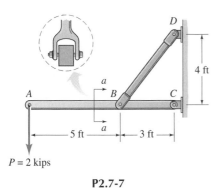

P2.7-6

Prob. 2.7-7. The horizontal beam AC in Fig. P2.7-7 supports a 2-kip load P at end A and is, in turn, supported by a bracket at C and a tie-rod BD. The shape of the tie-rod end at B is shown in the enlargement of View a-a. Determine the average shear stress in the bolt at B for the loading condition shown, if the bolt has a diameter $d_B = 0.75$ in.

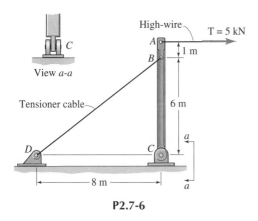

P2.7-7

Prob. 2.7-8. Two equal-leg angles are used to suspend a load P from a wood beam, as shown in side view in Fig. P2.7-8a and end view in Fig. P2.7-8b. If $P = 10$ kN and the diameter of the bolts passing through the two angles and the beam is 6.5 mm, what is the average shear stress in these bolts? (Neglect friction between the angles and the beam.)

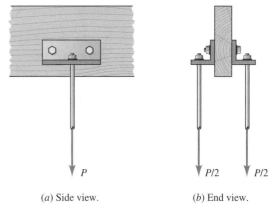

(a) Side view.   (b) End view.

P2.7-8

Prob. 2.7-9. Two $\frac{3}{4}$-in. nylon rods are spliced together by gluing a 4-in. section of plastic pipe over the rod ends, as shown in Fig. P2.7-9. If a tensile force of $P = 530$ lb is applied to the spliced nylon rod, what is the average shear stress in the glue joint between the pipe and the rods?

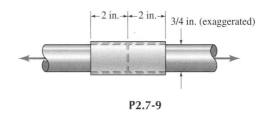

P2.7-9

Prob. 2.7-10. A pipe flange is attached by four bolts, whose effective diameter is 0.425 in., to a concrete base. The bolts are uniformly spaced around an 8-in.-diameter bolt circle, as shown in Fig. P2.7-10. If a twisting couple $T = 5000$ lb·in. is applied to the pipe flange, as shown in Fig. P2.7-10, what is the average shear stress in each of the four bolts? Neglect friction between the pipe flange and the concrete base.

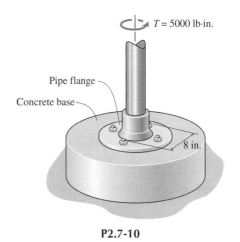

P2.7-10

Prob. 2.7-11. A rectangular plate (dashed lines show original configuration) is uniformly deformed into the shape of a parallelogram (shaded figure) as shown in Fig. P2.7-11. (a) Determine the average shear strain, call it $\gamma_{xy}(A)$, between lines in the directions x and y shown in the figure. (b) Determine the average shear strain, call it $\gamma_{x'y'}(B)$, between lines in the directions of x' and y' shown in the figure. (Hint: Don't forget that shear strain is a signed quantity, that is, it can be either positive or negative.)

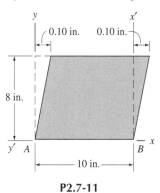

P2.7-11

Prob. 2.7-12. Shear stress τ produces a shear strain γ_{xy} (between lines in the x direction and lines in the y direction) of $\gamma_{xy} = 1200\ \mu$ (i.e., $\gamma = 0.0012\ \frac{m}{m}$). (a) Determine the horizontal displacement δ_A of point A. (b) Determine the shear strain $\gamma_{x'y'}$ between the lines in the x' direction and the y' direction, as shown on Fig. P2.7-12.

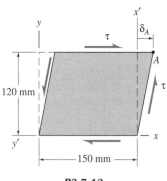

P2.7-12

Prob. 2.7-13. Two identical symmetrically placed rubber pads transmit load from a rectangular bar to a C-shaped bracket, as shown in Fig. P2.7-13. (a) Determine the average shear stress, τ, in the rubber pads on planes parallel to the top and bottom surfaces of the pads if $P = 250$ N and the dimensions of the rubber pads are: $b = 50$ mm, $w = 80$ mm, and $h = 25$ mm. (Although the load is transmitted predominately by shearing deformation, the pads are not undergoing pure shear. However, you can still calculate the *average* shear stress and *average* shear strain.) (b) If the shear modulus of elasticity of the rubber is $G_r = 0.6$ MPa, what is the average shear strain, γ, related to the average shear stress τ computed in Part (a)? (c) Based on the average shear strain determined in Part (b) what is the relative displacement, δ, between the rectangular bar and the C-shaped bracket when the load $P = 250$ N is applied?

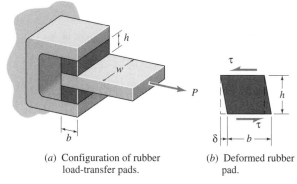

(a) Configuration of rubber load-transfer pads.

(b) Deformed rubber pad.

P2.7-13 and P2.7-14

Prob. 2.7-14. Two identical, symmetrically placed rubber pads transmit load from a rectangular bar to a C-shaped bracket, as shown in Fig. 2.7-14. The dimensions of the rubber pads are: $b = 3$ in., $w = 4$ in., and $h = 2$ in. The shear modulus of elasticity of the rubber is $G_r = 100$ psi. If the maximum relative displacement between the bar and the bracket is $\delta_{max} = 0.25$ in., what is the maximum value of load P that may be applied? (Use average shear strain and average shear stress in solving this problem.)

*Prob. 2.7-15. Vibration isolators like the one shown in Fig. P2.7-15 are used to support sensitive instruments. Each isolator consists of a hollow rubber cylinder of outer diameter D, inner diameter d, and height h. A steel center post of diameter d is bonded to the inner surface of the rubber cylinder, and the outer surface of the rubber cylinder is bonded to the inner surface of a steel-tube base. (a) Derive an expression for the average shear stress in the rubber as a function of the distance r from the center of the isolator. (b) Derive an expression relating the load P to the downward displacement of the center post, using G as the shear

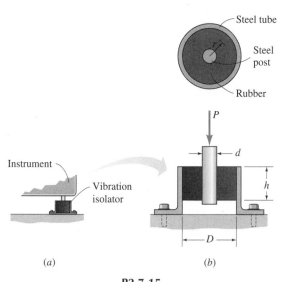

(a)

(b)

P2.7-15

modulus of the rubber, and assuming that the steel post and steel tube are rigid (compared with the rubber). (Hint: Since the shear strain varies with the distance r from the center, an integral is required.)

Prob. 2.8-1. A bar with rectangular cross section is subjected to an axial tensile load P, as shown in Fig. P2.8-1. (a) Determine the angle, call it θ_{na}, of the plane NN' on which $\tau_{nt} = 2\sigma_n$, that is, the plane on which the magnitude of the shear stress is twice the magnitude of the normal stress. (b) Determine the angle, call it θ_{nb}, of the plane on which $\sigma_n = 2\tau_{nt}$. (Hint: You can get approximate answers from Fig. 2.30.)

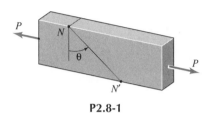

P2.8-1

Prob. 2.8-2. A wood cube that measures $3\frac{1}{2}$ in. on each side is tested in compression, as illustrated in Fig. P2.8-2. The direction of the grain of the wood is shown in the figure. Determine the normal stress σ_n and shear stress τ_{nt} on planes that are parallel to the grain of the wood.

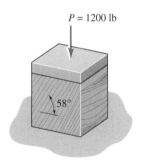

P2.8-2

Prob. 2.8-3. A prismatic bar in tension has a cross section that measures $20 \text{ mm} \times 65 \text{ mm}$ and supports a tensile load $P = 200$ kN, as illustrated in Fig. P2.8-3. Determine the normal and shear stresses on the n and t faces of an element oriented at angle $\theta = 30°$.

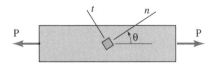

P2.8-3 and P2.8-4

Prob. 2.8-4. The prismatic bar in Fig. P2.8-4 is subjected to an axial compressive load $P = -70$ kips. The cross-sectional area of the bar is 2.0 in.^{2}. Determine the normal stress and the shear stress on the n face and on the t face of an element oriented at angle $\theta = 50°$.

Prob. 2.8-5. The tie rod BD that supports beam AC has a diameter of $\frac{7}{8}$ in. (a) Determine the normal stress on a typical cross section a-a of the tie rod. (b) Determine the normal stress σ and shear stress τ on a typical vertical section b-b through the tie rod.

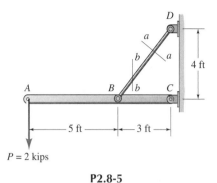

P2.8-5

Prob. 2.8-6. A 6-in.-diameter concrete test cylinder is subjected to a compressive load $P = 110$ kips, as shown in Fig. P2.8-6. The cylinder fails along a plane that makes an angle of 62° to the horizontal. (a) Determine the (compressive) axial stress in the cylinder when it reaches the failure load. (b) Determine the normal stress, σ, and shear stress, τ, on the failure plane at failure.

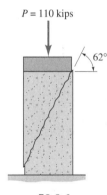

P2.8-6

Prob. 2.8-7. The plane NN' makes an angle $\theta = 26°$ with respect to the cross section of the prismatic bar shown in Fig. P2.8-7. The dimensions of the rectangular cross section of the bar are 1 in. $\times$ 2 in. Under the action of an axial tensile load, P, the normal stress on the NN' plane is $\sigma_n = 8$ ksi. (a) Determine the value of the axial load P. (b) Determine the shear stress τ_{nt} on the NN' plane. (c) Determine the maximum shear stress in the bar.

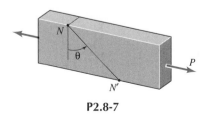

P2.8-7

Prob. 2.8-8. Determine the allowable tensile load P for the prismatic bar shown in Fig. P2.8-8 if the allowable tensile stress is $\sigma_{allow.} = 22$ ksi and the allowable shear stress is $\tau_{allow.} = 14$ ksi. The cross-sectional dimensions of the bar are $\frac{1}{4}$ in. $\times 2$ in.

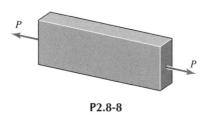

P2.8-8

Prob. 2.8-9. An aluminum bar with a square cross section of dimension b is subjected to a compressive load, $P = 50$ kN, as shown in Fig. P2.8-9. If the allowable compressive stress for the aluminum is $\sigma_{allow.} = -130$ MPa, and $\tau_{allow.} = 75$ MPa, what is the minimum value of the dimension b, to the nearest millimeter?

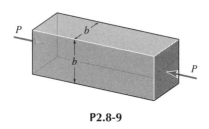

P2.8-9

Prob. 2.8-10. Either a finger-joint splice, Fig. P2.8-10a, or a diagonal lap-joint splice, Fig. P2.8-10b, may be used to glue two wood strips together to form a longer tension member. Determine the ratio of allowable loads, $(P_a)_{allow.}/(P_b)_{allow.}$, for the following two glue-strength cases: (a) the glue is twice as strong in tension as it is in shear, that is $\tau_{allow.} = 0.5\sigma_{allow.}$, and (b) the glue is twice as strong in shear as it is in tension, that is $\tau_{allow.} = 2\sigma_{allow.}$. (Hint: For each of the above cases, determine ($P_{allow.}$ in terms of $\sigma_{allow.}$, using the given glue-strength ratios.)

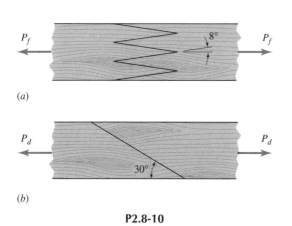

(a)

(b)

P2.8-10

Prob. 2.8-11. At room temperature (70°F) and with no axial load ($P = 0$) the extensional strain of the prismatic bar (Fig. P2.8-11) in the axial direction is zero, that is $\epsilon_x = 0$. Subsequently, the bar is heated to 120° F and a tensile load P is applied. The material properties for the bar are: $E = 10 \times 10^3$ ksi and $\alpha = 13 \times 10^{-6}/°F$, and the cross-sectional area of the bar is 1.8 in². For the latter load/temperature condition, the extensional strain is found to be $\epsilon_x = 900 \times 10^{-6} \frac{in.}{in.}$. (a) Determine the value of the axial tensile load P. (b) Determine the normal stress and the shear stress on the oblique plane NN'. (Note: The total strain is the sum of strain associated with normal stress σ_x (Eq. 2.14) and the strain due to change of temperature ΔT (Eq. 2.8).)

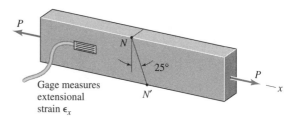

Gage measures extensional strain ϵ_x

P2.8-11

Prob. 2.8-12. The steel eyebar in Fig. P2.8-12 has cross-sectional dimensions 20 mm $\times$ 90 mm. To obtain the required length of bar, it is necessary to weld together two separate lengths along a diagonal joint that makes an angle θ with respect to the cross section. The stresses in the weld are limited to 100 MPa in tension and 70 MPa in shear. (a) Determine the optimum angle θ ($0 \leq \theta \leq 45°$) for the welded joint. (b) Determine the maximum safe load for the optimally spliced bar.

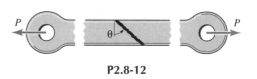

P2.8-12

Prob. 2.9-1. The normal stress on the rectangular cross section $ABCD$ in Fig. P2.9-1 varies linearly with respect to position (y, z) in the cross section. That is, σ_x has the form $\sigma_x = a + by + cz$. The values of σ_x at corners A, B, and C are: $\sigma_{xA} = 12$ ksi, $\sigma_{xB} = 8$ ksi, and $\sigma_{xC} = 12$ ksi. (a) Determine the value of σ_{xD}, the normal stress at corner D. (b) Determine the axial force, F_x. (c) Determine the bending moment M_y.

*Prob. 2.9-2. The normal stress, σ_x, on the trapezoidal cross section shown in Fig. P2.9-2b varies linearly with y, the distance from the bottom edge of the section. The values of σ_x at the bottom edge and top edge are $\sigma_{xA} = 20$ MPa and $\sigma_{xB} = 10$ MPa, respectively, as indicated in Fig. P2.9-2a. (a) Determine the value of the axial force, F_x. (b) Locate the point R in the cross section through which the resultant axial force, F_x, acts. (Note: F_x must have the same moment, M_z, about the z axis as the moment due to the distributed stress σ_x.)

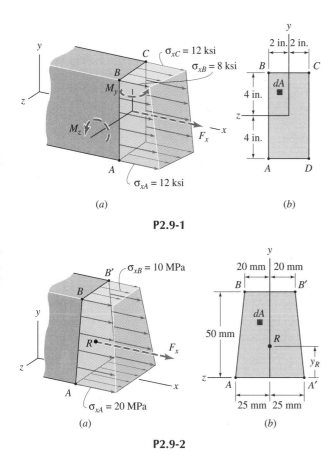

(a)

$\sigma_{xC} = 12$ ksi
$\sigma_{xB} = 8$ ksi

$\sigma_{xA} = 12$ ksi

(b)

P2.9-1

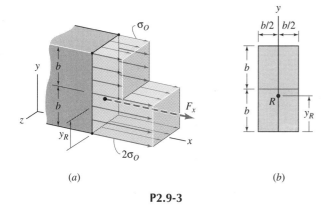

(a)

$\sigma_{xB} = 10$ MPa

$\sigma_{xA} = 20$ MPa

(b)

20 mm | 20 mm

50 mm

25 mm | 25 mm

P2.9-2

Prob. 2.9-3. The normal stress, σ_x, over the top half of the cross section of the rectangular bar in Fig. P2.9-3 is σ_0, while the normal stress acting on the bottom half of the cross section is $2\sigma_0$. (a) Determine the value of the resultant axial force, F_x. (b) Locate the point R in the cross section through which the resultant axial force, F_x, acts.

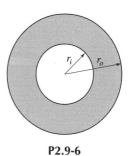

(a)

σ_0

$2\sigma_0$

(b)

$b/2$ | $b/2$

P2.9-3

Prob. 2.9-4. On a particular cross section of a rectangular bar, there is shear stress whose distribution has the form

$$\tau_{xy} = \tau_{max}\left[1 - \left(\frac{y}{4}\right)^2\right]$$

where y is measured from the centroid of the cross section (Fig. P2.9-4b). If the shear stress τ_{xy} may not exceed $\tau_{allow.} = 10$ ksi, what is the maximum shear force V_y that may be applied to the bar at this cross section?

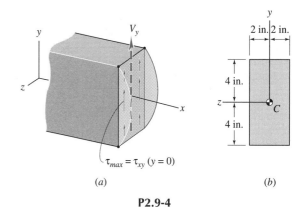

V_y

$\tau_{max} = \tau_{xy}$ ($y = 0$)

(a)

2 in. | 2 in.

4 in.

C

4 in.

(b)

P2.9-4

Prob. 2.9-5. On the cross section of a circular rod, the shear stress at a point acts in the circumferential direction at that point, as illustrated in Fig. P2.9-5. The shear stress magnitude varies linearly with distance from the center of the cross section, that is $\tau = \dfrac{\tau_{max}\rho}{r}$. Using the ring-shaped area in Fig. P2.9-5b, determine the formula that relates τ_{max} and the resultant torque, T.

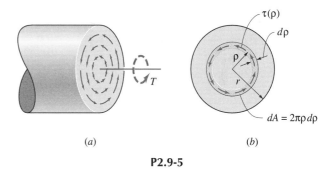

$\tau(\rho)$

$d\rho$

$dA = 2\pi\rho\,d\rho$

(a)

(b)

P2.9-5

Prob. 2.9-6. Determine the relationship between τ_{max} and T if, instead of acting on a solid circular bar, as in Fig. P2.9-5, the shear stress distribution $\tau = \dfrac{\tau_{max}\rho}{r_o}$ acts on a tubular cylinder with

r_i r_o

P2.9-6

outer radius r_o and inner radius r_i. (The cross-sectional dimensions are shown in Fig. P7.9-6. See Prob. 2.9-5 for an illustration of the shear stress distribution on a circular cross section and for the definitions of T and ρ.)

Prob. 2.9-7. If the magnitude of the shear stress on a solid, circular rod of radius r varies with radial position ρ as shown in Fig. P2.9-7, determine the formula that relates the resultant torque T to the maximum shear stress τ_Y. (See Prob. 2.9-5 for an illustration of the shear stress distribution on the cross section and for the definitions of T and ρ, and use the area shown in Fig. 2.9-5b.)

The undeformed and deformed plates are shown in Figs. P2.9-9a and P2.9-9b, respectively. Using the definition extensional strain in Eq. 2.29, determine an expression for $\epsilon_x(x, y)$, the extensional strain in the x direction.

*Prob. 2.9-10. Using the definition of shear strain, Eq. 2.30, and using the "before deformation" and "after deformation" sketches in Fig. P2.9-10, determine an expression for the shear strain γ_{xy} as a function of position in the plate, that is, determine $\gamma_{xy}(x, y)$.

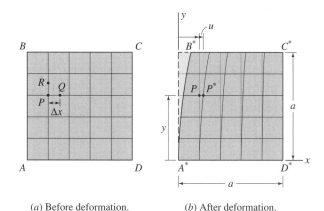

(a) Before deformation.　　　(b) After deformation.

P2.9-9 and P2.9-10

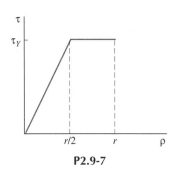

P2.9-7

Prob. 2.9-8. (a) Using Figs. P2.9-8 and the definition of extensional strain given in Eq. 2.29, show that the change in length, ΔL, of a thin wire whose original length is L is given by $\Delta L = \int_0^L \epsilon_x(x)\, dx$, where $\epsilon_x(x)$ is the extensional strain of the wire at x. (b) Determine the elongation of a 2-m-long wire if it has a coefficient of thermal expansion $\alpha = 20 \times 10^{-6}/°C$, and if the change in temperature along the wire is given by $\Delta T = 20x$ (°C).

Prob. 2.9-11. A thin, square plate $ABCD$ undergoes deformation in which no point in the plate moves in the y direction. Every horizontal line (except the bottom edge) is uniformly stretched as edge CD remains straight and rotates clockwise. Using the definition of extensional strain in Eq. 2.29, determine an expression for the extensional strain in the x direction, $\epsilon_x(x, y)$.

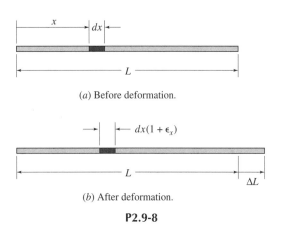

(a) Before deformation.

(b) After deformation.

P2.9-8

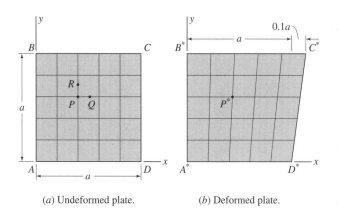

(a) Undeformed plate.　　　(b) Deformed plate.

P2.9-11 and P2.9-12

Prob. 2.9-9. A thin, square plate $ABCD$ undergoes deformation such that a typical point P with coordinates (x, y) moves horizontally an amount

$$u(x, y) \equiv \overline{PP^*} = \frac{1}{100}(a - x)\left(\frac{y}{a}\right)^2$$

Prob. 2.9-12. Using the definition of shear strain, Eq. 2.30, and using the "undeformed plate" and "deformed plate" sketches in Fig. P2.9-12, determine an expression for the shear strain γ_{xy} as a function of position in the plate, that is, $\gamma_{xy}(x, y)$.

Prob. 2.9-13. A thin, rectangular plate $ABCD$ undergoes deformation such that a typical point P at coordinates (x, y) moves vertically an amount

$$v(x, y) \equiv \overline{PP^*} = \frac{1}{100}(b - y)\left(\frac{x}{a}\right)^3$$

The undeformed and deformed plates are shown in Figs. P2.9-13a and P2.9-13b, respectively. Using the definition extensional strain in Eq. 2.29, determine an expression for $\epsilon_y(x, y)$, the extensional strain in the y direction.

*Prob. 2.9-14. Using the definition of shear strain, Eq. 2.30, and using the "undeformed plate" and "deformed plate" sketches in Fig. P2.9-14, determine an expression for the shear strain γ_{xy} as a function of position in the plate, that is, $\gamma_{xy}(x, y)$.

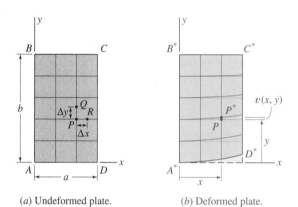

(a) Undeformed plate.　　　(b) Deformed plate.

P2.9-13 and P2.9-14

*Prob. 2.9-15. A typical point P at coordinates (x, y) in a flat plate moves through <u>small</u> displacements $u(x, y)$ and $v(x, y)$ in the x direction and the y direction, respectively. Using the definition of extensional strain, Eq. 2.29, and using the "before deformation" and "after deformation" sketches in Fig. 2.9-15,

show that the formula for the extensional strain in the x direction, $\epsilon_x(x, y)$, is the partial differential equation

$$\epsilon_x = \frac{\partial u}{\partial x}$$

Prob. 2.9-16. The thin, rectangular plate $ABCD$ shown in Fig. P2.9-16a undergoes uniform stretching in the x direction, so that it is elongated by an amount δ. Determine an expression for the (uniform) extensional strain ϵ_n of the diagonal AC. Express your answer in terms of δ, L, and the angle θ. Assume that $\delta \ll L$ and see Appendix A.2 for relevant approximations.

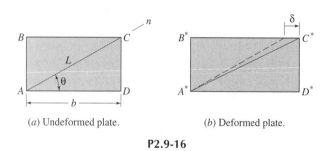

(a) Undeformed plate.　　　(b) Deformed plate.

P2.9-16

*Prob. 2.9-17. A rectangular plate $ABCD$ with base b and height h is uniformly stretched an amount δ_x in the x direction and δ_y in the y direction to become the enlarged rectangle $AB^*C^*D^*$ shown in Fig. P2.9-17. Determine an expression for the (uniform) extensional strain ϵ_n of the diagonal AC. Express your answer in terms of δ_x, δ_y, L, and θ, where $L = \sqrt{b^2 + h^2}$ and $\tan \theta = h/b$. Base your calculations on the small-displacement assumptions, that is, assume that $\delta_x \ll L$ and $\delta_y \ll L$. (See Appendix A.2 for relevant approximations.)

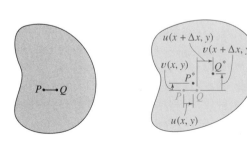

(a) Before deformation.　　　(b) After deformation.

P2.9-15

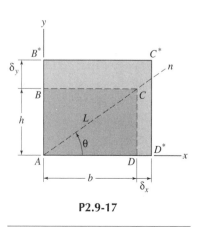

P2.9-17

Prob. 2.10-1. When thin sheets of material, like the top "skin" of the airplane wing in Fig. P2.10-1, are subjected to stress, they are said to be in a state of *plane stress,* with $\sigma_z = \tau_{xz} = \tau_{yz} = 0$. Starting with Eqs. 2.32, with $\Delta T = 0$, show that for the case of plane stress Hooke's Law can be written as

$$\sigma_x = \frac{E}{1 - \nu^2} (\epsilon_x + \nu\epsilon_y), \qquad \sigma_y = \frac{E}{1 - \nu^2} (\epsilon_y + \nu\epsilon_x)$$

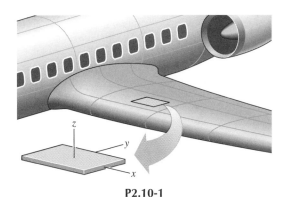

P2.10-1

Prob. 2.10-2. Figure P2.10-2 shows a small portion of a thin aluminum-alloy plate in plane stress ($\sigma_z = \tau_{xz} = \tau_{yz} = 0$). At a particular point in the plate $\epsilon_x = 600\mu$, $\epsilon_y = -200\mu$, and $\gamma_{xy} = 200\mu$. For the aluminum alloy, $E = 10 \times 10^3$ ksi and $\nu = 0.33$. Determine the stresses σ_x, σ_y, and τ_{xy} at this point in the plate. (Note: Start with Eqs. 2.32, not with Eqs. 2.34.)

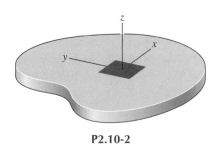

P2.10-2

Prob. 2.10-3. Determine the state of strain that corresponds to the following three-dimensional state of stress at a certain point in a steel machine component:

$$\sigma_x = 60 \text{ MPa}, \quad \sigma_y = 20 \text{ MPa}, \quad \sigma_z = 30 \text{ MPa}$$

$$\tau_{xy} = 20 \text{ MPa}, \quad \tau_{xz} = 15 \text{ MPa}, \quad \tau_{yz} = 10 \text{ MPa}$$

Use $E = 210$ GPa and $\nu = 0.30$ for the steel.

Prob. 2.10-4. The flat-bar plastic test specimen shown in Fig. P2.10-4 has a reduced-area "test section" that measures 0.5 in. × 1.0 in. Within the test section a strain gage oriented in the axial direction measures $\epsilon_x = 0.002 \frac{\text{in.}}{\text{in.}}$, while a strain gage

mounted in the transverse direction measures $\epsilon_y = -0.0008 \frac{\text{in.}}{\text{in.}}$, when the load on the specimen is $P = 300$ lb. (a) Determine the values of the modulus of elasticity, E, and Poisson's ratio, ν. (b) Determine the value of the dilatation, ϵ_V, within the test section.

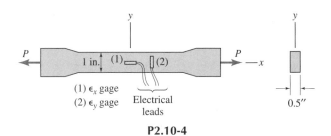

P2.10-4

Prob. 2.10-5. A titanium-alloy bar has the following original dimensions: $a = 10$ in., $b = 4$ in., and $c = 2$ in. The bar is subjected to stresses $\sigma_x = 14$ ksi and $\sigma_y = -6$ ksi, as indicated in Fig. P2.10-5. The remaining stresses—σ_z, τ_{xy}, τ_{xz}, and τ_{yz}— are all zero. Let $E = 16 \times 10^3$ ksi and $\nu = 0.33$ for the titanium alloy. (a) Determine the changes in the lengths: Δa, Δb, and Δc, where $a^* = a + \Delta a$, etc. (b) Determine the dilatation, ϵ_V.

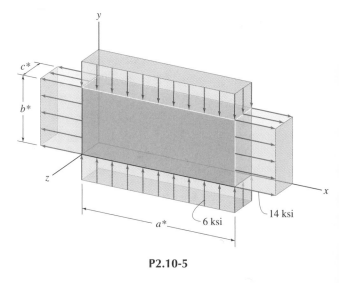

P2.10-5

Prob. 2.10-6. An aluminum-alloy plate is subjected to a biaxial state of stress, as illustrated in Fig. P2.10-6 ($\sigma_z = \tau_{xz} = \tau_{yz} =$

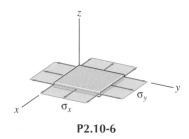

P2.10-6

$\tau_{xy} = 0$). For the aluminum alloy, $E = 72$ GPa and $\nu = 0.33$. Determine the stresses σ_x and σ_y if $\epsilon_x = 200\mu$ and $\epsilon_y = 140\mu$. (Note: Start with Eqs. 2.32, not with Eqs. 2.34.)

Prob. 2.10-7. A block of linearly elastic material (E, ν) is compressed between two rigid, perfectly smooth surfaces by an applied stress $\sigma_x = -\sigma_0$, as depicted in Fig. P2.10-7. The only other nonzero stress is the stress σ_y induced by the restraining surfaces at $y = 0$ and $y = b$. (a) Determine the value of the restraining stress σ_y. (b) Determine Δa, the change in the x dimension of the block. (c) Determine the change Δt in the thickness t in the z direction.

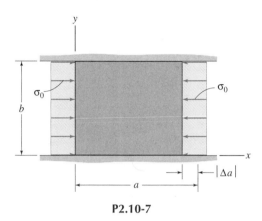

P2.10-7

Prob. 2.10-8. A thin, rectangular plate is subjected to a uniform biaxial state of stress (σ_x, σ_y). All other components of stress are zero. The initial dimensions of the plate are $L_x = 4$ in. and $L_y = 2$ in., but after the loading is applied, the dimensions are $L_x^* = 4.00176$ in., and $L_y^* = 2.00344$ in. If it is known that $\sigma_x = 10$ ksi and $E = 10 \times 10^3$ ksi, (a) What is the value of Poisson's ratio? (b) What is the value of σ_y?

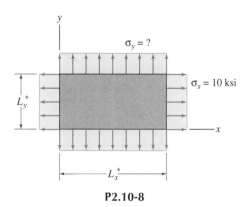

P2.10-8

Prob. 2.10-9. At a point in a thin steel plate in plane stress $(\sigma_z = \tau_{xz} = \tau_{yz} = 0)$, $\epsilon_x = 800\mu$, $\epsilon_y = -400\mu$, and $\gamma_{xy} = 200\mu$. For the steel plate, $E = 200$ GPa and $\nu = 0.30$. (a) Determine the extensional strain ϵ_z at this point. (b) Determine the stresses σ_x, σ_y and τ_{xy} at this point. (c) Determine the dilatation, ϵ_V, at this point.

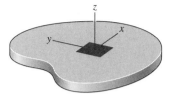

P2.10-9

*Prob. 2.10-10. A block of linearly elastic material (E, ν) is placed under hydrostatic pressure: $\sigma_x = \sigma_y = \sigma_z = -p$; $\tau_{xy} = \tau_{xz} = \tau_{yz} = 0$, as shown in Fig. P2.10-10. (a) Determine an expression for the extensional strain ϵ_x $(= \epsilon_y = \epsilon_z)$. (b) Determine an expression for the dilatation, ϵ_V. (c) The bulk modulus, k_b, of a material is defined as the ratio of the hydrostatic pressure, p, to the magnitude of the volume change per unit volume, $|\epsilon_V|$, that is, $k_b \equiv \dfrac{p}{|\epsilon_V|}$. Determine an expression for the bulk modulus of this block of linearly elastic material. Express your answer in terms of E and ν.

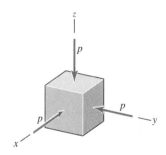

(Stresses on hidden faces not shown.)

P2.10-10

Prob. 2.12-1. Boom AC is supported by a rectangular steel bar BD, and it is attached to a bracket at C by a high-strength-steel pin. Assume that the pin at B is adequate to sustain the loading applied to it, and that the design-critical components are the bar BD and the pin at C. The factor of safety against failure of BD by yielding is $FS_\sigma = 3.0$, and the factor of safety against ultimate shear failure of the pin at C is $FS_\tau = 3.3$. (a) Determine the required thickness, t, of the rectangular bar BD, whose width is b. (b) Determine the required diameter, d, of the pin at C.

$$P = 2400 \text{ lb}, \quad L = 6 \text{ ft}, \quad h = 6 \text{ ft}$$

Bar BD: $b = 1$ in., $\sigma_Y = 36$ ksi, Pin C: $\tau_U = 60$ ksi

Prob. 2.12-2. Repeat Prob. 2.12-1 using the following data:

$$P = 10 \text{ kN}, \quad L = 3 \text{ m}, \quad h = 2 \text{ m}$$

Bar BD: $b = 25$ mm, $\sigma_Y = 250$ MPa, Pin C: $\tau_U = 400$ MPa

89

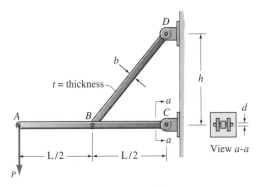

P2.12-1, P2.12-2, and P2.12-3

Prob. 2.12-3. Using the same basic figure and factors of safety as for Prob. 2.12-1, determine the allowable load, P_{allow}.

$$L = 60 \text{ in.,} \quad h = 40 \text{ in.}$$

Bar BD: $b = 2.0$ in., $t = 0.25$ in., $\sigma_Y = 36$ ksi

Pin C: $d = 0.375$ in., $\tau_U = 60$ ksi

Prob. 2.12-4. A load of $W = 2$ kips is suspended from a cable at end C of a rigid beam AC, whose length is $b = 8$ ft. Beam AC, in turn, is supported by a steel rod BD ($E = 30 \times 10^3$ ksi) of diameter d and length $L = 12$ ft. The rod BD is to be sized so that there will be a factor of safety with respect to yielding of $FS_\sigma = 4.0$ and a factor of safety with respect to deflection of $FS_\delta = 3.0$. The yield strength of rod BD is $\sigma_Y = 50$ ksi, and the maximum displacement at C is limited to $(\delta_C)_{max} = 0.25$ in. $\left[\text{i.e., } (\delta_C)_{allow.} = \dfrac{(\delta_C)_{max}}{FS_\delta} \right]$. Determine the required diameter, d, of rod BD to the nearest $\frac{1}{8}$ in.

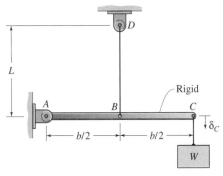

P2.12-4, P2.12-5, and P2.12-6

Prob. 2.12-5. Rework Prob. 2.12-4 if $\sigma_Y = 36$ ksi and the length of rod BD is $L = 8$ ft.

Prob. 2.12-6. A load W is to be suspended from a cable at end C of a rigid beam AC, whose length is $b = 3$ m. Beam AC, in turn, is supported by a steel rod of diameter $d = 25$ mm and length $L = 2.5$ m. Rod BD is made of steel with a yield point $\sigma_Y = 250$ MPa, and modulus of elasticity $E = 200$ GPa. If the maximum displacement at C is $(\delta_C)_{max} = 10$ mm, and there is

to be a factor of safety with respect to yielding of BD of $FS_\sigma = 3.3$ and with respect to displacement of $FS_\delta = 3.0$, what is the allowable weight that can be suspended from the beam at C?

Prob. 2.12-7. An angle bracket ABC is restrained by a high-strength steel wire CD, and it supports a load P at A, as shown in Fig. P2.12-7. The strength properties of the wire and the shear pin at B are $\sigma_Y = 80$ ksi (wire), and $\tau_U = 60$ ksi (pin at B). If the wire and pin are to be sized to provide a factor of safety against yielding of the wire of $FS_\sigma = 3.3$ and a factor of safety against ultimate shear failure of the pin of $FS_\tau = 3.5$, what are the required diameters of the wire (to the nearest $\frac{1}{32}$ in.) and the pin (to the nearest $\frac{1}{16}$ in.)?

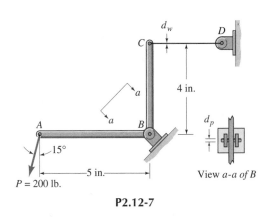

P2.12-7

Prob. 2.12-8. As shown in Fig. P2.12-8, a hydraulic cylinder at C is used to exert tension on a pump sucker-rod attached to the rocker-frame ABC at A. The design-critical elements are the diameter d_r of the sucker rod and the diameter d_p of the pin at B. Both are to be designed with a factor of safety of $FS = 3.0$, the sucker rod with respect to yielding of the rod, and the pin at B with respect to ultimate shear failure. The strength properties of the rod and pin are $\sigma_Y = 340$ MPa and $\tau_U = 400$ MPa, respectively. Determine the required diameters of the rod and pin (to the

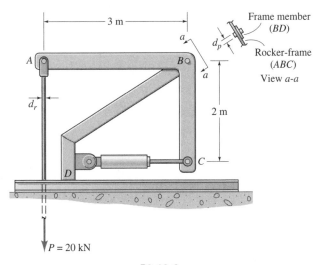

P2.12-8

nearest millimeter) if the maximum sucker-rod tension is P_{max} = $P_{allow.}$ = 20 kN.

Prob. 2.12-9. As illustrated in Fig. P2.12-9, a pump motor applies a force P_A at point A on a "rocking-horse" beam ABC. The beam, in turn, applies a tensile force P_C to the sucker-rod that is attached to the beam at C. This pumping system is to operate with a factor of safety FS_σ = 4.0 with respect to yielding and FS_τ = 3.0 with respect to shear failure of the pin at B. The strength properties of the rod and shear pin are σ_Y = 50 ksi and τ_U = 60 ksi, respectively. If the sucker-rod must be able to exert a pull of P_C = 4 kips, what is the required diameter, d_r, of the sucker-rod? (b) What is the required diameter, d_p, of the pin at B?

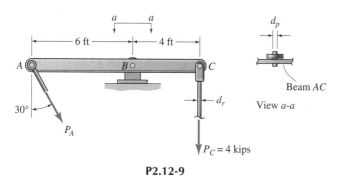

Beam AC
View a-a

P2.12-9

Prob. 2.12-10. The L-shaped loading-platform frame in Fig. P2.12-10 is supported by a high-strength steel shear pin at C and by a tie-rod AB. Both the tie rod and the pin are to be sized with a factor of safety of FS = 3.0, the tie-rod with respect to tensile yielding and the pin with respect to shear failure. The strength properties of the rod and pin are: σ_Y = 340 MPa and τ_U = 340 MPa; the respective lengths are: L_1 = 1.5 m and $L_2 = L_3$ = 2.0 m. (a) If the loading platform is to be able to handle loads W up to W_{max} = 10 kN, what is the required diameter, d_r, of the tie-rod (to the nearest millimeter)? (b) What is the required diameter, d_p, of the shear pin at C (to the nearest millimeter)?

Prob. 2.12-11. The L-shaped loading frame in Fig. P2.12-11 is supported by a high-strength shear pin (d_p = 0.5 in, τ_U = 50 ksi)

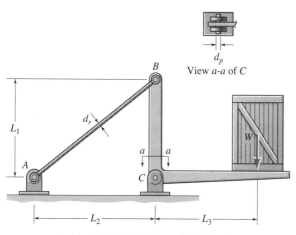

d_p
View a-a of C

P2.12-10, P2.12-11, and P2.12-12

and by a tie-rod AB (d_r = 0.625 in., σ_Y = 50 ksi). Both the tie-rod and the pin are to be sized with a factor of safety of FS = 3.0, the tie-rod with respect to tensile yielding, and the shear pin with respect to ultimate shear failure. Determine the allowable platform load, $W_{allow.}$. Let L_1 = 6 ft, $L_2 = L_3$ = 8 ft.

Prob. 2.12-12. Repeat Prob. 2.12-11 using σ_Y = 36 ksi and L_2 = 12 ft, with the remaining data unchanged.

Prob. 2.12-13. Repeat Example Problem 2.14, adding an equal downward load P at joint B of the planar truss. Determine $P_{allow.}$, the allowable downward load that can be applied simultaneously at joints A and B.

Prob. 2.12-14. The four-element, pin-jointed planar truss in Fig. P2.12-14 supports a single downward load P at joint C. All members have the same cross-sectional area, A_i = 0.80 in^2, and the same length, L. The allowable stress in tension in $(\sigma_T)_{allow.}$ = 20 ksi, and the allowable stress in compression is $(\sigma_C)_{allow.}$ = 12 ksi. Determine the allowable load, $P_{allow.}$.

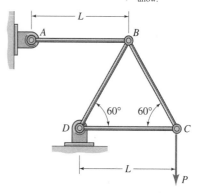

P2.12-14

Prob. 2.12-15. A three-bar, pin-jointed, planar truss supports a single vertical load P at joint B. Joint C is free to move vertically. The allowable stress in tension is $(\sigma_T)_{allow.}$ = 20 ksi, and the allowable stress in compression is $(\sigma_C)_{allow.}$ = 12 ksi. If the truss is to support a maximum load P_{max} = 12 kips, what are the required cross-sectional areas, A_i, of the three truss members?

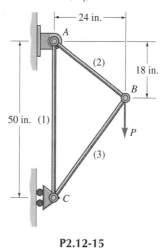

P2.12-15

COMPUTER EXERCISES—Section 2.12.

For **Problems C2.12-1 through C2.12-4** *you are to develop a computer program to generate the required graph(s) that will enable you to choose the "optimum design." You may use a spreadsheet program or other mathematical application program (e.g., TK Solver or Mathcad), or you may write a program in a computer language (e.g., BASIC or FORTRAN).*

*Prob. C2.12-1. The pin-jointed planar truss shown in Fig. PC2.12-1*a* is to be made of three steel two-force members and support a single vertical load $P = 1.2$ kips at joint B. Joint C is free to move vertically. For the steel truss members, the allowable stress in tension is $(\sigma_T)_{\text{allow.}} = 20$ ksi, the allowable stress in compression is $(\sigma_C)_{\text{allow.}} = 12$ ksi, and the weight density is 0.284 lb/in^3. You are to consider truss designs for which the vertical member AC has lengths varying from $L_1 = 18$ in. to $L_1 = 50$ in. (a) Show that, if each member has the minimum cross sectional area that meets the strength criteria stated above, the weight W of the truss can be expressed as a function of the length L_1 of member AC by the function that is plotted in Fig. PC2.12-1*b*. (b) What value of L_1 gives the minimum-weight truss, and what is the weight of that truss?

*Prob. C2.12-2. The pin-jointed planar truss shown in Fig. PC2.12-2*a* is to be made of two steel two-force members and support a single vertical load $P = 10$ kN at joint B. For the steel truss members, the allowable stress in tension is $(\sigma_T)_{\text{allow.}} = 150$ MPa, the allowable stress in compression is $(\sigma_C)_{\text{allow.}} = 100$ MPa, and the weight density is 77.0 kN/m^3. You are to consider truss designs for which joint B can be located at any point along the vertical line that is 1 m to the right of AC, with y_B varying from $y_B = 0$ to $y_B = 2$m. (a) Show that, if each member has the minimum cross sectional area that meets the strength criteria stated above, the weight W of the truss can be expressed as a function of y_B, the position of joint B, by the function that is plotted in Fig. PC2.12-2*b*. (b) What value of y_B gives the minimum-weight truss, and what is the weight of that truss?

*Prob. C2.12-3. The pin-jointed planar truss shown in Fig. PC2.12-3 is to be made of three steel two-force members and is to support vertical loads $P_B = 2$ kips at joint B and $P_C = 3$ kips at joint C. The lengths of members AB and BC are $L_1 = 30$ in. and $L_2 = 24$ in., respectively. Joint C is free to move vertically. For the steel truss members, the allowable stress in tension is $(\sigma_T)_{\text{allow.}} = 20$ ksi, the allowable stress in compression is

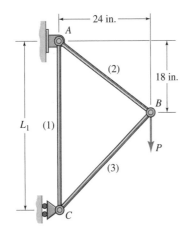

(*a*) A three-bar planar truss.

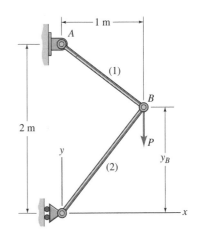

(*a*) A two-bar planar truss.

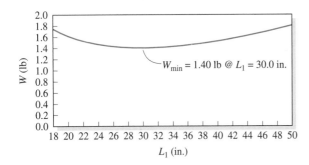

(*b*) Minimum-weight design for the three-bar planar truss.

PC2.12-1

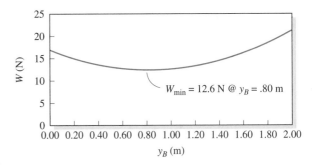

(*b*) Minimum-weight design for the two-bar planar truss.

PC2.12-2

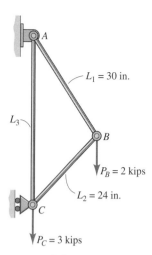

$L_1 = 30$ in.

L_3

B

$P_B = 2$ kips

$L_2 = 24$ in.

C

$P_C = 3$ kips

PC2.12-3

$(\sigma_C)_{\text{allow.}} = 12$ ksi, and the weight density is 0.284 lb/in³. You are to consider truss designs for which the vertical member AC has lengths varying from $L_3 = 18$ in. to $L_3 = 50$ in. (a) Show that, if each member has the minimum cross sectional area that meets the strength criteria stated above, the weight W of the truss can be expressed as a function of the length L_3 of member AC by a function that is similar to the one plotted in Fig. PC2.12-1b. (Hint: Use the *law of cosines* to obtain expressions for the angle at joint A and the angle at joint C.) (b) What value of L_3 gives the minimum-weight truss, and what is the weight of that truss?

*Prob. C2.12-4. The pin-jointed planar truss shown in Fig. PC2.12-4 is to be made of two aluminum two-force members and support a single horizontal load $P = 50$ kN at joint B. For the aluminum truss members, the allowable stress in tension is $(\sigma_T)_{\text{allow.}} = 200$ MPa, the allowable stress in compression is $(\sigma_C)_{\text{allow.}} = 130$ MPa, and the weight density is 28.0 kN/m³. You are to consider truss designs for which support C can be located at any point along the x axis, with x_C varying from $x_C = 1$ m to $x_C = 2.4$ m. (a) Show that, if each member has the minimum cross sectional area that meets the strength criteria stated above, the weight W of the truss can be expressed as a function of x_C, the position of support C, by a function that is similar to the one plotted in Fig. PC2.12-2b. (b) What value of x_C gives the minimum-weight truss, and what is the weight of that truss?

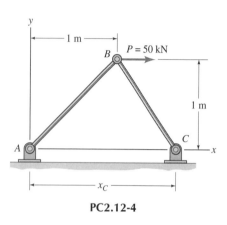

PC2.12-4

3 AXIAL DEFORMATION

3.1 INTRODUCTION

In Chapter 2 uniform axial deformation was used to introduce the concepts of normal stress and extensional strain and to describe the experiments required to determine the stress-strain behavior of materials. In this chapter we will pursue the topic of axial deformation in greater detail. In the process, we will employ the fundamental concepts of deformable-body mechanics from Section 1.2—*equilibrium, geometry of deformation,* and *material behavior*—to solve axial-deformation problems. We begin with a definition of *axial deformation.*

> *A structural member having a straight, longitudinal axis is said to undergo* **axial deformation** *if its axis remains straight and its cross sections remain plane, remain perpendicular to the axis, and do not rotate about the axis as the member deforms.*

There are many examples of axial-deformation members: columns in buildings, hoist cables, and truss members in space structures, to name just a few. The picture in Fig. 3.1 illustrates several stages in the construction of columns (piers) for a highway interchange. On the left is an example of the steel reinforcement for a column, and on the right is a completed column with reinforcement protruding from the top of the column. The columns of bridges like the ones shown in Fig. 3.1 act primarily as axial deformation members.

FIGURE 3.1 Some reinforced concrete columns for highway interchange bridges.

3.2 BASIC THEORY OF AXIAL DEFORMATION ■■■■■■■■■■■

Let us now develop the theory of axial deformation by applying the three fundamental concepts of deformable-body mechanics: *equilibrium, geometry of deformation,* and *material behavior.* We begin by considering the geometry of deformation.

Geometry of Deformation. The theory of axial deformation applies to straight, slender members with a cross section that is either constant or that changes slowly along the length of the member. Figure 3.2 shows such an element before and after it has undergone axial deformation.

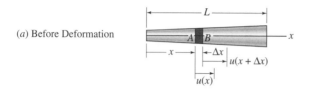
(a) Before Deformation

(b) After Deformation

FIGURE 3.2 The geometry of axial deformation.

The definition of *axial deformation* is characterized by two fundamental *kinematic assumptions*:

1. The axis of the member remains straight.
2. Cross sections, which are plane and are perpendicular to the axis before deformation, remain plane and perpendicular to the axis after deformation. And, the cross sections do not rotate about the axis.

These assumptions are illustrated in Fig. 3.2, where A and B designate cross sections at x and $(x + \Delta x)$ prior to deformation, and where A^* and B^* designate these same cross sections after deformation.

The distance that a cross section moves in the axial direction is called its *displacement.* The displacement of cross section A is labeled $u(x)$, while the neighboring section B displaces an amount $u(x + \Delta x)$. The displacement $u(x)$ is taken to be <u>positive in the $+x$ direction</u>. An expression for the *extensional strain* of any fiber[1] of infinitesimal length Δx parallel to the x-axis extending from section A to section B of the undeformed member may be determined from the fundamental definition of extensional strain, which is given in Eq. 2.29.

$$\epsilon_x = \lim_{\Delta x \to 0} \left(\frac{\Delta x^* - \Delta x}{\Delta x} \right) = \lim_{\Delta x \to 0} \left[\frac{u(x + \Delta x) - u(x)}{\Delta x} \right] = \frac{du}{dx}$$

Therefore, the extensional strain is the derivative (with respect to x) of the displacement, or

$$\epsilon_x(x) = \frac{du(x)}{dx} \tag{3.1}$$

---

[1]The word *fiber* signifies a line of material particles.

95

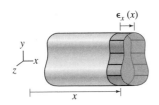

FIGURE 3.3 The extensional strain distribution for a member undergoing axial deformation.

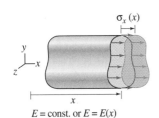

E = const. or $E = E(x)$

FIGURE 3.4 The stress distribution for a member undergoing axial deformation.

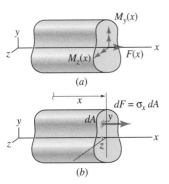

FIGURE 3.5 (*a*) Stress resultants. (*b*) Normal stress at a point.

The two fundamental kinematic assumptions stated above imply that the extensional strain ϵ_x may be a function of x, but that it is not a function of position in the cross section, that is, of y or z. To emphasize this point, a plot of the *strain distribution* at an arbitrary cross section of x is shown in Fig. 3.3 superimposed on a sketch of a portion of the member. To reiterate, **axial deformation is characterized by an extensional strain that is not a function of position in the cross section**.

Material Behavior. Having the strain distribution given by Eq. 3.1 and illustrated in Fig. 3.3, we can now employ the uniaxial stress-strain behavior of materials from Section 2.10 to determine the stress distribution in a member undergoing axial deformation. Let us take first the simplest case—linearly elastic behavior—and let us assume that the temperature remains constant (i.e., $\Delta T = 0$) and that $\sigma_y = \sigma_z = 0$. Then, from Eq. 2.32a we get the uniaxial stress-strain equation

$$\sigma_x = E\epsilon_x \tag{3.2}$$

By combining Eq. 3.2 with Eq. 3.1, we get

$$\sigma_x(x, y, z) = E(x, y, z)\epsilon_x(x) \tag{3.3}$$

where it is specifically noted that the extensional strain, ϵ_x, does not vary with y or z. If Young's modulus, E, is constant throughout the member, Eq. 3.3 reduces to the important special case of uniform axial deformation that we examined in Chapter 2.

If Young's modulus is constant throughout a cross section, that is $E = E(x)$ or $E = $ const., then the stress distribution, like the strain distribution, is constant on the cross section. Then,

$$\sigma_x(x) = E(x)\epsilon_x(x) \tag{3.4}$$

This stress distribution is illustrated in Fig. 3.4.

Stress Resultants and Equilibrium. Deformable-body mechanics problems are simplified by making assumptions, like the axial-deformation kinematic assumptions discussed earlier in this section, that reduce a basically three-dimensional problem to a one-dimensional problem. Thus far we have found that axial deformation is characterized by an extensional strain, ϵ_x, that is independent of position in the cross section. And when the material follows Hooke's law we can use Eq. 3.3 to determine the stress distribution on a cross section. As we did previously in Section 2.2, we now replace the stress, which is distributed over the cross section, by *stress resultants* or by a single *stress resultant*. Figure 3.5 shows the three stress resultants that are related to the normal stress σ_x and defines the sign convention for these stress resultants.

The *axial force, F*, on the cross section will always be taken to be positive in tension. The *bending moments* M_y and M_z are taken positive according to the right-hand rule. By summing up the contributions to these stress resultants of the forces dF on infinitesimal areas, dA, we get

$$
\begin{aligned}
F(x) &= \int_A \sigma_x dA \\
M_y(x) &= \int_A z\sigma_x dA \\
M_z(x) &= -\int_A y\sigma_x dA
\end{aligned}
\tag{3.5}
$$

These equations define the three stress resultants associated with σ_x.

Before continuing our discussion of the axial-deformation behavior of a homogeneous, linearly elastic member (Eq. 3.4 and Fig. 3.4), let us determine the resultant of the normal stresses on the cross section of a nonhomogeneous bar.

EXAMPLE 3.1

A bimetallic bar is made of two linearly elastic materials, material 1 and material 2, that are bonded together at their interface, as shown in Fig. 1. Assume that $E_2 > E_1$. Determine the distribution of normal stress that must be applied at each end if the bar is to undergo axial deformation, and determine the location of the point in the cross section where the resultant force P must act. Express your answers in terms of P, E_1, E_2, and the dimensions of the bar.

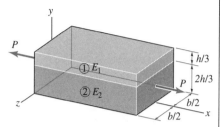

Fig. 1 A bimetallic bar.

Plan the Solution The bar is said to undergo axial deformation, so we know from Eq. 3.1 that the strain, $\epsilon_x(x)$, is constant on every cross section. And since the bar is prismatic (constant cross section) and is loaded only at its ends, we can assume that $\epsilon_x(x) = \epsilon = $ const for the entire bar. Since there are two values of E, Eq. 3.3 will lead to different values of stress in the two materials. We can use the stress resultant integrals of Eqs. 3.5 to relate these two stresses to the resultant force.

Solution

Strain Distribution: Because the bar is prismatic and is undergoing axial deformation, and because external forces are applied only at the ends of the bar, $\epsilon(x)$ will have a constant value everywhere. Let

$$\epsilon(x) \equiv \epsilon = \text{const} \tag{1}$$

Stress Distribution: From Eq. (1) and Eq. 3.3, the stresses in the two parts of the bar will be

$$\sigma_{x1} = E_1\epsilon, \qquad \sigma_{x2} = E_2\epsilon \tag{2}$$

as illustrated in Fig. 2.

Resultant Force: The resultant force and moments on the cross section are given by Eq. 3.5.

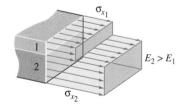

Fig. 2 The stress distribution in a bimetallic bar.

$$F(x) = \int_A \sigma_x dA$$

$$M_y(x) = \int_A z\sigma_x dA \tag{3}$$

$$M_z(x) = -\int_A y\sigma_x dA$$

But, from Example 2.1 we know that the resultant of a constant normal stress distribution is a force acting through the centroid of the area on which the constant stress acts. Therefore, we can replace the stress distribution of Fig. 2 by two axial forces, P_1 and P_2, acting at the centroids of their respective areas of the cross section, as shown in Fig. 3b.

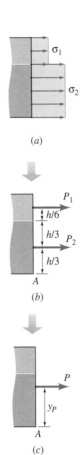

(a)

(b)

(c)

Fig. 3

From Eqs. (2) and the $F(x)$ equation in Eq. (3),

$$P_1 = E_1 \epsilon A_1 = E_1 \epsilon \left(\frac{bh}{3} \right), \quad P_2 = E_2 \epsilon A_2 = E_2 \epsilon \left(\frac{2bh}{3} \right) \quad (4)$$

Taking the summation of forces in the x direction, we get

$$\xrightarrow{+} \sum F_x = 0: \qquad P_1 + P_2 = P \qquad (5)$$

Thus, the resultant force acting on the ends of the bar (and at every cross section, for that matter) is related to the extensional strain ϵ by the equation

$$P = \frac{\epsilon bh}{3} (E_1 + 2E_2) \qquad (6)$$

Solving Eq. (6) for ϵ and inserting the result into the Eqs. (2), we get the stresses

$$\sigma_{x1} = \frac{3PE_1}{bh(E_1 + 2E_2)}, \quad \sigma_{x2} = \frac{3PE_2}{bh(E_1 + 2E_2)} \qquad \textbf{Ans.} \quad (7)$$

In Fig. 3b the forces P_1 and P_2 are shown acting at the centroids of the areas on which they act. The location of the single resultant force, P, in Fig. 3c can be determined by taking moments about a z axis passing through point A.

$$+\circlearrowleft \left(\sum M \right)_A = 0: \qquad P_1 \left(\frac{5h}{6} \right) + P_2 \left(\frac{h}{3} \right) = Py_P \qquad (8)$$

Thus, the resultant force acting on each end of the bar (and at every cross section, for that matter) is a force P located in the yz plane at

$$y_P = h \left(\frac{5E_1 + 4E_2}{6E_1 + 12E_2} \right), \quad z_P = 0 \qquad \textbf{Ans.} \quad (9)$$

Review the Solution   A good check on the results above is to let $E_1 = E_2 = E$. Then, we get

$$\sigma_{x1} = \sigma_{x2} = \frac{P}{bh}, \quad y_P = \frac{h}{2}$$

which is consistent with Example 2.1.

Let us return to the special case where Young's modulus is independent of position in the cross section, and where σ_x is given by Eq. 3.4 as illustrated in Fig. 3.4. Then, since $\int_A y\, dA \equiv \bar{y}A$ and $\int_A z\, dA \equiv \bar{z}A$,

$$F(x) = \sigma_x(x) \int_A dA = \sigma_x(x) A(x)$$

$$M_y(x) = \sigma_x(x) \int_A z\, dA = \sigma_x(x)(\bar{z}A(x)) \qquad (3.6)$$

$$M_z(x) = -\sigma_x(x) \int_A y\, dA = -\sigma_x(x)(\bar{y}A(x))$$

We can simplify the axial-deformation problem by choosing the x-axis such that it passes through the centroid of the cross section. Then, $\bar{y} = \bar{z} = 0$, and $M_y = M_z = 0$. Thus, **a member having $E = E(x)$ will undergo axial deformation if the applied load is an axial force $F(x)$ acting through the centroid of the cross section**, as shown in Fig. 3.6.

We have determined the stress resultant on the cross section when a member undergoes axial deformation. Equilibrium is satisfied by drawing a free-body diagram and relating this internal axial force, $F(x)$, to the external forces applied to the member.

Axial Deformation of a Homogeneous, Linearly Elastic Member.

Thus far in this section, "Basic Theory of Axial Deformation," we have derived the strain distribution (Eq. 3.1) for a member undergoing axial deformation and the stress distribution (Eq. 3.4) for a member with $E = E(x)$ undergoing axial deformation. We have also illustrated the procedure for determining the stress resultant on the cross section when a member undergoes axial deformation. Let us now consider the special case of a *homogeneous, linearly elastic, member undergoing axial deformation.* The relevant equations are repeated here for convenience.

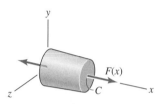

FIGURE 3.6 The resultant force on the cross section for axial deformation of a bar with $E = E(x)$.

$$\epsilon_x(x) = \frac{du}{dx}$$

$$\sigma_x(x) = E\epsilon_x(x) \tag{3.7}$$

$$F(x) = EA(x)\epsilon_x(x)$$

To apply the theory of axial deformation to a member with constant modulus of elasticity, we need to relate the stress σ_x and the displacement $u(x)$ to the external loads on a structure. This is done by combining Eqs. 3.7b and 3.7c to give

$$\sigma_x \equiv \sigma_x(x) = \frac{F(x)}{A(x)} \tag{3.8}$$

and by combining Eqs. 3.7a and 3.7c to give

$$\frac{du}{dx} = \frac{F(x)}{EA(x)} \tag{3.9}$$

If we need the displacement at an arbitrary section x, we can employ a dummy variable of integration in Eq. 3.9 and write

$$u(x) = u(0) + \int_0^x \frac{F(\xi)d\xi}{A(\xi)E} \tag{3.10}$$

Note that $u(L) - u(0)$ is just the *elongation* of the member of length L, that is,

$$e \equiv u(L) - u(0) = \int_0^L \frac{F(x)dx}{A(x)E} \tag{3.11}$$

Equation 3.9 (or Eq. 3.10 or Eq. 3.11) expresses the *force-deformation behavior* of a homogeneous member undergoing axial deformation. The next example problem illustrates the use of Eq. 3.11.

Fig. 1

Fig. 2. Free-body diagram.

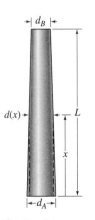

Fig. 3. Geometry.

The tapered column in Fig. 1 is subjected to a downward force P acting through the centroid of the top cross section at B. The column has a circular cross section, with a diameter that varies linearly from d_A at the bottom to d_B at the top. Determine an expression for the amount that the column shortens under the action of load P.

Plan the Solution   The change in length of the column (in this case, the *shortening* of the column) can be computed by using Eq. 3.11.

Solution   From Eq. 3.11, the shortening of the column, say δ, is

$$\delta = -\int_0^L \frac{F(x)dx}{A(x)E} \tag{1}$$

We need to use equilibrium to determine $F(x)$, and geometry to determine $A(x)$.

Equilibrium: From the free-body diagram in Fig. 2,

$$+\uparrow \sum F(x) = 0: \qquad F(x) = -P = \text{const} \tag{2}$$

Geometry: The cross-sectional area is

$$A(x) = (\pi/4)d^2(x)$$

By referring to Fig. 3 and employing similar triangles, we can determine $d(x)$ in terms of d_A and d_B.

$$\frac{d_A - d(x)}{x} = \frac{d_A - d_B}{L}$$

Thus,

$$d(x) = d_A - (x/L)(d_A - d_B)$$
$$A(x) = (\pi/4)[d_A - (x/L)(d_A - d_B)]^2 \tag{3}$$

Force-Deformation: Combining Eqs. (1) through (3), we get

$$\delta = \frac{4P}{\pi E} \int_0^L \frac{dx}{[d_A - (x/L)(d_A - d_B)]^2}$$

Finally, by evaluating this integral we obtain the answer

$$\delta = \frac{4PL}{\pi E d_A d_B} \qquad \text{Ans.} \quad (4)$$

Review the Solution   The right side of Eq. (4) has the dimension of length, as it should. Also note that an increase in P or L causes an increase in δ, while a larger E or ''fatter'' column decreases δ. These are reasonable effects. Finally, if $d_A = d_B \equiv d$, $\delta = PL/AE$, which is the expression we got in Example 2.5 (Section 2.6) for a uniform, linearly elastic member under axial loading.

3.3 SAINT-VENANT'S PRINCIPLE

Up to this point we have treated the distribution of normal stress σ_x on a cross section of a uniform member undergoing axial deformation as being uniform across the cross section. However, near points of application of load, plane sections do not remain plane, and the normal stress is not uniform.

In Section 12.2 we will take up the topic of *stress concentration,* the increase in stress caused by abrupt changes in cross section, holes, and so on. Here, however, let us briefly examine the stress distribution near points of application of concentrated loads. Consider the short compression bar, with cross-sectional area $A = bt,$ shown in Fig. 3.7.[2]

From Fig. 3.7 we can make the following three observations:

- The *average stress* is the same on all cross sections, namely $\sigma = -P/A$.
- Near the ends of the bar, where the concentrated load is applied (e.g., Fig. 3.7(*b*)), there is a *stress concentration,* with higher stress near the point of application of the load.
- At distances from the point of application of the load that are greater than the width of the compression bar (e.g., Fig. 3.7(*d*)), the stress distribution is essentially uniform.

The third observation above is referred to as *Saint-Venant's Principle.*[3] The significance of the principle can be stated as follows:

The stresses and strains in a body at points that are sufficiently remote from points of application of load depend only on the static resultant of the loads and not on the distribution of the loads.

Thus, the stress distribution in Fig. 3.7(*d*) would not be greatly altered if the single load *P* were replaced by two loads of magnitude *P*/2, shown as dashed arrows in the figure.

Throughout the remainder of this text on mechanics of deformable bodies we will obtain expressions for stress distributions in and deformations of various members under various types of loading. On the basis of Saint-Venant's Principle we can say that the expressions we derive are valid except very near to points of loading or support, or near to an abrupt change in cross section.

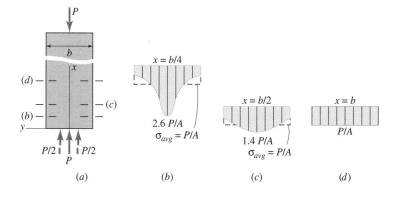

(a)   *(b)*   *(c)*   *(d)*

FIGURE 3.7 The effect of a concentrated load on the distribution of normal stress.

---

[2]Figure 3.7 is adapted from information in Section 24 of *Theory of Elasticity,* 3rd ed., [Ref. 3-1].

[3]Barré de Saint-Venant (1797–1886) is credited with many outstanding contributions to the theory of elasticity, especially his theories for torsion and bending of prismatic bars with various cross-sectional shapes. [Ref. 3-2]

3.4 ELASTIC BEHAVIOR OF A UNIFORM AXIALLY LOADED MEMBER

■■■■■■■■■■■■■

Many structures, for example the two structures in Fig. 3.8, incorporate one or more uniform axial-deformation members. Whereas the axial forces in the two-member truss of Example 1.1 (Section 1.4) could be determined from statics alone, to determine the member forces in the structures in Fig. 3.8 requires consideration of the deformation of the members, since these structures are statically indeterminate. To simplify the solution of problems involving uniform members (i.e., members that have uniform cross section and constant E), we will apply the fundamental equations of equilibrium, material behavior, and geometry of deformation to determine the basic force-deformation behavior of a typical *uniform axial-deformation element*.

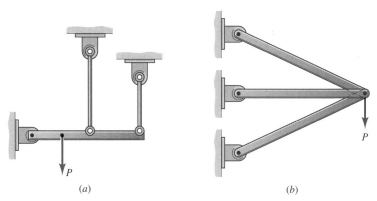

(a) (b)

FIGURE 3.8 Two structures that employ axial-deformation members.

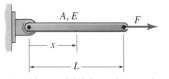

(a) A uniform axial-deformation member.

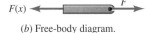

(b) Free-body diagram.

FIGURE 3.9 A typical uniform axial-deformation element.

Equilibrium. Using the free-body diagram in Fig. 3.9(*b*), we get

$$\overset{+}{\rightarrow} \sum F_x = 0: \qquad\qquad F(x) = F = \text{const}$$

Material Behavior and Geometry of Deformation. The derivation of Eq. 3.11 included both material behavior (linearly elastic) and geometry of deformation; hence, we can simply write

$$e = \int_0^L \frac{F(x)dx}{A(x)E} = \frac{F}{AE}\int_0^L dx = \frac{FL}{AE} \tag{3.12}$$

Sometimes it will be convenient to write Eq. 3.12 in the form

$$\boxed{e_i = f_i F_i, \text{ where } f_i \equiv (L/AE)_i} \tag{3.13}$$

where the subscript *i* refers to element *i*. Sometimes it will be more convenient to write it in the form

$$\boxed{F_i = k_i e_i, \text{ where } k_i \equiv (AE/L)_i} \tag{3.14}$$

The parameter *f* is called the *flexibility coefficient* for the axial-deformation member, or element, and *k* is called the *stiffness coefficient*.

103

Structures with Uniform Axial-Deformation Members

- The *flexibility coefficient, f,* is the elongation produced when a unit force is applied to the member. Its dimensions are L/F.
- The *stiffness coefficient, k,* is the force required to produce a unit elongation of a member. Its dimensions are F/L.

Either of these last two equations characterizes the *force-deformation behavior* of the linearly elastic uniform element.

3.5 STRUCTURES WITH UNIFORM AXIAL-DEFORMATION MEMBERS

This section is very important because in it we introduce the fundamental problem-solving strategy that is used extensively throughout the rest of the book. We solve a number of example problems using the fundamental equations: *equilibrium, element force-deformation behavior,* and *geometry of deformation.* Note carefully how each of these three fundamental types of equations enters into the solution of each problem. There are two systematic methods for combining these fundamental equations to obtain the final solution of the problem. One is called the *displacement method,* or *stiffness method;* the other is called the *force method,* or *flexibility method.* Several examples are used to illustrate these systematic problem-solving procedures.

EXAMPLE 3.3

Consider again the two-element axial-deformation member of Example 2.2 (Section 2.2), shown here in Fig. 1. Determine the total elongation of the member *ABC* if segment *AB* is steel ($E_1 = 200$ GPa) and segment *BC* is aluminum ($E_2 = 70$ Gpa).

Plan the Solution The total elongation is the sum of the elongations of the two elements. Equation 3.13 can be used to express the elongation of each element in terms of its axial force, and we have the axial force values in Example 2.2 (Section 2.2).

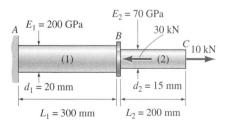

Fig. 1

Solution

Equilibrium: From the free-body diagrams and equilibrium equations of Example 2.2 we have

$$F_1 = -20 \text{ kN}, \qquad F_2 = 10 \text{ kN} \tag{1}$$

where F_i is the axial force in element i.

Element Force-Deformation Behavior: For the uniform element in Fig. 2, the force-deformation equation, Eq. 3.13, gives

$$e_i = \frac{F_i L_i}{A_i E_i}, \qquad i = 1, 2 \tag{2}$$

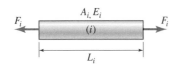

Fig. 2 A typical element.

Geometry of Deformation: Since the two elements are attached end-to-end, we have

$$e_{\text{total}} = e_1 + e_2 \tag{3}$$

Combining Eqs. (2) and (3) we get

$$e_{\text{total}} = \sum_{i=1}^{2} \frac{F_i L_i}{A_i E_i} \tag{4}$$

Then,

$$e_{\text{total}} = \frac{(-20 \text{ kN})(300 \text{ mm})}{(314.2 \text{ mm}^2)(200 \text{ kN/mm}^2)}$$

$$+ \frac{(10 \text{ kN})(200 \text{ mm})}{(176.7 \text{ mm}^2)(70 \text{ kN/mm}^2)}$$

$$= -0.0955 \text{ mm} + 0.1617 \text{ mm} = 0.0662 \text{ mm}$$

$$e_{\text{total}} = 6.62(10^{-2}) \text{ mm} \qquad \text{Ans.}$$

Review the Solution   Note how each of the quantities F_i, L_i, A_i, and E_i enters into the elongation of a uniform element. Since F_1 is negative, element AB is shortened. Likewise, since F_2 is positive, element BC is lengthened. Also note the effect of the relative values of $A_1 E_1$ and $A_2 E_2$ in the denominator of Eq. (4).

Example 3.3 is a statically determinate problem. We will now consider a more difficult statically indeterminate problem to illustrate the importance of carefully considering *equilibrium, force-deformation behavior,* and *geometry of deformation.* In the remaining examples in this chapter these fundamental equations will be highlighted by a rectangular shaded background, and the key solution step will be highlighted by a shaded background with red border.

A stepped rod is made up of three uniform elements, or members, as shown in Fig. 1. The rod exactly fits between rigid walls when no external forces are applied, and the ends of the rod are welded to the rigid walls. (a) Determine the displacements u_B and u_C of the two joints, or *nodes,* where the external loads P_B and P_C are applied. (b) Determine the internal force in each of the three elements. (c) Determine the reaction forces P_A and P_D at the fixed ends A and D.

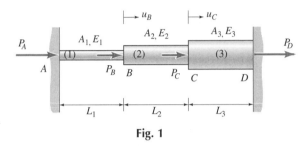

Fig. 1

Plan the Solution We can think of this structure as being composed of three uniform elements and two connecting joints. We can write equilibrium equations for the joints and force-deformation equations for the separate elements. Finally, we can relate element elongations to the displacements of the joints. We can then combine these three sets of equations to get the required answers.

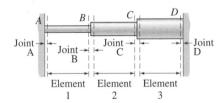

Fig. 2 A structure divided into elements and joints.

If the external forces P_B and P_C in Fig. 1 act to the right (i.e., if they are positive), we should find that element (1) is in tension and that element (3) is in compression.

The reasons for dividing the structure into joints and elements are as follows:

1. From Section 3.3 we know the force-deformation behavior of a uniform, axially loaded member, and each of the elements in this problem is such a member.

2. Each equilibrium equation written for one of the joints will relate the element forces acting on that joint to the external loads and/or reactions acting on the joint.

Solution

(a) To determine the nodal displacements u_B and u_C, we apply the three fundamentals of solid mechanics—*equilibrium, material behavior*, and *geometry of deformation*.

Equilibrium: We isolate each of the joints as a free body, and on each free body we show the external forces acting on the joint and the internal forces exerted by the elements that connect to the joint. We label the element forces and adopt the same sign convention that was used in Section 3.3, namely, that the element forces are taken to be positive in tension. The symbol F_i is used for the force in element i to distinguish element forces from the externally applied loads and reactions.

For Node (Joint) B:

$$\overset{+}{\rightarrow} \sum F_x = 0: \qquad -F_1 + F_2 + P_B = 0$$

For Node (Joint) C:

$$\overset{+}{\rightarrow} \sum F_x = 0: \qquad -F_2 + F_3 + P_C = 0$$

$$\boxed{\begin{aligned} F_1 - F_2 &= P_B \\ F_2 - F_3 &= P_c \end{aligned}} \qquad \textbf{Equilibrium} \quad (1)$$

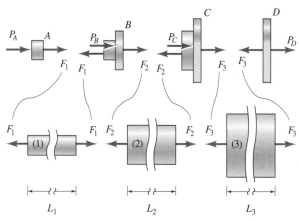

Fig. 3 Free-body diagrams.

Equations (1) relate the unknown internal element forces to the known external loads. Since there are three unknowns, but only two equilibrium equations, this problem is *statically indeterminate*; that is, the internal forces cannot be determined by statics alone. There is no way that we can find another equilibrium equation that will not be just a combination of these two joint-equilibrium equations. Thus, to find additional equations we must look to element force-deformation behavior and to the geometry of deformation.

Element Force-Deformation Behavior: We have three uniform, axial-deformation elements, and for each one we can write an element force-deformation equation like Eq. 3.13 or Eq. 3.14. We choose Eq. 3.14 for this example. We have called the element forces F_1, F_2, and F_3 (tension positive), so we have

$$
\begin{aligned}
F_1 &= k_1 e_1 \text{ where } k_1 = (A_1 E_1 / L_1) \\
F_2 &= k_2 e_2 \text{ where } k_2 = (A_2 E_2 / L_2) \\
F_3 &= k_3 e_3 \text{ where } k_3 = (A_3 E_3 / L_3)
\end{aligned}
\qquad
\text{Element Force-Deformation}
\qquad (2)
$$

In Eqs. (2) the e's are the element elongations. A positive F_i (tension) produces a positive e_i (element gets longer), since the k_i's are, by definition, positive.

Although we have three new equations, we have introduced three new unknowns, so we now have five equations in six unknowns—three F's and three e's. Therefore, we must look to the geometry of deformation for more equations.

Geometry of Deformation: The joint displacements are labeled u_B and u_C in Fig. 1. We can easily relate the elongation of each element to these two joint displacements by using the definition of elongation of an element, that is,

$$
e \equiv u(L) - u(0)
$$

So,

$$
\begin{aligned}
e_1 &= u_B - u_A = u_B \\
e_2 &= u_C - u_B \\
e_3 &= u_D - u_C = -u_C
\end{aligned}
\qquad
\text{Geometry of Deformation}
\qquad (3)
$$

Here we have used the fact that the displacements at joints A and D are zero. (Note that, since e is positive when the element gets longer, a displacement of joint C to the right by an amount u_C implies a shortening of element 3 by that amount.)

If we now count equations and unknowns, we find that we have eight equations and eight unknowns. Rather than just combine Eqs. (1) through (3) in some arbitrary order, we can note that, since there are two joints, there are two equilibrium equations, and also there are two joint displacements, u_B and u_C. By substituting Eqs. (3) into Eqs. (2), we will be able to write the three F's in terms of the two u's. If we then substitute these equations, call them Eqs. (2'), into Eqs. (1), we will have *two equilibrium equations expressed in terms of two unknown joint displacements.*

Substitute Eqs. (3) (deformation geometry) into Eqs. (2) (element force-deformation behavior) to obtain

$$
F_1 = k_1 u_B, \quad F_2 = k_2 (u_C - u_B), \quad F_3 = k_3 (-u_C) \qquad (2')
$$

Now substitute Eqs. (2') into Eqs. (1).

$$
k_1 u_B - k_2 (u_C - u_B) = P_B
$$

$$
k_2 (u_C - u_B) - k_3 (-u_C) = P_C
$$

or

$$(k_1 + k_2)u_B - k_2 u_C = P_B$$
$$-k_2 u_B + (k_2 + k_3)u_C = P_C$$

Joint Equilibrium in Terms of Joint Displacements (1′)

The solution of these two equations in two unknowns is

$$u_B = \frac{(k_2 + k_3)P_B + k_2 P_C}{k_1 k_2 + k_1 k_3 + k_2 k_3}$$

$$u_C = \frac{k_2 P_B + (k_1 + k_2)P_C}{k_1 k_2 + k_1 k_3 + k_2 k_3}$$

Ans. (a) (4)

The strategy of substituting deformation geometry equations (3) into element force-deformation equations (2) into equilibrium equations (1) is called the *displacement method* because the major step in the solution gives answers that are displacements (Eqs. (4)). It is also sometimes referred to as the *stiffness method* since stiffness coefficients appear in the final solution.

(b) Now we need to determine the element forces. This is simple to do, because we only need to substitute the nodal displacements into Eqs. (2′). Thus, we get

$$F_1 = k_1 u_B = \frac{(k_1 k_2 + k_1 k_3)P_B + k_1 k_2 P_C}{k_1 k_2 + k_1 k_3 + k_2 k_3}$$

$$F_2 = k_2(u_C - u_B) = \frac{-k_2 k_3 P_B + k_1 k_2 P_C}{k_1 k_2 + k_1 k_3 + k_2 k_3}$$

$$F_3 = k_3(-u_C) = \frac{-k_3[k_2 P_B + (k_1 + k_2)P_C]}{k_1 k_2 + k_1 k_3 + k_2 k_3}$$

Ans. (b) (5)

(c) To determine the reactions at A and D, we can note from the free-body diagrams of node A and node D in Fig. 3 that

$$\xrightarrow{+} \sum F_x = 0: \qquad P_A + F_1 = 0, \quad P_D - F_3 = 0 \tag{6}$$

Then,

$$P_A = -F_1, \qquad P_D = F_3 \qquad \textbf{Ans. (c)} \quad (7)$$

where F_1 and F_3 are given in Eqs. (5).

Review the Solution As one check of our work, we can substitute Eqs. (5) back into Eqs. (1) to see if equilibrium is satisfied.

Is $F_1 - F_2 = P_B$? Yes. Is $F_2 - F_3 = P_C$? Yes.

The fact that equilibrium is satisfied by our answers means that we have probably not made errors in our solution. Also, from Eqs. (5a) and (5c) we see that, when P_B and P_C are both positive, element (1) is in tension and element (3) is in compression. This is what we expected to find.

The Displacement Method.[4] The preceding solution by the displacement method
became quite lengthy, not because we included unnecessary steps, but primarily because
of the discussion of the solution as it progressed. Before illustrating a force-method solution, let us summarize the steps involved in a *displacement-method solution.*

109

**Structures with Uniform
Axial-Deformation Members**

DISPLACEMENT-METHOD SOLUTION PROCEDURE

1. Identify the independent *system displacement variables.* (In Example 3.4, u_B and u_C are the nodal displacements; they determine the element elongations.)

2. Draw a free-body diagram associated with each of the displacement variables identified in Step 1. Write *equations of equilibrium* for these free bodies to relate the external loads to the (unknown) element forces.

3. Write the element *force-deformation equation* for each axial-deformation element. Equation 3.14,

$$F_i = k_i e_i, \qquad k_i = (AE/L)_i$$

   is the most convenient form to use for this step.

4. Use *geometry of deformation* to relate the element elongations, e_i, to the system displacement variables that were identified in Step 1.

5. Substitute the deformation-geometry equations (Step 4) into the force-deformation equations (Step 3). This gives *element forces in terms of system displacements.*

6. Substitute the results of Step 5 into the equilibrium equations of Step 2. This gives *equilibrium equations written in terms of system displacements.*

7. Solve the equations obtained in Step 6. The answer will be the *system displacements.*

8. Substitute the system displacements of Step 7 into the equations obtained in Step 5. The answer will be the *element forces.* If element normal stresses are required, the element forces can be divided by the respective element cross-sectional areas, that is, $\sigma_i = F_i/A_i$.

9. Review the solution to make sure that all answers seem to be correct.

The Force Method.[5] The three fundamental types of equations are needed in a solution by the force method, just as they are in a displacement-method solution. The steps that are outlined below constitute a systematic *force-method solution.*

FORCE-METHOD SOLUTION PROCEDURE

1. Determine the number of *redundant internal forces* in the system. The number of redundants is equal to the total number of unknown internal forces minus number of independent equilibrium equations that are available, that is

$$N_R = N_U - N_E$$

   Select N_R internal forces as *redundant internal forces.* The other N_E forces will be the *determinate internal forces.* If the system is statically determinate, there are no redundants, and all of the internal forces can be obtained by using the equilibrium equations alone, as in Example 2.2.

2. Use *free-body diagrams* and N_E equilibrium equations to obtain expressions for the N_E determinate internal force(s) in terms of the N_R redundant internal force(s) and the external loads.

3. Write the element *force-deformation equation* for each axial deformation element. Equation 3.13,

$$e_i = f_i F_i, \qquad f_i = (L/AE)_i$$

   is the most convenient form to use for this step.

4. Use the *geometry of deformation* to write N_R geometric compatibility equations in terms of the element elongations.

5. Substitute the N_E determinate-force equations of Step 2 into the element force-deformation equations of Step 3. This produces force-deformation equations in terms of the N_R redundant internal force(s).

6. Substitute the results of Step 5 into the N_R geometric-compatibility equations of Step 4. This gives N_R geometric-compatibility equations in terms of the N_R redundant internal forces.

7. Solve the equations obtained in Step 6. The answer will be the N_R redundant internal forces.

8. To obtain system displacements, substitute the redundant internal forces into the equilibrium equations of Step 2. Then, substitute the element forces into the force-deformation equations of Step 3. Finally, use geometry of deformation to relate element elongations to system displacements.[6]

9. Review the solution to make sure that all answers seem to be correct.

---

[4]This method is also called the *stiffness method.*

[5]This method is also referred to as the *flexibility method.*

[6]If a system is statically indeterminate and displacements are required, in addition to element forces and/or reactions at the supports, the displacement method is generally easier to use than the force method.

We will now solve the problem in Example 3.4 again, but this time we will employ the force method instead of the displacement method. Notice that we must still begin the solution by considering the three fundamental types of equations: *equilibrium, element force-deformation,* and *geometry of deformation.*

■■■■■■■■■■■■■■■□ EXAMPLE 3.5 □□□□□□□■■■■■■■■■■■■

For the three-element stepped rod in Fig. 1, use the force method to solve the problem stated in Example 3.4.

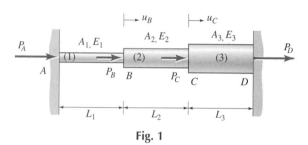

Fig. 1

(a) Determine the normal stress, σ_i, in each of the three elements. Let F_1, the internal force in member AB, be the redundant force. (b) Determine the reactions P_A and P_D at A and D. (c) Determine the joint displacements u_B and u_C.

Plan the Solution   A single cut anywhere along element (1) will leave the structure statically determinate. Therefore, $N_R = 1$. We can let F_1 be the redundant force and follow the steps outlined as the *Force Method.* The answers should agree with the answers in Example 3.4.

Solution
(a) We follow the force-method solution procedure to determine the redundant force F_1.

Equilibrium: Having selected F_1 as the redundant internal force (Step 1), we need to draw two ($N_E = N_U - N_R$) free-body diagrams that can be used to express F_2 and F_3 in terms of the redundant force F_1. By cutting on each side of the joint at B, we produce the free-body diagram of Fig. 2a, which relates F_2 to F_1. Similarly, by cutting through elements (1) and (3) we get the free-body diagram in Fig. 2b, which relates F_3 to the redundant force F_1.
   For Fig. 2a:

$$\overset{+}{\rightarrow} \sum F_x = 0: \qquad -F_1 + F_2 + P_B = 0$$

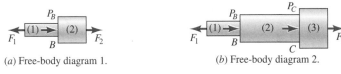

(a) Free-body diagram 1. (b) Free-body diagram 2.

Fig. 2 Free-body diagrams.

For Fig. 2b:

$$\xrightarrow{+} \sum F_x = 0: \qquad -F_1 + P_B + P_C + F_3 = 0$$

Writing F_2 and F_3 in terms of F_1 and the external loads, we get

$$F_2 = F_1 - P_B$$
$$F_3 = F_1 - P_B - P_C$$

Equilibrium (1)

The equilibrium equations in Eq. (1) constitute Step 2 of the force-method procedure.

Element Force-Deformation Behavior: The force-deformation equations for the three elements can be written in the form given by Eq. 3.13, namely

$$e_1 = f_1 F_1, \qquad f_1 = (L/AE)_1$$
$$e_2 = f_2 F_2, \qquad f_2 = (L/AE)_2$$
$$e_3 = f_3 F_3, \qquad f_3 = (L/AE)_3$$

Element Force-Deformation (2)

These equations constitute Step 3 of the force-method procedure. At this point we have five equations and six unknowns, so we must press on. Since we have completely exhausted the possible equilibrium equations and element force-deformation equations, we know that we must turn to the geometry of deformation.

Geometry of Deformation: Since we have one redundant force, F_1, we need one *equation of geometric compatibility* (Step 4). We can see (Fig. 1) that, because the total length of the three-element rod system remains constant, there is no elongation of the rod. The appropriate equation is

$$e_{\text{total}} = e_1 + e_2 + e_3 = 0$$

Geometry of Deformation (3)

Force-Method Solution: Following Steps 5 and 6 we first combine Eqs. (1) and (2) to get

$$e_1 = f_1 F_1, \quad e_2 = f_2(F_1 - P_B), \quad e_3 = f_3(F_1 - P_B - P_C) \qquad (2')$$

which are element force-deformation equations written in terms of the redundant force. Next we substitute Eqs. (2') into Eq. (3) to get

$$(f_1 + f_2 + f_3)F_1 = f_2 P_B + f_3(P_B + P_C) \qquad (3')$$

and, solving for the redundant force, we get

$$\boxed{F_1 = \frac{(f_2 + f_3)P_B + f_3 P_C}{f_1 + f_2 + f_3}} \qquad (4)$$

This is the key solution step. We have determined the redundant internal force in terms of the given external loads and properties of the structure. The approach we have used is called the *force method* because the major solution step gives us a force quantity. It is also sometimes referred to as the *flexibility method* because flexibility coefficients, f_i, appear in this key solution step. Knowing the redundant force F_1, we can now use the equilibrium equations, Eqs. (1), to determine the determinate forces F_2 and F_3. Then

$$F_2 = F_1 - P_B = \frac{-f_1 P_B + f_3 P_C}{f_1 + f_2 + f_3}$$

$$F_3 = F_1 - P_B - P_C = \frac{-f_1 P_B - (f_1 + f_2)P_C}{f_1 + f_2 + f_3} \qquad (5)$$

Since we were asked for the stresses in the three elements, we need to divide the forces in Eqs. (4) and (5) by their respective cross-sectional areas,

$$\sigma_1 = \frac{F_1}{A_1}, \quad \sigma_2 = \frac{F_2}{A_2}, \quad \sigma_3 = \frac{F_3}{A_3} \qquad \text{Ans. (a)} \quad (6)$$

(b) The reactions at A and D are determined from equilibrium, just as they were in part (c) of Example 3.4. The result is

$$P_A = -F_1, \qquad P_D = F_3 \qquad \text{Ans. (b)} \quad (7)$$

(c) To obtain the displacements u_B and u_C we need to use geometry of deformation equations in the form of Eqs. (3) of Example 3.4. We only need to use the first and last equations, writing

$$u_B = e_1, \qquad u_C = -e_3 \qquad (8)$$

Equations (2a) and (2c) may be substituted into these to give

$$u_B = f_1 F_1, \qquad u_C = -f_3 F_3 \qquad (9)$$

and forces F_1 and F_3 from Eqs. (4) and (5b), respectively, can be substituted into Eqs. (9) giving

$$
\left.
\begin{aligned}
u_B &= \frac{f_1[(f_2 + f_3)P_B + f_3 P_C]}{f_1 + f_2 + f_3} \\[2mm]
u_C &= \frac{f_3[f_1 P_B + (f_1 + f_2)P_C]}{f_1 + f_2 + f_3}
\end{aligned}
\right\} \qquad \textbf{Ans. (c)} \quad (10)
$$

Review the Solution   To show that the above force-method solution gives us the same results as the displacement method, let us compare the F_2 of this solution with the F_2 of Example 3.4, recalling that $f_i = 1/k_i$.

$$F_2 = \frac{-(1/k_1)P_B + (1/k_3)P_C}{(1/k_1) + (1/k_2) + (1/k_3)}$$

$$F_2 = \frac{-k_2 k_3 P_B + k_1 k_2 P_C}{k_1 k_2 + k_1 k_3 + k_2 k_3}$$

From Examples 3.4 and 3.5 it should be clear that **problem solving in deformable-body mechanics involves a systematic application of the three fundamental types of equations:** *equilibrium, element force-deformation behavior, and geometry of deformation*. One order of eliminating variables to get the final solution leads to the displacement method; the other leads to the force method. You may ask: Why bother with the displacement method and the force method? Why not just solve N equations in N unknowns without bothering about the order? One answer to these questions is that problem solving is just easier if it is approached in a systematic manner. A second answer is that these same two approaches are found to be useful not only in this introduction to the mechanics of deformable bodies but in all subsequent topics in solid mechanics, for instance in the theory of elasticity, in the theory of plates and shells, in matrix structural analysis, and in finite element analysis.

Either the displacement method or the force method can be used as long as we have all the necessary equations at hand. However, it may be more convenient to use one procedure rather than the other. The displacement method is more straightforward because we don't have to select redundants. However, there are circumstances where the force method might require less work. For example, if we were asked only to determine the reaction P_A at the left wall, but not for the nodal displacements, for the rod in Fig. 3.10,

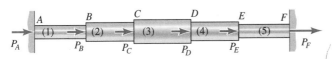

FIGURE 3.10   A sample statically indeterminate structure.

113

we could let F_1 be the redundant and, using the force method, solve the deformation compatibility equation.

$$e_1 + e_2 + e_3 + e_4 + e_5 = 0$$

once it has been written in terms of F_1. (See Homework Problem 3.5-14).

The displacement method was used as the basis for the development of AXIALDEF, a computer program to solve axial-deformation problems like the ones in Example Problems 3.4 and 3.5. This computer program, one of the **MechSOLID** collection of computer programs, is described in Appendix G.3. Computer homework exercises for Section 3.5 follow the regular Section 3.5 homework problems at the end of the chapter.

To reinforce your understanding of the basic problem-solving approach introduced in the preceding two examples, let us consider a problem of the type illustrated in Fig. 3.8a. In Example 3.6 we will employ the displacement-method procedure, and we will solve the same problem by the force method in Example 3.7. Then, you can compare the two solutions and note their similarities and differences. Subsequently, only one or the other of the two methods will be used to solve example problems.

■■■■■■■■■■■■■■■ E X A M P L E   3 . 6 ■■■■■■■■■■■■■■

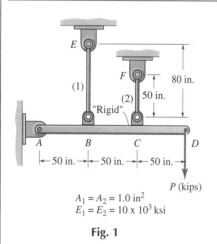

$A_1 = A_2 = 1.0$ in^2
$E_1 = E_2 = 10 \times 10^3$ ksi

Fig. 1

Consider the statically indeterminate system in Fig. 1. A load P is hung from the end of a rigid beam AD, which, in turn, is supported by two uniform rods. How much load can be applied without exceeding an allowable tensile stress of 30 ksi in either of the hanger rods? Assume small-angle rotation of AD.

Plan the Solution   This problem is not quite as straightforward as the previous one in Examples 3.4 and 3.5. Now the load is unknown, but the allowable internal stress is known. Since the internal stress is just the internal force divided by the area, it is clear that we must eventually get equations that relate F_1 and F_2, the forces in the two rods, to the load P. We can follow the steps outlined previously under the heading *Displacement Method Solution Procedure.*

Solution   Let us begin by writing down the three fundamental types of equations.

Equilibrium: We must ask ourselves the question: What free-body diagram(s) would lead to an equilibrium equation (or equations) that relate the external load P to the internal element forces F_1 and F_2? Clearly, a free-body diagram of the rigid beam should be used, since the *system displacement* in this case is the rotation of the beam about the pin at A (Step 1). Figure 2 shows the appropriate free-body diagram.

Since we are not specifically asked to determine the reactions A_x and A_y, we do not need to sum forces. A moment equation for the free-body diagram in Fig. 2 will directly relate F_1 and F_2 to P, so we write

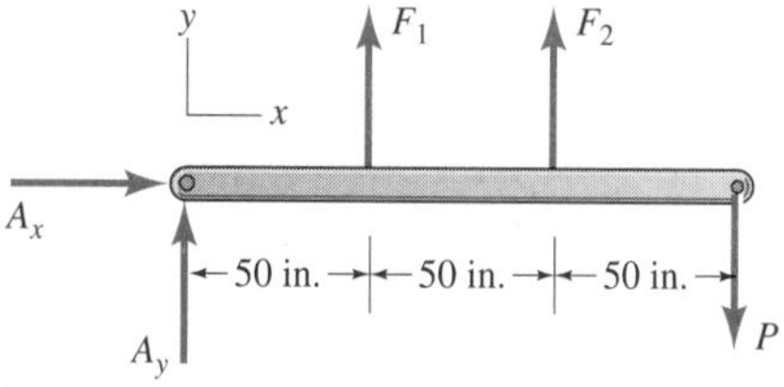

$$+\circlearrowleft \left(\sum M \right)_A = 0: \qquad 50F_1 + 100F_2 - 150P = 0$$

Fig. 2 Free-body diagram.

or

$$F_1 + 2F_2 = 3P \qquad \textbf{Equilibrium} \quad (1)$$

Element Force–Deformation Behavior: Equation 3.14 is the convenient form to use, so we first compute the stiffness coefficients for the two rods, labeled elements (1) and (2).

$$k_1 = \left(\frac{AE}{L}\right)_1 = \frac{(1.0 \text{ in}^2)(10 \times 10^3 \text{ ksi})}{80 \text{ in.}} = 125 \text{ kips/in.}$$

$$k_2 = \left(\frac{AE}{L}\right)_2 = \frac{(1.0 \text{ in}^2)(10 \times 10^3 \text{ ksi})}{50 \text{ in.}} = 200 \text{ kips/in.}$$

and, from Eq. 3.14,

$$F_1 = k_1 e_1 = 125\, e_1 \qquad \textbf{Element Force–}$$
$$\textbf{Deformation} \quad (2)$$
$$F_2 = k_2 e_2 = 200\, e_2 \qquad \textbf{Behavior}$$

Geometry of Deformation: Because *AD* is assumed to be rigid, we can sketch a *deformation diagram* that relates the elongations of the hanger rods to the rotation of the beam *AD* about *A* (Fig. 3). Since θ is assumed to be small, we can write

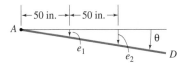

Fig. 3 Deformation diagram.

$$e_1 \doteq (50 \text{ in.})(\theta \text{ rad}) \qquad \textbf{Deformation}$$
$$\textbf{Geometry} \quad (3)$$
$$e_2 \doteq (100 \text{ in.})(\theta \text{ rad})$$

Displacement-Method Solution: The displacement method is to substitute in the following order: (3) → (2) → (1).

$$F_1 = 125\, e_1 = 125(50\,\theta) = 6250\,\theta \qquad (2')$$

$$F_2 = 200\, e_2 = 200(100\,\theta) = 20{,}000\,\theta$$

$$6250\,\theta + 2(20{,}000\,\theta) = 3P \qquad (1')$$

Finally, we can solve this equation for the unknown angular displacement θ.

$$\theta = \frac{3P}{46{,}250} \text{ rad} \qquad (4)$$

The stresses in the rods are given by

$$\sigma_1 = \frac{F_1}{A_1} = \frac{(F_1 \text{ kips})}{(1.0 \text{ in}^2)} \qquad (5)$$

$$\sigma_2 = \frac{F_2}{A_2} = \frac{(F_2 \text{ kips})}{(1.0 \text{ in}^2)}$$

Substituting Eq. (4) into Eqs. (2′), and substituting the result into Eqs. (5), we get

$$\sigma_1 = 6250 \ \theta = \left(\frac{15}{37}\right) P \text{ ksi}$$

$$\sigma_2 = 20{,}000 \ \theta = \left(\frac{48}{37}\right) P \text{ ksi} > \sigma_1$$

Since rod (2) is more highly stressed than rod (1), we set σ_2 equal to the allowable stress and get

$$P = \frac{37}{38}\sigma_{\text{allow.}} = \frac{37(30)}{48} = 23.1 \text{ kips} \qquad \qquad \textbf{Ans.}$$

Review the Solution   One way to verify the results is to check equilibrium.

$$F_1 = \left(\frac{15}{37}\right) P = 9.38 \text{ kips}, \qquad F_2 = \left(\frac{48}{37}\right) P = 30.0 \text{ kips}$$

Is $F_1 + 2F_2 = 3P$?   Yes.

■■■■■■■■■■■■■■■□ **EXAMPLE 3.7** □■■■■■■■■■■■■■■■■

The force method will now be used in solving the problem stated in Example 3.6.

Plan the Solution   We still need to employ equilibrium, element force-deformation behavior, and deformation geometry. However, this time we follow the steps outlined under the heading—*Force-Method Solution Procedure*. Since there are two unknown internal forces and only one equilibrium equation, we need to select one redundant force (Step 1).

Solution

Equilibrium: We can use the free-body diagram and equilibrium equation from Example 3.6.

$$F_1 + 2F_2 = 3P$$

Let F_2 be the redundant force, and use this equilibrium equation to express F_1 in terms of F_2 (Step 2).

$$F_1 = 3P - 2F_2 \qquad \qquad \textbf{Equilibrium} \quad (1)$$

Element Force–Deformation Behavior: The form given in Eq. 3.13 is used.

$$f_1 = \frac{1}{k_1} = \frac{1}{(125 \text{ kips/in.})} = 8(10^{-3}) \text{ in./kip}$$

$$f_2 = \frac{1}{k_2} = \frac{1}{(200 \text{ kips/in.})} = 5(10^{-3}) \text{ in./kip}$$

$$e_1 = f_1 F_1 = 8(10^{-3})F_1$$
$$e_2 = f_2 F_2 = 5(10^{-3})F_2$$

Element Force–Deformation Behavior (2)

Geometry of Deformation: Since we have one redundant force ($N_R = 1$) we need one equation of *deformation compatibility*. In Example 3.6 we used a deformation diagram to obtain

$$e_1 = 50\ \theta, \qquad e_2 = 100\ \theta$$

Therefore, by eliminating θ we get the compatibility equation

$$e_2 = 2e_1$$

Deformation Geometry (3)

Force-Method Solution: For the force-method approach, we substitute in the following order: (1) $\rightarrow$ (2) $\rightarrow$ (3).

Substitute Eq. (1) into Eq. (2) to get

$$e_1 = 8(10^{-3})(3P - 2F_2) \tag{2'}$$
$$e_2 = 5(10^{-3})F_2$$

Then substitute Eqs. (2') into the geometric compatibility equation, Eq. (3).

$$5(10^{-3})F_2 = 2(8)(10^{-3})(3P - 2F_2) \tag{3'}$$

or

$$F_2 = \left(\frac{48}{37}\right) P \tag{4}$$

This is the key solution step, but we still need to solve for F_1 and relate P to the allowable stress. Combining Eqs. (1) and (4), we get

$$F_1 = 3P - 2\left(\frac{48}{37}\right) P = \left(\frac{15}{37}\right) P$$

The remainder of the problem is identical to Example 3.4, that is,

$$\sigma_1 = \frac{F_1}{A_1} = \left(\frac{15}{37}\right) P, \qquad \sigma_2 = \frac{F_2}{A_2} = \left(\frac{48}{37}\right) P > \sigma_1$$

Therefore,

$$P = \frac{37}{48}\ \sigma_{\text{allow.}} = \frac{37(30)}{48} = 23.1 \text{ kips} \qquad \text{Ans.}$$

Review the Solution   We have obtained the same answers that were obtained by the displacement method. Notice that, since we did not really need to know the displacement, the force method may be a little more efficient than the displacement method in solving this particular problem.

Appendix G.3 describes how the displacement method can conveniently be used to develop a computer algorithm for solving axial-deformation problems. Examples G-3 and G-4 compare a "hand solution" of an axial-deformation problem with an AXIALDEF computer solution of the same problem. Computer Exercises for Section 3.5 follow the regular Section 3.5 homework exercises at the end of the chapter.

■■■■■■■■■■ 3.6 THERMAL STRESSES IN AXIAL DEFORMATION

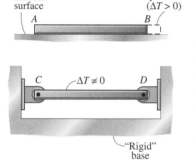

FIGURE 3.11 Restraint of thermal deformation.

Thermal strain was introduced in Section 2.3. In this section we examine temperature effects in greater detail. Two axial-deformation members are shown in Fig. 3.11. Member *AB* is supported on a smooth surface and, when heated or cooled, is free to expand or contract. No axial force (or stress) is induced in this member, since it can freely expand or contract. Member *CD* is assumed to be force free when the pins are installed to connect it to a rigid base. If it is subsequently heated, it will tend to expand, in accordance with Eq. 2.8 (Section 2.3). However, since it is prevented from expanding by the rigid walls, a compressive force will be induced in this member. Conversely, if it is cooled, it will tend to contract and pull away from the rigid walls, and in the process tension will be induced in this member. Even if the base to which member *CD* is attached is not completely rigid, the presence of some restraining structure will cause compressive stress to be induced in *CD* as a result of heating the bar, and tension will be induced by cooling the bar.

There is a straightforward way to determine the effect of heating or cooling on an axial-deformation member. From Eq. 2.8, the extensional strain due to temperature change is given by

$$(\epsilon_x)_t = \alpha \Delta T \tag{3.15}$$

where α is the coefficient of thermal expansion and ΔT is the temperature increase above the reference temperature (e.g., room temperature). ΔT is positive if the temperature increases, and a positive ΔT causes a tendency for the bar to expand.

If we consider only slender members, where it is reasonable to assume that the only significant normal stress is the axial stress σ_x (i.e., we assume that σ_y and σ_z are negligible), then Eq. 2.32a gives

$$\boxed{\epsilon_x = \frac{\sigma_x}{E} + \alpha \Delta T} \tag{3.16}$$

That is, ϵ_x is the sum of a strain due to the axial stress σ_x and a thermal strain due to the temperature change, or

$$\epsilon_x = (\epsilon_x)_\sigma + (\epsilon_x)_t$$

Equation 3.16 is the *stress-strain-temperature equation* for axial deformation of a slender, linearly elastic member.

To solve axial deformation problems that involve temperature change, we need to use the *stress-strain-temperature equation*, Eq. 3.16, but we also need to account for equilibrium and the geometry of deformation. With regard to the *geometry of deformation*, we know that the strain-displacement equation for axial deformation, Eq. 3.1, holds. That is, the strain ϵ_x is independent of position in a cross section and is given by

$$\epsilon = \frac{du}{dx}$$

(3.1)
repeated

With regard to equilibrium, we know that the axial stress σ_x is related to the axial force $F(x)$ by Eq. 3.5a, that is

$$F(x) = \int \sigma_x dA$$

(3.5a)
repeated

If we restrict our attention to situations where E, α, and ΔT may depend on x but are independent of position in the cross section, we can combine Eq. 3.16 with Eqs. 3.1 and 3.5a to get

$$\sigma_x(x) = \frac{F(x)}{A(x)}$$

(3.17)

and

$$\frac{du(x)}{dx} = \frac{F(x)}{A(x)E(x)} + \alpha(x)\Delta T(x)$$

(3.18)

In the same manner that was used to obtain Eq. 3.11, we can integrate Eq. 3.18 to get the following expression for the elongation of the member:

$$e \equiv u(L) - u(0) = \int_0^L \frac{F(x)dx}{A(x)E(x)} + \int_0^L \alpha(x)\Delta T(x)dx$$

(3.19)

Example 3.8 illustrates the application of Eqs. 3.16, 3.17, and 3.19.[7]

---

[7]Problems of this type are sometimes referred to as *thermal stress* problems.

■■■■■■■■■■■■■■■□ E X A M P L E 3 . 8 □■■■■■■■■■■■■■■

The slender, uniform rod in Fig. 1 is attached to rigid supports at A and B, and it is stress free when $\Delta T = 0$. It is surrounded by a heating element that is capable of producing the linearly varying temperature change $\Delta T(x)$ shown in Fig. 2. Determine the stress distribution $\sigma_x(x)$ that is induced by this nonuniform heating.

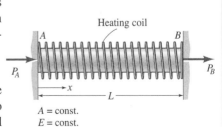

Plan the Solution   The only external forces applied to the rod AB are the reactions at A and B, as shown in Fig. 1. We can use a free-body diagram to relate $F(x)$ to P_A (or P_B). Since the supports at A and B are rigid, the total elongation of AB is zero. Since $\Delta T > 0$ all along the bar, we can expect a compressive stress to result.

Heating coil

$A = $ const.
$E = $ const.
$\alpha = $ const.

Fig. 1   A heated, uniform bar.

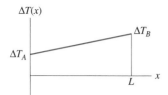

Fig. 2 The temperature change.

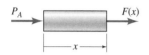

Fig. 3 Free-body diagram.

Solution

Equilibrium: To determine how $F(x)$ varies with x, we draw the free-body diagram shown in Fig. 3, with a cut at an arbitrary section x.

$$\overset{+}{\rightarrow} \sum F_x = 0: \qquad \boxed{F(x) = -P_A = \text{const.}} \qquad \textbf{Equilibrium} \quad (1)$$

Force–Temperature–Deformation Behavior: From Eq. 3.19

$$e = \int_0^L \left(\frac{F}{AE} + \alpha \Delta T(x) \right) dx$$

From Fig. 2, the temperature distribution is given by

$$\Delta T(x) = \Delta T_A + (\Delta T_B - \Delta T_A)\left(\frac{x}{L}\right)$$

Then, since $F(x) = \text{const}$,

$$e = \frac{FL}{AE} + \alpha \int_0^L \left[\Delta T_A + (\Delta T_B - \Delta T_A)\left(\frac{x}{L}\right) \right] dx$$

or

$$\boxed{e = \frac{FL}{AE} + \alpha L\left(\frac{\Delta T_A + \Delta T_B}{2}\right)} \quad \begin{matrix} \textbf{Force–} \\ \textbf{Temperature–} \\ \textbf{Deformation} \\ \textbf{Behavior} \end{matrix} \quad (2)$$

Geometry of Deformation: Since the ends of the rod cannot move,

$$\boxed{e = 0} \qquad \begin{matrix} \textbf{Geometry of} \\ \textbf{Deformation} \end{matrix} \quad (3)$$

Solution: Combine Eqs. (2) and (3) to get

$$\boxed{F = -AE\alpha\left(\frac{\Delta T_A + \Delta T_B}{2}\right)} \quad (4)$$

Then, since $\sigma_x = \dfrac{F}{A}$,

$$\sigma_x = -E\alpha\left(\frac{\Delta T_A + \Delta T_B}{2}\right) \qquad \textbf{Ans.} \quad (5)$$

Review the Solution The right-hand side of Eq. (5) has the dimensions of stress (F/L^2), as it should. The stress is negative, which is what we would expect to happen when the bar is heated.

It may surprise you that the stress is independent of x, even though ΔT varies with x. This is the direct consequence of the fact that the equilibrium equation, Eq. (1), states that $F(x)$ is a constant, not a function of x, like ΔT is!

Let us now consider the uniformly heated, uniform, axial-deformation element in Fig. 3.12. We will repeat the steps of Section 3.4 to determine the *force-temperature-deformation equation* for a *uniform member subjected to a uniform temperature change.*

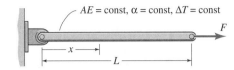

$AE = \text{const}, \alpha = \text{const}, \Delta T = \text{const}$

F

x

L

(*a*) A uniform element with axial force F and with $\Delta T = \text{const}$

$F(x)$ F

(*b*) Free-body diagram.

FIGURE 3.12 A heated (cooled) axial-deformation element.

Equilibrium. Draw a free-body diagram with the member cut at an arbitrary section x.

$\overset{+}{\rightarrow} \sum F_x = 0:$ $F(x) = F = \text{const}$

From Eq. 3.17,

$$\sigma_x = \frac{F(x)}{A(x)} = \frac{F}{A} = \text{const}$$

Stress-Strain-Temperature. From Eq. 3.19 and the preceding equation,

$$e = \int_0^L \epsilon_x dx = \frac{FL}{AE} + \alpha L \Delta T$$

Thus, **the total elongation, e, consists of the elongation due to the applied force, F, plus the elongation due to the uniform temperature change, ΔT.**

$$\boxed{e = fF + \alpha L \Delta T} \qquad (3.20)$$

In terms of the stiffness coefficient, we can write Eq. 3.20 as

$$\boxed{F = k(e - \alpha L \Delta T)} \qquad (3.21)$$

The only fundamental equations that are affected by temperature change are the force-deformation equations, Eqs. 3.13 and 3.14, which are replaced by Eqs. 3.20 and 3.21, respectively. **No changes are required in the equilibrium equations or geometry-of-deformation equations!** We will now apply Eqs. 3.20 and 3.21 in several example problems. As before, we can formulate the solution in the displacement-method format or the force-method format. Both methods are illustrated in the following examples.

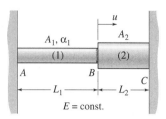

Fig. 1

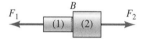

Fig. 2 Free-body diagram.

Two elements are stress-free when they are welded together at B and welded to rigid walls at A and C (Fig. 1). Subsequently, element (1) is heated by an amount ΔT, while element (2) is held at the reference temperature.

Using the *displacement method,* (a) determine the displacement, u, of the joint at B; and (b) determine the internal force induced in each element.

Plan the Solution This problem can be solved by using the same steps as those used in Example 3.4. To incorporate the thermal strain, we just need to use Eq. 3.21 instead of Eq. 3.14.

Solution
(a) Solve for the displacement u.

Equilibrium: Although there are no external loads, that is, no external forces other than the reactions at A and C, we still need an equilibrium equation of the joint at B in order to relate the element forces to each other. Note that tension is assumed positive on the free-body diagram, Fig. 2.

$$\overset{+}{\rightarrow} \sum F_x = 0: \qquad \boxed{-F_1 + F_2 = 0} \qquad \textbf{Equilibrium} \quad (1)$$

Element Force–Temperature–Deformation Behavior: These are the equations through which the temperature effect enters the solution. Since we are asked to use the displacement (stiffness) method, Eq. 3.21 is the appropriate equation to use. We write this equation for each element, noting that $\Delta T_1 = \Delta T$ and $\Delta T_2 = 0$.

$$\boxed{\begin{aligned} F_1 &= k_1(e_1 - \alpha_1 L_1 \Delta T) \\ F_2 &= k_2 e_2 \end{aligned}} \qquad \begin{aligned} &\textbf{Element Force–} \\ &\textbf{Temperature–} \\ &\textbf{Deformation} \\ &\textbf{Behavior} \end{aligned} \quad (2)$$

where $k_i = \left(\dfrac{AE}{L}\right)_i$.

Geometry of Deformation: In Eqs. (2), e_1 and e_2 are the <u>total elongations</u> of the respective elements, just as e represents total elongation in Sections 3.3 and 3.4. Therefore, it is a purely geometric exercise to relate the e's to the joint displacement u.

$$\boxed{e_1 = u, \qquad e_2 = -u} \qquad \begin{aligned} &\textbf{Geometry of} \\ &\textbf{Deformation} \end{aligned} \quad (3)$$

Displacement-Method Solution: Now we simply combine the above sets of equations in the order (3) → (2) → (1), which constitutes a displacement-method solution. Then, Eqs. (2) become

$$F_1 = k_1(u - \alpha_1 L_1 \Delta T), \qquad F_2 = k_2(-u) \qquad (2')$$

leading to the equilibrium equation in the form

$$-k_1(u - \alpha_1 L_1 \Delta T) + k_2(-u) = 0 \qquad (1')$$

Inserting expressions for the k_i's and solving for u, we get

$$u = \frac{A_1 E \alpha_1 \Delta T}{\left(\dfrac{A_1 E}{L_1}\right) + \left(\dfrac{A_2 E}{L_2}\right)}$$

or, finally,

$$u = \frac{A_1 L_1 L_2 \alpha_1 \Delta T}{A_1 L_2 + A_2 L_1} \qquad \text{Ans. (a)} \quad (4)$$

(b) Solve for the element forces F_1 and F_2. We can substitute the expression for u in Eq. (4) into Eqs. (2') to obtain F_1 and F_2.

$$F_1 = \left(\frac{A_1 E}{L_1}\right)\left(\frac{A_1 L_1 L_2 \alpha_1 \Delta T}{A_1 L_2 + A_2 L_1} - \alpha_1 L_1 \Delta T\right)$$

$$F_1 = \frac{A_1 A_2 E \alpha_1 L_1 \Delta T}{A_1 L_2 + A_2 L_1}$$

$$F_2 = k_2(-u) = \left(\frac{A_2 E}{L_2}\right)\left(\frac{-A_1 L_1 L_2 \alpha_1 \Delta T}{A_1 L_2 + A_2 L_1}\right)$$

Therefore,

$$F_1 = F_2 = \frac{-A_1 A_2 E \alpha_1 L_1 \Delta T}{A_1 L_2 + A_2 L_1} \qquad \text{Ans. (b)} \quad (5)$$

Review the Solution   As we would expect, heating element (1) causes a compression to be induced in both elements and, since there is no external force on the joint connecting the two elements, the same compressive force is induced in each element. The right-hand side of Eq. (5) has the proper dimensions of force.
   Homework Problem 3.6.18 is a force-method version of this problem.

EXAMPLE 3.10

In Fig. 1 a cylindrical aluminum core is surrounded by a titanium sleeve, and both are attached at each end to a rigid end-plate. Set up the fundamental equations and combine them in displacement-method order to solve for the following: (a) the elongation of the composite bar if the entire structure is heated by 100°F. (Assume that the core and the outer sleeve are both stress free at the reference temperature.) (b) the axial stress induced in each material.

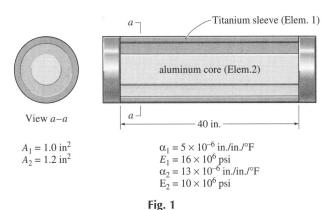

$A_1 = 1.0 \text{ in}^2$
$A_2 = 1.2 \text{ in}^2$

$\alpha_1 = 5 \times 10^{-6} \text{ in./in./°F}$
$E_1 = 16 \times 10^6 \text{ psi}$
$\alpha_2 = 13 \times 10^{-6} \text{ in./in./°F}$
$E_2 = 10 \times 10^6 \text{ psi}$

Fig. 1

Plan the Solution This two-element problem is simple enough that we can "reason" that, because $\alpha_{\text{alum.}} > \alpha_{\text{titan.}}$, the aluminum will tend to expand more than the titanium and will therefore be in compression if both are heated the same amount. However, in a more complicated situation we would essentially have to "solve" the entire problem just to determine which elements are in compression and which are in tension. Fortunately, we do not have to do this! We just <u>assume tension in each element</u> and let the final solution tell us which elements are in tension and which are in compression.

Solution
(a) Solve for the change in length.

Equilibrium: The basic question to ask in setting up the equilibrium equation is: What free-body diagram can I use that will relate the internal element forces to each other and to the external loads? (In this case, the external loads are zero.) The answer is, of course, one of the end-plates to which both the core and the sleeve are attached, as shown in Fig. 2.

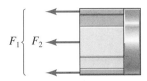

Fig. 2 Free-body diagram.

$$\xrightarrow{+} \sum F_x = 0: \qquad \boxed{-F_1 - F_2 = 0} \qquad \textbf{Equilibrium} \quad (1)$$

Element Force–Temperature–Deformation Behavior: Since we have been asked to set up equations that can be combined in the displacement-method order, we use the force-temperature-deformation (F-T-D) format of Eq. 3.21, which involves stiffness coefficients, k_i:

$$F_1 = k_1(e_1 - \alpha_1 L_1 \Delta T_1), \qquad F_2 = k_2(e_2 - \alpha_2 L_2 \Delta T_2)$$

where

$$k_1 = \left(\frac{AE}{L}\right)_1 = \frac{(1.0 \text{ in}^2)(16 \times 10^6 \text{ psi})}{(40 \text{ in.})} = 4(10^5) \text{ lb/in.}$$

$$k_2 = \left(\frac{AE}{L}\right)_2 = \frac{(1.2 \text{ in}^2)(10 \times 10^6 \text{ psi})}{(40 \text{ in.})} = 3(10^5) \text{ lb/in.}$$

$$\alpha_1 L_1 \Delta T_1 = (5 \times 10^{-6}/°F)(40 \text{ in.})(100°F) = 0.020 \text{ in.}$$

$$\alpha_2 L_2 \Delta T_2 = (13 \times 10^{-6}/°F)(40 \text{ in.})(100°F) = 0.052 \text{ in.}$$

Combining the above equations, we get

$$
\begin{aligned}
F_1 &= 4(10^5)(e_1 - 0.020) \\
F_2 &= 3(10^5)(e_2 - 0.052)
\end{aligned}
\qquad
\begin{array}{l}
\textbf{Element Force–} \\
\textbf{Temperature} \\
\textbf{Deformation}
\end{array}
\quad (2)
$$

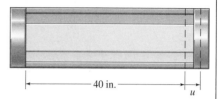

Geometry of Deformation: Let u be the displacement of the right-hand end-plate in the deformation diagram (Fig. 3), and recall that e_1 and e_2 are the <u>total elongations</u> of the respective elements. Then,

$$e_1 = u, \qquad e_2 = u \qquad \begin{array}{l}\textbf{Geometry of} \\ \textbf{Deformation}\end{array} \quad (3)$$

|←——————— 40 in. ———————→| u

Fig. 3 Deformation diagram.

Displacement-Method Solution: In Eqs. (1) through (3) we have five equations in five unknowns. Hence, we can use the displacement method to solve for u; then we can solve for internal forces. Substituting Eq. (3) into Eq. (2), we get

$$F_1 = 4(10^5)(u - 0.020), \qquad F_2 = 3(10^5)(u - 0.052) \qquad (2')$$

These expressions are then substituted into Eq. (1) to give

$$4(10^5)(u - 0.020) + 3(10^5)(u - 0.052) = 0$$

or,

$$7u = 4(0.020) + 3(0.052) = 0.236 \qquad (1')$$

Then,

$$\boxed{u = 0.0337 \text{ in.}} \qquad \textbf{Ans. (a)} \quad (4)$$

(b) Now, we can solve for the internal forces. Equations (4) and (2′) may now be combined to give

$$F_1 = 4(10^5)(0.0337 - 0.02) = 5486 \text{ lb}$$

$$F_2 = 3(10^5)(0.0337 - 0.052) = -5486 \text{ lb}$$

Rounding off these numbers, we get

$$\left.\begin{aligned}F_{\text{sleeve}} \equiv F_1 &= 5490 \text{ lb (5490 lb T)} \\ F_{\text{core}} \equiv F_2 &= -5490 \text{ lb (5490 lb C)}\end{aligned}\right\} \quad \textbf{Ans. (b)} \quad (10)$$

Review the Solution The final value of the elongation, u, is between the free-expansion values of aluminum and titanium, which is what we would expect. The signs of F_1 and F_2 agree with our ''Plan the Solution'' discussion.

(See Homework Problem 3.6-19 for a force-method version of this problem.)

The preceding examples clearly illustrate that **the solution of thermal deformation problems involving axial deformation of uniform elements only requires a modification of the element force-displacement equations, but absolutely no change to either the equilibrium equations or the deformation-geometry equations**.

As a final example of thermal deformation of systems with uniform axial-deformation members, consider the following modification of Example 3.6.

■■■■■■■■■■■■■■■■ E X A M P L E 3 . 1 1 ■■■■■■■■■■■■■■■

The structure of Example 3.6 is modified by replacing the external load, P, by a third rod element at D (Fig. 1). Determine the force in the three rods if element (2) is cooled by 100°F. Assume that the rotation of beam AD is small and that all rods are force-free when the system is assembled.

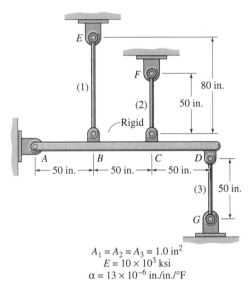

$$A_1 = A_2 = A_3 = 1.0 \text{ in}^2$$
$$E = 10 \times 10^3 \text{ ksi}$$
$$\alpha = 13 \times 10^{-6} \text{ in./in./°F}$$

Fig. 1

Plan the Solution   Although this problem could be solved by either the force method or the displacement method, since it is the element forces that are required, we will use the force (flexibility) method.

Since rod (2) is cooled, it will tend to shorten, and the force induced in it will be tension. This will cause the beam AD to rotate counterclockwise.

There are three unknown rod forces and, by taking moments about point A, we will get one equilibrium equation. Therefore, the number of redundant internal forces is $N_R = N_U - N_E = 3 - 1 = 2$.

Solution

Equilibrium: Since we want a free-body diagram that will permit us to relate the internal element forces to each other and to the external loads (actually, there are no external loads in this problem), we select the rigid beam AD, just as we did in Example 3.6. Figure 2 shows the resulting free-body diagram. As always, we select tension as positive for the force in each element.

$$+\circlearrowleft \left(\sum M \right)_A = 0: \qquad F_1(50 \text{ in.}) + F_2(100 \text{ in.}) - F_3(150 \text{ in.}) = 0$$

$$F_1 + 2F_2 - 3F_3 = 0 \qquad \textbf{Equilibrium} \quad (1)$$

Element Force–Temperature–Deformation: Since we have chosen to employ the force method, Eq. 3.20 is the most convenient form.

$$e_1 = f_1 F_1, \quad e_2 = f_2 F_2 + \alpha_2 L_2 \Delta T_2, \quad e_3 = f_3 F_3$$

where

$$f_1 = \left(\frac{L}{AE} \right)_1 = \frac{(80 \text{ in.})}{(1.0 \text{ in}^2)(10 \times 10^3 \text{ ksi})} = 0.0080 \text{ in./kip}$$

$$f_2 = \left(\frac{L}{AE} \right)_2 = \frac{(50 \text{ in.})}{(1.0 \text{ in}^2)(10 \times 10^3 \text{ ksi})} = 0.0050 \text{ in./kip}$$

$$f_3 = \left(\frac{L}{AE} \right)_3 = \frac{(50 \text{ in.})}{(1.0 \text{ in}^2)(10 \times 10^3 \text{ ksi})} = 0.0050 \text{ in./kip}$$

$$\alpha_2 L_2 \Delta T_2 = (13 \times 10^{-6}/°\text{F})(50 \text{ in.})(-100°\text{F}) = -0.0650 \text{ in.}$$

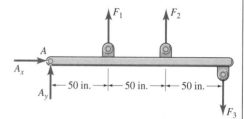

Fig. 2 Free-body diagram.

Thus,

$$
\begin{aligned}
e_1 &= 0.0080 F_1 \\
e_2 &= 0.0050 F_2 - 0.0650 \\
e_3 &= 0.0050 F_3
\end{aligned}
$$
Element Force–Temperature–Deformation Behavior (2)

Geometry of Deformation: We need to sketch a *deformation diagram.* We can essentially repeat the one in Example 3.6, adding, of course, the third element's elongation. (We could equally well assume the rotation of *AD* to be counterclockwise, rather than clockwise.)

In Eqs. (2) a positive *e* corresponds to a lengthening of the element. Thus, on the above deformation diagram (Fig. 3) a clockwise rotation of *AD* corresponds to elongations of rods (1) and (2) but a contraction of rod (3), that is, we need to use $(-e_3)$.

Before writing any deformation-geometry equations, let us see how many equations we need. It is stated in Step 4 of the *Force Method* procedure that N_R redundant internal forces must be selected and that there will be exactly N_R deformation compatibility equations relating the element elongations. From Eq. (1) we see that there are three unknowns and only one equilibrium equation; hence we need to select two element forces as redundants and to look for two deformation compatibility equations relating the *e*'s. Without even introducing the beam's rotation angle θ, we can make use of similar triangles (for small θ) and write

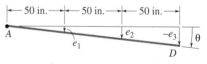

Fig. 3 Deformation diagram.

$$e_2 = 2e_1, \qquad -e_3 = 3e_1 \qquad \textbf{Deformation Compatibility} \quad (3)$$

127

Force-Method Solution: Now we use the force-method steps to combine Eqs. (1) through (3). First, we select two redundants and use Eq. (1) to write the third force, that is, the determinate force, in terms of the redundants. Let F_2 and F_3 be the redundants. (This choice is completely arbitrary.) Then, from Eq. (1)

$$F_1 = -2F_2 + 3F_3 \tag{1'}$$

This equation is now substituted into Eqs. (2).

$$e_1 = 0.0080(-2F_2 + 3F_3)$$
$$e_2 = 0.0050F_2 - 0.0650 \tag{2'}$$
$$e_3 = 0.0050F_3$$

Finally, these equations are substituted into the deformation compatibility equations, Eqs. (3), to obtain

$$0.0050F_2 - 0.0650 = 2(0.0080)(-2F_2 + 3F_3)$$
$$-0.0050F_3 = 3(0.0080)(-2F_2 + 3F_3)$$

or

$$0.037F_2 - 0.048F_3 = 0.0650 \tag{3'}$$
$$-0.048F_2 + 0.077F_3 = 0$$

The solution of these two simultaneous algebraic equations gives the two redundant forces

$$\boxed{F_2 = 9.1835 \text{ kips}, \quad F_3 = 5.7248 \text{ kips}} \tag{4}$$

Then, from Eq. (1'),

$$F_1 = -2F_2 + 3F_3 = -1.1927 \text{ kips}.$$

Rounding off these values to two decimal places, we have

$$\left. \begin{array}{l} F_1 = -1.19 \text{ kips (1.19 kips C)} \\ F_2 = 9.18 \text{ kips (9.18 kips T)} \\ F_3 = 5.72 \text{ kips (5.72 kips T)} \end{array} \right\} \qquad \text{Ans.} \quad (5)$$

Review the Solution   As we expected, cooling element (2) puts element (1) in compression, while elements (2) and (3) are put in tension. (Although an "equilibrium check" is frequently used as one means of verifying the correctness of a solution, an equilibrium check would not be of much help here, since the answer for F_1 was obtained directly from the equilibrium equation.)

(See Homework Problems 3.6-20 for a displacement-method version of this problem.)

The displacement method was used as the basis for the development of AXIALDEF, a computer program to solve axial-deformation problems, including ones that involve temperature change, like Example Problem 3.9. This computer program, one of the *MechSOLID* collection of computer programs, is described in Appendix G.3. Computer homework exercises for Section 3.6 follow the regular Section 3.6 homework problems at the end of the chapter.

3.7 GEOMETRIC "MISFITS"

In Section 3.4 a systematic procedure was introduced for solving problems involving structures having uniform axial-deformation elements subjected to external loads. You learned that three types of equations are required: *equilibrium, element force–deformation behavior,* and *deformation geometry.* The external loads enter the solution through the equilibrium equations, which relate the element axial forces to each other and to the external loads. Section 3.5 showed that when axial-deformation members are heated or cooled, that information enters the problem solution through a modification of the element force-deformation equations. In this section we will treat problems in which modification of the deformation-geometry equation(s) is required. Figure 3.13 shows two examples where this geometric misfit situation arises.[8]

In the structure on the left, one member is fabricated an amount δ too short. This misfit, or incompatibility, must be accounted for in relating the actual elongation of this member to the other geometric quantities (namely, the rotation of the beam and the elongations of the other two rods). Figure 3.13b represents a bolt surrounded by a sleeve (or pipe). If the nut on the bolt is just snugged up against the washer, there will be no axial force in the bolt or the sleeve. However, if the nut is tightened further against the washer, the portion of the bolt between the nut and bolt head is, in effect, made "too short." This misfit must be accounted for in the deformation-geometry equation(s). It is emphasized that **no change is made to either the equilibrium equations or the element force-deformation equations to account for misfits.** Geometric incompatibility is a purely geometric problem. Examples will now be given to illustrate this fact.

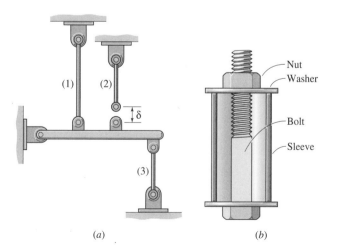

(a) *(b)*

FIGURE 3.13 Two structures that exhibit geometric "misfits."

---

[8]The topic of geometric "misfits," or geometric incompatibility, is sometimes referred to by the name *prestrain effects* or by the name *initial stresses.*

When the structure in Example 3.11 was fabricated, element (2) was made $\delta_2 = 0.065$ in. too short. Use the displacement method to: (a) determine the rotation of rigid beam AD that results if rod (2) is stretched, attached to the beam at C, and then released, and (b) determine the axial forces induced in the three rods.

Plan the Solution We will need to write equilibrium and force-deformation equations. The ''misfit'' of rod (2) can be accounted for in the deformation diagram.

When rod (2) is attached to the beam, it will tend to return to its unstretched length. In the process, it will pull upward on the beam, causing compression in rod (1) and tension in rod (3).

Solution
(a) Determine the rotation of the beam AD.

Equilibrium: Equilibrium is handled in exactly the same manner as in Example 3.11, that is, by use of the same free-body diagram and the same equilibrium equation for moments about A, giving

$$F_1 + 2F_2 - 3F_3 = 0 \qquad \textbf{Equilibrium} \quad (1)$$

Element Force–Deformation: Since use of the displacement method was specified in the problem statement, we will use Eq. 3.14.

$$F_1 = k_1 e_1, \qquad F_2 = k_2 e_2, \qquad F_3 = k_3 e_3$$

where

$$k_1 = \left(\frac{AE}{L}\right)_1 = \frac{(1.0 \text{ in}^2)(10 \times 10^3 \text{ ksi})}{(80 \text{ in.})} = 125 \text{ kips/in.}$$

$$k_2 = k_3 = \left(\frac{AE}{L}\right)_2 = \frac{(1.0 \text{ in}^2)(10 \times 10^3 \text{ ksi})}{(50 \text{ in.})} = 200 \text{ kips/in.}$$

(In computing k_2, we ignored the error in length L_2 because $\delta_2 \ll L_2$.) Then,

$$\begin{array}{l} F_1 = 125 \ e_1 \\ F_2 = 200 \ e_2 \\ F_3 = 200 \ e_3 \end{array} \qquad \begin{array}{l} \textbf{Element Force} \\ \textbf{Deformation} \end{array} \quad (2)$$

Note that the e's in Eqs. (2) are the <u>total elongations</u> of the elements from their undeformed lengths. Since this is our definition of e, the gap of δ_2 between the end of element (2) and the rigid beam does not enter directly into the element force-deformation equations!

Geometry of Deformation: As in Example 3.11, we draw a deformation diagram (Fig. 1) so that we can properly relate the element elongations that appear in Eqs. (2) to the final rotation of the beam. This will be similar to the defor-

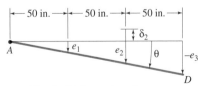

Fig. 1 Deformation diagram.

130

mation diagram of Example 3.11, but must account for the length error δ_2 of element (2). We can assume that the beam rotates clockwise or that it rotates counterclockwise. Our answer will tell us whether the beam actually rotates in the direction that we assumed. So, let clockwise be assumed positive, as in Example 3.11.

Using the deformation diagram in Fig. 1 and assuming the rotation angle θ to be small, we can relate the three e's to θ as follows:

$$
\begin{aligned}
e_1 &= 50\ \theta \\
e_2 &= 100\ \theta + \delta_2 = 100\ \theta + 0.065 \qquad \text{\textbf{Deformation}} \qquad (3) \\
-e_3 &= 150\ \theta \qquad\qquad\qquad\qquad\qquad\qquad \text{\textbf{Geometry}}
\end{aligned}
$$

Displacement-Method Solution: A displacement-method solution consists of substituting Eqs. (3) into Eq. (2), and then substituting the resulting expressions into Eq. (1).

$$ F_1 = 125(50)\ \theta = 6250\ \theta $$

$$ F_2 = 200(100\ \theta + 0.065) = 20{,}000\ \theta + 13.000 \qquad (2') $$

$$ F_3 = 200(-150\ \theta) = -30000\ \theta $$

$$ 6250\ \theta + 2(20{,}000\ \theta + 13.000) - 3(-30{,}000\ \theta) = 0 \qquad (1') $$

Finally, Eq. (1') is solved for the angular displacement

$$ \boxed{\theta = -1.908(10^{-4})\ \text{rad}} \qquad \text{Ans. (a)} \quad (4) $$

(b) Solve for the element forces. Substituting the preceding value of θ into Eq. (2'), we get

$$ F_1 = 6250\ \theta = -1.193\ \text{kips} $$

$$ F_2 = 20{,}000\ \theta + 13.000 = 9.183\ \text{kips} $$

$$ F_3 = -30{,}000\ \theta = 5.725\ \text{kips} $$

Rounded to two decimal places, the element forces are:

$$
\left.
\begin{aligned}
F_1 &= -1.19\ \text{kips (1.19 kips C)} \\
F_2 &= 9.18\ \text{kips (9.18 kips T)} \\
F_3 &= 5.72\ \text{kips (5.72 kips T)}
\end{aligned}
\right\} \qquad \text{Ans. (b)} \quad (5)
$$

Review the Solution   Note that these answers are the same as the answers in Example 3.11, but also note carefully the difference in the two problems. In Example 3.11 *element (2) was cooled* by 100°F and this led to an $\alpha_2 L_2 \Delta T_2$ term of -0.065 inches in the *element force-temperature-deformation equation* for element (2). In the present example we used the same numerical value, -0.065 in., as the amount by which *element (2) was manufactured too short.* This information was introduced through the *deformation-geometry equation* for element (2).

Now let us see how bolt-tightening problems, like the one illustrated in Fig. 3.13*b*, fit into the classification of geometric misfits.

■■■■■■■■■■■■■■■ E X A M P L E 3 . 1 3 ■■■■■■■■■■■■■■■■■

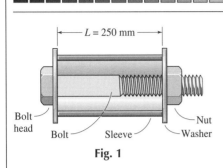

Fig. 1

A 14-mm-diameter steel bolt, and a steel pipe with 19-mm ID and 25-mm OD are arranged as shown in Fig. 1. The pitch of the (single) thread is 2 mm. What stresses will be produced in the steel bolt and sleeve if the nut is tightened by $\frac{1}{8}$ turn? (For a single thread the *pitch* is the distance the nut advances along the thread in one complete revolution of the nut.) Neglect the thickness of the washers, but assume them to be rigid. Let $E = 200$ GPa for the bolt and pipe.

Plan the Solution Let us use the displacement method so that we can more easily relate the amount the nut is advanced along the thread to the deformation this produces in the bolt and sleeve. As we discussed earlier in this section, the ''misfit'' due to tightening of the nut will enter the solution through the deformation-geometry equations. Tightening the nut should put the bolt in tension and the sleeve in compression.

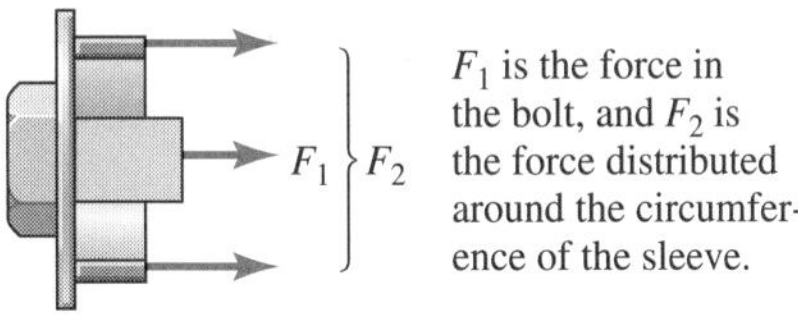

Fig. 2 Free-body diagram.

F_1 is the force in the bolt, and F_2 is the force distributed around the circumference of the sleeve.

Solution

Equilibrium: As always, we should ask the question: What free-body diagram can we draw that will enable us to relate the internal element forces to each other and to the external forces? (In this problem there are no external forces.) One answer is that a cut made just inside the washer at either end will expose the internal forces so that they can be included in the equilibrium equation, as illustrated in Fig. 2. Again, as always, we take tension in the elements to be positive. Then, we sum forces in the axial direction.

$$\stackrel{+}{\rightarrow} \sum F_x = 0: \qquad \boxed{F_1 + F_2 = 0} \qquad \textbf{Equilibrium} \quad (1)$$

Element Force–Deformation Behavior: Since we decided to use the displacement method, Eq. 3.14 is the more convenient form.

$$F_1 = k_1 e_1, \qquad F_2 = k_2 e_2$$

where, as always, e is the <u>total elongation</u> of an element and is positive when the element gets longer.

$$k_1 = \left(\frac{AE}{L}\right)_1 = \frac{\pi(7 \text{ mm})^2(200 \text{ kN/mm}^2)}{(250 \text{ mm})} = 123.2 \text{ kN/mm}$$

$$k_2 = \left(\frac{AE}{L}\right)_2 = \frac{\pi[(12.5 \text{ mm})^2 - (9.5 \text{ mm})^2](200 \text{ kN/mm}^2)}{(250 \text{ mm})}$$

$$= 165.9 \text{ kN/mm}$$

Therefore,

$$\boxed{\begin{array}{l} F_1 = 123.2 \, e_1 \\ F_2 = 165.9 \, e_2 \end{array}} \qquad \begin{array}{l} \textbf{Element Force–} \\ \textbf{Deformation} \quad (2) \\ \textbf{Behavior} \end{array}$$

Deformation Geometry: We need to assess the deformation that results when the nut is tightened on the bolt. To do this, let us suppose that the head of the bolt does not move, while the right-hand washer and nut move to the left an amount δ when the nut is tightened, as depicted in Fig. 3. Then, we need to relate the two elongations in Eqs. (2) to the displacement δ.

The sleeve is shortened by the amount the right-hand washer moves to the left, so

$$e_2 = -\delta$$

If the sleeve were to be removed and the nut advanced $\frac{1}{8}$ turn, the working length of the bolt would be $[250 \text{ mm} - (\frac{1}{8})(2 \text{ mm})] = 249.75 \text{ mm}$. Thus, there is a "misfit," and we would have to stretch the bolt by 0.25 mm to restore it to the 250-mm length, from which the displacement δ is measured. Thus,

$$\begin{aligned} e_1 &= 0.25 - \delta \\ e_2 &= -\delta \end{aligned} \qquad \begin{aligned} \textbf{Deformation} \\ \textbf{Geometry} \end{aligned} \qquad (3)$$

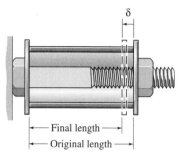

Fig. 3 Deformation diagram.

Displacement-Method Solution: Now, we simply combine Eqs. (1) through (3) in the displacement-method order, that is, $(3) \to (2) \to (1)$.

$$F_1 = 123.2(0.25 - \delta), \qquad F_2 = 165.9(-\delta) \qquad (2')$$

$$123.2(0.25 - \delta) + 165.9(-\delta) = 0 \qquad (1')$$

Finally, solving Eq. $(1')$ for the displacement u, we get

$$\delta = 0.1065 \text{ mm} \qquad (4)$$

$$F_1 = 123.2(0.25 - \delta) = 17.67 \text{ kN}$$

$$F_2 = 165.9(-\delta) = -17.67 \text{ kN}$$

and the element stresses are given by

$$\sigma_{\text{bolt}} \equiv \sigma_1 = \frac{F_1}{A_1} = \frac{17.67 \text{ kN}}{\pi(7.0 \text{ mm})^2} = 114.8 \text{ MPa} \qquad \textbf{Ans.}$$

$$\sigma_{\text{sleeve}} \equiv \sigma_2 = \frac{F_2}{A_2} = \frac{-17.67 \text{ kN}}{\pi[(12.5 \text{ mm})^2 - (9.5 \text{ mm})^2]}$$

$$= -85.2 \text{ MPa} \qquad \textbf{Ans.}$$

Thus, the bolt has a tensile stress of 115 MPa, and the sleeve has a compressive stress of 85 MPa as a result of tightening the nut by $\frac{1}{8}$ turn.

Review the Solution   We expected the bolt to be in tension and the sleeve to be in compression as a result of tightening the nut, and this is the result that we obtained. The yield strength of the steel is not stated in the problem, but assuming that it is greater than 300 MPa (see Appendix F), the values of σ_{bolt} and σ_{sleeve} are reasonable.

TABLE 3.1   A Summary of Problem-Solving Strategies

Input	Relevant Equation Set	
External forces	Equilibrium	(1)
Temperature changes	Element force-temperature-deformation	(2)
Geometric misfits	Deformation geometry	(3)
Displacement Method:	(3) → (2) → (1)	
Force Method:	(1) → (2) → (3)	

The displacement method was used as the basis for the development of AXIALDEF, a computer program to solve axial-deformation problems involving externally applied axial loads, element temperature changes, and "misfits." This computer program, one of the ***MechSOLID*** collection of computer programs, is described in Appendix G.3. Computer homework exercises for Section 3.7 follow the regular Section 3.7 homework problems at the end of the chapter.

From the examples in Sections 3.5 through 3.7, we can draw the conclusions that are summarized in Table 3.1.

■■■■■■■■■■■ *3.8 INTRODUCTION TO THE ANALYSIS OF PLANAR TRUSSES

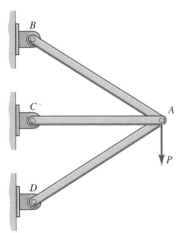

FIGURE 3.14 A planar truss.

Figure 3.14 shows a three-element planar truss. In this section we examine the fundamental equations for an inclined element that is pinned to a rigid base at one end and connected to another member or members at the other end, like the elements in the planar truss in Fig. 3.14.[9] We restrict our attention to uniform, linearly elastic elements.

Figure 3.15 shows a typical uniform planar truss element that is assumed to be subjected to an axial force F and uniformly heated by an amount ΔT.[10] Let u be the horizontal displacement of the "free" end, positive in the $+x$ direction, and let v be the vertical displacement of the free end, positive in the $+y$ direction. Let F_x be the x component of force applied to the element at the free end, and F_y be the y component. Then, F is the total axial force (tension is assumed positive). The <u>angle θ is always taken positive counterclockwise and is measured from the x axis</u>, as shown in Figs. 3.15 and 3.16.

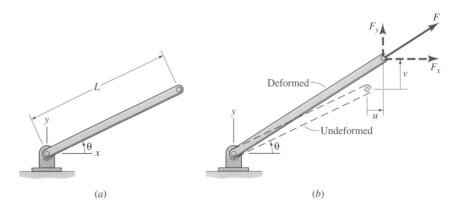

FIGURE 3.15 Planar truss element notation.

(a)   (b)

---

[9]This section exemplifies the systematic problem-solving procedure that is employed in *matrix structural analysis*. Since only elements that have one end pinned to a rigid base are considered here, the problems can be readily solved, even without the use of a computer.

[10]The temperature increment ΔT is measured from the *reference temperature,* the temperature at which the truss element is stress-free.

Equilibrium. When we show a free-body diagram of the joint to which the element in Fig. 3.15 is attached, we can show either the axial force F, or we can show the components F_x and F_y.

Element Force–Temperature–Deformation Behavior. Equations 3.20 and 3.21 relate the axial force F to the elongation of the element and to its (uniform) temperature change. That is,

$$e = fF + \alpha L \Delta T \qquad (3.20)\text{ repeated}$$

$$F = k(e - \alpha L \Delta T) \qquad (3.21)\text{ repeated}$$

Now we need to relate the elongation of e to the joint's displacement components u and v.

Geometry of Deformation. We will assume that u and v are very small in comparison with L. We need expressions that relate e to u and v. Let us take u and v separately, as shown in Fig. 3.16.

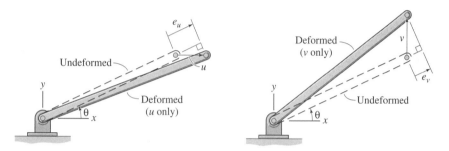

FIGURE 3.16 The contributions of displacements u and v to the elongation e.

Since u and v are assumed to be small, we can determine their contributions to e by projecting the displacements onto the original element direction, noting that the small triangles have a vertex angle θ.[11] Thus,

$$e = e_u + e_v = u\cos\theta + v\sin\theta \qquad (3.22)$$

It is beyond the scope of this text to treat trusses with more than one node (joint), since this is, properly, a principal topic treated in courses in matrix structural analysis. We will, however, consider one-node trusses in order to show how easily the procedures used in Sections 3.5 through 3.7 to solve axial-deformation problems can be extended to the solution of planar truss problems. If a one-node planar truss has only two elements, it is statically determinate. The Force-Method Procedure outlined below is a very efficient way to solve this type of problem. If the one-node truss has more than two elements, like the three-element planar truss in Fig. 3.14, the Displacement-Method Procedure is preferable.

FORCE-METHOD PROCEDURE FOR STATICALLY DETERMINATE TRUSS PROBLEMS

1. Draw a free-body diagram of the joint where the truss members are joined by a pin. Solve the two equilibrium equations for the unknown element forces, F_1 and F_2.
2. Use Eq. 3.20 to determine the elongations e_i due to forces F_i and temperature changes ΔT_i.
3. Write two geometry equations of the form given in Eq. 3.22, and solve these for the joint displacement components u and v.

[11]See Problem 3.8-26 for an exercise that illustrates the validity of this approximation.

In Example 3.14 we consider a statically determinate planar truss acted on by an external load and having one member cooled. In Example 3.15 we consider a statically indeterminate truss with one member that is not manufactured to the correct length.

■■■■■■■■■■■■■□□ **EXAMPLE 3.14** □□□□□□□□□□□□□■■■

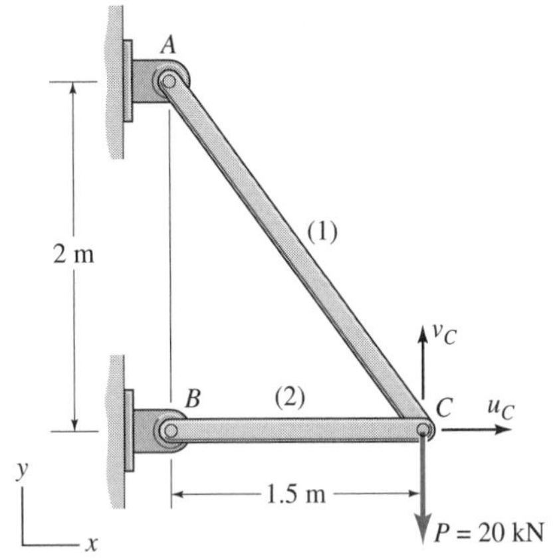

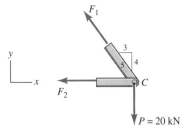

$A_1 = 1000$ mm^2, $E_1 = 200$ GPa, $\alpha_1 = 20(10^{-6})/°C$
$A_2 = 1000$ mm^2, $E_2 = 100$ GPa

Fig. 1

F_1

$\frac{3}{5}$ 4 C

F_2

$P = 20$ kN

Fig. 2 Free-body diagram of joint C.

A (statically determinate) two-bar planar truss has the configuration shown in Fig. 1 when it is assembled. If a downward load $P = 20$ kN is applied to the pin at C, and, at the same time, element (1) is <u>cooled</u> by 20° C, (a) what are the stresses σ_1 and σ_2 in elements (1) and (2), respectively? (b) What are the horizontal and vertical displacements, u_C and v_C, respectively?

Solution
(a) Determine the element stresses σ_1 and σ_2.

Equilibrium: We should use a free-body diagram of joint C shown in Fig. 2, summing forces in the x and y directions.

$$\xrightarrow{+} \sum F_x = 0: \qquad -\tfrac{3}{5}F_1 - F_2 = 0, \quad F_2 = -\tfrac{3}{5}F_1$$

$$+\uparrow \sum F_y = 0: \qquad \tfrac{4}{5}F_1 - 20 \text{ kN} = 0, \quad F_1 = 25 \text{ kN}$$

$$\boxed{\begin{array}{l} F_1 = 25 \text{ kN} \\ F_2 = -15 \text{ kN} \end{array}} \qquad \textbf{Equilibrium} \qquad (1)$$

Thus, the stresses are:

$$\sigma_1 = \frac{F_1}{A_1} = \frac{25\ 000\text{N}}{0.001\text{m}^2} = 25\text{MPa}$$

$$\sigma_2 = \frac{F_2}{A_2} = \frac{-15\ 000\text{N}}{0.001\text{m}^2} = -15\text{MPa}$$

$$\sigma_1 = 25\text{MPa (T)}, \qquad \sigma_2 = 15\text{MPa (C)} \qquad \textbf{Ans. (a)}$$

(b) Determine the joint displacement components u_C and v_C.

Element Force–Temperature–Deformation Behavior: From Eq. 3.20,

$$e_i = f_i F_i + \alpha_i L_i \Delta T_i \qquad \begin{array}{l} \textbf{Element} \\ \textbf{Force–Temp.} \\ \textbf{Deformation Behavior} \end{array} \qquad (2)$$

where

$$f_1 = \left(\frac{L}{AE}\right)_1 = \frac{2.5\text{m}}{(0.001\text{m}^2)(200 \times 10^9\text{N/m}^2)} = 1.250(10^{-8})\text{m/N}$$

$$f_2 = \left(\frac{L}{AE}\right)_2 = \frac{1.5\text{m}}{(0.001\text{m}^2)(100 \times 10^9\text{N/m}^2)} = 1.500(10^{-8})\text{m/N}$$

Then, from Eq. (2),

$$e_1 = (1.250 \times 10^{-8}\text{m/N})(25\ 000\text{N}) + (20 \times 10^{-6}/°\text{C})(2.5\text{m})(-20°\text{C})$$

$$= -0.000\ 688\text{m} = -0.688\text{mm}$$

$$e_2 = (1.500 \times 10^{-8}\text{m/N})(-15\ 000\text{N}) = -0.225\text{mm}$$

Geometry of Deformation: From Eq. 3.22,

$$e_i = u_C \cos \theta_i + v_C \sin \theta_i \qquad \text{Geometry of Deformation} \qquad (3)$$

$$\cos \theta_1 = \frac{3}{5}, \ \sin \theta_1 = -\frac{4}{5}, \ \cos \theta_2 = 1, \ \sin \theta_2 = 0$$

$$e_1 = \frac{3}{5} u_C - \frac{4}{5} v_C = -0.688\text{mm}$$

$$e_2 = u_C = -0.225\text{mm}$$

Then,

$$u_C = -0.225\text{mm}, \qquad v_C = 0.691\text{mm} \qquad \textbf{Ans. (b)}$$

Review the Solution   The displacement of joint C, upward and to the left, seems a bit strange, since the applied force P pulls downward. However, we must remember that since element (1) is cooled, it will tend to pull joint C upward and to the left, apparently overriding the downward displacement due to load P.

Note that Eq. (3), the geometry-of-deformation equation, is the only really "new" feature that distinguishes this truss problem from the axial-deformation problems treated in Sections 3.5 through 3.7.

If the truss is statically indeterminate, that is, if it has more than two members but only one joint, a displacement-method solution is the most straightforward. Of course, the displacement method could also be used to solve statically determinate truss problems.

DISPLACEMENT-METHOD PROCEDURE FOR TRUSS PROBLEMS

1. Draw a free-body diagram of the joint where the truss members are joined together by a pin. Write the two equilibrium equations in terms of the unknown element forces, F_i.

2. Write an element *force-temperature-deformation equation* for each element. Equation 3.21,

$$F_i = k_i(e_i - \alpha_i L_i \Delta T_i), \qquad k_i = (AE/L)_i$$

   is the most convenient form to use for this step.

3. Use *geometry of deformation*, Eq. 3.22, to relate the element elongations to the joint displacement components u and v.

4. Substitute the deformation-geometry equations (Step 3) into the force-temperature-deformation equations (Step 2). This gives *element forces in terms of system displacements.*

5. Substitute the results of Step 4 into the equilibrium equations of Step 1. This gives *equilibrium equations written in terms of system displacements.*

6. Solve the equations obtained in Step 5. The answer will be the *system displacements.*

7. Substitute the system displacements of Step 6 into the equations obtained in Step 4. The answer will be the *element forces.* If element normal stresses are required, the element forces can be divided by the respective element cross-sectional areas, that is, $\sigma_i = F_i/A_i$.

The following example problem illustrates the solution of truss problems using the displacement method.

137

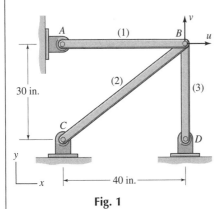

Fig. 1

A three-element truss has the configuration shown in Fig. 1. Each member has a cross-sectional area of 1.2 in^2, and all of them are made of aluminum, with $E = 10 \times 10^3$ ksi. When the truss members were fabricated, members (1) and (3) were manufactured correctly (i.e., correct lengths of $L_1 = 40$ in. and $L_3 = 30$ in.). However, member (2) has a distance between hole centers of $L_2 = 49.90$ in., rather than the correct value of 50.00 in. If member (2) is stretched so that a pin can be inserted to connect all three members together at B, and then the system is released, (a) what displacements, $u \equiv u_B$ and $v \equiv v_B$, will occur at node B? (b) What forces will be induced in the three members?

Plan the Solution As always, we need to formulate equations for equilibrium, element force–deformation behavior, and geometry of deformation. We must incorporate the "misfit" condition in the deformation-compatibility equations.

Because member (2) will tend to return to its original length but will be restrained by members (1) and (3), we should expect member (2) to be in tension and the other two members to be in compression.

Solution
(a) Use the displacement method to solve for the displacements u and v.

Equilibrium: The node at B (Fig. 2) is acted on only by the element forces, since there is no external load. As usual, the axial forces in the elements are taken positive in tension.

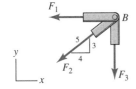

Fig. 2 Free-body diagram.

$$\xrightarrow{+} \sum F_x = 0: \qquad -F_1 - \frac{4}{5} F_2 = 0$$

$$+\uparrow \sum F_y = 0: \qquad -\frac{3}{5} F_2 - F_3 = 0$$

$$F_1 + \frac{4}{5} F_2 = 0 \qquad\qquad \textbf{Equilibrium} \quad (1)$$

$$\frac{3}{5} F_2 + F_3 = 0$$

Element Force-Deformation Behavior: We have three elements, so we need to write Eq. 3.21 for each of these elements. It will be helpful to compile a table of element properties first.

$$k_1 = \left(\frac{AE}{L}\right)_1 = \frac{(1.2 \text{ in}^2)(10 \times 10^3 \text{ ksi})}{(40 \text{ in.})} = 300 \text{ kips/in.}$$

$$k_2 = \left(\frac{AE}{L}\right)_2 = \frac{(1.2 \text{ in}^2)(10 \times 10^3 \text{ ksi})}{(50 \text{ in.})} = 240 \text{ kips/in.}$$

$$k_3 = \left(\frac{AE}{L}\right)_3 = \frac{(1.2 \text{ in}^2)(10 \times 10^3 \text{ ksi})}{(30 \text{ in.})} = 400 \text{ kips/in.}$$

138

Table of Elements Properties

Element, i	L_i (in.)	k_i (kips/in.)	$\cos \theta_i$	$\sin \theta_i$
1	40	300	1.0	0.0
2	50	240	0.8	0.6
3	30	400	0.0	1.0

Then, from Eq. 3.21,

$$F_1 = k_1 e_1 = 300e_1$$
$$F_2 = k_2 e_2 = 240e_2$$
$$F_3 = k_3 e_3 = 400e_3$$

Element Force–
Deformation   (2)
Behavior

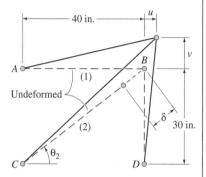

Fig. 3 Deformation diagram.

Geometry of Deformation: We sketch a *deformation diagram* (Fig. 1) so that we can see how to relate the element elongations to the system displacements u and v. We greatly exaggerate the gap between the end of element (2) and node B, where it is supposed to connect to elements (1) and (3).

First, we can use Eq. 3.22 to express the elongation of each element in terms of the displacement at its "free" end, that is, end B.

$$e_i = u_i \cos \theta_i + v_i \sin \theta_i \qquad i = 1, 2, 3$$

However, we will have to modify the equation for element (2) to account for the "misfit." Since all elements are pinned together at joint B,

$$u_1 = u_2 = u_3 \equiv u, \qquad v_1 = v_2 = v_3 \equiv v$$

Combining these equations, and referring to the Table of Element Properties, we get

$$e_1 = u \cos \theta_1 + v \sin \theta_1 = u$$
$$e_2 = u \cos \theta_2 + v \sin \theta_2 + \delta$$
$$\quad = 0.8u + 0.6v + \delta$$
$$e_3 = u \cos \theta_3 + v \sin \theta_3 = v$$

Deformation
Compatibility   (3)

Displacement-Method Solution: All we need to do is combine Eqs. (1) through (3) in the displacement-method order: (3) → (2) → (1).

$$F_1 = 300u, \quad F_2 = 192u + 144v + 24.00, \quad F_3 = 400v \qquad (2')$$

Then, from Eqs. (1) and (2'), we have

$$300u + 0.8(192u + 144v + 24.00) = 0 \qquad (1')$$
$$0.6(192u + 144v + 24.00) + 400v = 0$$

or

$$453.6u + 115.2v = -19.20$$
$$115.2u + 486.4v = -14.40$$

Nodal
Equilibrium
in terms of   (4)
Nodal
Displacements

139

Solving these simultaneous algebraic equations, we get

$$u = -3.70(10^{-2}) \text{ in.}$$
$$v = -2.08(10^{-2}) \text{ in.}$$

Ans. (a)   (5)

(b) Solve for the element forces. The displacements u and v can be substituted into Eqs. (2') to give

$$F_1 = 300u = -11.11 \text{ kips}$$
$$F_2 = 192u + 144v + 24.00 = 13.89 \text{ kips}$$
$$F_3 = 400v = -8.33 \text{ kips}$$

$$\left.\begin{array}{l} F_1 = 11.11 \text{ kips (C)} \\ F_2 = 13.89 \text{ kips (T)} \\ F_3 = 8.33 \text{ kips (C)} \end{array}\right\}$$

Ans. (b)   (6)

Review the Solution   One way to verify our results is to check equilibrium (using full calculator precision, not the reported rounded values).

Is $F_1 + 0.8F_2 = 0$?   Yes.

Is $0.6F_2 + F_3 = 0$ Yes.

Also, we note that all of the element forces have the signs that we expected them to have, that is, member (2) is in tension, while members (1) and (3) are in compression.

From the two truss examples of this section we can conclude that, by establishing a geometric relationship between the elongation of an element and the cartesian components of displacement at its ''free'' end (Fig. 3.16), we can solve planar truss problems involving elements attached to ''ground'' at one end by exactly the same procedures used in Sections 3.5 through 3.7. Specifically, the problem-solving strategies summarized in Table 3.1 at the end of Section 3.7 still apply.

*3.9 INELASTIC AXIAL DEFORMATION

Up to this point in our analysis of axial deformation we have considered only linearly elastic behavior. It is important, however, that we also study the inelastic behavior of deformable bodies whose stress has exceeded the proportional limit. If properly designed, structures and machines do not completely fail when the stress at one or more points reaches the proportional limit, or yield point. This is true for statically indeterminate structures, as you will discover in the elastic-plastic analysis that follows. In many cases present design codes acknowledge the fact that ''failure'' does not necessarily correspond to the first yielding of a structural element or machine component.

Fundamental Equations. Of the three fundamental concepts of deformable-body mechanics—*equilibrium*, *geometry of deformation*, and *material behavior*—**only material behavior is treated in a manner different from that in the previous analysis of**

linearly elastic behavior. Let us consider the axially loaded element in Fig. 3.17*a*. Assume that it is homogeneous (i.e., it has the same material properties throughout), and that it may be stressed beyond the proportional limit of the material.

(*a*) A slender member subjected to axial loading.

Geometry of Deformation: For inelastic axial deformation, just as for linearly elastic axial deformation, we assume that *the axis of the member remains straight and that plane sections remain plane.* Therefore, Eq. 3.1 holds, so the strain is independent of position in the cross section, and the *strain-displacement equation* for axial deformation is

$$\epsilon = \frac{du}{dx} \qquad (3.1)$$
repeated

The *elongation* of the axial-deformation member is obtained by integrating Eq. 3.1, that is, by summing the elongations of infinitesimal fibers of length *dx*.

$$e = u(L) - u(0) = \int_0^L \epsilon(x)dx \qquad (3.23)$$

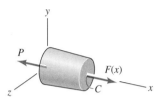

(*b*) A free-body diagram of a portion of the member.

FIGURE 3.17 An element undergoing axial deformation.

Equilibrium: The extensional strain may vary with *x*, but it is constant at any cross section. Since the material is assumed to be homogeneous, the stress will also be a constant on the cross section. Then Eqs. 3.5 become

$$F = \int_A \sigma dA = \sigma A$$

$$M = \int_A z\sigma dA = \bar{z}\sigma A = \bar{z}F = 0 \qquad (3.24)$$

$$M_z = -\int_A y\sigma dA = -\bar{y}\sigma A = -\bar{y}F = 0$$

Therefore, axial deformation occurs when the axial force *F(x)* acts through the centroid of the cross section of the member, the loading situation shown in Fig. 3.17*b*. The axial stress is related to the (internal) axial force *F(x)* by

$$\sigma = \frac{F(x)}{A(x)} \qquad (3.25)$$

Material Behavior: Figure 3.18 depicts three typical stress-strain curves that exhibit inelastic behavior beyond the proportional limit, σ_{PL}. Figure 3.18*a* is typical of metals like aluminum; Fig. 3.18*b* is an idealization of the behavior of a material with a definite yield point, like mild steel; and Fig. 3.18*c* is an idealization of a material that exhibits strain hardening after it yields. In each of the three cases illustrated in Fig 3.18 the material is

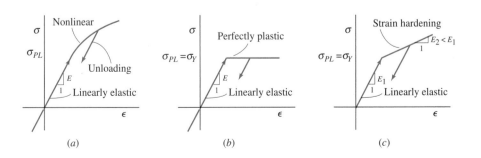

FIGURE 3.18 Three types of nonlinear stress-strain behavior: (a) elastic, nonlinearly plastic, (b) elastic, perfectly plastic, and (c) elastic, strain-hardening.

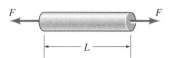

FIGURE 3.19 A prismatic bar with axial loading.

linearly elastic (and, therefore, it obeys Hooke's Law) up to the proportional limit. And, in each case, if unloading occurs from a point on the stress-strain diagram above the yield stress, the unloading will proceed along a path that is parallel to the original linearly elastic portion of the σ versus ϵ curve, as illustrated in Fig. 3.18.

Uniform End-Loaded Element. If the element is prismatic and is loaded only by axial forces at its ends, as in Fig. 3.19, then

$$\sigma = \frac{F}{A} = \text{const.}$$

$$e = \epsilon(\sigma)L$$

(3.26)

Elastic-Plastic Analysis of Statically Indeterminate Structures. An important illustration of inelastic axial deformation is the behavior of a statically indeterminate system whose elements exhibit *elastic, perfectly plastic* behavior, as represented by the stress-strain diagram in Fig. 3.18b. Consider the statically indeterminate rod system in Fig. 3.20a. Let both members be made of a material whose stress-strain curve exhibits the elastic, perfectly plastic behavior shown in Fig. 3.20b. (It is assumed that the compressive yield stress has the same magnitude, σ_Y, as the tensile yield stress.) We wish to determine a load-displacement curve that relates the force P at B to the displacement u of point B in Fig. 3.20a.

We handle equilibrium and geometry of deformation just as in the previous analysis of linearly elastic axial deformation.

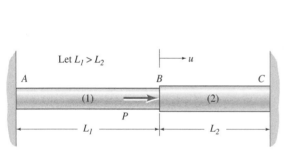

(*a*) A two-element rod system with axial loading.

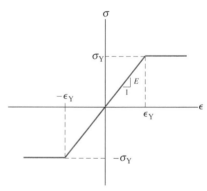

(*b*) Stress-strain diagram.

FIGURE 3.20 An elastic-plastic system with axial loading.

Equilibrium: From a free-body diagram of the node (joint) at B, we get

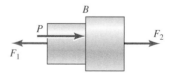

$$F_1 - F_2 = P$$ **Equilibrium** (3.27)

Geometry of Deformation: By examining Fig. 3.20a, we see that element 1 elongates an amount u and element 2 shortens the same amount u when node B moves to the right by an amount u. Therefore, from deformation compatibility, we have

$$e_1 = u, \qquad e_2 = -u \qquad \text{Geometric Compatibility} \qquad (3.28)$$

Material Behavior: Now let us consider the linear and nonlinear material behavior that occurs <u>as the load P is increased</u>. There are three cases: Case 1—both elements are linearly elastic; Case 2—one member has yielded, but the other is still linearly elastic; and Case 3—both elements have yielded.

Case 1—Both elements are linearly elastic. From Eq. 3.14 of Section 3.4,

$$F_i = k_i e_i = \left(\frac{A_i E_i}{L_i}\right) e_i \qquad \text{Material Behavior— Case 1} \qquad (3.29)$$

Combining Eqs. 3.27 through 3.29 in *displacement-method* fashion, we get

$$(k_1 + k_2)u = P \rightarrow \quad u = \frac{P}{\left(\dfrac{AE}{L}\right)_1 + \left(\dfrac{AE}{L}\right)_2} \qquad (a)$$

The element stresses are

$$\sigma_1 = \frac{F_1}{A_1} = \frac{k_1 u}{A_1} = \frac{P L_2}{A_1 L_2 + A_2 L_1}$$

$$\sigma_2 = \frac{F_2}{A_2} = \frac{-k_2 u}{A_2} = \frac{-P L_1}{A_1 L_2 + A_2 L_1} \qquad (b)$$

It was assumed in Fig. 3.20a that $L_1 > L_2$. Therefore, from Eqs. (b), $|\sigma_2| > |\sigma_1|$, so element 2 reaches yield in compression before element 1 reaches yield in tension. The *yield load, P_Y,* is therefore determined by setting $\sigma_2 = -\sigma_Y$, giving

$$P_Y = \sigma_Y\left(\frac{A_1 L_2 + A_2 L_1}{L_1}\right) \qquad (c)$$

We can also use the second of Eqs. (b) to obtain

$$u_Y = \frac{\sigma_Y L_2}{E} \qquad (d)$$

Case 2—One element has yielded. Since element 2 yields first, for this case we must replace the linear force-elongation equation of element 2 to reflect a constant yield stress of $(-\sigma_Y)$. Element 1, however, is still linearly elastic, so the material behavior for Case 2 is represented by the equations

$$F_1 = k_1 e_1$$
$$F_2 = -\sigma_Y A_2 \qquad \text{Material Behavior— Case 2} \qquad (3.30)$$

Combining Eqs. 3.27 (equilibrium), 3.28 (geometric compatibility), and 3.30 (material behavior) in displacement-method fashion, we get

$$k_1 u = P + F_2 = P - \sigma_Y A_2$$

or

$$u = \frac{(P - \sigma_Y A_2)L_1}{A_1 E} \qquad (e)$$

This phase of loading behavior extends from $P = P_Y$ to $P = P_P$, the *plastic load*, at which rod 1 also yields.

Case 3—Both elements have yielded. In this case both elements have yielded, and the appropriate material behavior equations (replacing Eqs. 3.29 and 3.30) are:

$$\begin{aligned} F_1 &= \sigma_Y A_1 \\ F_2 &= -\sigma_Y A_2 \end{aligned} \qquad \begin{array}{l} \textbf{Material} \\ \textbf{Behavior—} \\ \textbf{Case 3} \end{array} \qquad (3.31)$$

Combining these with the equilibrium equation, Eq. 3.27, we get the value of the plastic load, P_P.

$$P_P = \sigma_Y(A_1 + A_2) \qquad (f)$$

This plastic load is reached when member 1 also yields, so equating the strain in member 1 to σ_Y/E, we get

$$\epsilon_{1P} = \frac{u_P}{L_1} = \frac{\sigma_Y}{E} \qquad (g)$$

or

$$u_P = \frac{\sigma_Y L_1}{E} \qquad (h)$$

Note that

$$\frac{P_P}{P_Y} = \frac{A_1 L_1 + A_2 L_1}{A_1 L_2 + A_2 L_1}, \qquad \frac{u_P}{u_Y} = \frac{L_1}{L_2} \qquad (i, j)$$

The above elastic-plastic analysis is summarized on the load-displacement diagram in Fig. 3.21, with the ranges of applicability of the three cases indicated on the plot.

FIGURE 3.21 A load-displacement diagram for elastic-plastic axial deformation.

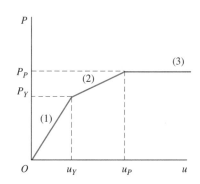

The two support rods in Example Problem 3.6 are made of structural steel that may be assumed to have a stress-strain diagram like Fig. 3.20b, with $E = 30(10^3)$ ksi and $\sigma_Y = 36$ ksi (see Fig. 1). (a) Construct a load-displacement diagram that relates the load P to the vertical displacement at D. Make the usual small-angle assumption for the rotation of the "rigid" beam AD. (b) Determine the allowable load if the applicable load factor is $\lambda = 1.5$, the resistance factor for first yielding is $\phi_Y = 0.90$, and the resistance factor for fully plastic behavior is $\phi_P = 0.70$. (See Section 2.12 for a discussion of load factors and resistance factors.)

Plan the Solution We can use the equilibrium analysis and the geometry-of-deformation analysis from Example Problem 3.6. As in the preceding discussion of elastic-plastic behavior, the material behavior here falls into three cases—(1) both support rods are linearly elastic, (2) one rod has yielded, and (3) both rods have yielded. The load-displacement diagram should resemble Fig. 3.21. The allowable load is determined by applying the load and resistance factors according to Section 2.12 (Eqs. 2.40 and 2.41).

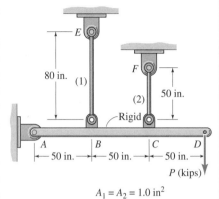

$A_1 = A_2 = 1.0$ in^2

Fig. 1

Solution
(a) Construct a load-displacement diagram relating load P to the vertical displacement at D.

Equilibrium: The free-body diagram for Example 3.6 is repeated here as Fig. 2, and the equation of equilibrium is

$$\left(\sum M\right)_A = 0: \qquad F_1 + 2F_2 = 3P \qquad \textbf{Equilibrium} \quad (1)$$

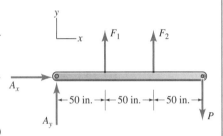

Fig. 2 Free-body diagram.

Geometry of Deformation: The deformation diagram of Example 3.6 is repeated here as Fig. 3. For small rotation of the rigid beam AD, the equations of geometric compatibility can be written as

$$e_1 = \frac{\delta}{3} \text{ in.}$$

$$e_2 = \frac{2\delta}{3} \text{ in.}$$

$$\textbf{Geometric Compatibility} \quad (2)$$

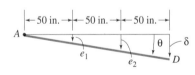

Fig. 3 Deformation diagram.

where δ is the vertical displacement at D.

Material Behavior: There are three cases to be considered: Case 1—both rods are linearly elastic (same as Example Problem 3.6), Case 2—one rod has yielded, and Case 3—both rods have yielded.

Case 1—Both rods are linearly elastic. For this case, from Example 3.6 we have

$$F_1 = k_1 e_1 = 375e_1 \text{ kips}$$

$$F_2 = k_2 e_2 = 600e_2 \text{ kips}$$

$$\textbf{Material Behavior—} \quad (3)$$
$$\textbf{Case 1}$$

145

Equations (1) through (3) may be combined to give the load-displacement equation

$$\delta = \frac{3}{925} P \quad \text{or} \quad P = \frac{925}{3}\delta \qquad \textbf{Ans. (a)} \quad (4)$$

The corresponding stresses in the two rods are obtained by combining Eqs. (2) through (4). Thus,

$$\sigma_1 = \frac{F_1}{A_1} = \frac{15}{37}P \text{ ksi}, \qquad \sigma_2 = \frac{F_2}{A_2} = \frac{48}{37}P \text{ ksi} \qquad (5)$$

Since $\sigma_2 > \sigma_1$, rod 2 will yield before rod 1 does. Therefore, the yield load P_Y is given by setting $\sigma_2 = \sigma_Y = 36$ ksi in Eq. (5b), the equation for σ_2. This gives

$$P_Y = 27.8 \text{ kips} \qquad \textbf{Ans. (a)} \quad (6)$$

The corresponding displacement at which first yield occurs is obtained by substituting Eq. (6) into Eq. (4a), the equation for δ, giving

$$\delta_Y = \frac{3}{925}(27.8) = 0.090 \text{ in.} \qquad \textbf{Ans. (a)} \quad (7)$$

The load-displacement formula for loads up to load P_Y is given by Eqs. (4).

Case 2—One rod has yielded; one is linearly elastic. After rod 2 yields and up until the load at which rod 1 yields, the material-behavior equations for rods 1 and 2 are:

$$\begin{aligned} F_1 &= 375e_1 \text{ kips} \\ F_2 &= \sigma_Y A_2 = 36 \text{ kips} \end{aligned} \qquad \begin{array}{l} \textbf{Material} \\ \textbf{Behavior—} \\ \textbf{Case 2} \end{array} \quad (8)$$

Combining Eqs. (1) (equilibrium equation), (2a) (deformation-geometry equation for e_1), and (8) (force-deformation equations), we get

$$375\left(\frac{1}{3}\delta\right) + 2(36) = 3P$$

or

$$P = \left(\frac{125}{3}\delta + 24\right) \text{ kips} \qquad \textbf{Ans. (a)} \quad (9)$$

This formula characterizes the load-displacement behavior from the yield load P_Y up to the fully plastic load P_P, which corresponds to the yielding of rod 1.

Case 3—Both rods have yielded. Once rod 1 also yields, the material-behavior equations become

$$\begin{aligned} F_1 &= \sigma_Y A_1 = 36 \text{ kips} \\ F_2 &= \sigma_Y A_2 = 36 \text{ kips} \end{aligned} \qquad \begin{array}{l} \textbf{Material} \\ \textbf{Behavior—} \\ \textbf{Case 3} \end{array} \quad (10)$$

These rod forces can be substituted into the equilibrium equation, Eq. (1), to give the plastic load P_P.

$$P_P = \frac{F_1 + 2F_2}{3} = \frac{36 + 2(36)}{3}$$

or

$$P_P = 36 \text{ kips} \qquad \textbf{Ans. (a)} \quad (11)$$

The corresponding displacement δ_P is obtained by substituting Eq. (11) into Eq. (9) to obtain

$$\delta_P = \frac{3}{125}(12) = 0.288 \text{ in.} \qquad \textbf{Ans. (a)} \quad (12)$$

We can now use Eqs. (6), (7), (11), and (12) to plot the load-displacement diagram, Fig. 4.

(b) Determine the allowable load. Based on first yield, the allowable load is given by

$$\lambda(P_{\text{allow.}})_Y = \phi_Y P_Y \qquad (13a)$$

$$(P_{\text{allow.}})_Y = \frac{0.90(27.75 \text{ kips})}{1.5} = 16.65 \text{ kips} \qquad (13b)$$

Based on the fully-plastic load, the allowable load is given by

$$\lambda(P_{\text{allow.}})_P = \phi_P P_P \qquad (14a)$$

So,

$$(P_{\text{allow.}})_P = \frac{0.70(36 \text{ kips})}{1.5} = 16.80 \text{ kips} \qquad (14b)$$

In this case, the *allowable load* is determined by the first-yield criterion. Therefore,

$$P_{\text{allow.}} = 16.6 \text{ kips} \qquad \textbf{Ans. (b)} \quad (15)$$

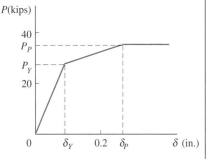

Fig. 4 A load-displacement diagram.

Review the Solution The key results in the preceding analysis are the "break points," (P_Y, δ_Y) and (P_P, δ_P), in the load-displacement diagram. Let us check the first of these, starting with $\delta_Y = 0.090$ in. The compatibility equations, Eqs. (2) are easily checked by referring to Fig. 3, and these give

$$e_{1Y} = 0.030 \text{ in.}, \qquad e_{2Y} = 0.060 \text{ in.}$$

The strain $\epsilon_{2Y} = e_{2Y}/L_2 = 0.0012$ is in agreement with the yield strain σ_Y/E, so we have the correct elongations. We can get the rod forces by using Eqs. (3). Thus, $F_{1Y} = 11.25$ kips and $F_{2Y} = 36$ kips. When these forces are substituted into the equilibrium equation, Eq. (1), we do get $P_Y = 27.75$ kips, so our solution appears to be correct.

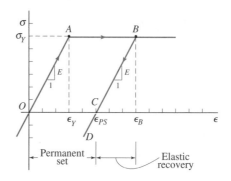

FIGURE 3.22 Loading and unloading paths for an elastic-plastic member.

Residual Stress. Let us now see what happens when the load is removed after yielding has occurred in one or more elements in an assemblage of axial deformation members made of an elastic-plastic material represented by the stress-strain curve in Fig. 3.22.

Statically Determinate Case. Consider a single axial-deformation member that has been stretched to a strain of ϵ_B. Upon removal of the load, the member will undergo *linearly elastic recovery* along the straight line *BD* to point *C*, where there is a residual strain, or *permanent set,* which is given by Fig. 3.22.

$$\epsilon_{PS} = \epsilon_B - \epsilon_Y$$

There is, however, no residual stress, because the load can be completely removed from the member, since the member is statically determinate. A similar situation would hold for a statically determinate assemblage of two or more members—there would be some permanent set, but there would not be any residual stress.

Statically Indeterminate Case. If the assemblage of axial-deformation members is statically indeterminate, it is possible to have self-equilibrating internal forces after all external loads have been removed. The corresponding member stresses are called *residual stresses,* because they are stresses that result from some prior loading, but they remain after removal of all external loads. The next example illustrates how residual stresses may result from inelastic material behavior.

EXAMPLE 3.17

Let us take a special case of the statically indeterminate system in Fig. 3.20. This special case is shown in Fig. 1a. If the load P is increased until $\epsilon_2 = -1.5\epsilon_Y$ and is then removed, what residual stresses will be left in the two elements? What is the permanent deformation? Let $A_1 = A_2 = A$, $L_1 = 2L_2 = 2L$, and let both elements be made of the elastic-plastic material depicted in Fig. 1b.

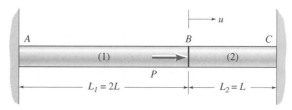

Fig. 1a A statically indeterminate assemblage.

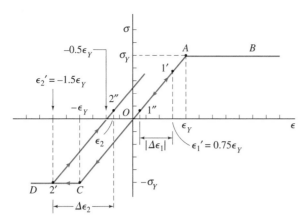

Fig. 1b Elastic-plastic material behavior.

Plan the Solution We can divide the analysis into a loading phase and an unloading phase. The analysis of the loading phase can be taken directly from the discussion of elastic-plastic analysis of statically indeterminate members at the beginning of Sect. 3.9. For the unloading phase we must set $P = 0$ in the equilibrium equation and use linearly elastic material behavior to represent the unloading $\sigma - \epsilon$ path.

Solution

Geometry of Deformation: Let us consider the geometry of deformation first, since this is the same for both loading and unloading. The member elongations are related to the displacement of node B by

$$e_1 = u, \qquad e_2 = -u \tag{1}$$

Since the strain is uniform along each member, the geometric compatibility equations can be written in terms of strains as

$$\epsilon_1 = \frac{e_1}{2L} = \frac{u}{2L}$$

$$\epsilon_2 = \frac{e_2}{L} = \frac{-u}{L}$$

Geometric Compatibility (2)

These strain equations will prove to be more convenient than Eqs. (1) since we have to carefully examine the loading and unloading paths on the stress-strain diagram.

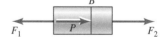

Fig. 2 Free-body diagram.

Equilibrium: Equilibrium of node B (Fig. 2) gives

$$F_1 - F_2 = P \qquad \textbf{Equilibrium} \tag{3}$$

(At the end of the unloading phase, $P = 0$.)

Loading Phase. You should follow the loading and unloading strain paths on Fig. 1b. We are given that the load is increased until load P produces a

149

strain $\epsilon_2' = -1.5\epsilon_Y$. For member 2 this corresponds to a path $O \to C \to 2'$ in Fig. 1b. Combining this information with Eq. (2b) we get

$$\epsilon_2' = -1.5\epsilon_Y = \frac{-u'}{L} \tag{4}$$

where the primes denote quantities when the load has been increased until it produces the given strain in member 2. Point $2'$ on Fig. 1b characterizes the state for member 2. From Eq. (4)

$$u' = 1.5\epsilon_Y L \tag{5}$$

Combining Eqs. (2a) and (5) gives

$$\epsilon_1' = \frac{1.5\epsilon_Y L}{2L} = 0.75\epsilon_Y \tag{6}$$

Therefore, member 1 is still elastic at the point designated $1'$ on Fig. 1b.

Material Behavior—Loading Phase: Since $|\epsilon_2'| > \epsilon_Y = |\epsilon_1'| < \epsilon_Y$ the appropriate material-behavior equations are given by Case 2 of the earlier elastic-plastic analysis, namely

$$
\begin{aligned}
F_1 &= A\sigma_1 = AE\epsilon_1 \\
F_2 &= A\sigma_2 = -\sigma_Y A
\end{aligned}
\qquad
\begin{aligned}
&\textbf{Material} \\
&\textbf{Behavior—Loading}
\end{aligned}
\tag{7}
$$

To determine the external load P' that corresponds to the given state (i.e., that makes $\epsilon_2' = -1.5\epsilon_Y$), we can combine Eqs. (7), (2a), and (5) to obtain the member forces

$$
\begin{aligned}
F_1' &= \frac{AE}{2L}(1.5\epsilon_Y L) = 0.75\sigma_Y A \\
F_2' &= -\sigma_Y A
\end{aligned}
\tag{8}
$$

These may be substituted into the equilibrium equation, Eq. (3), giving the external load

$$P' = 1.75\sigma_Y A \tag{9}$$

From Eq. (5) the corresponding displacement of node B is

$$u' = 1.5\frac{\sigma_Y L}{E} \tag{10}$$

Unloading Phase. When the load is removed, members 1 and 2 unload to points $1''$ and $2''$, respectively, on Fig. 1b. Since $P'' = 0$, the equilibrium equation, Eq. (1), becomes

$$F_1'' - F_2'' = 0 \tag{11}$$

The geometry of deformation is still described by Eqs. (1) and (2), but the material-behavior equations must now correspond to the unloading paths indicated in Fig. 1b. Point $1''$ lies on the original linear portion, AC, of the $\sigma - \epsilon$ diagram, but point $2''$ lies along the unloading path $2' - 2''$ that intersects the ϵ axis at $\epsilon_2' + \epsilon_Y = -0.5\epsilon_Y$. Hence, along these two respective paths,

$$F_1 = A\sigma_1 = AE\epsilon_1$$
$$F_2 = A\sigma_2 = AE(\epsilon_2 + 0.5\epsilon_Y)$$

Material Behavior—Unloading (12)

Substituting Eqs. (2) into Eqs. (12), we get the following expressions for the internal forces:

$$F_1 = AE\left(\frac{u}{2L}\right)$$
$$F_2 = AE\left(-\frac{u}{L} + \frac{\sigma_Y}{2E}\right)$$

(13)

The *permanent displacement of point B, u'',* at which these forces satisfy the equilibrium equation, Eq. (11), is

$$u'' = \frac{1}{3}\frac{\sigma_Y L}{E}$$

Ans. (14)

with corresponding element forces

$$F_1'' = F_2'' = \frac{1}{6}\sigma_Y A$$

(15)

Therefore, the two members are left with a *residual stress* of

$$\sigma_1'' = \sigma_2'' = \frac{1}{6}\sigma_Y$$

Ans. (16)

Review the Solution   A key equation is the geometric-compatibility equation, Eq. (2). It tells us that for any increment of displacement Δu, $\Delta\epsilon_1 = \Delta u/2L$ and $\Delta\epsilon_2 = -\Delta u/L$. Therefore, we can easily check points $2'$ and $1'$ on Fig. 1b. From the corresponding stresses, $\sigma_1 = \frac{3}{4}\sigma_Y$ and $\sigma_2 = -\sigma_Y$, and the nodal equilibrium equation we can verify that the force P' given in Eq. (9) is correct.

Next, to determine the residual stresses we know that $F_1'' = F_2''$, so $\sigma_1'' = \sigma_2''$. We also know that $\Delta\epsilon_1 = -\frac{1}{2}\Delta\epsilon_2$. We see in Fig. 1b that points $1''$ and $2''$ have equal stress and that the strain increments, $\Delta\epsilon_1 \equiv (\epsilon_{1''} - \epsilon_{1'})$ and $\Delta\epsilon_2 \equiv (\epsilon_{2''} - \epsilon_{2'})$, satisfy this strain-increment equation. Therefore, our residual stress answers in Eqs. (16) appear to be correct.

Following the elastic-plastic loading and unloading behavior of a statically indeterminate structure, as in the above analysis, is obviously not a very easy task. However, note that by considering separately the roles of *equilibrium,* of *geometry of deformation,* and of *material behavior,* we have been able to make the analysis both straightforward and tractable.

Prob. 3.2-1. A bimetallic bar is made by bonding together two homogenous rectangular bars, each having a width b, thickness t, and length L. The moduli of elasticity of the bars are E_1 and E_2, respectively. An axial force P is applied to the ends of the bimetallic bar at location ($y = y_P$, $z = 0$) such that the bar undergoes axial deformation. Let $L = 50$ in., $b = 2$ in., $t = 1$ in., $E_1 = 10 \times 10^3$ ksi, $E_2 = 30 \times 10^3$ ksi, $P = 12$ kips. (a) Determine the normal stress in each material; that is, determine σ_{x1} and σ_{x2}. (b) Determine the value of y_P. (c) Determine the elongation of the bar.

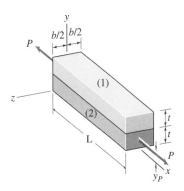

P3.2-1 and P3.2-2

Prob. 3.2-2. Solve Prob. 3.2-1 using the following dimensions, material constants, and load: $L = 2$ m, $b = 50$ mm, $t = 25$ mm, $E_1 = 70$ GPa, $E_2 = 210$ GPa, $P = 48$ kN.

Prob. 3.2-3. A magnesium-alloy rod ($E_m = 6.5 \times 10^3$ ksi) of diameter $d_m = 1$ in. is encased by a brass tube ($E_b = 15 \times 10^3$ ksi) with outer diameter $d_b = 1.5$ in.; both have length $L = 20$ in. An axial load $P = 10$ kips is applied to the resulting bimetallic rod. (a) Determine the normal stresses σ_{xm} and σ_{xb} in each material, assuming that the magnesium rod and brass tube are securely bonded to each other. (b) Determine the elongation of the bimetallic rod.

Prob. 3.2-4. A magnesium-alloy rod ($E_m = 45$ GPa) of diameter $d_m = 30$ mm is encased by a brass tube ($E_b = 100$ GPa) with outer diameter d_b; both have length $L = 600$ mm. An axial load $P = 40$ kN is applied to the resulting bimetallic rod. Assume that the magnesium rod and the brass tube are securely bonded to each other. (a) If half of the load P is carried by the magnesium rod and half by the brass tube, what is the outer diameter, d_b, of the tube? (b) What is the elongation of the bimetallic tube?

Prob. 3.2-5. A steel pipe is filled with concrete, and the resulting column is subjected to a compressive load $P = 80$ kips. The pipe has an outer diameter of 12.75 in. and an inside diameter of 12.00 in. The elastic moduli of the steel and concrete are: $E_s = 30 \times 10^3$ ksi and $E_c = 3.6 \times 10^3$ ksi. (a) Determine the stress in the steel and the concrete due to this loading. (b) If the initial length of the column is $L = 12$ ft, how much does the column shorten when the load is applied? (Ignore radial expansion of the concrete and steel due to Poisson's ratio effect.)

P3.2-5 and P3.5-20

Prob. 3.2-6. A homogenous rod of length L and modulus E is a conical frustum with diameter $d(x)$ that varies linearly from d_0 at

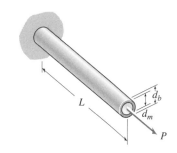

P3.2-3, P3.2-4, P3.5-18, and P3.5-19

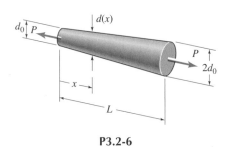

P3.2-6

one end to $2d_0$ at the other end, with $d_0 \ll L$. An axial load P is applied to the rod, as shown in Fig. P3.2-6. (a) Determine an expression for the stress distribution on an arbitrary cross section, $\sigma_x(x)$. (b) Determine an expression for the elongation of the rod.

Prob. 3.2-7. The tapered stone pier in Fig. P3.2-7 is 16 ft high, and it has a square cross section with side dimension that varies linearly from 24 in. at the top to 36 in. at the bottom. Assume that the stone is linearly elastic with modulus of elasticity $E = 4.0 \times 10^3$ ksi. Determine the shortening of the pier under a compressive load of $P = 150$ kips. (Neglect the weight of the stone.)

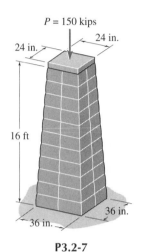

P3.2-7

Prob. 3.2-8. A flat bar with rectangular cross section has a constant thickness t and an unstretched length L. The width of the bar varies linearly from b_1 at one end to b_2 at the other end, and its modulus of elasticity is E. (a) Derive a formula for the elongation, e, of the bar when it is subjected to an axial tensile load P, as shown in Fig. P3.2-8. (b) Calculate the elongation for the following case: $b_1 = 50$ mm, $b_2 = 100$ mm, $t = 25$ mm, $L = 2$ m, $P = 250$ kN, $E = 70$ GPa.

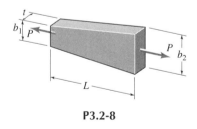

P3.2-8

Prob. 3.2-9. A bar of unstretched length L and cross-sectional area A is made of material with modulus of elasticity E and specific weight γ (weight per unit volume). (a) Determine the elongation of the bar due to its own weight (i.e., with $P = 0$). (b) Determine the compressive axial force P that would be required to return the bar to its original length L. Express your answer in terms of γ, A, E, and L.

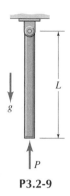

P3.2-9

Prob. 3.2-10. To help support a vertical retaining wall, a metal rod is pushed into the soil embankment. The soil surrounding the rod exerts a distributed axial force on the rod that varies linearly along the rod as shown. The rod has an initial length L and cross-sectional area A, and it is made of material whose modulus of elasticity is E. (a) What force P is required just as the rod becomes fully embedded in the soil and end B contacts the retaining wall? (b) What is the total shortening of the rod caused by the force P and the distributed resisting force? Express your answers in terms of p_{max}, A, E, and L.

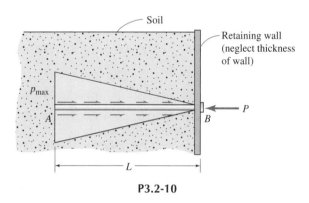

P3.2-10

Prob. 3.2-11. The stiffeners in airplane wings may be analyzed as uniform rods subjected to distributed loading. The stiffener shown below has a cross-sectional area of 0.60 in^2 and is made of aluminum ($E_a = 10 \times 10^6$ psi). Determine the elongation of section AB of the stiffener, whose original length is 20 in., if the

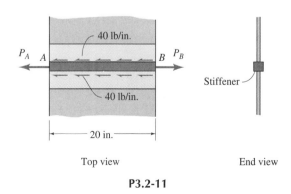

P3.2-11

stress in the stiffener at end A is $\sigma_A = 5{,}000$ psi, and a uniform distributed loading of 40 lb/in. is applied on either side of the stiffener over the 20-in. length from A to B.

*Prob. 3.2-12. Determine an expression for the elongation of the conical-frustum-shaped bar in Fig. P3.2-12 due to its own weight. The unstretched length of the bar is L, and its diameter varies from $2d_0$ at the top to d_0 at the bottom. The specific weight of the material is γ, and its modulus of elasticity is E. (Assume that $d_0 \ll L$, so that the normal stress may be assumed to be uniform over a cross section.)

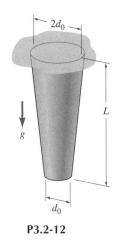

P3.2-12

*Prob. 3.2-13. A flat bar with rectangular cross section and constant thickness t has a width that varies linearly from $2b_0$ at the top end to b_0 at the bottom end. The unstretched length of the tapered bar is L, and it is made of material with specific weight γ and modulus of elasticity E. (a) Derive a formula for the elongation of the bar when loaded only by its own weight. (b) Calculate the elongation of the bar for the following case: $b_0 = 25$ mm, $L = 10$ m, $E = 200$ GPa, $\gamma = 77$ kN/m^3.

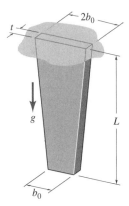

P3.2-13

Prob. 3.2-14. A carnival ride consists of passenger cars at each end of a connecting tubular member. The cars are supported by wheels that roll on a track, so the only force exerted on the connecting tube when the ride is up to its full speed of $\omega = 20$ rpm is the centrifugal force that the cars exert on the tube. (Neglect the mass of the tube in comparison with the mass of a car, neglect air resistance, and neglect the distance from A to the center of rotation and the distance from B to the center of gravity of the car.) (a) If each car weighs 1000 lb, including the weight of passengers, what cross-sectional area of connecting tube is required if the allowable tensile stress in the tube is 10 ksi? (b) What is the elongation of the segment AB of the tube if it has the cross-sectional area determined in Part (a)?

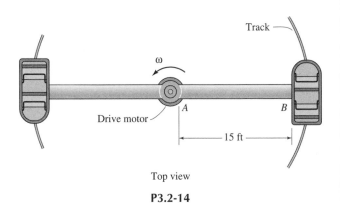

P3.2-14

Prob. 3.2-15. Determine an expression for the total elongation, e, of the bimetallic bar in Example 3.1.

Prob. 3.2-16. For the bimetallic bar in Fig. P3.2-16, $E_1 = 30 \times 10^3$ ksi, $L = 100$ in., $b = 2$ in., and $h_1 = h_2 = 0.6$ in. (a) If the bimetallic bar undergoes *axial deformation* under the action of load P at $y_P = 0.4$ in., what is the value of the modulus of elasticity, E_2, of material (2)? (b) Determine the total elongation of the bar for a load of $P = 2$ kips.

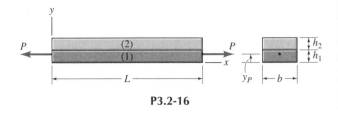

P3.2-16

Prob. 3.2-17. A uniform circular cylinder of diameter d and length L is made of material with modulus of elasticity E and specific weight γ. It hangs from a rigid "ceiling" at A, as shown in Fig. P3.2-17. Using Eq. 3.10, determine an expression for the (downward) vertical displacement, $u(x)$, of the cross section at distance x from end A.

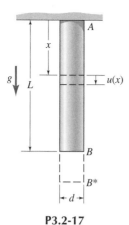

P3.2-17

*Prob. 3.2-18.** The solid conical frustum in Fig. P3.2-18 tapers gradually from a diameter d_0 at its left end, A, to a diameter $2d_0$ at its right end, B. The right end is fixed to a rigid wall, and a compressive axial load of magnitude P is uniformly distributed over end A. The modulus of elasticity of the bar is E and its length is L. Using Eq. 3.10, determine an expression for the horizontal displacement, $u(x)$, of the cross section at distance x from end A.

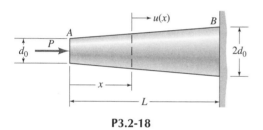

P3.2-18

Prob. 3.2-19. A uniform circular cylinder of diameter d and length L is made of material with modulus of elasticity E. It is fixed to a rigid wall at end A and subjected to distributed external axial loading of magnitude $p(x)$ per unit length, as shown in Fig. P3.2-19a. The axial stress, $\sigma_x(x)$, varies linearly with x as shown in Fig. P3.2-19b. (a) Determine an expression for the distributed loading, $p(x)$. (Hint: Draw a free-body diagram of the bar from section x to section $(x + \Delta x)$.) (b) Using Eq. 3.10, determine an expression for the horizontal displacement, $u(x)$, of the cross section at distance x from end A.

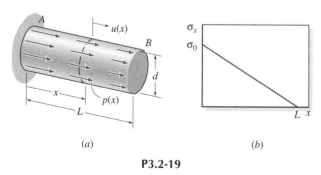

(a)   (b)

P3.2-19

*Prob. 3.2-20.** A uniform circular cylinder of diameter d and length L is made of material with modulus of elasticity E. It is

fixed to a rigid wall at end B and subjected to distributed external axial loading of magnitude $p(x)$ per unit length, as shown in Fig. P3.2-20a. The axial stress, $\sigma_x(x)$, varies quadratically with x as shown in Fig. P3.2-20b. (a) Determine an expression for the distributed loading, $p(x)$. (Hint: Draw a free-body diagram of the bar from section x to section $(x + \Delta x)$.) (b) Using Eq. 3.10, determine an expression for the horizontal displacement, $u(x)$, of the cross section at distance x from end A.

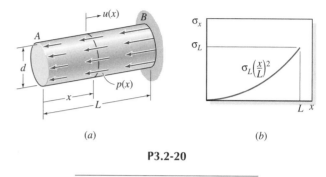

(a)   (b)

P3.2-20

Problems 3.5-1 through 3.5-8 *are statically determinate problems. The appropriate solution procedure for these problems is the* **force method***, since the force in any member can be determined directly from equilibrium equations. Therefore, the appropriate order of solution is*

> **Equilibrium equations**
> **Force-deformation relationships**
> **Geometry-of-deformation equations**

Prob. 3.5-1. A solid brass rod AB and a solid aluminum rod BC are connected together by a coupler at B. Assume that the coupler is rigid, and neglect its length. The diameters of the two segments are $d_1 = 2.5$ in. and $d_2 = 2.0$ in., and the moduli for the brass and aluminum are $E_1 = 15 \times 10^3$ ksi and $E_2 = 10 \times 10^3$ ksi, respectively. Determine the displacements u_B at the coupler and u_C at the right end, due to the two loads shown in Fig. P3.5-1.

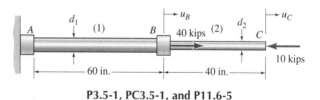

P3.5-1, PC3.5-1, and P11.6-5

Prob. 3.5-2. The diameter of the central one-third of a 50-mm-diameter steel rod is reduced to 20 mm, forming a three-segment rod, as shown in Fig. P3.5-2. For the loading shown, determine

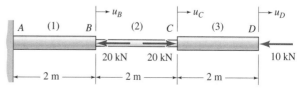

P3.5-2 and PC3.5-2

155

the displacements u_B, u_C, and u_D at nodes B, C, and D, respectively. Let $E = 200$ GPa.

Prob. 3.5-3. A three-segment stepped aluminum-alloy column is subjected to the vertical axial loads shown in Fig. P3.5-3. The cross-sectional areas of the segments are $A_1 = 6.0$ in², $A_2 = 9.0$ in², and $A_3 = 14.0$ in²; and the modulus is $E = 10 \times 10^3$ ksi. Determine the vertical displacements u_A, u_B, and u_C under the given loading of the column.

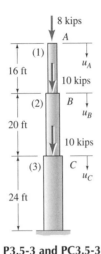

P3.5-3 and PC3.5-3

Prob. 3.5-4. A uniform rod of diameter d is subjected to axial loads at the three cross sections, as illustrated in Fig. P3.5-4. If the displacement at the right end, u_D, cannot exceed 5 mm, and the maximum axial stress in the rod cannot exceed 80 MPa, what is the minimum allowable diameter of the (circular) cylindrical rod? Use $E = 70$ GPa.

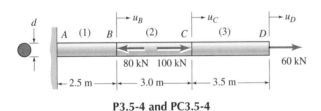

P3.5-4 and PC3.5-4

Prob. 3.5-5. A 1.2-in.-diameter aluminum-alloy hangar rod is supported by a steel pipe with an inside diameter of 3.0 in. Determine the minimum thickness of the steel pipe if the maximum displacement, u_C, at the lower end of the hanger rod is 0.10 in. Let $E_1 \equiv E_{al} = 10 \times 10^3$ ksi and $E_2 \equiv E_{st} = 30 \times 10^3$ ksi.

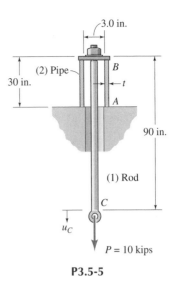

P3.5-5

Prob. 3.5-6. A hanger rod CD is supported by a rigid beam AB that is, in turn, supported by identical vertical rods at A and B. If the three rods all have the same cross-sectional area, A, and modulus of elasticity, E, determine an expression that relates the load P to the vertical displacement u_D.

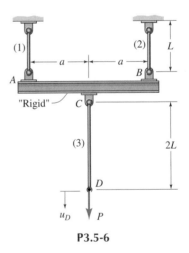

P3.5-6

Prob. 3.5-7. A rigid beam AB is supported by vertical rods at its ends, and it supports a downward load at C of $P = 60$ kN as shown in Fig. P.3.5-7. The diameter of the support rod at A is $d_1 = 25$ mm. Both hanger rods are made of steel ($E = 210$ GPa). Neglect the weight of beam AB. (a) If it is found that $u_B = 2u_A$, what is the diameter, d_2 of the hanger rod at B? (b) What is the corresponding displacement at the load point, C?

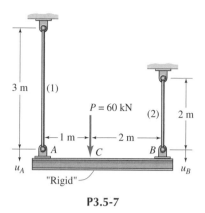

P3.5-7

Prob. 3.5-8. The pin-jointed planar truss in Fig. P3.5-8 consists of three aluminum-alloy members, each having a cross-sectional area of 1.5 in^2 and modulus $E = 10 \times 10^3$ ksi. If a vertical load $P = 10$ kips hangs from the pin at B, what is the vertical displacement, u_C, of the roller at C?

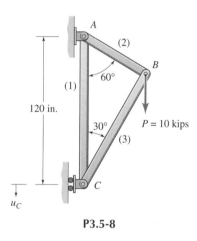

P3.5-8

Problems 3.5-9 through 3.5-30 *are statically indeterminate problems. In all cases, it is important that the* **equilibrium equations**, *the* **element force-deformation behavior**, *and the* **geometry of deformation** *be accounted for. Where a particular "method" is specified—the* **displacement method** *or the* **force method**—*it is expected that that method will be used; otherwise, either method may be used.*

Prob. 3.5-9. Two uniform, linearly elastic members are joined together at B, and the resulting two-segment rod is attached to rigid supports at ends A and C. A single external force, P, is applied to the joint at B. Member (1) is steel with modulus $E_1 = 30 \times 10^3$ ksi, cross-sectional area $A_1 = 2.0$ in^2 and length $L_1 = 80$ in.; member (2) is made of aluminum alloy with $E_2 = 10 \times 10^3$ ksi, $A_2 = 3.6$ in^2, and $L_2 = 60$ in. Let $P = 16$ kips, and use the *force method*: (a) to solve for the axial stresses σ_1 and σ_2 in

the two elements, and (b) to solve for the horizontal displacement, u, of joint B.

Prob. 3.5-10. For the two-segment rod in Fig. P3.5-10, member (1) is steel with $E_1 = 210$ GPa, $A_1 = 1000$ mm^2, and $L_1 = 2$ m; member (2) is a titanium alloy with $E_2 = 120$ GPa, $A_2 = 1200$ mm^2, and $L_2 = 1.8$ m. Let $P = 40$ kN, and use the *displacement method*: (a) to solve for the horizontal displacement, u, of joint B, and (b) to solve for the axial stresses σ_1 and σ_2 in the two elements.

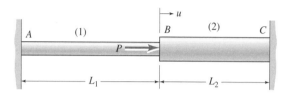

P3.5-9, P3.5-10, P3.5-11, and PC3.5-5

Prob. 3.5-11. For the rod in Fig. P3.5-11, let $E_1 = E_2 = E$ and $A_1 = A_2 = A$, and let an external horizontal force P be applied to the rod at distance L_1 from end A and L_2 from end B. If the stresses are found to satisfy the equation $|\sigma_2| = 3\sigma_1$, what is the ratio L_1/L_2?

Prob. 3.5-12. A three-segment rod is attached to rigid supports at ends A and D and is subjected to equal and opposite external loads P at nodes (joints) B and C, as shown in Fig. P3.5-12. The rod is homogeneous and linearly elastic, with modulus of elasticity E. Let $A_1 = A_3 = A$, and $A_2 = 2A$; let $L_1 = 2L$ and $L_2 = L_3 = L$. (a) Let the force in member (2) be the *redundant force*, and use the *force method* to determine the axial stresses σ_1, σ_2, and σ_3 in the elements. (b) Determine expressions for the horizontal displacements u_B and u_C at nodes B and C, respectively.

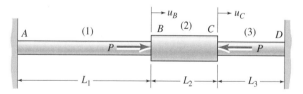

P3.5-12, P3.5-13, and PC3.5-10

Prob. 3.5-13. For the three-segment stepped rod in Prob. 3.5-12, (a) use the *displacement method* to solve for the horizontal displacements u_B and u_C at nodes B and C, respectively, and (b) solve for the axial stresses σ_1, σ_2, and σ_3 in the three elements that comprise the stepped rod.

Prob. 3.5-14. A five-segment stepped rod is stress free when it is attached to rigid supports at ends A and F. Subsequently, it is subjected to two external loads of magnitude P at nodes B and D, as shown in Fig. P3.5-14. The bar is homogeneous with modulus of elasticity E, and the cross-sectional areas are $A_1 = A_5 = A$, and $A_2 = A_4 = 2A$, and $A_3 = 3A$. Use the *force method* to solve for the reaction P_F at end F. (Hint: Let $P_F = F_5$ be the redundant force, and use the force method to solve directly for this force.)

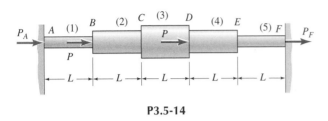

P3.5-14

Prob. 3.5-15. A steel pipe with outer diameter d_o and inner diameter d_i and a solid aluminum-alloy rod of diameter d form a three-segment system that undergoes axial deformation due to a single load P acting on a collar at point C, as shown in Fig. P3.5-15. (a) Using the *displacement method,* determine the displacements u_B and u_C, respectively, and (b) calculate the axial stresses σ_1, σ_2, and σ_3 in the three elements.

$$d_o = 2 \text{ in., } d_i = 1.5 \text{ in., } d = 0.75 \text{ in.}$$

$$L_1 = L_2 = 30 \text{ in., } L_3 = 50 \text{ in.}$$

$$P = 12 \text{ kips, } E_1 = 30 \times 10^3 \text{ ksi, } E_2 = E_3 = 10 \times 10^3 \text{ ksi}$$

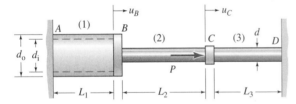

P3.5-15, P3.5-16, PC3.5-6, PC3.5-7, PC3.5-8, PC3.5-9, and P11.6-11

Prob. 3.5-16. Consider the three-segment system shown in Fig. P3.5-16. Using the dimensions, material properties, and load given below, and using the *force method,* (a) determine the element axial forces F_1, F_2, and F_3, and (b) calculate the two nodal displacements u_B and u_C.

$$d_o = 50 \text{ mm, } d_i = 36 \text{ mm, } d = 20 \text{ mm}$$

$$L_1 = L_2 = 1 \text{ m, } L_3 = 2 \text{ m, } P = 50 \text{ kN}$$

$$E_1 = 210 \text{ GPa, } E_2 = E_3 = 70 \text{ GPa}$$

Prob. 3.5-17. Three members that are attached together and supported by two rigid walls undergo axial deformation due to a single applied load P, as shown in Fig. P3.5-17. Member AB is

a solid brass rod with diameter $d_1 = 1.0$ in. and modulus $E_1 = 15 \times 10^3$ ksi; member BC is a steel pipe with outer diameter $(d_o)_2 = 1.0$ in., inner diameter $(d_i)_2 = 0.75$ in., and modulus $E_2 = 30 \times 10^3$ ksi; and member BD is a solid aluminum-alloy rod with diameter $d_3 = 0.50$ in. and modulus $E_3 = 10 \times 10^3$ ksi. $L_1 = L_2 = 20$ in., $L_3 = 40$ in. (a) Determine the displacement u_D of the point D where the load $P = 2$ kips is applied, and (b) calculate the axial stress, σ_2, in the pipe section, BC.

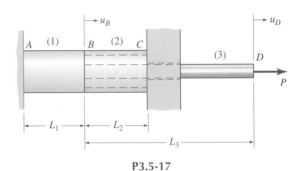

P3.5-17

Prob. 3.5-18. Repeat Prob. 3.2-3, treating the magnesium-alloy rod as element (1) and the brass tube as element (2). Use either the *force method* or the *displacement method* to solve for the requested information.

Prob. 3.5-19. Repeat Prob. 3.2-4, treating the magnesium-alloy rod as element (1) and the brass tube as element (2). Use either the *force method* or the *displacement method* to solve for the requested information.

Prob. 3.5-20. With the figure and data of Prob. 3.2-5, (a) use the *displacement method* to solve for the shortening of the 12-ft, concrete-filled pipe column due to application of the 80-kip load. (b) Calculate the stress in the steel and the stress in the concrete under the given loading. (c) The load, P, and the shortening of the column, u, are related by the equation $P = k_{\text{eff}} u$. Calculate the effective stiffness, k_{eff}, of this 2-material column.

Prob. 3.5-21. A 15-in.-diameter, reinforced-concrete column has six $\frac{9}{8}$-in.-diameter steel reinforcing rods uniformly spaced around

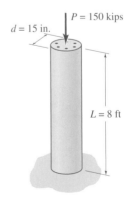

P3.5-21

a circle. $E_c = 3 \times 10^3$ ksi and $E_s = 30 \times 10^3$ ksi. (a) Determine the axial stress in the concrete and the axial stress in the steel under an axial load $P = 150$ kips. (Note: Treat the six reinforcing bars as a single element, and treat the concrete as a second element. Use the net area of concrete.) (b) How much does the column shorten under the 150-kip load?

Prob. 3.5-22. A steel pipe ($E_1 = 200$ GPa) surrounds a solid aluminum-alloy rod ($E_2 = 70$ GPa), and together they are subjected to a compressive force of 200 kN acting on rigid end caps. Determine the shortening of this bimetallic compression member, and determine the normal stresses in the pipe and in the rod.

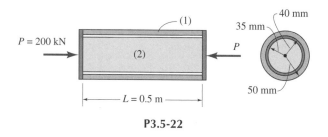

P3.5-22

Prob. 3.5-23. A rigid beam AD is supported by a smooth pin at D and by vertical steel rods attached to the beam at points A and C. Neglect the weight of the beam, and assume that the support rods are stress-free when $P = 0$. (a) Use the *displacement method* to solve for the rotation, θ, of the beam AD when load P is applied. (b) Determine the axial stresses in support rods (1) and (2) when load P is applied.

$$A_1 = A_2 = 1.0 \text{ in}^2, \quad L_1 = 40 \text{ in.}, \quad L_2 = 60 \text{ in.}$$
$$E_1 = E_2 = 30 \times 10^3 \text{ ksi}, \quad P = 10 \text{ kips}$$
$$a = 20 \text{ in.}, \quad b = 60 \text{ in.}$$

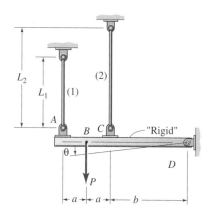

P3.5-23, P3.5-24, and P11.6-14

Prob. 3.5-24. With Fig. P3.5-24 and the data below, (a) use the *force method* to solve for the forces F_1 and F_2 in the support rods.

Use force F_1 as the redundant force. (b) Determine the elongation of support rod (1).

$$A_1 = A_2 = 500 \text{ mm}^2, \quad L_1 = 1 \text{ m}, \quad L_2 = 2 \text{ m}$$
$$E_1 = E_2 = 210 \text{ GPa}, \quad P = 50 \text{ kN}$$
$$a = 0.50 \text{ m}, \quad b = 1.5 \text{ m}$$

Prob. 3.5-25. A rigid beam AD is supported by a smooth pin at B and by vertical rods attached to the beam at points A and C. Neglect the weight of the beam, and assume that the rods are stress-free when $P = 0$. (a) Use the *displacement method* to solve for the rotation, θ, of the beam AD when load P is applied. (b) Determine the axial stresses in support rods (1) and (2) when load P is applied.

$$A_1 = A_2 = 200 \text{ mm}^2, \quad L_1 = L_2 = 2 \text{ m}$$
$$E_1 = 70 \text{ GPa}, \quad E_2 = 100 \text{ GPa}$$
$$a = 1 \text{ m}, b = 1.5 \text{ m}, \quad P = 20 \text{ kN}$$

Prob. 3.5-26. With Fig. P3.5-26 and the data below, (a) use the *force method* to solve for the forces F_1 and F_2 in the support rods. Use force F_1 as the redundant force. (b) Determine the axial stress in each support rod. (c) Calculate the elongation of support rod (1).

$$A_1 = 1.0 \text{ in}^2, \quad A_2 = 0.50 \text{ in}^2, \quad L_1 = L_2 = 5 \text{ ft}$$
$$E_1 = E_2 = 10 \times 10^3 \text{ ksi}$$
$$a = 2 \text{ ft}, \quad b = 4 \text{ ft}, \quad P = 5 \text{ kips}$$

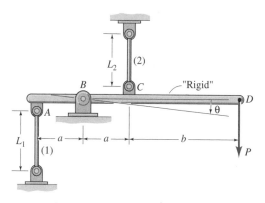

P3.5-25, P3.5-26, and P11.6-13

*Prob. 3.5-27. A rigid beam AD is supported by three identical vertical rods that are attached to the beam at points A, C, and D. When the rods are initially attached to the beam and $P = 0$, the rods are stress-free. (a) Use the *displacement method* to solve for the vertical displacements u_A and u_D of points A and D when the load P is applied to the beam at point B. (b) Calculate the axial stress in each of the three support rods when the load is applied.

$$A = 1.0 \text{ in}^2, \quad L = 60 \text{ in.}, \quad E = 30 \times 10^3 \text{ ksi}$$
$$a = 20 \text{ in.}, \quad b = 40 \text{ in.}, \quad c = 60 \text{ in.}, \quad P = 10 \text{ kips}$$

159

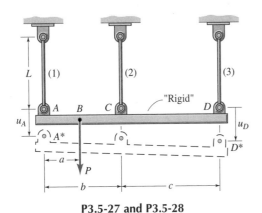

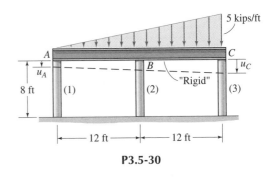

P3.5-30

P3.5-27 and P3.5-28

Computer Exercises—Section 3.5. *Use the* ***MechSOLID*** *computer program AXIALDEF to solve the homework exercises C3.5-1 through C3.5-10. Refer to Appendix G.3 for a description of this computer program and for examples of its use. Hand in computer printouts of the following* ***MechSOLID*** *"screens": AXIALDEF: Model, AXIALDEF: Boundary Conditions, AXIALDEF: Results, AXIALDEF: Displacement Plot, AXIALDEF: Force Plot. Samples of these screens are illustrated in Example G-4.*

*Prob. 3.5-28. With Fig. P3.5-28 and the data below, (a) use the *force method* to solve for the axial forces F_1, F_2, and F_3 in the support bars. Use F_2 as the redundant force. (Hint: Take moments about A and D to relate F_3 and F_1 to F_2.) (b) Determine the vertical displacement of point B, the point of application of the load.

$$A = 500 \text{ mm}^2, \quad L = 2 \text{ m}, \quad E = 100 \text{ GPa}$$

$$a = 0.5 \text{ m}, \quad b = 1.0 \text{ m}, \quad c = 1.5 \text{ m}, \quad P = 20 \text{ kN}$$

*Prob. 3.5-29. A rigid beam AE is supported by three vertical rods, as shown in Fig. P3.5-29. (a) Determine the vertical displacements of points A and E when vertical forces are applied at points B and D as shown in the figure. (b) Determine the axial stress in each of the three support rods.

$$A_1 = A, \quad A_2 = A_3 = 2A, \quad E_1 = E_2 = E_3 = E$$

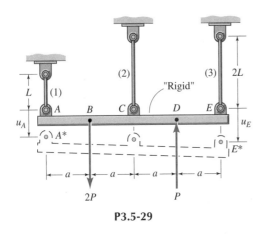

P3.5-29

*Prob. 3.5-30. Three steel pipe columns ($E = 29 \times 10^3$ ksi) support a rigid beam AC, which, in turn, supports a linearly varying distributed load. Neglect the weight of beam AC and determine the compressive normal stress in each of the columns.

$$A_1 = A_2 = 2.2 \text{ in}^2, \quad A_3 = 4.3 \text{ in}^2$$

Prob. C3.5-1. Use AXIALDEF to solve Prob. 3.5-1. Let node A in Fig. P3.5-1 be relabeled as node ①, node B be relabeled as node ②, and node C be relabeled as node ③.

Prob. C3.5-2. Use AXIALDEF to solve Prob. 3.5-2. Let node A in Fig. P3.5-2 be relabeled as node ①, node B be relabeled as node ②, and node C be relabeled as node ③.

Prob. C3.5-3. Use AXIALDEF to solve Prob. 3.5-3. Let node A in Fig. P3.5-3 be relabeled as node ①, node B be relabeled as node ②, and so on.

Prob. C3.5-4. Use AXIALDEF to solve Prob. 3.5-4. Let node A in Fig. P3.5-4 be relabeled as node ①, node B be relabeled as node ②, and so on.

Prob. C3.5-5. Use AXIALDEF to solve Prob. 3.5-10. Let node A in Fig. P3.5-10 be relabeled as node ①, node B be relabeled as node ②, and node C be relabeled as node ③.

Prob. C3.5-6. Use AXIALDEF to solve Prob. 3.5-15. Let node A in Fig. P3.5-15 be relabeled as node ①, node B be relabeled as node ②, and so on.

Prob. C3.5-7. Use AXIALDEF to solve Prob. 3.5-16. Let node A in Fig. P3.5-16 be relabeled as node ①, node B be relabeled as node ②, etc. (Since AXIALDEF uses the displacement method, you may ignore the request that the force method be used.)

*Prob. C3.5-8. (a) Use AXIALDEF to solve Prob. 3.5-15 as it is stated. Call this "Solution A." (b) Keep the problem parameters the same, except replace d_o and d_i by the following: $d_o = 1.75$ in., $d_i = 1.5$ in. Use AXIALDEF to solve this revised problem, and call this "Solution B." (c) Again, keep all parameters the same as in the original problem, except replace d_o, and d_i by the following: $d_o = 1.5$ in., $d_i = 1.25$ in. Use AXIALDEF to solve this revised problem, and call this "Solution C." (d) Create a table of nodal displacements and a table of element axial stresses. Discuss the effects that changing A_1 has on the nodal displacements and on the element axial stresses. Which design best minimizes the element stresses?

*Prob. C3.5-9. (a) Use AXIALDEF to solve Prob. 3.5-16 as it is stated. Call this "Solution A." (Since AXIALDEF is based on the displacement method, you can ignore the request that you use the force method to solve this problem.) (b) Keep the problem parameters the same, except replace d_o and d_i by the following: $d_o = 44$ mm, $d_i = 36$ mm. Use AXIALDEF to solve this revised problem, and call this "Solution B." (c) Again, keep all parameters the same as in the original problem, except replace d_o and d_i by the following: $d_o = 38$ mm, $d_i = 30$ mm. Use AXIALDEF to solve this revised problem, and call this "Solution C." (d) Create a table of nodal displacements and a table of element axial stresses. Discuss the effects that changing A_1 has on the nodal displacements and on the element axial stresses. Which design best minimizes the element stresses?

*Prob. C3.5-10. The AXIALDEF program can be used to solve nonnumeric problems like Probs. 3.5-12 and 3.5-13 if you remember that nodal displacements will have the form $u_i = \bar{u}_i(PL/AE)$, where $\bar{u}_i$ is a nondimensional factor that is obtained by AXIALDEF if you input the following data: $P = L = A = E = 1$. Likewise, element axial forces will have the form $F_i = \bar{F}_i P$, and element axial stresses will have the form $\sigma_i = \bar{\sigma}_i(P/A)$, where $\bar{F}_i$ and $\bar{\sigma}_i$ are the internal forces and internal stresses output by AXIALDEF.

Use AXIALDEF to solve Prob. 3.5-13. Let the nodes in Fig. P3.5-13 be relabeled as follows: $A \rightarrow \boxed{1}$, $B \rightarrow \boxed{2}$, and so on.

---

*All the problems in **Section 3.6** are statically indeterminate problems. In all cases, it is important that the **equilibrium equations**, the **element force-temperature-deformation behavior**, and the **geometry of deformation** be accounted for. Where a particular "method" is specified—the **displacement method** or the **force method**—it is expected that that method will be used; otherwise, either method may be used.*

Prob. 3.6-1. A square aluminum-alloy bar is attached to rigid walls at ends A and B when the temperature of the bar is 90°F. If the bar is cooled to 40°F, what is the maximum normal stress in the bar and what is the maximum <u>shear</u> stress? (Recall Section 2.8.) Let $\alpha = 13 \times 10^{-6}/°F$ and $E = 10 \times 10^3$ ksi, and give your answers in kips/in².

Prob. 3.6-2. If the square steel bar in Fig. P3.6-2 is stress free when it is attached to the rigid walls at A and B, how much must the temperature of the bar be raised, in °F, to cause the maximum <u>shear</u> stress in the bar to be 5 ksi? (Recall Section 2.8). Let $\alpha = 6.5 \times 10^{-6}/°F$ and $E = 30 \times 10^3$ ksi.

P3.6-1 and P3.6-2

Prob. 3.6-3. Two uniform, linearly elastic rods are joined together at B, and the resulting two-segment rod is attached to rigid supports at A and C. Element (1) is steel with modulus $E_1 = 30 \times 10^3$ ksi, cross-sectional area $A_1 = 2.0$ in², length $L_1 = 80$ in., and coefficient of thermal expansion $\alpha_1 = 7 \times 10^{-6}/°F$; the corresponding values for the aluminum element (2) are: $E_2 = 10 \times 10^3$ ksi, $A_2 = 3.6$ in², $L_2 = 60$ in., and $\alpha_2 = 13 \times 10^{-6}/°F$. (a) Use the *displacement method* to determine the axial stresses σ_1 and σ_2 in the rods if the temperature of both is raised by $\Delta T_1 = \Delta T_2 = 60°F$. (b) Does node B move to the right, or does it move to the left? How much does it move?

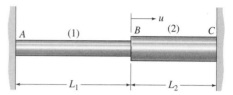

P3.6-3, P3.6-4, P3.6-5, PC3.6-1, and PC3.6-2

Prob. 3.6-4. For the two-element rod system in Fig. P3.6-4, element (1) is steel with modulus $E_1 = 210$ GPa, area $A_1 = 1000$ mm², length $L_1 = 2$ m, and coefficient of thermal expansion $\alpha_1 = 12 \times 10^{-6}/°C$; element (2) is of titanium alloy with $E_2 = 120$ GPa, $A_2 = 1200$ mm², $L_2 = 1.8$ m, and $\alpha_2 = 8 \times 10^{-6}/°C$. Use the *force method* to determine the axial stresses σ_1 and σ_2 in the rods if the temperature of both is raised by $\Delta T_1 = \Delta T_2 = 30°C$.

Prob. 3.6-5. The two rod elements in Fig. P3.6-5 are stress free when they are assembled together and attached to the rigid walls at A and C. Determine expressions for the stresses $\sigma_1(\Delta T)$ and $\sigma_2(\Delta T)$ that would result from a uniform temperature increase $\Delta T_1 = \Delta T_2 = \Delta T$. Express your answers in terms of $E_1, E_2, A_1, A_2, L_1, L_2, \alpha_1, \alpha_2$, and ΔT.

Prob. 3.6-6. The pipe-sleeve element (1) and cylindrical-rod core element (2) in Fig. P3.6-6 are both stress free after they are attached to rigid end-plates. (a) Use the *displacement method* to determine an expression for the total elongation, e, of the pipe/rod system if the temperature of the system is uniformly increased by ΔT. Express your answer in terms of $E_1, E_2, A_1, A_2, L_1 = L_2 \equiv L, \alpha_1, \alpha_2$, and ΔT. (b) Determine expressions for the stresses σ_1 and σ_2 in the pipe and rod, respectively, due to the uniform temperature increase ΔT.

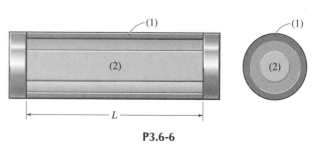

P3.6-6

Prob. 3.6-7. A three-segment rod is attached to rigid supports at ends A and D. When the rod is initially attached to the supports, it is stress free. Subsequently, the middle segment is heated by an amount ΔT_2, while segments (1) and (3) are kept at their original temperature (i.e., $\Delta T_1 = \Delta T_3 = 0$). There are no external forces acting on the rod, other than the reactions at ends A and D. Let the force in segment (2) be the redundant force, and use the *force method* to determine the axial stresses σ_1, σ_2, and σ_3 in the elements. Let $A_1 = A_3 = A$, $A_2 = 2A$, $L_2 = L_3 = L$, $L_1 = 2L$, $E = $ const., $\alpha = $ const.

Prob. 3.6-8. The three-element, stepped-rod system in Fig. P3.6-8 is stress free when ends A and D are attached to rigid walls. Subsequently, the middle segment is heated by an amount ΔT_2, while $\Delta T_1 = \Delta T_3 = 0$. (a) Use the *displacement method* to solve for the horizontal displacements u_B and u_C at joints B and C, respectively, and (b) solve for the axial stresses σ_1, σ_2, and σ_3 induced in the three segments.

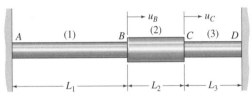

P3.6-7, P3.6-8, and PC3.6-7

Prob. 3.6-9. A steel pipe with outer diameter d_o and inner diameter d_i, and a solid aluminum-alloy rod of diameter d form a three-segment system. When the system is welded to rigid supports at A and D, it is stress free. Subsequently, the steel pipe is cooled by 100°F (i.e., $\Delta T_1 = -100°F$), while $\Delta T_2 = \Delta T_3 = 0$. Let the force in segment (2) be the redundant force, and use the *force method* to determine the axial stresses σ_1, σ_2, and σ_3 in the elements.

$$d_o = 2 \text{ in.}, \quad d_i = 1.5 \text{ in.}, \quad d = 0.75 \text{ in.}$$

$$L_1 = L_2 = 30 \text{ in.}, \quad L_3 = 50 \text{ in.}$$

$$E_1 = 30 \times 10^3 \text{ ksi}, \quad E_2 = E_3 = 10 \times 10^3 \text{ ksi}$$

$$\alpha_1 = 6.5 \times 10^{-6}/°F$$

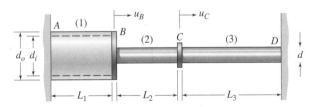

P3.6-9, P3.6-10, PC3.6-3, PC3.6-4, PC3.6-5, and PC3.6-6

Prob. 3.6-10. The three-element rod system in Fig. P3.6-10 is stress free when it is welded to rigid supports at A and D. Sub-

sequently, segment (2) is cooled by an amount $\Delta T_2 = -20°C$, while $\Delta T_1 = \Delta T_3 = 0$. The dimensions and material properties of the system are

$$d_o = 50 \text{ mm}, \quad d_i = 36 \text{ mm}, \quad d = 20 \text{ mm},$$

$$L_1 = L_2 = 1 \text{ m}, \quad L_3 = 2 \text{ m}$$

$$E_1 = 210 \text{ GPa}, \quad E_2 = E_3 = 70 \text{ GPa},$$

$$\alpha_2 = 23 \times 10^{-6}/°C$$

(a) Use the *displacement method* to solve for the horizontal displacements u_B and u_C at joints B and C, respectively, and (b) solve for the stresses σ_1, σ_2, and σ_3 that are induced in the three segments.

Prob. 3.6-11. A "rigid" beam AD is supported by a smooth pin at D and by vertical steel rods attached to the beam at points A, B, and C. Neglect the weight of the beam, and assume that the support rods are stress-free when the system is assembled. Use the *displacement method* to solve this problem. (a) Solve for the rotation, θ, of the beam AD when element (3) is cooled by 80°F (i.e., $\Delta T_3 = -80°F$). (b) Solve for the axial stresses σ_1, σ_2, and σ_3 induced in the three rods by the cooling of rod (3).

$$A_1 = A_2 = A_3 = 1.0 \text{ in.}^2, \quad L_1 = L_3 = 40 \text{ in.}, \quad L_2 = 60 \text{ in.}$$

$$a = 20 \text{ in.}, \quad b = 60 \text{ in.},$$

$$E_1 = E_2 = E_3 = 30 \times 10^3 \text{ ksi}, \quad \alpha_3 = 6.5 \times 10^{-6}/°F$$

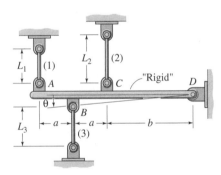

P3.6-11 and P3.6-12

*Prob. 3.6-12. The "rigid" beam AD in Fig. P3.6-12 is supported by a smooth pin D and is attached to vertical steel rods at points A, B, and C. Neglect the weight of the beam, and assume that the steel rods are stress-free when the system is assembled.

Use the *force method* to solve for the stress induced in each of the steel rods if rod (2) is cooled by 10°C, that is, $\Delta T_1 = \Delta T_3 = 0$; $\Delta T_2 = -10°C$. Let F_1 and F_3 be the redundant forces.

$$A_1 = A_2 = A_3 = 500 \text{ mm}^2, \quad L_1 = L_3 = 1 \text{ m}, \quad L_2 = 2 \text{ m}$$

$$a = 0.5 \text{ m}, \quad b = 1.5 \text{ m}, \quad E_1 = E_2 = E_3 = 210 \text{ GPa},$$

$$\alpha_2 = 12 \times 10^{-6}/°C$$

*Prob. 3.6-13. A "rigid" beam AD is supported by a smooth pin at B and by vertical aluminum-alloy rods attached to the beam at points A, C, and D. Neglect the weight of the beam, and assume that the rods are stress free when the system is assembled.

If rods (1) and (2) are kept at the assembly temperature (i.e., $\Delta T_1 = \Delta T_2 = 0$), and rod (3) is cooled by an amount ΔT (i.e., $\Delta T_3 = -\Delta T$), how much must rod (3) be cooled to induce a tensile stress of 5 ksi in rod (1)? Let F_1 and F_2 be redundant forces, and use the *force method* to solve this problem.

$$A_1 = A_3 = 1.0 \text{ in}^2, \quad A_2 = 0.50 \text{ in}^2, \quad L_1 = L_2 = L_3 = 5 \text{ ft}$$

$$E_1 = E_2 = E_3 = 10 \times 10^3 \text{ ksi}, \quad \alpha_3 = 23 \times 10^{-6}/°F,$$

$$a = 2 \text{ ft}, \quad b = 4 \text{ ft}$$

Prob. 3.6-14. Use the *displacement method* to solve for the temperature $\Delta T_3 \equiv -\Delta T$ in Prob. 3.6-13.

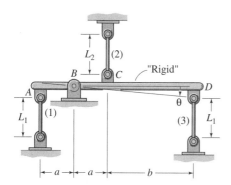

P3.6-13, P3.6-14, and P3.6-15

Prob. 3.6-15. A "rigid" beam AD is supported by a smooth pin at B and by vertical rods at A, C, and D. Neglect the weight of the beam, and assume that the rods are stress free when the system is assembled. Rods (1) and (2) are kept at the assembly temperature (i.e., $\Delta T_1 = \Delta T_2 = 0$), and rod (3) is cooled by an amount $\Delta T_3 = -50°C$.

$$A_1 = A_2 = A_3 = 200 \text{ mm}^2, \quad L_1 = L_2 = L_3 = 2 \text{ m}$$

$$a = 1 \text{ m}, \quad b = 1.5 \text{ m}, \quad E_1 = E_3 = 70 \text{ GPa},$$

$$E_2 = 100 \text{ GPa}, \quad \alpha_3 = 20 \times 10^{-6}/°C$$

(a) Use the *displacement method* to solve for the angle of rotation, θ, of the beam caused by cooling rod (3). (b) Determine the stresses σ_1, σ_2, and σ_3 induced in the three rods by the cooling of rod (3).

Prob. 3.6-16. A "rigid" beam AC is supported by three vertical rods, as shown in Fig. P3.6-16. When originally assembled, all of the rods are stress free and beam AC is horizontal. Subsequently, rod (1) is maintained at the assembly temperature (i.e., $\Delta T_1 = 0$), and rods (2) and (3) are heated the same amount ($\Delta T_2 = \Delta T_3 = \Delta T$). (a) Use the *displacement method* to solve for the displacements u_A and u_C of the ends of the beam, and (b) solve for the stresses induced in the three support rods.

$$A_1 = A, \quad A_2 = A_3 = 2A, \quad E_1 = E_2 = E_3 = E$$

Prob. 3.6-17. A "rigid" beam AC is supported by three vertical

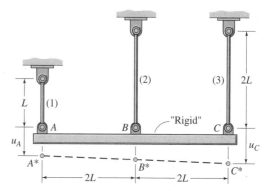

P3.6-16, P3.6-17, P11.6-15, and P11.6-16

rods, as shown in Fig. P3.6-17. When originally assembled, the rods are all stress free and beam AC is horizontal. If rod (1) is maintained at the assembly temperature (i.e., $\Delta T_1 = 0$), and rods (2) and (3) are heated the same amount ($\Delta T_2 = \Delta T_3 = \Delta T$), determine the stresses induced in the three support rods. Use the *force method* to solve this problem, letting F_1 be the redundant force. (Hint: Take $(\Sigma M)_B$ and $(\Sigma M)_C$ for the equilibrium equations.)

$$A_1 = A, \quad A_2 = A_3 = 2A, \quad E_1 = E_2 = E_3 = E$$

Prob. 3.6-18. For the two-element system described in the problem statement of Example 3.9, use the *force method*: (a) to determine the internal force induced in each element when element (1) is heated by an amount ΔT, and (b) to determine the displacement, u, of the joint at B.

Prob. 3.6-19. The entire titanium-sleeve/aluminum-core system described in the problem statement of Example 3.10 is heated by 100°F. Use the *force method*: (a) to determine the axial stress induced in each material, and (b) to determine the total elongation, e, of the sleeve/core system. (Assume that the core and the sleeve are both stress free at the reference temperature.)

Prob. 3.6-20. For the statically indeterminate system described in the problem statement of Example 3.11, use the *displacement method*: (a) to determine the (small) angle through which the "rigid" beam AD rotates when rod (2) is cooled by 100°F, and (b) to determine the forces F_1, F_2, and F_3 induced in the rods when rod (2) is cooled by 100°F.

Computer Exercises—Section 3.6. *Use the MechSOLID computer program AXIALDEF to solve the homework exercises C3.6-1 through C3.6-7. Refer to Appendix G.3 for a description of this computer program and for examples of its use. Hand in computer printouts of the following MechSOLID "screens": AXIALDEF: Model, AXIALDEF: Boundary Conditions, AXIALDEF: Results, AXIALDEF: Displacement Plot, AXIALDEF: Force Plot. Samples of these screens are illustrated in Example G-4.*

Prob. C3.6-1. Use AXIALDEF to solve Prob. 3.6-3. Let node A in Fig. P3.6-3 be relabeled as node ①, node B be relabeled as node ②, and node C be relabeled as node ③.

Prob. C3.6-2. Use AXIALDEF to solve Prob. 3.6-4. Let node *A* in Fig. P3.6-4 be relabeled as node ①, node *B* be relabeled as node ②, and node *C* be relabeled as node ③.

Prob. C3.6-3. Use AXIALDEF to solve Prob. 3.6-9 (instead of using the force method, as prescribed in the problem statement). Let node *A* in Fig. P3.6-9 be relabeled as node ①, node *B* be relabeled as node ②, and so on.

Prob. C3.6-4. Use AXIALDEF to solve Prob. 3.6-10. (Since AXIALDEF is based on the displacement method, you will, in effect, be using the displacement method as prescribed in the problem statement.) Let node *A* in Fig. P3.6-10 be relabeled as node ①, node *B* be relabeled as node ②, and so on.

*Prob. C3.6-5. (a) Use AXIALDEF to solve Prob. 3.6-9 (instead of using the force method, as prescribed in the problem statement). Call this "Solution *A*." (b) Keep all of the problem parameters the same, except replace d_o and d_i by the following: $d_o = 1.75$ in., $d_i = 1.5$ in. Use AXIALDEF to solve this revised problem, and call this "Solution *B*." (c) Again, keep all parameters the same as in the original problem, except replace d_o, and d_i by the following: $d_o = 1.5$ in., $d_i = 1.25$ in. Use AXIALDEF to solve this revised problem, and call this "Solution *C*." (d) Create a table of nodal displacements and a table of element axial stresses. Discuss the effect that changing A_1 has on the nodal displacements and on the element axial stresses. Which design best minimizes the element stresses?

*Prob. C3.6-6. (a) Use AXIALDEF to solve Prob. 3.6-10. (Since AXIALDEF is based on the displacement method, you will, in effect, be using the displacement method as prescribed in the problem statement.) Call this "Solution *A*." (b) Keep all of the problem parameters the same, except replace d_o and d_i by the following: $d_o = 44$ mm, $d_i = 36$ mm. Use AXIALDEF to solve this revised problem, and call this "Solution *B*." (c) Again, keep all parameters the same as in the original problem, except replace d_o and d_i by the following: $d_o = 38$ mm, $d_i = 30$ mm. Use AXIALDEF to solve this revised problem, and call this "Solution *C*." (d) Create a table of nodal displacements and a table of element axial stresses. Discuss the effect that changing A_1 has on the nodal displacements and on the element axial stresses. Which design best minimizes the element stresses?

*Prob. C3.6-7. The AXIALDEF program can be used to solve nonnumeric problems like Probs. 3.6-7 and 3.6-8 if you remember that nodal displacements will have the form $u_i = \bar{u}_i(\alpha L \Delta T)$, where $\bar{u}_i$ is a nondimensional factor that is obtained by AXIALDEF if you input the following data: $\alpha = L = \Delta T = A = E = 1$. Likewise, element axial forces will have the form $F_i = \bar{F}_i(AE\alpha\Delta T)$, and element axial stresses will have the form $\sigma_i = \bar{\sigma}_i(E\alpha\Delta T)$, where $\bar{F}_i$ and $\bar{\sigma}_i$ are the internal forces and internal stresses output by AXIALDEF.

Use AXIALDEF to solve Prob. 3.6-7 (instead of using the force method as prescribed in the problem statement). Let the nodes in Fig. P3.6-7 be relabeled as follows: $A \rightarrow$ ①, $B \rightarrow$ ②, and so on.

All of the problems in **Section 3.7** *are statically indeterminate problems. In all cases, it is important that the* **equilibrium equations,** *the* **element force-temperature-deformation behavior,** *and the* **geometry of deformation** *be accounted for. Where a particular "method" is specified—the* **displacement method** *or the* **force method**—*it is expected that that method will be used; otherwise, either method may be used.*

Prob. 3.7-1. A two-rod system with flanges at ends *A* and *C* was supposed to exactly fit between two rigid walls, as shown in Fig. P3.7-1. Element (1) is a steel pipe with outer diameter $d_o = 2$ in. and inner diameter $d_i = 1.5$ in.; element (2) is a solid steel rod with diameter $d = 0.75$ in. Bolts are installed in the flange at *C* and are tightened until the gap of $\delta = 0.1$ in. is closed. $E_1 = E_2 = E_{\text{steel}} = 30 \times 10^3$ ksi. (a) Using the *displacement method,* determine the displacement u_B of the joint *B* when the gap at *C* has been closed. (b) Determine the stress induced in each element as a result of closing the gap δ.

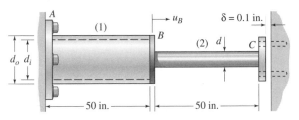

P3.7-1, P3.7-2, P3.7-18, P3.7-19, and PC3.7-1

Prob. 3.7-2. For the two-rod system of Prob. 3.7-1, use the *force method* to determine the stresses σ_1 and σ_2 induced in elements (1) and (2), respectively, when the gap at *C* is closed.

Prob. 3.7-3. Elements (1) and (2) were supposed to exactly fit between the rigid walls at *A* and *C* in Fig. P3.7-3. However, element (1) was manufactured a small amount, $\delta \ll L$, too long. However, an ingenious technician was able to compress element (1) and insert it in the space between *A* and *B* (indicated by dashed lines). (a) Use the *displacement method* to determine the amount u_B by which element (2) is shortened when element (1) is forced

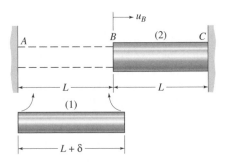

P3.7-3, P3.7-4, and PC3.7-5

into the dashed-line position. (b) Determine the stress σ_1 induced in element (1) and the stress σ_2 induced in element (2).

$$L_1 = L_2 = L, \quad A_1 = A, \quad A_2 = 1.5A, \quad E_1 = E_2 = E$$

Assume $\delta \ll L$ in calculating the stiffness coefficient k_1.

Prob. 3.7-4. When element (1) in Prob. 3.7-3 is forced to fit into the dashed-line position, axial stress is induced in each of the two elements. Use the *force method* to determine the stress induced in element (2).

Prob. 3.7-5. Two flat bars are supposed to be connected together by a pin at B, and the system is supposed to be stress free after the pin is inserted (Fig. P3.7-5a). However, the distance between the rigid walls at A and C was measured incorrectly so that, when the bars are manufactured, there is a misalignment of the holes in the two bars of $\delta = 1$ mm (exaggerated in Fig. P3.7-5b). Determine the stress σ_1 induced in bar (1) and the stress σ_2 induced in bar (2) if they are stretched to align the holes, the pin is inserted at B, and the stretching force is then removed.

$$E_1 = E_2 = 210 \text{ GPa}, \quad A_1 = 25 \text{ mm}^2, \quad A_2 = 50 \text{ mm}^2$$

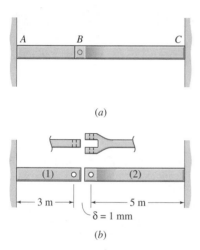

(a)

(b)

P3.7-5 and PC3.7-2

Prob. 3.7-6. Two uniform, linearly elastic members are joined together at B, and the resulting two-segment rod is attached to a rigid support at end A. When there is no load on the 2-element bar (i.e., when $P = 0$) there is a gap of 0.2 mm between the end of element (2) and the rigid wall at C. Element (1) is steel with modulus $E_1 = 210$ GPa, cross-sectional area $A_1 = 1000$ mm^2, and length $L_1 = 2.1$ m; element (2) is titanium alloy with $E_2 = 120$ GPa, $A_2 = 1000$ mm^2, and $L_2 = 1.8$ m. A single external force $P = 50$ kN is applied at node B. (a) Use the *displacement method* to calculate the horizontal displacement, u, of joint B when load P is applied. (b) Determine the axial stresses, σ_1 and σ_2, in the two elements.

Prob. 3.7-7. With the data of Prob. 3.7-6, use the *force method* to solve for the axial stresses, σ_1 and σ_2, in the two support rods

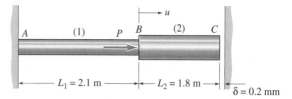

P3.7-6, P3.7-7, P3.7-16, P3.7-17, PC3.7-3, and PC3.7-4

when the load $P = 50$ kN is applied. Let F_1 be the redundant force.

*Prob. 3.7-8. A "rigid" beam AD is supported by a smooth pin at D and by vertical steel rods attached to the beam at points A and C. Neglect the weight of the beam. Element (1) was fabricated the correct length, and, when only rod (1) is attached to beam AD, the beam is horizontal (i.e., $\theta = 0$). Element (2), on the other hand, was fabricated 0.2 in. too short, and it has to be stretched manually in order to connect it to the beam by a pin at C. (a) Use the *displacement method* to solve for the rotation, θ, of the beam AD when rod (2) is connected and then the load P is applied. (b) Solve for the axial stresses, σ_1 and σ_2, in the two support rods when load P is applied.

$$A_1 = A_2 = 1.0 \text{ in}^2, \quad L_1 = 40 \text{ in.}, \quad L_2 = 60 \text{ in.}$$

$$E_1 = E_2 = 30 \times 10^3 \text{ ksi}, \quad a = 20 \text{ in.}$$

$$b = 60 \text{ in.}, \quad P = 2 \text{ kips}$$

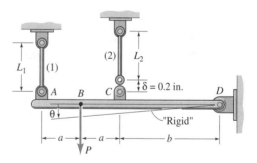

P3.7-8, P3.7-9, and P3.7-20

Prob. 3.7-9. With the data of Prob. 3.7-8, use the *force method* to solve for the axial stresses, σ_1 and σ_2, in the two support rods when the load $P = 2$ kips is applied. Let F_1 be the redundant force.

Prob. 3.7-10. A "rigid" beam AD is supported by a smooth pin B and by vertical rods that are attached to the beam at points A and C. Neglect the weight of the beam. The beam is horizontal when only rod (1) is attached and $P = 0$, but rod (2) was manufactured $\delta = 0.5$ mm too short. (a) Use the *displacement method* to solve for the rotation, θ, of the beam AD after rod (2) is attached to the bracket at its top end and, in addition, the load $P = 20$ kN

165

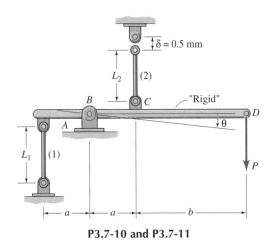

P3.7-10 and P3.7-11

is applied. (b) Determine the final axial stress in each of the support rods.

$$A_1 = A_2 = 200 \text{ mm}^2, \quad L_1 = L_2 = 2 \text{ m},$$

$$E_1 = 70 \text{ GPa}, \quad E_2 = 100 \text{ GPa}$$

$$P = 20 \text{ kN}, \quad a = 1 \text{ m}, \quad b = 1.5 \text{ m}$$

Prob. 3.7-11. With the data of Prob. 3.7-10, use the *force method* to solve for the axial stresses, σ_1 and σ_2, in the two support rods after rod (2) is attached to the upper bracket and the load $P = 20$ kN is applied. Let F_1 be the redundant force.

Prob. 3.7-12. The mechanical system of Fig. P3.7-12 consists of a brass rod ($E_1 = 15 \times 10^3$ ksi) with cross-sectional area $A_1 = 1.5$ in.^{2}, a structural steel rod ($E_2 = 30 \times 10^3$ ksi) with cross-sectional area $A_2 = 0.75$ in.^{2}, and a "rigid" beam AC. The beam is supported by a smooth pin at B. The nuts at A and C are initially tightened just enough to remove the slack, leaving the two rods stress free and $\theta = 0$. (a) Use the *displacement method* to determine the (small) angle θ through which the beam AB would rotate if the nut at C were to be advanced (i.e., tightened) by one turn (0.1 in. of thread-length). (b) Determine the axial stress induced in rod (1) by advancing the nut at C one turn.

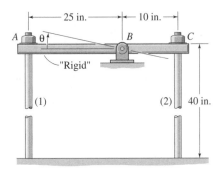

P3.7-12 and P3.7-13

Prob. 3.7-13. With the data of Prob. 3.7-12, use the *force method* to solve for the axial stress induced in rod (1) when the nut at C is advanced (i.e., tightened) one turn (0.1 in. of thread-length).

166

Prob. 3.7-14. A brass pipe sleeve ($E_1 = 15 \times 10^3$ ksi) with outer diameter $d_o = 3$ in. and inner diameter $d_i = 2.5$ in. is held in compression against a rigid machine wall by a high-strength steel bolt ($E_2 = 30 \times 10^3$ ksi) with diameter $d = 1$ in. The head of the bolt bears on a 0.25-in.-thick (rigid) washer, which, in turn, bears on the brass sleeve. The bolt is initially advanced until there is no slack (i.e., the sleeve is held in contact with the rigid machine wall, but there is no stress induced in the bolt or the sleeve). (a) Use the *displacement method* to determine the amount, u, by which the sleeve is shortened when the bolt is advanced $\frac{1}{4}$ turn (0.01 in. of thread-length). (b) Determine the resulting stress, σ_1, in the sleeve and the stress σ_2 in the bolt.

Prob. 3.7-15. With the data of Prob. 3.7-14, use the *force method* to solve for the axial stresses induced in the sleeve (σ_1) and the bolt (σ_2) when the bolt is advanced an additional $\frac{1}{4}$ turn beyond the just-tight position.

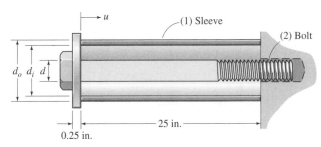

P3.7-14 and P3.7-15

Prob. 3.7-16. Solve Prob. 3.7-6 if, <u>instead of</u> having a force P applied at B, the entire 2-segment rod is heated by 20°C. Let $\alpha_1 = 12 \times 10^{-6}/°$C and $\alpha_2 = 10 \times 10^{-6}/°$C.

Prob. 3.7-17. Solve Prob. 3.7-7 if, <u>instead of</u> having a force P applied at B, the entire 2-segment rod is heated by 20°C. Let $\alpha_1 = 12 \times 10^{-6}/°$C and $\alpha_2 = 10 \times 10^{-6}/°$C.

Prob. 3.7-18. Solve Prob. 3.7-1 if, <u>in addition to</u> having the flange tightened against the wall at C, the pipe-rod system ABC is heated uniformly by 50°F. Use $\alpha = 6.5 \times 10^{-6}/°$F.

Prob. 3.7-19. Solve Prob. 3.7-2 if, <u>in addition to</u> having the flange tightened against the wall at C, the pipe-rod system ABC is heated uniformly by 50°F. Use $\alpha = 6.5 \times 10^{-6}/°$F.

*Prob. 3.7-20. Consider the statically indeterminate rod/beam system described in the problem statement of Prob. 3.7-8. After element (2) is connected to the beam at C and the downward load $P = 2$ kips is applied to the beam at B, rod (1) is <u>heated</u> by $\Delta T_1 = 50$°F. (a) Use the *displacement method* to solve for the rotation, θ, of the beam AD when rod (2) is connected, external load P is applied, and rod (1) is heated. Let $\alpha = 6.5 \times 10^{-6}/°$F for both rods. (b) Solve for the axial stresses, σ_1 and σ_2, in the two support rods under the conditions stated in Part (a).

Computer Exercises—Section 3.7 *Use the MechSOLID computer program AXIALDEF to solve the homework exercises C3.7-*

*1 through C3.7-5. Refer to Appendix G.3 for a description of this computer program and for examples of its use. Hand in computer printouts of the following **MechSOLID** "screens": AXIALDEF: Model, AXIALDEF: Boundary Conditions, AXIALDEF: Results, AXIALDEF: Displacement Plot, AXIALDEF: Force Plot. Samples of these screens are illustrated in Example G-4.*

Prob. C3.7-1. Use AXIALDEF to solve Prob. 3.7-1. (Since AXIALDEF is based on the displacement method, you will, in effect, be using the displacement method as prescribed in the problem statement.) Let node A in Fig. P3.7-1 be relabeled as node ①, node B be relabeled as node ②, and node C be relabeled as node ③. (a) Assume that both elements were fabricated the correct lengths, but that node ③ (the right-hand flange) must undergo a prescribed displacement of $u_3 = 0.1$ in. (b) Re-solve the problem, assuming this time that element (2) was fabricated 0.1 in. too short (i.e., it should have been fabricated 50.1 in. long). In this case node ③ will be the right-hand wall where the flange on element (2) is to be attached. That is, for this case $u_3 = 0$.

Prob. C3.7-2. Use AXIALDEF to solve Prob. 3.7-5. Let node A in Fig. P3.7-5 be relabeled as node ①, node B be relabeled as node ②, and node C be relabeled as node ③. Assume that element (2) was fabricated 1 mm too short.

Prob. C3.7-3. Use AXIALDEF to solve Prob. 3.7-6. (Since AXIALDEF is based on the displacement method, you will, in effect, be using the displacement method as prescribed in the problem statement.) Let node A in Fig. P3.7-6 be relabeled as node ①, node B be relabeled as node ②, and so on. Assume that both elements were fabricated the correct lengths, but that node ③, the right-hand end of element (2), must undergo a prescribed displacement of $u_3 = 0.2$ mm.

Prob. C3.7-4. Use AXIALDEF to solve Prob. 3.7-6, with one addition: in addition to the applied load P and the gap δ, let element (1) be <u>cooled</u> by 50°C, with $\alpha_1 = 12 \times 10^{-6}/°C$. (Since AXIALDEF is based on the displacement method, you will, in effect, be using the displacement method as prescribed in the problem statement.) Let node A in Fig. P3.7-6 be relabeled as node ①, node B be relabeled as node ②, and so on. Assume that both elements were fabricated the correct lengths, but that node ③, the right-hand end of element (2), must undergo a prescribed displacement of $u_3 = 0.2$ mm.

*Prob. C3.7-5. The AXIALDEF program can be used to solve nonnumeric problems like Probs. 3.7-3 if you remember that nodal displacements will have the form $u_i = \bar{u}_i \delta$, where $\bar{u}_i$ is a nondimensional factor that is obtained by AXIALDEF if you input the following data: $A = E = L = \delta = 1$. Likewise, element axial forces will have the form $F_i = \bar{F}_i \left(\dfrac{AE\delta}{L} \right)$, and element axial stresses will have the form $\sigma_i = \bar{\sigma}_i \left(\dfrac{E\delta}{L} \right)$, where $\bar{F}_i$ and σ_i are the internal forces and internal stresses output by AXIALDEF.

Use AXIALDEF to solve Prob. 3.7-3. (Since AXIALDEF is based on the displacement method, you will, in effect, be using the displacement method as prescribed in the problem statement.)

Let the nodes in Fig. P3.7-3 be relabeled as follows: $A \rightarrow$ ①, $B \rightarrow$ ②, and so on.

Problems 3.8-1 through 3.8-12 *are statically-determinate planar truss problems. The appropriate solution procedure for these problems is the* **force method**, *since the internal resisting force in any member can be determined directly from equilibrium equations. Therefore, the appropriate order of solution is:*

> **Equilibrium equations**
> **Force-deformation relationships**
> **Geometry-of-deformation equations**

Prob. 3.8-1. Bar AC is $L_1 = 8$ ft long, has a cross sectional area $A_1 = 1.0$ in², and is made of steel with a modulus of elasticity $E_1 = 30 \times 10^3$ ksi. Member BC has the following properties: $L_2 = 10$ ft, $A_2 = 1.4$ in², and $E_2 = 10 \times 10^3$ ksi. A load $P = 10$ kips acts downward on the pin at joint C. (a) Determine the axial stresses in rods (1) and (2). (b) Determine the horizontal and vertical displacements of joint C, u_C and v_C, respectively.

Prob. 3.8-2. Repeat Prob. 3.8-1 if member (1) is heated by 80°F and, at the same time, the load $P = 10$ kips is applied. Member (2) is held at the reference temperature. Use $\alpha_1 = 8 \times 10^{-6}/°F$.

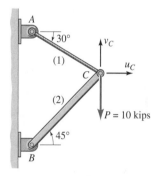

P3.8-1 and P3.8-2

Prob. 3.8-3. A two-bar planar truss has the geometry shown in Fig. P3.8-3. A force $P = 200$ kN is applied to the truss at joint C. (a) What are the resulting stresses in elements (1) and (2)? (b) What are the horizontal and vertical displacements of joint C, u_C and v_C, respectively?

$$A_1 = A_2 = 1500 \text{ mm}^2, \qquad E_1 = E_2 = 70 \text{ GPa}$$

Prob. 3.8-4. Repeat Prob. 3.8-3 if member (2) is heated by 25°C and, at the same time, the load $P = 200$ kN is applied to the truss at C. Member (1) is held at the reference temperature, that is, $\Delta T_1 = 0$. Use $\alpha_2 = 23 \times 10^{-6}/°C$.

Prob. 3.8-5. A force $P = 400$ kN is applied at joint C of the planar truss in Fig. P3.8-5. (a) Determine the required cross-sectional areas of members (1) and (2) if the allowable stresses

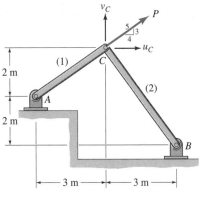

P3.8-3, P3.8-4, and P3.8-5

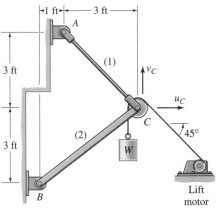

P3.8-7

are 100 MPa in tension and 75 MPa in compression. (b) For the truss with areas determined in Part (a), determine the displacements u_C and v_C of joint C. $E_1 = E_2 = 70$ GPa.

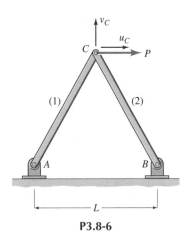

P3.8-6

Prob. 3.8-6. Each of the two bars of the planar truss in Fig. P3.8-6 has length L, and they are made of a material with modulus of elasticity E. If $A_1 = A$, and $A_2 = 2A$, and a horizontal load P is applied at joint C, determine: (a) the stresses σ_1 and σ_2 in the two bars, and (b) the horizontal displacement, u_C, and vertical displacement, v_C, of the pin joint at C.

Prob. 3.8-7. A planar truss ACB is part of an apparatus used to lift a weight W, as shown in Fig. P3.8-7. If $W = 4$ kips, determine: (a) the stresses σ_1 and σ_2 in the two truss members, and (b) the horizontal and vertical displacements of joint C, u_C and v_C, respectively.

$$A_1 = 0.8 \text{ in}^2, \quad A_2 = 1.0 \text{ in}^2$$

$$E_1 = 10 \times 10^3 \text{ ksi}, \quad E_2 = 30 \times 10^3 \text{ ksi}$$

Prob. 3.8-8. The two tie rods in Fig. P3.8-8 support a maximum horizontal load $P = 10$ kips at joint B. Rod (1) is made of structural steel with a modulus of elasticity $E_1 = 29 \times 10^3$ ksi and a yield point of $\sigma_{Y1} = 36$ ksi. The corresponding properties for the aluminum rod (2) are $E_2 = 10 \times 10^3$ ksi and $\sigma_{Y2} = 60$ ksi. (a) If the two-bar truss is to have a factor of safety of 2.5 with respect to failure by yielding, what are the required areas A_1 and A_2? (b) If the truss is sized according to the requirements stated in Part (a), and loaded by a force $P = 10$ kips, what will be the values of u_B and v_B, the horizontal and vertical displacements at joint B, at the maximum-load condition?

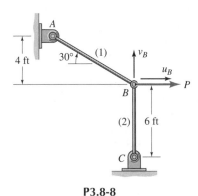

P3.8-8

Prob. 3.8-9. The rigid beam BD in Fig. P3.8-9 is supported by a steel rod that is connected to the beam by a smooth pin at C. The beam is horizontal when there is no weight W hanging at D. The properties of rod AC are: $A = 500$ mm^2 and $E = 200$ GPa. (a) If a weight $W = 10$ kN is suspended from the beam at D, what is the stress in rod AC? Neglect the weight of the beam. (b) What is the elongation of rod AC? (c) What is the vertical displacement of D? (Make the assumption that beam BD rotates through an angle that is small, so points C and D can be assumed to move only vertically.)

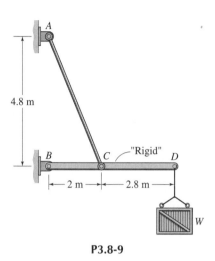

P3.8-9

*Prob. 3.8-10. The crane hoist in Fig. P3.8-10 consists of a "rigid," "weightless" boom BC supported by a tie-rod AB. The rod has a cross-sectional area $A = 0.75$ in^2 and is made of steel with modulus of elasticity $E = 29 \times 10^3$ ksi. (a) If the yield point of the steel is $\sigma_Y = 50$ ksi and there is to be a factor of safety with respect to yielding of $FS = 2.5$, what is the maximum weight W that can safely be hoisted by the cable suspended from the pin at B? (b) If a load $W = 2$ kips is suspended from the pin

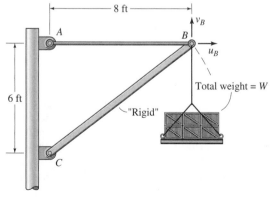

P3.8-10

at B, what will be the vertical displacement v_B? (Note: Assume that the angle through which boom BC rotates is small, so point B moves along the dashed line perpendicular to the original line BC. That is, $v_B = -\frac{4}{3}u_B$.)

*Prob. 3.8-11. Truss members (1) and (2) in Fig. P3.8-11 have the same modulus of elasticity E. The cross-sectional areas of the members are $A_1 = A$ and $A_2 = 2A$. At what angle θ must load P be applied in order to make $v_B = u_B$, that is, to make joint B move upward to the right at 45°?

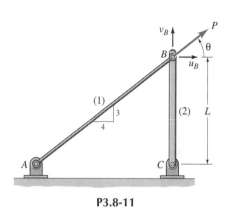

P3.8-11

*Prob. 3.8-12. The cross-sectional area of tie rod AB is $A_1 = A$, and the cross-sectional area of post BC is $A_2 = 8A$. Both are made of the same material with modulus of elasticity E. A load P is applied horizontally at B. If the length, $L_2 = L$, of the post is fixed, but the length L_1 and position of the bracket at A are both variable, determine the angle θ of the tie rod that will minimize the horizontal displacement u_B at B. (Hint: Obtain an expression for u_B as a function of θ; differentiate this expression with respect to θ; and set $du_B/d\theta$ equal to 0. You will obtain an equation involving $\sin \theta$ and $\cos \theta$ that can be solved by trial and error or by computer.)

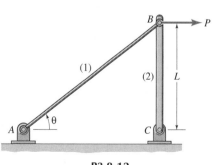

P3.8-12

169

Prob. 3.8-13. The three truss members in Fig. P3.8-13 all have cross-sectional area A and modulus of elasticity E. A horizontal load P is applied to the truss at joint A. (a) Determine expressions for the horizontal and vertical displacements of joint A, that is, u_A and v_A. (b) Determine expressions for the member axial forces F_1, F_2, and F_3.

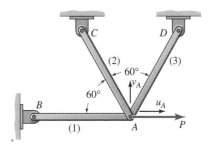

P3.8-16, P3.8-17, P3.8-18, and P11.8-54

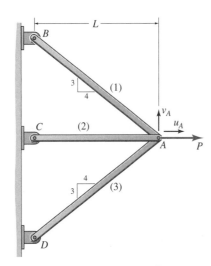

P3.8-13, P3.8-14, P3.8-15, and P11.8-53

Prob. 3.8-14. Repeat Prob. 3.8-13 with $A_1 = 2A$, $A_2 = A_3 = A$.

Prob. 3.8-15. Repeat Prob. 3.8-13 with $A_1 = 2A$, $A_2 = A_3 = A$. Let $P = 0$, but let member (2), whose coefficient of thermal expansion is α_2, be heated by an amount $\Delta T_2 = \Delta T$, with $\Delta T_1 = \Delta T_3 = 0$.

Prob. 3.8-16. Each of the three truss members in Fig. P3.8-16 has a length L and modulus of elasticity E. The cross-sectional areas of the members are $A_1 = A_2 = A$ and $A_3 = 2A$. (a) Determine expressions for the horizontal and vertical displacements, u_A and v_A, at joint A when a horizontal load P is applied to the right to the pin at joint A. (b) Determine expressions for the axial forces—F_1, F_2, and F_3—in the three truss members when load P is applied.

Prob. 3.8-17. For the truss in Fig. P3.8-17: $L_1 = L_2 = L_3 = 2.1$ m, $E_1 = E_2 = E_3 = 70$ GPa, $A_1 = A_2 = 600$ mm^2, and $A_3 = 900$ mm^2. (a) If the truss is subjected to a horizontal force $P = 60$ kN at joint A, what are the resulting horizontal displace-

ment, u_A, and vertical displacement, v_A, at that joint? (b) What are the resulting axial forces in the members—F_1, F_2, and F_3?

Prob. 3.8-18. Re-solve Problem 3.8-17, but this time let $P = 0$, $\Delta T_1 = \Delta T_3 = 0$, and $\Delta T_2 = -20°$C. Let $\alpha_2 = 23 \times 10^{-6}/°$C.

Prob. 3.8-19. For the truss in Fig. P3.8-19: $A_1 = A_2 = A_3 = 1.0$ in^2, $E_1 = E_2 = E_3 = 30 \times 10^3$ ksi, and $P = 15$ kips. (a) Determine the horizontal displacement, u_A, and vertical displacement, v_A, at joint A. (b) Determine the member axial forces—F_1, F_2, and F_3.

Prob. 3.8-20. If $P = 0$, how much must member (2) in Prob. 3.8-19 be cooled to cause joint A to move upward by 0.05 in. (i.e., $v_A = 0.05$ in.)? Let $\alpha_2 = 6.5 \times 10^{-6}/°$F?

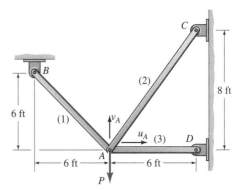

P3.8-19, P3.8-20, P3.8-21, P3.8-22(a), and P11.8-55

Prob. 3.8-21. Consider the truss in Prob. 3.8-19. Suppose that member (3) was originally fabricated 0.05 in. too short (i.e., $\Delta L_3 = -0.05$ in.) so that it had to be temporarily stretched enough to insert the pin at A. (a) If $P = 0$, what displacements u_A and v_A will result when the truss is forcefully assembled as described above? (b) What "initial stresses"—σ_1, σ_2, and σ_3—will be induced in the members when the truss is forcefully assembled?

*Prob. 3.8-22. Re-solve Prob. 3.8-19, this time assuming that there is a clearance of 0.02 in. in the hole at A in member (2) so that the pin does not act on member (2) at A until this clearance (gap) has been closed. [See Prob. 3.8-19 for Fig. P3.8-22(a).]

170

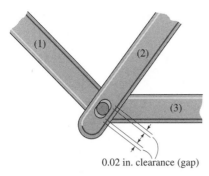

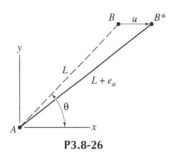

P3.8-26

P3.8-22(b)

Prob. 3.8-23. Three members form the pin-jointed truss in Fig. P3.8-23. The joint at A is constrained by a slider block to move only in the horizontal direction (i.e., $v_A = 0$). All members have the same cross-sectional area $A = 400$ mm^2, and the modulus of elasticity is $E = 70$ GPa for all three members. (a) Determine the horizontal load P that would be required to move joint A to the right by 5 mm (i.e., $u_A = 5$ mm). (b) Determine the member forces—F_1, F_2, and F_3—corresponding to the load P determined in Part (a). (c) Determine the vertical reaction between the slider block at A and the track.

stresses''—σ_1, σ_2, and σ_3—will be induced in the truss members due to the error in the original length of member (2)?

Prob. 3.8-26. Using the Pythagorean Theorem, show that if u/L is ''small'' (i.e., $u/L \ll 1$), the change in length, e_u, of member AB, of initial length L, is given by

$$e_u = u \cos \theta$$

if end B moves in the x-direction from B to B^* by an amount u, while end A does not move. (Note: This problem relates to Fig. 3.16 and Eq. 3.22.)

*Prob. 3.9-1. The two-segment bar in Fig. P3.9-1a has a constant cross section and is homogeneous, with bilinear stress-strain behavior given by the diagram in Fig. P3.9-1b. External loads of magnitude P are applied at B and C as shown. (a) If $A = 1.0$ in^2, determine the value of P that causes first yielding in the bar AC. (Call this load P_{Y1}.) (b) Determine the total elongation, u_C, of the bar AC if $P = 20$ kips and $L = 20$ in.

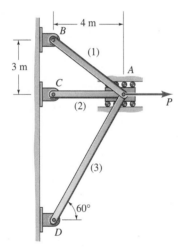

P3.8-23, P3.8-24, and P3.8-25

Prob. 3.8-24. The coefficient of thermal expansion for all members of the truss in Prob. 3.8-23 is $\alpha = 23 \times 10^{-6}/°C$. (a) If $P = 0$, but all three members of the truss are uniformly heated by $\Delta T = 20°C$, how much will joint A move? That is, what horizontal displacement, u_A, will occur as a result of the heating of all members? (b) What stresses—σ_1, σ_2, and σ_3—will be induced in the truss members due to the uniform heating?

Prob. 3.8-25. Member (2) of the truss in Prob. 3.8-23 was manufactured 2 mm too short (i.e., $\delta_2 = -2$ mm), so that members of the truss have to be forcefully assembled. (a) What will be the ''initial displacement'' of joint A, that is, what is the value of u_A prior to the application of any external load P? (b) What ''initial

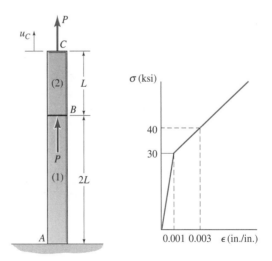

(a) A two-segment bar.   (b) A bilinear stress-strain curve.

P3.9-1 and P3.9-2

*Prob. 3.9-2. For the two-segment bar in Prob. 3.9-1, (a) determine the value, P_{Y1}, of load P that causes first yielding in the two-segment bar AC. (b) Determine the value, P_{Y2}, of load P that causes the other segment to yield also. (c) Sketch a load-elongation diagram (i.e., P vs. u_C) for loads up to $2P_{Y2}$.

Prob. 3.9-3. The two-segment statically indeterminate bar in Fig. P3.9-3a has a constant cross-sectional area $A = 0.8$ in^2. It is made entirely of material that has an elastic, perfectly-plastic stress-strain behavior illustrated in Fig. P3.9-3b, with $\sigma_Y = 36$ ksi and $E = 30 \times 10^3$ ksi. (a) Determine the load P_Y at which first yielding occurs, and determine the corresponding displacement u_Y of section B where the load P is applied. (b) Determine the load P_U at which yielding occurs in the remaining segment of the bar, and determine the corresponding displacement u_U of section B. (c) Sketch a load-displacement diagram, that is, sketch a diagram of P versus u up to P_U.

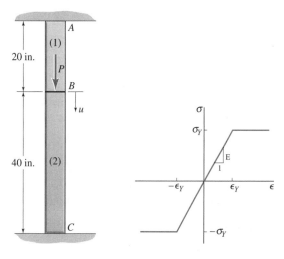

(a) A two-segment bar.   (b) A stress-strain diagram for an elastic, perfectly-plastic material.

P3.9-3 and P3.9-7

Prob. 3.9-4. Repeat Prob. 3.9-3 for the two-segment bar in Fig. 3.9-4. Let the bar have a stress-strain diagram of the form illustrated in Fig. P3.9-3b, with $\sigma_Y = 250$ MPa and $E = 200$ GPa. The cross-sectional areas of the respective segments of the bar are $A_1 = 500$ mm^2 and $A_2 = 800$ mm^2.

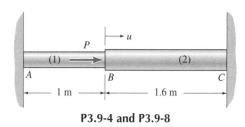

P3.9-4 and P3.9-8

*Prob. 3.9-5. A symmetric three-bar planar truss is loaded by a single horizontal force P at joint A. The members of the truss are all made of the same linearly elastic, perfectly-plastic material (see Fig. P3.9-3b) with $\sigma_Y = 36$ ksi and $E = 30 \times 10^3$ ksi, and all have a cross-sectional area $A = 2.0$ in^2. (a) Determine the load P_Y at which first yielding occurs, and determine the corresponding displacement u_Y of joint A, where the load P is applied. (b) Determine the load P_U at which yielding occurs in the remaining members of the truss, and determine the corresponding

displacement u_U of joint A. (c) Sketch a load-displacement diagram, that is, sketch a diagram of P versus u up to P_U.

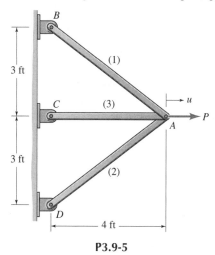

P3.9-5

*Prob. 3.9-6. When there is no load on the two-segment bar in Fig. P3.9-6, there is a 1-mm gap between the bar and the rigid

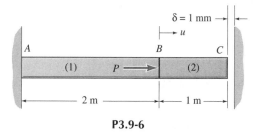

P3.9-6

wall at C. The two-segment rod is made of linearly elastic, perfectly plastic material with a stress-strain diagram like the one in Fig. P3.9-3b, with $\sigma_Y = 250$ MPa and $E = 200$ GPa. The bar has a constant cross-sectional area $A = 500$ mm^2. (a) Determine the value of P, say P_R, where the right end of the bar initially makes contact with the rigid wall at C. Determine the displacement u_R of node B that corresponds to the load P_R. (b) Determine the load P_Y at which first yielding occurs, and determine the corresponding displacement u_Y of section B where the load P is applied. (c) Determine the load P_U at which yielding occurs in the remaining segment of the bar, and determine the corresponding displacement, u_U, of section B. (d) Sketch a load-displacement diagram, that is, sketch a diagram of P versus u up to P_U.

Problems 3.9-7 and 3.9-8 *are statically indeterminate, and they involve unloading as well as loading. You are strongly urged to follow the procedure used in Example Problem 3.17 of tracking the loading and unloading stress-strain behavior of each element on a plot like the one in Fig. 1b of that example problem.*

*Prob. 3.9-7. Determine the residual stresses in segments (1) and (2) of the bar AC in Prob. 3.9-3 if a force $P = 50$ kips is applied and then completely removed.

*Prob. 3.9-8. Determine the residual stresses in segments (1) and (2) of the bar AC in Prob. 3.9-4 if a force $P = 300$ kN is applied and then completely removed.

TORSION

4

4.1 INTRODUCTION

In Chapter 3 we considered the behavior of slender members subjected to axial loading, that is, to forces applied along the longitudinal axis of the member. In this chapter we will concentrate on the behavior of slender members subjected to torsional loading, that is, loading by couples that produce twisting of the member about its axis.

Figure 4.1 shows a common example of torsional loading and indicates the shear stresses and the stress resultant associated with torsion. A torque (couple) of magnitude $2Pb$ is applied to the lug-wrench shaft AB by the application of equal and opposite forces of magnitude P at the ends of arm CD (Fig. 4.1a,b). We say that shaft AB is a *torsion member*. As indicated in Fig. 4.1c, the shaft AB is subjected to equal and opposite *torques* of magnitude T that twist one end relative to the other, and, as shown in Fig. 4.1d, the torque T acting on a cross section between A and B is the resultant of distributed shear stresses.

Another common example of a torsion member is a power transmission shaft, like the drive shaft of an automobile or a truck. Several names are applied to torsion members, depending on the application: *shaft*, *torque tube*, *torsion rod*, *torsion bar*, or simply *torsion member*. This chapter deals primarily with torsion of slender members with circular cross sections, such as solid or tubular circular cylinders. However, torsion of noncircular thin-wall tubular members and torsion of noncircular prismatic bars are treated in Sections 4.7 and 4.8, respectively.

FIGURE 4.1 An example of torsion.

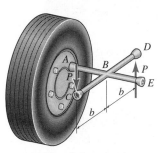

(a) Use of a lug-wrench.

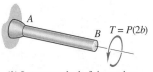

(b) Lug-wrench shaft in torsion.

(c) A torsion rod.

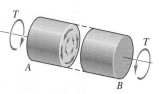

(d) The shear-stress distribution.

TABLE 4.1 Analogy Between Axial Deformation and Torsion

Axial Deformation	Torsion
Axial Force (F)	Torque (T)
Elongation (e)	Twist angle (ϕ)
Normal stress (σ)	Shear stress (τ)
Extensional strain (ϵ)	Shear strain (γ)
Modulus of elasticity (E)	Shear modulus (G)

You will find that the key concepts of *equilibrium, material behavior,* and *geometry of deformation,* which were stressed in Chapter 3, have their counterparts in the theory of torsion. There is a direct analogy between axial deformation and torsion, as indicated by the entries in Table 4.1. Thus, although there is new ''theory'' to be learned in Chapter 4, particularly in Section 4.2, you should quickly feel at home solving problems in the same systematic manner used to solve the axial-deformation problems of Chapter 3.

4.2 BASIC THEORY OF TORSION OF CIRCULAR BARS

As noted in the previous section, most of this chapter concerns torsion of members with circular cross sections. Figure 4.2 shows the deformation patterns of a circular cylinder and a uniform square bar that have each been subjected to torsion. When a circular shaft, whether solid or tubular, is subjected to torsion, each cross section remains plane and simply rotates about the axis of the member. On the other hand, as can be seen in Fig. 4.2*b*, cross sections of the square bar become warped. Because of the mathematical simplicity of the theory of torsion of members with circular cross section, and because of the widespread application of such members, we will now develop the *theory of torsion for circular members.*

Geometry of Deformation of Circular Bars.
As in our previous discussion of axial deformation, we begin with an analysis of the geometry of torsional deformation and

FIGURE 4.2 Examples of torsional deformation of steel bars.

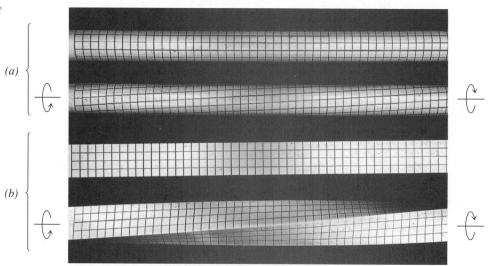

(a)

(b)

develop a *strain-displacement equation*. Figure 4.3 shows a circular cylinder before and after the application of equal and opposite torques to the ends of the member. Note that the circumferential lines, which represent plane cross sections before deformation, remain in a plane after deformation, but that longitudinal lines, which are parallel to the axis of the member before deformation, become helical as a result of torsional deformation. Furthermore, right angles, such as the before-deformation angle *ABC*, are no longer right angles after deformation. This angle change is evidence of shear deformation due to torsion. We will now develop a mathematical expression for the torsional *strain-displacement relationship*.

Strain-Displacement Analysis.

Torsion of circular members is characterized by two fundamental *torsional-deformation assumptions*:

1. The axis remains straight and remains inextensible.
2. Every cross section remains plane, remains perpendicular to the axis, and remains undistorted as it rotates about the axis.

Our task now is to establish a relationship between torsional deformation (angle of rotation) and the resulting shear strain.

Consider the torsion bar in Fig. 4.4*a*. The circular cylinder of radius r and length L is attached to a rigid base at the left end and loaded by a torque T_L at the right end. The *torque* T_L causes the torsion bar to twist through an *angle of twist* ϕ_L.

A *sign convention* for torsion is defined as follows:

- The *longitudinal axis* of the bar is labeled the x axis, with one end of the member being taken as the origin. (In Fig. 4.4*a* the left end is taken as the origin, with the x axis going from left to right.)

- A *positive internal (resisting) torque*, $T(x)$, is a couple that acts on the cross section at x in a right-hand-rule sense about the outer normal to the cross section. On a cross-sectional cut at x there will be equal and opposite torques $T(x)$, as indicated in Fig. 4.4*b*.

- A *positive angle of rotation*, $\phi(x)$, is a rotation of the cross section at x in a right-hand-rule sense about the x axis, as illustrated in Fig. 4.4*c*.

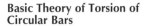
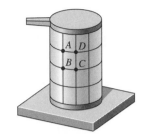

Before deformation.

(a)

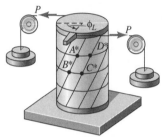

After deformation.

(b)

FIGURE 4.3 The deformation of a circular cylinder.

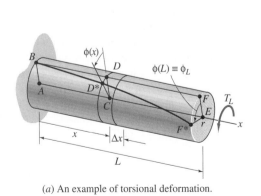

(a) An example of torsional deformation.

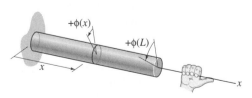

(b) Sign convention for internal (resisting) torque $T(x)$.

(c) Sign convention for angle of rotation $\phi(x)$.

FIGURE 4.4 Torsional deformation; sign convention for torsion.

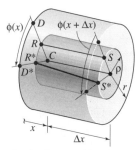

(a) An element of length Δx.

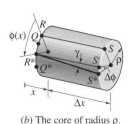

(b) The core of radius ρ.

FIGURE 4.5 Torsional-deformation details.

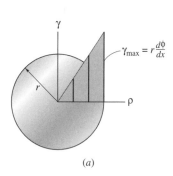

(a)

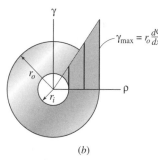

(b)

FIGURE 4.6 Examples of torsional shear-strain distribution.

The torsional-deformation assumptions are illustrated in Fig. 4.4a where, before deformation, *ABFE* is a radial plane. When torque T_L is applied to the end $x = L$, the radial lines *CD* and *EF* rotate to positions *CD** and *EF**, respectively, as the cross sections rotate through angles $\phi(x)$ and $\phi_L \equiv \phi(L)$, respectively. To determine the shear strain associated with this twisting of a circular cylinder, we use Fig. 4.5a, where we have redrawn the portion of the shaft between the cross sections at x and $(x + \Delta x)$. In Fig. 4.5b we concentrate on the central core of radius ρ and length Δx.

The angle *QRS* in Fig. 4.5b is a right angle. However, as a result of torsional deformation the angle *QRS* becomes angle *Q*R*S*, which is no longer a right angle, but is smaller by the shear strain angle

$$\gamma \equiv \gamma(x, \rho) = \frac{\pi}{2} - \angle Q^*R^*S = \angle S'R^*S^*$$

as seen in Fig. 4.5b. Since γ is small, we can approximate the angle by its tangent, at the same time taking the limit as $\Delta x \to 0$, to get

$$\gamma = \lim_{\Delta x \to 0} \frac{\overline{S^*S'}}{\overline{R^*S'}} = \lim_{\Delta x \to 0} \frac{\rho\Delta\phi}{\Delta x} = \rho\frac{d\phi}{dx}$$

Therefore, the *strain-displacement relationship* for torsional deformation of a circular member is

$$\boxed{\gamma = \rho\frac{d\phi}{dx}} \tag{4.1}$$

The derivative $d\phi/dx$ is called the *twist rate*. Figure 4.6 shows plots of this *shear-strain distribution* at a typical cross section of a solid circular cylinder (Fig. 4.6a) and of a tubular circular cylinder (Fig. 4.6b). Note that, in each case, γ varies linearly with ρ, the distance from the axis.

Stress-Strain Behavior. The twisting of a shaft with circular cross sections produces shear strains throughout the shaft that are related to the twist rate by Eq. 4.1. This torsional deformation, which is illustrated again in Fig. 4.7a, results in shear stresses in the shaft. The shear stress on a typical cross section is illustrated in Fig. 4.7b, and Fig. 4.7c shows the shear stress, τ, and the corresponding shear strain, γ, at typical points. As indicated by Fig. 4.7d, the shear stress has the same distribution, $\tau(x, \rho)$, along every radial line in the cross section at x. Also note that there is shear stress not only on cross sections, but there is always an accompanying shear stress that acts on radial planes (from Eq. 2.31).

At this point in our development of a theory of torsion of circular bars we need a relationship between the shear strain, γ, and the shear stress, τ. Chapter 2 describes how a tension test may be used to determine the uniaxial stress-strain behavior of materials, that is, σ versus ϵ. We will defer discussion of torsion testing until later in this chapter (Section 4.5) and will here confine our attention to the simplest case—linearly elastic material behavior.[1] For this case, Eq. 2.22 gives Hooke's Law for shear as

$$\boxed{\tau = G\gamma} \tag{4.2}$$

where G is called the shear modulus of elasticity, or simply the *shear modulus*.

---

[1] In Section 4.9 we will consider nonlinear stress-strain behavior.

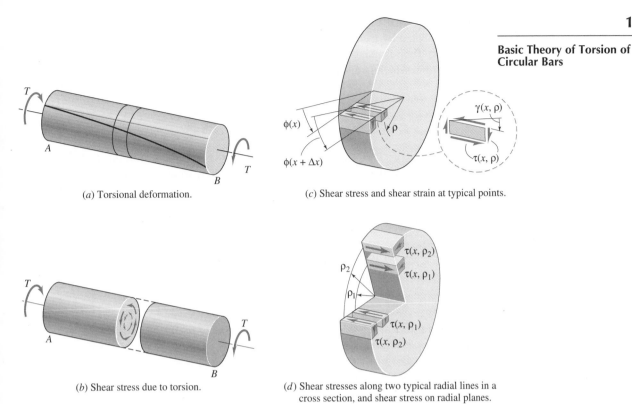

(a) Torsional deformation.

(c) Shear stress and shear strain at typical points.

(b) Shear stress due to torsion.

(d) Shear stresses along two typical radial lines in a cross section, and shear stress on radial planes.

FIGURE 4.7 Relationship of shear strain to shear stress in torsion of a circular shaft.

By combining Eqs. 4.1 and 4.2 we obtain

$$\tau = G\rho\frac{d\phi}{dx} \tag{4.3}$$

This equation gives the *shear-stress distribution* at a typical cross section. If the torsion bar is homogeneous (i.e., $G =$ const), then the shear stress varies linearly with the distance ρ from the center of the shaft, with the maximum shear stress acting at the outer edge of the cross section, as illustrated in Fig. 4.8. As indicated in Fig. 4.9a, this same shear stress distribution acts along each radial line in the cross section.

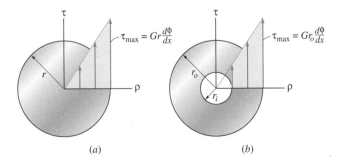

(a)

(b)

FIGURE 4.8 Torsional shear-stress distribution along a typical radial line (homogeneous, linearly elastic case).

Stress Resultant and Equilibrium. Figure 4.9 shows the increment of shear force, dF_s, contributed by a shear stress τ acting on an incremental area dA at distance ρ from the center. The resultant of the incremental torques $dT = \rho dF_s$ is the torque $T(x)$ at the cross section, which is given by

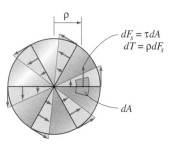

$$T = \int_A \rho \, dF_s = \int_A \rho \, \tau dA \tag{4.4}$$

Equation 4.4 is general, since the distribution of τ is not specified in this equation.

For the particular case of linearly elastic behavior, where Eq. 4.3 holds, we get

(a) The shear stress distribution.

$$T = \int_A \rho \left(G\rho \frac{d\phi}{dx} \right) dA$$

Furthermore, if G is independent of ρ we get

$$T = G\frac{d\phi}{dx} \int_A \rho^2 \, dA \tag{4.5}$$

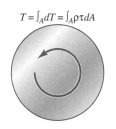

$T = \int_A dT = \int_A \rho\tau dA$

The integral in Eq. 4.5 is called the *polar moment of inertia,* for which we will use the symbol I_p.[2]

$$I_p = \int_A \rho^2 \, dA \tag{4.6}$$

(b) The resultant torque.

FIGURE 4.9 The relationship of torque to shear stress.

It is shown in Appendix C.2 that, for a solid cross section of radius r

$$I_p = \frac{\pi r^4}{2} \tag{4.7}$$

and for a tubular shaft of outer radius r_o and inner radius r_i

$$I_p = \frac{\pi}{2}(r_o^4 - r_i^4) \tag{4.8}$$

Then, Eq. 4.5 becomes

$$T = GI_p \frac{d\phi}{dx} \tag{4.9}$$

This is called the *torque-twist equation.* Note its similarity to the axial-deformation analogue, Eq. 3.9. As indicated in footnote 2, this equation is frequently written as

$$T = GJ \frac{d\phi}{dx} \tag{4.10}$$

---

[2]Many texts use the symbol J. This sometimes leads to confusion, because J is also used in the discussion of torsion of noncircular members (e.g., Sections 4.7 and 4.8), where it does <u>not</u> refer to the polar moment of inertia.

and the combined symbol *GJ* is referred to as the *torsional rigidity*. While Eq. 4.9 applies only to circular members, Eq. 4.10 is also used in more general cases to define the torsional rigidity of nonhomogeneous and noncircular members. (See, for example, Sections 4.7 and 4.8.)

When applying Eq. 4.9 (or Eq. 4.10), it is important to remember that **a sign convention is associated with both *T* and ϕ—a positive torque *T* produces a positive twist rate $d\phi/dx$.** We have chosen to let the right-hand-rule establish the positive sense of each (Figs. 4.4*b* and 4.4*c*).

By combining Eqs. 4.3 and 4.9 we get the *torsion formula* (or elastic torsion formula)

$$\tau = \frac{T\rho}{I_p} \tag{4.11}$$

Summary of Torsion Theory. Let us summarize the theory of torsion as applied to the special case of a linearly elastic member with circular cross section and with a modulus *G* that is independent of radial position ρ, that is, where $G = G(x)$ or $G =$ constant.[3] While the derivation of the torsion theory summarized in Eqs. 4.9 and 4.11 was based on deformation of a circular cylinder, it can also be applied to approximate the behavior of a torsion member whose radius varies slowly with *x*. Figure 4.10 shows such a member with distributed and concentrated external torques applied to the member. Thus, we can write the *torsion formula*, Eq. 4.11, in the form

$$\tau \equiv \tau(x, \rho) = \frac{T(x)\rho}{I_p(x)} \tag{4.12}$$

We are usually most interested in the maximum shear stress at a cross section, in which case the torsion formula becomes[4]

$$\tau_{\text{max}} = \frac{T_{\text{max}}r}{I_p} \tag{4.13}$$

where *r* is the radius of the shaft if it is a solid shaft, or the outer radius if the shaft is tubular.

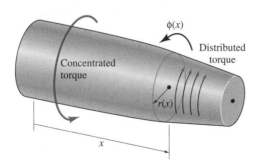

FIGURE 4.10 A torsion member with varying radius and with torsional loading along its length.

---

[3]Actually, there are very few applications where *G* varies continuously with *x*. However, in Section 4.4 there are examples of torsion bars that have different segments made of different linearly elastic materials.

[4]In this equation, T_{max} stands for the maximum absolute value of *T*, since τ_{max} means the <u>magnitude</u> of the maximum shear stress.

The *torque-twist equation,* Eq. 4.9, can be written in the general form

$$\frac{d\phi}{dx} = \frac{T(x)}{G(x)I_p(x)} \tag{4.14}$$

Finally, Eq. 4.14 may be integrated to give the *angle of rotation, $\phi(x)$*, at cross section x

$$\phi(x) = \phi(0) + \int_0^x \frac{T(\xi)\,d\xi}{G(\xi)\,I_p(\xi)} \tag{4.15}$$

or the *angle of twist, ϕ.*[5]

$$\phi \equiv \phi(L) - \phi(0) = \int_0^L \frac{T(x)\,dx}{G(x)\,I_p(x)} \tag{4.16}$$

Note the similarity of Eqs. 4.15 and 4.16 to the axial deformation analogues, Eqs. 3.10 and 3.11, respectively.

---

[5]Rather than pick another symbol (e.g., $\Delta\phi$) for the twist angle (i.e., the relative angle of rotation between the two ends), we will follow convention and use the same symbol ϕ, but without parentheses.

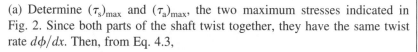

EXAMPLE 4.1

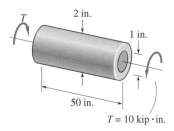

Fig. 1

A bimetallic torsion bar consists of an aluminum shell ($G_a = 4 \times 10^3$ ksi) bonded to the outside of a steel core ($G_s = 11 \times 10^3$ ksi). The shaft has the dimensions shown in Fig. 1 and is loaded by end torques of magnitude $T = 10$ kip · in. (a) Determine the maximum shear stress in the steel core and the maximum shear stress in the aluminum shell. (b) Determine the total twist angle of the composite torsion bar.

Plan the Solution Since the shear modulus varies with radial position, we must use Eq. 4.3 to express the shear stress distribution in the composite shaft; then we can determine the corresponding torque from Eq. 4.4.

Solution
(a) Determine $(\tau_s)_{max}$ and $(\tau_a)_{max}$, the two maximum stresses indicated in Fig. 2. Since both parts of the shaft twist together, they have the same twist rate $d\phi/dx$. Then, from Eq. 4.3,

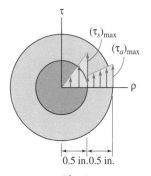

Fig. 2

$$\tau_s = G_s \rho \frac{d\phi}{dx}, \quad 0 \le \rho \le 0.5 \text{ in.} \tag{1a}$$

$$\tau_a = G_a \rho \frac{d\phi}{dx}, \quad 0.5 \text{ in.} < \rho \le 1.0 \text{ in.} \tag{1b}$$

These shear stress distributions are plotted on the sketch in Fig. 2. Substituting these expressions for τ into Eq. 4.4, we get

$$T = \int_A \rho \tau dA = G_s \frac{d\phi}{dx} \int_{As} \rho^2 dA + G_a \frac{d\phi}{dx} \int_{Aa} \rho^2 dA$$

$$T = \frac{d\phi}{dx} [G_s I_{ps} + G_a I_{pa}]$$

Then,

$$\frac{d\phi}{dx} = \frac{T}{[G_s I_{ps} + G_a I_{pa}]} \tag{3}$$

So,

$$\frac{d\phi}{dx} = \frac{10 \text{ kip} \cdot \text{in.}}{[(11 \times 10^3 \text{ ksi})(\frac{\pi}{2})(0.5 \text{ in.})^4 + (4 \times 10^3 \text{ ksi})(\frac{\pi}{2})[(1.0 \text{ in})^4 - (0.5 \text{ in})^4]]} \tag{4}$$

$$= 1.4346(10^{-3}) \text{ rad/in.}$$

Evaluating Eq. (1a) at $\rho = 0.5$ in., we get

$$(\tau_s)_{max} = (11 \times 10^3 \text{ ksi})(0.5 \text{ in.})(1.4346(10^{-3}) \text{ rad/in.})$$

$$(\tau_s)_{max} = 7.89 \text{ ksi} \qquad\qquad \text{Ans. (a)} \quad (5a)$$

and evaluating Eq. (1b) at $\rho = 1.0$ in., we get

$$(\tau_a)_{max} = (4 \times 10^3 \text{ ksi})(1.0 \text{ in.})(1.4346(10^{-3}) \text{ rad/in.})$$

$$(\tau_a)_{max} = 5.74 \text{ ksi} \qquad\qquad \text{Ans. (a)} \quad (5b)$$

As is indicated in Fig. 2, the maximum shear stress in the steel core is larger than the maximum shear stress in the aluminum shell for the given dimensions of this composite torsion bar.

(b) Determine the total angle of twist of the torsion bar. We cannot use the torque-twist equation, Eq. 4.14, since it is for a homogeneous shaft. For this composite shaft we obtained a different equation, Eq. (3), for the twist rate. Since the twist rate $d\phi/dx$ is constant along the shaft, we get

$$\phi = \int_0^L \frac{d\phi}{dx} dx = \frac{d\phi}{dx} L = (1.4346 \times 10^{-3} \text{ rad/in.})(50 \text{ in.})$$

$$\phi = 7.17 (10^{-2}) \text{ rad} \qquad\qquad \text{Ans. (b)} \quad (6)$$

Review the Solution   If the torsion bar was homogeneous (either aluminum or steel), its maximum shear stress would be at the outer radius, $\rho = 1.0$ in., and would be given by the elastic torsion formula, Eq. 4.13.

$$\tau_{max} = \frac{Tr}{I_p} = \frac{(10 \text{ kip} \cdot \text{in.})(1.0 \text{ in.})}{\frac{\pi}{2}(1.0 \text{ in.})^4} = 6.37 \text{ ksi}$$

Because the steel core is stiffer than the aluminum shell, it is reasonable that $(\tau_a)_{max}$ is somewhat less than this value.

EXAMPLE 4.2

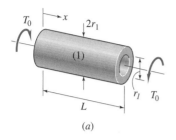

(a)

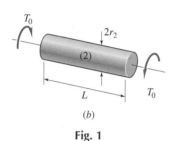

(b)

Fig. 1

A tubular shaft of length L has an outer radius r_1 and an inner radius $r_1/2$, and it is made of material with shear modulus G. It is subjected to end torques of magnitude T_0, as shown in Fig. 1a. (a) Determine an expression for τ_{max} for the shaft in Fig. 1a. (b) Determine an expression for the twist angle, ϕ, for this shaft. (c) If the same torque T_0 is applied to the solid circular shaft in Fig. 1b producing the same maximum shear stress τ_{max} as in the tubular shaft, what is the radius, r_2, of the solid shaft? What is the ratio of the weight W_2 of the solid shaft of the weight W_1 of the tubular shaft?

Plan the Solution   We can use the torsion formula, Eq. 4.13, to determine an expression for τ_{max}, and the torque-twist equation, Eq. 4.16, to determine the total twist angle, ϕ. Since there are only end torques, from equilibrium we get $T_{max1} = T_{max2} = T_0$.

Solution
(a) Determine an expression for τ_{max} for shaft (1). From the torsion formula, Eq. 4.13,

$$\tau_{max} = \frac{T_{max}r}{I_p} \qquad (1)$$

From Eq. 4.8, the polar moment of inertia of the tubular shaft is

$$I_{p1} = \frac{\pi}{2}\left[r_1^4 - \left(\frac{r_1}{2}\right)^4\right] = \frac{15}{32}\pi r_1^4$$

Therefore,

$$\tau_{max1} = \frac{T_0 r_1}{\frac{15}{32}(\pi r_1^4)} = \frac{32T_0}{15\pi r_1^3} \qquad \text{Ans. (a)} \quad (2)$$

(b) Determine an expression for the twist angle ϕ for shaft (1). From the torque-twist equation, Eq. 4.16,

$$\phi_1 = \int_0^L \frac{T(x)dx}{G(x)I_{p_1}(x)} = \frac{T_0 L}{GI_{p1}} = \frac{32T_0 L}{15\pi G r_1^4} \qquad \text{Ans. (b)} \quad (3)$$

(c) Determine the radius r_2 such that $\tau_{max2} = \tau_{max1}$ when $T_2 = T_1 = T_0$. From the torsion formula, Eq. 4.13,

$$\tau_{max2} = \frac{T_{max2}r_2}{I_{p2}} = \frac{T_0 r_2}{\frac{\pi}{2}(r_2)^4} = \frac{2T_0}{\pi r_2^3} \qquad (4)$$

Equating τ_{max1} from Eq. (2) with τ_{max2} from Eq. (4) we get

$$\frac{32T_0}{15\pi r_1^3} = \frac{2T_0}{\pi r_2^3}$$

so

$$r_2 = \left(\frac{15}{16}\right)^{1/3} r_1 = 0.979 r_1 \qquad \text{Ans. (c)} \quad (5)$$

Therefore, the ratio of the weights of the two shafts is

$$\frac{W_2}{W_1} = \frac{A_2 L_2}{A_1 L_1} = \frac{\pi r_2^2 L}{\pi [r_1^2 - (r_1/2)^2] L} = \frac{4}{3}\left(\frac{r_2}{r_1}\right)^2 = 1.277 \qquad \text{Ans. (c)} \quad (6)$$

Review the Solution The answer to Part (c) seems questionable, since it says that the radius of the solid shaft must be almost equal to the radius of the tubular shaft even though both have the same "strength." However, a recheck of Eqs. (2) and (4) shows that both are correct. From answers (5) and (6) we can conclude that the tubular shaft has a definite strength-to-weight advantage over the solid shaft, since the solid shaft weighs 28% more than the tubular shaft, even though both have the same strength.

The strength-to-weight advantage of tubular shafts is the reason that they are frequently used as torsion members in applications where weight is an important consideration.

■■■■■■■■■■■■■■□ **EXAMPLE 4.3** □■■■■■■■■■■■■■■■

A uniform shaft of radius r and length L is subjected to a uniform distributed external torque t_0 (moment per unit length). (See Fig. 1.) (a) Determine an expression for the maximum shear stress τ_{max}. (b) Determine an expression for the total twist angle $\phi \equiv \phi_L$.

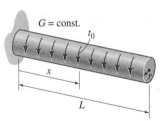

Fig. 1

Plan the Solution We need to determine $T(x)$ from equilibrium so that we can apply Eqs. 4.13 to determine the maximum shear stress and Eq. 4.16 to determine the twist angle. The maximum internal torque occurs at the wall at the left end, so the maximum shear stress occurs there also.

Solution

(a) Determine τ_{max}.

Equilibrium: We need to draw a free-body diagram and write the appropriate moment equilibrium equation. An appropriate free-body diagram is shown in Fig. 2. On the section at x we show the internal torque $T(x)$ in the positive sense according to the right-hand rule.

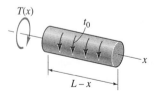

Fig. 2 Free-body diagram.

$$\sum M_x = 0: \qquad T(x) = t_0(L - x) \qquad (1)$$

Shear Stress: The maximum shear stress occurs at $x = 0$, where $T(0) = T_{max} = t_0 L$. Then, from the torsion formula, Eq. 4.13,

$$\tau_{max} = \frac{T_{max} r}{I_p} = \frac{t_0 L r}{I_p} = \frac{t_0 L r}{\frac{\pi}{2}(r^4)}$$

or

$$\tau_{\max} = \frac{2t_0 L}{\pi r^3} \qquad \text{Ans. (a)} \quad (2)$$

(b) Solve for the twist angle $\phi \equiv \phi_L$.

Torque-Twist: From the torque-twist relationship, Eq. 4.16,

$$\phi_L = \phi(0) + \int_0^L \frac{T(x)dx}{GI_p} = \int_0^L \frac{t_0(L-x)\,dx}{GI_p}$$

$$= \frac{t_0}{GI_p}\left[\left(Lx - \frac{x^2}{2}\right)\right]_0^L$$

or

$$\phi_L = \frac{t_0 L^2}{2GI_p} = \frac{t_0 L^2}{\pi r^4 G} \qquad \text{Ans. (b)} \quad (3)$$

Review the Solution   In this problem we can check to see that the answers are dimensionally correct, that is, F/L^2 in Eq. (2) and dimensionless in Eq. (3). We can also observe that t_0, L, r, and G have the proper effect on the answers (e.g., a longer bar will have a larger twist angle).

4.3   LINEARLY ELASTIC BEHAVIOR OF A UNIFORM TORSION MEMBER

In Section 3.4, Eqs. 3.13 and 3.14 were derived to describe the elastic-deformation be-havior of a uniform, linearly elastic, axially end-loaded element. This led to the definition of stiffness and flexibility coefficients for relating the axial force to the elongation of a uniform member. In the present section we examine an analogous theory for the linearly elastic behavior of a uniform torsion bar, or *torsion element,* and, in Section 4.4 we apply the displacement method and force method to solve problems for assemblages, like the ones shown in Fig. 4.11.

Figure 4.12 shows a typical uniform torsion member and defines the sign convention. Let the element have constant G and constant I_p along its length, and let the member be

FIGURE 4.11 Examples of assemblages of torsion bars.

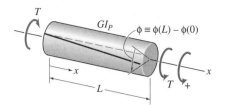

FIGURE 4.12 A typical end-loaded uniform torsion element.

subjected to equilibrating end torques, T, as shown. Finally, let

$$\phi \equiv \phi(L) - \phi(0) \qquad (4.17)$$

be the *twist angle* (i.e., the rotation of the end $x = L$ relative to the end $x = 0$), with positive sense as shown in Fig. 4.12.

From Eq. 4.16,

$$\phi = \int_0^L \frac{T\,dx}{GI_p} = \frac{TL}{GI_p}$$

As in Section 3.3, we can express this equation two ways. For a torsion member, or *torsion element,* designated as element i,

$$\phi_i = f_{ti}T_i, \text{ where } f_{ti} = \left(\frac{L}{GI_p}\right)_i \qquad (4.18)$$

or

$$T_i = k_{ti}\phi_i, \text{ where } k_{ti} = \left(\frac{GI_p}{L}\right)_i \qquad (4.19)$$

where the subscript t stands for torsion and subscript i identifies the particular torsion element. We call f_t the *torsional flexibility coefficient* and k_t the *torsional stiffness coefficient.* Note the similarity between Eqs. 3.13 and 4.18 and between Eqs. 3.14 and 4.19.

Once the torque T_i in an individual member has been determined, the maximum shear stress for that member can be computed using Eq. 4.13.

4.4 ASSEMBLAGES OF UNIFORM TORSION MEMBERS ■■■■■■■■■■■■

In Fig. 4.11 two configurations of torsion-bar assemblages are illustrated. To solve torsion problems we can employ the same problem-solving strategies used for axial deformation in Section 3.5. To establish a systematic problem-solving procedure we first define a *notation convention* and a *sign convention.* Then we set up the three fundamental types of equations: *equilibrium, element force-deformation* behavior (here *torque-twist* behavior), and *geometry of deformation.* Finally, we combine these equations employing either the *displacement-method* sequence or the *force-method* sequence.

- *Notation Conventions:*[6]
    - (a) Member-identification subscripts will always be either numbers or the generic subscript i, as in Eqs. 4.18 and 4.19. For example, T_i is the torque at every

---

[6]These ''notation conventions'' are necessitated by the fact that the symbol T is conventionally used for both externally applied load and reaction torques and also for internal resisting torques. Similarly for the symbol ϕ. In Appendix G.4 the symbol $\overline{T}$ is used for externally applied torque, and a nodal rotation angle is denoted by the symbol $\overline{\phi}$

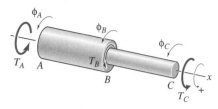

FIGURE 4.13 Sign convention for external torques and rotation angles.

cross section of element i, and specifically at each end of element i; ϕ_i is the relative twist angle between the two ends of element i.

(b) Externally applied torques and nodal (joint) rotation angles will always be denoted by uppercase letter subscripts or will be given numerical values.

- *Sign Conventions*:

(a) Figure 4.12 establishes the right-hand-rule sign convention for the torque T_i acting on an element, and the sign convention for the corresponding twist angle ϕ_i.

(b) When several elements form an assemblage, a sign convention is needed for external torques and nodal rotation angles. This sign convention is illustrated in Fig. 4.13. First, a direction for $+x$ is selected; then positive external torques and positive nodal rotations follow the right-hand rule with respect to the $+x$ axis.

■■■■■■■■■■■■■■□ EXAMPLE 4.4 □■■■■■■■■■■■■■■

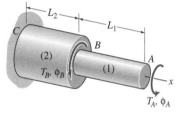

Fig. 1

A statically determinate two-element torsion bar is shown in Fig. 1. T_A and T_B are external torques applied at nodes A and B, respectively, and ϕ_A and ϕ_B are the rotation angles at these two nodes. An x axis has been established, and these T's and ϕ's have been taken to be positive in the right-hand-rule sense about this x axis. (a) Determine the internal (resisting) torques in members (1) and (2). (b) Determine the rotation angles at A and B.

Plan the Solution To determine the internal torques T_1 and T_2 in elements (1) and (2), respectively, we can draw free-body diagrams and write the corresponding equilibrium equations. Then we can use the element torque-twist equation, Eq. 4.18, and deformation compatibility to determine the two rotation angles.

Solution
(a) Determine the internal torques in members (1) and (2).

Equilibrium: Free-body diagrams that expose the unknown element torques T_1 and T_2 are shown in Fig. 2. We can write a moment-equilibrium equation for each of these. (Note that the sense of T_1 and of T_2 is established by the

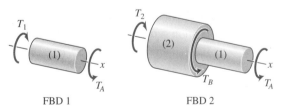

FBD 1   FBD 2

Fig. 2 Free-body diagrams.

right-hand-rule sign convention in Fig. 4.12.)

For FBD 1: $\qquad \sum M_x = 0: \qquad T_A - T_1 = 0$

For FBD 2: $\qquad \sum M_x = 0: \qquad T_A + T_B - T_2 = 0$

Hence,

$$\left. \begin{array}{l} T_1 = T_A \\ T_2 = T_A + T_B \end{array} \right. \qquad \text{Equilibrium} \quad \text{Ans. (a)} \quad (1)$$

(b) Determine the rotation angles at A and B.

Element Torque-Twist: Since we now know T_1 and T_2, Eq. 4.18 is the appropriate torque-twist equation to use.

$$\phi_i = f_{ti}T_i, \qquad i = 1, 2 \qquad \begin{array}{l} \textbf{Torque-Twist} \\ \textbf{Behavior} \end{array} \quad (2)$$

where

$$f_{ti} = \left(\frac{L}{GI_p} \right)_i \qquad (3)$$

Geometry of Deformation: We can relate the nodal rotation angles ϕ_A and ϕ_B to the element twist angles ϕ_i as follows:

$$\phi_1 = \phi_A - \phi_B$$
$$\phi_2 = \phi_B - \phi_C = \phi_B$$

or

$$\left. \begin{array}{l} \phi_A = \phi_1 + \phi_2 \\ \phi_B = \phi_2 \end{array} \right. \qquad \begin{array}{l} \textbf{Geometry} \\ \textbf{of} \\ \textbf{Deformation} \end{array} \quad (4)$$

Force-Method Solution: To determine ϕ_A and ϕ_B we can substitute $(1) \rightarrow (2) \rightarrow (4)$. Finally,

$$\phi_A = f_{t1}T_A + f_{t2}(T_A + T_B)$$
$$\phi_B = f_{t2}(T_A + T_B)$$

or

$$\left. \begin{array}{l} \phi_A = \dfrac{T_A L_1}{G_1 I_{p1}} + \dfrac{(T_A + T_B)L_2}{G_2 I_{p2}} \\[4mm] \phi_B = \dfrac{(T_A + T_B)L_2}{G_2 I_{p2}} \end{array} \right\} \qquad \text{Ans. (b)} \quad (5)$$

We will now solve several statically indeterminate problems. The displacement method will be used in solving Example 4.5.

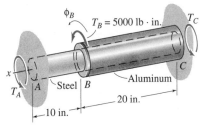

Fig. 1

A high-strength steel shaft ($G = 11.5 \times 10^6$ psi) of radius $r = 1.0$ in. and length $L = 30$ in. is sheathed over 20 in. of its length by an aluminum-alloy tube ($G = 3.9 \times 10^6$ psi) of outer radius $r_0 = 1.5$ in., as shown in Fig. 1. Ends (nodes) A and C are fixed. An external torque $T_B = 5000$ lb·in. is applied at node B as shown. Use the displacement (stiffness) method: (a) to determine the rotation angle ϕ_B; (b) to determine the reaction torques T_A and T_C; and (c) to determine the maximum shear stress in the steel and the maximum shear stress in the aluminum.

Plan the Solution   This is a statically indeterminate problem. There is only one unknown nodal displacement, ϕ_B. Since the section BC of the shaft is composed of an inner steel core and an outer aluminum shell, we can call these portions separate elements, namely elements (2) and (3). Figure 2 shows the nodes A, B, and C and elements (1), (2), and (3). (Note that both external and internal torques act on the nodes.) For a displacement-method solution we will solve an equilibrium equation that is eventually written in terms of the unknown nodal rotation ϕ_B and the three element stiffness coefficients, k_{ti}.

Solution
(a) Determine ϕ_B. Use the displacement method.

Equilibrium: The torsion-bar assemblage is separated into nodes and elements in Fig. 2.

The nodal equilibrium equations are:

For Node A: $\sum M_x = 0$: $T_A - T_1 = 0$

For Node B: $\sum M_x = 0$: $T_B + T_1 - T_2 - T_3 = 0$

For Node C: $\sum M_x = 0$: $T_C + T_2 + T_3 = 0$

$$T_1 = T_A$$
$$-T_1 + T_2 + T_3 = T_B \qquad \textbf{Equilibrium} \quad (1)$$
$$-T_2 - T_3 = T_C$$

(Equations (1a) and (1c) are needed only because the reaction torques T_A and T_C are to be determined. Equation (1b) is the "active" equilibrium equation, that is, the one for a node that has nonzero rotation.)

Fig. 2 Free-body diagrams.

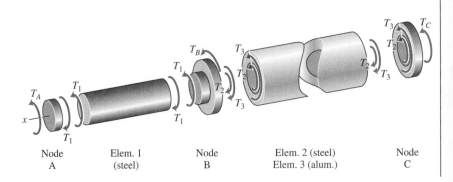

| Node A | Elem. 1 (steel) | Node B | Elem. 2 (steel) Elem. 3 (alum.) | Node C |

Element Torque-Twist Behavior: For a displacement-method solution, Eq. 4.19 is the appropriate form of the torque-twist equation to use, that is,

$$T_i = k_{ti}\phi_i, \qquad e = 1, 2, 3$$

where

$$k_{ti} = \left(\frac{GI_p}{L}\right)_i$$

$$I_{p1} = I_{p2} = \frac{\pi(1.0 \text{ in})^4}{2} = 1.571 \text{ in}^4$$

$$I_{p3} = \frac{\pi(1.5 \text{ in})^4}{2} - \frac{\pi(1.0 \text{ in})^4}{2} = 6.381 \text{ in}^4$$

$$k_{t1} = \left(\frac{GI_p}{L}\right)_1 = \frac{(11.5 \times 10^6 \text{ psi})(1.571 \text{ in}^4)}{(10 \text{ in.})} = 1.806(10^6) \frac{\text{lb} \cdot \text{in.}}{\text{rad}}$$

$$k_{t2} = \left(\frac{GI_p}{L}\right)_2 = \frac{(11.5 \times 10^6 \text{ psi})(1.571 \text{ in}^4)}{(20 \text{ in.})} = 9.032(10^5) \frac{\text{lb} \cdot \text{in.}}{\text{rad}}$$

$$k_{t3} = \left(\frac{GI_p}{L}\right)_3 = \frac{(3.9 \times 10^6 \text{ psi})(6.381 \text{ in}^4)}{(20 \text{ in.})} = 1.244(10^6) \frac{\text{lb} \cdot \text{in.}}{\text{rad}}$$

Thus, the element torque-twist equations are

$$\begin{aligned} T_1 &= 1.806(10^6)\phi_1 \\ T_2 &= 9.032(10^5)\phi_2 \\ T_3 &= 1.244(10^6)\phi_3 \end{aligned}$$

Torque-Twist Behavior (2)

Geometry of Deformation: The element relative-twist angles are

$$\phi_1 = \phi_A - \phi_B = -\phi_B$$
$$\phi_2 = \phi_B - \phi_C = \phi_B$$
$$\phi_3 = \phi_B - \phi_C = \phi_B$$

or

$$\begin{aligned} \phi_1 &= -\phi_B \\ \phi_2 &= \phi_3 = \phi_B \end{aligned}$$

Geometry of Deformation (3)

Displacement-Method Solution: The displacement method for determining ϕ_B consists of substituting Eqs. (3) into (2) into (1b).

$$T_1 = 1.806(10^6)(-\phi_B)$$
$$T_2 = 9.032(10^5)(\phi_B) \qquad (2')$$
$$T_3 = 1.244(10^6)(\phi_B)$$

$$[1.806(10^6) + 9.032(10^5) + 1.244(10^6)]\phi_B = 5000 \text{ lb} \cdot \text{in.} \qquad (1')$$

Therefore, the unknown nodal displacement ϕ_B is

$$\phi_B = 1.265(10^{-3}) \text{ rad} \qquad \textbf{Ans. (a)} \quad (4)$$

(b) Determine the reactions T_A and T_C. Combining Eqs. (2') and (4), we get

$$T_1 = 1.806(10^6)(-1.265)(10^{-3}) = -2284 \text{ lb·in.}$$
$$T_2 = 9.032(10^5)(1.265)(10^{-3}) = 1142 \text{ lb·in.}$$
$$T_3 = 1.244(10^6)(1.265)(10^{-3}) = 1574 \text{ lb·in.}$$

Combining these answers with Eqs. (1a) and (1c), and rounding to three significant figures, we get

$$\left.\begin{array}{l} T_A = T_1 = -2280 \text{ lb·in.} \\ T_C = -T_2 - T_3 = -2720 \text{ lb·in.} \end{array}\right\} \qquad \textbf{Ans. (b)} \quad (5)$$

(c) Determine the maximum shear stress in each part. Since elements (1) and (2) are both steel, and both have the same radius, and since $|T_2| < |T_1|$, the maximum shear stress in the steel will occur in the outer fibers of section AB.

$$(\tau_{\max})_{\text{steel}} = \frac{|T_1| r_1}{I_{p1}} = \frac{2284(1.0)}{1.571} = 1454 \text{ psi}$$

$$(\tau_{\max})_{\text{alum.}} = \frac{|T_3| r_3}{I_{p3}} = \frac{1574(1.5)}{6.381} = 369.9 \text{ psi}$$

Rounding to three places, we get

$$\left.\begin{array}{l} (\tau_{\max})_{\text{steel}} = 1450 \text{ psi} \\ (\tau_{\max})_{\text{alum.}} = 370 \text{ psi} \end{array}\right\} \qquad \textbf{Ans. (c)} \quad (6)$$

Review the Solution One way to check results is to check the overall equilibrium of the shaft. Because of the negative signs of T_A and T_C, their magnitudes, 2280 lb·in. and 2720 lb·in., respectively, are shown, along with couple-arrows indicating the proper sense, in Fig. 3.

$$\sum M_x = 0: \quad \textbf{Is } (-2280 + 5000 - 2720) = 0? \quad \textbf{Yes}$$

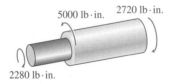

5000 lb·in.　2720 lb·in.

2280 lb·in.

Fig. 3　A free-body diagram.

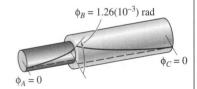

$\phi_B = 1.26(10^{-3})$ rad

$\phi_C = 0$

$\phi_A = 0$

Fig. 4　The deformation diagram.

We can also examine the deformations. Since the ends are restrained, $\phi_A = \phi_C = 0$. The sketch in Fig. 4 shows the deformation due to the positive torque $T_B = 5000$ lb·in. (ϕ_B is exaggerated.)

The displacement method was used as the basis for the development of TORSDEF, a computer program to solve torsional-deformation problems like Example Problem 4.5. This computer program, one of the *MechSOLID* collection of computer programs, is described in Appendix G.4. Computer homework exercises for Section 4.4 follow the regular Section 4.4 homework problems at the end of the chapter.

In Examples 4.4 and 4.5 we considered coaxial assemblages of uniform torsion elements. Torsion members may also be coupled together through gears or through belts and pulleys. The next example shows how such problems can be solved using the fundamental equations of *equilibrium*, *torque-twist behavior*, and *deformation compatibility*.

■■■■■■■■■■■■■■■ E X A M P L E 4 . 6 ■■■■■■■■■■■■■■■

Shafts *AB* and *CE* in Fig. 1 have the same diameter and are made of the same material. A torque T_E is applied to the shaft-gear system at *E*. Assume that torque is transmitted from shaft *CE* to shaft *AB* by a single gear-tooth contact force, and neglect the thickness of the gears.

Using the displacement (stiffness) method: (a) determine an expression for the rotation of gear *B*; (b) determine an expression for the shaft rotation at *E*; and (c) determine the torque transmitted to the base at *C*.

Plan the Solution We have three uniform elements, which we can number as shown in Fig. 2. There is a relationship between torques applied to the two gears (equilibrium), and also a relationship between their angles and directions of rotation (deformation compatibility). Let x and x' axes be designated for the shafts so that we can adopt a consistent sign convention for the various torques and rotation angles that enter into the solution.

The torque in element (3) can be determined from statics (i.e., equilibrium) alone, but the problem is statically indeterminate, since there are restraints at both *A* and *C*.

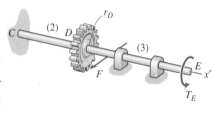

Fig. 1

Solution
(a) Determine ϕ_B, the rotation angle of gear *B*.

Equilibrium: We can separate the system into shaft elements and nodes, and we can then write a moment-equilibrium equation for each node. The gear-tooth force *F* is assumed to be acting normal to the radius at the point of contact. Note that all torques in Fig. 3 are shown in the positive sense according to the sign convention stated at the beginning of this section. The sense assumed for the gear-tooth contact force is arbitrary, so long as Newton's third law of "action and reaction" is observed. (Only torques and gear-tooth forces are shown on the "free-body diagrams." Other forces, such as reaction forces at the bearings and at the wall, are omitted to reduce the complexity of the diagrams.) We now write the moment-equilibrium equation for each node.

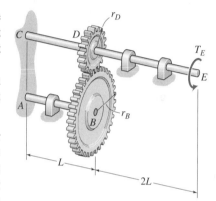

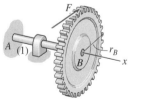

Fig. 2

For Node A:	$\sum M_x = 0$:	$T_1 - T_A = 0$
For Node B:	$\sum M_x = 0$:	$-T_1 - Fr_B = 0$
For Node C:	$\sum M_{x'} = 0$:	$T_2 - T_C = 0$
For Node D:	$\sum M_{x'} = 0$:	$-T_2 + T_3 - Fr_D = 0$
For Node E:	$\sum M_{x'} = 0$:	$-T_3 + T_E = 0$

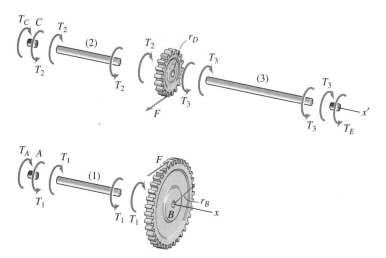

Note: All torques are shown, but force reactions
(except for the gear-tooth force) are omitted for clarity.

Fig. 3 "Free-body diagrams."

Summarizing the nodal equilibrium equations, we have

$$
\begin{aligned}
T_1 &= T_A \\
-T_1 - Fr_B &= 0 \\
T_2 &= T_C \qquad\qquad \textbf{Equilibrium} \quad (1)\\
-T_2 + T_3 - Fr_D &= 0 \\
T_3 &= T_E
\end{aligned}
$$

Equation (1e) enables us to determine T_3 from the given torque T_E. Equations (1a) and (1c) allow us to determine the reaction torques T_A and T_C once T_1 and T_2 have been determined. This leaves Eqs. (1b) and (1d) as the primary equilibrium equations to be used in determining T_1, T_2, and F. That is, we have two equations in three unknowns, so the problem is statically indeterminate.

Element Torque-Twist Behavior: Since we are to use the displacement method, Eq. 4.19 is the appropriate form for the torque-twist behavior. The sign convention for T_i and ϕ_i is shown in Fig. 4.

$$
T_i = k_{ti}\phi_i, \qquad i = 1, 2, 3
$$

where

$$
k_{ti} = \left(\frac{GI_p}{L}\right)_i
$$

All members have the same GI_p, and the length of element (3) is twice that of the other two elements. Let $k_t \equiv GI_p/L$. Then,

$$
k_{t1} = k_{t2} \equiv k_t, \qquad k_{t3} = \frac{GI_p}{2L} \equiv \frac{k_t}{2}
$$

Fig. 4 An element.

192

So,

$$T_1 = k_t\phi_1$$

$$T_2 = k_t\phi_2$$

$$T_3 = \left(\frac{k_t}{2}\right)\phi_3$$

Element Torque-Twist Behavior (2)

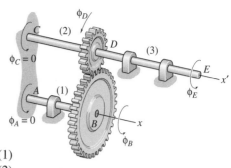

Fig. 5 Deformation diagram.

Deformation Geometry: We need two types of displacement information: (1) the relationships of the element twist angles ϕ_i to the nodal rotations, and (2) the relationship of the rotation of gear B to the rotation of gear D. The sign convention for nodal rotation angles is shown on the deformation sketch in Fig. 5.

From Fig. 5,

$$\phi_1 = \phi_B - \phi_A = \phi_B$$

$$\phi_2 = \phi_D - \phi_C = \phi_D$$

$$\phi_3 = \phi_E - \phi_D$$

Element Twist-Angle Definitions (3)

In Fig. 6 the gear rotation angles ϕ_B and ϕ_D are shown in the positive sense. When gear B rotates through a positive angle ϕ_B, gear D will rotate in a negative sense by an amount such that the circumferential contact lengths will be equal. Thus,

$$r_B\phi_B = -r_D\phi_D$$

Rotational-Displacement Compatibility (4)

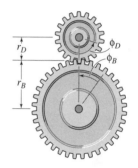

Fig. 6 Gear compatibility.

Since we are to use the displacement method to determine ϕ_B, we can use Eq. (4) to eliminate ϕ_D and then substitute (3) → (2) → (1).

$$\phi_1 = \phi_B$$

$$\phi_2 = -(r_B/r_D)\phi_B \qquad (3')$$

$$\phi_3 = \phi_E + (r_B/r_D)\phi_B$$

$$T_1 = k_t\phi_B$$

$$T_2 = -k_t(r_B/r_D)\phi_B \qquad (2')$$

$$T_3 = (k_t/2)[\phi_E + (r_B/r_D)\phi_B]$$

We can eliminate T_3 and F from Eqs. (1b), (1d), and (1e) to get

$$-T_2 + (r_D/r_B)T_1 = -T_E$$

and then we can combine this equation with Eqs. (2'a) and (2'b) to get

$$k_t(r_B/r_D)\phi_B + (r_D/r_B)(k_t\phi_B) = -T_E \qquad (1')$$

193

from which we can solve for the required angle of rotation ϕ_B.

$$\phi_B = \frac{-T_E/k_t}{(r_B/r_D) + (r_D/r_B)} \qquad \text{Ans. (a)} \quad (5)$$

(b) Determine the rotation angle ϕ_E. We can use Eqs. (1e), (2'c) and (5) to obtain an expression for ϕ_E.

$$\phi_E = (2/k_t)T_E - (r_B/r_D)\phi_B$$

or

$$\phi_E = (T_E/k_t)\left[2 + \frac{(r_B/r_D)}{(r_B/r_D) + (r_D/r_B)} \right] \qquad \text{Ans. (b)} \quad (6)$$

(c) Determine the reaction torque T_C. T_C may be obtained by combining Eqs. (1c), (2'b), and (5) to get

$$T_C = -T_2 = \frac{-(r_B/r_D)T_E}{(r_B/r_D) + (r_D/r_B)} \qquad \text{Ans. (c)} \quad (7)$$

Review the Solution Because this is a fairly complex problem, we should first check each answer to see if it has the proper sign and proper dimensions. The answers (5) through (7) satisfy these two requirements.

Next, we can observe that the magnitude of the torque transmitted from T_E to the base at C should be less than T_E, since part of the reaction to T_E is via the gear at B to the base at A. From Eq. (7) we can verify that $|T_C| < T_E$ as expected.

4.5 STRESS DISTRIBUTION IN CIRCULAR TORSION BARS; TORSION TESTING

Stress Distribution. In Section 2.8 we found that associated with pure normal stresses on the cross section of an axial-deformation member there are shear stresses and normal stresses on inclined cuts. Figure 4.7d shows the shear-stress distribution on a cross section due to torque and the corresponding shear stress on radial planes, and Fig. 4.14 shows that this leads to pure shear on an element between cross sections. We will now examine the stress distribution on faces that are inclined to the axis of a torsion bar.

Figure 4.14a shows a torsion bar and indicates the orientation of a "cut" inclined at

FIGURE 4.14 (a) A bar in torsion; (b) a pure-shear element.

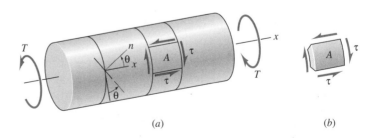

(a)   (b)

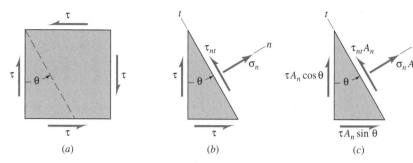

FIGURE 4.15 The state of stress for a pure-shear element.

angle θ with respect to the cross section. To determine the stresses on the inclined cut, we can take a triangular "wedge," as shown in Fig. 4.15. The stresses are shown on Fig. 4.15b, while Fig. 4.15c is a free-body diagram of the element; that is, it shows the forces that act on the respective faces of the element.

To determine expressions for σ_n and τ_{nt}, the normal stress and the shear stress on an inclined cut, we can write equilibrium equations for the free body in Fig. 4.15c. The n and t axes are taken normal to the inclined cut and tangential to the inclined cut, respectively.

$$+\nearrow \sum F_n = 0: \quad \sigma_n A_n + \tau A_n \cos \theta \, (\sin \theta) + \tau A_n \sin \theta \, (\cos \theta) = 0$$

$$\sigma_n = -2\tau \sin \theta \cos \theta$$

$$\nwarrow \sum F_t = 0: \quad \tau_{nt} A_n + \tau A_n \cos^2 \theta - \tau A_n \sin^2 \theta = 0$$

$$\tau_{nt} = -\tau(\cos^2 \theta - \sin^2 \theta)$$

The expressions for σ_n and τ_{nt} can be simplified by introducing the trigonometric identities

$$\sin (2\theta) = 2 \sin \theta \cos \theta, \quad \cos (2\theta) = \cos^2 \theta - \sin^2 \theta$$

Therefore,

$$\boxed{\begin{aligned} \sigma_n &= -\tau \sin(2\theta) \\ \tau_{nt} &= -\tau \cos(2\theta) \end{aligned}} \tag{4.20}$$

From Eqs. 4.20 it is clear that τ_{nt} has a maximum magnitude of τ for $\theta = 0°$ or $\theta = 90°$. On the other hand, σ_n has a maximum magnitude for $\theta = \pm 45°$. From Eqs. 4.20, $\sigma_{45°} = -\tau$, $\sigma_{-45°} = \tau$, and $\tau_{nt} = 0$ for $\theta = \pm 45°$. Figure 4.16 shows the original pure-shear element and an element rotated at 45° to this element.

Strain Distribution. Hooke's Law relating shear stress to shear strain, Eq. 2.22, gives

$$\gamma = \frac{\tau}{G} \tag{4.21}$$

Thus, the shear stress due to torsion, depicted in Fig. 4.16a, produces the shear deformation illustrated in Fig. 4.17a.

Along the $\pm 45°$ directions, which are the directions of maximum compression and maximum tension, respectively, as depicted in Fig. 4.16b, the extensional strains may be

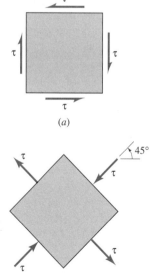

FIGURE 4.16 A pure-shear element, and the associated maximum-normal-stress element.

obtained by applying Hooke's Law in the form of Eqs. 2.32, namely

$$\epsilon_{45°} = \frac{1}{E}[-\tau - \nu(\tau)] = \frac{-\tau}{E}(1 + \nu)$$

$$\epsilon_{-45°} = \frac{1}{E}[\tau - \nu(-\tau)] = \frac{\tau}{E}(1 + \nu)$$

Since E is related to G by Eq. 2.35, the above equations can be written in the form

$$\epsilon_{\text{max.comp.}} = \epsilon_{45°} = -\frac{\tau}{2G}$$

$$\epsilon_{\text{max.tens.}} = \epsilon_{-45°} = \frac{\tau}{2G}$$

(4.22)

The deformation of a $\pm 45°$ element is illustrated in Fig. 4.17b.

Torsion Testing. To determine the shear modulus of a material, to determine its shear strength properties, or to determine other properties associated with torsion, bars may be tested in a torsion testing machine like the one shown in Fig. 4.18. The torque and twist angle are recorded as the torsion bar is twisted.

Figure 4.19 shows specimens tested to failure in a torsion testing machine. This figure illustrates the importance of Eq. 4.20 and the fact that σ_n and τ_{nt} vary with orientation. Both shafts in Fig. 4.19 were tested to failure in pure torsion. The failure of the mild-steel bar in Fig. 4.19a is quite different from that of the cast-iron bar in Fig. 4.19b. By comparing the failure planes in Figs. 4.19 with the stress elements in Fig. 4.16, we can conclude that the mild steel bar failed to shear, while the cast iron fracture occurred on a maximum-tensile-stress surface. This is consistent with the types of tensile-test failures discussed in Section 2.4, leading to the conclusion that brittle members are weaker in tension than in shear, while ductile members are weaker in shear.

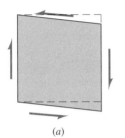

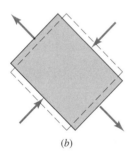

(a)

(b)

FIGURE 4.17 Deformation of a pure-shear element, and the associated maximum-normal-stress element.

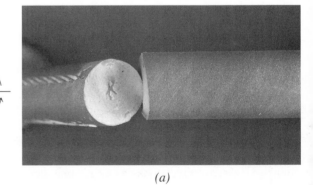

(a)

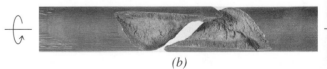

(b)

FIGURE 4.19 The failure surfaces of mild-steel and cast-iron torsion bars.

FIGURE 4.18 A torsion testing machine.

The torque tube in Fig. 1, with outer diameter $d_o = 1.25$ in. and inner diameter 1.00 in., is subjected to a torque $T = 1000$ lb·in. A strain gage oriented at an angle $\theta = -45°$ with respect to the axis, which measures the extensional strain along this direction, gives a reading of $\epsilon_{-45°} = 190$ μin./in. (a) Determine the value of the maximum shear stress, τ_{max}. (b) Determine the shear modulus of elasticity, G. (c) Determine the angle of twist in a section of the tube of length $L = 30$ in.

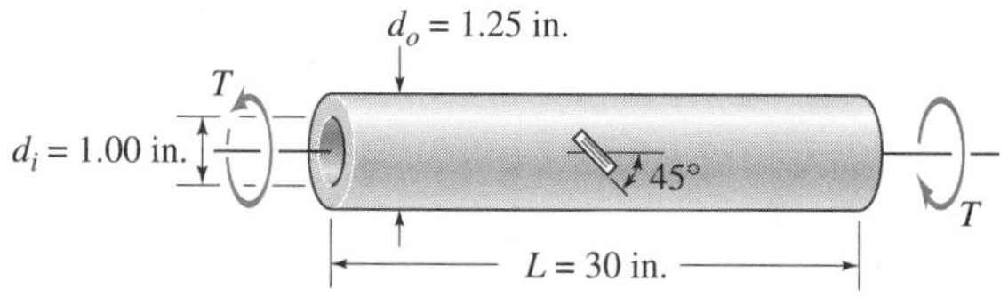

Fig. 1

Plan the Solution   The maximum shear stress, which occurs at the outer surface, can be determined directly from Eq. 4.13. Since the extensional strain along the direction $\theta = -45°$ is given, Eq. 4.22b can be used to compute G, the shear modulus. Finally, the angle of twist can be calculated from Eq. 4.16.

Solution
(a) The maximum shear stress occurs at the outer surface and is given by

$$\tau_{max} = \frac{Tr_o}{I_p}$$

where

$$I_p = \frac{\pi(r_o^4 - r_i^4)}{2} = \frac{\pi[(0.625 \text{ in.})^4 - (0.5 \text{ in.})^4]}{2} = 0.1415 \text{ in}^4$$

Then,

$$\tau_{max} = \frac{(1000 \text{ lb}\cdot\text{in.})(0.625 \text{ in.})}{0.1415 \text{ in}^4} = 4417 \text{ psi}$$

or, rounding to three significant figures,

$$\tau_{max} = 4420 \text{ psi} \qquad \textbf{Ans. (a)}$$

(b) From Eq. 4.22b,

$$G = \frac{\tau_{max}}{2\epsilon_{-45°}} = \frac{4417 \text{ psi}}{2(0.000190 \text{ in./in.})} = 11.6(10^6) \text{ psi} \qquad \textbf{Ans. (b)}$$

(c) From Eq. 4.16,

$$\phi = \frac{TL}{GI_p} = \frac{(1000 \text{ lb}\cdot\text{in.})(30 \text{ in.})}{(11.6(10^6) \text{ psi})(0.1415 \text{ in}^4)} = 0.0182 \text{ rad} \qquad \textbf{Ans. (c)}$$

Solid circular shafts and tubular circular shafts are frequently employed to transmit power from one device to another, for example from a turbine to a generator in a power plant or from the motor to the wheels of a car, truck, or other vehicle. Figure 4.20 shows a motor, a pulley, and the circular shaft connecting them. The motor at A supplies power to the pulley at D through the shaft BC, which is rotating at constant speed ω and which exerts a torque T at C, where the shaft is attached to the pulley D.

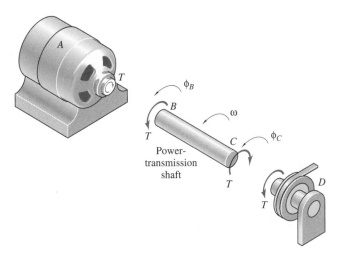

FIGURE 4.20 A power-transmission shaft.

The *work* that the shaft does on the pulley at D is the torque times the angle (in radians) through which the shaft rotates, that is,

$$\mathcal{W}_{onD} = T\phi_C$$

The *power* delivered at C is given by

$$\mathcal{P} = \frac{d\mathcal{W}}{dt} = T\frac{d\phi_C}{dt}$$

or

$$\boxed{\mathcal{P} = T\omega} \tag{4.23}$$

where ω is the rotational speed of the shaft in radians per second. The rotational speed is frequently expressed in revolutions per second or in revolutions per minute; the conversions are:

$$\omega(\text{rad/s}) = 2\pi f(\text{rev/s}) = \frac{2\pi n\ (\text{rpm})}{60}$$

where f is the rotational speed in revolutions per second (rev/s) and n is the speed in revolutions per minute (rpm).

When U.S. Customary units are used, the power is usually expressed in *horsepower* (hp), where

$$1 \text{ hp} = 550 \text{ lb} \cdot \text{ft/s} = 6600 \text{ lb} \cdot \text{in/s}$$

When SI units are used and T is expressed in N · m, the power will be in N · m/s = J/s, or watts (W). The conversion factor is

$$1 \text{ hp} = 745.7 \text{ W}$$

Once the torque has been determined, the angle of twist of a shaft and the complete stress distribution in the shaft can be determined. Two examples will now be given to illustrate how Eq. 4.23 is employed in designing power-transmission shafts.

■■■■■■■■■■■■■■■■□ **E X A M P L E   4 . 8** □□□□□□□□□□□□□□■■■

An electric motor delivers 10 hp to a pump through a solid circular shaft that is rotating at 875 rpm. If the shaft has an allowable shear stress of $\tau_{\text{allow.}} = 20$ ksi, what is the minimum required diameter of the shaft as a multiple of $\frac{1}{16}$ in.?

Solution

Torque: Equation 4.23 relates power to torque.

$$\mathcal{P} = T\omega \tag{1}$$

To get the torque T in units of lb · in. we will have to introduce consistent units for $\mathcal{P}$ and ω. Thus,

$$\mathcal{P} = (10 \text{ hp})\left(\frac{6600 \text{ lb} \cdot \text{in./s}}{1 \text{ hp}}\right) = 66{,}000 \text{ lb} \cdot \text{in./s}$$

$$\omega = (875 \text{ rpm})\left(\frac{2\pi \text{ rad}}{1 \text{ rev}}\right)\left(\frac{1 \text{ min}}{60 \text{ sec}}\right) = 91.63 \text{ rad/s}$$

Therefore,

$$T = \frac{\mathcal{P}}{\omega} = \frac{66{,}000 \text{ lb} \cdot \text{in./s}}{91.63 \text{ rad/s}} = 720.3 \text{ lb} \cdot \text{in.} \tag{2}$$

Allowable-Stress Design: Using the torsion formula for a solid shaft, Eq. 4.13, we can now determine the required diameter of the shaft.

$$\tau_{\text{allow.}} \geq \tau_{\text{max}} = \frac{Tr}{I_p}: \quad d^3_{\text{min.}} = \frac{16T}{\pi\tau_{\text{allow.}}} = \frac{16(720.3 \text{ lb} \cdot \text{in.})}{\pi(20 \times 10^3)\text{psi}} \tag{4}$$

So, the required diameter is $d_{\text{min.}} = 0.568$ in., or, to the next greater $\frac{1}{16}$ in.,

$$d_{\text{min}} = \frac{5}{8} \text{ in.} \qquad\qquad \textbf{Ans.} \quad (5)$$

A truck driveshaft is to be a tube with an outer diameter of 46 mm. It is to be made of steel ($\tau_{\text{allow.}} = 80$ MPa), and it must transmit 120 kW of power at an angular speed of 40 rev/s. Determine, to the nearest even millimeter (that satisfies the design allowable), the maximum inner diameter that the shaft may have.

Solution   The key equations are the torque-power equation, Eq. 4.23, and the torsion formula, Eq. 4.13.

Torque:

$$\mathcal{P} = T\omega = T(2\pi f) \tag{1}$$

so

$$T = \frac{\mathcal{P}}{2\pi f} = \frac{120 \text{ kW}}{2\pi(40 \text{ rev/s})} = 477.5 \text{ N} \cdot \text{m} \tag{2}$$

Allowable-Stress Design:

$$\tau_{\text{allow.}} \geq \tau_{\text{max}} = \frac{Tr_o}{I_p} \tag{3}$$

For a circular tube,

$$I_p = \frac{\pi}{32}(d_o^4 - d_i^4) \tag{4}$$

where d_o is the outer diameter of the shaft, and d_i is the inner diameter.
Combining Eqs. (3) and (4) we get

$$d_i^4 \leq d_o^4 - \frac{16 d_o T}{\pi \tau_{\text{allow.}}} \tag{5}$$

Finally, from Eqs. (2) and (5),

$$d_i^4 \leq (46 \times 10^{-3}\text{m})^4 - \frac{16(46 \times 10^{-3}\text{m})}{\pi}\left(\frac{477.5 \text{ N} \cdot \text{m}}{80 \times 10^6 \text{N/m}^2}\right) \tag{6}$$

from which, $(d_i)_{\text{max}} = 41.89$ mm. Although this value for $(d_i)_{\text{max}}$ is very close to 42 mm, we must pick an inner diameter of

$$d_i = 40 \text{ mm} \qquad \text{Ans.} \tag{7}$$

to keep $\tau_{\text{max}} \leq \tau_{\text{allow.}}$.

Although power-transmission shafts and many other torsion members are solid circular cylinders or tubular members with circular cross section, not every important torsion member has a circular cross section. For example, airplane fuselages, wings, and tails are all thin-wall members that are subjected to loading that includes torsion. Figure 4.21 depicts the structure of the wing of a light aircraft. The part of the wing structure that resists torsion is referred to as the *torque box*. The wing in Fig. 4.21 has a *single-cell torque box*, which is outlined with dashed lines.[7]

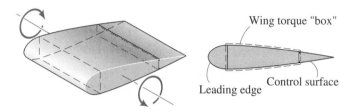

Wing torque "box"

Leading edge   Control surface

FIGURE 4.21 An airplane wing with torque box outlined.

Under certain conditions, the shear stress distribution on the cross section of a thin-wall torsion member can be determined easily. These simplifying conditions are:

- The member is cylindrical, that is, the cross section does not vary along the length of the member.
- The cross section is ''closed,'' that is, there is no longitudinal slit in the member.
- The wall thickness is small compared with the cross-sectional dimensions of the member.
- The member is subjected to end torques only.
- The ends are not restrained from warping.

Shear Flow. Figure 4.22 shows a portion of a typical closed, single-cell thin-wall torsion member. The thickness, t, may vary with circumferential position. The key assumption that is made in order to simplify the analysis of the stress distribution in <u>closed, thin-wall</u> members is—*the shear stress is constant through the thickness and is parallel to the median curve defining the cross section.* Figures 4.22b and 4.22c illustrate the

FIGURE 4.22 A closed, thin-wall, single-cell torsion member.

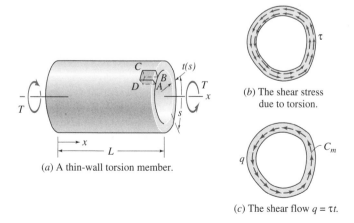

(a) A thin-wall torsion member.

(b) The shear stress due to torsion.

(c) The shear flow $q = \tau t$.

---

[7]Analysis of multi-cell thin-wall torsion members is beyond the scope of this book. See, for example, [Ref. 4-1].

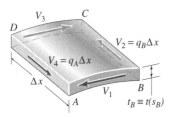

FIGURE 4.23 Free-body diagram of a thin-tube element.

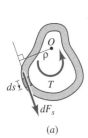

(a)

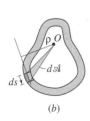

(b)

(c)

FIGURE 4.24 The torque due to shear flow.

significance of this assumption, which permits us to define a quantity called *shear flow*, q, by the equation

$$q = \tau t \tag{4.24}$$

where the wall thickness t can be a function of circumferential location s. Now we can proceed to determine the shear flow, q, and thus determine the shear stress τ.

The small element $ABCD$ in Fig. 4.22*a* is now removed and shown in Fig. 4.23 as a free body with shear forces acting on its faces. The shear stress on the face AB at A is τ_A. The same shear stress must also be acting on the face AD. Hence, the shear force on face AD is

$$V_4 = \tau_A t_A \Delta x = q_A \Delta x \tag{4.25}$$

Similarly, the shear force on face BC is

$$V_2 = \tau_B t_B \Delta x = q_B \Delta x \tag{4.26}$$

Since it was assumed that the member is subjected to torque only (thus, there is no axial stress σ_x), the equation of axial equilibrium is

$$\searrow \sum F_x = 0: \qquad V_4 - V_2 = 0, \qquad (q_A - q_B)\Delta x = 0$$

or

$$q_A = q_B \tag{4.27}$$

Since A and B are arbitrary points in the cross section, Eq. 4.27 implies that q is independent of location in the cross section; that is, q is independent of s. Hence,

$$q = \tau t = \text{const.} \tag{4.28}$$

Shear Stress Resultant; Torque. Equilibrium of forces in the x-direction leads to a shear flow that is constant (Eq. 4.28). Now we need to determine the relationship between the shear flow, q, and the torque T. The incremental force, dF_s, due to the shear flow, q, on a differential element of area of the cross section is illustrated in Fig. 4.24*a*. The force on this element of cross section,

$$dF_s = q\, ds \tag{4.29}$$

acts tangent to the *median curve*, C_m, as indicated in Fig. 4.24*a*, and the moment of this force dF_s about an arbitrary point O in the cross section is

$$dT = \rho dF_s = q\rho ds \tag{4.30}$$

where ρ is the perpendicular distance from point O to the line of action of dF_s. To get the torque on the cross section, we sum the contributions around the curve C_m, that is,

$$T = \oint_{C_m} q\rho ds \tag{4.31}$$

Since q is constant, we are left with evaluating $\oint \rho ds$ around C_m. This integral is a purely geometrical quantity. Figure 4.24*b* illustrates the fact that the origin O and base ds form a triangle whose area is given by

$$dA_m = \frac{1}{2}(\rho ds) \tag{4.32}$$

Hence,

$$\oint_{C_m} \rho ds = 2A_m \tag{4.33}$$

where A_m is the shaded area enclosed by the median curve C_m, as illustrated in Fig. 4.24c. (Note: The area A_m is not the area of the material cross section!) Finally, combining Eqs. 4.31 and 4.33, we get

$$T = 2qA_m \tag{4.34}$$

When combined with Eq. 4.24, Eq. 4.34 gives the following expression for the shear stress acting on the cross section of a closed, single-cell, thin-wall torsion member:

$$\tau = \frac{T}{2tA_m} \tag{4.35}$$

where $\tau \equiv \tau(s)$ is the average shear stress at location s in the cross section, and where $t \equiv t(s)$ is the local thickness at the point where the shear stress is evaluated.

Angle of Twist. To determine the angle of twist of a closed, thin-wall torsion member, like the one in Fig. 4.22, we can employ basic strain displacement and stress-strain relationships, or we can employ energy methods. The latter approach is discussed in Section 11.4. The result is

$$\phi = \frac{TL}{4 A_m^2 G} \oint_{C_m} \frac{ds}{t(s)} \tag{4.36}$$

where G is the shear modulus, and where the integral is evaluated around the median curve C_m shown in Fig. 4.24c.

In the discussion of Eqs. 4.9 and 4.10 it was indicated that the relationship between torque and twist rate for a circular shaft is

$$T = GI_p \frac{d\phi}{dx} \tag{4.9 repeated}$$

where I_p is the polar moment of inertia. Furthermore, it was indicated that the relationship between the torque T and the twist rate $d\phi/dx$ is frequently written as

$$T = GJ \frac{d\phi}{dx} \tag{4.10 repeated}$$

where GJ is called the *torsional rigidity* and where, for noncircular cross sections, J is <u>not</u> the polar moment of inertia. From Eqs. 4.10 and 4.36 we can see that, for a closed, single-cell, thin-wall torsion member, the torsional rigidity, GJ, is given by

$$GJ = \frac{4 A_m^2 G}{\oint_{C_m} \dfrac{ds}{t(s)}} \tag{4.37}$$

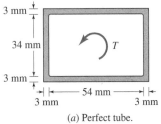

(a) Perfect tube.

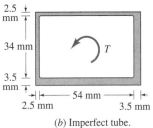

(b) Imperfect tube.

Fig. 1

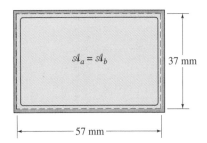

Fig. 2 Area enclosed by the median curve for tubes (a) and (b).

(a) Determine the maximum torque that may be applied to an aluminum-alloy tube if the allowable shear stress is $\tau_{\text{allow.}} = 100$ MPa. The tube has the cross section shown in Fig. 1a. (b) When the aluminum tube was extruded, the hole was not perfectly centered and the tube was found to have the actual dimensions shown in Fig. 1b. If the torque determined in Part (a) is applied to the imperfect tube, what will the maximum shear stress be? Where will this maximum shear stress occur?

Plan the Solution   Equation 4.35 can be used to determine the allowable torque, given the allowable shear stress and the cross-sectional dimensions.

Solution
(a) Determine the allowable torque for the perfect tube. Equation 4.35 can be written in the form

$$T_{\text{allow.}} = 2\mathcal{A}_m t_{\text{min}} \tau_{\text{allow.}} \tag{1}$$

Figure 2 shows the median curve used in calculating $\mathcal{A}_a = \mathcal{A}_b$.

$$\mathcal{A}_a = \mathcal{A}_b = (37\text{mm})(57\text{mm}) = 2.109(10^{-3})\text{m}^2 \tag{2}$$

The minimum wall thickness is 3 mm. Therefore, from Eqs. (1) and (2) we get

$$(T_a)_{\text{allow.}} = 2\mathcal{A}_a(t_a)_{\text{min}}\tau_{\text{allow.}}$$
$$= 2[2.109(10^{-3})\ \text{m}^2](0.003\ \text{m})[100(10^6)\ \text{N/m}^2]$$
$$(T_a)_{\text{allow.}} = 1265\ \text{N} \cdot \text{m} \qquad\qquad \textbf{Ans. (a)} \tag{3}$$

(b) Determine the maximum shear stress in the imperfect tube. In this case, Eq. 4.35 can be written in the form

$$T_b = 2\mathcal{A}_b(t_b)_{\text{min}}(\tau_b)_{\text{max}} \tag{4}$$

Since $\mathcal{A}_a = \mathcal{A}_b$ and the two torque tubes are to have the same torque, Eqs. (3) and (4) can be combined to give

$$(\tau_b)_{\text{max}} = \tau_{\text{allow.}}\left[\frac{(t_a)_{\text{min}}}{(t_b)_{\text{min}}}\right]$$
$$= 100\ \text{MPa}\left(\frac{0.003\ \text{m}}{0.0025\ \text{m}}\right)$$

Therefore, the maximum shear stress in the imperfect tube is in the minimum-thickness section of the tube wall and has the value

$$(\tau_b)_{\text{max}} = 120\ \text{MPa} \qquad\qquad \textbf{Ans. (b)}$$

Note that the shear stresses in the two tubes are inversely proportional to their minimum wall thicknesses.

Figure 4.22a indicates how shear stresses "flow" in the thin wall of a torque tube. Note that the rectangular tubes in Fig. 1 of the preceding example problem have filleted inner corners to minimize the stress concentration that occurs due to the 90° change in direction of the shear stress at each corner of the rectangular cross section. (See Section 12.2 for a discussion of stress concentrations.)

■■■■■■■■■■■■■■■ E X A M P L E 4 . 1 1 ■■■■■■■■■■■■■■■

Determine the torsional rigidity, GJ, for the thin-wall tubular member whose cross section is shown in Fig. 1. The shear modulus is G.

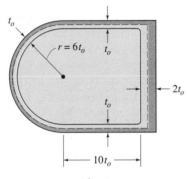

Fig. 1

Solution This is a straightforward application of Eq. 4.37, but, since the thickness of the wall of the torque tube is piecewise-constant, we can write Eq. 4.38 in the form

$$GJ = \frac{4\,\mathcal{A}_m^2 G}{\displaystyle\sum_i \left(\frac{\Delta s_i}{t_i}\right)} \tag{1}$$

The area $\mathcal{A}_m$ enclosed by the dashed median curve in Fig. 1 is

$$\mathcal{A}_m = \frac{\pi}{2}(6.5t_0)^2 + (11t_0)(13t_0) = 209.4t_0^2 \tag{2}$$

Referring to Fig. 1, we can evaluate the sum in the denominator of Eq. (1) as

$$\sum_i \left(\frac{\Delta s_i}{t_i}\right) = \frac{\pi(6.5t_0)}{t_0} + \frac{2(11t_0)}{t_0} + \frac{13t_0}{2t_0} = 48.92 \tag{3}$$

Combining Eqs. (1) through (3), we get

$$GJ = \frac{4(209.4t_0^2)^2 G}{48.92} = 3584Gt_0^4$$

or, rounded to three significant figures,

$$GJ = 3580Gt_0^4 \qquad\qquad \textbf{Ans.}$$

In Section 4.2 the deformation of a circular cylinder twisted by equal and opposite torques applied at its ends was described. That discussion, supported by the photos in Fig. 4.2, pointed out that, for torsion members with circular cross sections, plane sections remain plane and simply rotate around the axis of the member. It is clear from the photo of the deformed square torsion bar in Fig. 4.2b that plane sections do not remain plane when a member with noncircular cross section is subjected to torsional loading. An important feature of the torsional deformation of noncircular prismatic bars is the *warping* of the cross sections.

The theory of elasticity may be used to relate the torque applied to such noncircular prismatic members to the resulting stress distribution and angle of twist.[8]

Stress Distribution and Angle of Twist.

The shear-stress distribution in noncircular torsion bars is quite different than the shear stress distribution in circular torsion members. Figure 4.25 compares the stress distribution in a circular bar with that in a rectangular bar. The shear stress on the circular cross section varies linearly with distance from the center and reaches its maximum at the outer surface (Eq. 4.13). In contrast, the shear stress at the corners of the rectangular torsion member in Fig. 4.25b must be zero. (Recall that $\tau_{xy} = \tau_{yx}$.) In fact, the maximum shear stress on a rectangular cross section occurs at the middle of the longer edge, which is the point on the periphery of the cross section that is nearest the center!

The maximum shear stress in a rectangular prismatic bar subjected to torsion may be expressed in the form

$$\tau_{\max} = \frac{T}{\alpha d t^2} \tag{4.38}$$

where α is a dimensionless constant obtained by a theory of elasticity solution and listed in Table 4.2, and where the dimensions d and t satisfy $d/t \geq 1$. The angle of twist of a

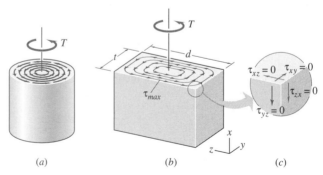

(a)   (b)   (c)

FIGURE 4.25 Torsion of circular and rectangular members.

---

[8]Saint-Venant (see Section 3.3) developed the theory of torsion for noncircular bars. He presented his famous memoir on torsion to the French Academy of Sciences in 1853. See, for example, Chapter 10 of [Ref. 4-2] for a discussion of Saint-Venant's theory of torsion.

TABLE 4.2   Torsion Constants for Rectangular Bars

d/t	1.00	1.50	1.75	2.00	2.50	3.00	4	6	8	10	∞
α	0.208	0.231	0.239	0.246	0.258	0.267	0.282	0.298	0.307	0.312	0.333
β	0.141	0.196	0.214	0.229	0.249	0.263	0.281	0.298	0.307	0.312	0.333

bar of length L can be expressed by

$$\phi = \frac{TL}{\beta dt^3 G} \qquad (4.39)$$

where β is a dimensionless constant with value as listed in Table 4.2.

The shear stress distribution on the cross section of a shaft with elliptical cross section is illustrated in Fig. 4.26. The maximum shear stress occurs at the boundary at the two ends of the minor axis of the ellipse and is given by

$$\tau_{max} = \frac{2T}{\pi a b^2} \qquad (4.40)$$

The angle of twist for an elliptical shaft of length L is given by

$$\phi = TL\left(\frac{a^2 + b^2}{\pi a^3 b^3 G}\right) \qquad (4.41)$$

Finally, the area of an ellipse is

$$A = \pi a b$$

It is important to note that the torsional behavior of circular bars is very special. If a channel section or a wide-flange section or any other noncircular cross section is subjected to torsional loading, its behavior must be analyzed by analytical methods like the Saint-Venant solution that produced Eqs. 4.38 through 4.41. Finite-element analysis may also be used to solve specific torsion problems.

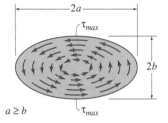

FIGURE 4.26   Torsion of an elliptical shaft.

■■■■■■■■■■■■■■■□ E X A M P L E   4 . 1 2 □■■■■■■■■■■■■■■■■

If torsion members having the cross sections shown in Fig. 1 have the same cross-sectional area and are subjected to torques that produce the same maximum shear stress, τ_{max}, in each, what is the torque carried by each?

Solution   Since the areas are to be the same, that is, $A_a = A_b = A_c = a^2$, the radius of the circular bar is given by

$$\pi c^2 = a^2, \qquad c = 0.5642a$$

(a)

(b)

$t = a$

$d = a$

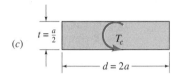

(c)

$t = \frac{a}{2}$

$d = 2a$

Fig. 1

For the circular bar $J = \frac{1}{2}\pi c^4$, and

$$\tau_{\max} = \frac{T_a c}{J} \tag{1}$$

Therefore,

$$T_a = \frac{\tau_{\max} J}{c} = \frac{\pi \tau_{\max} c^3}{2} = \frac{\pi \tau_{\max}(0.5642a)^2}{2} \tag{2}$$

or

$$T_{\text{circle}} \equiv T_a = 0.282\tau_{\max}a^3 \tag{3}$$

For rectangular bars, Eq. 4.38 gives

$$\tau_{\max} = \frac{T}{\alpha d t^2} \tag{4}$$

For the square bar in Fig. 1b, $d/t = 1$, and Table 4.2 gives $\alpha = 0.208$. Therefore,

$$T_{\text{square}} \equiv T_b = 0.208\tau_{\max}a^3 \tag{5}$$

Finally, for the rectangle in Fig. 1c, $d/t = 4$, so Table 4.2 gives $\alpha = 0.282$. Therefore, Eq. (2) gives

$$T_{4:1\text{rect.}} \equiv T_c = 0.282\tau_{\max}(2a)(a/2)^2 = 0.141\tau_{\max}a^3 \tag{6}$$

Summarizing the above results, we get

$$\frac{T_{\text{circle}}}{T_{\text{square}}} = 1.36, \qquad \frac{T_{\text{circle}}}{T_{4:1\text{rect.}}} = 2.00 \qquad \textbf{Ans.} \quad (7a,b)$$

That is, the circular bar can support 36% more torque than a square bar of equal area; the circular bar can support 100% higher torque than can a rectangle with a 4 : 1 ratio of sides.

*4.9 INELASTIC TORSION OF CIRCULAR RODS

In the preceding sections of Chapter 4, we have considered torsion of linearly elastic members, the simplest case being the torsion of rods with circular cross section. Now we will examine the behavior of circular rods that are subjected to torques that produce shear stresses beyond the proportional limit. Inelastic torsion is similar in many respects to the inelastic axial deformation discussed in Section 3.9, with one very important difference. In the case of axial deformation, the strain and stress are uniform over the entire cross section of the axial-deformation member, but in the case of torsion, both shear strain and shear stress vary with distance from the center of the torsion rod.

Fundamental Equations. Of the three fundamentals of deformable-body mechanics—*equilibrium, geometry of deformation,* and *material behavior*—only the material behavior differs when we consider inelastic torsion rather than the linearly elastic behavior treated so far in Chapter 4.

Geometry of Deformation: The *strain-displacement equation,* Eq. 4.1, holds for inelastic as well as for linearly elastic torsion.

$$\gamma(x, \rho) = \rho\frac{d\phi}{dx} \equiv \rho\theta \qquad (4.42)$$

where

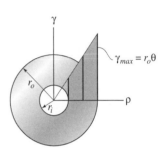

γ = the shear strain due to torsion,

ρ = the distance from the center of the rod to the point in the cross section where the strain is to be determined,

ϕ = the angle of twist at section x; $\theta \equiv \dfrac{d\phi}{dx}$ is the twist rate.

This linear strain distribution is sketched in Fig. 4.27 (repeat of Fig. 4.6).

FIGURE 4.27 Torsional shear-strain distribution.

Equilibrium: The shear stress $\tau(x, \rho)$ is related to the resultant torque $T(x)$ by the *equilibrium equation,* Eq. 4.4.

$$T(x) = \int_A \rho\tau dA \qquad \begin{matrix}(4.4)\\ \text{repeated}\end{matrix}$$

Figure 4.28 illustrates how this integral can be evaluated for a solid shaft by using

$$T(x) = 2\pi \int_0^r \tau\rho^2 d\rho \qquad (4.43)$$

This internal torque $T(x)$ is related to the external load and reaction torques through free-body diagrams and equations of equilibrium.

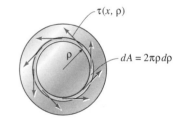

(*a*) Shear stress τ on dA ring at radius ρ.

Material Behavior: So far in Chapter 4 we have only considered linearly elastic behavior characterized by Hooke's Law for shear, $\tau = G\gamma$. Now, however, we will consider stresses beyond the proportional limit τ_{PL} for materials that have stress-strain curves in shear like those in Fig. 4.29. For the general case, therefore, the stress-strain diagram must be employed to establish the appropriate *material behavior* in one of the following two forms:

$$\tau = \tau(\gamma), \quad \text{or} \quad \gamma = \gamma(\tau) \qquad (4.44)$$

(Later, in the discussion of residual stresses, we will discuss how shear stress and shear strain are related during unloading.)

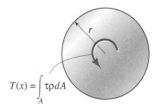

(*b*) Resultant internal torque at section x.

FIGURE 4.28 The relationship of torque to shear stress.

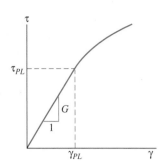

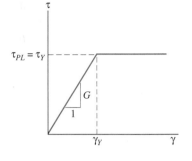

(a) Linearly elastic, nonlinearly plastic material.   *(b)* Linearly elastic, perfectly plastic material.

FIGURE 4.29 Typical diagrams of shear stress versus shear strain.

Elastic-Plastic Torque-Twist Analysis. Equations 4.42 (strain-displacement), 4.43 (equilibrium), and 4.44 (material behavior) permit a complete solution of any problem of torsion of circular rods. The procedure will be illustrated here with the analysis of a rod made of elastic-plastic material having a stress-strain curve like that in Fig. 4.29*b*. As in the case of inelastic axial deformation (Section 3.9), there are also three cases to be considered for torsion: Case 1—linearly elastic behavior, Case 2—partially plastic behavior, and Case 3—fully plastic behavior.

Case 1—Linearly elastic behavior. In the linearly elastic range, that is, up to $\tau = \tau_Y$, the material obeys Hooke's Law,

$$\tau = G\gamma \tag{4.45}$$

In Section 4.2 the following torque-stress (Eq. 4.12) and torque-twist (Eq. 4.14) equations were derived for a homogeneous, linearly elastic rod.

$$\tau = \frac{T\rho}{I_p} \tag{4.46}$$

$$\theta = \frac{T}{GI_p} \tag{4.47}$$

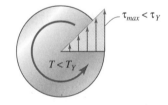

(a) Linearly elastic case.

where I_p is the polar moment of inertia of the cross section. The maximum elastic torque, or *yield torque* (Fig. 4.30*b*), is obtained by setting $\tau = \tau_Y$ and $\rho = r$ in Eq. 4.46. Thus,

$$T_Y = \frac{\tau_Y I_p}{r} \tag{4.48}$$

Then, from Eq. 4.47

$$\theta_Y = \frac{T_Y}{GI_p} = \frac{\tau_Y}{Gr} \tag{4.49}$$

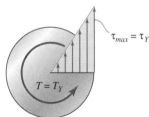

(b) Maximum elastic torque case.

FIGURE 4.30 Shear-stress distribution—elastic behavior.

Case 2—Partially plastic behavior. When the torque exceeds T_Y, the shear strain over a portion of the cross section exceeds the yield shear strain γ_Y. The shear-stress distribution then exhibits an *elastic core* of radius r_Y and a *plastic annulus,* as depicted in Fig. 4.31*a*. This stress-strain behavior for the partially plastic condition is given by the equations

$$\begin{aligned}
\tau &= \tau_Y\left(\frac{\rho}{r_Y}\right) & 0 \le \rho \le r_Y \\
\tau &= \tau_Y & r_Y \le \rho \le r
\end{aligned} \tag{4.50}$$

The corresponding torque is obtained by substituting Eqs. 4.50 into Eq. 4.43 giving

$$T = 2\pi \int_0^{r_Y} \left(\frac{\tau_Y \rho}{r_Y} \right) \rho^2 d\rho + 2\pi \int_{r_Y}^r \tau_Y \rho^2 d\rho$$

$$T = \frac{\pi \tau_Y}{6} \left(4r^3 - r_Y^3 \right) \tag{4.51}$$

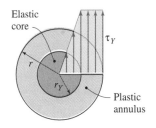

To determine the twist rate, θ, for this value of torque we can set the shear strain at $\rho = r_Y$ equal to γ_Y. Therefore, from Eq. 4.42,

$$\theta = \frac{\tau_Y}{Gr_Y} \tag{4.52}$$

(a) Partially-plastic-torsion stress distribution.

Selecting a value of r_Y that satisfies $0 < r_Y < r$, we can eliminate r_Y from Eqs. 4.51 and 4.52 and thereby get a torque-twist relationship.

Case 3—Fully plastic torque. An elastic-plastic material, like mild steel, can experience very large shear strain while the shear stress remains constant at τ_Y. Therefore, it is possible to have $\gamma_{\max} \equiv \gamma(x, r) \gg \gamma_Y$. Then, $r_Y \to 0$ and we get the *fully plastic stress distribution* depicted in Fig. 4.31b. The material equation for the fully plastic case is

$$\tau = \tau_Y \qquad 0 < \rho \le r \tag{4.53}$$

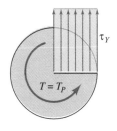

The *plastic torque, T_P,* can be obtained by setting $r_Y = 0$ in Eq. 4.51, giving

$$T_P = \frac{2\pi \tau_Y r^3}{3} \tag{4.54}$$

(b) Fully-plastic-torsion stress distribution.

FIGURE 4.31 Partially plastic torsion and fully plastic torsion.

This value can be approached, but not actually reached, since the strain-displacement equation, Eq. 4.42, does not permit the shear strain to actually reach γ_Y at $\rho = 0$ for a finite twist angle. Comparing Eqs. 4.48 and 4.54 we see that, for a solid circular shaft,

$$T_P = \frac{4}{3} T_Y \tag{4.55}$$

The above elastic-plastic torque-twist analysis is summarized in the *torque-twist curve* in Fig. 4.32.

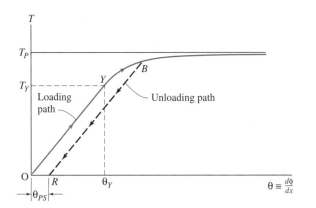

FIGURE 4.32 The torque-twist curve for an elastic-plastic circular torsion bar.

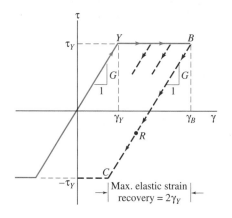

FIGURE 4.33 A stress-strain diagram illustrating elastic recovery.

Unloading; Residual Stress. If the torque is allowed to exceed T_Y, say to T_B in Fig. 4.32, and then the torque is removed, the unloading curve will parallel the initial linearly elastic portion, OY, of the torque-twist curve. When the torque is completely removed, there will be *residual stresses* and a residual angle of twist, or *permanent set,* left in the rod (See Fig. 4.2).

Suppose that the maximum shear strain (at $\rho = r$) at torque T_B (Fig. 4.32) is γ_B, as indicated in Fig. 4.33. A subsequent decrease in torque would cause unloading at $\rho = r$ along the path BC indicated in Fig. 4.33. At all radii, whether in the elastic core or in the plastic annulus, unloading will occur in a linearly elastic manner along lines satisfying Hooke's Law in the form

$$\Delta\tau = G\Delta\gamma \tag{4.56}$$

where the Δ's indicate increments of stress and strain.

Since unloading takes place elastically, we can solve for the residual stresses in a torsion rod by superposing the plastic stress distribution due to torque T_B and an elastic stress distribution due to an equal and opposite torque $(-T_B)$. For the fully plastic case, we get the stress distributions shown in Fig. 4.34. Then, the maximum elastic-recovery stress $(\tau_{er})_{\text{max}}$ in Fig. 4.34b is given by setting $T = T_P$ in the elastic-shear-stress formula, Eq. 4.46. Thus,

$$(\tau_{er})_{\text{max}} = \frac{T_P r}{I_p} = \frac{4}{3}\tau_Y \tag{4.57}$$

The clockwise-acting shear stress at $\rho = r$ in Fig. 4.34c is the shear stress at point R on the elastic recovery curve BC in Fig. 4.33. Its magnitude is $\tau_Y/3$.

FIGURE 4.34
Determination of residual shear stresses in torsion by superposition.

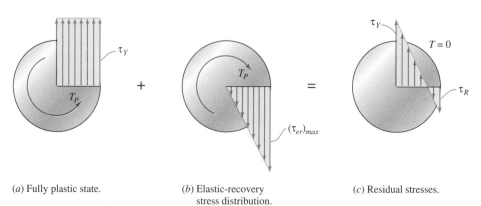

(*a*) Fully plastic state.

(*b*) Elastic-recovery stress distribution.

(*c*) Residual stresses.

A tubular shaft with an outer diameter of 120 mm and an inner diameter of 100 mm is made of elastic-plastic material whose τ vs. γ curve is shown in Fig. 1b. The shear modulus is $G = 200$ GPa, and the yield stress in shear is $\tau_Y = 100$ MPa. (a) Determine the yield torque, T_Y, for this tubular shaft. (b) Determine the fully plastic torque, T_P. (c) Determine the distribution of residual shear stress if the torque is completely removed following loading to T_P.

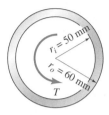

(a) Torque tube.　　　(b) Elastic-plastic material behavior.

Fig. 1

Plan the Solution　We can use the results of the above discussion of elastic-plastic torsion, but it will be necessary to incorporate the correct expression for I_p for a tubular shaft, to integrate only over $r_i \leq \rho \leq r_o$, and to note that the shaft becomes fully plastic when $\tau(r_i) = \tau_Y$.

Solution
(a) The yield torque is the torque that makes $\tau(r_o) = \tau_Y$. From Eq. 4.26,

$$T_Y = \frac{\tau_Y I_p}{r_o} = \frac{\tau_Y (\pi/2)(r_o^4 - r_i^4)}{r_o} \quad (1)$$

$$= \frac{[100(10^6)\ \text{N/m}^2](\pi/2)[(0.06\ \text{m})^4 - (0.05\ \text{m})^4]}{(0.06\ \text{m})}$$

$$T_Y = 17.57\ \text{kN} \cdot \text{m} \qquad \textbf{Ans. (a)} \quad (2)$$

The stress distribution produced by the yield torque is shown in Fig. 2a.

(b) The shear stress distribution produced by the fully plastic torque, T_P, is shown in Fig. 2b. The fully plastic torque may be obtained by setting $\tau = \tau_Y$ in Eq. 4.43 but integrating from r_i to r_o.

$$T_P = 2\pi\tau_Y \int_{r_i}^{r_o} \rho^2 d\rho \quad (3)$$

$$= \frac{2\pi[100(10^6)\ \text{N/m}^2]}{3}[(0.06\ \text{m})^3 - (0.05\ \text{m})^3]$$

$$T_P = 19.06\ \text{kN} \cdot \text{m} \qquad \textbf{Ans. (b)} \quad (4)$$

(Note that $T_P = 1.08T_Y$, whereas $T_P = 1.33T_Y$ for a solid shaft.)

(c) We can use the superposition of the linearly elastic recovery stress distri-

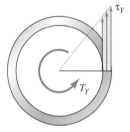

(a) Stress distribution for yield torque T_Y.

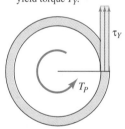

(b) Stress distribution for fully plastic torque T_P.

Fig. 2

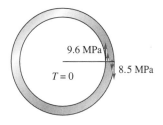

(a) Elastic recovery stresses.

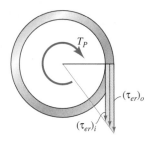

(b) Residual stress distribution.

Fig. 3

bution of Fig. 3a and the fully plastic stress distribution of Fig. 2b to obtain the residual stress distribution in the torque tube when T_P is completely removed. Because the recovery is linearly elastic, we can set $T = T_P$ and use Eq. 4.46 to evaluate $(\tau_{er})_i$ and $(\tau_{er})_o$, the elastic recovery shear stresses at the r_i and r_o, respectively.

$$(\tau_{er}) = \frac{T_P\rho}{I_p} = \frac{(19.06 \text{ kN} \cdot \text{m})\rho}{(\pi/2)[(0.06 \text{ m})^4 - (0.05 \text{ m})^4]} \quad (5)$$

Therefore,

$$(\tau_{er})_i = (\tau_{er})_{\rho=0.05 \text{ m}} = 90.4 \text{ MPa} \quad (6)$$
$$(\tau_{er})_o = (\tau_{er})_{\rho=0.06 \text{ m}} = 108.5 \text{ MPa}$$

Finally, the maximum residual stresses are:

$$\left.\begin{array}{l}(\tau_r)_o = \tau_Y - (\tau_{er})_o = -8.5 \text{ MPa} \\ (\tau_r)_i = \tau_Y - (\tau_{er})_i = 9.6 \text{ MPa}\end{array}\right\} \quad \text{Ans. (c)}$$

The residual stress distribution is shown in Fig. 3b.

Review the Solution   Figures 2 and 3 are very helpful in checking our solution. The calculations in Eq. (1) should be rechecked to make sure that T_Y is correct. Because the tube is relatively thin-walled, the stresses in the fully plastic state (Fig. 2b) are only a small amount larger than the stresses produced by the yield torque (Fig. 2a). Therefore, the ratio of $T_P/T_Y = 1.08$ is reasonable.

Finally, for Part (c), we know that the resultant torque is zero. Therefore, some shear stresses must be acting counterclockwise and some acting clockwise. The stresses nearer to the center of the shaft must be larger than those near the outer surface, because the latter stresses have a longer moment arm ρ. Therefore, the values of $(\tau_r)_i = 9.6$ MPa (ccw) and $(\tau_r)_o = 8.5$ MPa (cw) seem reasonable.

4.10 PROBLEMS ■■

Prob. 4.2-1. The tubular shaft in Fig. P4.2-1 has an outside diameter $d_o = 2.0$ in. and an inside diameter $d_i = 1.2$ in. The maximum shear stress in the shaft is $\tau_{max} = 8$ ksi. (a) Sketch the shear stress distribution on an arbitrary cross section, indicating the values of the maximum and minimum shear stresses. (b) By integrating over the cross section (Eq. 4.4), determine the value of the torque T. (c) Use the *torsion formula* to determine T. (d) What percent additional torque would be required to cause the same maximum shear stress, $\tau_{max} = 8$ ksi, in a solid shaft of

the same outer diameter d_o, and by what percentage would the weight of the solid shaft exceed the weight of the tabular shaft?

Prob. 4.2-2. Repeat Prob. 4.2-1 for a shaft with $d_o = 60$ mm, $d_i = 40$ mm, and $\tau_{max} = 90$ MPa.

Prob. 4.2-3. A solid, homogeneous shaft of diameter $d = 4.0$ in. is subjected to a torque $T = 120$ kip · in. (Fig. P4.2-3). (a) Determine the maximum shear stress in the shaft. (b) Determine the percentage of the total torque that is carried by the inner core of diameter $d_c = 2.0$ in. (shown darkly shaded in the figure).

P4.2-1 and P4.2-2

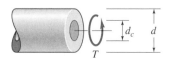

P4.2-3 and P4.2-4

(c) Determine the percentage of the total weight of the shaft that lies within this inner core.

Prob. 4.2-4. The solid, homogeneous shaft in Fig. P4.2-4 has a diameter of $d = 80$ mm and a maximum shear stress of $\tau_{max} = 40$ MPa. (a) Determine the torque T acting on the shaft. (b) Determine the percentage of the total torque that is carried by the inner core of diameter $d_c = 40$ mm (shown darkly shaded in the figure).

Prob. 4.2-5. The shaft in Fig. P4.2-5 has an outside diameter $d_o = 4.0$ in. and is made of a steel alloy that has an allowable shear stress of $\tau_{allow.} = 12$ ksi. The shaft is to be subjected to a torque $T = 60$ kip $\cdot$ in. Relative to the weight of a solid shaft, by what percentage could the weight of the shaft be reduced by removing a core of diameter d_i?

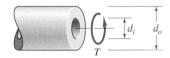

P4.2-5 and P4.2-6

Prob. 4.2-6. Repeat Prob. 4.2-5 with $d_o = 60$ mm, $\tau_{allow.} = 80$ MPa, and $T = 2$ kN $\cdot$ m.

Prob. 4.2-7. The solid shaft in Fig. P4.2-7 is made of an aluminum alloy that has an allowable shear stress $\tau_{allow.} = 10$ ksi and a shear modulus of elasticity $G = 3800$ ksi. The diameter of the shaft is $d = 1.6$ in., and its length is $L = 32$ in. If the allowable angle of twist over the 32-in. length of the shaft is $\phi_{allow.} = 0.10$ rad, what is the value of the allowable torque, $T_{allow.}$?

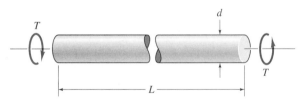

P4.2-7 and P4.2-8

Prob. 4.2-8. The solid shaft in Fig. P4.2-8 is made of brass that has an allowable shear stress $\tau_{allow.} = 120$ MPa and a shear modulus of elasticity $G = 39$ GPa. The length of the shaft is $L = 2$ m, and over this length the allowable angle of twist is $\phi_{allow.} = 0.10$ rad. If the shaft is to be subjected to a maximum torque of $T = 25$ kN $\cdot$ m, what is the required diameter of the shaft?

Prob. 4.2-9. The solid circular shaft of diameter d in Fig. P4.2-9a has a maximum shear stress $\tau_{max} = \tau_a$ under the action of a torque T_a. If the solid shaft is replaced by a tubular shaft with a ratio of outside diameter to inside diameter of $d_o/d_i = 1.2$ but weighing

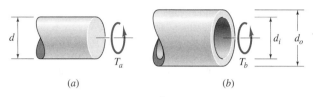

P4.2-9 and P4.2-10

the same as the solid shaft Fig. P4.2-9b, by what percentage would the torque have to be increased in order to produce the same maximum shear stress?

Prob. 4.2-10. The solid circular shaft of diameter d in Fig. P4.2-10a has a twist rate of $\dfrac{d\phi}{dx} = \theta_a$ under the action of the torque T_a. If the solid shaft is replaced by a tubular shaft with a ratio of outside diameter to inside diameter of $d_o/d_i = 1.2$ but weighing the same as the solid shaft (Fig. P4.2-10b), by what percentage would the torque have to be increased in order to produce the same twist rate θ_a in the tubular shaft?

Prob. 4.2-11. The solid shaft in Fig. P4.2-11 has a diameter that varies linearly from d_o at $x = 0$ to $2d_o$ at $x = L$. It is subjected to a torque T_0 at $x = 0$, and is attached to a rigid wall at $x = L$. (a) Determine an expression for the maximum (cross-sectional) shear stress in the tapered shaft as a function of the distance x from the left end. (b) Determine an expression for the total angle of twist of the shaft, ϕ_0. The shear modulus of elasticity is G.

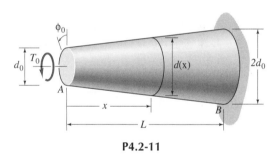

P4.2-11

*__Prob. 4.2-12.__ Repeat Prob. P4.2-11 if a hole of constant diameter $d_i = d_o/2$ is drilled out along the axis of the shaft shown in Fig. P4.2-11, giving the resulting tubular shaft a linearly varying wall thickness as shown in Fig. P4.2-12.

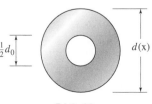

P4.2-12

215

Prob. 4.2-13. The solid circular shaft in Fig. P4.2-13 is subjected to a distributed external torque that varies linearly from intensity of t_0 per unit of length at $x = 0$ to zero at $x = L$. The shaft has a diameter d and shear modulus G and is fixed to a rigid wall at $x = 0$. (a) Determine an expression for the maximum (cross-sectional) shear stress in the shaft as a function of the distance x from the left end. (b) Determine an expression for the total angle of twist, ϕ_B, at the free end. The shear modulus of elasticity is G.

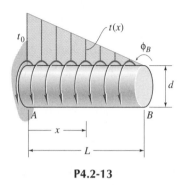

P4.2-13

Prob. 4.2-14. Solve Prob. 4.2-13 if the stated externally-applied torque distribution is replaced by the quadratically-varying torque shown in Fig. P4.2-14.

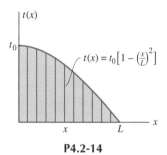

P4.2-14

Prob. 4.2-15. Solve Prob. 4.2-13 if the stated externally applied torque distribution is replaced by the torque distribution shown in Fig. P4.2-15.

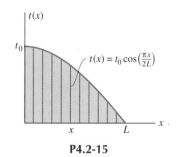

P4.2-15

Problems 4.4-1 through 4.4-18 *are multisegment, statically determinate problems. The appropriate solution procedure for these problems is the* **force method**, *since the internal resisting torque in any member can be determined directly from equilibrium equations. Therefore, the appropriate order of solution is:*

> **Equilibrium equations**
> **Torque-twist relationships**
> **Geometry-of-deformation (e.g., compatibility) equations**

Prob. 4.4-1. A stepped steel shaft AC ($G = 12 \times 10^6$ psi) is subjected to torsional loads at sections B and C as shown in Fig. P4.4-1. The diameters are: $d_1 = 2.0$ in. and $d_2 = 1.0$ in. (a) Determine the maximum shear stress in the shaft. Identify the location(s) where this maximum shear stress occurs. (b) Determine the angle of rotation at C, ϕ_C.

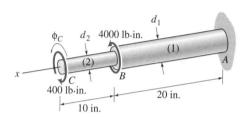

P4.4-1, P4.4-9, and PC4.4-1

Prob. 4.4-2. The aluminum-alloy shaft AC in Fig. P4.4-2 ($G = 26$ GPa) has an 800-mm-long solid segment AB and a 400-mm-long tubular segment BC. The shaft is subjected to the torsional loading shown in the figure. The diameters are $d_1 = (d_o)_2 = 60$ mm and $(d_i)_2 = 30$ mm. (a) Determine the maximum shear stress in the shaft, and identify the location(s) where this maximum shear stress occurs. (b) Determine the angle of rotation at B, ϕ_B, and the angle of rotation at C, ϕ_C.

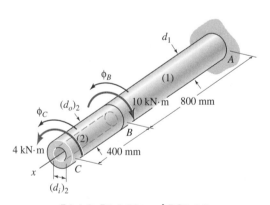

P4.4-2, P4.4-10, and PC4.4-2

Prob. 4.4-3. The aluminum-alloy shaft AD in Fig. P4.4-3 ($G = 3.8 \times 10^3$ ksi) has a 30-in.-long solid segment AB and a 40-in.-long tubular segment BD. The shaft is subjected to the torsional loads as shown in the figure. The respective diameters are: $d_1 =$

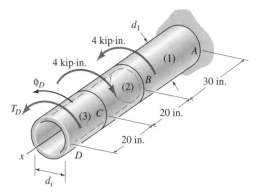

P4.4-3, P4.4-11, and PC4.4-3

$(d_o)_2 = (d_o)_3 = 2.5$ in., and $(d_i)_2 = (d_i)_3 = 2.0$ in. (a) Determine the value of the torque T_D added at D that would make $\phi_D = 0$. (b) For the loading described in Part (a), determine the maximum shear stress in the shaft AD.

Prob. 4.4-4. The steel shaft AD in Fig. P4.4-4 ($G = 80$ GPa) is subjected to torsional loads at sections B and D, as shown in the figure. The diameters are: $d_1 = d_2 = 40$ mm., and $d_3 = 25$ mm. (a) Determine the value of the torque T_D added at D that would make the rotation at C equal to zero, that is, make $\phi_C = 0$. (b) For the loading as determined in Part (a), determine the maximum shear stress in each of the three rod segments.

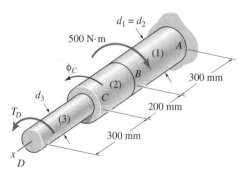

P4.4-4, P4.4-12, and PC4.4-4

Prob. 4.4-5. A plumber is cutting threads on the end of a 4-ft.-long section of 1-in.-diameter* pipe. This section, AB, has already been attached to a 2-in.-diameter* section, BC, by a reducing coupler and, in turn, to a 3-in.-diameter* section CD. The shear modulus of the steel pipe is $G = 11.5 \times 10^3$ ksi. Neglecting the dimensions and flexibility of the couplers, (a) determine the maximum shear stress in the assembly, (b) determine the angle of rotation at end A, ϕ_A, and (c) determine the torsional stiffness, k_t, of the assembly. That is, determine the torque that must be applied at A to produce a unit rotation (i.e., one radian rotation) at A. (*The stated diameters are nominal pipe diameters. See Table D.7 for section properties of the pipe.)

Prob. 4.4-6. The maximum shearing stress in the stepped steel shaft shown in Fig. P4.4-6 is 120 MPa. The shear modulus is

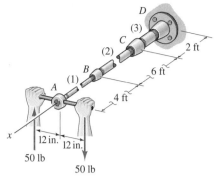

P4.4-5

$G = 80$ GPa, and the diameters of the elements are $d_1 = 50$ mm, $d_2 = 100$ mm, and $d_3 = 150$ mm. (a) The externally applied torques at A and B are shown on the figure. Determine the third applied torque, T_C, at C. (b) Determine the rotation of end A, ϕ_A, under the action of all three external torques.

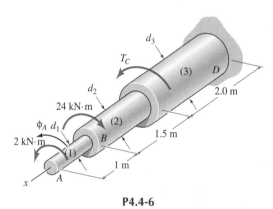

P4.4-6

Prob. 4.4-7. A uniform 1-in.-diameter steel shaft ($G = 12 \times 10^6$ psi) is supported by frictionless bearings and is used to transmit torques from gear B to gears at A and C as shown in Fig. P4.4-7. (a) Determine the maximum shear stress in element (1), (i.e., in segment AB), and the maximum shear stress in element (2). (b) Determine the relative rotation between the two ends; that is, determine $\phi_{C/A} \equiv \phi_C - \phi_A$.

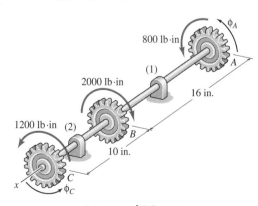

P4.4-7 and PC4.4-5

Prob. 4.4-8. A torque is applied to gear A of a two-shaft system and is transmitted through gears at B and C to a fixed end at D. The shafts are made of steel ($G = 80$ GPa). Each shaft has a diameter, $d = 32$ mm, and they are supported by frictionless bearings as shown in Fig. P4.4-8. If the torque applied to gear A is 400 N · m, and D is restrained, (a) determine the maximum shear stress in each shaft, and (b) determine the angle of rotation of the gear A relative to its no-load position.

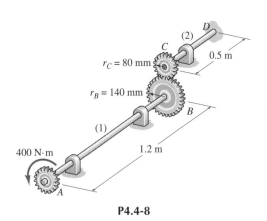

P4.4-8

Prob. 4.4-9. Determine the torsional stiffness, k_t, of the stepped shaft AC in Fig. P4.4-1.

Prob. 4.4-10. Determine the torsional stiffness, k_t, of the shaft AC in Fig. P4.4-2.

Prob. 4.4-11. Determine the torsional stiffness, k_t, of the shaft AD in Fig. P4.4-3.

Prob. 4.4-12. Determine the torsional stiffness, k_t, of the shaft AD in Fig. P4.4-4.

Prob. 4.4-13. A stepped steel shaft AC ($G = 11.5 \times 10^3$ ksi) is subjected to external torques at sections B and C as shown in Fig. P4.4-13. The diameter of element (1) is $d_1 = 2.0$ in. Diameter d_2 is to be determined such that the maximum shear stress in the stepped shaft does not exceed the allowable shear stress $\tau_{\text{allow.}} = 8$ ksi and the total angle of twist does not exceed $\phi_{\text{allow.}} = 0.06$ rad. Express your answer to the nearest ¼ in. that satisfies these two criteria.

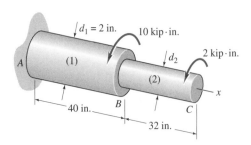

P4.4-13

Prob. 4.4-14. A two-element aluminum shaft AC ($G = 27$ GPa) is subjected to external torques at sections B and C as shown in Fig. P4.4-14. The outer diameter of the shaft is $d_1 = d_2 = 50$ mm. The inner diameter of element (2), d_{i2}, is to be determined such that the maximum shear stress in the shaft does not exceed the allowable shear stress $\tau_{\text{allow.}} = 35$ MPa and the total angle of twist does not exceed $\phi_{\text{allow.}} = 0.08$ rad. Express your answer, d_{i2}, to the nearest 2 mm that satisfies these two criteria.

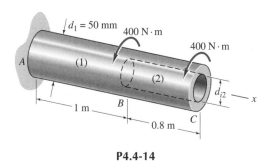

P4.4-14

Prob. 4.4-15. The aluminum-alloy shaft AD in Fig. P4.4-15 ($G = 3.8 \times 10^3$ ksi) has a 40-in.-long tubular section AC and a 30-in.-long solid section CD. The diameters are $(d_o)_1 = (d_o)_2 = d_3 = 2.0$ in. and $(d_i)_1 = (d_i)_2 = 1.75$ in. A 2 kip · in. torque acts at section B. What is the allowable torque $T_A \geq 0$ that may be added at end A if the angle of rotation at A is not to exceed $(\phi_a)_{\text{allow.}} = 0.10$ rad, and the magnitude of the shear stress is not to exceed $\tau_{\text{allow.}} = 4$ ksi anywhere in the shaft?

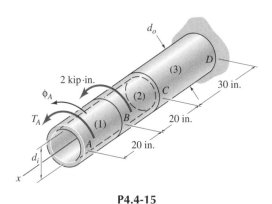

P4.4-15

*Prob. 4.4-16. The solid steel shaft AD in Fig. P4.4-16 is subjected to applied torques at sections A and C, as shown in the figure. The shear modulus is $G = 80$ GPa, and the diameters of the respective segments of the shaft are $d_1 = 25$ mm and $d_2 = d_3 = 40$ mm. If the angle of rotation at A is not to exceed

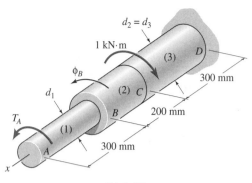

P4.4-16

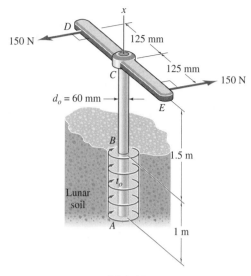

P4.4-18

$|\phi_A|_{\text{allow.}} = 0.0125$ rad in either direction, and the magnitude of the shear stress is not to exceed $\tau_{\text{allow.}} = 70$ MPa anywhere in the shaft, (a) what is the <u>minimum</u> torque T_A that <u>must</u> be applied at end A, and (b) what is the <u>maximum</u> torque T_A that <u>may</u> be applied at end A?

Prob. 4.4-17. A uniform shaft of diameter d, length L, and shear modulus G, is subjected to a uniformly distributed torque t_0 over half of its length, as shown in Fig. P4.4-17. (a) Determine an expression for τ_{max}, and indicate the location(s) where τ_{max} occurs. (b) Determine the angle of rotation at A, ϕ_A, and the angle of rotation at B, ϕ_B.

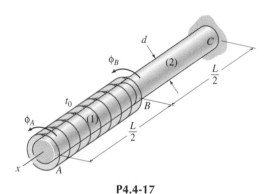

P4.4-17

Prob. 4.4-18. A lunar soil sampler consists of a tubular shaft that has an inside diameter of 45 mm and an outside diameter of 60 mm. The shaft is made of stainless steel, with a shear modulus $G = 80$ GPa. In Fig. P4.4-18, the sampler is being removed from the sampling hole by two 150-N forces that the astronaut exerts at right angles to arm DE and parallel to the lunar surface. Assume that the external torque that is exerted by the soil from A to B is uniformly distributed, with magnitude t_0. (a) Determine the maximum shear stress in the shaft of the sampler tube AC, and indicate the location(s) where it acts. (b) Determine the angle of twist of end C with respect to end A, that is, determine $\phi_{A/C} \equiv \phi_A - \phi_C$.

Problems **4.4-19 through 4.4-33** *are statically indeterminate problems. In some cases, the solution method to be used (i.e., the* **force method***, or the* **displacement method***) is specified, and you are expected to use that particular method. If the method of solution is not specified, you should mentally "walk through" the solution and pick the approach that will most effectively lead from the given information to the required answer(s). Remember that the basic types of equations to use are the following:* **equilibrium, torque-twist***, and* **geometry of deformation***.*

*Prob. 4.4-19. A uniform shaft of diameter d and length L is subjected to a distributed external loading

$$t(x) = t_0[1 - (x/L)^2]$$

Determine the reactions, T_A and T_B, at the fixed supports at A and B.

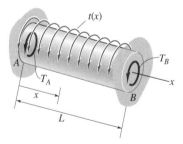

P4.4-19 and P4.4-20

*Prob. 4.4-20. Repeat Prob. 4.4-19 for a distributed torsional loading

$$t(x) = t_0 \cos\left(\frac{\pi x}{2L}\right)$$

219

Prob. 4.4-21. A stepped steel shaft AC ($G = 12 \times 10^6$ psi) is subjected to a torsional load of 4000 lb·in. at B and is fixed to rigid supports at A and C, as shown in Fig. P4.4-21. The diameters of two segments are $d_1 = 1.0$ in. and $d_2 = 2.0$ in. (a) Using the *force method,* and treating the (internal) torque in segment (1) as the redundant, determine both T_1 and T_2, the internal torques in the two segments. (b) Determine the maximum shear stress in each segment. (c) Determine ϕ_B, the angle of rotation of the shaft at joint B.

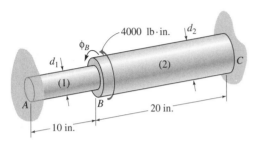

P4.4-21 and P4.4-22

Prob. 4.4-22. Consider the stepped steel shaft with torsional loading as given in Prob. 4.4-21. (a) Use the *displacement method* to determine ϕ_B, the rotation of the shaft at joint B. (b) Determine the internal torques, T_1 and T_2, in the two segments of the shaft. (c) Determine the maximum shear stress in each segment of the shaft.

Prob. 4.4-23. The aluminum-alloy shaft AC in Fig. P4.4-23 ($G = 26$ GPa) has a 400-mm-long tubular section AB and an 800-mm-long solid segment BC. The shaft is subjected to a 10 kN · m torsional load at section B, and it is fixed to rigid supports at ends A and C. The diameters are $(d_o)_1 = d_2 = 60$ mm and $(d_i)_1 = 30$ mm. (a) Using the *force method,* and treating the (internal) torque in segment (1) as the redundant, determine both T_1 and T_2, the internal torques in the two segments. (b) Determine the maximum shear stress in each segment. (c) Determine ϕ_B, the angle of rotation of the shaft at joint B.

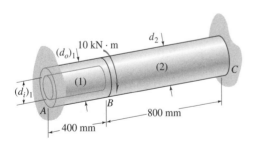

P4.4-23, P4.4-24, and PC4.4-6

Prob. 4.4-24. Consider the aluminum-alloy shaft AC and torsional loading given in Prob. P4.4-23. (a) Use the *displacement*

220

method to determine ϕ_B, the rotation of the shaft at joint B. (b) Determine the internal torques, T_1 and T_2, in the two segments of the shaft. (c) Determine the maximum shear stress in each segment of the shaft.

Prob. 4.4-25. The uniform 1-in.-diameter steel shaft in Fig. P4.4-25 ($G = 12 \times 10^6$ psi) is supported by frictionless bearings and is used to transmit torques from the gear at B to the gears at A and C, as shown. Using the *force method,* determine the maximum shear stress in each of the two segments of the shaft: (a) if the relative rotation between gears A and C is $\phi_{A/C} \equiv \phi_A - \phi_C = 0$, and (b) if the relative rotation between gears A and C is $\phi_{A/C} \equiv \phi_A - \phi_C = 0.01$ rad.

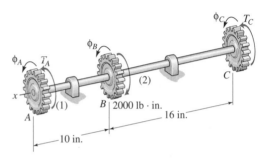

P4.4-25

Prob. 4.4-26. A torque T_A is applied to gear A of the two-shaft system in Fig. P4.4-26, producing a rotation $\phi_A = 0.05$ rad. The shafts are made of steel ($G = 80$ GPa), and each has a diameter of $d = 32$ mm. The shafts are supported by frictionless bearings, and end D of shaft CD is restrained. (a) Using the *displacement method,* determine the angle of rotation of gear C and the angle of rotation at gear B. (b) Determine the internal torques in shafts (1) and (2). (c) Determine the maximum shear stress in the two-shaft system.

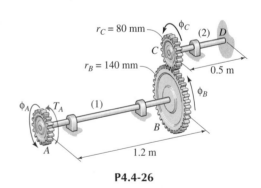

P4.4-26

Prob. 4.4-27. A solid circular shaft AC of total length $L = L_1 + L_2$ is fixed against rotation at ends A and C and is loaded by a torque T_B at joint B, as shown in Fig. P4.4-27. If the diameters of the two segments of the shaft are in the ratio $d_2/d_1 = 1.2$, what ratio of lengths, L_2/L_1, will cause the maximum shear stress to be the same in the two segments, that is, will make $(\tau_{max})_1 = (\tau_{max})_2$?

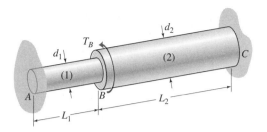

P4.4-27 and P4.4-28

Prob. 4.4-28. A solid circular shaft AC of total length $L = L_1 + L_2$ is fixed against rotation at ends A and C and is loaded by a torque T_B at joint B, as shown in Fig. P4.4-28. If the lengths of the two segments of the shaft are in the ratio $L_2/L_1 = 2.0$, what ratio of diameters, d_2/d_1, will cause the maximum shear stress to be the same in both segments, that is, will make $(\tau_{max})_1 = (\tau_{max})_2$?

Prob. 4.4-29. A composite shaft of length $L = 6$ ft is made by shrink-fitting a titanium-alloy sleeve (element 1) over an aluminum-alloy core (element 2). The two-component shaft is fixed against rotation at end B. The shear moduli are $G_1 = 6 \times 10^3$ ksi and $G_2 = 4 \times 10^3$ ksi, respectively, and the diameters are $(d_o)_1 = 2.25$ in. and $(d_i)_1 = d_2 = 2.0$ in. (a) Use the *displacement method* to determine the angle of rotation at A, ϕ_A, produced by an applied torque $T_A = 20$ kip · in. (b) Determine the maximum shear stresses $(\tau_{max})_1$ in the titanium sleeve and $(\tau_{max})_2$ in the aluminum core.

Prob. 4.4-30. As illustrated in Fig. P4.4-30, a composite shaft of length $L =$ is made by shrink-fitting a sleeve (element 1) over a solid core (element 2). Thus, $d_2 = (d_i)_1$. A torsional load T_A is applied to the shaft at A, and end B is fixed against rotation. Let $G_2 > G_1$, and determine an expression for the ratio of diameters $(d_o/d_i)_1$: (a) if the torque T_A is evenly divided between the sleeve and the core, and (b) if the maximum shear stress in the sleeve, $(\tau_{max})_1$, is the same as the maximum shear stress in the core, $(\tau_{max})_2$. Express your answers in terms of the shear-modulus ratio G_2/G_1.

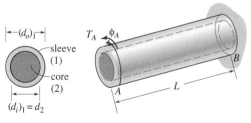

P4.4-29, P4.4-30, and P4.4-31

Prob. 4.4-31. A composite shaft is made by shrink-fitting a brass sleeve ($G_1 = 39$ GPa) over an aluminum-alloy core ($G_2 = 26$ GPa), and the shaft is subjected to an end-torque T_A. The diameter of the aluminum core is 2.0 in. What thickness is required for the

brass sleeve in order to reduce the angle of twist of the composite shaft to one-half the angle of twist of the aluminum core alone when subjected to the same end torque T_A?

Prob. 4.4-32. A uniform shaft with fixed ends at A and D is subjected to external torques of magnitude T_0 and $2T_0$, as shown in Fig. P4.4-32. The diameter of the shaft is d, and its shear modulus is G. (a) Determine expressions for the maximum shear stress in each of the three segments of the shaft: $(\tau_{max})_1$, $(\tau_{max})_2$, and $(\tau_{max})_3$. (b) Determine an expression for the angle of rotation at B, ϕ_B.

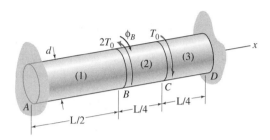

P4.4-32 and PC4.4-9

Prob. 4.4-33. The aluminum-alloy shaft AD in Fig. P4.4-33 ($G = 3.8 \times 10^3$ ksi) has a 40-in.-long tubular segment AC and a 30-in.-long solid segment CD, and it is fixed against rotation at ends A and D. The shaft, whose respective diameters are $(d_o)_1 = (d_o)_2 = d_3 = 2.5$ in. and $(d_i)_1 = (d_i)_2 = 2.0$ in., is subjected to torsional loads at sections B and C, as shown in the figure. (a) Determine the reaction torques at ends A and D. (b) Determine the maximum shear stress in the shaft. (c) Determine the angle of rotation at section C, ϕ_C.

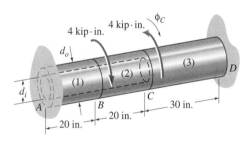

P4.4-33, PC4.4-7, and PC4.4-8

Computer Exercises—Sect. 4.4 *Use the MechSOLID computer program TORSDEF to solve the homework exercises C4.4-1 through C4.4-9. Refer to Appendix G.4 for a description of this computer program and for an example of its use. Hand in computer printouts of the following MechSOLID "screens": TORSDEF: Model, TORSDEF: Boundary Conditions, TORSDEF: Results, TORSDEF: Rotation Angle Plot, TORSDEF: Torque Plot. Samples of these screens are illustrated in Example G-5.*

Prob. C4.4-1. Use TORSDEF to solve Prob. 4.4-1. Let node A in Fig. P4.4-1 be relabeled as node $\boxed{1}$, node B be relabeled as node $\boxed{2}$, and node C be relabeled as node $\boxed{3}$.

Prob. C4.4-2. Use TORSDEF to solve Prob. 4.4-2. Let node A in Fig. P4.4-2 be relabeled as node ①, node B be relabeled as node ②, and node C be relabeled as node ③.

Prob. C4.4-3. Use TORSDEF to solve Prob. 4.4-3. Let node A in Fig. P4.4-3 be relabeled as node ①, node B be relabeled as node ②, and so on.

Prob. C4.4-4. Use TORSDEF to solve Prob. 4.4-4. Let node A in Fig. P4.4-4 be relabeled as node ①, node B be relabeled as node ②, and so on.

Prob. C4.4-5. Use TORSDEF to solve Prob. 4.4-7. Let node A in Fig. P4.4-7 be relabeled as node ①, node B be relabeled as node ②, and node C be relabeled as node ③. Let $\phi_A = 0$. (Note: This means that the torque at A cannot be specified, but will be obtained as a reaction.)

Prob. C4.4-6. Use TORSDEF to solve Prob. 4.4-24. Let node A in Fig. P4.4-24 be relabeled as node ①, node B be relabeled as node ②, etc.

Prob. C4.4-7. Use TORSDEF to solve Prob. 4.4-33. Let node A in Fig. P4.4-33 be relabeled as node ①, node B be relabeled as node ②, etc.

*Prob. C4.4-8. (a) Use TORSDEF to solve Prob. 4.4-33 as it is stated. Call this ''Solution A.'' (b) Keep all of the problem parameters the same, except replace d_i by $d_i = 1.5$ in. Use TORSDEF to solve this revised problem, and call this ''Solution B.'' (c) Again, keep all parameters the same as in the original problem, except replace d_i by $d_i = 1.00$ in. Use TORSDEF to solve this revised problem, and call this ''Solution C.'' (d) Create a table of nodal angular displacements and a table of element maximum shear stresses. Discuss the effect that changing $I_{p1} = I_{p2}$ has on the nodal displacements and on the element shear stresses.

*Prob. C4.4-9. The TORSDEF program can be used to solve non-numeric problems like Prob. 4.4-32 if you remember that nodal displacements will have the form $\overline{\phi}_i = \hat{\phi}_i(T_0 L / G I_p)$, where $\hat{\phi}_i$ is a nondimensional factor that is obtained by TORSDEF if you input the following data: $T_0 = L = G = I_p = 1$. Likewise, element internal torques will have the form $T_i = \hat{T}_i T_0$ and element shear stresses will have the form $\tau = \hat{\tau}(T_0 d / I_p)$, where $\hat{T}_i$ is an internal torque output by TORSDEF.

Use TORSDEF to solve Prob. 4.4-32. Let the nodes in Fig. P4.4-32 be relabeled as follows: $A \rightarrow$ ①, $B \rightarrow$ ②, etc.

Prob. 4.5-1. A round wood dowel is made of oak having an allowable stress in shear, parallel to the grain, of $\tau_{\text{allow.}} = 200$ psi. (a) If the diameter of the dowel is 2 in., what is the maximum permissible torque T that can be applied? (b) Why would the

wood dowel fail in shear, rather than in tension or compression? Describe the shear failure surface(s).

This figure of a **generic torsion member** *applies to all of the remaining problems for Section 4.5.*

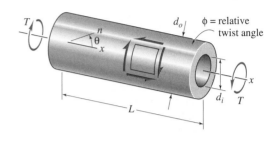

Problems 4.5-2 through 4.5-5. *For these four problems, use the generic torsion member shown above. For each problem: (a) determine the maximum tensile stress, the maximum compressive stress, and the maximum shear stress; and (b) show the stresses determined in Part (a) on properly oriented stress elements (e.g., see Fig. 4.16).*

In solving these problems, make the sense of the stresses consistent with the sense of the torque T *shown on the generic torsion member figure.*

Prob. 4.5-2. (See the ''generic torsion member'' figure.) $T = 5$ kip · in., $d_o = 1.2$ in., and $d_i = 0$ in.

Prob. 4.5-3. (See the ''generic torsion member'' figure.) $T = 200$ N · m, $d_o = 30$ mm, and $d_i = 0$ mm.

Prob. 4.5-4. (See the ''generic torsion member'' figure.) $T = 2.5$ kip · in., $d_o = 2.0$ in., and $d_i = 1.8$ in.

Prob. 4.5-5. (See the ''generic torsion member'' figure.) $T = 500$ N · m, $d_o = 40$ mm, and $d_i = 30$ mm.

Problems 4.5-6 through 4.5-9. *For these four problems, use the generic torsion member and determine (a) the value of the applied torque,* T, *and (b) the value of the maximum tensile stress,* σ_{maxT}.

Prob. 4.5-6. $L = 30$ in., $d_o = 0.75$ in., $d_i = 0$ in., $\phi = 0.09$ rad, and $G = 11 \times 10^3$ ksi.

Prob. 4.5-7. $L = 1$ m, $d_o = 20$ mm, $d_i = 0$ mm, $\phi = 0.10$ rad, and $G = 80$ GPa.

Prob. 4.5-8. $L = 10$ ft, $d_o = 3.5$ in., $d_i = 3.0$ in., $\phi = 0.04$ rad, and $G = 11 \times 10^3$ ksi.

Prob. 4.5-9. $L = 4$ m, $d_o = 40$ mm, $d_i = 30$ mm, $\phi = 0.60$ rad, and $G = 28$ GPa.

Prob. 4.5-10. For the generic torsion member, $T = 800$ lb · in., $G = 4 \times 10^3$ ksi, $d_o = 0.875$ in., and $d_i = 0$ in. Determine (a) the maximum shear strain, $\gamma_{\text{max,}}$ and (b) the maximum tensile strain, ϵ_{maxT}.

Prob. 4.5-11. For the generic torsion member, $T = 600$ N · m, $G = 28$ GPa, $d_o = 50$ mm, and $d_i = 40$ mm. Determine (a) the

P4.5-1

222

maximum shear strain, γ_{max}; and (b) the maximum tensile strain, ϵ_{maxT}.

Prob. 4.5-12. For the generic torsion member, $T = 500$ lb·in., $d_o = 0.625$ in., and $d_i = 0$ in. The extensional strain for $\theta = 45°$ is $\epsilon_{45°} = -1300\ \mu$. (a) Determine the value of the maximum shear stress, τ_{max}. (b) Determine the shear modulus of elasticity, G.

Prob. 4.5-13. For the generic torsion member, $T = 90$ N·m, $d_o = 20$ mm, and $d_i = 16$ mm. The extensional strain for $\theta = -45°$ is $\epsilon_{-45°} = 1200\ \mu$. (a) Determine the value of the maximum shear stress, τ_{max}. (b) Determine the shear modulus of elasticity, G.

Prob. 4.5-14. A steel shaft ($G = 11.8 \times 10^3$ psi) has the form of the generic torsion member pictured on page 222. The outside diameter of the shaft is $d_o = 1.0$ in., and it is required to carry torques up to $T_{max} = 1200$ lb·in. If the maximum allowable extensional strain for this shaft is $400\ \mu$, what is the minimum wall thickness of the shaft to the nearest $\frac{1}{32}$ in.

Prob. 4.6-1. A solid circular shaft having a diameter $d = 2$ in. rotates at an angular speed of $\omega = 95$ rpm. If the shaft is made of steel with an allowable shear stress of $\tau_{allow.} = 8$ ksi, what is the maximum power, $\mathcal{P}$ (hp), that can be delivered by the shaft?

Prob. 4.6-2. A hollow bronze shaft ($\tau_{allow.} = 100$ MPa) has an outside diameter of $d_o = 60$ mm and an inside diameter of $d_i = 50$ mm. How much power, $\mathcal{P}$ (kW), can be delivered by this shaft rotating at a speed of $5\frac{rev}{s}$?

Prob. 4.6-3. A solid turbine shaft delivers 10,000 hp at 60 rpm. The diameter of the shaft is 24 in., its length is 8 ft, and it is made of steel with a shear modulus of elasticity $G = 12 \times 10^3$ ksi. (a) What is the value of the maximum shear stress under the above operating conditions? (b) What is the angle of twist between the two ends of the shaft?

Prob. 4.6-4. A hollow drive shaft delivers 150 kW of power to a rock crusher at a rotating speed of $5\frac{rev}{s}$. The shaft has an outer diameter of $d_o = 120$ mm and an inside diameter of $d_i = 100$ mm, and it is made of steel with a shear modulus of elasticity $G = 80$ GPa. (a) Determine the maximum shear stress in the shaft under the above operating conditions. (b) If the shaft is 3 m long, what is the angle of twist of the shaft?

Prob. 4.6-5. A solid shaft is to be designed to deliver 500 kW of power at a rotational speed of $30\frac{rev}{s}$. If the shaft is to be made of bronze with an allowable shear stress $\tau_{allow.} = 50$ MPa, what is the required diameter of the shaft (to the nearest mm)?

Prob. 4.6-6. The drive shaft of the outboard motor in Fig. P4.6-6 delivers 6 hp to the propeller gear box at 4500 rpm. (a) If the shaft is solid brass, ($G = 5.6 \times 10^3$ ksi) with an allowable shear stress $\tau_{allow.} = 10$ ksi, what is the required diameter of the shaft (to the nearest $\frac{1}{16}$ in.)? (b) If the shaft is 25 in. long, and if it has

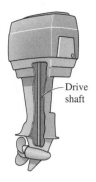

P4.6-6

the diameter determined in Part (a), what is the angle of twist of the shaft?

Prob. 4.6-7. The drive shaft of an inboard motor boat (Fig. P4.6-7) is required to deliver 100 hp at 300 rpm. A shaft is to be selected from the following group of available shafts: A($d_{oA} = 1.900$ in., $d_{iA} = 1.610$ in.), B($d_{oB} = 1.900$ in., $d_{iB} = 1.500$ in.), C($d_{oC} = 1.900$ in., $d_{iC} = 1.202$ in.). Select the lightest shaft that meets an allowable shear stress requirement of $\tau_{allow.} = 20$ ksi.

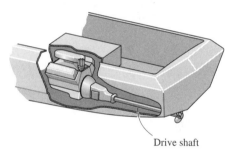

Drive shaft

P4.6-7

Prob. 4.6-8. A hollow circular shaft is to be designed to deliver 100 kW of power at a rotational speed of $50\frac{rev}{s}$. The allowable shear stress is 50 MPa, and the shaft is to be designed so that the outside diameter and the inside diameter have the ratio $d_o/d_i = 1.25$. Calculate the minimum allowable inside diameter, and express your answer to the nearest millimeter.

Prob. 4.6-9. An automobile engine delivers 145 hp to the drive shaft at 5500 rpm. If the allowable shaft stress in the drive shaft is 6 ksi, determine (a) the minimum diameter required for a solid drive shaft, (b) the minimum outside diameter of a hollow shaft with an inner diameter of 1.5 in., and (c) the percent savings in weight if the hollow shaft is used rather than the solid shaft.

Prob. 4.6-10. A solid circular drive shaft 2 m long transmits 560 kW of power from the turbine engine of a helicopter to its rotor at $15\frac{rev}{s}$. If the shaft is steel, with shear modulus $G = 80$ GPa and allowable shear stress $\tau_{allow.} = 60$ MPa, and if the allowable angle of twist of the shaft is 0.03 rad, determine the minimum permissible diameter of the drive shaft.

223

Prob. 4.6-11. A solid circular drive shaft 20 in. long transmits 60 hp from a motorcycle's transmission to its rear wheel at 4600 rpm. If the shaft is made of steel with shear modulus $G = 11 \times 10^3$ ksi and allowable shear stress $\tau_{allow.} = 10$ ksi, and if the allowable angle of twist of the drive shaft is 0.04 rad, determine the minimum permissible diameter of the drive shaft. Express your answer to the nearest $\frac{1}{16}$ in.

Prob. 4.6-12. A tubular shaft of outer diameter $d_o = 50$ mm and inner diameter $d_i = 40$ mm drives a wind turbine that is producing 7 kW of power. (Assume 100% efficiency of the turbine.) If the allowable shear stress in the steel shaft is $\tau_{allow.} = 50$ MPa, what is the maximum speed, ω_{max}, at which the blades may be allowed to rotate? (Neglect stresses in the shaft other than ones that are directly due to torsion.)

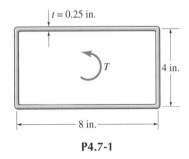

Turbine Shaft ω

P4.6-12

Problems 4.7-1 through 4.7-8. *For the thin-wall tubular sections shown in Figs. P4.7-1 through P4.7-8; (a) determine the maximum shear stress in the cross section, and (b) determine the value of the torsion constant, J.*

Prob. 4.7-1. See the problem statement above, and use $T = 100$ kip · in.

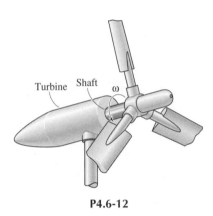

$t = 0.25$ in.

T

4 in.

8 in.

P4.7-1

Prob. 4.7-2. See the problem statement preceding Prob. 4.7-1, and use $T = 5$ kN · m.

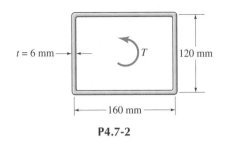

$t = 6$ mm T 120 mm

160 mm

P4.7-2

Prob. 4.7-3. See the problem statement preceding Prob. 4.7-1, and use $T = 800$ kip · in.

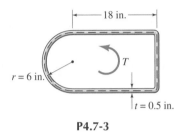

18 in.

$r = 6$ in. T

$t = 0.5$ in.

P4.7-3

Prob. 4.7-4. The tube in Fig. P4.7-4 has an elliptical cross section. See the problem statement preceding Prob. 4.7-1, and use $T = 90$ KN · m, $a = 250$ mm, $b = 150$ mm, $t = 10$ mm. (See the Table of Geometric Properties of Plane Areas inside the back cover.)

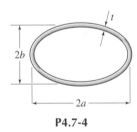

t

$2b$

$2a$

P4.7-4

Prob. 4.7-5. The "nose" of the cross section in Fig. P4.7-5 is semielliptical. See the problem statement preceding Prob. 4.7-1 and use $T = 800$ N · m.

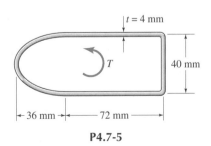

$t = 4$ mm

T 40 mm

36 mm 72 mm

P4.7-5

Prob. 4.7-6. See the problem statement preceding Prob. 4.7-1 and use $T = 25$ kip $\cdot$ in.

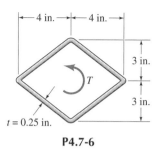

- 4 in. - 4 in. -
3 in.
3 in.
$t = 0.25$ in.
T

P4.7-6

Prob. 4.7-7. See the problem preceding Prob. 4.7-1 and use $T = 4$ kN $\cdot$ m, $a = 90$ mm, $b = 120$ mm, $c = 150$ mm, $t_1 = 3$ mm, and $t_2 = 5$ mm.

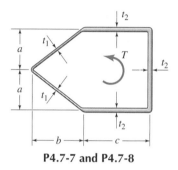

t_2
t_1
a
T
t_2
a
t_1
t_2
b c

P4.7-7 and P4.7-8

Prob. 4.7-8. Repeat Prob. 4.7-7 using $T = 500$ kip $\cdot$ in., $a = 9$ in., $b = 12$ in., $c = 15$ in., $t_1 = 0.375$ in., and $t_2 = 0.625$ in.

Prob. 4.7-9. A tubular shaft having an inside diameter of $d_i = 2.0$ in. and a wall thickness of $t = 0.20$ in., is subjected to a torque $T = 5$ kip $\cdot$ in. (Note: The wall thickness is one-tenth of the inner diameter so that you can examine the range of validity of thin-wall torsion theory.) Determine the maximum shear stress in the tube: (a) using the exact theory for torsion of circular shafts, and (b) using the (approximate) thin-wall torsion theory of this section.

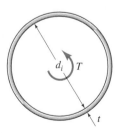

d_i T
t

P4.7-9 and P4.7-10

Prob. 4.7-10. For the same tubular shaft of Prob. 4.7-9, determine the angle of twist per unit length: (a) using the exact theory for

torsion of circular shafts, and (b) using the (approximate) thin-wall torsion theory of this section. Let $T = 5$ kip $\cdot$ in., as before, and let $G = 4 \times 10^3$ ksi.

*Prob. 4.7-11. A thin-wall tube has uniform thickness t and cross-sectional dimensions b and h, measured to the median line of the cross section. Let the length of the median curve, $L_m = 2b + 2h$, and the thickness t be constant (hence, the weight will be constant), but let the ratio $\alpha \equiv \frac{b}{h}$ ($b \geq h$) vary. (a) Determine an expression that relates the maximum shear stress in a rectangular tube to the ratio α. (b) From your result in (a), show the maximum shear stress will be smallest when the tube is square (i.e., when $\alpha = 1$).

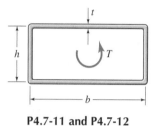

t
h
T
b

P4.7-11 and P4.7-12

*Prob. 4.7-12. Repeat Prob. 4.7-11, but consider the angle of twist, rather than maximum shear stress. That is, (a) determine an expression that relates the angle of twist in a rectangular tube to the ratio α and (b) show that the angle of twist per unit length is smallest for a square tube.

Prob. 4.7-13. A 4-in.-square steel tube has a wall thickness $t = 0.25$ in. You are to compare the torsion behavior of the square tube to that of a tube with circular cross section having the same median-curve length ($L_m = 16$ in.). Use thin-wall torsion theory to (a) determine the shear-stress ratio τ_c/τ_s, where τ_c and τ_s are the maximum shear stresses in the circular tube and the square tube, respectively, when both members are subjected to a torque $T = 80$ kip $\cdot$ in. (b) Determine the ratio J_c/J_s, where J_c and J_s are the respective circle and square area properties defined by the torsional stiffness equation, Eq. 4.37.

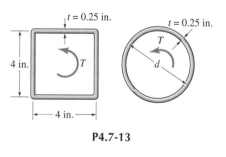

$t = 0.25$ in.
$t = 0.25$ in.
4 in.
T
T
d
4 in.

P4.7-13

Prob. 4.8-1. A torsion member with square cross section is subjected to end torques, as shown in Fig. P4.8-1. Let $d = 1.0$ in., $L = 20$ in., $G = 11 \times 10^3$ ksi, and $T = 50$ lb $\cdot$ in. (a) Determine

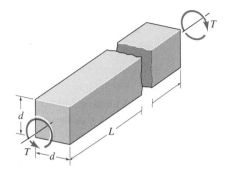

P4.8-1 and P4.8-2

the maximum shear stress in the shaft. (b) Determine the angle of twist of the shaft. (c) Determine the cross-sectional area of a circular shaft that would carry the same torque as this square shaft without any increase in maximum shear stress.

Prob. 4.8-2. Repeat Prob. 4.8-1 using $d = 50$ mm, $L = 2$ m, $G = 80$ GPa, and $T = 500$ N · m.

Prob. 4.8-3. A torsion member with elliptical cross section is subjected to end torques, as illustrated in Fig. P4.8-3. Let $a = 1.0$ in., $b = 0.5$ in., $L = 24$ in., $G = 11 \times 10^3$ ksi, and $T = 3000$ lb · in. (a) Determine the maximum shear stress in the shaft. (b) Determine the angle of twist of the shaft. (c) Determine the cross-sectional area of the elliptical cross section of this shaft. If the same torque T were to be applied to a shaft with circular cross section having the same cross-sectional area as this elliptical shaft, by what percent would the maximum shear stress decrease?

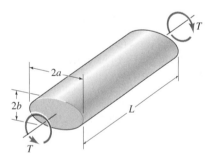

P4.8-3 and P4.8-4

Prob. 4.8-4. Repeat Prob. 4.8-3 using $a = 40$ mm, $b = 30$ mm, $L = 2$ m, $G = 80$ GPa, and $T = 2.5$ kN · m.

Prob. 4.8-5. (a) Determine the ratio of the maximum shear stress in a rectangular cross section to the maximum shear stress in an elliptical cross section,

$$\frac{(\tau_{max})_{rect.}}{(\tau_{max})_{ell.}}$$

if

$$\left(\frac{d}{t}\right)_{rect.} = \left(\frac{a}{b}\right)_{ell.} = 2$$

and if both torsion bars have the same cross-sectional area and are subjected to the same torque. (b) Determine the ratio of twist rates, $(\phi/L)_r/(\phi/L)_e$, for the same two torsion bars.

Prob. 4.8-6. The torsion of thin-wall prismatic members, like angle sections and channel sections, can be analyzed by using Eqs. 4.38 and 4.39, taking the dashed centerline length, as in Fig. P4.8-6, as the dimension d in the formulas. For the cross section shown in Fig. P4.8-6, (a) determine the maximum shear stress due to torsion, and (b) determine the angle of twist of a torsion bar of length L. Use $G = 11 \times 10^3$ ksi. (You will have to interpolate the values given in Table 4.2 to get α and β for the d/t ratio of this cross section.)

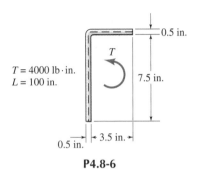

P4.8-6

Prob. 4.8-7. Repeat Prob. 4.8-6 for a channel-shaped torsion member whose cross section is shown in Fig. P4.8-7. Use $G = 80$ GPa.

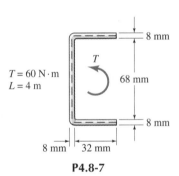

P4.8-7

***Prob. 4.8-8.** A thin-wall-torsion member with square cross section has a wall thickness equal to one-tenth of its outer cross-sectional dimension, d. The member is subjected to pure torsion by torques of magnitude T. (a) Use the thin-wall torsion theory of Section 4.7 to estimate the maximum shear stress on the cross section. (b) Use the prismatic-torsion-bar theory of Section 4.8

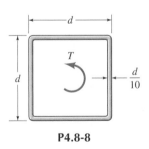

P4.8-8

to determine an expression for the maximum shear stress. (Hint: Subtract a solution for the inner hole from a solution for a solid bar, forcing both to have the same twist rate.) Discuss your results in Parts (a) and (b).

*Prob. 4.8-9. A thin-wall-torsion member with rectangular cross section has a wall thickness equal to one-sixth of its shorter cross-sectional dimension. The member is subjected to pure torsion by torques of magnitude T. The shear modulus is G, and the length is L. (a) Use the thin-wall torsion theory of Section 4.7 to estimate the angle of twist of the cross section. (b) Use the prismatic-torsion-bar theory of Section 4.8 to determine an expression for the angle of twist. (Hint: Subtract a solution for the inner hole from a solution for a solid bar, forcing both to have the same twist rate.) Would you say that thin-wall torsion theory is appropriate for analyzing a torsion member with these dimensions?

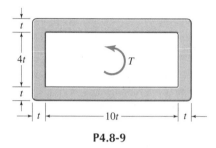

P4.8-9

*Prob. 4.8-10. A thin-wall-torsion member with elliptical cross section has a wall thickness equal to one-twelfth of its outer major principal axis; one sixth of its outer minor principal axis. The member is subjected to pure torsion by torques of magnitude T. (a) Use the thin-wall torsion theory of Section 4.7 to estimate the maximum shear stress on the cross section. (b) Use the prismatic-torsion-bar theory of Section 4.8 to determine an expression for the maximum shear stress. (Hint: Subtract a solution for the inner hole from a solution for a solid bar, forcing both to have the same twist rate.) Compare your results for Parts (a) and (b).

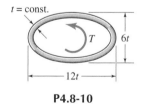

P4.8-10

Prob. 4.9-1. A 3-in.-diameter solid circular shaft made of elastic, perfectly plastic material is subjected to a torque T that produces partially plastic deformation with an elastic core of radius $r_Y = 0.5$ in. (a) Sketch the shear-stress distribution in the shaft. (b) Determine the torque required to produce this partially plastic stress distribution. (c) Determine the angle of twist of the shaft, which is 4 ft. long.

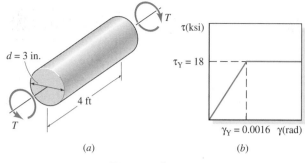

P4.9-1 and P4.9-2

Prob. 4.9-2. For the shaft in Fig. P4.9-2, (a) determine the value of the yield torque T_Y. (b) Determine the angle of twist, ϕ, for the following values of torque, T: T_Y, 1.1 T_Y, 1.2 T_Y, and 1.3 T_Y. Sketch the T versus ϕ curve for $0 \le T \le 1.3T_Y$. (c) Determine the fully plastic torque, T_P, for this shaft.

Prob. 4.9-3. A 50-mm-diameter shaft is made of elastic, perfectly plastic material and is subjected to a torque $T = 4$ kN · m. (a) Verify that this torque produces a partially-plastic stress distribution, that is, show that $T_Y \le 4$ kN · m $\le T_P$. (b) Determine the radius, r_Y, of the elastic core produced by the 4 kN · m torque. (c) Determine the angle of twist of this 2-m-long shaft.

Prob. 4.9-4. For the shaft in Fig. P4.9-4, (a) determine the value of the yield torque, T_Y. (b) Determine the angle of twist, ϕ, for the following values of torque, T: T_Y, 1.1 T_Y, 1.2 T_Y, and 1.3 T_Y. Sketch the T versus ϕ curve for $0 \le T \le 1.3T_Y$. (c) Determine the fully-plastic torque, T_P, for this shaft.

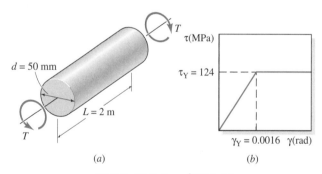

P4.9-3, P4.9-4, and P4.9-10

*Prob. 4.9-5. A tubular shaft is made of elastic, perfectly plastic material with shear modulus $G = 6 \times 10^3$ ksi and yield stress $\tau_Y = 4$ ksi. The dimensions of the shaft are: $d_o = 1$ in., $d_i = 0.6$ in., and $L = 30$ in. (a) Determine the yield torque, T_Y, and the fully plastic torque, T_P, for this shaft. (b) What percent of the cross-sectional area of the shaft has yielded when $T = 1.1 T_Y$? (c) Determine the angle of twist of the shaft when the torque is just enough to cause yielding at the inner surface of the tubular shaft.

*Prob. 4.9-6. A tubular shaft is made of elastic, perfectly-plastic material with shear modulus $G = 40$ GPa and yield stress $\tau_Y =$

227

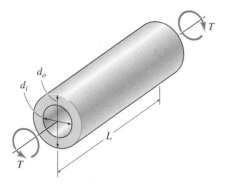

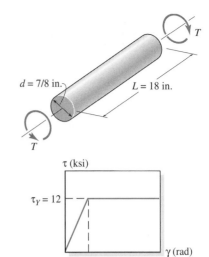

P4.9-5, P4.9-6, P4.9-11, and P4.9-12

30 MPa. The dimensions of the shaft are: $d_o = 40$ mm, $d_i = 25$ mm, and $L = 1$ m. (a) Determine the yield torque, T_Y, and the fully plastic torque, T_P, for this shaft. (b) What percent of the cross-sectional area of the shaft has yielded when $T = 1.1\ T_Y$? (c) Determine the angle of twist of the shaft when the torque is just enough to cause yielding at the inner surface of the tubular shaft.

Prob. 4.9-7. Derive the formula, similar to Eq. 4.51, that relates the partially plastic torque, T, to the outer radius, r_Y, of the elastic core of an elastic, perfectly plastic tubular shaft (Fig. P4.9-7) having outer radius r_o, inner radius r_i, and yield stress in shear, τ_Y.

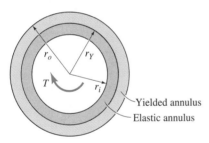

Yielded annulus
Elastic annulus

P4.9-7

Prob. 4.9-8. The two ends of a solid circular shaft of diameter $d = \frac{7}{8}$ in. and length $L = 18$ in. are rotated with respect to each other by exactly one revolution (i.e., $\phi = 2\pi$ rad). The shaft is made of elastic, perfectly plastic material with τ versus γ curve as shown in Fig. P4.9-8b. (a) Determine the maximum shear strain for this loading condition. (b) Determine the radius of the elastic core for this loading condition.

*Prob. 4.9-9. The two ends of the solid circular shaft in Fig. P4.9-9a are rotated to each other by an angle of $\phi = 6°$. The shaft is made of elastic, perfectly-plastic material with τ versus γ curve as shown in Fig. P4.9-9b. (a) Determine the torque required to produce this twist angle. (b) If the torque determined in Part (a) is completely removed, what is the permanent angle of twist left in the shaft? (c) What is the residual stress distribution in the shaft after the torque is removed?

P4.9-8 and P4.9-9

*Prob. 4.9-10. The solid circular shaft in Fig. P4.9-10 (see Prob. 4.9-3) is subjected to a torque T that produces a twist angle $\phi = 10°$ between the two ends of the shaft. (a) Determine the torque required to produce this twist angle. (b) If the torque determined in Part (a) is completely removed, what is the permanent angle of twist left in the shaft? (c) What is the residual stress distribution in the shaft after the torque is removed?

*Prob. 4.9-11. A tubular shaft with dimensions $d_o = 1$ in., $d_i = 0.6$ in., and $L = 30$ in. (see Prob. 4.9-5) is made of elastic, perfectly plastic material with shear modulus $G = 6 \times 10^3$ ksi and yield stress $\tau_Y = 4$ ksi. The shaft is subjected to a torque T that is sufficient to cause yielding from $r_Y = 0.4$ in. to the outer surface. (a) Determine the value of the torque T required to produce this state of stress in the shaft. (b) What is the shear strain at the outer surface for this loading? (c) If the torque determined in Part (a) is completely removed, what is the permanent angle of twist of the shaft? (d) Sketch the residual stress distribution after removal of the original torque. Indicate the value of the shear stress at the outer surface of the tubular shaft and the shear stress at the inner surface.

*Prob. 4.9-12. A tubular shaft with dimensions $d_o = 40$ mm, $d_i = 20$ mm, and $L = 1$ m (see Prob. 4.9-5) is made of elastic, perfectly plastic material with shear modulus $G = 40$ GPa and yield stress $\tau_Y = 30$ MPa. The shaft is subjected to a torque T that is sufficient to cause yielding from $r_Y = 15$ mm to the outer surface. (a) Determine the value of the torque, T, required to produce this state of stress in the shaft. (b) What is the shear strain at the outer surface for this loading? (c) If the torque determined in Part (a) is completely removed, what is the permanent angle of twist of the shaft? (d) Sketch the residual stress distribution after removal of the original torque. Indicate the value of the shear stress at the outer surface of the tubular shaft and the shear stress at the inner surface.

EQUILIBRIUM OF BEAMS

5

5.1 INTRODUCTION

The behavior of slender members subjected to axial loading and to torsional loading was discussed in Chapter 3 and Chapter 4, respectively. Now we turn our attention to the problem of determining the stress distribution in, and the deflection of, beams.

> A **beam** is a structural member that is designed to support transverse loads, that is, loads that act perpendicular to the axis of the beam. A beam resists applied loads by a combination of internal transverse shear force and bending moment.

Figure 5.1 shows steel beams (lower right) and concrete beams (center) that will support bridge decking (roadway) and the vehicles that pass over the bridges.

FIGURE 5.1 Several bridge beams during bridge construction.

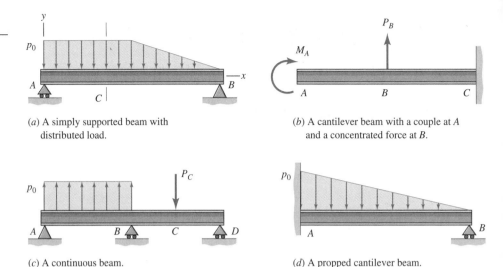

(a) A simply supported beam with distributed load.

(b) A cantilever beam with a couple at A and a concentrated force at B.

(c) A continuous beam.

(d) A propped cantilever beam.

FIGURE 5.2 Examples of beams with various types of loads and supports.

The loads applied to beams are distributed transverse loads and/or concentrated forces and couples. Figure 5.2a illustrates a *simply supported beam* with *distributed loading*; a *cantilever beam* with *concentrated loads* is illustrated in Fig. 5.2b. A *continuous beam* and a *propped cantilever beam* are shown in Figs. 5.2c and 5.2d, respectively.

The symbols used for idealized supports in Fig. 5.2 are to be interpreted, on free-body diagrams for example, as indicated in Table 1.1 in Section 1.4. The force and moment components at a support depend on the type of support; they are called *reactions* since they react to the applied loads. The roller symbol in Table 1.1(1) depicts an actual support that can either push or pull on the beam; that is, A_y can be either negative or positive, depending on the loads on the beam. Likewise, the pin-support symbol in Table 1.1(3) stands for a frictionless pin at the axis of the beam. The cantilever support in Table 1.1(4) is capable of exerting a moment M_A as well as the force components A_x and A_y.

Consider now the simply supported beam in Fig. 5.2a. The downward distributed load gives rise to upward reactions at the supports at A and B. The roller symbol at A implies that the reaction force can have no horizontal component. If we pass an imaginary cutting plane at C, as indicated in Fig. 5.2a, and we draw separate free-body diagrams of AC and CB (Fig. 5.3), we see that a transverse shear force V_C and a bending moment M_C must act at section C to maintain the force equilibrium and moment equilibrium of these two adjoining free bodies. The principle of action and reaction determines the relationship of the directions of V_C and M_C on the two free-body diagrams.

The *sign conventions for internal stress resultants* are illustrated in Fig. 5.4. The sign

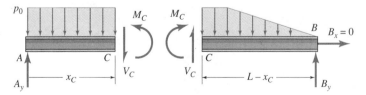

FIGURE 5.3 The transverse shear force (V) and bending moment (M) at a cross section.

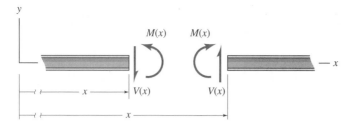

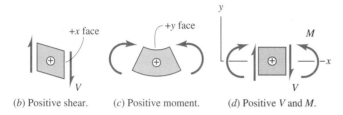

(a) Positive V and M on section "x."

(b) Positive shear. (c) Positive moment. (d) Positive V and M.

FIGURE 5.4 The sign convention for internal stress resultants $V(x)$ and $M(x)$.

conventions may be stated in words as follows:

- A *positive shear force*, V, acts in the $-y$ direction on a $+x$ face.
- A *positive bending moment*, M, makes the $+y$ face of the beam concave.

Figures 5.4b and 5.4c illustrate the physical meaning of positive shear force and positive bending moment, while Fig. 5.4d summarizes the sign conventions for the internal stress resultants in beams. **It is very important to observe these sign conventions for V and M**, because equations will be developed to relate the stress distribution in beams and the deflection of beams to these two stress resultants.

The discussion of beams is divided into three chapters. In the present chapter we concentrate on *equilibrium of beams*, and we solve for the shear force and bending moment in various types of beams subjected to various loading conditions. In Chapter 6 we introduce displacement assumptions that permit us to determine the *normal-stress distribution* associated with the bending moment M; then we determine the *shear-stress distribution* associated with the transverse shear force V. Finally, in Chapter 7 we solve for the *deflection of beams*, including statically indeterminate beams.

5.2 EQUILIBRIUM OF BEAMS USING FINITE FREE-BODY DIAGRAMS

To determine the stress distribution in a beam or to determine the deflected shape of a beam under load, we need to consider *equilibrium, material behavior,* and *geometry of deformation.* In the remainder of Chapter 5 we will concentrate on equilibrium of beams. That is, we will draw free-body diagrams and write equilibrium equations in order to relate the shear-force and bending-moment stress resultants to the applied loads. Several examples that illustrate the use of finite-length free-body diagrams are given in this section. In Section 5.3 we will employ infinitesimal free-body diagrams, and in Section 5.4 shear and moment diagrams are discussed. Finally, in Section 5.5 discontinuity functions are used to represent loads, shear, and moment.

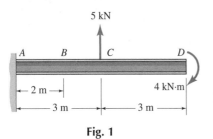

Fig. 1

The cantilever beam AD in Fig. 1 is subjected to a concentrated force of 5 kN at C and a couple of 4 kN·m at D. Determine the shear V_B and bending moment M_B at a section 2 m to the right of the support A.

Plan the Solution   We can use either a free-body diagram of AB or a free-body diagram of BD. Since the former would require us to compute the support reactions at A, we will, instead, use a free-body diagram of BD.

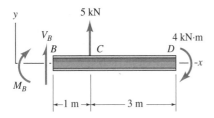

Fig. 2   A free-body diagram.

Solution   Let us draw a free-body diagram of BD and show the shear force V_B and bending moment M_B in the positive sense according to the sign convention in Fig. 5.4 (Figure 2).

$$+\uparrow \ \sum F_y = 0: \qquad\qquad V_B + 5 \text{ kN} = 0$$

$$V_B = -5 \text{ kN} \qquad\qquad \textbf{Ans.}$$

$$+\circlearrowleft \left(\sum M\right)_B = 0: \quad M_B - (5 \text{ kN})(1 \text{ m}) + 4 \text{ kN·m} = 0$$

$$M_B = 1 \text{ kN·m} \qquad\qquad \textbf{Ans.}$$

Review the Solution   To satisfy force and moment equilibrium of the cantilever beam, there are internal stress resultants at B as shown in Fig. 3. As a check, we should be able to satisfy $\sum M = 0$ about any point, for example point C.

$$\text{Is} \left(\sum M\right)_C = 1 \text{ kN·m} - (5 \text{ kN})(1 \text{ m}) + 4 \text{ kN·m} = 0? \qquad \text{Yes}$$

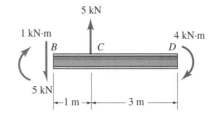

Fig. 3   A free-body diagram.

In the preceding example, the shear force and bending moment were required at a specific cross section. Using a finite free body permits these values to be determined directly from the corresponding equilibrium equations. The finite-free-body approach is also useful when expressions for $V(x)$ and $M(x)$ are required over some portion of the beam. Example 5.2 illustrates this type of problem and also illustrates a way to handle distributed loads.

The simply supported beam AC in Fig. 1 is subjected to a distributed downward loading as shown. The load varies linearly between B and C. (a) Determine the reactions at A and C, (b) determine expressions for $V(x)$ and $M(x)$ for $0 < x \leq 6$ ft, and (c) determine expressions for $V(x)$ and $M(x)$ for 6 ft $\leq x < 12$ ft.

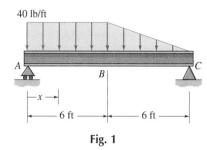

Fig. 1

Plan the Solution By using a free-body diagram of the entire beam, AC, we can determine the reactions at A and C. We will need to make a "cut" between A and B to determine the shear force and bending moment required in Part (b), and we will need to make a "cut" between B and C to answer Part (c). The distributed loads can be replaced, on each free-body diagram, by their resultants.

Solution

(a) Determine the reactions. On the free-body diagram in Fig. 2, the resultants of the uniform distributed load on AB and the linearly varying distributed load on BC are shown with dashed-arrow symbols. The reactions at A and C can now be determined from three equilibrium equations.

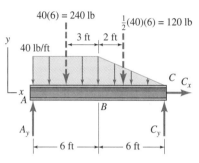

Fig. 2 A free-body diagram of beam AC.

$$+\circlearrowleft \left(\sum M \right)_A = 0: \quad (240 \text{ lb})(3 \text{ ft}) + (120 \text{ lb})(8 \text{ ft}) - C_y(12 \text{ ft}) = 0$$

$$C_y = 140 \text{ lb} \qquad \text{Ans. (a)}$$

$$+\circlearrowleft \left(\sum M \right)_C = 0: \quad A_y(12 \text{ ft}) - (240 \text{ lb})(9 \text{ ft}) - (120 \text{ lb})(4 \text{ ft}) = 0$$

$$A_y = 220 \text{ lb} \qquad \text{Ans. (a)}$$

$$\xrightarrow{+} \sum F_x = 0: \quad C_x = 0 \qquad \text{Ans. (a)}$$

(Note: From now on we will ignore axial reactions and internal axial forces on beams that have no axial applied loads.)

Before going on to parts (b) and (c), we should check the above answers.

$$+\uparrow \sum F_y = 0: \quad \text{Is} \quad 220 + 140 - 240 - 120 = 0? \quad \text{Yes.}$$

(b) Determine expressions for $V(x)$ and $M(x)$ for $0 < x \leq 6$ ft. To do this, we can make a cut between A and B and designate this portion "Interval 1," giving the free-body diagram in Fig. 3. Let us use the symbols V_1 and M_1 for expressions that are valid in this interval. The resultant of the uniform distributed load is shown as a dashed arrow. The shear and moment are shown in the positive sense according to the sign conventions of Fig. 5.4.

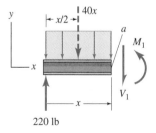

Fig. 3 A free-body diagram for Interval 1.

$$+\uparrow \sum F_y = 0: \quad 220 \text{ lb} - (40x)\text{lb} - V_1 = 0$$

$$V_1 = (220 - 40x)\text{lb}, \ 0 < x \leq 6 \text{ ft} \qquad \text{Ans. (b)}$$

$$+\circlearrowleft \left(\sum M \right)_a = 0: \quad (220 \text{ lb})(x \text{ ft}) - (40x \text{ lb})(\tfrac{x}{2} \text{ ft}) - M_1 = 0$$

$$M_1 = (220x - 20x^2) \text{ lb·ft}, \ 0 < x \leq 6 \text{ ft} \qquad \text{Ans. (b)}$$

233

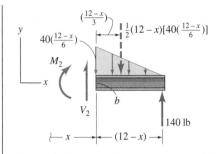

Fig. 4 A free-body diagram for Interval 2.

(c) Determine expressions for $V(x)$ and $M(x)$ for 6 ft $\leq x < 12$ ft. To do this, we can take either a free-body diagram from A to a cut between B and C, or a free-body diagram from the cut to the right end, C. The latter, which is shown in Fig. 4, will be easier because we will then only have to deal with a single triangular load.

$$+\uparrow \sum F_y = 0: \qquad V_2 - \frac{40}{12}(12 - x)^2 \text{ lb} + 140 \text{ lb} = 0$$

$$V_2 = \left[-140 + \frac{10}{3}(12 - x)^2 \right] \text{ lb, } 6 \text{ ft} \leq x < 12 \text{ ft} \qquad \textbf{Ans. (c)}$$

$$+\circlearrowleft \left(\sum M \right)_b = 0:$$

$$M_2 + \left[\frac{40}{12}(12 - x)^2 \text{ lb} \right]\left[\left(\frac{12 - x}{3} \right) \text{ ft} \right] - (140 \text{ lb})[(12 - x) \text{ ft}] = 0$$

$$M_2 = \left[140(12 - x) - \frac{10}{9}(12 - x)^3 \right] \text{ lb}\cdot\text{ft, } 6 \text{ ft} \leq x < 12 \text{ ft} \quad \textbf{Ans. (c)}$$

Review the Solution We have already, in Part (a), performed an equilibrium check on the reactions. Since there is no concentrated transverse load or couple at B, we should have $V_1(x = 6 \text{ ft}) = V_2(6 \text{ ft})$ and $M_1(6 \text{ ft}) = M_2(6 \text{ ft})$.

$$V_1(6 \text{ ft}) = 220 - 40(6) = -20 \text{ lb}$$

$$V_2(6 \text{ ft}) = -140 + \frac{10}{3}(6)^2 = -20 \text{ lb} = V_1(6 \text{ ft})$$

$$M_1(6 \text{ ft}) = 220(6) - 20(6)^2 = 600 \text{ lb}\cdot\text{ft}$$

$$M_2(6 \text{ ft}) = 140(6) - \frac{10}{9}(6)^3 = 600 \text{ lb}\cdot\text{ft} = M_1(6 \text{ ft})$$

In Section 5.4 shear and moment diagrams will be used to graphically represent $V(x)$ and $M(x)$.

In both Example 5.1 and Example 5.2 it was possible for us to solve the equilibrium equations and to determine values of (or expressions for) shear and moment, since in each case the beam is statically determinate. The next example illustrates the type of equilibrium results that are obtained for statically indeterminate beams.

███████████████████ E X A M P L E 5 . 3 ███████████████████

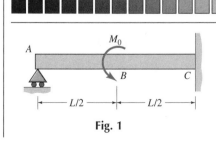

Fig. 1

The propped cantilever beam AC in Fig. 1 has a couple applied at its center B. Determine expressions for the reactions (i.e., the shear force and bending moment) at C in terms of the applied couple M_0 and the reaction at A.

Plan the Solution A free-body diagram of the whole beam will permit us to relate the reactions at C to the reaction at A.

Solution

$$+\!\uparrow \; \sum F_y = 0: \qquad A_y + C_y = 0$$

$$C_y = -A_y \qquad \text{Ans.}$$

$$+\circlearrowleft \left(\sum M\right)_C = 0: \quad A_y L - M_0 - M_C = 0$$

$$M_C = A_y L - M_0 \qquad \text{Ans.}$$

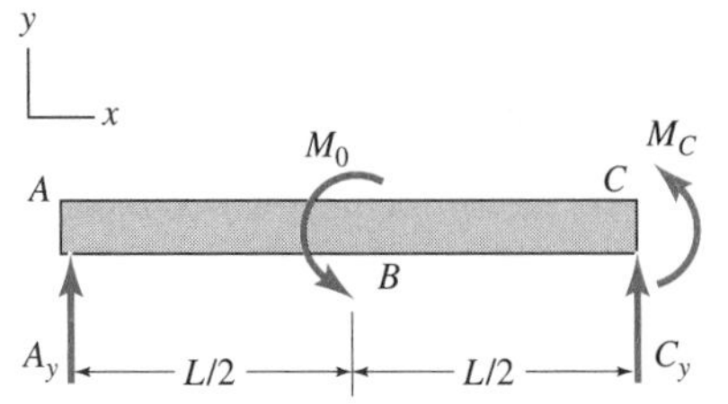

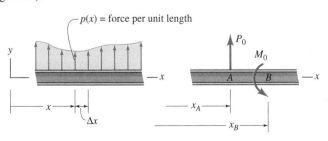

Fig. 2 A free-body diagram of beam AC.

Review the Solution The results above are typical of *statically indeterminate problems,* that is, problems where the equations of statics (i.e., equilibrium) are not sufficient to determine all of the unknowns. Here we have three unknowns, but only two equilibrium equations, other than the trivial one for horizontal equilibrium. Therefore, two reactions can be written in terms of the third reaction (here A_y). This one is called a *redundant.* That is, the support at A is not essential to prevent collapse of the beam.

Just as for statically indeterminate axial deformation and torsion problems, in order to solve statically indeterminate beam problems we must consider the deformation of the beam. Statically indeterminate beam problems are examined in Chapter 7, *Deflection of Beams.*

5.3 EQUILIBRIUM RELATIONSHIPS AMONG LOADS, SHEAR FORCE, AND BENDING MOMENT

In the previous section we used finite free-body diagrams to determine values of shear force and bending moment at specific cross sections, and to determine expressions for $V(x)$ and $M(x)$ over specified ranges of x. Here we use *infinitesimal free-body diagrams* to obtain equations that relate the external loading to the internal shear force and bending moment. These expressions will be especially helpful in Section 5.4, where we discuss shear and moment diagrams. In addition to the sign conventions for shear force and bending moment, given in Fig. 5.4, we need to adopt a *sign convention for external loads* (Fig. 5.5).

- *Positive distributed loads* and *positive concentrated loads* act in the $+y$ direction (e.g., loads $p(x)$ and P_0 in Fig. 5.5).
- A *positive external couple* acts in a right-hand-rule sense with respect to the z axis, that is, counterclockwise as viewed in the xy plane (e.g., the external couple M_0 in Fig. 5.5*b*).

FIGURE 5.5 The sign convention for external loads on a beam.

(*a*) Distributed load.

(*b*) Concentrated force and couple.

235

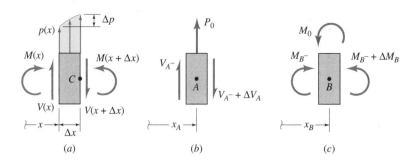

FIGURE 5.6 Infinitesimal free-body diagrams.

First, let us consider a portion of the beam where there are no concentrated external loads, and let us establish equilibrium equations for an infinitesimal free-body diagram. Take the segment of beam from x to $(x + \Delta x)$ in Fig. 5.5a, as redrawn in Fig. 5.6a. For equilibrium of the free body in Fig. 5.6a.

$$+\uparrow \sum F_y = 0: \qquad V(x) + p(x)\,\Delta x + \mathcal{O}(\Delta p \cdot \Delta x) - V(x + \Delta x) = 0$$

where $\mathcal{O}(\cdots)$ means "of the order of." Collecting terms and dividing by Δx, we get

$$\frac{V(x + \Delta x) - V(x)}{\Delta x} = p(x) + \frac{\mathcal{O}(\Delta p \cdot \Delta x)}{\Delta x}$$

By taking the limit as $\Delta x \to 0$, we get

$$\boxed{\frac{dV}{dx} = p(x)} \tag{5.1}$$

since the limit of the $\mathcal{O}(\cdot)$ term is zero. To satisfy moment equilibrium for the free body in Fig. 5.6a, we can take moments about point C at $(x + \Delta x)$.

$$+\zeta \left(\sum M\right)_C = 0:$$

$$M(x) - M(x + \Delta x) + p(x)\frac{(\Delta x)^2}{2} + \mathcal{O}(\Delta p \cdot \Delta x^2) + V(x)\Delta x = 0$$

Dividing through by Δx and taking the limit as $\Delta x \to 0$, we obtain

$$\boxed{\frac{dM}{dx} = V(x)} \tag{5.2}$$

Wherever there is an external concentrated force, such as P_0 in Fig. 5.5b, or a concentrated couple, such as M_0 in Fig. 5.5c, there will be a step change in shear or moment, respectively. From the partial free-body diagram in Fig. 5.6b (moments have been omitted for clarity),

$$+\uparrow \sum F_y = 0: \qquad V_{A^-} - (V_{A^-} + \Delta V_A) + P_0 = 0$$

$$\boxed{\Delta V_A = P_0} \tag{5.3}$$

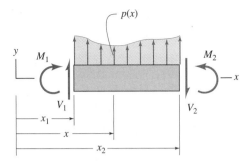

FIGURE 5.7 A free-body diagram of a finite portion of a beam.

where V_{A^-} is the (internal) shear force just to the left of the point x_A where P_0 is applied. That is, a concentrated force P_0 at coordinate x_A will cause a step change ΔV_A in shear having the same sign as P_0.

An external couple M_0 at coordinate x_B causes a step change in the moment at x_B. From Fig. 5.6c (shear forces have been omitted for clarity),

$$+\circlearrowleft \left(\sum M\right)_B = 0: \qquad M_{B^-} - (M_{B^-} + \Delta M_B) - M_0 = 0$$

or

$$\boxed{\Delta M_B = -M_0} \tag{5.4}$$

Equations 5.1 and 5.2 are differential equations relating the distributed load $p(x)$ to the shear force $V(x)$, and the shear force $V(x)$ to the bending moment $M(x)$. Let $x_1 \leq x \leq x_2$ be a portion of the beam that is free of concentrated forces or couples (Fig. 5.7). We can integrate Eqs. 5.1 and 5.2 over this portion of the beam to get

$$\int_{x_1}^{x_2} \frac{dV}{dx}dx = V_2 - V_1 = \int_{x_1}^{x_2} p(x)dx$$

or

$$\boxed{V_2 - V_1 = \int_{x_1}^{x_2} p(x)dx} \tag{5.5}$$

Similarly, from Eq. 5.2,

$$\boxed{M_2 - M_1 = \int_{x_1}^{x_2} V(x)dx} \tag{5.6}$$

Equations 5.5 and 5.6 can be stated in words as follows:

- The *change in shear* from Section 1 to Section 2 is equal to the *area under the load diagram* from 1 to 2. (The "area" that results from negative $p(x)$ is negative.)
- The *change in moment* from Section 1 to Section 2 is equal to the *area under the shear diagram* from 1 to 2. (The "area" that results from negative $V(x)$ is negative.)

Equations 5.1 through 5.6 will be very useful to us in Sections 5.4 and 5.5, where we draw shear and moment diagrams. And we can employ modifications of Eqs. 5.5 and 5.6

to determine expressions for $V(x)$ and $M(x)$. Thus,

$$V(x) = V_1 + \int_{x_1}^{x} p(\xi)d\xi \qquad (5.7)$$

and

$$M(x) = M_1 + \int_{x_1}^{x} V(\xi)d\xi \qquad (5.8)$$

As a simple example of the application of Eqs. 5.1 through 5.8, let us apply Eq. 5.1 to the expression for $V_1(x)$ derived in Example 5.2.

$$\frac{dV_1}{dx} = p_1(x), \qquad 0 < x \le 6 \text{ ft}$$

Since $V_1 = 220 - 40x$,

$$p_1(x) = \frac{dV_1}{dx} = -40 \text{ lb/ft}$$

which is, of course, the value of the distributed load in this interval (Interval 1) of the beam. The negative sign agrees with the downward direction of the distributed load as shown in Fig. 1 of Example 5.2.

Returning to Example 5.2, we can use Eq. 5.8 to determine $M_1(x)$ from $V_1(x)$.

$$M_1(x) - M(0) = \int_{0}^{x} V_1(\xi)d\xi = \int_{0}^{x} (220 - 40\xi)d\xi$$

Since $M(0) = 0$, due to the roller support at A,

$$M_1(x) = (220x - 20x^2) \text{ lb} \cdot \text{ft}$$

as was obtained in Example 5.2 by using a finite free-body diagram.

5.4 SHEAR FORCE AND BENDING MOMENT DIAGRAMS

In Example 5.2 we obtained expressions for $V(x)$ and $M(x)$ for a simply supported beam with distributed loading. However, to *design* a beam (i.e., to select a beam of appropriate material and cross section) we need to ask questions like "What is the maximum value of the shear force, and where does it occur?" and "What is the maximum value of the bending moment, and where does it occur?" These questions are much more readily answered if we have a plot of $V(x)$ and a plot of $M(x)$. These plots are called the *shear diagram* and the *moment diagram*, respectively:

In this section two methods for constructing shear and moment diagrams are described:[1]

- Method 1 (*"Equilibrium Method"*)—Use finite free-body diagrams to obtain expressions for $V(x)$ and $M(x)$, and then plot these expressions; and

---

[1]A third method, which uses Eqs. 5.7 and 5.8 in a very systematic manner, is presented in Section 5.5.

- Method 2 (*"Graphical Method"*)—Make use of Eqs. 5.1 through 5.6 to sketch $V(x)$ and $M(x)$ diagrams.[2]

The following examples illustrate both procedures. Example 5.4 illustrates the first method. Examples 5.5 through 5.7 illustrate the second method.

---

[2]The so-called graphical method is most useful when the "areas" in Eqs. 5.5 and 5.6 are simple rectangles or triangles, that is, when the loads on the beam are either concentrated loads or uniform distributed loads. The graphical method is also useful in interpreting the results of an equilibrium-method solution.

■■■■■■■■■■■■■■□□□ EXAMPLE 5.4 □□□□□□□□□□□□□□■■

Derive expressions for $V(x)$ and $M(x)$ for the cantilever beam with linearly varying load shown in Fig. 1. Use these expressions to plot shear and moment diagrams for this beam.

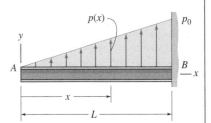

Fig. 1

Plan the Solution   We can use a finite free-body diagram to determine the required expressions for $V(x)$ and $M(x)$.

Solution   The triangle in the free-body diagram is similar to the triangle in the problem statement; so, by similar triangles,

$$\frac{p(x)}{x} = \frac{p_0}{L} \quad \rightarrow \quad p(x) = \left(\frac{x}{L}\right) p_0$$

$+\uparrow \sum F_y = 0:$

$$\left(\frac{x}{2}\right) p(x) - V(x) = 0$$

$$V(x) = \frac{p_0 x^2}{2L}$$

$+\circlearrowleft \left(\sum M\right)_a = 0:$

$$[p(x)]\left(\frac{x}{2}\right)\left(\frac{x}{3}\right) - M(x) = 0$$

$$M(x) = \frac{p_0 x^3}{6L}$$

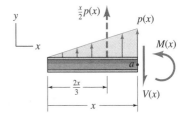

Fig. 2   A free-body diagram.

To plot these expressions for the shear force $V(x)$ and bending moment $M(x)$, we first calculate $V(L)$ and $M(L)$. Figure 3 shows the plots of $V(x)$ and $M(x)$.

$$V(L) = \frac{p_0 L}{2}, \quad M(L) = \frac{p_0 L^2}{6}$$

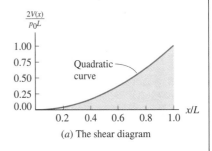

(*a*) The shear diagram

Review the Solution   The shear is positive everywhere, as we expect from the free-body diagram. The maximum shear occurs at the cantilever support at B and is equal to the total area under the load curve in the problem statement. The upward load will bend the beam upward, so it will be concave upward everywhere. This is consistent with the fact that the bending moment is positive for the entire length of the beam. The maximum bending moment occurs at B. Finally, the shear and moment have the proper dimensions, (F) and $(F \cdot L)$, respectively.

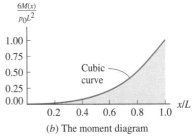

(*b*) The moment diagram

Fig. 3   The shear diagram and the moment diagram.

Examples 5.5 through 5.7 illustrate the graphical method of using Eqs. 5.1 through 5.6 to construct shear and moment diagrams, proceeding from the left end to the right end of the beam. If you study carefully the numbered steps that are used in these examples, you should be able to construct shear and moment diagrams for any beam with simple loading, once you have determined the loads and reactions acting on the beam.

■■■■■■■■■■■■■■□□ E X A M P L E 5 . 5 □□□□□□□□□□■■■■■

Use Eqs. 5.1 through 5.6 to sketch shear and moment diagrams for the simply-supported beam shown in Fig. 1.

Plan the Solution We can use a free-body diagram of the beam AC to determine the reactions at A and C. Since there is no distributed load on the beam, $p(x) = 0$ everywhere. Because of the concentrated load at B, we need to consider two spans, $0 < x < a$ and $a < x < L$.

Solution

Equilibrium—Reactions: To determine the reactions A_y and C_y, we first draw the free-body diagram of the entire beam AC (Fig. 2).

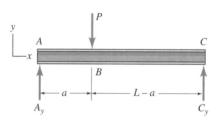

Fig. 2 A free-body diagram.

$$+\circlearrowleft \left(\sum M\right)_A = 0: \quad Pa - C_yL = 0, \quad C_y = P\left(\frac{a}{L}\right)$$

$$+\circlearrowleft \left(\sum M\right)_C = 0: \quad A_yL - P(L - a) = 0, \quad A_y = \frac{P(L - a)}{L}$$

Shear Diagram: Equations 5.1, 5.3, and 5.5 involve the shear. Using these equations, we can sketch $V(x)$ progressively from $x = 0$ to $x = L$. It is convenient to sketch the shear and moment diagrams directly below the loading diagram (Fig. 3a). Each step involved in sketching $V(x)$ is numbered in Fig. 3b.

1. The shear at $x = 0^-$ is zero.
2. The shear at $x = 0^+$ is determined from Eq. 5.3, that is, $\Delta V_A = A_y = \frac{P(L - a)}{L}$. Note that, because of the sign convention for shear, *an upward concentrated force causes an upward jump in the shear diagram.*
3. For $0 < x < a$, $p(x) = 0$. Therefore, from Eq. 5.1, $dV/dx = 0$.
4. At $x = a$ there is a downward force P, so $\Delta V_B = -P$.
5. For $a < x < L$, $p(x) = 0$, so $\frac{dV}{dx} = 0$.
6. The reaction at C causes $\Delta V_C = \frac{Pa}{L}$, which closes the shear diagram back to zero at $x = L^+$.

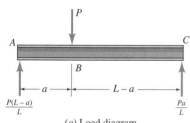

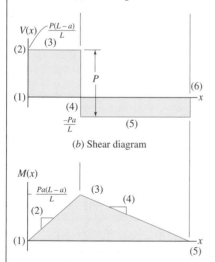

Fig. 1

(a) Load diagram

(b) Shear diagram

(c) Moment diagram

Fig. 3 Shear and moment diagrams.

Moment Diagram: Equations 5.2, 5.4, and 5.6 relate to $M(x)$ and can be used to sketch the moment diagram in Fig. 3c. Steps in the construction of the moment diagram are explained and numbered.

1. The moment at $x = 0$ is zero [simply supported beam].

2. For $0 < x < a$, Eq. 5.2 gives $\dfrac{dM}{dx} = V(x) = \dfrac{P(L - a)}{L} = $ constant.

3. At $x = a$, $M(a)$ can be determined from Eq. 5.6 as the area of the rectangle under the shear curve from $x = 0$ to $x = a$. Therefore $M(a) - 0 = \displaystyle\int_0^a V(x)dx = \dfrac{P(L - a)}{L}(a)$.

4. For $a < x < L$, Eq. 5.2 gives $\dfrac{dM}{dx} = V(x) = \dfrac{-Pa}{L} = $ constant.

5. Equation 5.6 gives $M(L) - M(a) = \displaystyle\int_a^L V(x)dx = -\dfrac{Pa}{L}(L - a)$, which closes the moment diagram back to zero at $x = L$, as it should [simple support at C].

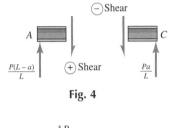

Fig. 4

Review the Solution The dimensions on the shear diagram (F) and the moment diagram ($F \cdot L$) are correct. If we draw finite free-body diagrams of the ends of the beam, we get Fig. 4. Therefore, the shear diagram in Fig. 3b has the correct signs according to the free-body sketches in Fig. 4 and the sign convention in Fig. 5.4. The downward force will bend the beam as shown in Fig. 5, which is consistent with the fact that the bending moment is positive everywhere.

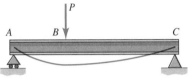

Fig. 5

◼◼◼◼◼◼◼◼◼◼◼◼◻◻◻◻ E X A M P L E 5 . 6 ◻◻◻◼◼◼◼◼◼◼◼◼◼◼◼◼

Determine the reactions and sketch the shear and moment diagrams for the beam shown in Fig. 1. (This beam is said to have an *overhang BC.*) Show all significant values (that is, maxima, minima, positions of maxima and minima, etc.) on the diagrams.

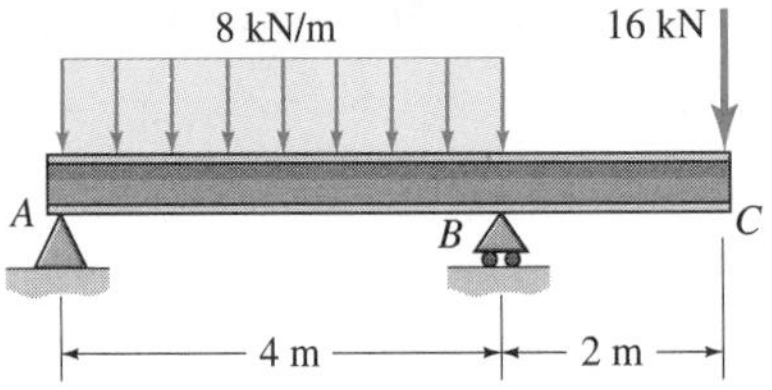

Plan the Solution We can use a free-body diagram of the whole beam to compute the reactions. Then we can use Eqs. 5.1 through 5.6 to sketch the $V(x)$ and $M(x)$ diagrams, as we did in Example 5.5.

Fig. 1

Solution

Equilibrium—Reactions: The reactions must be determined first. Figure 2 shows the appropriate free-body diagram.

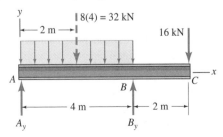

Fig. 2 A free-body diagram.

241

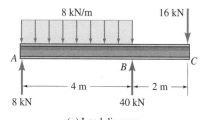

8 kN/m ... 16 kN

A ... B ... C

4 m — 2 m

8 kN ... 40 kN

(a) Load diagram

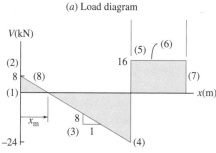

V(kN)

(2)

8 (8)

(1)

(5) (6)

16

(7)

x(m)

x_m

8

(3) 1

−24

(4)

(b) Shear diagram

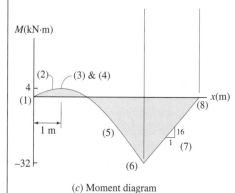

M(kN·m)

(2) (3) & (4)

4

(1)

x(m)

(8)

1 m

(5)

16

1

(7)

−32

(6)

(c) Moment diagram

Fig. 3 Shear and moment diagrams.

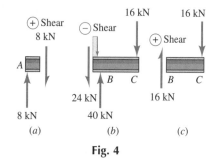

(+) Shear

8 kN

A

8 kN

(a)

(−) Shear

16 kN

B ... C

24 kN

40 kN

(b)

(+) Shear

16 kN

B ... C

16 kN

16 kN

(c)

Fig. 4

242

$+\circlearrowleft \left(\sum M \right)_A = 0:$

$$(8 \text{ kN/m})(4 \text{ m})(2 \text{ m}) + (16 \text{ kN})(6 \text{ m}) - B_y(4 \text{ m}) = 0$$

$$B_y = 40 \text{ kN} \qquad \text{Ans.}$$

$+\circlearrowleft \left(\sum M \right)_B = 0:$

$$A_y(4 \text{ m}) - (8 \text{ kN/m})(4 \text{ m})(2 \text{ m}) + (16 \text{ kN})(2 \text{ m}) = 0$$

$$A_y = 8 \text{ kN} \qquad \text{Ans.}$$

Check: Is $\sum F_y = 0$? $8 - 32 + 40 - 16 = 0$? Yes

It is convenient to sketch the V (Fig. 3b) and M (Fig. 3c) diagrams directly below a sketch of the beam that has all of the loads and reactions shown (Fig. 3a).

Shear Diagram: The following steps are used in sketching the shear diagram (Fig. 3b).

1. $V(0^-) = 0$ [no shear at end of beam].
2. $V(0^+) = 8$ kN [Eq. 5.3].
3. $dV/dx = -8$ kN/m [Eq. 5.1].
4. $V(4^-) = V(0^+) + (-8 \text{ kN/m})(4 \text{ m}) = 8 - 32 = -24$ kN [Eq. 5.5].
5. $V(4^+) - V(4^-) = 40$ kN [Eq. 5.3].
6. $dV/dx = 0$ [Eq. 5.1].
7. $V(6^+) = V(6^-) - 16 = 0$.
8. Since $dV/dx = -8$ for $0 < x < 4$ m, and since $V(0^+) = 8$ kN, $V = 0$ at $x_m = 1$ m [Eq. 5.5].

Moment Diagram: The steps employed in constructing the moment diagram (Fig. 3c) using Eqs. 5.1 through 5.6 will now be described:

1. $M(0)$ [no moment at end of beam].
2. From $dM/dx = V(x)$ we have the slope of $M(x)$ going from $+8$ kN·m/m at $x = 0^+$ to zero at $x = x_m = 1$ m. Therefore, the moment diagram for $0 < x < 1$ m must have the general shape ⌒ [Eq. 5.2].
3. $M(x)$ is maximum where $V(x) = 0$ [Eq. 5.2].
4. $M(1) = M(0) + \int_0^1 V(x)dx = \frac{1}{2}(8 \text{ kN})(1 \text{ m}) = 4$ kN·m [Eq. 5.6; area of triangle].
5. From $x = 1$ m to $x = 4$ m, $V(x)$ gets progressively more negative. Therefore, $M(x)$ must have the general shape ⌒ here [Eq. 5.2].
6. $M(4) = M(0) + \int_0^4 V(x)dx = 0 + \frac{1}{2}(8 \text{ kN})(1 \text{ m}) + \frac{1}{2}(-24 \text{ kN})(3 \text{ m}) = -32$ kN·m [Eq. 5.6; net of areas of triangles]
7. $dM/dx = V(x) = 16$ kN [Eq. 5.2].
8. $M(6) = M(4) + \int_4^6 V(x)dx = -32(\text{kN·m}) + (16 \text{ kN})(2 \text{ m}) = 0$ [Eq. 5.6; no moment at end of beam].

The maximum shear occurs just to the left of the support at B and has a magnitude of 24 kN. The maximum positive moment occurs at $x = 1$ m and has a magnitude of 4 kN·m; and the maximum negative moment occurs at B and has a magnitude of 32 kN·m.

Review the Solution   By imagining cuts just to the right of A (Fig. 4*a*), just to the left of B (Fig. 4*b*), and just to the right of B (Fig. 4*c*), we can check the sign of the shear at these points.

The moment diagram is best checked by seeing if the sign of the moment diagram corresponds to a reasonable deflected shape, that is, concave upward where $M(x)$ is positive and concave downward where $M(x)$ is negative, as illustrated in Fig. 5.4*c*. Where $M(x) = 0$, the beam is locally straight, that is, it is neither concave upward nor concave downward.

We are able to sketch a plausible deflection curve that passes over the supports at A and B and that is concave upward where $M(x)$ is positive and concave downward where M is negative. The distributed load between A and B and the concentrated load at C could, indeed, cause the beam to deflect as sketched.

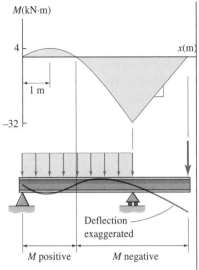

Fig. 5   A sketch showing the deflection of beam AC.

The idea that a positive bending moment makes a beam concave toward the $+y$ side, whereas a negative bending moment causes the beam to be concave toward the $-y$ side, is in accord with the definition of positive bending moment in Fig. 5.4*c*. This fact was used in Examples 5.5 and 5.6 above to check the bending moment diagrams. In the next chapter we derive a mathematical relationship between bending moment and curvature, and in Chapter 7 we use this relationship to obtain expressions for the deflection of beams.

The VM-DIAG computer program, one of the ***MechSOLID*** collection of computer programs, may be used to plot shear and bending moment diagrams. This program, which is based on Section 5.5, is described in Appendix G.5. Computer exercises utilizing VM-DIAG follow the regular homework exercises for Section 5.5.

*5.5   DISCONTINUITY FUNCTIONS TO REPRESENT LOADS, SHEAR, AND MOMENT

Wherever there is a discontinuity in the loading on a beam or where there is a support, there will be a discontinuity in integrals that involve the loads and reactions. For example, the beam in Fig. 5.8 has four intervals—*AB, BC, CD,* and *DE.* Consequently, four separate expressions $V_i(x)$ and four expressions $M_i(x)$ would be required to specify $V(x)$ and $M(x)$ for this beam. The introduction of discontinuity functions simplifies the process of determining expressions for shear and bending moment, and it greatly simplifies the process of solving for the slope and deflection of a beam.[3] For example, for the beam in Fig. 5.8,

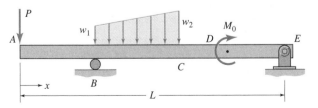

FIGURE 5.8   A beam with several applied loads.

---

[3]Discontinuity functions provide a very systematic way to apply Eqs. 5.7 and 5.8 to plot shear and moment diagrams. They will also be employed in Section 7.7 to solve beam-deflection problems.

243

$V(x)$ can be written as a single compact expression, valid for $0 \le x \le L$, rather than as four separate expressions.

Singularity functions will be used to represent concentrated external forces and concentrated couples, and *Macauley functions* will be used to represent distributed loads. Together, they are referred to as *discontinuity functions*.[4]

Macaulay Functions.

Macaulay functions are useful in expressing functions that are zero up to some particular value of the independent variable and nonzero for larger values of the independent variable. For example, Fig. 5.9 shows the *unit step function* $F_0(x - a)$ and the *unit ramp function* $F_1(x - a)$, where a is the value of the independent variable x at which the discontinuity occurs.

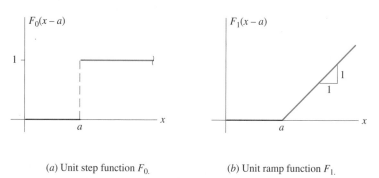

(a) Unit step function F_0.   (b) Unit ramp function F_1.

FIGURE 5.9   Examples of Macaulay Functions.

For integer values of $n \ge 1$, the Macaulay functions F_n, are defined by the following expressions.[5]

$$F_n(x - a) \equiv \langle x - a \rangle^n = \begin{cases} 0 & \text{for } x \le a \\ (x - a)^n & \text{for } x \ge a \end{cases} \quad n = 1, 2, \dots \tag{5.9}$$

As illustrated by the function F_1 in Fig. 5.9, these functions have the value zero for $x \le a$ and the value $(x - a)^n$ for $x \ge a$.[6] The units of F_n are the units of x^n (e.g., ft^n, m^n, etc.).

The *unit step function* $F_0(x - a)$, which is illustrated in Fig. 5.9a, is a special case.[7] It is defined by the expression[8]

$$F_0(x - a) \equiv \langle x - a \rangle^0 = \begin{cases} 0 & \text{for } x < a \\ 1 & \text{for } x \ge a \end{cases} \tag{5.10}$$

---

[4]Sometimes both types of discontinuity functions are called singularity functions.

[5]The English mathematician W. H. Macaulay (1857–1936) introduced the use of special brackets to represent these discontinuity functions. It has been common practice to use angle brackets for this purpose and to refer to them as *Macaulay brackets*.

[6]The definition here allows negative values of x to $x = -\infty$, but since we will use discontinuity functions in conjunction with beams of finite length with their origin at $x = 0$, we need not be concerned with values of x less than $x = 0$, as indicated in Fig. 5.9. The origin, $x = 0$ is assumed to be located such that, for $x \le 0^-$, $F_n \equiv 0$ for all n.

[7]The unit step function is also known as the *Heaviside step function* after the English physicist and electrical engineer Oliver Heaviside (1850–1925).

[8]No purpose is served here by leaving the value of F_0 ambiguous at $x = a$. Therefore, the above definition sets $F_0 = 0$ for $x < 0$ and $F_0 = 1$ for $x \ge a$, that is, $F_0 = 1$ at $x = a$.

From the definition of the higher-order Macaulay functions in Eq. 5.9 and the definition of the unit step function in Eq. 5.10, it can be seen that Macaulay functions can be expressed in terms of the unit step function by the following equation:

$$F_n(x - a) \equiv \langle x - a \rangle^n = (x - a)^n \langle x - a \rangle^0 \quad n = 0, 1, 2, \ldots \quad (5.11)$$

Using the definitions in Eqs. 5.9, we can obtain expressions for derivatives and integrals of Macaulay functions for $n \geq 1$. Thus,

$$\frac{dF_n}{dx} = nF_{n-1} \quad n = 1, 2, \ldots$$

$$\int_{0-}^{x} F_n dx = \frac{1}{n+1} F_{n+1} \quad n = 1, 2, \ldots \quad (5.12)$$

Based on the definition of the unit step function F_0 in Eq. 5.10, the derivative of F_0 does not exist at $x = a$, but

$$\int_{0-}^{x} F_0 dx = F_1 \quad (5.13)$$

Expressions for Macaulay functions and their derivatives and integrals are summarized in Table 5.1.

TABLE 5.1 Discontinuity Functions

	Name	Definition of $F_n(x - a)$	Graph	Derivative and Integral
Singularity Functions	Unit doublet function	$F_{-2} \equiv \langle x - a \rangle^{-2} = \begin{cases} 0 & x \neq a \\ +\infty & x = a^- \\ -\infty & x = a^+ \end{cases}$		No derivative $\int_{0-}^{x} F_{-2} dx = F_{-1}$
	Unit impulse function	$F_{-1} \equiv \langle x - a \rangle^{-1} = \begin{cases} 0 & x \neq a \\ +\infty & x = a \end{cases}$		No derivative $\int_{0-}^{x} F_{-1} dx = F_0$
Macaulay Functions	Unit step function	$F_0 \equiv \langle x - a \rangle^0 = \begin{cases} 0 & x < a \\ 1 & x \geq a \end{cases}$		No derivative $\int_{0-}^{x} F_0 dx = F_1$
	Unit ramp function	$F_1 \equiv \langle x - a \rangle^1 = \begin{cases} 0 & x \leq a \\ x - a & x \geq a \end{cases}$		$\dfrac{d}{dx} F_1 = F_0$ $\int_{0-}^{x} F_1 dx = \dfrac{F_2}{2}$
	Unit Macaulay functions	$F_n \equiv \langle x - a \rangle^n = \begin{cases} 0 & x \leq a \\ (x - a)^n & x \geq a \end{cases}$ $n = 1, 2, 3, \ldots$		$\dfrac{d}{dx} F_n = nF_{n-1}$ $n = 1, 2, 3, \ldots$ $\int_{0-}^{x} F_n dx = \dfrac{F_{n+1}}{n + 1}$ $n = 0, 1, 2, 3, \ldots$

FIGURE 5.10 Distributed loads represented by Macaulay functions.

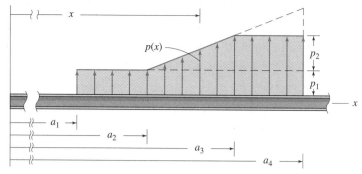

$$p(x) = p_1\langle x - a_1\rangle^0 + \left(\frac{p_2}{a_3 - a_2}\right)\langle x - a_2\rangle^1 - \left(\frac{p_2}{a_3 - a_2}\right)\langle x - a_3\rangle^1 - (p_1 + p_2)\langle x - a_4\rangle^0$$

Figure 5.10 illustrates how the Macaulay functions can be used to represent distributed loads on beams. It must be remembered that Macaulay functions continue indefinitely for $x > a$. Therefore, when a particular load pattern terminates at some value of x, a new Macaulay function must be introduced to cancel out the effect of that previous Macaulay function. Note how the dimensions of each term in $p(x)$ are preserved even though the term $\langle x - a\rangle^0$ is dimensionless whereas $\langle x - a\rangle^1$ has the dimensions of length.

Singularity Functions. The two singularity functions of interest here are the *unit impulse function*, F_{-1}, which can be used to represent a concentrated force, and the *unit doublet function*, F_{-2}, which can be used to represent a concentrated couple. Singularity functions are similar in some respects to the Macaulay functions, but they are defined for negative values of n. They become singular (i.e., infinite) at $x = a$, and they are zero for $x \neq a$. By definition, singularity functions satisfy the equation

$$\int_{0-}^{x} F_n dx = F_{n+1} \quad n = -2, -1 \tag{5.14}$$

(Note how this differs from Eq. 5.12*b*.) The units of singularity functions are the same as the units of x^n, even though n is negative.

Use of discontinuity functions to represent loads, shear, and moment. The sign conventions for loads and for internal shear force and bending moment are given in Figs. 5.5 and 5.4, respectively, and in Section 5.3 relationships among loads, shear force, and bending moment were presented. By way of review,

$$\Delta V = P_0 \tag{5.3}$$
repeated

$$\Delta M = -M_0 \tag{5.4}$$
repeated

$$V(x) = V_1 + \int_{x_1}^{x} p(\xi)d\xi \tag{5.7}$$
repeated

$$M(x) = M_1 + \int_{x_1}^{x} V(\xi)d\xi \tag{5.8}$$
repeated

To apply the latter two when $p(x)$ and $V(x)$ are represented by discontinuity functions, we let $x_1 = 0^-$ and let $V_1(0^-) = M_1(0^-) = 0$. Therefore, Eqs. 5.7 and 5.8 can be written as

$$V(x) = \int_{0-}^{x} p(\xi)d\xi, \qquad M(x) = \int_{0-}^{x} V(\xi)d\xi \tag{5.15}$$

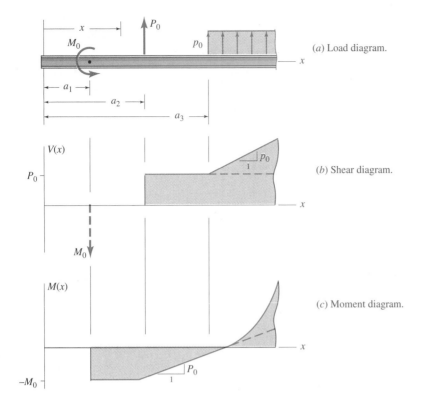

(a) Load diagram.

(b) Shear diagram.

(c) Moment diagram.

FIGURE 5.11 Examples of the use of discontinuity functions to represent load, shear, and moment.

The load function for the beam in Fig. 5.11 can be written in terms of discontinuity functions by referring to Table 5.1; and we can use Eqs. 5.12 through 5.15 to integrate these discontinuity functions to get

$$p(x) = -\underline{M_0\langle x - a_1 \rangle^{-2}} + \underline{P_0\langle x - a_2 \rangle^{-1}} + p_0\langle x - a_3 \rangle^0$$

$$V(x) = \int_{0-}^{x} p(\xi)d\xi = -\underline{M_0\langle x - a_1 \rangle^{-1}} + P_0\langle x - a_2 \rangle^0 + p_0\langle x - a_3 \rangle^1 \quad (5.16)$$

$$M(x) = \int_{0-}^{x} V(\xi)d\xi = -M_0\langle x - a_1 \rangle^0 + P_0\langle x - a_2 \rangle^1 + \frac{p_0}{2}\langle x - a_3 \rangle^2$$

Note that the beam in Fig. 5.11 is shown to extend on the left to $x = 0^-$. Singularity functions are underscored in the above equations for $p(x)$ and $V(x)$.

Table 5.2 summarizes load, shear, and moment relationships represented by discontinuity functions. The terms in this table that represent singularity functions are underscored to emphasize the fact that, strictly speaking, singularities have infinite values at the point $x = a_i$ and zero values everywhere else. Whereas it is common practice to represent concentrated couples and forces in the manner indicated in Case 1 and Case 2 of the "Load" column of Table 5.2, the moment M_0 in the "Shear" column of Case 1 is not a true concentrated shear force. Its effect is to cause the jump in moment at $x = a$ in the "Moment" column of Case 1, but otherwise it can be ignored. Therefore, it is shown dashed in Table 5.2, Case 1. Note how the amplitude of the load function is defined in Cases 4 through 6 to assure proper dimensionality.

In this section, only statically determinate problems are considered. Therefore, the reactions are treated as known quantities that have been obtained by the use of equilibrium equations. In Section 7.7, both statically determinate problems and statically indeterminate problems are examined.

TABLE 5.2   A Summary of Loads, Shear, and Moment Represented By Discontinuity Functions

Case	Load	Shear	Moment
1	$-M_0 \langle x-a \rangle^{-2}$	$-M_0 \langle x-a \rangle^{-1}$	$-M_0 \langle x-a \rangle^{0}$
2	$P_0 \langle x-a \rangle^{-1}$	$P_0 \langle x-a \rangle^{0}$	$P_0 \langle x-a \rangle^{1}$
3	$p_0 \langle x-a \rangle^{0}$	$p_0 \langle x-a \rangle^{1}$	$\dfrac{p_0}{2} \langle x-a \rangle^{2}$
4	$\dfrac{p_1}{b} \langle x-a \rangle^{1}$	$\dfrac{p_1}{2b} \langle x-a \rangle^{2}$	$\dfrac{p_1}{6b} \langle x-a \rangle^{3}$
5	$\dfrac{p_2}{b^2} \langle x-a \rangle^{2}$	$\dfrac{p_2}{3b^2} \langle x-a \rangle^{3}$	$\dfrac{p_2}{12b^2} \langle x-a \rangle^{4}$
6	$\dfrac{p_n}{b^n} \langle x-a \rangle^{n}$	$\dfrac{p_n}{b^n (n+1)} \langle x-a \rangle^{n+1}$	$\dfrac{p_n}{b^n (n+1)(n+2)} \langle x-a \rangle^{n+2}$

For the beam in Fig. 1 of Example Problem 5.6, (a) use discontinuity functions to obtain expressions for $p(x)$, $V(x)$, and $M(x)$, and (b) use the discontinuity functions from Part (a) to construct shear and moment diagrams for the beam, indicating the contribution of each term in the discontinuity-function expressions.

The loads and reactions from Example Problem 5.6 are given in Fig. 1.

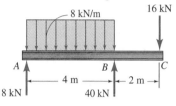

Fig. 1

Plan the Solution We can refer to Cases 2 and 3 of Table 5.2 to construct the load function $p(x)$ and then to perform the required integrations to get $V(x)$ and $M(x)$.

Solution

(a) By referring to the "Load" column of Cases 2 and 3 of Table 5.2, we can write

$$p(x) = (8 \text{ kN})\langle x - 0 \text{ m}\rangle^{-1} - (8 \text{ kN/m})[\langle x - 0 \text{ m}\rangle^0 - \langle x - 4 \text{ m}\rangle^0]$$

$$+ (40 \text{ kN})\langle x - 4 \text{ m}\rangle^{-1} - (16 \text{ kN})\langle x - 6 \text{ m}\rangle^{-1} \qquad \text{Ans.} \quad (1)$$

Integrating Eq. 1 by referring to Eqs. 5.12 through 5.14 or to the "Shear" column of Table 5.2, we get

$$V(x) = \int_{0-}^{x} p(\xi)d\xi:$$

$$\overset{(a)}{V(x)} = (8 \text{ kN})\langle x\rangle^0 \overset{(b)}{-} (8 \text{ kN/m})[\langle x\rangle^1 \overset{(c)}{-} \langle x - 4 \text{ m}\rangle^1]$$

$$\overset{(d)}{+ (40 \text{ kN})\langle x - 4 \text{ m}\rangle^0} \overset{(e)}{- (16 \text{ kN})\langle x - 6 \text{ m}\rangle^0}$$

$$\text{Ans.} \quad (2)$$

and, from the "Moment" column of Table 5.2, we get

$$M(x) = \int_{0-}^{x} V(\xi)d\xi:$$

$$\overset{(a')}{M(x)} = (8 \text{ kN})\langle x\rangle^1 \overset{(b')}{-} (8 \text{ kN/m}) [\tfrac{1}{2}\langle x\rangle^2 \overset{(c')}{-} \tfrac{1}{2}\langle x - 4 \text{ m}\rangle^2]$$

$$\overset{(d')}{+ (40 \text{ kN})\langle x - 4 \text{ m}\rangle^1} \overset{(e')}{- (16 \text{ kN})\langle x - 6 \text{ m}\rangle^1}$$

$$\text{Ans.} \quad (3)$$

(b) In Fig. 2, Eqs. (2) and (3) are plotted term-by-term; the separate terms are then summed so the results can be compared with the shear diagram and the moment diagrams obtained in Example Problem 5.6.

Review the Solution Since the shear diagram in Fig. 2b and the moment diagram in Fig. 2c both close to zero at the right end, our results are probably correct. For this problem we could use Eqs. 5.1 through 5.6 to check the shear and moment diagrams above. That is, the procedure used to construct the shear diagram and the moment diagram in Example Problem 5.6 can be used to check the results obtained by the discontinuity-function method.

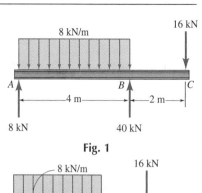

(a)

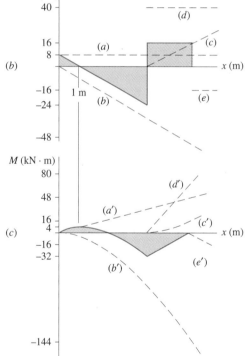

(b)

(c)

Fig. 2 Load, shear, and moment diagrams.

249

Observe that it is a very straightforward procedure to obtain the discontinuity-function expressions for $V(x)$ and $M(x)$. Each term in $p(x)$, $V(x)$, and $M(x)$ can be evaluated separately and the results summed to get the final discontinuity-function expressions. Likewise, graphs of $p(x)$, $V(x)$, and $M(x)$ can be easily constructed from the discontinuity-function expressions. This makes the discontinuity-function method an ideal one to serve as a basis for a computer program to evaluate and plot shear diagrams and moment diagrams. The VM-DIAG computer program, one of the ***MechSOLID*** collection of computer programs, is described in Appendix G.5. As noted there, VM-DIAG is based directly on discontinuity-function expressions. Computer homework exercises for Section 5.5 follow the regular Section 5.5 homework problems at the end of the chapter.

In Section 7.7 discontinuity functions will be used to solve beam deflections problems, including statically indeterminate problems.

5.6 PROBLEMS ■■

In **Problems 5.2-1 through 5.2-16**, *you are to determine the internal resultants (transverse shear force V and bending moment M) at specific cross sections of beams. Use the sign conventions for V and M given in Fig. 5.4, p. 231, and* ***always draw complete and correct free-body diagrams***. *The minus-sign superscript signifies "just to the left of" the referenced point; the plus-sign superscript means "just to the right of" the referenced point.*

Prob. 5.2-1. For the beam AE in Fig. P5.2-1, (a) determine the transverse shear force V_{B^-} and the bending moment M_{B^-} at a section just to the left of the support at B, and (b) determine V_C and M_C at section C. Express your answers in terms of P and a.

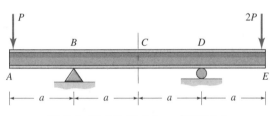

P5.2-1, P5.4-8, P5.5-1, and PC5.5-5

Prob. 5.2-2. Transverse loads are applied to the beam in Fig. P5.2-2 at A and C, and a 3 kip · ft concentrated couple is applied to the beam at E. Determine (a) the transverse shear force V_{C^-} and the bending moment M_{C^-} at a section just to the left of the

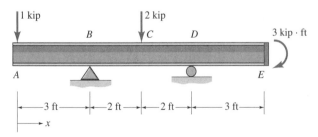

P5.2-2, P5.2-18, P5.4-1, P5.5-2, and PC5.5-2

2-kip load at C, and (b) the shear, V_{D^-} and moment M_{D^-} just to the left of the support at D.

Prob. 5.2-3. The shaft AD in Fig. P5.2-3 is supported by bearings at A and C. Assume that these bearings produce concentrated reaction forces that are normal to the shaft. Determine (a) the reactions at A and C; (b) the transverse shear force V_{B^+} and the bending moment M_{B^+} just to the right of B; and (c) the shear force V_{C^-} and bending moment M_{C^-} just to the left of C.

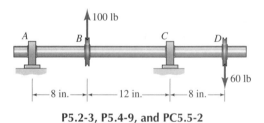

P5.2-3, P5.4-9, and PC5.5-2

Prob. 5.2-4. For the simply supported beam AE in Fig. P5.2-4, (a) determine V_{C^-} and M_{C^-}, the internal resultants just to the left of the 20-kN load at C, and (b) determine V_{D^-} and M_{D^-}, the internal resultants just to the left of the 20-kN load at D.

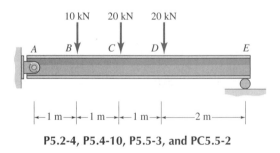

P5.2-4, P5.4-10, P5.5-3, and PC5.5-2

Prob. 5.2-5. Concentrated couples are applied to beam AE in Fig. P5.2-5 at B and E. Determine V_C and M_C, the internal resultants at section C, the middle of the beam. Express your answers in terms of M_0 and a.

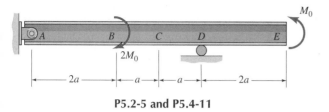

P5.2-5 and P5.4-11

Prob. 5.2-6. The shaft in Fig. P5.2-6 is supported by bearings at B and D that can only exert forces normal to the shaft. Belts that pass over pulleys at A and E exert parallel forces of 100 N and 200 N, respectively, as shown. Determine the transverse shear force V_C and the bending moment M_C at section C, midway between the two supports.

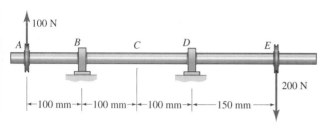

P5.2-6 and PC5.5-3

Prob. 5.2-7. Two transverse forces and a couple are applied as external loads to the cantilever beam AC in Fig. P5.2-7. Determine the transverse shear force V_A and the bending moment M_A at the fixed end A.

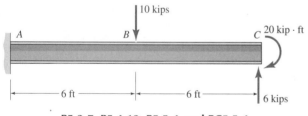

P5.2-7, P5.4-12, P5.5-4, and PC5.5-4

Prob. 5.2-8. For the cantilever beam AD in Fig. P5.2-8, determine the reactions at D; that is, determine V_D and M_D. Express your answers in terms of P and a.

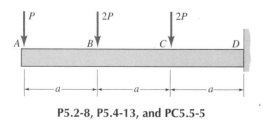

P5.2-8, P5.4-13, and PC5.5-5

Prob. 5.2-9. A uniformly distributed load of intensity p_0 per unit length is applied to beam AD, as shown in Fig. P5.2-9. Determine the internal resultants V_B and M_B at section B, midway between the supports.

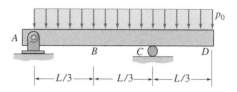

P5.2-9, P5.4-14, P5.5-5, and PC5.5-6

Prob. 5.2-10. A uniformly distributed load of 1 kip/ft and a concentrated transverse load of 8 kips are applied to the simply supported beam in Fig. P5.2-10. Determine the transverse shear force V_C and the bending moment M_C at section C, midway between the supports.

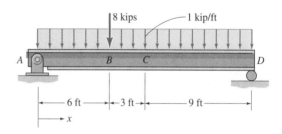

P5.2-10, P5.2-19, P5.4-2, P5.4-15, P5.5-6, and PC5.5-3

Prob. 5.2-11. A drilling engineer wishes to support pipes of length L on blocks so that the magnitude of the bending moment in the pipes directly over the supports is equal to the magnitude of the bending moment at the center of the pipes (see Fig. P5.2-11). Let w be the weight of the pipes per unit length, and neglect the width of the support blocks. (a) Determine the distance from each end, aL, at which the engineer should place the supports. (b) Determine the corresponding magnitude of the bending moments at the supports and at the center.

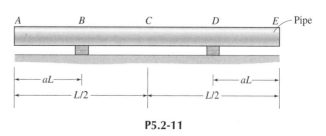

P5.2-11

Prob. 5.2-12. A concentrated couple of 4 kN · m and a uniformly distributed load of 1 kN/m are applied to beam AE, as shown in Fig. P5.2-12. Determine the transverse shear force V_C and the bending moment M_C at section C, the middle of the beam.

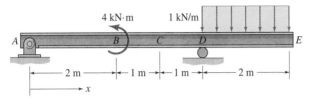

P5.2-12, P5.2-20, P5.4-3, P5.4-16, P5.5-7, and PC5.5-4

Prob. 5.2-13. The simply supported beam AD supports a concentrated load of 50 lb and a linearly varying load of maximum intensity 20 lb/in., as shown in Fig. P5.2-13. Determine the transverse shear force V_E and the bending moment M_E at section E, which is 3 in. to the right of the support at A.

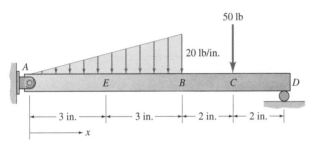

P5.2-13, P5.2-21, P5.4-4, and P5.4-17, P5.5-9, and PC5.5-3

Prob. 5.2-14. Two beam segments, AC and CE, are connected together at C by a frictionless pin. Segment CE is cantilevered from a rigid support at E, and segment AC has a roller support at A. (a) Determine the reactions at A and E. (b) Determine V_D and M_D, the internal resultants at section D, which is 5 ft to the left of the support E.

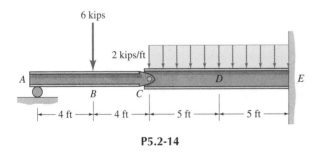

P5.2-14

Prob. 5.2-15. An L-shaped frame BC is welded to the beam AD in Fig. P5.2-15. A downward vertical load of 5 kips is applied at C as shown. Neglecting the width of the connection at B, determine the following: (a) the transverse shear force V_{B-} and bending moment M_{B-} just to the left of B, and (b) the transverse shear force V_{B+} and bending moment M_{B+} just to the right of B.

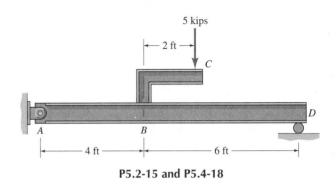

P5.2-15 and P5.4-18

Prob. 5.2-16. A freight-handling system consists of a "traveling beam," CD, supported by a simply supported beam AB. A drive system at C and a frictionless roller at E enable beam CD to move (i.e., to travel) along beam AB. Determine the following: (a) the roller force at E, (b) the reaction force at B, and (c) V_{E-} and M_{E-}, the internal resultants in beam AB at a section just to the left of the roller at E. Express your answers in terms of P, L_1, L_2, and a, where the distance aL_1 locates beam CD relative to beam AB.

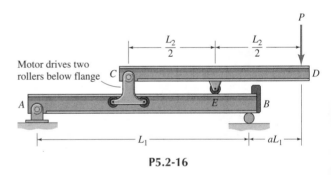

P5.2-16

In Problems 5.2-17 through 5.2-33, *you are to determine expressions for the internal resultants as functions of position x. That is, expressions for shear force V(x), bending moment M(x), and, occasionally, axial force F(x) are to be determined.* **Always draw correct free-body diagrams.** *Use the sign conventions for V and M given in Fig. 5.4 on p. 231, and let F be assumed positive in tension.*

Prob. 5.2-17. A linearly varying load of maximum intensity p_0 is applied to the simply supported beam AB in Fig. P5.2-17. Determine expressions for the transverse shear force $V(x)$ and the bending moment $M(x)$ at an arbitrary section x.

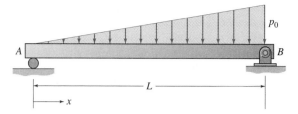

P5.2-17, P5.4-19, P5.5-10, and PC5.5-6

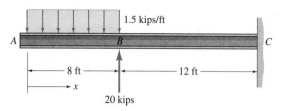

P5.2-23, P5.4-6, P5.4-21, P5.5-8, and PC5.5-4

Prob. 5.2-18. For the beam in Fig. P5.2-2, determine the following: (a) $V_2(x)$ and $M_2(x)$, the internal resultants at an arbitrary section in "Interval 2" between B and C, that is, for (3 ft $< x <$ 5 ft), and (b) $V_3(x)$ and $M_3(x)$, the internal resultants at an arbitrary section in "Interval 3" between C and D.

Prob. 5.2-19. For the beam in Fig. 5.2-10, determine the following: (a) $V_1(x)$ and $M_1(x)$, the internal resultants at an arbitrary section in "Interval 1" between A and B, that is, for (0 $< x < 6$ ft), and (b) $V_2(x)$ and $M_2(x)$, the internal resultants at an arbitrary section in "Interval 2" between B and D.

Prob. 5.2-20. For the beam in Fig. P5.2-12, determine the following: (a) $V_1(x)$ and $M_1(x)$, the internal resultants at an arbitrary section between A and B, that is, for (0 $< x < 2$ m), and (b) $V_2(x)$ and $M_2(x)$, the internal resultants at an arbitrary section between B and D.

Prob. 5.2-21. For the beam AD in Fig. P5.2-13, determine the following: (a) $V_1(x)$ and $M_1(x)$, the internal resultants at an arbitrary section between A and B, that is, for (0 $< x < 6$ in.), and (b) $V_2(x)$ and $M_2(x)$, the internal resultants at an arbitrary section between B and C.

Prob. 5.2-22. For the beam in Fig. P5.2-22, determine the following: (a) V_A and M_A, the reactions at the cantilever end A; (b) $V_1(x)$ and $M_1(x)$, the internal resultants at an arbitrary section between A and B, that is, for (0 $< x \leq L/2$); and (c) $V_2(x)$ and $M_2(x)$, the internal resultants at an arbitrary section between B and C, that is, for ($L/2 \leq x \leq L$).

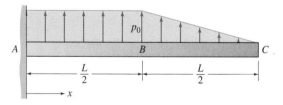

P5.2-22, P5.4-5, P5.4-20, P5.5-11, and PC5.5-7

Prob. 5.2-23. For the cantilever beam AC in Fig. P5.2-23, (a) determine $V_1(x)$ and $M_1(x)$, the internal resultants at an arbitrary section between A and B, that is, for (0 $< x < 8$ ft); (b) determine $V_2(x)$ and $M_2(x)$, the internal resultants at an arbitrary section between B and C; and (c) evaluate the shear force V_C and the bending moment M_C at the cantilever end C.

Prob. 5.2-24. The beam AB in Fig. P5.2-24 has a distributed load that varies linearly from p_0 at $x = 0$ to $2p_0$ at $x = L$. (a) Determine the reactions at A and B, and (b) determine $V(x)$ and $M(x)$, the internal resultants at an arbitrary section between A and B.

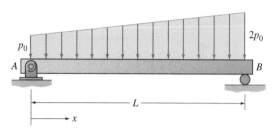

P5.2-24, P5.4-7, P5.4-22, P5.5-12, and PC5.5-7

Prob. 5.2-25. A boy and girl position themselves as shown in Fig. P5.2-25, so that their canoe remains level. Assume that the canoe behaves like a beam, and assume that the buoyant force of the water is uniformly distributed over the 16-ft waterline length. (a) If the weight of the boy is $W_b = 200$ lb, determine the weight, W_g, of the girl. (b) Determine the buoyant force per unit length, p_0. (c) Determine expressions for $V(x)$ and $M(x)$ at an arbitrary cross section of the canoe between A and B, that is, for (-3 ft $< x < 5$ ft).

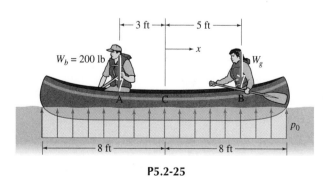

P5.2-25

Prob. 5.2-26. A hanger bar supports a load of 450 N and is, in turn, supported by a pin AF that passes through the hanger bar and a support bracket, as shown in Fig. P5.2-26a. Assume that the pin is a beam subjected to distributed loading from the bracket and the hanger rod as indicated in Fig. P5.2-26b. (a) Determine the maximum distributed-load intensities, p_t and p_b, on the top and bottom of the pin, respectively. (b) Determine expressions for $V_3(x)$ and $M_3(x)$, the shear and moment in the pin in the span between C and D, that is, for (25 mm $\leq x \leq$ 45 mm).

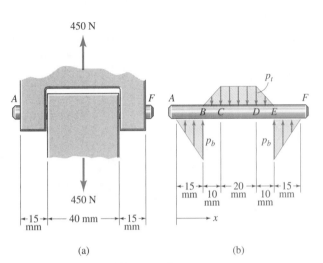

(a)   (b)

P5.2-26 and P5.4-23

Prob. 5.2-27. A crane-hoist frame is supported by a frictionless pin at A and a smooth roller at D. (a) Determine the reactions at A and D. (b) Determine expressions for internal resultants $F_2(x)$, $V_2(x)$, and $M_2(x)$, at an arbitrary cross section of the frame between B and C, that is, for (24 in. $< x <$ 64 in.). Note the orientation of the x and y axes for this problem.

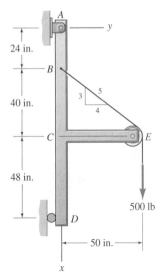

P5.2-27 and P5.4-24

Prob. 5.2-28. The boom of the shop crane shown in Fig. P5.2-28 is supported by a frictionless pin at A, and its elevation angle is controlled by the hydraulic cylinder between pins at B and D. (a) Determine the reactions at A and B when the boom is horizontal and supports a load of 900 N at C. (b) For the same boom angle and loading, determine expressions for internal resultants $F_1(x)$, $V_1(x)$, and $M_1(x)$ at an arbitrary cross section of the boom between A and B, that is, for (0 $< x <$ 0.75 m).

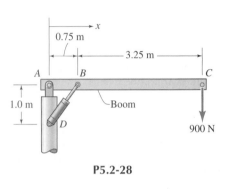

P5.2-28

Prob. 5.2-29. The boom of the shop crane shown in Fig. P5.2-29 is supported by a frictionless pin at A, and its angle is controlled by the hydraulic cylinder between pins at B and D. (a) If the boom is lowered until the hydraulic cylinder length, BD, is 1 m, determine the reactions at A and B when a 900-N load is picked up at C. (b) For the same boom angle and loading, determine expressions for internal resultants $F_1(x)$, $V_1(x)$, and $M_1(x)$ at an arbitrary cross section of the boom between A and B, that is, for (0 $< x <$ 0.75 m).

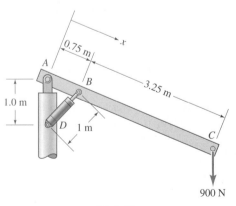

P5.2-29

254

Problems 5.2-30 through 5.2-33 *are statically indeterminate problems. In each case draw a correct free-body diagram and write the equilibrium equations for the internal resultants V(x) and M(x) in terms of the specified redundant reaction.*

Prob. 5.2-30. For the propped-cantilever beam *AB* in Fig. P5.2-30, determine expressions for the internal resultants $V(x)$ and $M(x)$ in terms of the reaction force at *A*.

Prob. 5.2-31. For the propped-cantilever beam *AB* in Fig. P5.2-31, determine expressions for the internal resultants $V(x)$ and $M(x)$ in terms of the moment at *B*.

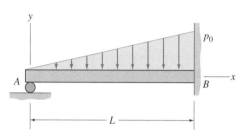

P5.2-30 and P5.2-31

Prob. 5.2-32. For the continuous beam *ABC* in Fig. P5.2-32 determine expressions for the internal resultants in each span in terms of the reaction force at *A*. That is, determine $V_1(x)$ and $M_1(x)$ in terms of A_y for the span $(0 < x < L/2)$, and, similarly, determine $V_2(x)$ and $M_2(x)$ for span *BC*.

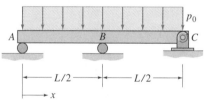

P5.2-32

Prob. 5.2-33. For the continuous beam *ABC* shown in Fig. P5.2-33, determine expressions for the internal resultants in each span in terms of the reaction force at *A*.

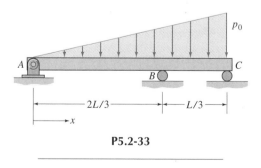

P5.2-33

When solving the following problems for Section 5.4, draw the requested shear-force and bending-moment diagrams approximately to scale. Label all critical ordinates, including the maximum and minimum values, and indicate the sections at which these occur.

Use the Equilibrium Equation Method *to solve* **Problems 5.4-1 through 5.4-7.** *Obtain and plot expressions for V(x) and M(x), as illustrated in Example 5.4. Most of the problems require several expressions $V_i(x)$ for shear and $M_i(x)$ for moment. Therefore, you will need to draw several free-body diagrams and solve the necessary equilibrium equations to obtain these expressions.*

Prob. 5.4-1. Plot the shear diagram and the moment diagram for beam *AE* in Prob. 5.2-2.

Prob. 5.4-2. Plot the shear diagram and the moment diagram for beam *AD* in Prob. 5.2-10.

Prob. 5.4-3. Plot the shear diagram and the moment diagram for beam *AE* in Prob. 5.2-12.

Prob. 5.4-4. Plot the shear diagram and the moment diagram for beam *AD* in Prob. 5.2-13.

Prob. 5.4-5. Plot the shear diagram and the moment diagram for beam *AC* in Prob. 5.2-22.

Prob. 5.4-6. Plot the shear diagram and the moment diagram for beam *AC* in Prob. 5.2-23.

Prob. 5.4-7. Plot the shear diagram and the moment diagram for beam *AB* in Prob. 5.2-24.

When solving **Problems 5.4-8 through 5.4-24** *use the* Graphical Method, *as illustrated in Examples 5.5 and 5.6. That is, use Eqs. 5.1, 5.3, and 5.5 in sketching shear diagrams, and use Eqs. 5.2, 5.4, and 5.6 in sketching moment diagrams.*

Prob. 5.4-8. Sketch the shear diagram and the moment diagram for beam *AE* in Prob. 5.2-1.

Prob. 5.4-9. Sketch the shear diagram and the moment diagram for beam *AD* in Prob. 5.2-3.

Prob. 5.4-10. Sketch the shear diagram and the moment diagram for beam *AE* in Prob. 5.2-4.

Prob. 5.4-11. Sketch the shear diagram and the moment diagram for beam *AE* in Prob. 5.2-5.

Prob. 5.4-12. Sketch the shear diagram and the moment diagram for beam *AC* in Prob. 5.2-7.

Prob. 5.4-13. Sketch the shear diagram and the moment diagram for beam *AD* in Prob. 5.2-8.

Prob. 5.4-14. Sketch the shear diagram and the moment diagram for beam *AD* in Prob. 5.2-9.

Prob. 5.4-15. Sketch the shear diagram and the moment diagram for beam *AD* in Prob. 5.2-10.

Prob. 5.4-16. Sketch the shear diagram and the moment diagram for beam *AE* in Prob. 5.2-12.

Prob. 5.4-17. Sketch the shear diagram and the moment diagram for beam AD in Prob. 5.2-13.

Prob. 5.4-18. Sketch the shear diagram and the moment diagram for beam AD in Prob. 5.2-15.

Prob. 5.4-19. Sketch the shear diagram and the moment diagram for beam AB in Prob. 5.2-17.

Prob. 5.4-20. Sketch the shear diagram and the moment diagram for cantilever beam AC in Prob. 5.2-22.

Prob. 5.4-21. Sketch the shear diagram and the moment diagram for cantilever beam AC in Prob. 5.2-23.

Prob. 5.4-22. Sketch the shear diagram and the moment diagram for the simply supported beam AB in Prob. 5.2-24.

Prob. 5.4-23. Sketch the shear diagram and the moment diagram for the pin AF in Prob. 5.2-26.

Prob. 5.4-24. Sketch the shear diagram and the moment diagram for the vertical frame member AD in Prob. 5.2-27.

In **Problems 5.5-1 through 5.5-8** *a beam and its loading are shown in the referenced figure. For each of these problems:*

(a)   Use equilibrium to verify the reactions.

(b)   Using discontinuity functions from the Load *column of Table 5.2, write an expression for the intensity p(x) of the equivalent distributed load. Include the reactions in your expression for the equivalent load.*

(c)   Perform a term-by-term integration of the load expression to obtain a discontinuity-function expression for the shear force V(x), and sketch a shear diagram like the one in Fig. 2b of Example Prob. 5.7.

(d)   Perform a term-by-term integration of the shear expression obtained in Part (c) to obtain a discontinuity-function expression for the moment M(x). Sketch a moment diagram like the one in Fig. 2c of Example Prob. 5.7.

Prob. 5.5-1. Use Fig. P5.2-1. $B_y = P/2$ and $D_y = 5P/2$.

Prob. 5.5-2. Use Fig. P5.2-2. $B_y = 2$ kips and $D_y = 1$ kip.

Prob. 5.5-3. Use Fig. P5.2-4. $A_y = 28$ kN and $E_y = 22$ kN.

Prob. 5.5-4. Use Fig. P5.2-7. $V_A = 4$ kips and $M_A = -8$ kip·ft.

Prob. 5.5-5. Use Fig. P5.2-9. $A_y = p_0L/4$ and $C_y = 3p_0L/4$.

Prob. 5.5-6. Use Fig. P5.2-10. $A_y = 14.33$ kips and $D_y = 11.67$ kips.

Prob. 5.5-7. Use Fig. P5.2-12. $A_y = 0.5$ kN and $D_y = 1.5$ kN.

Prob. 5.5-8. Use Fig. P5.2-23. $V_C = 8$ kips and $M_C = 48$ kip·ft.

For **Problems 5.5-9 through 5.5-15,** *carry out the same steps outlined above for Probs. 5.5-1 through 5.5-8, with the exception of omitting Part (d).*

Prob. 5.5-9. Use Fig. P5.2-13. $A_y = 46$ lb and $D_y = 64$ lb.

Prob. 5.5-10. Use Fig. P5.2-17. $A_y = p_0L/6$ and $B_y = p_0L/3$.

Prob. 5.5-11. Use Fig. P5.2-22. $V_A = -3p_0L/4$ and $M_A = 7p_0L^2/24$.

Prob. 5.5-12. Use Fig. P5.2-24. $A_y = 2p_0L/3$ and $B_y = 5p_0L/6$.

Prob. 5.5-13. Use Fig. P5.5-13. $V_B = -320$ lb and $M_B = -1440$ lb·ft.

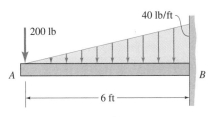

P5.5-13 and PC5.5-1

Prob. 5.5-14. Use Fig. P5.5-14. $A_y = 5.75$ kN and $C_y = 1.75$ kN.

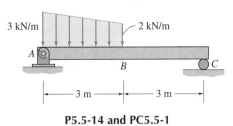

P5.5-14 and PC5.5-1

Prob. 5.5-15. Use Fig. P5.5-15. $A_y = B_y = 6.75$ kips.

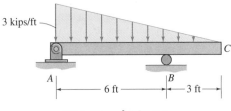

P5.5-15 and PC5.5-1

Computer Exercises—Sect. 5.5 *Use the MechSOLID computer program V/M-DIAG to solve the homework exercises C5.5-1 through C5.5-9. Refer to Appendix G.5 for a description of this computer program and for an example of its use. Hand in computer printouts of the following MechSOLID "screens": V/M DIAG: Results, V/M DIAG: Shear Diagram, V/M DIAG: Moment Diagram. Samples of these screens are illustrated in Example G-6.*

Prob. C5.5-1. Use V/M-DIAG to plot shear and moment diagrams for the beams described in the following homework problems: (a) Prob. 5.5-13, (b) Prob. 5.5-14, and (c) Prob. 5.5-15.

Prob. C5.5-2. Use V/M-DIAG to plot shear and moment diagrams for the beams described in the following homework problems: (a) Prob. 5.2-2, (b) Prob. 5.2-3, and (c) Prob. 5.2-4.

Prob. C5.5-3. Use V/M-DIAG to plot shear and moment diagrams for the beams described in the following homework problems: (a) Prob. 5.2-6, (b) Prob. 5.2-10, and (c) Prob. 5.2-13.

Prob. C5.5-4. Use V/M-DIAG to plot shear and moment diagrams for the beams described in the following homework problems: (a) Prob. 5.2-7, (b) Prob. 5.2-12, and (c) Prob. 5.2-23.

Prob. C5.5-5. The V/M-DIAG computer program can be used to solve nonnumeric problems like Probs. 5.2-1 and 5.2-8 if you remember that the transverse shear force will have the form $V = \hat{V}P$, where $\hat{V}$ is a nondimensional factor that is obtained by V/M-DIAG if you input the following data: $P = a = 1$. Likewise, the bending moment will have the form $M = \hat{M}Pa$, where $\hat{M}$ is the bending moment output by V/M-DIAG. (a) Use V/M-DIAG to plot shear and moment diagrams for the beam in Prob. 5.2-1. (b) Use V/M-DIAG to plot shear and moment diagrams for the cantilever beam in Prob. 5.2-8.

Prob. C5.5-6. The V/M-DIAG computer program can be used to solve nonnumeric problems like Probs. 5.2-9 and 5.2-17 if you remember that the transverse shear force will have the form $V = \hat{V}(p_0L)$, where $\hat{V}$ is a nondimensional factor that is obtained by V/M-DIAG if you input the following data: $p_0 = L = 1$. Likewise, the bending moment will have the form $M = \hat{M}(p_0L^2)$, where $\hat{M}$ is the moment output by V/M-DIAG. (a) Use V/M-DIAG to plot shear and moment diagrams for the beam in Prob. 5.2-9. (b) Use V/M-DIAG to plot shear and moment diagrams for the simply supported beam in Prob. 5.2-17.

Prob. C5.5-7. The V/M-DIAG computer program can be used to solve nonnumeric problems like Probs. 5.2-22 and 5.2-24 if you remember that the transverse shear force will have the form $V = \hat{V}(p_0L)$, where $\hat{V}$ is a nondimensional factor that is obtained by V/M-DIAG if you input the following data: $p_0 = L = 1$. Likewise, the bending moment will have the form $M = \hat{M}(p_0L^2)$, where $\hat{M}$ is the moment output by V/M-DIAG. (a) Use V/M-DIAG to plot shear and moment diagrams for the cantilever beam in Prob. 5.2-22. (b) Use V/M-DIAG to plot shear and moment diagrams for the simply supported beam in Prob. 5.2-24.

Probs. C5.5-8. Use V/M-DIAG computer program to plot shear and bending moment diagrams for the two beams shown.

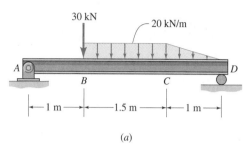

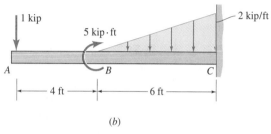

(a)

(b)

PC5.5-8

Probs. C5.5-9. Use V/M-DIAG computer program to plot shear and bending moment diagrams for the two beams shown.

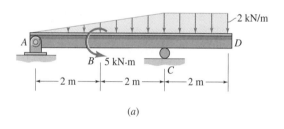

(a)

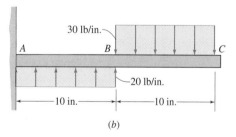

(b)

PC5.5-9

STRESSES IN BEAMS

6

6.1 INTRODUCTION

In this chapter we continue our study of beams by determining how the bending moment $M(x)$ and the transverse shear force $V(x)$ are related to the normal stress and the shear stress at section x. *Loads* (transverse forces or couples) applied to a beam cause it to deflect laterally, as illustrated in Fig. 6.1. The *deflection curve,* shown dashed in Fig. 6.1, may be used to characterize the beam's deformation. And, by relating the curvature of the deflection curve to the bending moment M, we can determine the distribution of the normal stress σ_x. You will discover that this derivation includes all three fundamental concepts of deformable-body mechanics: *geometry of deformation* (in the strain-displacement analysis), *material behavior* (in the stress-strain relations), and *equilibrium* (in the definition of stress resultants and in relating stress resultants to the external loads and reactions).

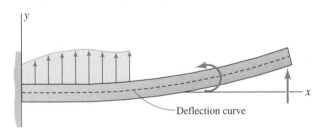

FIGURE 6.1 The deflection of a beam.

Beam-Deformation Terminology. To simplify the study of beams, we initially consider only straight beams that have a *longitudinal plane of symmetry* (LPS), and for which the loading and support are symmetric with respect to this LPS, as illustrated in Fig. 6.2. As indicated in Fig. 6.2a, we let the LPS, or *plane of bending,* be labeled the xy plane. (In Section 6.6 we consider a more general case of unsymmetric bending.)

To investigate the distribution of stresses in a beam, like the one in Fig. 6.1, it is convenient to imagine the beam to be a bundle of *longitudinal fibers* parallel to the x axis. Figure 6.3a depicts a few of these imaginary "fibers." Under the action of an applied bending moment M, there is a shortening of the upper fibers and a stretching of the lower fibers, causing the beam segment shown in Fig. 6.3 to be curved upward. But some lon-

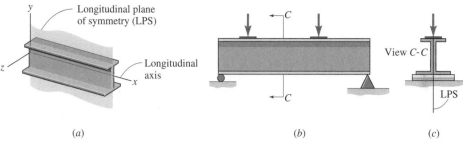

(a)   *(b)*   *(c)*

FIGURE 6.2   A beam with loading and support in its longitudinal plane of symmetry.

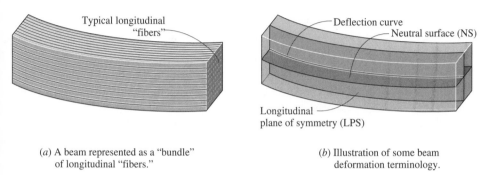

(a) A beam represented as a "bundle"   *(b)* Illustration of some beam
       of longitudinal "fibers."   deformation terminology.

FIGURE 6.3   A portion of a deformed beam.

gitudinal fibers retain their original length. These are said to form the *neutral surface* (NS), which is illustrated in Fig. 6.3*b*. For convenience, we position the undeformed beam in the *xyz* coordinate frame with the LPS lying in the *xy* plane, and with the *xz* plane corresponding to the neutral surface.[1] The line of intersection of the LPS and the NS (i.e., the *x* axis in the undeformed beam) is the *deflection curve* of the deformed beam, as indicated in Figs. 6.1 and 6.3*b*.

6.2   STRAIN-DISPLACEMENT ANALYSIS

Pure Bending.   Let us begin our analysis of beams by examining the deformation of a beam segment subjected to *pure bending*, that is, a segment for which $M(x)$ is constant. If equal couples M_0 are applied to the ends of a segment of beam, as in Fig. 6.4*b*, the moment is constant along the segment and the segment is said to be in pure bending. Lines *AB* and *DE* in Fig. 6.4*a* represent the edges of typical cross sections in the undeformed beam; lines *A*B** and *D*E** in Fig. 6.4*b* represent these same cross sections after deformation.

Since $M(x) = M_0 =$ constant, pure-bending deformation has a symmetry and uniformity such that the axis of the beam becomes a circular arc, and the *cross sections remain plane and remain perpendicular to the deformed axis.* Hence, lines like *A*B** and *D*E**, which represent cross sections after deformation, all intersect at a common point *C**, called the *center of curvature.*

---

[1]In Section 6.3 the position of the NS (the *xz* plane) with respect to the beam is determined.

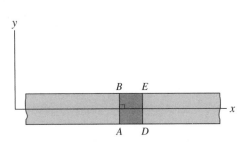

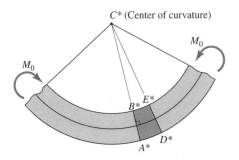

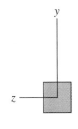

(a) The undeformed beam.

(b) The deformed beam.

(c) The cross section before deformation.

FIGURE 6.4 A beam segment undergoing pure bending.

The above description of beam deformation applies rigorously only to the case of pure bending, that is, when $dM/dx = V(x) = 0$. However, even in the case of nonuniform bending, the assumption that cross sections remain plane and remain perpendicular to the deformed axis leads to expressions for extensional strain ϵ_x and for normal stress σ_x that are quite accurate if the beam is long compared with its cross-sectional dimensions; that is, if the beam is "slender."

Kinematic Assumptions of Bernoulli-Euler Beam Theory. The previous discussion can be summarized in the following four deformation assumptions of Bernoulli-Euler beam theory:[2]

1. The beam possesses a *longitudinal plane of symmetry* (LPS), and is loaded and supported symmetrically with respect to this LPS.
2. There is a longitudinal plane perpendicular to the LPS that remains free of strain (i.e., $\epsilon_x = 0$) as the beam deforms. This plane is called the *neutral surface* (NS). The intersection of the neutral surface with a cross section is called the *neutral axis* (NA) of the cross section. The intersection of the NS with the LPS is called the *axis* of the beam; it forms the *deflection curve* of the deformed beam.
3. **Cross sections, which are plane and are perpendicular to the axis of the undeformed beam, remain plane and remain perpendicular to the axis of the deformed beam, that is, to the deflection curve.**
4. Deformation in the plane of a cross section (e.g., transverse strains ϵ_y and ϵ_z) may be neglected in deriving an expression for the longitudinal strain ϵ_x.

The third assumption above is the key assumption of Bernoulli-Euler beam theory, and it leads to a practical theory of bending of beams that is comparable to the theories of axial deformation and torsion covered previously.

Strain–Displacement Analysis; Longitudinal Strain. Because of Assumptions 1 and 4, the fibers in any plane parallel to the xy plane behave identically to the corresponding fibers that lie in the xy plane (i.e., the LPS). Therefore, bending deformation

---

[2]The development of the beam theory presented here is attributed principally to the work of Jacob Bernoulli (1654–1705) and Leonard Euler (1707–1783). Jacob Bernoulli, a prominent member of the famous Bernoulli family of mathematicians and physicists, studied the deflection of beams, finding that the curvature of an elastic beam at any point is proportional to the bending moment at that point. The work of Jacob Bernoulli and his nephew Daniel Bernoulli (1700–1782) led Euler to his discovery of the differential equation of the elastic curve of a beam. [Ref. 6-1]

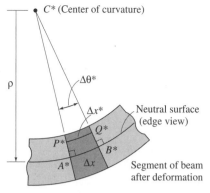

(a) The deformed beam segment.

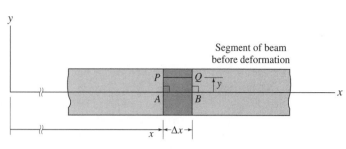

(b) The undeformed beam segment.

FIGURE 6.5 The geometry of deformation of a beam segment.

is independent of the coordinate z, so the drawing in Fig. 6.5 represents the deformation of any plane in the beam parallel to the LPS. Using Fig. 6.5 and the preceding four deformation assumptions, we can develop an expression for the extensional strain ϵ_x in a longitudinal fiber at coordinates (x,y) in the beam. In Fig. 6.5b, points A and P lie in the cross-sectional plane at coordinate x in the undeformed beam; similarly, points B and Q lie in the cross-sectional plane at $(x + \Delta x)$ in the undeformed beam. Line segment PQ is parallel to the x axis and lies at distance $+y$ above the NS (xz plane). Therefore, in the undeformed beam the infinitesimal fibers AB and PQ are both of length Δx. From Assumption 3, points A^* and P^* lie in a plane that is perpendicular to the neutral surface of the deformed beam, and points B^* and Q^* lie in a plane that also is perpendicular to the deformed neutral surface. According to Assumption 2, however, the length of A^*B^*, a fiber lying in the neutral surface, is unchanged; that is, the length of $\overline{A^*B^*}$ is still Δx, as indicated in Fig. 6.5a. Finally, by virtue of Assumption 4, $\overline{A^*P^*} = \overline{AP} = y$, and $\overline{B^*Q^*} = \overline{BQ} = y$.

Figure 6.5, therefore, embodies all four deformation assumptions of Bernoulli-Euler beam theory, so we can use it in deriving an expression for the extensional strain of a longitudinal fiber.

From the general definition of extensional strain in Eq. 2.29, we can express the extensional strain in the longitudinal fiber PQ as

$$\epsilon_x \equiv \epsilon_x(x,\ y) = \lim_{Q \to P}\left[\frac{(\overline{P^*Q^*} - \overline{PQ})}{\overline{PQ}}\right] = \lim_{\Delta x \to 0}\left[\frac{(\Delta x^* - \Delta x)}{\Delta x}\right] \qquad (6.1)$$

Considering A^*B^* to be the arc of a circle of radius $\rho(x) \equiv \rho$ subtending an angle $\Delta\theta^*$, we get

$$\overline{A^*B^*} = \Delta x = \rho\,\Delta\theta^*$$

Similarly,

$$\overline{P^*Q^*} = \Delta x^* = (\rho - y)\Delta\theta^*$$

Combining these two equations with Eq. 6.1, we get

$$\epsilon_x = \lim_{\Delta x \to 0}\left[\frac{[(\rho - y)\Delta\theta^* - \rho\,\Delta\theta^*]}{\rho\,\Delta\theta^*}\right] = -\frac{y}{\rho} \qquad (6.2)$$

That is, the extensional strain ϵ_x at point $(x,\ y,\ z)$ in the beam is independent of z and is

261

related to the local *radius of curvature*, $\rho(x)$, by the following *strain-displacement equation*:

$$\epsilon_x(x,\,y) = -\frac{y}{\rho(x)}$$

The reciprocal of the radius of curvature, $\kappa \equiv 1/\rho$, is called the *longitudinal curvature*, or just the *curvature*.

As noted earlier, the assumptions that lead to this *strain-displacement equation* are strictly valid for the case of pure bending, that is, when $V \equiv 0$. However, Eq. 6.3 can also be used for analyzing the bending of beams with $V \neq 0$ [i.e., $M = M(x)$] if the beam is slender.

Transverse Strains. The longitudinal strain at a point in the beam is given by Eq. 6.3. In Section 6.3 we will show that the transverse stresses σ_y and σ_z are negligible in comparison with the normal stress σ_x. Therefore, from the generalized Hooke's Law, Eq. 2.32, there is a *Poisson's-ratio effect* that produces transverse strains

$$\epsilon_y = \epsilon_z = -\nu\epsilon_x \tag{6.4}$$

The deformed shape of a rectangular beam segment undergoing pure bending is shown in Fig. 6.6. Consider the transverse strain ϵ_z for this beam. Above the neutral axis (i.e., where y is positive), the beam is compressed axially, so ϵ_x is negative there. Then, from Eq. 6.4, ϵ_z will be positive. Therefore, since ϵ_z is positive, fibers oriented in the z direction and lying above the NA elongate. Correspondingly, fibers oriented in the z direction and lying below the NA contract, since ϵ_z is negative there. This produces the *anticlastic curvature* of the beam that is illustrated in Fig. 6.6. In Homework Problem 6.2-5 you will be asked to prove that the *transverse curvature*, $\kappa' \equiv 1/\rho'$, is related to the longitudinal curvature $\kappa \equiv 1/\rho$ by the equation

$$\kappa' = \nu\kappa \tag{6.5}$$

You can easily demonstrate this anticlastic curvature if you take a soft eraser with rectangular cross section (e.g., a ''pink pearl'' eraser) and bend it between your thumb and index finger.

As noted previously in Assumption 4, deformation in the plane of a cross section may be neglected in the derivation of the strain-displacement equation, Eq. 6.3.

Summary of Strain-Displacement Analysis. Before we go on to discuss the normal stress σ_x and its relationship to $\rho(x)$ and $M(x)$, let us examine thoroughly the expression for extensional strain, Eq. 6.3, which we obtained from the four Bernoulli-Euler

FIGURE 6.6 Transverse deformation of a beam segment in pure bending.

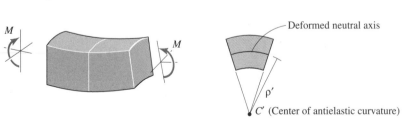

(a) The deformed beam segment and its cross section.

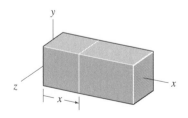

(b) The undeformed beam segment and its cross section.

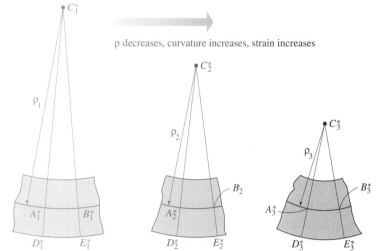

ρ decreases, curvature increases, strain increases

(a) (b) (c)

FIGURE 6.7 The relationship of the curvature of a beam to the extensional strain, ϵ_x.

deformation assumptions and the fundamental definition of extensional strain. We observe the following:

1. The extensional strain ϵ_x is independent of z, the thickness coordinate of the beam.

2. The extensional strain ϵ_x is inversely proportional to the radius of curvature at cross section x. Figure 6.7 illustrates the fact that as ρ decreases, the curvature κ increases (i.e., the beam becomes more curved), and the strain at any given fiber increases. For example, the length $\overline{A^*B^*}$ (which is equal to $\overline{AB}$ since this is identified as lying in the neutral surface) is the same for all three figures in Fig. 6.7. But, since $\overline{D_1^*E_1^*} < \overline{D_2^*E_2^*} < \overline{D_3^*E_3^*}$, the length change in the bottom fiber as a result of bending increases from left to right in Fig. 6.7, and therefore the strain increases from the least-curved beam in Fig. 6.7a to the more-curved beams in Figs. 6.7b and 6.7c.

3. The signs of $\rho(x)$ and y govern the sign of ϵ_x. If $\rho(x)$ is positive, the *center of curvature* lies above the beam, that is, on the $+y$ side of the beam. Therefore, when $\rho(x)$ is positive, the deformed beam is concave upward. Because of the minus sign in Eq. 6.3, the fibers above the neutral surface (i.e., fibers having positive y) are in compression, while the fibers below the neutral surface are in tension. This is the case for the beam segment in Fig. 6.8a. If $\rho(x)$ is negative, the center of

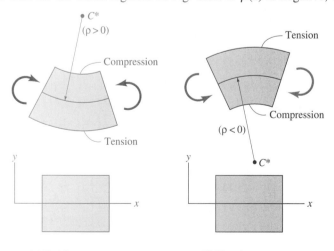

(a) Positive curvature. (b) Negative curvature.

FIGURE 6.8 Illustrations of positive and negative curvature.

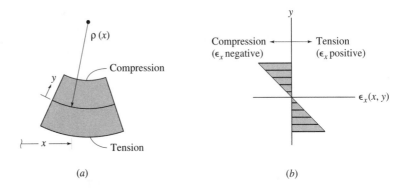

FIGURE 6.9 The strain distribution at a cross section where $\rho(x)$ is positive.

(a)　　　　　*(b)*

curvature lies below the beam, and the beam is curved concave downward, as shown as Fig. 6.8*b*.

4. Finally, **the strain ϵ_x is proportional to the distance y from the neutral surface**. This linear *strain distribution* is illustrated in Fig. 6.9.

The fact that ϵ_x varies linearly with y is completely independent of the material properties of the beam. For example, the beam could consist of two or more different materials, like concrete and steel (Section 6.5), or the beam could be partly elastic and partly plastic (Section 6.7). As long as the deformation can be characterized by the four assumptions listed in this section, the *extensional strain ϵ_x is proportional to y,* as given by Eq. 6.3 and illustrated in Fig. 6.9*b*.

■■■■■■■■■■■■■■■■□□□ **EXAMPLE 6.1** □□□□□□□□□□□□□□□■■

A couple M_0 acts on the end of a slender cantilever beam as shown in Fig. 1. Take $\frac{L}{c} = 30$, where L is the original length of the beam and $2c$ is the depth of the beam. (a) Determine an expression for the normalized radius of curvature (ρ/c) if the bottom fiber (at $y = -c$) is at the tensile yield strain of the material, ϵ_Y. (b) Determine the ratio (δ_{max}/c) for this loading condition. (c) Determine ρ and δ_{max} if $L = 15$ ft and $\epsilon_Y = 0.001$.

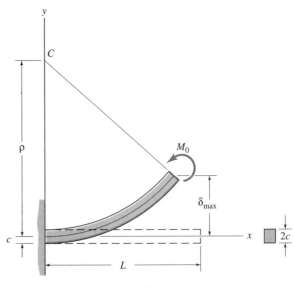

Fig. 1

Plan the Solution Part (a) is a straightforward application of Eq. 6.3, which relates extensional strain ϵ_x to the radius of curvature of the deflection curve. To determine δ_{max} in Part (b), we need, in addition, to use the geometric properties of a circle.

Solution
(a) From the strain-displacement equation, Eq. 6.3,

$$\epsilon_x = -\frac{y}{\rho}$$

The bottom fiber is in tension. Therefore,

$$\rho = -\frac{(-c)}{\epsilon_Y}$$

so

$$\frac{\rho}{c} = \frac{1}{\epsilon_Y} \qquad \text{Ans. (a)}$$

(b) From the sketch, in Fig. 2, of the deflection curve,

$$L = \rho\theta$$

and

$$\delta_{max} = \rho(1 - \cos\theta)$$

Therefore, for this particular beam,

$$\theta = \frac{L}{\rho} = \frac{L}{c}\frac{c}{\rho} = 30\,\epsilon_Y$$

or,

$$\frac{\delta_{max}}{c} = \frac{1}{\epsilon_Y}[1 - \cos(30\,\epsilon_Y)] \qquad \text{Ans. (b)}$$

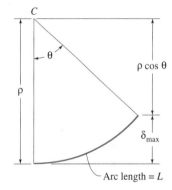

Fig. 2 The geometry of a circular arc.

(c) For $L = 15$ ft and $\epsilon_Y = 0.001$, we get

$$\theta = 30\,\epsilon_Y = 0.03 \text{ rad}$$

Finally,

$$\rho = \frac{L}{30\,\epsilon_Y} = 500 \text{ ft} = 6000 \text{ in.} \qquad \text{Ans. (c)}$$

$$\delta_{max} = \rho(1 - \cos\theta) = 0.225 \text{ ft} = 2.70 \text{ in.} \qquad \text{Ans. (c)}$$

Review the Solution Note that with the numerical value of $L = 15$ ft and the given ratio $L/c = 30$, $c = L/30 = 6$ in. Hence, for this 12-in.-depth beam, a maximum deflection that is less than one-fourth of the depth of the beam causes the extreme fiber of the beam to reach the yield strain $\epsilon_Y = 0.001$. (Figures 1 and 2 obviously exaggerate the deflection.)

In the previous section, assumptions were made about the geometry of deformation of slender beams, and an expression for the resulting extensional strain ϵ_x was derived, Eq. 6.3. The corresponding normal stress in beams, σ_x, is often called the *flexural stress*. To obtain an expression for the flexural stress, we need to consider the material behavior, that is, the stress-strain-temperature behavior of the material. To simplify our initial study of stresses in beams, let us assume that the material is linearly elastic and isotropic, and that the temperature remains constant. Then, the following two assumptions permit us to determine the flexural stress σ_x:

1. The material obeys Hooke's law, Eq. 2.32a, with $\Delta T = 0$.
2. The transverse normal stresses, σ_y and σ_z, may be neglected in comparison with the primary normal stress, σ_x.

By combining these two assumptions, we find that the uniaxial stress-strain equation

$$\sigma_x = E\epsilon_x \tag{6.6}$$

applies to bending of linearly elastic beams. When Eqs. 6.3 and 6.6 are combined, we obtain the following expression

$$\sigma_x = \frac{-Ey}{\rho} \tag{6.7}$$

If $E = $ const, or if $E = E(x)$, the normal stress on a cross section is linear in y, as given by Eq. 6.8 and indicated in Fig. 6.10.[3]

$$\sigma_x(x, y) = \frac{-E(x)y}{\rho(x)} \tag{6.8}$$

FIGURE 6.10 The flexural stress distribution at a cross section where $\rho(x)$ is positive.

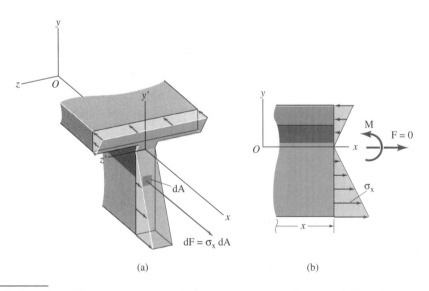

(a)   (b)

---

[3]In Section 6.5 we will consider stresses in nonhomogeneous beams, that is, stresses in beams that are made of more than one material.

266

As indicated in Fig. 6.10, the stress resultants that are related to the normal stress σ_x acting on the cross section are:

$$F(x) = \int_A \sigma_x \, dA, \qquad M(x) = -\int_A y\sigma_x \, dA \qquad (6.9)$$

A positive moment produces compression in the $+y$ fibers of the beam.

In Section 9.4 we will consider axial deformation combined with bending, but, for the present discussion of bending alone, let $F \equiv 0$. Therefore, substituting Eq. 6.8 into Eqs. 6.9, we get

$$F = -\frac{E}{\rho} \int_A y \, dA = 0, \qquad M = \frac{E}{\rho} \int_A y^2 \, dA \qquad (6.10)$$

The integrals appearing in Eqs. 6.10 are section properties that are defined in Appendix C:

$$\int_A dA = A, \qquad \int_A y \, dA = \bar{y}A, \qquad \int_A y^2 \, dA = I_z \qquad (6.11)$$

where A is the cross-sectional area, $\bar{y}$ is the y coordinate of the *centroid* of the cross section, and I_z is the *area moment of inertia* about the z axis of the cross section.

In order to satisfy the condition $F = 0$, we must make $\bar{y} = 0$. That is, the z axis of the cross section (labeled the z' axis in Figs. 6.10a and 6.11a) must pass through the centroid of the cross section. Thus, **the x axis passes through the centroid of each cross section of the undeformed beam**. The z' axis is called the *neutral axis of the cross section,* or simply the *neutral axis (NA)*, because it is the boundary between the portion of the cross section that is in compression and the portion that is in tension, as indicated in Fig. 6.11.[4]

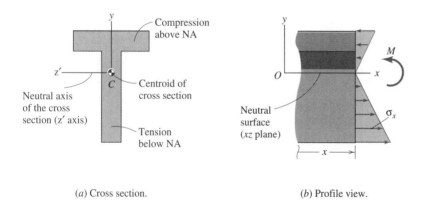

(a) Cross section.   (b) Profile view.

FIGURE 6.11   (a) The location of the neutral axis of the cross section, and (b) the flexural stress distribution for a homogeneous beam in bending.

---

[4]In the future, the ''z axis of the cross section'' will just be labeled z, not z', even though the true z axis does not lie in the particular cross section under consideration.

Combining Eqs. 6.10b and 6.11c, we obtain the *moment-curvature equation* of Bernoulli-Euler beam theory, namely

$$M = \frac{EI}{\rho} = EI\,\kappa \qquad \text{Moment-curvature equation} \qquad (6.12)$$

The *curvature* $\kappa(x)$ is related to the *radius of curvature* $\rho(x)$ by $\kappa(x) = \dfrac{1}{\rho(x)}$. The product EI is called the *flexural rigidity* of the beam. (In Eq. 6.12 the subscript has been dropped from I_z to simplify the remainder of the discussion of bending of symmetric beams. Subscripts will be needed again in the Section 6.6 on Unsymmetric Bending.)

We can relate the moment-curvature equation, Eq. 6.12, to the deformed-beam segments in Fig. 6.8 by noting that a positive bending moment, $M(x)$, leads to a positive value of $\rho(x)$, which means that the beam is concave upward, as shown in Fig. 6.8a. Conversely, a negative moment produces a negative curvature, which means that the center of curvature lies in the $-y$ direction, as shown in Fig. 6.8b.

Finally, Eqs. 6.8 and 6.12 may be combined to give the important *flexure formula* of Bernoulli-Euler beam theory.[5]

$$\sigma_x = \frac{-My}{I} \qquad \text{Flexure formula} \qquad (6.13)$$

By making the assumptions that plane sections remain plane and that the material is linearly elastic with $E = E(x)$, we have obtained an expression for the stress distribution on a cross section subjected to bending moment $M(x)$. This is the linear stress distribution illustrated in Fig. 6.12.[6]

An assumption made in the derivation of the flexure formula, Eq. 6.13, is that σ_x is much greater than either σ_y or σ_z. It is left as an exercise for the reader to show that this is a reasonable assumption if the beam is long in comparison with its cross-sectional dimensions. (Homework Problem 6.3-36)

FIGURE 6.12 The flexural stress distribution in a linearly elastic beam.

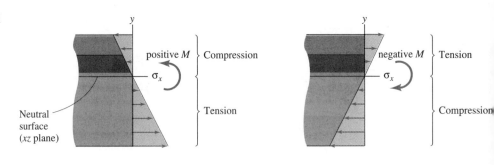

(a) Positive moment.　　　　(b) Negative moment.

---

[5]According to the sign convention adopted in this text and illustrated in Fig. 5.4, a positive moment produces compression in the $+y$ fibers of the beam. This results in a minus sign in Eq. 6.13. Some textbooks adopt a different sign convention that leads to a plus sign in the flexure formula.

[6]Compressive stresses as well as tensile stresses may be shown acting on the cross section, as in Fig. 6.11b. However, to emphasize here that σ_x is linear in y, compressive stresses are shown in Fig. 6.12 as a continuation of the straight-line plot that depicts tensile stresses.

The cross section of a beam is a T with the dimensions shown in Fig. 1. The moment at the section is $M = 4$ kip·ft. Determine (a) the location of the neutral axis of the cross section, (b) the moment of inertia with respect to the neutral axis, and (c) the maximum tensile stress and the maximum compressive stress on the cross section.

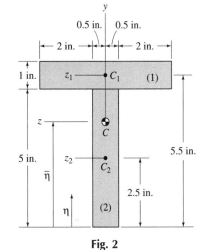

Fig. 1

Plan the Solution To use the flexure formula, Eq. 6.13, we need first to locate the centroid of the cross section and then compute the moment of inertia about an axis through the centroid (Appendix C). Since M is positive, the maximum compressive stress occurs in the top fibers, and the maximum tensile stress occurs in the bottom fibers.

Solution

(a) As indicated in Fig. 2, we can pick an arbitrary origin at the bottom and let a coordinate in the y-direction be called η. Then, by summing area contributions to the first moment, we have

$$\int_A \eta \, dA = \int_{A_1} \eta \, dA + \int_{A_2} \eta \, dA$$

or

$$\bar{\eta}A = \bar{\eta}_1 A_1 + \bar{\eta}_2 A_2 = (5.5 \text{ in.})(5 \text{ in}^2) + (2.5 \text{ in.})(5 \text{ in}^2) = 40 \text{ in}^3$$

where

$$A = A_1 + A_2 = (5 \text{ in}^2) + (5 \text{ in}^2) = 10 \text{ in}^2$$

Then,

$$\bar{\eta} = \frac{40 \text{ in}^3}{10 \text{ in}^2} = 4.0 \text{ in.} \qquad \textbf{Ans. (a)}$$

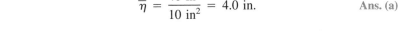

Fig. 2

(b) The moment of inertia of a rectangle about an axis through its own centroid is

$$I_C = \frac{1}{12}bh^3$$

and, from the parallel-axis theorem, the moment of inertia about an axis through C' parallel to the axis through the centroid C is

$$I_{C'} = I_C + Ad_{CC'}^2$$

Therefore,

$$I = (I_{C_1} + A_1 d_{C_1 C}^2) + (I_{C_2} + A_2 d_{C_2 C}^2)$$

$$= \left[\frac{1}{12}(5 \text{ in.})(1 \text{ in.})^3 + (5 \text{ in.})(1 \text{ in.})(1.5 \text{ in.})^2 \right]$$

$$+ \left[\frac{1}{12}(1 \text{ in.})(5 \text{ in.})^3 + (1 \text{ in.})(5 \text{ in.})(1.5 \text{ in.})^2 \right]$$

269

so,

$$I = 33.3 \text{ in}^4 \qquad \text{Ans. (b)}$$

(c) The maximum compression occurs at the top of the beam, and the maximum tension occurs at the bottom of the beam. From Eq. 6.13,

$$\sigma_x = \frac{-My}{I}:$$

$$\sigma_{\text{max C}} = \sigma(x, 2 \text{ in.}) = \frac{-(4 \text{ kip} \cdot \text{ft})(12 \text{ in./ft})(2 \text{ in.})}{33.3 \text{ in}^4}$$

$$= -2.88 \text{ ksi } (2.88 \text{ ksi C})$$

$$\sigma_{\text{max T}} = \sigma(x, -4 \text{ in.}) = \frac{-(4 \text{ kip} \cdot \text{ft})(12 \text{ in./ft})(-4 \text{ in.})}{33.3 \text{ in}^4}$$

$$= 5.76 \text{ ksi } (5.76 \text{ ksi T})$$

$$\sigma_{\text{max T}} = 5.76 \text{ ksi}, \qquad \sigma_{\text{max C}} = -2.88 \text{ ksi} \qquad \text{Ans. (c)}$$

Review the Solution The centroid must lie between the centroids of the two areas A_1 and A_2. Furthermore, since the areas are equal, the combined centroid C lies midway between the individual centroids, as we obtained above.

To see if the order of magnitude of I is reasonable, we can compare our answer with the moment of inertia of a 1 in. $\times$ 6 in. web about its own centroid ($\frac{1}{12}(1)(6)^3 = 18 \text{ in}^4$). Thus, the value of $I = 33.3 \text{ in}^4$ appears reasonable.

Finally, since the T-section is not symmetric about the neutral axis, the maximum tension and maximum compression are not equal in magnitude, but, since σ is linear in y, their magnitudes are in the ratio of the distances to the top and bottom fibers.

(See Homework Problem 6.6-7, where the effect of misalignment of the load is examined.)

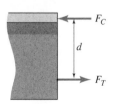

Fig. 1

For the beam of Example 6.2, (a) determine the resultant compressive force F_C, the resultant tensile force F_T, and the distance that separates them; and (b) show that F_C and F_T form a couple of magnitude $M = 4 \text{ kip} \cdot \text{ft}$.

Plan the Solution We can sketch the stress distribution on a figure like Fig. 6.11b. Then we can apply the standard procedures for locating the resultant of distributed forces to get the magnitude and the location of F_C and F_T.

Solution

(a) We first represent the area under various portions of the stress distribution by forces F_{C1}, F_{C2}, and so forth, as indicated in Fig. 2b. For example, from the fact that F_T is the resultant of the tensile stress, we have

$$F_T = \frac{1}{2}(5.76 \text{ kips/in}^2)(1 \text{ in.})(4 \text{ in.}) = 11.52 \text{ kips}$$

Similarly,

$$F_{C1} = \frac{1}{2}(1.44 \text{ kips/in}^2)(1 \text{ in.})(1 \text{ in.}) = 0.72 \text{ kips}$$

$$F_{C2} = (1.44 \text{ kips/in}^2)(5 \text{ in.})(1 \text{ in.}) = 7.20 \text{ kips}$$

$$F_{C3} = \frac{1}{2}(1.44 \text{ kips/in}^2)(5 \text{ in.})(1 \text{ in.}) = 3.60 \text{ kips}$$

$$F_C = F_{C1} + F_{C2} + F_{C3} = 11.52 \text{ kips}$$

Since F_C is the resultant of F_{C1}, F_{C2}, and F_{C3}, we determine its location by computing first moments:

$$F_{C1}(d_{C1}) + F_{C2}(d_{C2}) + F_{C3}(d_{C3}) = F_C d_C$$

So,

$$d_C = \frac{(0.72 \text{ kips})(0.67 \text{ in.}) + (7.20 \text{ kips})(1.50 \text{ in.}) + (3.60 \text{ kips})(1.67 \text{ in.})}{11.52 \text{ kips}}$$

or

$$d_C = 1.50 \text{ in.}$$

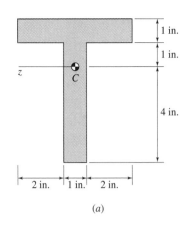

(a)

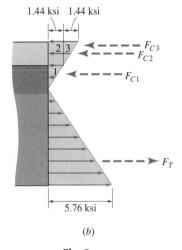

(b)

Fig. 2

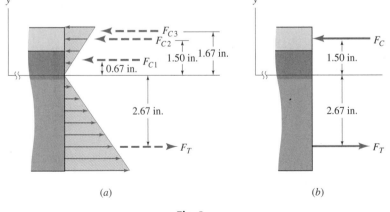

(a)   (b)

Fig. 3

271

Therefore,

$$d = d_C + d_T = 1.50 \text{ in.} + 2.67 \text{ in.} = 4.17 \text{ in.}$$

$$F_C = F_T = 11.52 \text{ kips,} \qquad d = 4.17 \text{ in.} \qquad \textbf{Ans. (a)}$$

(b) Since $F_C = F_T$, the resultant force on the cross section is zero. The couple formed by F_C and F_T is given by

$$F_C d = F_T d = (11.52 \text{ kips})(4.17 \text{ in.}) = 48.0 \text{ kip} \cdot \text{in.} \qquad \textbf{Ans. (b)}$$

which is the value of the applied moment.

Review the Solution The results of this example agree with our solution in Example 6.2. We can conclude that the stresses on a cross section due to a bending moment acting on the cross section are equivalent to a couple consisting of equal tensile and compressive forces acting on their respective portions of the cross section.

While the previous two examples illustrate the calculation of flexural stresses on a particular cross section with prescribed moment, it is also important for the absolute maximum tensile stress and absolute maximum compressive stress in a beam under given support and loading conditions to be determined. This is where a bending-moment diagram is very useful, as you will see in the following example.

■■■■■■■■■■■■■■■■■■■■ EXAMPLE 6.4 ■■■■■■■■■■■■■■■■■■■■

A beam whose cross section is the T section of Example Problem 6.2 is subjected to the loading shown in Fig. 1a. The shear diagram that corresponds to this loading is given in Fig. 1b. (a) Using the procedure illustrated in Examples 5.5 through 5.7, sketch the moment diagram for this beam. (b) Compute the maximum compressive flexural stress and the maximum tensile flexural stress in this beam.

Plan the Solution Once we have constructed the moment diagram, primarily using Eqs. 5.2 and 5.6, we can determine the cross sections at which we must

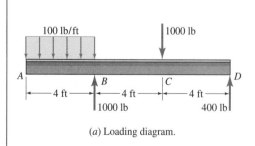

(a) Loading diagram.

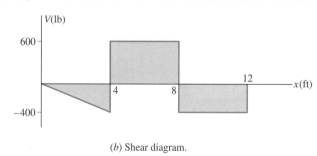

(b) Shear diagram.

Fig. 1

272

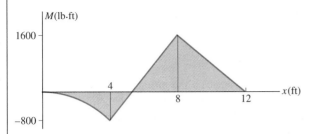

Fig. 2 Moment diagram.

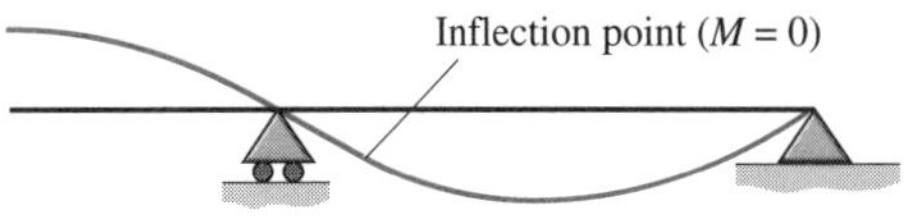

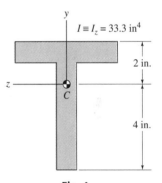

Fig. 3 Deflection curve.

calculate maximum and minimum flexural stresses. The flexure formula, Eq. 6.13, may then be used to calculate the maximum compressive stress and the maximum tensile stress.

Solution
(a) Use

$$V = \frac{dM}{dx}$$

and

$$M_2 - M_1 = \int_{x_1}^{x_2} V(x)dx$$

to construct the moment diagram (Fig. 2). From the diagram, the maximum negative moment is $M_B \equiv M(4 \text{ ft}) = -800 \text{ lb} \cdot \text{ft}$, and the maximum positive moment is $M_C \equiv M(8 \text{ ft}) = 1600 \text{ lb} \cdot \text{ft}$.

(b) In Example Problem 6.2, the location of the centroid and the value of $I \equiv I_z$ were calculated. These are shown in Fig. 4. At section B, the top of the beam is in tension and the bottom is in compression; at section C the top of the beam is in compression and the bottom is in tension. Just to be safe, we can compute σ_x at these four points and then select the maxima. We use the flexure formula, Eq. 6.13.

$$\sigma_x = \frac{-My}{I}:$$

$$\sigma_x(4 \text{ ft}, 2 \text{ in.}) = \frac{-(-800 \text{ lb} \cdot \text{ft})(12 \text{ in./ft})(2 \text{ in.})}{33.3 \text{ in}^4} = 576 \text{ psi}$$

$$\sigma_x(4 \text{ ft}, -4 \text{ in.}) = \frac{-(-800 \text{ lb} \cdot \text{ft})(12 \text{ in./ft})(-4 \text{ in.})}{33.3 \text{ in}^4} = -1152 \text{ psi}$$

$$\sigma_x(8 \text{ ft}, 2 \text{ in.}) = \frac{-(1600 \text{ lb} \cdot \text{ft})(12 \text{ in./ft})(2 \text{ in.})}{33.3 \text{ in}^4} = -1152 \text{ psi}$$

$$\sigma_x(8 \text{ ft}, -4 \text{ in.}) = \frac{-(1600 \text{ lb} \cdot \text{ft})(12 \text{ in./ft})(-4 \text{ in.})}{33.3 \text{ in}^4} = 2304 \text{ psi}$$

Fig. 4

Therefore, the maximum tensile flexural stress occurs at the bottom of the beam at section C. By coincidence, the particular loading and cross section of this beam produce equal maxima of compressive flexural stress at the bottom

273

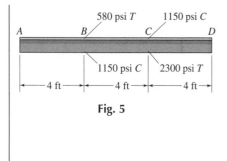

580 psi *T*   1150 psi *C*

1150 psi *C*   2300 psi *T*

|← 4 ft →|← 4 ft →|← 4 ft →|

Fig. 5

of the beam at *B* and at the top of the beam at *C*. The flexural stresses at sections *B* and *C*, rounded to the nearest 10 psi, are indicated on Fig. 5.

$$\sigma_{\text{max T}} = 2300 \text{ psi}, \qquad \sigma_{\text{max C}} = -1150 \text{ psi} \qquad \textbf{Ans. (c)}$$

Review the Solution   First, as a check on the moment diagram, we can see if the moment distribution corresponds to a reasonable shape of the deflected beam, according to Fig. 6.8 and Eq. 6.12. A sketch of the deflection curve (Fig. 3) is drawn adjacent to the moment diagram. The shape of the deflection curve does seem reasonable. The values of bending moments M_B and M_C can be checked easily by using free-body diagrams based on Fig. 1*a*. Each does have the correct magnitude and sign. The magnitudes and signs of the four calculated stresses are easily checked.

In Example Problems 6.2 and 6.4, both the loading and the dimensions, including the shape of the cross section, were given. In Section 6.4 we will consider a method for selecting an appropriate cross section when the loading on the beam is given.

6.4 DESIGN OF BEAMS FOR STRENGTH

The design of beams for specific applications is usually governed by detailed specifications and codes involving design requirements and procedures that are beyond the scope of this text.[7] In this section, however, we will consider the design of beams based on *allowable flexural stress*. That is, given the bending-moment distribution in a beam, and given the allowable tensile stress and allowable compressive stress of the material to be used, an appropriate beam cross section is to be selected. We will return to the topic of design of beams in Chapter 9, where the combined effect of flexural stress (σ_x) and transverse shear stress (τ_{xy}) will be considered.

Before we consider the actual process of selecting the cross section of a beam, let us first look at the types of beams available for selection. Standard sizes are available for beams made of wood and for beams made of steel, aluminum, and other metals. Appendices D.1 through D.10 give the properties of selected steel, aluminum, and wood structural shapes. More extensive tables may be found, for example, in publications of the American Institute of Steel Construction (AISC) [Ref. 6-2] and the Aluminum Association [Ref. 6-3]. Figure 6.13 illustrates five structural steel shapes.

Shapes like the ones illustrated in Fig. 6.13 are produced by passing a hot billet of metal between sets of rollers that, after several passes, produce the desired shape. The most commonly used shape is the wide-flange section illustrated in Fig. 6.13*a*. The American Standard beam, commonly called the I-beam (Fig. 6-13*b*), is less frequently used because it tends to have excessive material in the web, and its flanges are generally too narrow to provide adequate lateral stiffness. A **W**-shape, **S**-shape, or **C**-shape is designated by its symbol, followed by its *nominal depth* in inches and its weight per foot in pounds (e.g., **W**12×96, **S**24×100, **C**10×25). In SI units, the *nominal depth* is given in mm and the *mass* is given in kg/m. Angles are designated by the symbol **L**, followed by the leg lengths (longer leg first), followed by the thickness (e.g., **L**8×8×3/4, **L**5×3×1/2).

---

[7]For example, the *Manual of Steel Construction* [Ref. 6-2], published by the American Institute of Steel Construction, describes how to design steel beams for structural applications.

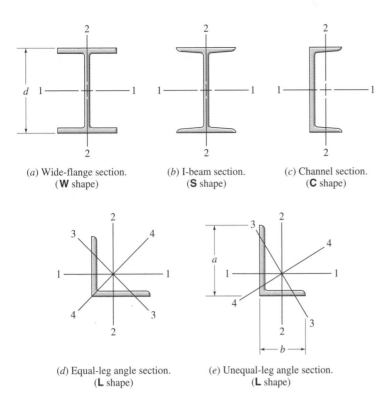

(*a*) Wide-flange section.
(**W** shape)

(*b*) I-beam section.
(**S** shape)

(*c*) Channel section.
(**C** shape)

(*d*) Equal-leg angle section.
(**L** shape)

(*e*) Unequal-leg angle section.
(**L** shape)

FIGURE 6.13 Several standard rolled structural steel shapes.

In the case of wood beams, it is important to note that the quoted dimensions of lumber are nominal, rough-cut, dimensions. The actual net dimensions, or *finish dimensions,* which are given in Appendix D.8, are smaller. Thus, the finish dimensions should be used in all structural calculations.

The flexure formula, Eq. 6.13, forms the basis for beam design based on flexural stress. Consider a general situation like Example Problem 6.4, where the beam is not doubly symmetric and where the maximum positive bending moment and the maximum negative bending moment have different magnitudes. From Eq. 6.13 the fiber stresses at the extreme (i.e., top and bottom) fibers may be written, respectively, as

$$\sigma_{top} \equiv \sigma_1 = \frac{-M(c_1)}{I} = \frac{-Mc_1}{I} \equiv -\frac{M}{S_1}$$

(6.14)

$$\sigma_{bot.} \equiv \sigma_2 = \frac{-M(-c_2)}{I} = \frac{Mc_2}{I} \equiv \frac{M}{S_2}$$

where

$$S = \frac{I}{c}$$

(6.15)

In this expression, I is the moment of inertia about the neutral axis, and c is the distance to an extreme fiber (see Fig. 6.14*a*). The quantity S, called the *elastic section modulus,* is a property of the cross-sectional dimensions. For most structural shapes, the values of S are tabulated, along with the location of the centroid of the cross section and the moment of inertia values (e.g., see Appendices D.1 through D.6).

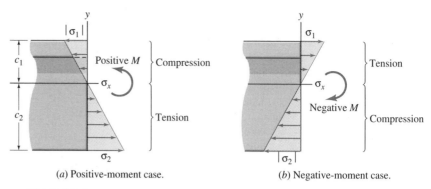

(a) Positive-moment case. (b) Negative-moment case.

FIGURE 6.14 Examples of the stress distribution in an unsymmetric beam.

Allowable-Stress Design.[8] Consider first the simple case of selecting a beam with doubly symmetric cross section made of material whose *allowable stress* (with the same magnitude in both tension and compression) is σ_{allow}. The value of σ_{allow} is determined by applying a *factor of safety* to the value of the yield strength of the material, as given in a table of material properties (see Appendix F). Let M_{max} be the maximum absolute value of the bending moment $M(x)$, that is, let

$$M_{max} \equiv \max_x |M(x)| \tag{6.16}$$

For design purposes, it is convenient to combine Eqs. 6.14, 6.15, and 6.16 and write the resulting *allowable-stress design equation* in the form

$$S_{design} = \frac{M_{max}}{\sigma_{allow}} \tag{6.17}$$

By selecting a section with $S \geq S_{design}$, we guarantee that the magnitude of the flexural stress will not exceed σ_{allow} anywhere in the beam.

Although there are a number of factors that must be considered in any design process in order to minimize the initial cost and the operating cost, it is usually desirable to select the lightest-weight structural member that satisfies the strength requirement (and all other requirements that are applicable). The following example problem illustrates this design process.

FIGURE 6.15 An example bending-moment diagram.

---

[8]Although the *Load and Resistance Factor Design Method,* discussed briefly in Section 2.12, is now widely used in designing beams (e.g., the *AISC Manual of Steel Construction, Load and Resistance Factor Design,* [Ref. 6-2], application of that method is limited, in this textbook, to Example 3.16.

■■■■■■■■■■■■■■■■■ EXAMPLE 6.5 ■■■■■■■■■■■■■■■■■■

From Appendix D.1, select a wide-flange steel beam to support the load distribution shown in Fig. 1a. Include weight of the beam in your calculations. The moment diagram for the beam, neglecting weight of the beam, is shown in Fig. 1c. Let $\sigma_{\text{allow}} = 19$ ksi.

Plan the Solution We can use Eq. 6.17 to compute the required section modulus. We need to pick the lightest section, with some margin in S so that we can accommodate the maximum moment with the beam weight included.

Solution

Preliminary Shape Selection: From Eq. 6.17,

$$S_{\text{design}} = \frac{M_{\text{max}}}{\sigma_{\text{allow}}} = \frac{(50 \text{ kip}\cdot\text{ft})(12 \text{ in./ft})}{(19 \text{ kips/in}^2)} = 31.6 \text{ in}^3$$

From Appendix D.1, there are two candidate sections, a **W**10×30 with $S = 32.4$ in³ and a **W**14×26 with $S = 35.3$ in³.[9] The **W**10×30 weighs 4 lb/ft more than the **W**14×26, so it seems that the best choice would be the **W**14×26. However, since this beam is deeper than the **W**10×30 (13.91 in. vs 10.47 in.), there may be justification for choosing the **W**10×30 as long as it meets the strength requirement. So, let us see if the **W**10×30 meets the design requirements when the weight of the beam is added to the loads in Fig. 1a.

Equilibrium Check: We could construct new shear and moment diagrams, but it may be quicker to use free-body diagrams. The maximum positive moment will occur near 5 ft (Fig. 1b) at the point where $V(x) = 0$. We first need to compute the reaction at A, which is labeled A_y in Fig. 2.

$$+\curvearrowleft \left(\sum M\right)_B = 0:$$

$$(4 \text{ kips/ft})(12 \text{ ft})(6 \text{ ft}) - (8 \text{ kips})(6 \text{ ft})$$
$$+ (w \text{ kips/ft})(18 \text{ ft})(3 \text{ ft}) - A_y(12 \text{ ft}) = 0$$
$$A_y = (20 + 4.5w) \text{ kips}$$

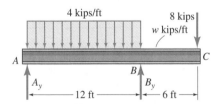

Fig. 2 Free-body diagram.

We can now determine the maximum positive moment using Fig. 3.

$$V(x_m) = (20 + 4.5w) - (4 + w)x_m = 0$$

---

[9]There are additional candidate sections in the complete AISC tables. [Ref. 6-2]

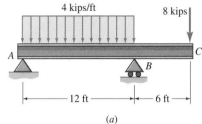

(a)

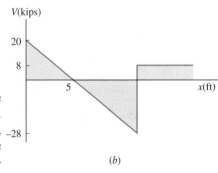

(b)

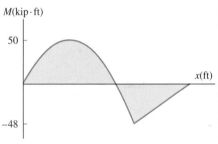

(c)

Fig. 1

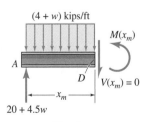

Fig. 3

277

Therefore, the maximum positive moment occurs at

$$x_m = \frac{20 + 4.5w}{4 + w}$$

$$\left(\sum M\right)_D = 0:$$

$$M(x_m) = (20 + 4.5w)\left(\frac{20 + 4.5w}{4 + w}\right) - (4 + w)\left(\frac{20 + 4.5w}{4 + w}\right)^2\left(\frac{1}{2}\right)$$

$$M(x_m) = \frac{(20 + 4.5w)^2}{2(4 + w)} \text{ kip} \cdot \text{ft}$$

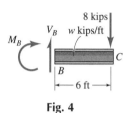

Fig. 4

The maximum negative moment will still occur at B, and its value can be computed using Fig. 4.

$$\left(\sum M\right)_B = 0: \qquad M_B = (-48 - 18w) \text{ kip} \cdot \text{ft}$$

For the **W**10×30 beam, $w = 30$ lb/ft $= 0.030$ kips/ft. Therefore,

$$M(x_m) = 50.3 \text{ kip} \cdot \text{ft}, \qquad M_B = -48.5 \text{ kip} \cdot \text{ft}$$

So $M_{\text{max}} = 50.3$ kip $\cdot$ ft, which would require

$$S_{\text{design}} = \frac{(50.3 \text{ kip} \cdot \text{ft})(12 \text{ in./ft})}{19 \text{ ksi}} = 31.8 \text{ in}^3$$

Final Shape Selection: Since the **W**14×26 beam is lighter than the **W**10×30, and since its section modulus is greater (35.3 in^3 vs 32.4 in^3), the **W**14×26 would be the best design, unless its added depth is, for some reason, undesirable. In that case, the **W**10×30 would be a perfectly satisfactory design choice.

Review the Solution It is interesting to note that an 18-ft **W**14×26 beam, weighing 468 lb is able, in this case, to support a total load of 56,000 lb. (We will again consider the design of the beam in Fig. 1 in Section 9.3, after we have taken up the topics of shear stress in beams and the state of stress at a point.)

The previous discussion and Example Problem 6.5 assumed the same allowable stress in tension and compression and assumed that the choice of beam cross section was to be made among candidate doubly symmetric sections. If the allowable stresses in tension and compression are different, this must be taken into account, and stresses must be computed where $M(x)$ is a maximum and also where it is a minimum. Also, if the cross section is not symmetric about the neutral axis, Eqs. 6.14 must be applied at the sections of maximum positive moment and maximum negative moment.

In Example Problem 6.5 you learned that it is desirable to select a beam that maximizes S, the elastic section modulus, and minimizes the weight. The next example problem indicates how the shape of the cross section affects these two quantities.

278

The **W**12×50 section, the **S**12×50 section, and the **S**15×50 section all weigh 50 lb/ft and have a cross-sectional area of 14.7 in.² Compare the **W**12×50 section, the **S**12×50 section, the **S**15×50 section, and the three "compact" sections in Fig. 1 on the basis of section modulus, S. Let $A = 14.7$ in² in each case.

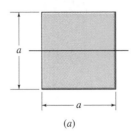

Plan the Solution Equation 6.15 gives the formula for the elastic section modulus, S, and Appendix C.2 and inside back covers give formulas for the area moments of inertia, I, for the compact sections. Appendices D.1 and D.3 give S values for the structural steel shapes.

Solution From Eq. 6.15,

$$S = \frac{I}{c}$$

and, from Appendix C.2,

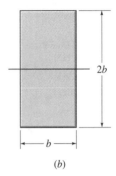

Square: $S = \dfrac{I}{c} = \left(\dfrac{a^4}{12}\right)\left(\dfrac{2}{a}\right) = \dfrac{a^3}{6}$

Rectangle: $S = \dfrac{I}{c} = \left[\dfrac{b(2b)^3}{12}\right]\left(\dfrac{1}{b}\right) = \dfrac{2b^3}{3}$

Circle: $S = \dfrac{I}{c} = \left(\dfrac{\pi r^4}{4}\right)\left(\dfrac{1}{r}\right) = \dfrac{\pi r^3}{4}$

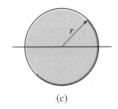

Since all shapes are to have an area of 14.7 in.²,

$$a = 3.834 \text{ in}, \quad b = 2.711 \text{ in.}, \quad r = 2.163 \text{ in.}$$

Fig. 1 Three "compact" shapes.

Then,

Shape	S (in³)
W12×50	64.7
S12×50	50.8
S15×50	64.8
Rectangle	13.3
Square	9.39
Circle	7.95

Review the Solution Two conclusions should be apparent from the preceding table: (a) "compact" shapes are not as efficient as beams that have a web connecting two flanges, and (b) wide-flange beams are well proportioned with regard to accommodating flexural stresses. For most applications the **W**12×50 shape would be more desirable than the **S**15×50 shape because it has essentially the same value of S but it is not as deep as the **S** shape, and it has wider flanges than the **S** shape.

(a) Reinforced concrete beam

(b) Sandwich beam

(c) Bimetallic beam

(d) Composite beam

FIGURE 6.16 Some nonhomogeneous beams.

Many structural applications of beams involve *nonhomogeneous beams*,[10] that is, beams made of two or more materials. Some significant examples, illustrated in Fig. 6.16, are: (a) reinforced concrete beams, (b) steel-wood "sandwich" beams, (c) bimetallic beams, and (d) fiber-reinforced-composite beams. The kinematic assumptions of Bernoulli-Euler beam theory apply to nonhomogeneous beams as well as to homogeneous ones, so the strain-displacement relationship of Eq. 6.3 applies. However, since the material properties are not constant throughout the cross-section, we must modify the analysis of Section 6.3 to accommodate the nonuniform material properties.

Direct-Method. Consider the simplest type of nonhomogeneous beam, namely, a rectangular beam consisting of two linearly elastic materials, as shown in Fig. 6.17. The two materials are labeled 1 and 2. We will assume, for purpose of illustration, that $E_1 > E_2$.

Strain Distribution: The strain distribution in Fig. 6.17b is given by Eq. 6.3, that is

$$\epsilon_x = \frac{-y}{\rho}$$

(6.3)
repeated

Stress Distribution: The flexural stress, which is illustrated in Fig. 6.17c, is given by

$$\sigma_{x1} = E_1\epsilon_x = \frac{-E_1 y}{\rho}$$

$$\sigma_{x2} = E_2\epsilon_x = \frac{-E_2 y}{\rho}$$

(6.18)

Location of the Neutral Axis: As in Section 6.3, we must first locate the neutral surface by setting $F(x) = 0$. From Eqs. 6.9a and 6.18, we get

$$F(x) = \int_A \sigma_x \, dA = -\frac{E_1}{\rho}\int_{A_1} y \, dA - \frac{E_2}{\rho}\int_{A_2} y \, dA = 0$$

(6.19)

This is a generalization of the equation defining the centroid of a cross section. In fact, we

FIGURE 6.17 The strain distribution and stress distribution in a nonhomogeneous, linearly elastic beam.

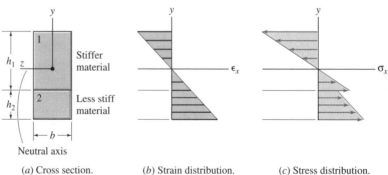

(a) Cross section.

(b) Strain distribution.

(c) Stress distribution.

---

[10]Beams of this type are sometimes called *composite beams*. The term *nonhomogeneous beams* is used here to avoid confusion with structures made of composite materials, which were discussed briefly in Section 2.4.

can write the above equation in the form

$$E_1\bar{y}_1A_1 + E_2\bar{y}_2A_2 = 0 \qquad (6.20)$$

This equation is used to establish the position of the neutral surface (i.e., the xz plane) in the beam.

Moment-Curvature Equation: Once the neutral axis has been located in the cross section, Eq. 6.9b is combined with Eqs. 6.18 to give

$$M = -\int_A y\sigma_x \, dA = \frac{1}{\rho}\left[E_1 \int_{A_1} y^2 \, dA + E_2 \int_{A_2} y^2 \, dA\right] \qquad (6.21)$$

or

$$M = \frac{1}{\rho}(E_1 I_1 + E_2 I_2) \equiv \frac{\overline{EI}}{\rho} \qquad (6.22)$$

where $\overline{EI}$ is the *weighted flexural rigidity*. It should be carefully noted that I_1 and I_2 are the moments of inertia of areas 1 and 2 <u>about the neutral axis</u>, defined by Eq. 6.20, not about their respective centroidal axes. Equation 6.22 is the *moment-curvature equation* for a beam that is nonhomogeneous within the cross section. Equations 6.20 and 6.22 can readily be extended to accommodate additional materials, but such situations are rare.

Flexure Formulas: Finally, we substitute Eq. 6.22 into Eqs. 6.18 to get the following expressions for the flexural stresses in materials 1 and 2.

$$\sigma_{x_1} = \frac{-M E_1 y}{\overline{EI}} \qquad \sigma_{x_2} = \frac{-M E_2 y}{\overline{EI}} \qquad (6.23)$$

■■■■■■■■■■■■■■□ **E X A M P L E   6 . 7** □■■■■■■■■■■■■■■

A nonhomogeneous beam having the dimensions shown in Fig. 1a is constructed by gluing a thin aluminum plate to the top side of a square wood beam. Take $E_1 \equiv E_{\text{alum.}} = 70$ GPa and $E_2 \equiv E_{\text{wood}} = 12$ GPa.

Determine the maximum flexural stress in the aluminum and the maximum flexural stress in the wood when a moment $M = 3$ kN·m is applied in the manner indicated in Fig. 1b.

Plan the Solution We must use Eq. 6.20 to locate the neutral axis, and we can then use Eqs. 6.23 to compute the flexural stresses in the two materials.

Solution

Neutral-Axis Location: The neutral axis of the beam is located at distance c_1, from the top of the beam, as indicated in Fig. 2. From Eq. 6.20,

$$E_1\bar{y}_1A_1 + E_2\bar{y}_2A_2 = 0$$

(a)

(b)

Fig. 1 A beam made of two materials.

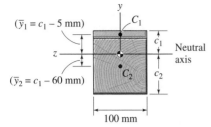

$(\bar{y}_1 = c_1 - 5 \text{ mm})$

$(\bar{y}_2 = c_1 - 60 \text{ mm})$

C_1

C_2

c_1

c_2

Neutral axis

y

z

100 mm

Fig. 2 Location of neutral axis.

$(70 \text{ GPa})(1000 \text{ mm}^2)(c_1 - 5 \text{ mm})$

$$+ (12 \text{ GPa})(10\,000 \text{ mm}^2)(c_1 - 60 \text{ mm}) = 0$$

Then,

$$c_1 = 39.74 \text{ mm}, \qquad c_2 = 70.26 \text{ mm}$$

and

$$\bar{y}_1 = 34.75 \text{ mm}, \qquad \bar{y}_2 = -20.26 \text{ mm}$$

Moment of Inertia: Since the moments of inertia I_1 and I_2 are to be taken about the z axis (neutral axis), we must use the parallel-axis theorem.

$$
\begin{aligned}
I_1 &= \frac{b_1 h_1^3}{12} + A_1 \bar{y}_1^2 \\[4pt]
&= \frac{(100 \text{ mm})(10 \text{ mm})^3}{12} + (100 \text{ mm})(10 \text{ mm})(34.74 \text{ mm})^2 \\[4pt]
&= 1.215(10^6) \text{ mm}^4 = 1.215(10^6) \text{ mm}^4 \, (10^3 \text{ mm/m})^{-4} \\[4pt]
I_1 &= 1.215(10^{-6}) \text{ m}^4
\end{aligned}
$$

Similarly,

$$I_2 = 12.439(10^{-6}) \text{ m}^4$$

Before calculating the flexural stresses we can check these moments of inertia by using the equation

$$I_z = I_1 + I_2 = 13.65(10^6) \text{ mm}^4$$

But,

$$
\begin{aligned}
I_z &= (I_z)_{\substack{\text{area} \\ \text{above}}} + (I_z)_{\substack{\text{area} \\ \text{below}}} \\[8pt]
&= \frac{1}{3}(100 \text{ mm})(39.74 \text{ mm})^3 + \frac{1}{3}(100 \text{ mm})(70.26 \text{ mm})^3 \\[4pt]
&= 13.65(10^6) \text{ mm}^4
\end{aligned}
$$

Flexural Stresses: The flexural stresses in materials 1 and 2 are given by Eqs. 6.23, with $\overline{EI}$ defined by Eq. 6.22,

$$
\begin{aligned}
\overline{EI} &= E_1 I_1 + E_2 I_2 = [70(10^9) \text{ N/m}^2][1.215(10^{-6}) \text{ m}^4] \\[4pt]
&\quad + [12(10^9) \text{ N/m}^2][12.439(10^{-6}) \text{ m}^4] \\[4pt]
&= 234.3(10^3) \text{ N} \cdot \text{m}^2 = 234.3 \text{ kN} \cdot \text{m}^2
\end{aligned}
$$

282

Finally, the maximum stresses in materials 1 and 2, which occur at $y = c_1$ and $y = -c_2$, respectively, are

$$(\sigma_{x1})_{max} = \frac{-M\,E_1(c_1)}{\overline{EI}}$$

$$= \frac{-(3\ kN \cdot m)(70 \times 10^6\ kN/m^2)(39.74 \times 10^{-3}\ m)}{234.3\ kN \cdot m^2}$$

$$= -35.6(10^3)\ kN/m^2 = -35.6\ MPa$$

$$(\sigma_{x2})_{max} = \frac{-M\,E_2(-c_2)}{\overline{EI}}$$

$$= \frac{-(3\ kN \cdot m)(12 \times 10^6\ kN/m^2)(-70.26 \times 10^{-3}\ m)}{234.3\ kN \cdot m^2}$$

$$= 10.79(10^3)\ kN/m^2 = 10.8\ MPa$$

$$(\sigma_{x1})_{max} = -35.6\ MPa, \qquad (\sigma_{x2})_{max} = 10.8\ MPa \qquad \textbf{Ans.}$$

Review the Solution This is the type of problem where we should check our result at each major step, as we did for the moment of inertia. We should expect the stiffer aluminum to experience higher stresses than those in the wood, and this is the case here.

This same problem is solved, in Example Problem 6.8, by the *transformed-section method*.

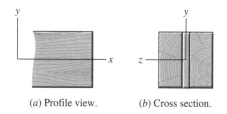

(a) Profile view. (b) Cross section.

FIGURE 6.18 A form of nonhomogeneous beam.

A nonhomogeneous beam may also be constructed by forming a sandwich beam that is symmetric about the plane of loading, as illustrated in Fig. 6.18. For example, beams of this type may be constructed by bonding wood beams to a steel-plate core. In this case Eqs. 6.22 and 6.23 are used, but it is unnecessary to use Eq. 6.20 since the location of the neutral axis is obvious.

Transformed-Section Method. Equations 6.18, 6.20, and 6.22 are the key equations in the analysis of nonhomogeneous beams since they completely relate the flexural stress distribution to the stress resultants $F(x) = 0$ and $M(x)$. By creating a *transformed section* it is possible to treat a nonhomogeneous beam essentially the same way as a homogeneous beam. That is, the neutral axis passes through the centroid of the transformed section, and the flexural stress is determined by a simple flexural formula of the form

$$\sigma = \frac{-My}{I} \qquad (6.13)$$
$$\text{repeated}$$

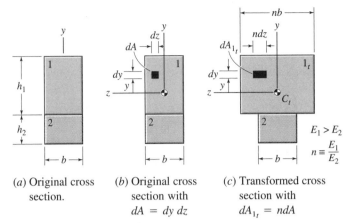

FIGURE 6.19 Basic geometry of the transformed section representing a nonhomogeneous beam.

Figure 6.19 shows the original two-material cross section (Figs. 6.19a,b) and the transformed section with *material 2 as the reference material* (Fig. 6.19c). (Material 1 is taken to be the stiffer material, as illustrated in Fig. 6.17.) Note that the transformed material is "widened" (since $E_1 > E_2$) only in the z direction.

Neutral Axis of the Transformed Section: We want y-distances to be the same in the original and the transformed section so that the distance y in the flexure formula will be unaltered. Let the ratio of moduli, with E_2 as the reference, be designated

$$n \equiv \frac{E_1}{E_2} \tag{6.24}$$

Then, Eq. 6.19, which locates the neutral axis, can be written as

$$\int_{A_1} y(n\,dA) + \int_{A_2} y\,dA = 0$$

or, for the transformed cross section A_t, simply

$$\boxed{\int_{A_t} y\,dA_t = 0} \tag{6.25}$$

Therefore, the *neutral axis passes through the centroid of the transformed section,* just as it would pass through the centroid of a homogeneous beam.

Moment-Curvature Equation: Equation 6.21 is the moment-curvature equation for a two-material, nonhomogeneous beam, and we want this equation to hold for the transformed section. Introducing the *modulus ratio n,* we can write Eq. 6.21 as

$$M = \frac{E_2}{\rho}\left[\int_{A_1} y^2(n\,dA) + \int_{A_2} y^2\,dA\right] \tag{6.26}$$

It is clear that the term in brackets in Eq. 6.26 is just the moment of inertia of the transformed section about its neutral axis (centroid). Hence, the moment-curvature equation can be written as

$$M = \frac{E_2 I_t}{\rho}$$ (6.27)

where I_t is given by

$$I_t = \int_{A_t} y^2 \, dA_t$$ (6.28)

Flexure Formulas: Finally, we can substitute Eq. 6.27 into Eqs. 6.18 to determine the stress distribution in each material. We get

$$\sigma_{x1} = -\frac{E_1 y}{\rho} = n\left(\frac{-M\,y}{I_t}\right)$$
$$\sigma_{x2} = -\frac{E_2 y}{\rho} = -\frac{M\,y}{I_t}$$ (6.29)

Thus, the stress in the reference material is computed using the standard flexure formula, but the stress in the transformed material must be multiplied by the modulus ratio, n.

Material 2 was taken as the reference material for the transformed-section analysis in Eqs. 6.24 through 6.29. Usually the less-stiff material is taken as the reference material, so that $n > 1$, but this is not essential. However, since the labeling of materials as "material 1" and "material 2" is completely arbitrary, the less-stiff material can always be labeled as "material 2."

■■■■■■■■■■■■■■■□ EXAMPLE 6.8 □■■■■■■■■■■■■■■■■

Solve the problem stated in Example Problem 6.7 using the transformed-section method, with material 2 as the reference material.

Plan the Solution The following four steps are required to solve this problem:

1. Sketch the transformed section using the modulus ratio n as the "z-stretch factor."
2. Locate the neutral axis (the centroid of the transformed section).
3. Compute I_t for the transformed section.
4. Use Eqs. 6.29 to compute the stresses.

Solution

Transformed Section: The modulus ratio (Eq. 6.24) is

$$n = \frac{E_1}{E_2} = \frac{70 \text{ GPa}}{12 \text{ GPa}} = 5.833$$

Therefore, the aluminum part of the cross section must be stretched by a factor $n = 5.833$.

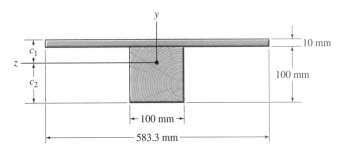

Fig. 1 The transformed section.

Neutral Axis: Since the transformed beam is "homogeneous" with $E_t = E_2$, the neutral axis passes through the centroid of the transformed section. Therefore,

$$(583.3 \text{ mm})(10 \text{ mm})(c_1 - 5 \text{ mm}) + (100 \text{ mm})(100 \text{ mm})(c_1 - 60 \text{ mm}) = 0$$

$$c_1 = 39.74 \text{ mm}, \qquad c_2 = 70.26 \text{ mm}$$

and

$$\bar{y}_1 = 34.74 \text{ mm}, \qquad \bar{y}_2 = -20.26 \text{ mm}$$

Moment of Inertia:

$$I_t = I_{1t} + I_2 = \sum\left(\frac{1}{12}b_t h^3 + A_t d^2\right)$$

$$= \frac{1}{12}(583.3 \text{ mm})(10 \text{ mm})^3 + (583.3 \text{ mm})(10 \text{ mm})(34.74 \text{ mm})^2$$

$$+ \frac{1}{12}(100 \text{ mm})(100 \text{ mm})^3 + (100 \text{ mm})(100 \text{ mm})(-20.26 \text{ mm})^2$$

$$= 19.53(10^6) \text{ mm}^4 = 19.53(10^{-6}) \text{ m}^4$$

Flexural Stresses: Using Eqs. 6.29, we get

$$\sigma_{x1} = n\left(\frac{-M c_1}{I_t}\right) = \frac{5.833(-3 \text{ kN}\cdot\text{m})(39.74 \times 10^{-3} \text{ m})}{(19.53 \times 10^{-6} \text{ m}^4)}$$

$$= -35.6 \text{ MPa}$$

$$\sigma_{x2} = \frac{-M(-c_2)}{I_t} = \frac{(-3 \text{ kN}\cdot\text{m})(-70.26 \times 10^{-3} \text{ m})}{(19.53 \times 10^{-6} \text{ m}^4)} = 10.8 \text{ MPa}$$

$$\sigma_{x1} = -35.6 \text{ MPa}, \qquad \sigma_{x2} = 10.8 \text{ MPa} \qquad \qquad \textbf{Ans.}$$

Review the Solution We have obtained the same answers as in Example Problem 6.7, so we can assume that they are probably correct.

Since the only difference between solving a transformed-section nonhomogeneous beam problem and solving a homogeneous beam problem is the multiplication of stress in the transformed area by n, this method for solving nonhomogeneous beam problems is preferred over the direct method used in Example Problem 6.7.

Thus far in Chapter 6, we have been considering flexural stress and strain in beams whose cross-sectional shape and whose loading and support conditions produce bending that is confined to a longitudinal plane of symmetry (LPS) of the beam. This simplifies the analysis in two important respects. First, the deflection of the beam can be characterized by a deflection curve in the LPS (e.g., Fig. 6.1); second, there is no tendency of the beam to twist. However, we also need to be able to analyze the behavior of beams that are not loaded and supported in this simple manner.

In Chapter 9 we will consider *combined bending and torsion,* as would be experienced, for example, by a beam that is loaded parallel to, but not along, an axis of symmetry, like the doubly symmetric box beam in Fig. 6.20. Here, however, we will consider loading that does not produce bending in a single longitudinal plane of symmetry. Two examples of this are a channel beam loaded parallel to its web (Fig. 6.21*a*), and a Z section (Fig. 6.21*b*). The former beam has a longitudinal plane of symmetry, but the loading is perpendicular to, not in or parallel to, the plane of symmetry. The latter beam has no longitudinal plane of symmetry.

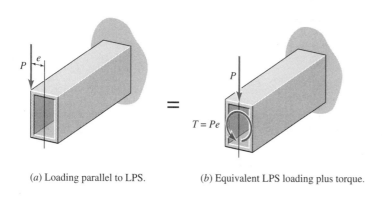

(*a*) Loading parallel to LPS.   (*b*) Equivalent LPS loading plus torque.

FIGURE 6.20 A beam with transverse loading parallel to an axis of symmetry.

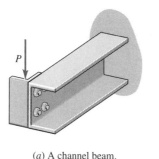
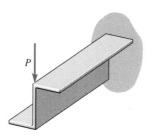

(*a*) A channel beam.   (*b*) A Z-section beam.

FIGURE 6.21 Two beams that are not loaded in a plane of symmetry.

Doubly Symmetric Beams with Inclined Loads. Before we study the general case of unsymmetric bending, let us generalize the results of Section 6.3 to the case of a doubly symmetric beam whose loading does not lie in either longitudinal plane of symmetry. Figure 6.22 shows a doubly symmetric beam with such an *inclined load,* that is, a load that simultaneously produces bending about both axes of symmetry in the cross section, as indicated in Fig. 6.22.

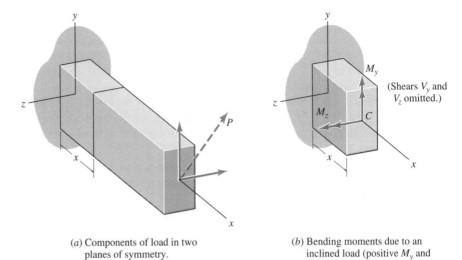

(a) Components of load in two planes of symmetry.

(b) Bending moments due to an inclined load (positive M_y and M_z shown).

FIGURE 6.22 A doubly-symmetric beam with inclined loading.

For a doubly symmetric, linearly elastic beam with $E = E(x)$, the flexure formula, Eq. 6.13, can be applied separately for M_y and M_z, and the two expressions can then be added to give

$$\sigma_x = \frac{M_y z}{I_y} - \frac{M_z y}{I_z} \qquad (6.30)$$

The second term on the right corresponds to bending about the z axis of symmetry and comes directly from Eq. 6.13. The first term is a modification of Eq. 6.13 for bending about the y axis of symmetry. Since the y axis and z axis are both axes of symmetry of the cross section, their origin is the centroid of the cross section, as indicated in Fig. 6.22b. Figure 6.23 illustrates the superposition of the stresses due to moment components M_y and M_z acting on a rectangular cross section. Due to the combined stresses, the beam will bend about the inclined NA indicated in Fig. 6.23c.

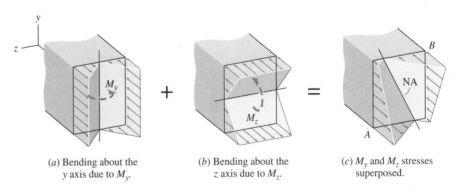

(a) Bending about the y axis due to M_y.

(b) Bending about the z axis due to M_z.

(c) M_y and M_z stresses superposed.

FIGURE 6.23 Flexural stresses due to inclined loading of a doubly symmetric beam.

Principal Axes of Inertia. In studying unsymmetric bending we will need to make use of several geometric properties of plane areas. (These are discussed in greater detail in Appendix C.) The *product of inertia* of an area A (Fig. 6.24a) with respect to a yz

reference frame in the plane of the area is defined by

$$I_{yz} \equiv \int_A yz \, dA \qquad (6.31)$$

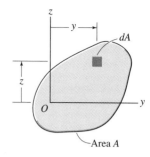

(a) A plane area.

If $I_{yz} = 0$, the y and z axes are said to be *principal axes of inertia* of the area. If either of the axes is an axis of symmetry, like the y axis in Fig. 6.24b, then $I_{yz} = 0$, since the contributions of symmetrically located dA's cancel in the integral for I_{yz}. Therefore, the axis of symmetry is one principal axis and any axis perpendicular to it is also a principal axis.

Finally, as discussed in Appendix C.3, for any planar area and for any origin in the plane, it is possible to orient a pair of orthogonal axes such that they are principal axes. For example, Fig. 6.24c shows the principal axes of inertia of an unequal-leg angle section with respect to an origin at the centroid. These are called *centroidal principal axes*. Appendix D.6 gives the angle of inclination of the centroidal principal axes of several unequal-leg angle sections.

Unsymmetric Bending: Principal-Axis Method.[11]

Equation 6.30 applies to bending of beams with doubly-symmetric cross section. Let us now determine an expression for the flexural stress in a beam with unsymmetric cross section (like the beams in Figs. 6.21b and 6.24c).

Consider the beam in Fig. 6.25a, which has no net axial force (i.e., $F = 0$), but which has both M_y and M_z; that is, consider an arbitrarily oriented bending moment acting on an arbitrarily shaped cross section. The contribution of the flexural stress σ_x on an elemental area dA is shown in Fig. 6.25b. The stress resultants on the cross section are obtained by summing the contributions of elemental forces $dF = \sigma_x \, dA$ over the area of the cross section. This gives

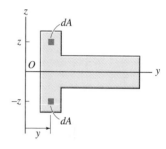

(b) An area with one axis of symmetry.

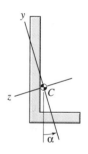

(c) Centroidal principal axes.

FIGURE 6.24 Illustrations for use in defining product of inertia and principal axes.

$$F(x) = \int_A \sigma_x \, dA$$

$$M_y(x) = \int z\sigma_x \, dA \qquad (6.32)$$

$$M_z(x) = -\int_A y\sigma_x \, dA$$

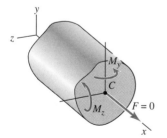

(a) The stress resultants related to σ_x.

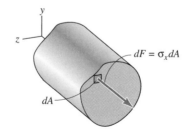

(b) The normal force on an elemental area.

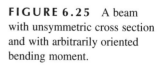

FIGURE 6.25 A beam with unsymmetric cross section and with arbitrarily oriented bending moment.

---

[11]An analysis of unsymmetric bending that does not directly employ the principal axes may be found in Ref. 6-4, Section 4.3 and in Ref. 6-5, Section 4.3.2.1.

Let us assume that σ_x has the bilinear form

$$\sigma_x = a_0 + a_1 y + a_2 z \tag{6.33}$$

Then, combining Eq. 6.32 with Eq. 6.33, we get

$$F(x) = a_0 \int_A dA + a_1 \int_A y\, dA + a_2 \int_A z\, dA$$

$$M_y(x) = a_0 \int_A z\, dA + a_1 \int_A zy\, dA + a_2 \int_A z^2\, dA \tag{6.34}$$

$$M_z(x) = -a_0 \int_A y\, dA - a_1 \int_A y^2\, dA - a_2 \int_A yz\, dA$$

Equations 6.34 can be simplified if we select the centroid of the cross section as the origin of the yz reference frame and if we orient the y and z axes along the centroidal principal axes. Then, $\bar{y} = \bar{z} = I_{yz} = 0$. Since $F = 0$, we get

$$a_0 = 0, \quad M_y = a_2 I_y, \quad M_z = -a_1 I_z \tag{6.35}$$

So, combining Eqs. 6.33 and 6.35, we get

$$\boxed{\sigma_x = \frac{M_y z}{I_y} - \frac{M_z y}{I_z}} \tag{6.36}$$

which is the same as Eq. 6.30. Therefore, we can conclude that the **the flexural stress σ_x in a beam with arbitrary cross section and with arbitrarily oriented bending moment is given by Eq. 6.36, provided that the y and z axes are centroidal principal axes**. Loading in a longitudinal plane of symmetry (LPS) is just a special case of this.

Orientation of the Neutral Axis: The orientation of the neutral axis in a cross section may be determined by setting $\sigma_x = 0$ in Eq. 6.36. Then, if (y^*, z^*) are the coordinates of points that lie on the neutral axis,

$$\left(\frac{M_y}{I_y}\right) z^* - \left(\frac{M_z}{I_z}\right) y^* = 0 \tag{6.37}$$

is the equation of the neutral axis in the yz plane.[12] Let the angles θ and β be defined by

$$\tan\theta = \frac{M_y}{M_z}, \qquad \tan\beta = \frac{y^*}{z^*} \tag{6.38}$$

as illustrated in Fig. 6.26a. That is, the moment vector is oriented at angle θ <u>measured clockwise from the positive z axis</u>, and the neutral axis is oriented at angle β <u>measured clockwise from the positive z axis</u>. Then Eq. 6.37 may be conveniently expressed as

$$\boxed{\tan\beta = \left(\frac{I_z}{I_y}\right)\tan\theta} \tag{6.39}$$

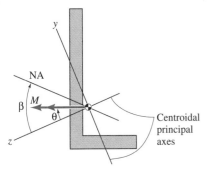

(a) Orientation of the couple vector and the NA with respect to the centroidal principal axes.

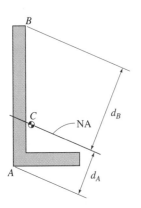

(b) Location of points of maximum tension and maximum compression.

FIGURE 6.26 A typical unsymmetric cross section.

---

[12]Recall that the y and z axes must be <u>centroidal principal axes</u>. It is arbitrary which centroidal principal axis is labeled the y axis and which is labeled the z axis, but the y and z axes must form a right-handed coordinate system with the x axis.

From Eq. 6.39 it is clear that β will lie in the same quadrant as θ, and that the relative orientation of the *NA* and **M** will depend on whether $I_z > I_y$ or $I_z < I_y$.

Figure 6.23c illustrates how flexural stress changes from tension to compression at the neutral axis, where $\sigma_x = 0$. Unlike the case of loading in a single principal plane, inclined loading produces a neutral-axis orientation that depends, at each cross section, on the orientation of the bending-moment vector at the particular cross section, that is, on the ratio of $M_y(x)$ to $M_z(x)$.

Maximum Tensile and Compressive Stresses: Figure 6.23c also illustrates the fact that the points of maximum tension and maximum compression on the cross section are the two points that are farthest from the neutral axis. These points are labeled A and B on Fig. 6.23c and on Fig. 6.26b. We need the coordinates of these points with respect to centroidal principal axes in order to find the maximum stresses from Eq. 6.36. The coordinates (y, z) in the principal-axis reference frame can be related to coordinates (y', z') by referring to Fig. 6.27. Using the shaded triangles, we get

$$y = y' \cos \phi + z' \sin \phi$$
$$z = -y' \sin \phi + z' \cos \phi \qquad (6.40)$$

The angle ϕ is measured counterclockwise from the y' axis to the y axis. Example Problem 6.9 illustrates the effect of load inclination on the bending of a beam with doubly symmetric cross section.

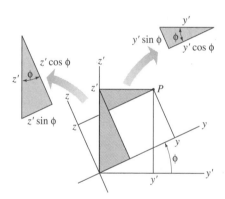

FIGURE 6.27
Determination of coordinates (y, z) of point P from coordinates (y', z').

■■■■■■■■■■■■■ □ EXAMPLE 6.9 □□□□□□□□□□□□■■

The beam in Fig. 1, an **S**12×50 I-beam, is subjected to a moment $M = 150$ kip·in. that is supposed to lie along the z axis (e.g., due to loading in the xy plane). (a) Determine the effect of a "load misalignment" of $\theta = 2°$ on the orientation of the neutral axis. Show the neutral axis on a sketch of the cross section; and (b) determine the maximum tensile stress and the maximum compressive stress on the section. Note the large difference in I_y and I_z, and note how this affects the solution.

Plan the Solution We can use Eq. 6.39 (or Eq. 6.37) to determine the orientation of the neutral axis. Then we can identify the two points farthest from the NA and use Eq. 6.36 to compute the flexural stress at these points.

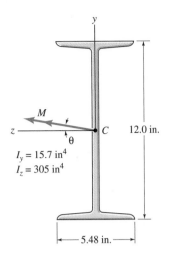

Fig. 1 An I-beam with inclined loading.

$I_y = 15.7 \text{ in}^4$
$I_z = 305 \text{ in}^4$

12.0 in.

5.48 in.

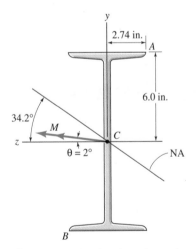

Fig. 2 Neutral axis orientation and points of maximum and minimum flexural stress.

Solution
(a) From Eq. 6.39,

$$\tan \beta = \left(\frac{I_z}{I_y}\right) \tan \theta \qquad (1)$$

where, from Fig. 1,

$$\tan \theta = \frac{M_y}{M_z} = \frac{M \sin \theta}{M \cos \theta} \qquad (2)$$

Therefore,

$$\tan \beta = \left(\frac{305 \text{ in}^4}{15.7 \text{ in}^4}\right) \tan \theta \qquad (3)$$

When $\theta = 0°$, $\beta = 0°$, and the z axis is the neutral axis, but when $\theta = 2°$ we get

$$\tan \beta = \left(\frac{305 \text{ in}^4}{15.7 \text{ in}^4}\right)(0.0349) = 0.678$$

So, for $\theta = 2°$,

$$\beta = 34.2° \qquad \text{Ans. (a)} \quad (4)$$

The orientation of the NA at this cross section is shown in Fig. 2.

(b) The maximum stresses will occur at the extreme points A and B. From Eq. 6.36,

$$\sigma_x = \frac{M_y z}{I_y} - \frac{M_z y}{I_z} \qquad (5)$$

$$\sigma_{x_A} \equiv \sigma_x(6.0, -2.74) = \frac{(150 \text{ kip} \cdot \text{in.})(\sin 2°)(-2.74 \text{ in.})}{15.7 \text{ in}^4}$$

$$- \frac{(150 \text{ kip} \cdot \text{in.})(\cos 2°)(6.0 \text{ in.})}{305 \text{ in}^4}$$

$$= -3.86 \text{ ksi}$$

From symmetry,

$$\sigma_{x_B} \equiv \sigma_x(-6.0, 2.74) = 3.86 \text{ ksi}$$

Had there been no misalignment, that is, for $\theta = 0$

$$\sigma_x(y, z) = \frac{-M_z y}{I_z} \qquad (6)$$

so, the maximum compression is at $y = 6.0$ in., or

$$(\sigma_x)_{\text{max C}} = \frac{-(150 \text{ kip} \cdot \text{in.})(6.0 \text{ in.})}{305 \text{ in}^4} = -2.95 \text{ ksi}$$

and

$$(\sigma_x)_{\text{max T}} = 2.95 \text{ ksi}$$

In summary,

θ	β	$(\sigma_x)_{max\ C}$	$(\sigma_x)_{max\ T}$
0°	0°	−2.95 ksi	2.95 ksi
2°	34.2°	−3.86 ksi	3.86 ksi

Thus, a small *misalignment of 2° increases the maximum stresses by 31%.* This is due to the large ratio of I_z/I_y.

Review the Solution   The answers look reasonable, since a nonzero value of M_y, together with a large ratio of I_z/I_y, will cause a significant rotation of the neutral axis, that is, a large value of β. The values of the maximum stresses are not unreasonable.

(See Homework Problem 6.6-5 where the same loading is applied to a very similar sized beam, a **W**12 × 50.)

Example Problem 6.10 illustrates the calculation of flexural stresses on an unsymmetric cross section.

■■■■■■■■■■■■■■■■■□ **E X A M P L E   6 . 1 0** ■□■■■■■■■■■■■■■■■

An unequal-leg angle section has the dimensions shown in Fig. 1.[13] At this cross section the moment is $M = 10\ kN \cdot m$ and is oriented parallel to the short leg of the angle, as shown. (a) Determine the orientation of the neutral axis of the cross section, and show this orientation on a sketch. (b) Determine the maximum tensile stress and the maximum compressive stress on the cross section.

Plan the Solution   We can use Eq. 6.39 to determine the orientation of the neutral axis. Then, we can locate the points in the cross section that are farthest from the NA and use Eq. 6.36 to compute the flexural stress at these points.

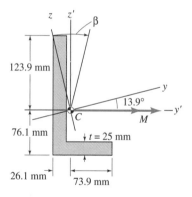

$I_y = 28.7(10^{-6})\ m^4$
$I_z = 3.08(10^{-6})\ m^4$

Fig. 1   An unequal-leg angle.

Solution
(a) From Eq. 6.39,[14]

$$\tan \beta = \left(\frac{I_z}{I_y}\right) \tan \theta \tag{1}$$

where

$$\tan \theta = \frac{M_y}{M_z} \tag{2}$$

But, from Fig. 1

$$M_y = M \cos(13.9°), \qquad M_z = -M \sin(13.9°)$$

---

[13]See Example Problems C.1 through C.4 for calculation of the section properties of an L-shaped cross section.

[14]Remember that β is measured from the z axis in the direction toward the y axis, as shown in Fig. 6.26a.

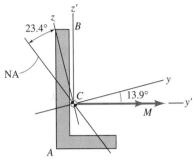

Fig. 2 The orientation of the neutral axis (NA).

Therefore,

$$\tan \beta = \left[\frac{(3.08 \times 10^{-6} \text{ m}^4)}{(28.7 \times 10^{-6} \text{ m}^4)} \right] [-\cot(13.9°)] = -0.434$$

$$\beta = -23.4° \qquad \textbf{Ans. (a)} \quad (3)$$

The correct orientation of the NA is shown in Fig. 2.

(b) Points A and B in Fig. 2 are the two points that are farthest from the neutral axis. To compute σ_x at these points we need their coordinates in the centroidal principal-coordinate reference frame. From Fig. 1 we get

$$(y_A', z_A') = (-26.1 \text{ mm}, -76.1 \text{ mm})$$

$$(y_B', z_B') = (-1.1 \text{ mm}, 123.9 \text{ mm})$$

Using Eqs. 6.40, we get

$$y_A = y_A' \cos \phi + z_A' \sin \phi$$
$$= (-26.1 \text{ mm})(\cos 13.9°) + (-76.1 \text{ mm})(\sin 13.9°)$$
$$y_A = -43.6 \text{ mm}$$
$$z_A = -y_A' \sin \phi + z_A' \cos \phi$$
$$= -(-26.1 \text{ mm})(\sin 13.9°) + (-76.1 \text{ mm})(\cos 13.9°)$$
$$z_A = -67.6 \text{ mm}$$

Similarly,

$$y_B = 28.7 \text{mm}, \qquad z_B = 120.5 \text{ mm}$$

From Eq. 6.36

$$\sigma_x = \frac{M_y z}{I_y} - \frac{M_z y}{I_z} \qquad (4)$$

so,

$$\sigma_{xA} \equiv \sigma_x(-43.6, -67.6)$$

$$= \frac{(9.71 \times 10^3 \text{ N} \cdot \text{m})(-67.6 \times 10^{-3} \text{ m})}{(28.7 \times 10^{-6} \text{ m}^4)}$$

$$- \frac{(-2.40 \times 10^3 \text{ N} \cdot \text{m})(-43.6 \times 10^{-3} \text{ m})}{(3.08 \times 10^{-6} \text{ m}^4)}$$

$$\sigma_{xA} = -56.9 \text{ MPa}$$

$$\sigma_{xB} \equiv \sigma_x(28.7, 120.5)$$

$$= \frac{(9.71 \times 10^3 \text{ N} \cdot \text{m})(120.5 \times 10^{-3}\text{m})}{(28.7 \times 10^{-6}\text{m}^4)}$$

$$- \frac{(-2.40 \times 10^3 \text{ N} \cdot \text{m})(28.7 \times 10^{-3} \text{ m})}{(3.08 \times 10^{-6} \text{ m}^4)}$$

$$\sigma_{xB} = 63.2 \text{ MPa}$$

In summary,

$$\left.\begin{array}{l} \sigma_{\text{max T}} = \sigma_{xB} = 63.2 \text{ MPa} \\ \sigma_{\text{max C}} = \sigma_{xA} = -56.9 \text{ MPa} \end{array}\right\} \quad \text{Ans. (b)} \quad (5)$$

Review the Solution   The best way to check the results of Part (a) and Part (b) is to draw the cross section to scale in order to verify the coordinates (y_A, z_A) and (y_B, z_B). Then, we can recheck the computations leading from Eq. (4) to the maximum stresses in Eq. (5). Since these have opposite signs, as they are supposed to, and since the ratio of their magnitudes is in agreement with their perpendicular distances from the NA, our results are probably correct.

*6.7 INELASTIC BENDING OF BEAMS

If the loads on a beam are large enough to cause the stress to exceed the yield strength, the beam is said to undergo *inelastic bending*. A beam does not totally collapse when its maximum stress reaches the yield strength. In fact, since the ultimate load that a beam can support may be much greater than the load that produces first yielding of the outer fibers, design codes now employ ultimate-load design concepts. (See Section 2.12.) Therefore, it is important for designers of structures and machines to understand the inelastic-bending behavior of beams so that this additional strength margin can be properly accounted for.

Fundamental Equations. In this section we will only consider pure bending of beams that are symmetrical about a longitudinal plane of symmetry (Fig. 6.28a,b). As in Sections 6.2 and 6.3, our analysis of pure bending will involve the three essentials of deformable-body mechanics—*geometry of deformation, equilibrium,* and *material behavior.*

Geometry of Deformation: The assumption that *plane sections remain plane* is valid for inelastic pure bending of beams as well as for linearly elastic bending (Section 6.2). This assumption leads to the *stain-displacement equation* (Eq. 6.3 repeated).

$$\boxed{\epsilon = -\frac{y}{\rho} = -y\,\kappa} \quad\quad (6.41)$$

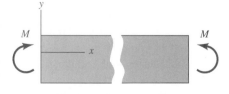 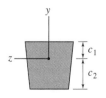 

FIGURE 6.28 Pure bending of a beam.

(a) Pure bending in longitudinal plane of symmetry.

(b) Cross section.

(c) Strain distribution.

295

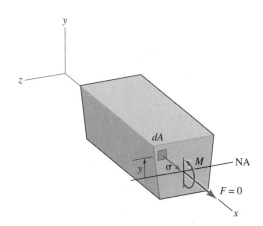

FIGURE 6.29 The stress resultants for pure bending.

where ϵ is the extensional strain of fibers at distance y from the neutral surface, ρ is the radius of curvature of the deformed axis of the beam, and $\kappa = 1/\rho$ is the curvature. This linear strain distribution is shown in Fig. 6.28c

Equilibrium: The stress resultants that are related to the bending of beams are the axial force F and the bending moment M, shown in Fig. 6.29. From Eqs. 6.9,

$$F = \int_A \sigma \, dA = 0$$

$$M \equiv M_z = -\int_A y \, \sigma \, dA$$

(6.42)

As in the case of linearly elastic behavior, Eq. 6.42a is used to locate the neutral axis in the cross section. As long as the flexural stress does not exceed the proportional limit, the neutral axis passes through the centroid of the cross section (Section 6.3). For inelastic bending, the location of the neutral axis depends on the shape of the cross section, on the stress-strain ($\sigma - \epsilon$) relationship, and on the magnitude of the applied moment, as will be illustrated in this section.

Material Behavior: Three types of nonlinear stress-strain behavior are illustrated in Fig. 3.18. Equations 6.43 are general expressions of the relationship of flexural stress σ to extensional strain ϵ for such nonlinear materials.

$$\sigma = \sigma(\epsilon), \quad \text{or} \quad \epsilon = \epsilon(\sigma)$$

(6.43)

To illustrate inelastic bending of beams we will examine only the case of elastic-plastic bending.

Elastic-Plastic Bending. Consider pure bending of a beam made of elastic-plastic material whose stress-strain curve is given in Fig. 6.30. The material follows Hooke's Law up to the proportional limit, which is assumed to be the same stress as the yield point (σ_Y). It is assumed that the material has a distinct yield point in compression at ($-\sigma_Y$), as indicated in Fig. 6.30. The corresponding yield strains in tension and compression are $\epsilon_Y = \sigma_Y/E$ and ($-\epsilon_Y$), respectively.

Figure 6.31 illustrates the strain distribution and the stress distribution for three levels of loading of a beam with a singly symmetric cross section (e.g., Fig. 6.28b).[15]

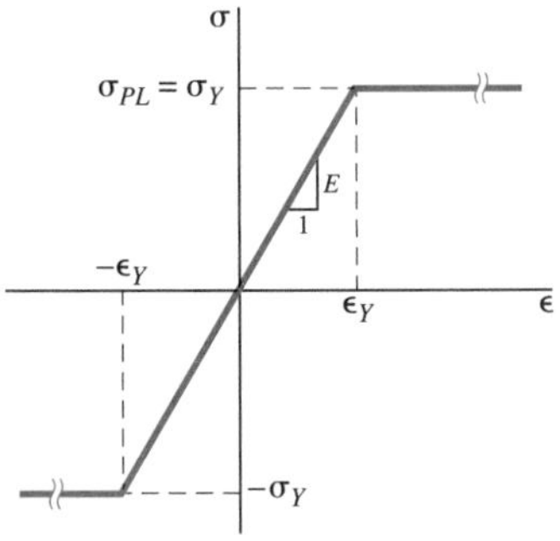

FIGURE 6.30 The stress-strain curve for an elastic-plastic material.

---

[15]Unloading behavior is discussed later under the topic of *residual stresses.*

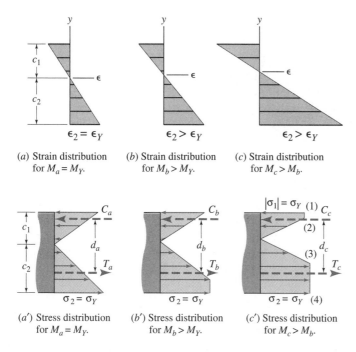

FIGURE 6.31 Strain distribution and stress distribution for an elastic-plastic beam.

Location of the Neutral Axis: Since, for pure bending, the axial force F is zero, the equation that determines the location of the neutral axis is Eq. 6.42a, which can be written as

$$\int_A \sigma \, dA = 0 \quad \rightarrow \quad T = C \tag{6.44}$$

where T is the tension on the cross section and C is the compression, as illustrated in Fig. 6.31. As the moment changes, the stress blocks that represent T and C change shape, and this change of shape leads to a shift in the location of the neutral axis, unless the cross section has two axes of symmetry. (Note in Figs. 6.31a through 6.31c and Figs. 6.31a′ through 6.31c′ that the neutral axis moves upward slightly with each increase in the applied moment.)

Maximum Elastic Moment: The maximum elastic moment, or *yield moment, M_Y*, is the moment that causes the fiber that is farthest from the elastic neutral axis to yield. Since the behavior is linearly elastic up the yield moment, we can use the flexural formula, Eq. 6.13, in the form

$$M_Y = \frac{\sigma_Y I}{c} \tag{6.45}$$

where c, the larger of the values c_1 and c_2, is measured from the *elastic neutral axis* (which passes through the centroid of the cross section).

Fully Plastic Moment: Materials like mild steel, that exhibit elastic-plastic behavior similar to that depicted in Fig. 6.30, are able to undergo extensional strains that are an order of magnitude or more greater than the yield strain ϵ_Y. In Fig. 6.31c′ there is an *elastic core*

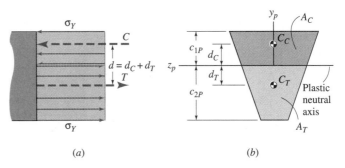

FIGURE 6.32 Fully plastic bending.

(2)–(3), a *compressive plastic zone* (1)–(2), and a *tensile plastic zone* (3)–(4). With increasing bending moment, the magnitude of the strain in the outer fibers increases, and the elastic zone shrinks in depth until, in the limit, there is a fully plastic compression zone above the (fully plastic) neutral axis and a fully plastic tensile zone below the neutral axis, as illustrated in Fig. 6.32a. Since the neutral axis is located by setting $T = C$ (Eq. 6.44), and since both tensile and compressive stress blocks have constant stress of magnitude σ_Y, the *plastic neutral axis* is determined by the purely geometric condition that

$$A_C = A_T = \frac{1}{2}A \tag{6.46}$$

The resultant forces C and T act at the centroids of the compression zone and tensile zone, respectively, so the *plastic moment, M_P,* is given by

$$\boxed{M_P = \frac{\sigma_Y A}{2}(d_C + d_T)} \tag{6.47}$$

where d_C and d_T are the distances from the plastic neutral axis to the centroids of the compression zone and the tensile zone, respectively.

The value of the plastic moment can be expressed in the compact form

$$\boxed{M_P = \sigma_Y Z}$$

where Z is the *plastic section modulus* for the cross section. Values of Z are tabulated in Appendix D for selected structural shapes. Recall from Section 6.4 that the yield moment can be expressed in terms of the *elastic section modulus S* by the formula

$$\boxed{M_Y = \sigma_Y S}$$

The ratio of the plastic moment M_P to the yield moment M_Y is called the *shape factor, f*.

$$\boxed{f = \frac{M_P}{M_Y} = \frac{Z}{S}} \tag{6.48}$$

This factor indicates the additional moment capacity of the beam beyond the moment that causes first yielding. The value of f for wide flange beams is typically in the range 1.1 to 1.2.

Determine expressions for the yield moment, the plastic moment, and the shape factor for a beam with rectangular cross section (Fig. 1*a*).

Solution   Because the rectangular cross section has two axes of symmetry, the neutral axis passes through the centroid for all values of applied moment. The stress state corresponding to the yield moment is shown in Fig. 1*b*, and the stress state for the plastic-moment case is shown in Fig. 1*c*.

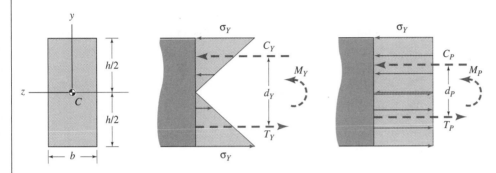

| (*a*) Cross section. | (*b*) Yield-moment case. | (*c*) Plastic-moment case. |

Fig. 1

Yield Moment: From Eq. 6.45,

$$M_Y = \frac{\sigma_Y I}{c} = \frac{\sigma_Y\left(\dfrac{bh^3}{12}\right)}{\left(\dfrac{h}{2}\right)} \qquad (1)$$

or

$$M_Y = \frac{1}{6}\sigma_Y bh^2 \qquad \text{Ans.} \quad (2)$$

Plastic Moment: From Eq. 6.47,

$$M_P = \frac{\sigma_Y A}{2}(d_C + d_T) = \frac{\sigma_Y(bh)}{2}\left(\frac{h}{4} + \frac{h}{4}\right) \qquad (3)$$

or

$$M_P = \frac{1}{4}\sigma_Y bh^2 \qquad \text{Ans.} \quad (4)$$

Shape Factor: From Eq. 6.48,

$$f = \frac{M_P}{M_Y} = \frac{6}{4} = 1.5 \qquad \text{Ans.} \quad (5)$$

Thus, for a rectangular beam, the plastic moment is 50% higher than the yield moment.

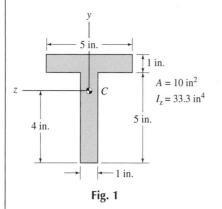

y

5 in.

1 in.

z —— C

$A = 10$ in^2
$I_z = 33.3$ in^4

4 in.

5 in.

1 in.

Fig. 1

The tee beam of Example Problem 6.2 is made of elastic-plastic material having a stress-strain diagram like Fig. 6.30 with $E = 29(10^3)$ ksi and $\sigma_Y = 40$ ksi. Determine the following: (a) the yield moment M_Y; (b) the location of the plastic neutral axis, and the value of the plastic moment M_P; and (c) the shape factor f.

The section properties calculated in Example Problem 6.2 are shown in Fig. 1.

Solution
(a) From Eq. 6.45, the yield moment is given by

$$M_Y = \frac{\sigma_Y I}{c} \tag{1}$$

The bottom fibers are farthest from the neutral axis, so

$$M_Y = \frac{(40 \text{ ksi})(33.3 \text{ in}^4)}{(4 \text{ in.})} = 333 \text{ kip} \cdot \text{in}$$

$$M_Y = 333 \text{ kip} \cdot \text{in} \qquad \text{Ans. (a)} \tag{2}$$

(b) The plastic neutral axis divides the cross section into two equal areas (Eq. 6.46). Therefore, for this problem, the plastic neutral axis falls at the flange-web interface, as illustrated in Fig. 2. From Eq. 6.47,

$$M_P = \frac{\sigma_Y A}{2}(d_C + d_T) = \frac{(40 \text{ ksi})(10 \text{ in}^2)}{2}(0.5 \text{ in.} + 2.5 \text{ in.})$$

or

$$M_P = 600 \text{ kip} \cdot \text{in} \qquad \text{Ans. (b)} \tag{3}$$

(c) The shape factor is given by Eq. 6.48.

$$f = \frac{M_P}{M_Y} = \frac{600 \text{ kip} \cdot \text{in}}{333 \text{ kip} \cdot \text{in}} = 1.80$$

$$f = 1.8 \qquad \text{Ans. (c)} \tag{4}$$

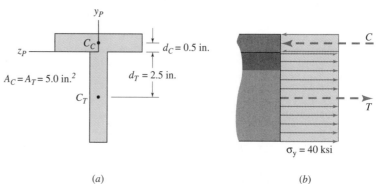

y_P

z_P

C_C

$d_C = 0.5$ in.

$A_C = A_T = 5.0$ in.^{2}

$d_T = 2.5$ in.

C_T

C

T

$\sigma_y = 40$ ksi

(a) (b)

Fig. 2

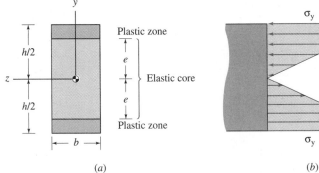
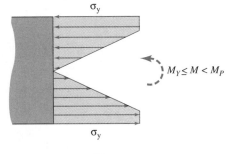

(a)

(b)

$M_Y \leq M < M_P$

FIGURE 6.33 Partially plastic bending of a rectangular beam.

Moment-Curvature Formulas. The moment-curvature equation for linearly elastic bending was derived in Section 6.3 and stated in Eq. 6.12, which we write here in terms of the curvature κ rather than the radius of curvature ρ.

$$M = EI\kappa \tag{6.49}$$

The yield moment M_Y is, therefore, related to the yield curvature κ_Y by the equation

$$M_Y = EI\kappa_Y \tag{6.50}$$

so the *moment-curvature equation* for linearly elastic behavior is

$$\frac{M}{M_Y} = \frac{\kappa}{\kappa_Y}, \quad 0 \leq M \leq M_Y \tag{6.51}$$

Let us now determine an expression that relates M/M_Y to κ/κ_Y for inelastic bending, and plot a curve of M/M_Y versus κ/κ_Y. The simplest case to consider is the rectangular beam, since the location of its neutral axis does not vary with the value of the applied moment and since it has a constant width. Figure 6.33 shows the stress distribution for the partially plastic case, with $2e$ being the depth of the elastic core. Summing the moment contributions of the elastic core and the two plastic zones we get

$$M = \sigma_Y b\left[\left(\frac{h}{2}\right)^2 - \frac{1}{3}e^2\right] \tag{6.52}$$

When $e = h/2$, we get the value of the *yield moment M_Y* as

$$M_Y = \frac{1}{6}\sigma_Y bh^2 \tag{6.53}$$

just as we previously obtained in Example Problem 6.11. When $e = 0$, Eq. 6.52 reduces to the formula for the *plastic moment, M_P*.

$$M_P = \frac{1}{4}\sigma_Y bh^2 \tag{6.54}$$

In order to determine a moment-curvature equation corresponding to the moment expression in Eq. 6.52, we need to relate the curvature κ to the value e. From the strain-displacement equation, Eq. 6.41 (Eq. 6.3), we get

$$\epsilon_Y = \frac{h}{2}\kappa_Y = e\kappa \tag{6.55}$$

since the yielding occurs at the outer fiber for M_Y and at $y = e$ for $M > M_Y$. (Note that $\kappa \geq \kappa_Y$.) Therefore,

$$e = \frac{h}{2}\left(\frac{\kappa_Y}{\kappa}\right) \tag{6.56}$$

Finally, we can combine Eqs. 6.52, 6.53, and 6.56 to get the following *moment-curvature equation*

$$\frac{M}{M_Y} = \frac{3}{2} - \frac{1}{2}\left(\frac{\kappa_Y}{\kappa}\right)^2, \qquad M_Y \leq M \leq M_P \tag{6.57}$$

Moment-curvature formulas for the rectangular beam, Eq. 6.51 and 6.57, are plotted as the solid-line curve in Fig. 6.34; the dashed-line curve is the moment-curvature curve for the T-beam of Example 6.12.

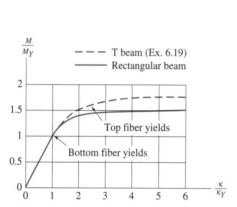

FIGURE 6.34 Moment-curvature plot for elastic-plastic beams.

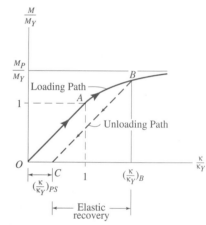

FIGURE 6.35 Loading-unloading curve of an elastic-plastic beam.

Residual Stresses. The elastic-plastic analysis summarized in the moment-curvature plots of Fig. 6.34 applies only when the load continues to increase. If the applied moment is reduced after exceeding the yield moment M_Y, unloading takes place along a moment-curvature path that is parallel to the original linear portion of the moment-curvature diagram, as illustrated in Fig. 6.35. This unloading behavior is very similar to the unloading behavior of circular torsion rods that was described in Section 4.9. Therefore, we can use the same procedure of adding the elastic-recovery stresses to the stresses due to plastic loading.

Figure 6.36 illustrates the superposition of stresses for a rectangular beam that has been loaded to the fully plastic state and then completely unloaded. The maximum elastic recovery stress for the rectangular beam in Fig. 6.36 is obtained by setting $M = -M_P$ and $y = h/2$ in the flexure formula, Eq. 6.13, giving[16]

$$(\sigma_{er})_{max} = \frac{-(-M_P)\left(\dfrac{h}{2}\right)}{\left(\dfrac{bh^3}{12}\right)} = \frac{\left(\dfrac{1}{4}\sigma_Y bh^2\right)\left(\dfrac{h}{2}\right)}{\left(\dfrac{bh^3}{12}\right)} = \frac{3}{2}\sigma_Y$$

---

[16]The maximum elastic recovery stress is sometimes called the *modulus of rupture*.

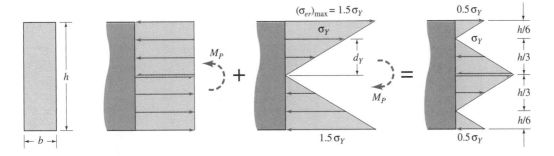

(a) Cross section. (b) Fully plastic stress state. (c) Elastic recovery stress state. (d) Residual stress state.

FIGURE 6.36 Residual stresses determined by superposition.

The point of zero stress in Fig. 6.36d may be determined by examining similar triangles in Fig. 6.36c. Thus,

$$\frac{\sigma_Y}{d_Y} = \frac{(\sigma_{er})_{\max}}{(h/2)} \rightarrow d_Y = \left(\frac{\sigma_Y}{1.5\sigma_Y}\right)\left(\frac{h}{2}\right) = \frac{h}{3}$$

In spite of the fact that all externally applied moment has been removed from the beam in Fig. 6.36d, it is obvious that the beam has self-equilibrating *residual stresses*. Any reloading would begin from this state, rather than from the stress-free state. After being completely unloaded, the beam would be left with a residual curvature called the *permanent set*. Figure 6.35b illustrates the permanent set that would result from unloading from point B on the moment-curvature diagram, a point well short of the (infinite) curvature required to provide the fully plastic state.

The next example problem illustrates the calculation of residual stresses.

■■■■■■■■■■■■■■■■ EXAMPLE 6.13 ■■■■■■■■■■■■■■■■■■

A wide-flange beam is subjected to a fully plastic moment, and then the moment is completely removed. Determine the distribution of residual stresses. The beam is made of (elastic-plastic) steel with a yield stress of $\sigma_Y = 200$ MPa.

Solution We can follow the superposition-of-stresses procedure that was illustrated in Fig. 6.36 for a rectangular beam. We first calculate M_P using Figs. 2a and 2d.

$$C_1 = T_1 = (200 \text{ MPa})(0.150 \text{ m})(0.010 \text{ m}) = 300 \text{ kN} \tag{1}$$

$$C_2 = T_2 = (200 \text{ MPa})(0.010 \text{ m})(0.075 \text{ m}) = 150 \text{ kN}$$

$$M_P = (300 \text{ kN})(0.160 \text{ m}) + (150 \text{ kN})(0.075 \text{ m}) = 59.25 \text{ kN} \cdot \text{m} \tag{2}$$

The maximum elastic recovery stress, or modulus of rupture, is obtained by

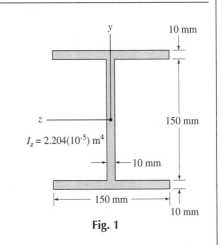

Fig. 1

303

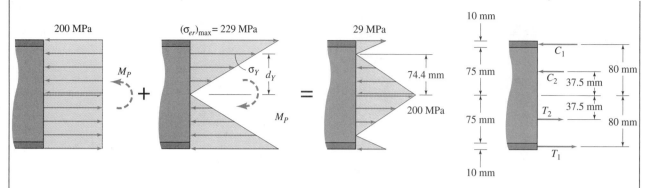

(a) Fully plastic stress state.   (b) Elastic recovery stress state.   (c) Residual stress state.   (d) Resultants of fully plastic stresses.

Fig. 2 Stress distributions and stress resultants.

setting $M = M_P$ in the flexure formula, Eq. 6.13, giving

$$(\sigma_{er})_{max} = \frac{M_P(\frac{h}{2})}{I} = \frac{(59.25 \text{ kN} \cdot \text{m})(0.085 \text{ m})}{2.204(10^{-5}) \text{ m}^4} = 229 \text{ MPa} \qquad (3)$$

To sketch the residual stress state in Fig. 2c we add the stress blocks in Figs. 2a and 2b. The point of zero stress in Fig. 2c may be determined by using similar triangles.

$$\frac{\sigma_Y}{d_Y} = \frac{(\sigma_{er})_{max}}{(h/2)} \rightarrow d_Y = \left(\frac{200}{229}\right)(85 \text{ mm}) = 74.4 \text{ mm} \qquad (4)$$

Therefore, the residual stress distribution has the form shown in Fig. 2c.

■■■■■■■■■■■ **6.8** SHEAR STRESS AND SHEAR FLOW IN BEAMS

Recall from Chapter 5 that transverse loads on a beam give rise not only to a bending moment $M(x)$ but also to a *transverse shear force* $V(x)$. The bending moment on a cross section is related to the flexural stress σ (Fig. 6.37b); the transverse shear force $V(x)$ is the resultant of distributed transverse shear stresses (Fig. 6.37c). We need to determine how the *transverse shear stress* τ varies with position in the cross section, that is, with respect to y and z. We will also find that surfaces other than cross sections (e.g., the neutral surface) also experience shear stress.

Basic Assumption. In developing the theories of axial deformation (Section 3.2), torsion (Section 4.2), and flexure (Section 6.3), we began by making kinematic assumptions and performing a strain-displacement analysis. Then we introduced the constitutive (stress-strain-temperature) equations to determine the respective stress distributions. However, when we seek to determine the distribution of shear stress in a beam and to relate the shear stress to the transverse shear force $V(x)$, we are immediately faced with a dilemma. Shear stress produces a change in angle, but in Section 6.3 we assumed that **plane sections remain plane and remain perpendicular to the deformed axis of the beam**

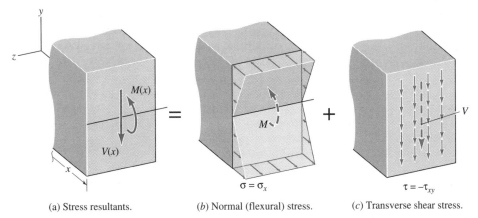

(a) Stress resultants.   (b) Normal (flexural) stress.   (c) Transverse shear stress.

FIGURE 6.37   The stress resultants and stresses on a cross section.

(i.e., without change of angle). Therefore, in order to derive an expression for the distribution of transverse shear stress, we make the following assumption:[17]

- *The distribution of flexural stress on a given cross section is not affected by the deformation due to shear.*

Thus, for example, we can analyze the shear stress in homogeneous, linearly elastic beams, using the flexural stress given by Eq. 6.13.

$$\sigma = \frac{-My}{I}$$

(6.13)
repeated

Because we have no convenient way to characterize the displacement due to shear (like the "plane sections" assumption that was used to characterize displacement due to flexure) we must follow a different approach. Fortunately, we can employ an equilibrium analysis to develop a shear-stress theory for beams.[18]

Shear-Stress Distribution. Before developing formulas for the distribution of shear stress in a beam, let us examine physical descriptions of shear stress and shear strain in beams. Figure 6.38 shows a cantilever beam subjected to a transverse load P at $x = 0$. (To simplify the drawings, a rectangular cross section is used in Fig. 6.38.) On an arbitrary cross section at x, the load P produces a positive bending moment and a positive shear force (as in Fig. 6.37a).

With linearly elastic behavior, the bending moment $M(x)$ produces the flexural stress distribution at section x as shown in Fig. 6.38b. Imagine now that a horizontal sectioning plane separates the portion of the beam up to section x into two parts, one from $y = y_1$ to

---

[17]There are some limitations on the theory that is based on this assumption. These are discussed in Section 6.9.

[18]A Russian engineer, D. J. Jourawski (1829–1891) was the first person to develop the elementary shear stress theory presented here. He developed this theory, which is based on equilibrium, while designing timber railroad bridges [Ref. 6-1].

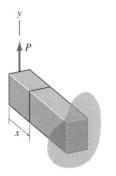

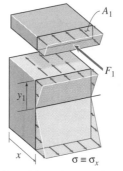

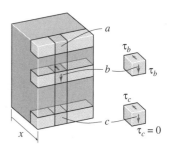

(a) A cantilever beam.

(b) Flexural stress distribution (linearly elastic).

(c) Unbalanced flexural stresses lead to shear stresses on a longitudinal section.

(d) Transverse and longitudinal shear stresses.

FIGURE 6.38 Shear stress distribution in a beam.

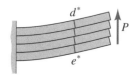

(a) A beam made of separate "planks."

(b) Slip between non-bonded "planks."

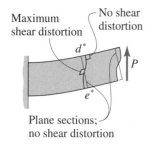

(c) Shear deformation of a uniform beam (or bonded-layer beam).

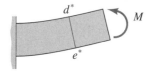

(d) Pure bending of a cantilever beam.

FIGURE 6.39 Some illustrations of shear deformation in beams.

the top of the beam, the other from $y = y_1$ to the bottom of the beam, as illustrated in Fig. 6.38c. The resultant of the flexural stress on A_1, the cross-sectional area above $y = y_1$, is designated as F_1. It is clear from Fig. 6.38c that the only way to satisfy $\Sigma F_x = 0$ for either the portion of beam above $y = y_1$ or the portion below $y = y_1$ is for there to be *longitudinal shear stress* on the horizontal sectioning plane, as indicated in Fig. 6.38c. The resultant of this longitudinal shear stress balances the normal force F_1. But this says that the unbalanced flexural stresses on this beam free body lead to *longitudinal* shear stresses on the free body. What about *transverse* shear stress on the cross section of the beam? Consider Fig. 6.38d, and recall from Section 2.10 that $\tau_{yx} = \tau_{xy}$; that is, the shear stress on a vertical plane must be equal to the shear stress on a horizontal plane at the intersection of the two planes. Thus, at point b in Fig. 6.38d the longitudinal shear stress and the vertical shear stress can both be labeled τ_b, because they are equal. This leads to two very interesting conclusions.

1. The *transverse shear stress* at an arbitrary level y in the cross section can be calculated by determining the *longitudinal shear stress* at this level.
2. The transverse shear stress must vanish at points a and c, since there is no horizontal shear stress on the top surface or on the bottom surface of the beam.

Shear-Strain Distribution. From the preceding discussion of beams subjected to transverse loads, and especially from Fig. 6.38c, it should be clear that planes parallel to the neutral surface (i.e., horizontal planes) must be able to transmit shear. Consider a cantilever beam like the one in Fig. 6.39a. In Fig. 6.39b the cantilever beam is made up of four "planks" that are not bonded together at levels $a - a'$, $b - b'$, and $c - c'$, but are free to slip along the surfaces of contact at these levels. Clearly, there is slip along the plank interfaces, and plane sections, like de, do not remain plane through the entire thickness of the beam in Fig. 6.39b. If the planks are bonded together to form a single beam (or if the beam is originally homogeneous throughout its depth), the beam will undergo shear deformation as illustrated in Fig. 6.39c. Because of shear deformation, plane sections do not remain plane, as they do in the case of pure bending (Fig. 6.39d). However, as noted earlier, the shear deformation has little effect on the distribution of flexural stress as long as the beam is slender (length greater than ten times depth).

Shear Flow in Beams. To derive an expression relating the shear stress τ to the resultant shear force V we will first define and derive an expression for *shear flow*. Consider the segment from x to $(x + \Delta x)$ of a beam subjected to transverse loads (Fig. 6.40a). The

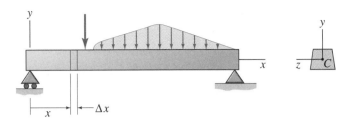

(*a*) Profile view.   (*b*) Cross section.

FIGURE 6.40 A beam subjected to transverse loading.

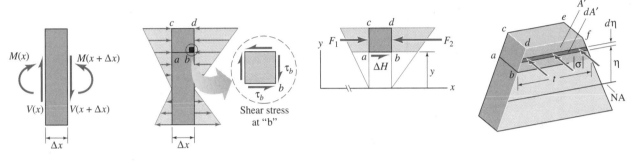

(*a*) An element of length Δx.   (*b*) The distribution of flexural stress.   (*c*) A free body diagram (minus vertical shear on *ac* and *bd*).   (*d*) The flexural stress contributing to F_2.

FIGURE 6.41 The stress resultants and stresses on segments of a beam.

segment from x to $(x + \Delta x)$ is enlarged in Fig. 6.41*a*, and the normal stresses acting on this segment are shown in Fig. 6.41*b*. (Note that $V(x)$ is taken in the positive sense according to the sign convention of Chapter 5.) From Eq. 5.2, repeated here,

$$V = \frac{dM}{dx}$$

(5.2)
repeated

Thus, when $V > 0$, $M(x + \Delta x) > M(x)$. Consequently, the flexural stresses at $x + \Delta x$ in Fig. 6.41*b* are shown larger in magnitude than the corresponding stresses at x.

As was suggested earlier, the key to the derivation of the distribution of transverse shear stress on a cross section is the fact that **the transverse shear stress is equal to the longitudinal shear stress at the same point** (Figs. 6.38*d* and 6.41*b*). Figures 6.41*c* and 6.41*d* focus on the portion above the level y, where the shear flow and the shear stress are to be determined. In Fig. 6.41*c* the resultant of the normal stresses on the area A' above level y at cross section x is labeled F_1; the corresponding resultant at $(x + \Delta x)$ is labeled F_2. The total shear force on the horizontal section at level y is labeled ΔH. Equilibrium requires that $\Sigma F_x = 0$, and since there is no x-component of force on the sides or top of the beam,

$$\Delta H = F_2 - F_1$$

(6.58)

The *shear flow, q,* which is the *shear force per unit length,* can be defined by taking the limit

$$q \equiv \lim_{\Delta x \to 0} \frac{\Delta H}{\Delta x}$$

(6.59)

which, from Eq. 6.58, can be calculated by using the expression

$$q \equiv q(x, y) = \lim_{\Delta x \to 0} \frac{(F_2 - F_1)}{\Delta x} \tag{6.60}$$

The forces F_1 and F_2 are the resultants of the normal stresses at x and $(x + \Delta x)$, respectively, over the area labeled A', as illustrated in Fig. 6.41d. Thus,[19]

$$F_1 = \int_{A'} |\sigma(x, \eta)| \, dA = \frac{M(x)}{I} \int_{A'} \eta \, dA$$

$$F_2 = \int_{A'} |\sigma(x + \Delta x, \eta)| \, dA = \frac{M(x + \Delta x)}{I} \int_{A'} \eta \, dA \tag{6.61}$$

where the area of integration, A', is from level y, where the shear flow is to be calculated, to the top (free surface) of the beam. The integral in Eq. 6.61 is just the *first moment of the area A' with respect to the neutral axis*. It is given the symbol Q, that is

$$Q(y) \equiv \int_{A'} \eta \, dA = A'\bar{y}' \tag{6.62}$$

where $\bar{y}'$ is the y coordinate of the centroid of area A'. Combining Eqs. 6.60 through 6.62, and recalling from Eq. 5.2 that $V = dM/dx$, we get

$$q = \lim_{\Delta x \to 0} \left[\frac{[M(x + \Delta x) - M(x)]}{\Delta x} \right] \frac{Q}{I}$$

or

$$q = \frac{VQ}{I} \qquad \textbf{Shear-flow formula} \tag{6.63}$$

We will now use this expression for shear flow to obtain an expression for the distribution of transverse shear stress. Later, we will also make use of Eq. 6.63 in Sections 6.10 through 6.12, where we examine shear stresses in thin-wall beams and built-up beams.

Shear-Stress Formula. Note that the shear flow, q, in Eq. 6.63 was derived by dividing the shear force, ΔH, by the length Δx over which it acts. If, instead, we divide ΔH by the area over which it acts, we get an average shear stress on the longitudinal plane at level y. From the equality of longitudinal and transverse shear, we get

$$\tau_{\text{avg}}(x, y) = \lim_{\Delta x \to 0} \frac{\Delta H}{t\Delta x} = \frac{V(x)Q(x, y)}{I(x)t(x, y)} \tag{6.64}$$

where $t(x, y)$ is the thickness, or width, of the beam at level y in section x, as indicated in Fig. 6.41d. The *transverse-shear-stress formula* is generally written without its xy qualifiers as, simply

$$\tau = \frac{VQ}{It} \qquad \textbf{Shear-stress formula} \tag{6.65}$$

---

[19]The variable of integration, η, locates a differential strip within area A'.

where

τ = the *average* transverse shear stress at level y in section x,
$Q = A'\bar{y}'$ = the first moment, with respect to the neutral axis, of the cross sectional area *above* level y,[20]
I = the moment of inertia <u>of the entire cross section</u>, taken with respect to the neutral axis, and
t = the width of the cross section at level y.

The *sign convention* implied in Eq. 6.65 is that the shear stress τ acts in the same direction as the resultant shear force V. Sometimes the absolute values of V and Q are used in the shear-stress formula to compute the *magnitude* of the shear stress, and the direction is simply assigned as that of the resultant shear force.

After illustrating how to use the transverse-shear-stress formula (in Example Problem 6.14) we will discuss the limitations of the formula.

---

[20]This derivation has been based on calculating Q using the area A' *above* level y. Since the first moment of the total area of the cross section, taken about the neutral axis, is zero (by definition of the neutral axis) the first moment of the area *below* level y is just the negative of the value Q in Eq. 6.62.

■■■■■■■■■■■■■■■■■■ E X A M P L E   6 . 1 4 ■■■■■■■■■■■■■■■■■■■

The rectangular beam of width b and height h (Fig. 1) is subjected to a transverse shear force V. (a) Determine the average shear stress as a function of y, (b) sketch the shear stress distribution, and (c) determine the maximum shear stress on the cross section.

Plan the Solution This is a straightforward application of the shear-stress formula, Eq. 6.65. Since V, I, and b are constant, $\tau_{\max}$ will occur at the neutral axis, where Q has its maximum value.

Solution
(a) The shear-stress formula, Eq. 6.65, is

$$\tau = \frac{VQ}{It}$$

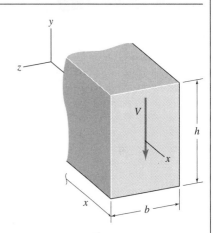

Fig. 1

As indicated in Fig. 2, the neutral axis of the rectangular cross section is at its mid-depth, and the area to be used in calculating Q is the area above level y.
Since $Q = A'\bar{y}'$, $t = b$, and $I = bh^3/12$,

$$\tau = \frac{V A'\bar{y}'}{Ib} = \frac{V\left[b\left(\dfrac{h}{2} - y\right)\right]\left[\dfrac{1}{2}\left(\dfrac{h}{2} + y\right)\right]}{\left(\dfrac{bh^3}{12}\right)b}$$

or

$$\tau = \frac{6V}{bh^3}\left(\frac{h^2}{4} - y^2\right) = \frac{3V}{2A}\left(\frac{c^2 - y^2}{c^2}\right) \qquad \textbf{Ans.} \text{(a)} \quad (1)$$

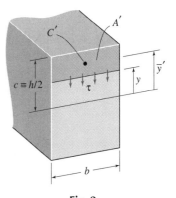

Fig. 2

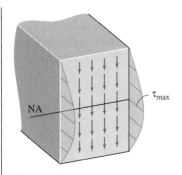

Fig. 3 The shear-stress distribution on a rectangular cross section.

(b) From Eq. 1 we see that the distribution of τ is parabolic. As we noted earlier when discussing Fig. 6.38d, τ vanishes at the top edge ($y = +c$) and at the bottom edge ($y = -c$). The parabolic distribution of τ is illustrated in Fig. 3.

(c) From Eq. 1 and Fig. 3 it is obvious that τ_{max} **occurs at the neutral axis**, as expected. Therefore,

$$\tau_{max} = \frac{3}{2}\frac{V}{A} \qquad \text{Ans. (c)}$$

Review the Solution   Unlike the case of direct shear (Section 2.7), where the shear stress is just the shear force divided by the area on which the shear stress acts, we have found a parabolic distribution of shear stress on the cross section of a rectangular beam. Since we know (Fig. 6.38d) that the shear stress must vanish at the top and bottom surfaces ($y = \pm c$), it is reasonable for the maximum shear stress to be 50% greater than the overall average shear stress, V/A.

6.9 LIMITATIONS ON THE SHEAR-STRESS FORMULA

The shear-flow formula and the shear-stress formula, Eqs. 6.63 and 6.65, respectively, may be applied to calculate the distribution of shear in a wide variety of beam shapes under a wide variety of loading conditions. These formulas were based on the flexure formula. Therefore, the limitations on the applicability of the flexure formula (slender beam, linearly elastic behavior, etc.) apply to these shear formulas also. However, there are some additional limitations on the shape of the beam and on the load distribution.

Effect of Cross-Sectional Shape. Consider first a beam with rectangular cross section. Figure 6.42 contrasts a "wide beam" ($b = 4h$ in Fig. 6.42a) and a "narrow beam" ($b = 0.5h$ in Fig. 6.42b). The corresponding distributions of shear stress at the neutral axis of the two rectangular shapes come from a *theory of elasticity* solution [Ref. 6-6, Section 124]. It is clear that the elementary shear stress theory represented by Eq. 6.65 is applicable only to narrow beams (e.g., $h \geq 2b$).

In mechanical applications, circular shafts often act as beams (Fig. 6.43). Although Fig. 6.41d and Eq. 6.64 deal with beams whose width may vary with y, the shear stress formula does not apply where the width, $t(x, y)$, varies rapidly. For example, the shear stresses at a line across a solid circular cross section, like line AA' in Fig. 6.43b, are not parallel to the y axis and cannot be determined by the shear stress formula.[21] However, along the neutral axis (diameter) of the circular cross section, the theory of elasticity [Ref. 6-6, Section 124] shows that the elementary solution is quite accurate, as indicated in Fig. 6.43b.[22]

Effect of the Load Distribution. Consider first a cantilever beam with end load P, as illustrated in Fig. 6.44a. For this loading, $V(x) = P = $ constant. The shear deformation of a central portion of the beam, of length $2\Delta x$, is shown in Fig. 6.44b. Since $V = $

---

[21]Beams having a circular tubular cross section are considered in Section 6.10.

[22]See Homework Problem 6.8-14 for the elementary solution based on Eq. 6.65.

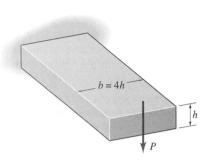

(a) A "wide beam," or plate.

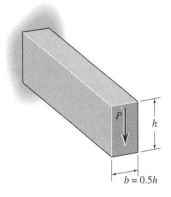

(b) A "narrow beam."

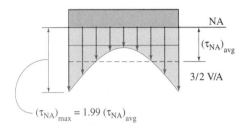

(c) Shear-stress distribution in the "wide beam" of Fig. 6.42a.

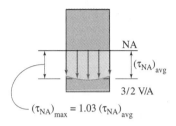

(d) Shear-stress distribution in the "narrow beam" of Fig. 6.42b.

FIGURE 6.42 The shear-stress distribution in "beams" with rectangular cross section.

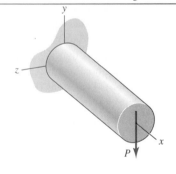

(a) A beam with circular cross section.

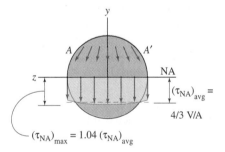

(b) The shear-stress distribution on a circular cross section.

FIGURE 6.43 The shear-stress distribution in a solid circular beam.

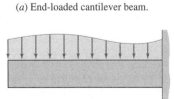

(a) End-loaded cantilever beam.

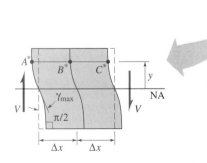

(b) Shear deformation when $V(x)$ = constant.

(c) A cantilever beam with distributed load.

FIGURE 6.44 An illustration of the effect of shear deformation.

constant, the shear stresses on cross sections away from the loaded end and away from the supported end produce the same warped shape at neighboring cross sections. Hence, considering shear deformation alone, $A^*B^* = B^*C^* = \Delta x$. Therefore, although the cross sections do not remain plane, the extensional strain based on the "plane-sections" assumption is not altered by the presence of shear. For loading conditions other than pure bending or constant shear, like the distributed loading in Fig. 6.44c, error in the flexural stress calculated by the flexure formula is introduced by the presence of shear deformation, but this error is small if the span of the beam is large in comparison with its depth.[23]

Effect of Length of Beam. The elementary expressions for flexural stress and transverse shear stress are accurate to within about 3% for beams whose length-to-depth ratio, L/h, is greater than 4.[24] A more elaborate analytical theory should be used to calculate the stresses in shorter spans, or the stresses should be computed by using a detailed finite-element solution.[25]

6.10 STRESSES IN THIN-WALL BEAMS

Many of the structural shapes used as beams can be classified as having thin-wall sections. That is, the wall thickness, t, is significantly smaller than the overall dimensions of the cross section, like the outer diameter, d_o, of a tubular beam (Fig. 6.45a) or the depth, h, of a wide-flange beam (Fig. 6.45b). The dashed curve in Fig. 6.45c is called the *centerline,* or *midline,* of the cross section. A *centerline coordinate s,* measured from an origin on the centerline, is used to locate points in the cross section at x. The wall thickness at location (x, s), identified as $t(x, s)$, is usually, but not always, independent of both x and s.

Some examples of thin-wall beams are ones with cross sections that are symmetric about two axes, like **W**-shapes (Fig. 6.46a), **S**-shapes (Fig. 6.46b), box beams (Fig. 6.46c), and circular pipe beams (Fig. 6.46d); channel sections (Fig. 6.46e), T-sections (Fig. 6.46f), and equal-leg angles (Fig. 6.46g), which have only one axis of symmetry in the cross section; and unequal-leg angle sections (Fig. 6.46h) and Z-sections (Fig. 6.46i), which have no axis of symmetry in the cross section.[26]

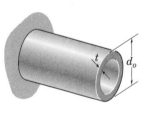 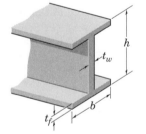 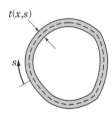

FIGURE 6.45 Dimensions of thin-wall beams.

(a) A circular tublar beam   (b) A wide-flange beam   (c) The cross section of a thin-wall beam

---

[23]See Ref. 6-6, Section 22 for a solution based on the theory of elasticity.

[24]This error estimate is based on an elasticity solution for bending of a beam by a uniform distributed load. (See Section 22 of Ref. 6-6).

[25]See Ref. 6-7, Section 7.5 for illustrations based on finite-element solutions.

[26]See Appendix D, *Section Properties of Selected Structural Shapes.*

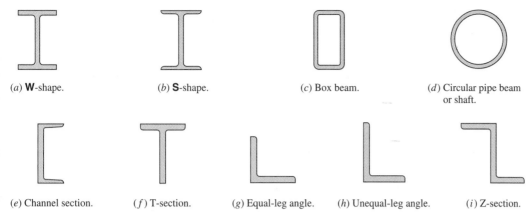

(a) **W**-shape.　　　(b) **S**-shape.　　　(c) Box beam.　　　(d) Circular pipe beam or shaft.

(e) Channel section.　　(f) T-section.　　(g) Equal-leg angle.　　(h) Unequal-leg angle.　　(i) Z-section.

FIGURE 6.46　Some thin-wall structural shapes.

The principal feature that distinguishes the theory of thin-wall beams is the fact that the *shear flow q(x, s)* can be computed at the location (x, s), and, because the section is thin, it can be assumed that:

- *the shear flow, q(x, s), is locally tangent to the centerline of the section,*
- *the average shear stress is given by*

$$\tau(x,\ s) = \frac{q(x,\ s)}{t(x,\ s)} \qquad (6.66)$$

and

- *this average shear stress accurately reflects the shear stress distribution through the thickness at location s in cross section x.*

Figure 6.47 illustrates shear flow along a thin-wall section and the corresponding average shear stress.

In this section we will illustrate the analysis of thin-wall beams by considering the stress distributions in wide-flange beams, in box beams, and in thin-wall beams with circular cross section.

Shear Stress in Web-Flange Beams.
Let us begin our analysis of stresses in thin-wall beams by considering the important case of doubly symmetric, open beams[27] that consist of a *web* lying in the longitudinal plane of symmetry and two equal *flanges,* as illustrated in Fig. 6.48. Wide-flange beams and I-beams fall in this category, as do steel plate girders used, for example, in buildings and bridges.

In Example Problem 6.6 it was shown that thin-wall beams, like **W**-sections and **S**-sections, are much more efficient than compact sections, such as rectangular sections and circular sections, in their resistance to bending, as expressed by the ratio S of bending

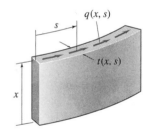

(a) Shear flow.

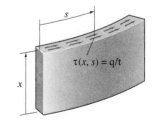

(b) Average shear stress.

FIGURE 6.47　An illustration of shear in a thin-wall beam.

---

[27]An *open cross section* is distinguished from *closed*, thin-wall sections like the box beam in Fig. 6.46c and the circular pipe beam in Fig. 6.46d.

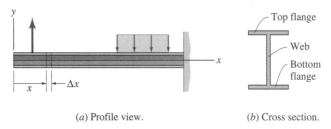

(*a*) Profile view.   (*b*) Cross section.

FIGURE 6.48 A doubly symmetric, thin-wall, web-flange beam.

moment to maximum flexural stress. Let us now consider the distribution of shear stress in beams of this type. To determine the shear-stress distribution in the flanges and in the web, we can follow the steps of the equilibrium analysis given in Section 6.8. In each case, that is, for flange shear and for web shear, we need to select the appropriate free-body diagram.[28] The segment of wide-flange beam in Fig. 6.48*a* from x to $(x + \Delta x)$ is enlarged in Fig. 6.49*a*, and the flexural stresses are shown on the cross sections at x and $(x + \Delta x)$.

Shear Flow in Flanges: Let us first determine the shear flow q_f, and associated shear stress, τ_f, in a flange.[29] Figure 6.49*b* shows that location of the cutting plane that isolates the free body to use in determining q_f. Figure 6.49*c* shows the appropriate free-body diagram. Note carefully how the shear force ΔF_f on the cutting plane is required to balance the flexural stresses acting on the flange outboard of the cutting plane. Note also that the direction of q_f in Fig. 6.49*b* is dictated by the direction of ΔF_f on the free-body diagram in Fig. 6.49*c*.

Having used 3-D sketches to clearly identify the free body that serves as a basis for this shear-flow analysis, we can now resort to a profile view (Fig. 6.50*a*) and a cross-sectional view (Fig. 6.50*b*) as we use Eq. 6.63 to calculate the flange shear flow q_f. From Eq. 6.63, the flange shear flow is given by

FIGURE 6.49 Illustrations of stresses in a web-flange beam.

$$q_f = \frac{VQ_f}{I} \tag{6.67}$$

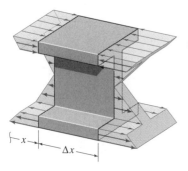

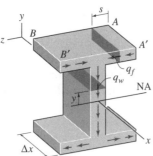

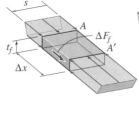

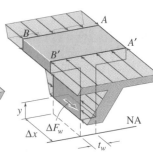

(*a*) A beam segment showing flexural stresses due to $M(x)$ and due to $M(x + \Delta x)$.

(*b*) Cutting planes used to determine the flange shear flow q_f and the web shear flow q_w.

(*c*) A free-body diagram for determining flange shear flow.

(*d*) A free-body diagram for determining web shear flow.

---

[28]Drawing 3-D free-body diagrams, like the ones in Figs. 6.49*c* and 6.49*d*, can be tedious, but the secret to developing a thorough understanding of shear stress in beams lies in the success with which you are able to visualize the manner in which unbalanced flexural stresses lead to shear stresses.

[29]Here we are considering shear flow that is parallel to the flange (i.e., parallel to the z axis). This is much more important than the flange shear that is parallel to the web.

The moment of inertia of the entire cross section, taken about the neutral axis, is

$$I = \frac{bh^3}{12} - \frac{(b - t_w)h_w^3}{12}$$

or

$$I = \frac{1}{12}(bh^3 - bh_w^3 + t_w h_w^3) \tag{6.68}$$

where $h_w = h - 2t_f$. From Eq. 6.62,

$$Q_f = A_f' \bar{y}_f' \tag{6.69}$$

where A_f' and $\bar{y}_f'$ are defined in Fig. 6.50b. Thus,

$$Q_f = s t_f \left(\frac{h}{2} - \frac{t_f}{2}\right)$$

or, in terms of h_w,

$$Q_f = \frac{s}{8}\left(h^2 - h_w^2\right) \tag{6.70}$$

Therefore, combining Eqs. 6.67, 6.68, and 6.70, we get the following expression for the *flange shear flow*:

$$q_f = \frac{3}{2}(V\,s)\frac{h^2 - h_w^2}{bh^3 - bh_w^3 + t_w h_w^3} \tag{6.71}$$

If the flange is thin, say $t_f \leq \frac{b}{5}$, the *flange shear stress* can be computed by using Eq. 6.66, that is

$$\tau_f = \frac{q_f}{t_f} = \frac{3}{2}\left(\frac{V\,s}{t_f}\right)\frac{h^2 - h_w^2}{bh^3 - bh_w^3 + t_w h_w^3} \tag{6.72}$$

After deriving an expression for shear flow and shear stress in the web, we will plot and discuss both flange shear and web shear.

Shear Flow in the Web: Starting with the free-body diagram in Fig. 6.49d, also shown in profile view in Fig. 6.51a, and following the same steps just used to derive expressions for flange shear flow and flange shear stress, we will now derive expressions for web shear flow and web shear stress. From Eq. 6.63,

$$q_w = \frac{VQ_w}{I} \tag{6.73}$$

where, from Eq. 6.62 and Fig. 6.51b,

$$Q_w = A_w' \bar{y}_w' = A_1' \bar{y}_1' + A_2' \bar{y}_2'$$

So, in terms of h_w,

$$Q_w = b\left(\frac{h}{2} - \frac{h_w}{2}\right)\left(\frac{1}{2}\right)\left(\frac{h}{2} + \frac{h_w}{2}\right) + t_w\left(\frac{h_w}{2} - y\right)\left(\frac{1}{2}\right)\left(\frac{h_w}{2} + y\right)$$

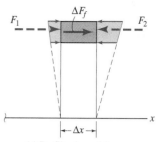

(a) Profile view of flange free-body diagram.

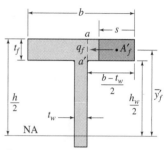

(b) The area on which the unbalanced flexural stresses act.

FIGURE 6.50 Information for use in calculating flange shear flow.

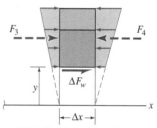

(a) Profile view of web-flange free-body diagram.

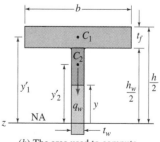

(b) The area used to compute shear flow in the web.

FIGURE 6.51 Information for use in calculating web shear flow.

315

or,

$$Q_w = \frac{b}{2}\left[\left(\frac{h}{2}\right)^2 - \left(\frac{h_w}{2}\right)^2\right] + \frac{t_w}{2}\left[\left(\frac{h_w}{2}\right)^2 - y^2\right] \tag{6.74}$$

Therefore, combining Eqs. 6.68, 6.73, and 6.74, we get the following expression for the *web shear flow*:

$$q_w = \frac{3V}{2}\left(\frac{bh^2 - bh_w^2 + t_wh_w^2 - 4t_wy^2}{bh^3 - bh_w^3 + t_wh_w^3}\right) \tag{6.75}$$

If the web is thin, then the *web shear stress* may be computed by using Eq. 6.66. So,

$$\tau_w = \frac{3V}{2t_w}\left(\frac{bh^2 - bh_w^2 + t_wh_w^2 - 4t_wy^2}{bh^3 - bh_w^3 + t_wh_w^3}\right) \tag{6.76}$$

Summary of Shear in Web-Flange Beams: From Eqs. 6.71 and 6.72 we see that:

1. The flange shear flow q_f and shear stress τ_f vary linearly with distances from the outer edge of the flange.
2. The maximum flange shear flow (and shear stress) occurs at the web-flange intersection (section $a - a'$ in Fig. 6.50b).
3. For the flange free body analyzed, shear flow is directed from the outer edge of the flange toward the web (Fig. 6.50b).
4. The maximum flange shear stress is given by[30]

$$(\tau_f)_{max} = \frac{3V(b - t_w)(h^2 - h_w^2)}{4t_f(bh^3 - bh_w^3 + t_wh_w^3)} \tag{6.77}$$

From similar flange shear analyses for the other three flange areas we can conclude that:

5. Equations 6.71 and 6.72 are valid for any of the four flange areas.
6. The shear flow in each of the four flange areas "flows" (in the sense of fluid flowing in a pipe) in the directions indicated on Fig. 6.49b and Fig. 6.52, and the shear flow in the web is in the direction of the resultant transverse shear force, V.

From Eqs. 6.75 and 6.76 we can conclude that:

7. The web shear flow and shear stress are parabolic, that is, they are quadratic functions of y.
8. The maxima of q_w and τ_w occur at the neutral axis ($y = 0$), and the minima occur at the web-flange junction ($y = \pm h_w/2$). (Because actual beam cross sections have fillets at the web-flange junctions to reduce stress concentration there, the minima are only theoretical minimum values.)
9. Expressions for the maximum and minimum web shear stresses are[31]

$$(\tau_w)_{max} = \frac{3V}{2t_w}\left(\frac{bh^2 - bh_w^2 + t_wh_w^2}{bh^3 - bh_w^3 + t_wh_w^3}\right) \tag{6.78}$$

[30]If V is negative, $(\tau_f)_{max}$ will be the magnitude of this quantity.

[31]Absolute values are taken if V is negative.

$$(\tau_w)_{\min} = \frac{3V\,b}{2t_w}\left(\frac{h^2 - h_w^2}{bh^3 - bh_w^3 + t_wh_w^3}\right) \qquad (6.79)$$

The magnitudes and directions of the web and flange shear stresses are indicated on Fig. 6.52.

From Eqs. 6.78 and 6.79 we note that, as $t_w/b \to 0$, $(\tau_w)_{\min} \to (\tau_w)_{\max}$, that is, the web shear stress approaches a constant shear stress. For beams of practical dimensions, the maximum web shear stress typically exceeds the minimum web shear stress by 10%–50%.

It is of interest to determine what percent of the vertical shear force V is carried by the web. We can determine this value by integrating the shear flow over the depth of the web, that is

$$V_w = \int_{-h_w/2}^{h_w/2} q_w(y)dy \qquad (6.80)$$

but we can also determine this value by multiplying the area under the τ_w curve in Fig. 6.52 by t_w, noting that the area consists of a rectangle and a parabola. Thus,

$$V_w = h_wt_w\left\{(\tau_w)_{\min} + \frac{2}{3}[(\tau_w)_{\max} - (\tau_w)_{\min}]\right\}$$

or

$$V_w = \frac{h_wt_w}{3}[2(\tau_w)_{\max} + (\tau_w)_{\min}] \qquad (6.81)$$

For practical beams, V_w typically accounts for more than 90% of the total vertical shear force. The remainder is carried by vertical shear in the flanges. This vertical shear in the flanges is quite small and cannot be computed by the shear flow method above.

The following Example Problem illustrates the calculation of shear stresses in wide-flange beams.

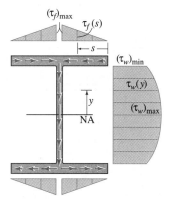

FIGURE 6.52 The shear stresses in a wide-flange beam.

(a) Determine the shear stress distribution in the flanges and the web of a **W**14×26 beam[32] subjected to the shear force $V = -28$ kips, as shown in Fig. 1.[33] (b) Determine the percent of the vertical shear that is carried by the web of this cross section.

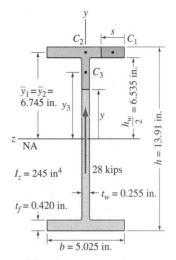

Fig. 1 A **W**14×26 wide-flange beam section.

Plan the Solution Although it would be possible to solve Part (a) by using the formulas in Eqs. 6.72 and 6.76, it will be more instructive to use the basic shear flow formula, Eq. 6.63, and the average shear stress formula, Eq. 6.66. We can determine the total web shear force by integrating the shear flow, as indicated in Eq. 6.80. Since the shear on the cross section is negative (i.e., it points in $+y$ direction in Fig. 1) the shear flows and shear stresses will act in the opposite sense to those shown in Fig. 6.49b.

Solution

(a) Equations 6.63 and 6.66 can be used to determine the shear flow and the shear stress in the flanges and the web.

Flange Shear Flow and Shear Stress: The flange shear flow is given by Eq. 6.63:

$$q_f = \frac{VQ_f}{I} = \frac{V A_f' \bar{y}_f'}{I} = \frac{V A_1 \bar{y}_1}{I} \tag{1}$$

$$q_f = \frac{(-28 \text{ kips})(0.420\, s \text{ in}^2)(6.745 \text{ in.})}{245 \text{ in}^4} = -0.324\, s \text{ kips/in.} \tag{2}$$

$$\tau_f = \frac{q_f}{t_f} = \frac{-0.324\, s \text{ kips/in.}}{0.420 \text{ in.}} = -0.771\, s \text{ ksi} \tag{3}$$

$$(\tau_f)_{max} = |(\tau_f)_{s=2.385 \text{ in.}}| = 1.84 \text{ ksi} \tag{4}$$

$$\tau_f = -0.771\, s \text{ ksi}, \qquad (\tau_f)_{max} = 1.84 \text{ ksi} \qquad \textbf{Ans. (a)}$$

---

[32]The properties of the **W**14×26 section are given in Appendix D.1.

[33]This is the shear force just to the left of the support at B in Example Problem 6.5.

Web Shear Flow and Web Shear Stress: The web shear flow is given by Eq. 6.63:

$$q_w = \frac{VQ_w}{I} = \frac{V A'_w \bar{y}'_w}{I} = \frac{V}{I}(A_2 \bar{y}_2 + A_3 \bar{y}_3) \qquad (5)$$

where A_2 is the area of the entire top flange. Therefore,

$$q_w = \left(\frac{-28 \text{ kips}}{245 \text{ in}^4}\right)[(5.025 \text{ in.})(0.420 \text{ in.})(6.745 \text{ in.})$$
$$+ (6.535 \text{ in.} - y)(0.255 \text{ in.})(0.5)(6.535 \text{ in.} + y)]$$

or

$$q_w = \frac{(-28)[14.235 + 0.1275(42.71 - y^2)]}{245}$$

$$= -(2.249 - 0.0146y^2) \text{ kips/in.}$$

$$\left.\begin{aligned}
\tau_w &= \frac{q_w}{t_w} = -(8.82 - 0.0571y^2) \text{ ksi} && (6)\\[4pt]
(\tau_w)_{\text{max}} &= |(\tau_w)_{y=0}| = 8.82 \text{ ksi} && (7a)\\[4pt]
(\tau_w)_{\text{min}} &= |(\tau_w)_{y=6.535 \text{ in.}}| = 6.38 \text{ ksi} && (7b)
\end{aligned}\right\}\text{Ans. (a)}$$

The results of the flange shear analysis and web shear analysis are summarized on Fig. 2. The negative shear ($V = -28$ kips) gives shear stresses in the opposite direction to those in Fig. 6.49*b*.

(b) To determine the percent of vertical shear that is carried by the web, let us use Eq. 6.80.

$$V_w = \int_{-h_w/2}^{h_w/2} q_w(y)dy = -\int_{-6.535}^{6.535} (2.249 - 0.0146y^2)\,dy$$

$$= -2.249(6.535)(2) + \frac{2}{3}(0.0146)(6.535)^3$$

$$= -29.40 + 2.71 = -26.69 \text{ kips}$$

Comparing V_w with V, we see that *95% of the vertical shear is carried by the web of this beam.*

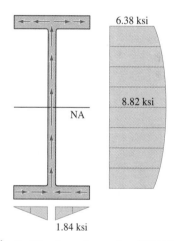

Fig. 2 The shear stresses in a **W**14×26 beam.

Review the Solution We can spot-check the above results by using Eq. 6.78 to compute $(\tau_w)_{\text{max}}$. When we do so, we get $(\tau_w)_{\text{max}} = 9.02$ ksi, which does not agree very well with the value of 8.82 ksi that we got in Eq. (7a). So let us compare the value of $I = 245$ in^4 from Appendix D.1 with the value calculated from Eq. 6.68. For the latter we get $I = 240$ in^4. The difference in I values accounts for the difference in $(\tau_w)_{\text{max}}$ values. This difference stems from the fact that the actual cross section has web-to-flange fillets that provide a slight increase in the value of the moment of inertia. This was not accounted for in Eq. 6.68.

The analysis procedure employed above to determine the shear stress in the web and the flanges of a doubly symmetric, thin-wall, open beam can also be applied to thin-wall open beams with one axis of symmetry and with loading in the plane of symmetry. For example, Fig. 6.53 shows a T-section and an equal-leg V-section. Since these sections have only one axis of symmetry (the longitudinal plane of symmetry) it is necessary to locate the neutral axis first before proceeding with a shear flow/shear stress analysis based on Eq. 6.63. Here, as before, **the maximum shear flow occurs at the neutral axis**, since the imbalance of normal forces is a maximum at the neutral axis.

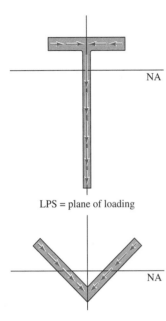

LPS = plane of loading

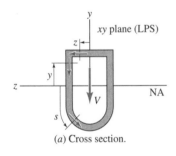

FIGURE 6.53 Shear in open beams with one axis of symmetry in the cross section.

Stresses in Closed, Thin-Wall Beams.

Box beams and circular pipe beams, as illustrated in Figs. 6.46c and 6.46d, may be classified as *closed, thin-wall sections.* Let us determine the shear-stress distribution in this type of beam, assuming that the beam has at least one plane of symmetry and that the loads lie in this LPS, as illustrated by the generic closed, thin-wall section in Fig. 6.54a.

The analysis of shear stress in closed, thin-wall beams differs from the analysis of open thin-wall beams, like the wide-flange beam illustrated in Fig. 6.48, in only one major respect, namely, the location of a point in the cross section where the shear vanishes. For the flange free-body diagram in Fig. 6.49c, the outboard edge AA' is free of shear, and the plot of $\tau_f(s)$ in Fig. 6.52 reflects this fact. Likewise, the free-body diagram in Fig. 6.49d includes the portion of beam from the web sectioning-plane at level y to the free surfaces of the flange. For a closed section we need to be able to determine the shear flow (and shear stress) at any arbitrary location in the cross section, say the locations identified by coordinates $z, y,$ and s in Fig. 6.54a. But, how do we choose an appropriate free body? One answer is that we can make use of symmetry about the xy-plane and always *select a free body that is symmetric in z.* Because of symmetry, there can be no longitudinal shear on surfaces exposed by a cutting plane in the LPS, and thus there can be no shear on the cross section at $z = 0$, as indicated in Fig. 6.55. (Can you explain why symmetry about the LPS makes $\tau = 0$ at $z = 0$?) The shear flow will be directed so that the resultant shear force on the cross section is V, as indicated in Fig. 6.54a. Finally, the sectioning cut should be made normal to the midline of the thin-wall section, like the cuts at $z, y,$ and s in Fig. 6.54a. In the next example problem we will illustrate how to select the appropriate free-body diagram and how to calculate shear flow and shear stress on a closed, thin-wall section.

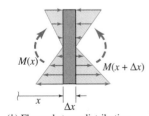

(a) Cross section.

(b) Flexural stress distribution.

FIGURE 6.54 A closed, thin-wall section.

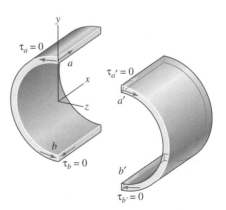

FIGURE 6.55 Shear stress on a sectioning plane in the xy-plane.

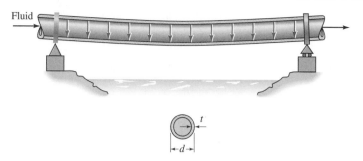

Fig. 1 A pipe transporting fluid and acting as a beam.

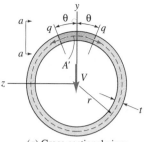

(a) Cross-sectional view.

A pipe conveying fluid over a narrow stream crossing must act as a beam as well as a conduit, as indicated in Fig. 1.

Assuming that the ratio of mean diameter to wall thickness satisfies the requirement $d/t \gg 1$, determine the shear flow and shear stress distribution at a section due to the transverse shear force, V, at that section. Also, determine the maximum shear stress on the cross section, and express τ_{max} in terms of V/A.

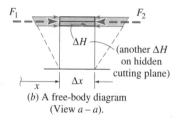

(b) A free-body diagram (View $a-a$).

Fig. 2 Selection of a symmetric free body.

Plan the Solution As noted in the preceding discussion of shear in closed, thin-wall beams, we should pick a free-body diagram that is symmetric about the xy-plane. Then we can apply the shear flow formula, Eq. 6.63, and the average shear formula, Eq. 6.66, to this free-body diagram.

Solution Because this is a circular cross section, it is convenient to select a free-body diagram that is defined by radial cutting planes at angle θ either side of the xy-plane, as indicated in Fig. 2. In Fig. 2b there are two surfaces that have equal shear forces ΔH which serve to balance the net force $(F_2 - F_1)$ due to the flexural stresses. Therefore, we get

$$2\Delta H = F_2 - F_1 \tag{1}$$

where, from the definition of shear flow

$$\Delta H = q\Delta x \tag{2}$$

Comparing Eq. (1) to Eq. 6.58, and following the same steps that were used to arrive at Eq. 6.63, we get

$$2q = \frac{VQ}{I} \tag{3}$$

for the symmetric free-body diagram in Fig. 2. Here, as in Eq. 6.63,

$$Q = \int_{A'} y\, dA = A'\overline{y}' \tag{4}$$

is the first moment, about the neutral axis, of the area acted on by the unbalanced flexural stresses, and I is the moment of inertia of the entire cross section about its neutral axis. The geometric properties for a thin ring and for a sector of a thin ring are shown in Fig. 3. Using the formulas in Fig. 3b, we have, for $t \ll r$,

$$Q = A'\overline{y}' \doteq 2r^2 t \sin\theta \tag{5}$$

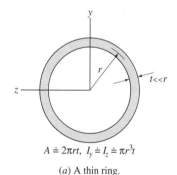

$A \doteq 2\pi rt, \quad I_y \doteq I_z \doteq \pi r^3 t$

(a) A thin ring.

$A \doteq 2rt\theta, \quad \overline{y} \doteq \dfrac{r\sin\theta}{\theta}$

(b) A sector of a thin ring.

Fig. 3 Some geometric properties of plane areas.

321

Combining the moment of inertia from Fig. 3a with Eqs. (3) and (5), we get

$$q = \frac{V(2r^2t\sin\theta)}{2(\pi r^3 t)} = \frac{V\sin\theta}{\pi r} \qquad \text{Ans.} \quad (6)$$

The average shear stress is obtained from the shear flow by using Eq. 6.66. Therefore,

$$\tau = \frac{q}{t} = \frac{V\sin\theta}{\pi r t} \qquad \text{Ans.} \quad (7)$$

Figure 4 illustrates the distribution of shear flow (and shear stress). The maximum shear stress is given by

$$\tau_{max} = \tau_{(\theta=\frac{\pi}{2})} = \frac{V}{\pi r t}$$

Since the area of the cross section can be approximated by

$$A \doteq 2\pi r t$$

for a *thin-wall pipe beam with $t \ll r$*, the maximum shear stress is[34]

$$\tau_{max} = \frac{2V}{A} \qquad \text{Ans.} \quad (8)$$

Review the Solution   As a check on the expression that we obtained for q, Eq. (6), we can see if the resultant of this distribution is V, as it is supposed to be. The force on the elemental area highlighted in Fig. 4 is

$$dF = qds = qrd\theta$$

The vertical component, as shown on the insert in Fig. 4, is, therefore,

$$dF_y = qr\sin\theta\,d\theta$$

So, due to symmetry, we have

$$F_y = 2\int_0^\pi qr\sin\theta\,d\theta$$

$$= \frac{2V}{\pi}\int_0^\pi \sin^2\theta\,d\theta$$

$$= \frac{2V}{\pi}\left[\frac{\theta}{2} - \frac{1}{4}\sin 2\theta\right]_0^\pi = V$$

Since the distribution looks ''reasonable'' (the shear flow has its maximum at the neutral axis and is zero on the plane of symmetry), and since it gives the correct resultant, we can assume that our answer is correct.

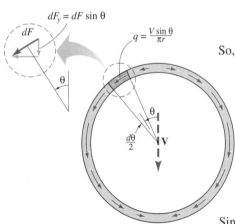

Fig. 4 The distribution of shear flow corresponding to a resultant shear force V.

---

[34]See Homework Problem 6.8-15 for shear in a ''thick-wall'' pipe beam.

6.11 SHEAR IN BUILT-UP BEAMS ■■■■■■■■■■

In Sections 6.8 and 6.10 we assumed the beams to be homogeneous, so it was appropriate for us to derive formulas for the *average shear stress* on a given surface. However, there are a number of important applications of beams where welds, glue, rivets, bolts, or nails are used to join two or more structural components to form a single beam. Several examples of such *built-up beams* are shown in Fig. 6.56. They are: (a) a glued-laminated timber beam, (b) a welded plate girder, (c) a wooden box beam constructed of several planks nailed together, and (d) a beam with reinforcing flange plates bolted to the basic beam. In such cases it is assumed that the beam acts as a unit (e.g., plane sections remain plane) and that the joining medium (nails, welds, bolts, etc.) is capable of transferring shear across the longitudinal junctions between component parts of the beam.

Shear may be transferred between adjoining parts of a built-up beam in three ways:[35]

1. by *shear distributed over the interface areas,* as when surfaces are glued together (Fig. 6.56a).
2. by *shear distributed along a line,* as when two metal parts are jointed by weld beads (Fig. 6.56b), and
3. by *discrete shear connectors,* as when two parts are nailed, riveted, or bolted together (Figs. 6.56c,d).

As was pointed out earlier in the discussion of Fig. 6.39, if a beam is composed of several pieces, it will only behave as a single beam (i.e., satisfy the ''plane sections'' hypothesis) if some provision is made to transfer shear between adjacent parts. When shear is distributed over a bond area, like the glued joints in Fig. 6.56a, the analysis is based on the *shear stress* in the bond layer. This has already been treated in Section 6.8 (e.g., Homework Problems 6.8-11 through 6.8-13). In the case of a linear shear connector (e.g., the weld beads in Fig. 6.56b), and the case of discrete shear connectors (e.g., the nails in Fig. 6.56c and the bolts in Fig. 6.56d), an analysis based on shear flow is appropriate. We will now consider the latter two cases.

The shear flow equation, Eq. 6.63, is used to compute the <u>required</u> shear flow, that is, the shear force per unit length that must be transferred from one part of the beam to the adjacent part under the given loading condition.

$$q_r = \frac{VQ}{I} = \frac{V A' \bar{y}'}{I} \tag{6.82}$$

where, as before, the area A' is the area on which the unbalanced flexural stresses responsible for the shear flow q_r are exerted.

Linear Shear Connectors. In the case of a linear shear connector, Eq. 6.82 determines the actual shear force per unit length that must be transferred through the linear shear connector(s). The shear connector (e.g., a weld bead) can exert an <u>allowable</u> shear force per unit length, which we will designate as q_a. Therefore, the joint will fail locally unless

$$q_a \geq q_r \tag{6.83}$$

Example Problem 6.14 illustrates this type of problem.

(a) A glued, laminated beam.

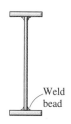

(b) A welded plate girder.

Weld bead

Nail

(c) A box beam.

(d) A reinforced beam.

FIGURE 6.56 Several examples of built-up beams.

---

[35]The analysis of the transfer of shear from steel reinforcing bars to the surrounding concrete is similar to the analysis of shear presented in this section, but is beyond the scope of this text.

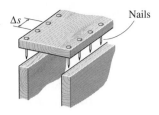

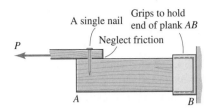

FIGURE 6.57 A nailed shear joint.

(a) Nails on a uniform spacing of Δs.

(b) A direct-shear test to determine the shear capacity of a single nail.

Discrete Shear Connectors. If discrete connectors, each having a shear-force capacity V_a, are spaced along a joint at a spacing Δs, as illustrated in Fig. 6.57a, then the joint will fail unless

$$V_a \geq q_r \Delta s \tag{6.84}$$

The <u>allowable</u> shear-force capacity of a single discrete shear connector, like the nail in Fig. 6.57b, is designated as V_a, whether this available strength represents shearing the nail in two (which is highly unlikely) or whether the joint fails by pull-out of the nail or crushing of the adjacent wood. Direct-shear tests, as illustrated in Fig. 6.57b, may be performed to determine the ultimate shear capacity of a certain size and type of nail used to connect two pieces of wood of a certain type.

Example Problem 6.15 illustrates this discrete-shear-connector type of problem.

■■■■■■■■■■■■■■□□ EXAMPLE 6.17 ■■■■■■■■■■■■■■■■

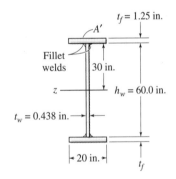

Fig. 1 (Drawing not to scale.)

A built-up plate girder has the dimensions shown in Fig. 1. If the shear force at this cross section is $V = 300$ kips, what is the shear flow in each fillet weld bead? $I_z = 54{,}800$ in⁴.

Plan the Solution Since there are two weld beads that attach each flange to the web, we can use Eq. 6.82 modified by a factor of two.

Solution From Eq. 6.82 (modified),

$$2q_r = \frac{VQ}{I} = \frac{V A' \overline{y}'}{I}$$

where, as indicated, Q is the first moment of the entire flange area, taken about the NA.

$$q_r = \frac{(300 \text{ kips})(20 \text{ in.})(1.25 \text{ in.})(30.625 \text{ in.})}{2(54{,}800 \text{ in}^4)}$$

$$q_r = 2.10 \text{ kips/in.} \qquad \text{Ans.}$$

Review the Solution A simple recheck of the calculations is all that is necessary here.

Interesting design information on steel girders and weld beads can be found in Ref. 6.2, but we will not pursue the design of welds further in this text.

A box beam is made by nailing together four boards in the configuration shown in Fig. 1a. The beam supports a concentrated load of 1000 lb at its midspan, and it rests on simple supports (Fig. 1b). If each nail can withstand an allowable shear force of 150 lb, what is the maximum spacing that can be used?

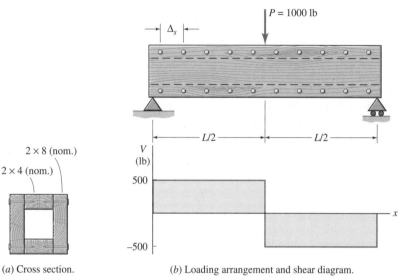

(a) Cross section. (b) Loading arrangement and shear diagram.

Fig. 1

Plan the Solution Since the shear force V is constant over each half of the beam, we can use a constant nail spacing Δs. The shear flow along each row of nails can be computed using the shaded area between A and B in Fig. 2. Equations 6.82 and 6.84 can be combined to compute the maximum spacing. The finish dimensions of the boards are given in Appendix D.8.

Solution Since there are two rows of nails transferring shear between the shaded area in Fig. 2 and the vertical boards, we need to modify Eq. 6.82 to read

$$2q_r = \frac{VQ}{I} = \frac{V A' \overline{y}'}{I} \qquad (1)$$

in order to calculate the shear flow along a single line of nails. The spacing along each row of nails is then given by Eq. 6.84.

$$\Delta s_{\max} = \frac{V_a}{q_r} \qquad (2)$$

The moment of inertia can be computed by taking a 6.5 in. × 7.25 in. rectangle and subtracting a 3.5 in. × 4.25 in. rectangle.

$$I = \frac{6.5(7.25)^3}{12} - \frac{3.5(4.25)^3}{12} = 184.0 \ \text{in}^4$$

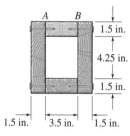

Fig. 2 Surfaced lumber design dimensions.

325

The shear flow is given by Eq. 1.

$$q_r = \frac{V A' \overline{y}'}{2I} = \frac{(500 \text{ lb})[(3.5 \text{ in.})(1.5 \text{ in.})](2.875 \text{ in.})}{2(184.0 \text{ in}^4)}$$

$$q_r = 20.5 \text{ lb/in.}$$

Then, from Eq. 2,

$$\Delta s_{\max} = \frac{V_a}{q_r} = \frac{150 \text{ lb}}{20.5 \text{ lb /in.}} = 7.32 \text{ in.}$$

Use $\Delta s = 7$ in. **Ans.**

Review the Solution The answer seems reasonable, and the calculations are easily rechecked and found to be correct.

Of course, factors other than shear strength should be considered in the design of joints. For example, nails cannot be driven too close to the edge of a piece of timber, and nails cannot be spaced too close together, or the wood is likely to split.

*6.12 SHEAR CENTER

In Section 6.6 "Unsymmetric Bending," you discovered that *loading of a beam, even an unsymmetric beam, parallel to a principal axis of inertia of the cross section produces bending that is confined to the corresponding longitudinal principal plane.* We considered the *direction of loading,* but we did not establish the *line of action of loading* that would produce bending without simultaneous twisting. Consider bending of a beam that has a thin-wall channel cross section and is loaded normal to the axis of symmetry of the cross section. Figure 6.58 shows this situation, with the load placed in three different positions. In Fig. 6.58*a* the load is too far to the left, so it produces both bending (which is not really visible) and a counterclockwise twisting, as indicated by the position of the pointer. In Fig. 6.58*b* a load that acts to the right of the centroid of the cross section clearly produces

 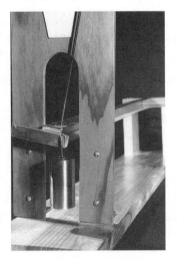 

(a) Bending plus counter-clockwise twisting.   *(b) Bending plus clockwise twisting.*   *(c) Bending without twisting.*

FIGURE 6.58 Loading of a thin-wall, open beam normal to its plane of symmetry.

twisting in a clockwise sense. Finally, by trial-and-error, a location is found where the load produces bending (again, too small to see) without any twisting (Fig. 6.58c).

*The **shear center** of a cross section is defined as that point in the cross section through which the resultant shear force V must pass if the beam is to bend without twisting.*

It is particularly important that the shear center of open, thin-wall beams be located, since, as Fig. 6.58 shows, such members are very flexible in torsion (See Section 4.8). There are other situations, as well, where locating the shear center of the cross section of a beam is very important. For example, the shear center of an airplane wing, which may in some cases be analyzed as a thin-wall closed beam, must be properly located with respect to the imposed aerodynamic loads (Fig. 6.59).

Shear-center location is important for all beams, but especially for thin-wall open beams. Therefore, we will begin our analysis of shear-center location by considering *singly symmetric, thin-wall beams,* similar to the channel section in Fig. 6.58. If a cross section has an axis of symmetry, the shear center will be located somewhere along the axis of symmetry of the section. An inclined load may be resolved into one component lying in the plane of symmetry, which contains one principal axis, and one component in the perpendicular (principal) direction, as illustrated in Fig. 6.60. Therefore, to locate the shear center, S, we need only consider shear that is perpendicular to the axis of symmetry of the cross section.

Consider the open, thin-wall, singly symmetric section in Fig. 6.61. At an arbitrary cross section there is a resultant shear force V, as shown in the pictorial view of Fig. 6.61a and in the cross-sectional view in Fig. 6.61b. The single shear force V is the resultant of a shear force V_2 in the web and the shear forces V_1 and V_3 in the two inclined flanges (Fig. 6.61c). Finally, these shear forces are the resultants of shear flows (Fig. 6.61d) that we can compute by the method of Section 6.8. The problem can therefore be stated as follows:

- *Determine the location of the shear center, S, so that the shear force acting through S is the resultant of shear stresses due to bending only (i.e., bending without twisting).*

First, we can relate the *eccentricity, e,* of the shear center to the shear resultants in Fig. 6.61c by setting the moment about O in Fig. 6.61b equal to the moment about O in Fig. 6.61c.

FIGURE 6.59 The shear center of a wing cross section.

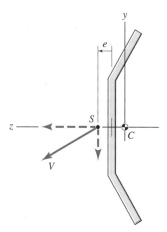

FIGURE 6.60 The shear center is on the axis of symmetry.

$$+\zeta \left(\sum M \right)_O = Ve = (V_1 \cos \phi)\left(\frac{h_w}{2}\right) + (V_3 \cos \phi)\left(\frac{h_w}{2}\right) \qquad (6.85)$$
$$\text{(b)} \qquad\qquad\qquad \text{(c)}$$

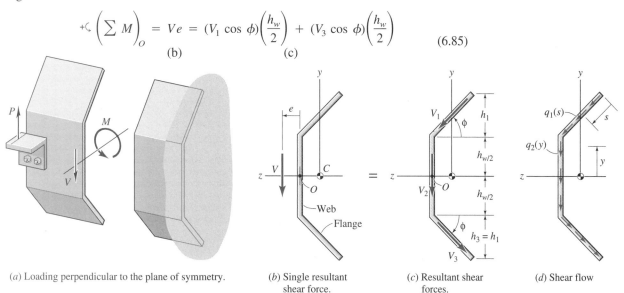

(a) Loading perpendicular to the plane of symmetry. (b) Single resultant shear force. (c) Resultant shear forces. (d) Shear flow

FIGURE 6.61 Steps in locating the shear center of a thin-wall, open beam.

We also must satisfy the force-equilibrium equation

$$\overset{+}{\leftarrow} \sum F_z = 0 = \underset{(b)}{V_1} \cos \phi - \underset{(c)}{V_3} \cos \phi \tag{6.86}$$

Combining these two equations, we get

$$e = \frac{V_1 h_w \cos \phi}{V} \tag{6.87}$$

Therefore, to locate the shear center for this cross section we need to determine the shear force V_1 in the flange. From Eq. 6.63, the shear flow in the inclined flange is given by

$$q_1(s) = \frac{V Q_1(s)}{I} \tag{6.88}$$

Figure 6.62 can be used in the determination of the cross-sectional properties that are required in the calculation of the shear flow, $q_1(s)$. We assume that the thicknesses of the flanges and of the web are small in comparison with other dimensions. First, we need the moment of inertia of the entire cross section, taken with respect to the z axis.

$$I \equiv I_z = (I_z)_{\text{web}} + 2(I_z)_{\text{flange}} \tag{6.89}$$

We can use the definition of moment of inertia, together with Fig. 6.62b, to determine an expression for $(I_z)_{\text{flange}}$.

$$(I_z)_{\text{flange}} = \int_{A_{\text{flange}}} y^2 \, dA = \int_0^b \left[\frac{h_w}{2} + (b - s) \sin \phi \right]^2 (t_f ds)$$

Carrying out the integration, we get[36]

$$(I_z)_{\text{flange}} = b t_f \left[\left(\frac{h_w}{2} \right)^2 + \left(\frac{b h_w}{2} \right) \sin \phi + \left(\frac{b^2}{3} \right) \sin^2 \phi \right]$$

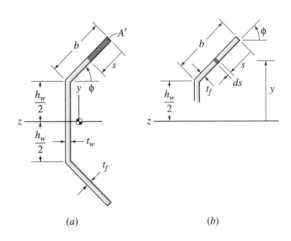

FIGURE 6.62 Cross-sectional dimensions used in calculating the shear flow in the flange.

(a)　　　　　(b)

---

[36]Check this expression by evaluating it for $\phi = 0$ and $\phi = 90°$.

Therefore, for the complete cross section,

$$I = \frac{t_w h_w^3}{12} + 2bt_f\left[\left(\frac{h_w}{2}\right)^2 + \left(\frac{bh_w}{2}\right)\sin\phi + \left(\frac{b^2}{3}\right)\sin^2\phi\right] \qquad (6.90)$$

We can use Fig. 6.62a to determine $Q_1(s)$.

$$Q_1(s) = \int_{A'} y \, dA = A'\overline{y}'$$

or

$$Q_1(s) = (t_f s)\left[\frac{h_w}{2} + \left(b - \frac{s}{2}\right)\sin\phi\right] \qquad (6.91)$$

Therefore, the flange shear flow is

$$q_1(s) = \frac{V t_f s}{I}\left[\frac{h_w}{2} + \left(b - \frac{s}{2}\right)\sin\phi\right] \qquad (6.92)$$

This distribution of shear flow in the flange is illustrated in Fig. 6.63. The web shear flow distribution is also illustrated.

The total shear force in the flange is given by the integral

$$V_1 = \int_0^b q_1(s)ds = \frac{V t_f b^2}{I}\left(\frac{h_w}{4} + \frac{b\sin\phi}{3}\right) \qquad (6.93)$$

Combining Eqs. 6.87 and 6.93 we finally obtain an expression for the location of the shear center.

$$e = \frac{b^2 t_f h_w \cos\phi}{I}\left(\frac{h_w}{4} + \frac{b\sin\phi}{3}\right) \qquad (6.94)$$

where the moment of inertia, I, is given by Eq. 6.90.

In Example Problems 6.16 and 6.17 we determine two special cases of the above analysis, a channel cross section and an equal-leg angle cross section, respectively.

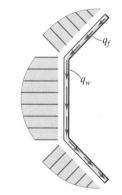

FIGURE 6.63 The shear-flow distribution in an open, thin-wall beam.

■■■■■■■■■■■■■■■□ EXAMPLE 6.19 □■■■■■■■■■■■■■■■

Use the fundamental equation of shear flow analysis for beams, Eq. 6.63, to determine the eccentricity of the shear center for the channel section in Fig. 1. Neglect the thickness in comparison with the other cross-sectional dimensions. For dimensional purposes, let V be in Newtons.

Plan the Solution We can follow the basic procedure that was followed in deriving Eq. 6.94, that is, we can derive an expression for e, like Eq. 6.87, based on resultant moments about O. This will involve the flange shear force. Next we can determine an expression for the flange shear flow, and then use the flange shear flow to determine the flange shear force needed in the expression for e. The shear center should lie "outside" the cross section.

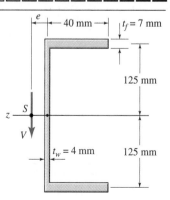

Fig. 1 (Drawing not to scale.)

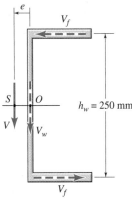

Fig. 2

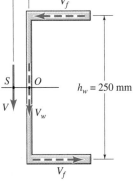

$t_f = 7$ mm

$\bar{y}' = 125$ mm

Fig. 3

$(q_f)_{max}$

Fig. 4

Solution Neglecting a small flange contribution ($\frac{1}{12}bt_f^3$), we get the following value for the moment of inertia.

$$I = \frac{t_w h_w^3}{12} + 2A_f d_f^2$$

$$I = \frac{(4 \text{ mm})(250 \text{ mm})^3}{12} + 2(40 \text{ mm})(7 \text{ mm})(125 \text{ mm})^2$$

$$I = 13.96(10^6) \text{ mm}^4 = 13.96(10^{-6}) \text{ m}^4$$

From resultant moments about the point O (through which V_w passes), (Fig. 2),

$$+\circlearrowleft \left(\sum M\right)_O = Ve = V_f h_w$$

so,

$$e = \frac{V_f h_w}{V} \tag{1}$$

From Eq. 6.63,

$$q_f = \frac{VQ_f}{I} \tag{2}$$

with Q_f based on the flange area A' in Fig. 3, that is,

$$Q_f = A'\bar{y}' = (7 \text{ mm})(s)(125 \text{ mm})$$

$$Q_f = 875s \text{ mm}^3$$

So,

$$q_f = \frac{(V \text{ N})(875s \text{ mm}^3)}{13.96(10^6) \text{ mm}^4} \tag{3}$$

$$(q_f)_{max} = (q_f)_{s=40 \text{ mm}}$$

$$= 2.51(10^{-3})V \text{ N/mm}$$

Since the shear flow distribution is triangular, as illustrated in Fig. 4, the flange shear force V_f is

$$V_f = \frac{1}{2}[2.507(10^{-3}) \ V \text{ N/mm}](40 \text{ mm})$$

$$V_f = 0.0501 \ V \text{ N} \tag{4}$$

Combining Eqs. (1) and (4) we get

$$e = 0.0501h_w = 12.5 \text{ mm} \qquad \textbf{Ans.}$$

Review the Solution The eccentricity e should be positive so that the shear center lies outside the cross section. Our answer has the correct sign, and the magnitude seems reasonable in comparison with the dimensions of the beam. Many commercial channel sections have an e/h_w ratio in the 0.05–0.15 range.

(a) Prove that the shear center of an equal-leg angle (Fig. 1) is at the corner of the angle, and prove that the resultant of the shear flow in the legs of the angle is equal to the total shear force on the section. The shear force acts normal to the axis of symmetry of the cross section. Assume that $t \ll b$. (b) Determine the maximum shear stress in the cross section in Fig. 1.

Plan the Solution   There is a resultant shear force along each leg of the angle. Since the lines of action of these two forces pass through the corner O, this point must be the shear center. We could develop an expression for this using basic shear-flow concepts (i.e., Eq. 6.63), or we can make use of Eq. 6.93, setting $h_w = 0$ and $\phi = 45°$. The maximum shear stress should occur at the neutral axis.

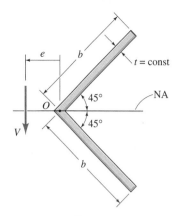

Fig. 1

Solution
(a) Since V_1 and V_2 both pass through point O, as indicated in Fig. 2, the resultant, V, must pass through this point. Therefore, the shear center lies at the intersection of the legs of the angle.[37] The flange shear is given by Eq. 6.93, with $h_w = 0$.

$$V_f = \frac{Vt_f b^3 \sin 45°}{3I} \tag{1}$$

where the moment of inertia is obtained by specializing Eq. 6.90 to give

$$I = \frac{2}{3} b^3 t_f \sin^2 45° = \frac{b^3 t_f}{3} \tag{2}$$

Combining Eqs. (1) and (2) we get

$$V_f = \frac{V}{\sqrt{2}} \tag{3}$$

Since

$$+\downarrow \sum F_y = 2V_f \sin 45° = V$$

we have shown that combining Eqs. 6.90 and 6.93 leads to the correct resultant shear force.

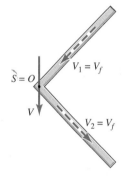

Fig. 2

(b) The shear flow distribution is given by Eq. 6.92, which specializes to

$$q_f(s) = \frac{Vt_f s}{I}\left(b - \frac{s}{2}\right) \sin 45° \tag{4}$$

where I is given by Eq. (2). This shear-flow distribution is depicted in Fig. 3. Since $\tau_f = q_f/t_f$,

$$(\tau_f)_{max} = \frac{(q_f)_{s=b}}{t_f}$$

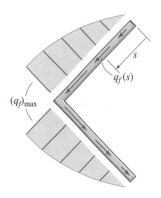

Fig. 3   Shear-flow distribution.

---

[37]This argument also holds for unequal-leg angles, but the shear-flow distribution in an unequal-leg angle must be obtained using the unsymmmetric bending theory of Section 6.6, so this problem, which asks for the shear flow, only treats equal-leg angles.

or

$$(\tau_f)_{max} = \left(\frac{3\sqrt{2}}{4}\right)\frac{V}{bt_f} \qquad\qquad \text{Ans.} \quad (5)$$

Review the Solution   As a rough check on our answer for τ_{max}, we can compare the preceding result, Eq. (5), with τ_{max} for a rectangle. From Example Problem 6.11,

$$(\tau_{max})_{rectangle} = \frac{3}{2}\frac{V}{A}$$

Since the area of the equal-leg angle is $2bt_f$, Eq. (5) can be expressed as

$$(\tau_{fmax})_{angle} = \frac{3\sqrt{2}}{2}\frac{V}{A}$$

Since the shear flows in the angle are not parallel with the resultant V, it is reasonable that the shear stress in the angle section is greater than that in a rectangle by a factor of $\sqrt{2}$.

See Homework Problem 6.12-16 for a chance to use the basic shear flow formula, Eq. 6.63, to obtain the above results.

6.13   PROBLEMS ■■■

Prob. 6.2-1. A steel strap with rectangular cross section is bent around a solid cylinder of radius $r = 30$ in. Assume that the neutral surface passes through the center of the strap's cross section. Determine the maximum strain, ϵ_{max}, in the steel strap (a) if the thickness of the strap is $h = \frac{1}{16}$ in., and (b) if the thickness of the strap is $h = \frac{1}{32}$ in.

the thickness of the strap is $h = 1$ mm, and (b) if the thickness of the strap is $h = 2$ mm.

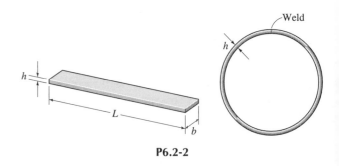

P6.2-2

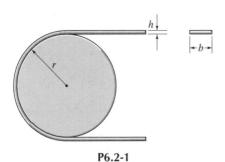

P6.2-1

Prob. 6.2-2. A steel strap of length L (meters) is bent to form a circle, and its ends are then butt-welded together. Determine the shortest length of strap that can be used if the maximum permissible strain in the strap is $\epsilon_{max} = 3.0 \times 10^{-3}$ mm/mm and (a) if

Prob. 6.2-3. Couples M_0 are applied to a steel strap of length $L = 30$ in. and thickness $h = \frac{1}{8}$ in. to bend it into the form of a circular arc, as shown in Fig. P6.2-3. The deflection δ at the midpoint of the arc, measured from a line joining the tips, is 3.5 in. (a) Determine the radius of curvature, ρ, of the strap, measured from the center of curvature to the neutral axis of the strap (which passes through the center of the cross section of the strap). (b) Calculate the maximum extensional strain ϵ_{max} in the curved strap.

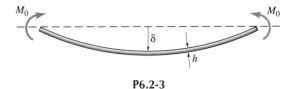

P6.2-3

Prob. 6.2-4. A couple M_0 acts on the end of a slender cantilever beam to bend it into the arc of a circle of radius ρ, as shown in Fig. P6.2-4. If the depth of the beam is $h = 2c = 6$ in., the length of the beam is $L = 5$ ft, and the extensional strain in the bottom fibers is $\epsilon(y = -3$ in.$) = 6(10^{-4})$ in./in., (a) determine the radius of curvature ρ, and (b) determine the tip deflection δ_{max}.

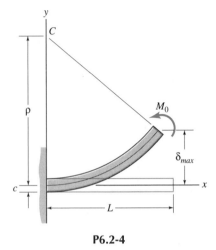

P6.2-4

Prob. 6.2-5. Prove that the transverse curvature, $\kappa' \equiv 1/\rho'$, is related to the longitudinal curvature, $\kappa \equiv 1/\rho$, by the equation $\kappa' = \nu\kappa$, that is, derive Eq. 6.5.

Prob. 6.3-1. A metal strap (Young's modulus $= E$) with rectangular cross section is bent around a solid circular cylinder of radius r, as shown in Fig. P6.3-1. Determine an expression for the maximum tensile stress in the curved portion of the strap. Express your answer in terms of the dimensions of the strap, the radius r, and the modulus E.

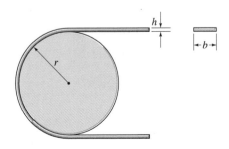

P6.3-1, P6.3-2, and P6.3-3

Prob. 6.3-2. A steel strap ($E = 210$ GPa) with rectangular cross section of dimensions $b = 6$ mm and $h = 1$ mm is bent around a solid cylinder of radius r as shown in Fig. P6.3-2. If the maximum permissible tensile stress in the steel strap is $\sigma_{max} = 200$ MPa, what is the minimum radius, r, of cylinder that can be used?

Prob. 6.3-3. A steel strap ($E = 29 \times 10^3$ ksi) with rectangular cross section is bent around a solid circular cylinder of radius $r = 20$ in., as shown in Fig. P6.3-3. If the maximum allowable flexural stress is not to exceed the yield strength ($\sigma_Y = 36$ ksi), what is the maximum thickness h that the strap can have?

Prob. 6.3-4. A rectangular aluminum strap ($E = 10 \times 10^3$ ksi) of length L (in.) is bent to form a circle, and its ends are then butt-welded together. Determine the shortest length of strap that can be used if the maximum permissible tensile stress in the strap is $\sigma_{max} = 50$ ksi. The dimensions of the strap are $b = 1$ in. and $h = \dfrac{1}{32}$ in.

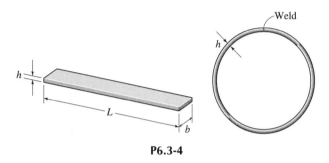

P6.3-4

Prob. 6.3-5. A couple $M_0 = 30$ kN·m acts on the end of a slender steel cantilever beam ($E = 210$ GPa) to bend it into the arc of a circle of radius ρ, as shown in Fig. P6.3-4. The cross section of the beam is a square of dimension $d = 100$ mm. (a) Calculate the maximum tensile stress in the beam. (b) Calculate the radius of curvature of the beam. (c) Calculate the maximum deflection, δ_{max}, of the beam if its length is $L = 4$ m.

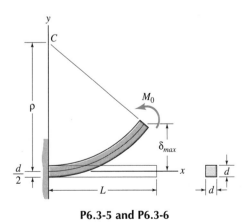

P6.3-5 and P6.3-6

Prob. 6.3-6. Solve Prob. 6.3-5 if $E = 30 \times 10^3$ ksi, $d = 4$ in., $L = 10$ ft, and $M_0 = 20$ kip·ft.

333

Prob. 6.3-7. A beam is made from three boards that are glued together to form a single beam, as shown in Fig. P6.3-7. The moment acting at this cross section is $M = 32$ kip·in. (a) Calculate the maximum tensile stress, and (b) calculate the total force acting on the top board.

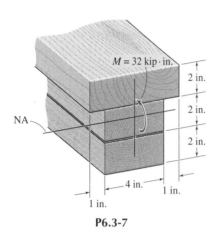

P6.3-7

Prob. 6.3-8. An extruded aluminum machine part has the cross section in Fig. 6.3-8. Determine the maximum moment M that can be applied to the member if the allowable flexural stress is $\sigma_{allow} = 200$ MPa.

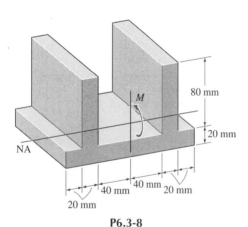

P6.3-8

Prob. 6.3-9. The structural steel wide-flange beam shown in Fig. P6.3-9 is a **W**10×60 section (See Table D.1). The allowable flexural stress in the beam is $\sigma_{allow} = 28$ ksi. (a) Calculate the maximum moment M_z, that can be applied (producing deflection in the xy-plane); and (b) calculate the maximum moment M_y that can be applied (producing deflection in the xz-plane).

Prob. 6.3-10. Repeat Prob. 6.3-9 if the wide-flange beam is a **W**250×89 section (see Table D.2) and the allowable flexural stress is $\sigma_{allow} = 200$ MPa.

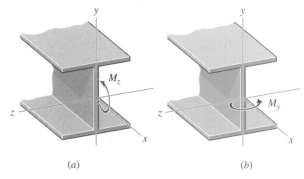

(a)   (b)

P6.3-9 and P6.3-10

Prob. 6.3-11. Determine the flexural stresses at points A and B in the cross section in Fig. P6.3-10 if the bending moment at this section is $M = 10$ kip·ft. The dimensions of the cross section are $b_f = 8$ in., $t_f = 2$ in., $h_w = 6$ in., and $t_w = 2$ in.

Prob. 6.3-12. Repeat Prob. 6.3-11 if $M = 15$ kN·m and the dimensions of the beam are $b_f = 200$ mm, $t_f = 50$ mm, $h_w = 150$ mm, and $t_w = 50$ mm.

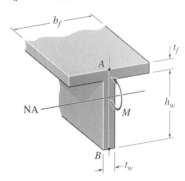

P6.3-11 and P6.3-12

Prob. 6.3-13. A timber beam consists of four planks fastened together with screws to form a box section 5.5 in. wide and 8.5 in. deep, as shown in Fig. P6.3-13. If the flexural stress at point B in the cross section is 900 psi (T), (a) determine the flexural stress at point A in the cross section; (b) determine the stress at point C in the cross section; and (c) determine the total force on the top plank.

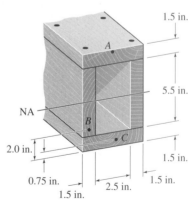

P6.3-13

Prob. 6.3-14. A tee-beam has the cross-sectional dimensions shown in Fig. P6.3-14. The bending moment at the given section has a magnitude of 40 kN·m and acts in the sense indicated. Determine the maximum tensile flexural stress and the maximum compressive flexural stress at this cross section.

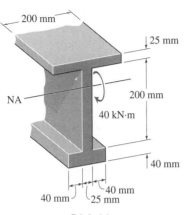

P6.3-14

Prob. 6.3-15. A simply-supported 4×8 in. (nominal) timber beam has the (finished) cross-sectional dimensions and loading shown in Fig. P6.3-15. Determine the flexural stress distribution on the cross section at the center of the beam, C, and make a three-dimensional sketch of this stress distribution.

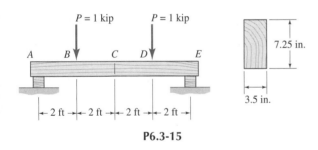

P6.3-15

Prob. 6.3-16. A rectangular timber beam AE has the cross-sectional dimensions and loading shown in Fig. P6.3-16. Determine the flexural stress distribution on the cross section at C, and make a three-dimensional sketch of this stress distribution.

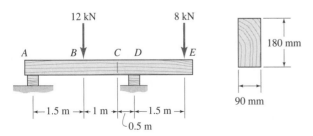

P6.3-16 and P6.3-27

Prob. 6.3-17. A wide-flange beam with overhangs is shown in Fig. P6.3-17 (see also Prob. 5.2-2). Determine the maximum tensile stress on the cross section at C. Neglect the weight of the beam.

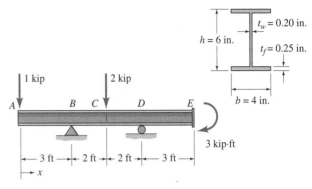

P6.3-17 and P6.3-28

Prob. 6.3-18. A simply supported wide-flange beam has the loading and cross-sectional dimensions shown in Fig. P6.3-18 (see also Prob. 5.2-4). Determine the maximum tensile stress on the cross section at C. Neglect the weight of the beam.

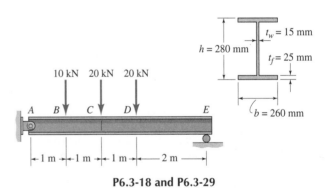

P6.3-18 and P6.3-29

Prob. 6.3-19. A W14×53 wide-flange beam supports the distributed load and concentrated load shown in Fig. P6.3-19 (see also Prob. 5.2-10). See Table D.1 for the cross-sectional properties of the beam. (a) Determine the maximum tensile stress on the cross section just to the left of the 8-kip load at B, and (b) determine the maximum compressive stress on the cross section at C.

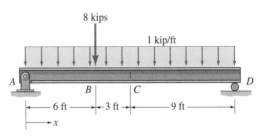

P6.3-19 and P6.3-30

Prob. 6.3-20. A **W**150×24 wide-flange beam supports the distributed load and concentrated load shown in Fig. P6.3-20 (see also Prob. 5.2-12). See Table D.2 for the cross-sectional properties of the beam. (a) Determine the maximum tensile stress on a cross section just to the left of *B*, where the 4 kN · m couple acts, and (b) determine the maximum tensile stress on a cross section just to right of *B*.

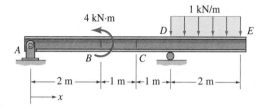

P6.3-20 and P6.3-31

Prob. 6.3-21. A channel section is used as a cantilever beam to support a uniformly distributed load of intensity $w = 25$ lb/ft and a concentrated load $P = 50$ lb, as shown in Fig. P6.3-21. Determine the maximum tensile flexural stress and the maximum compressive flexural stress at end *A*. The relevant dimensions of the cross section are shown below, and the moment of inertia about the neutral axis (NA) is $I = 5.14$ in⁴.

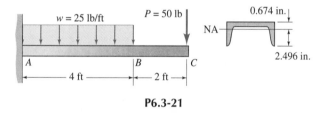

P6.3-21

Prob. 6.3-22. A structural tee section is used as a cantilever beam to support a triangular distributed load of maximum intensity $w = 40$ lb/ft and a concentrated load $P = 200$ lb, as shown in Fig. P6.3-22. Determine the maximum tensile flexural stress and the maximum compressive flexural stress at end *B*. The relevant dimensions of the cross section are shown below, and the moment of inertia about the neutral axis (NA) is $I = 33.1$ in⁴.

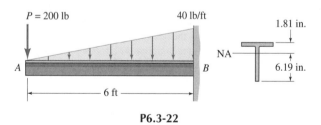

P6.3-22

Prob. 6.3-23. The shaft *AD* in Fig. P6.3-23 is supported by bearings at *A* and *C* (see also Prob. 5.2-3). Assume that the bearings produce concentrated reaction forces that are normal to the shaft. (a) Sketch shear and moment diagrams for the shaft, and (b) determine the maximum flexural stress in the shaft if its diameter is $d = \frac{3}{4}$ in.

336

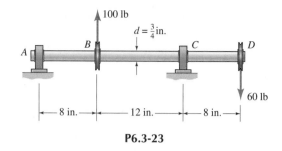

P6.3-23

Prob. 6.3-24. A cantilever beam is subjected to two concentrated loads, and its cross section has the dimensions shown in Fig. P6.3-24. (a) Sketch shear and moment diagrams for the beam, expressing the shear values in terms of *P* and the bending moment values in terms of *PL*. (b) If the allowable flexural stress (tension or compression) is $\sigma_{\text{allow}} = 180$ MPa, determine the maximum value of load *P* (Newtons) that can be applied to the beam. The moment of inertia about the neutral axis (NA) is $I = 122(10^6)$ mm⁴.

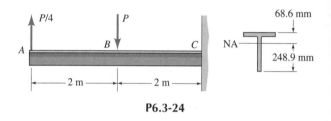

P6.3-24

Prob. 6.3-25. A **W**24×94 wide-flange member is used as a cantilever beam to support a uniformly distributed load of intensity $w = 1.5$ kips/ft and a concentrated load $P = 20$ kips as shown in Fig. P6.3-25 (see also Prob. 5.2-23). (a) Sketch shear and moment diagrams for the beam *AC*, and (b) determine the maximum flexural stress in the beam. Include the weight of the beam in your calculation of the flexural stress.

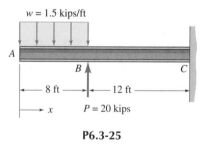

P6.3-25

Prob. 6.3-26. To limit the downward motion of *D*, a microswitch is installed as shown in Fig. P6.3-26a. When fully depressed, the switch arm *AC* can be modeled as a cantilever beam with a concentrated force $P = 0.02$ oz applied at end *C* and a reaction of $2P = 0.04$ oz at the contact point *B*, as shown in Fig. P6.3-26b. (a) Sketch shear and moment diagrams for the beam *AC*, and (b) determine the maximum flexural stress in the beam when the switch is fully activated. Express the stress in psi. The cross section of the beam is shown in Fig. P6.3-26c.

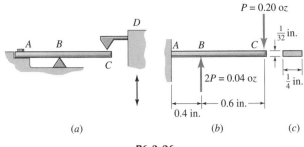

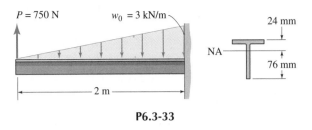

P = 0.20 oz

D

A B

C

$\frac{1}{32}$ in.

A B C

2P = 0.04 oz

$\frac{1}{4}$ in.

0.6 in.

0.4 in.

(a)   (b)   (c)

P6.3-26

P = 750 N w_0 = 3 kN/m

24 mm

NA

76 mm

2 m

P6.3-33

Prob. 6.3-27. For beam AE in Fig. P6.3-16, (a) sketch shear and moment diagrams, and (b) determine the maximum flexural stress in the beam.

Prob. 6.3-28. For beam AE in Fig. P6.3-17, (a) sketch shear and moment diagrams, and (b) determine the maximum flexural stress in the beam.

Prob. 6.3-29. For beam AE in Fig. P6.3-18, (a) sketch shear and moment diagrams, and (b) determine the maximum flexural stress in the beam.

Prob. 6.3-30. For beam AD in Fig. P6.3-19, (a) sketch shear and moment diagrams, and (b) determine the maximum flexural stress in the beam.

Prob. 6.3-31. For beam AE in Fig. P6.3-20, (a) sketch shear and moment diagrams, and (b) determine the maximum flexural stress in the beam.

Prob. 6.3-32. A channel section is used as a cantilever beam to support a uniformly distributed load of intensity $w = 100$ lb/ft and a concentrated load of $P = 100$ lb, as shown in Fig. P6.3-32. (a) Sketch shear and moment diagrams for the beam, and (b) determine the maximum tensile flexural stress and the maximum compressive flexural stress in the beam. The relevant dimensions of the cross section are shown below, and the moment of inertia about the neutral axis (NA) is $I = 5.14$ in⁴.

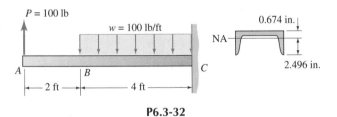

P = 100 lb

w = 100 lb/ft

0.674 in.

NA

B   C

2.496 in.

A

2 ft   4 ft

P6.3-32

Prob. 6.3-33. A structural tee section is used as a cantilever beam to support a triangularly distributed load of maximum intensity $w_0 = 3$ kN/m and a concentrated load $P = 750$ N as shown in Fig. P6.3-33. (a) Sketch shear and moment diagrams for the beam, and (b) determine the maximum tensile flexural stress and the maximum compressive flexural stress in the beam. The relevant dimensions of the cross section are shown below, and the moment of inertia about the beam's neutral axis (NA) is $I = 895(10^3)$ mm⁴.

Prob. 6.3-34. Determine the maximum uniform distributed load, w, that can be applied to a cantilever beam with tee section as shown in Fig. P6.3-34. The allowable tensile stress is $(\sigma_{\text{allow}})_T = 20$ ksi; and the allowable compressive stress is $(\sigma_{\text{allow}})_C = 16$ ksi.

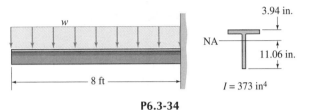

w

3.94 in.

NA

11.06 in.

8 ft

$I = 373$ in⁴

P6.3-34

Prob. 6.3-35. Determine the maximum uniform distributed load, w, that can be applied to the beam with overhang shown in Fig. P6.3-35. The allowable stress (magnitude) in tension or compression is 150 MPa, and the beam is a **W**310×97 (see Table D.2 of Appendix D).

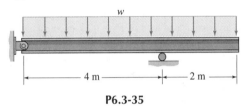

w

4 m   2 m

P6.3-35

Prob. 6.3-36. One of the assumptions made in deriving the flexure formula, Eq. 6-13, was that σ_y and σ_z are much smaller than σ_x. (a) Using the cantilever beam with uniformly distributed load, shown in Fig. P6.3-36, derive an expression for the ratio of the magnitude of the maximum flexural stress at the top of the beam, $\sigma_{xm} \equiv |\sigma_x(x, y = h/2)|$, at an arbitrary cross section x to the maximum transverse normal stress in the beam, $\sigma_{ym} = p/b$. (b) Use your results from Part (a) to show that the stated assumption is satisfied practically everywhere in this cantilever beam with uniform distributed load.

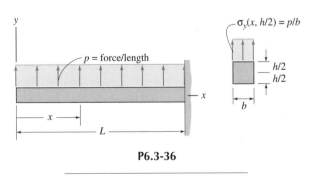

y

p = force/length

$\sigma_y(x, h/2) = p/b$

h/2

h/2

x

x

L

b

P6.3-36

In **Problems. 6.4-1 through 6.4-15** *ignore the weight of the beam in comparison with the applied loading. It is suggested that you use shear and moment diagrams to determine the locations of critical sections.*

Prob. 6.4-1. The simply-supported beam in Fig. P6.4-1 is subjected to a uniform downward load of $w = 8$ kips/ft on a span of $L = 8$ ft. The allowable stress (magnitude) in tension or compression is $\sigma_{allow} = 20$ ksi. From Table D.1 of Appendix D, select the lightest wide-flange steel beam that may be used for this application.

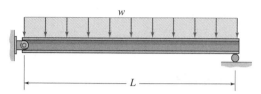

P6.4-1 and P6.4-2

Prob. 6.4-2. The simply supported beam in Fig. 6.4-2 is subjected to a uniform downward load of $w = 90$ kN/m on a span of $L = 3$ m. The allowable stress (magnitude) in tension or compression is $\sigma_{allow} = 150$ MPa. From Table D.2 of Appendix D, select the lightest wide-flange steel beam that may be used for this application.

Prob. 6.4-3. The simply supported beam in Fig. P6.4-3 is subjected to a concentrated load of $P = 40$ kips at the center of its span of $L = 8$ ft. The allowable stress (magnitude) in tension or compression is $\sigma_{allow} = 20$ ksi. From Table D.1 of Appendix D select the lightest wide-flange steel beam that may be used for this application.

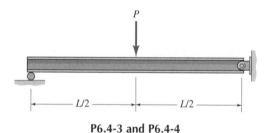

P6.4-3 and P6.4-4

Prob. 6.4-4. The simply supported beam in Fig. P6.4-4 is subjected to a concentrated load of $P = 200$ kN at the center of its span of $L = 3$ m. The allowable stress (magnitude) in tension or compression is $\sigma_{allow} = 150$ MPa. From Table D.2 of Appendix D select the lightest wide-flange steel beam that may be used for this application.

Prob. 6.4-5. A simply supported timber beam supports a triangularly distributed load, as shown in Fig. P6.4-5. The magnitude of the allowable stress in tension or compression is $\sigma_{allow} = 800$ psi. From Table D.8 of Appendix D select the lightest timber beam that may be used for this application if the nominal depth of the beam may not exceed twice its nominal width.

338

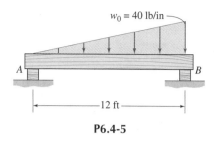

P6.4-5

Prob. 6.4-6. A simply supported timber beam supports a linearly varying load, as shown in Fig. P6.4-6. The magnitude of the allowable stress in tension or compression is $\sigma_{allow} = 6$ MPa. Determine, to the nearest 10 mm, the dimension b of the lightest timber beam with square cross section that can be used for this application.

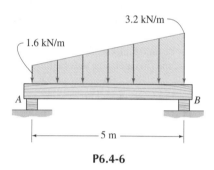

P6.4-6

Prob. 6.4-7. From Table D.1 of Appendix D, select the lightest wide-flange steel beam that may be used for the application shown in Fig. P6.4-7. The allowable stress (magnitude) in tension or compression is $\sigma_{allow} = 20$ ksi.

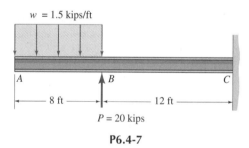

P6.4-7

Prob. 6.4-8. From Table D.2 of Appendix D, select the lightest wide-flange steel beam that may be used for the application shown in Fig. P6.4-8. The allowable stress (magnitude) in tension or compression is $\sigma_{allow} = 150$ MPa.

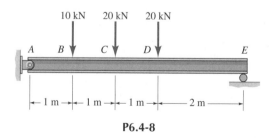

P6.4-8

Prob. 6.4-9. Two angle sections with equal legs (from Table D.5 of Appendix D) are to be welded together to form a beam with a channel cross section as illustrated in Fig. P6.4-9b. The loads are $w = 1.5$ kips/ft and $P = 20$ kips. If the allowable stress (magnitude) in tension or compression is $\sigma_{allow} = 20$ ksi, what is the lightest angle that could be used to make the beam shown in Fig. P6.4-9a? (Note: Assume that $t/b \ll 1$ in obtaining expressions for the moment of inertia, etc.)

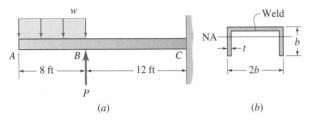

(a)   (b)

P6.4-9 and P6.4-10

Prob. 6.4-10. Two angle sections with equal legs (from Table D.5 of Appendix D) are to be welded together to form a beam with a channel cross section as illustrated in Fig. P6.4-10b. The loads are $w = 1.5$ kips/ft and $P = 20$ kips. If the allowable stress in tension is $(\sigma_{allow})_T = 20$ ksi and the allowable stress magnitude in compression is $(\sigma_{allow})_C = 15$ ksi, what is the lightest angle that could be used to make the beam shown in Fig. P6.4-10a? (Note: Assume that $t/b \ll 1$ in obtaining expressions for the moment of inertia, etc.)

Prob. 6.4-11. A sawmill cuts rectangular timber beams from circular logs. If two beams of width b and depth h are to be cut from a single log of diameter d, what b and h would give the strongest beams, that is, the beams with maximum value of S? (Neglect the width of saw cuts.)

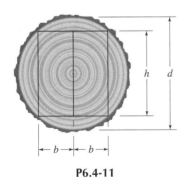

P6.4-11

Prob. 6.4-12. Beam AE rests on supports at B and D, with overhangs AB and DE. It is subjected to two concentrated loads and a concentrated couple, as shown in Fig. P6.4-12. If the allowable stress for this application is $\sigma_{allow} = 14$ ksi (in tension or compression), select a structural steel wide-flange section from those listed in Table D.1.

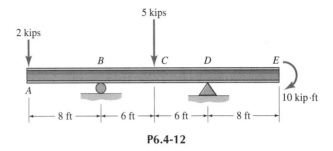

P6.4-12

Prob. 6.4-13. An 18-ft.-long timber beam is subjected to two concentrated loads, as shown in Fig. P6.4-13. The allowable tensile (and compressive) stress for the wood is $\sigma_{allow} = 960$ psi. Select the best timber beam for this application from those listed in Table D.8.

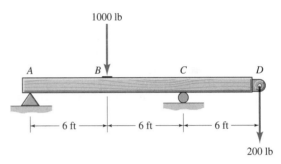

P6.4-13

Prob. 6.4-14. A structural designer wishes to locate the roller support B such that the beam AD will have the same maximum magnitude of positive moment and negative moment when subjected to concentrated loads P_A, and P_C acting on the beam as shown. Once the overhang length b has been determined, the designer must choose an appropriate beam to satisfy a flexural stress allowable of σ_{allow} (tension or compression). (a) Determine the overhang length b. (b) From Table D.2, select the best wide-flange section for this application.

Let $P_A = 20$ kN, $P_C = 40$ kN, $L = 2$ m, $\sigma_{allow} = 80$ MPa.

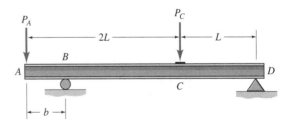

P6.4-14 and P6.4-15

Prob. 6.4-15. Repeat Prob. 6.4-14 for the following parameters: $P_A = 4$ kips, $P_C = 12$ kips, $L = 4$ ft, $\sigma_{allow} = 16$ ksi.

Prob. 6.5-1. A timber beam (nominal 6 in. × 12 in.) is reinforced by bonding a steel splice plate on either side, as shown in Fig. P6.5-1. If the maximum moment resisted by the beam is $M_{max} = 20$ kip·ft., and the ratio of elastic moduli is $E_s/E_w = 20$, what are the maximum normal stresses in the steel and the timber?

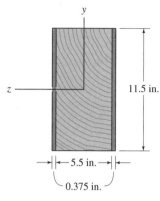

5.5 in.

0.375 in.

P6.5-1

Prob. 6.5-2. A steel/timber sandwich beam is fabricated from two 90 mm × 180 mm timber beams and a 12 mm × 180 mm steel plate, as illustrated in Fig. P6.5-2. The ratio of elastic moduli is $E_s/E_w = 20$. If the allowable stresses in the steel and wood are $(\sigma_{allow})_s = 124$ MPa and $(\sigma_{allow})_w = 8$ MPa, respectively, what is the maximum moment that can be safely applied to this sandwich beam?

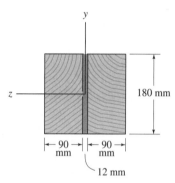

180 mm

90 mm   90 mm

12 mm

P6.5-2

Prob. 6.5-3. A timber beam (nominal 6 in. × 10 in.) is reinforced by attaching $\frac{3}{8}$-in.-thick steel plates to its top and bottom surfaces.

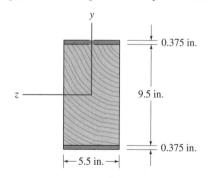

0.375 in.

9.5 in.

0.375 in.

5.5 in.

P6.5-3 and P6.5-4a

340

The ratio of elastic moduli of the steel and wood is $E_s/E_w = 20$. If the beam is subjected to a bending moment $M = 200$ kip·in, what are the maximum stresses in the steel and wood, $(\sigma_{max})_s$ and $(\sigma_{max})_w$?

Prob. 6.5-4. A steel-reinforced timber beam has the cross section shown in Fig. P6.5-4a (see Prob. 6.5-3) and is used as an 18-ft-long simply supported beam to carry a uniformly distributed load $w = 500$ lb/ft, as shown in Fig. P6.5-4b. Determine the maximum normal stresses in the steel and the wood, $(\sigma_{max})_s$ and $(\sigma_{max})_w$, under this loading condition. The ratio of elastic moduli of the steel and wood is $E_s/E_w = 20$.

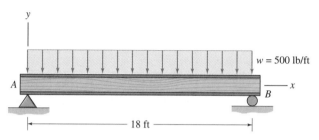

$w = 500$ lb/ft

18 ft

P6.5-4b

Prob. 6.5-5. The reinforced timber beam shown in Fig. P6.5-5a is simply supported and is subjected to quarter-point loading, as shown in Fig. P6.5-5b. If the allowable stresses in the steel and wood are $(\sigma_{allow})_s = 124$ MPa and $(\sigma_{allow})_w = 8$ MPa, respectively, and the ratio of elastic moduli is $E_s/E_w = 20$, determine the (maximum) allowable load P.

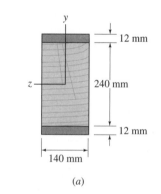

12 mm

240 mm

12 mm

140 mm

(a)

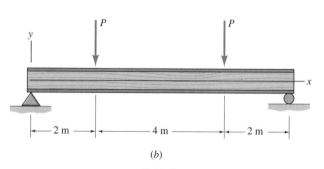

P P

2 m   4 m   2 m

(b)

P6.5-5

Prob. 6.5-6. The cross section of a bimetallic beam is shown in Fig. P6.5-6. The elastic moduli of the two metallic components are $E_1 = 15,000$ ksi and $E_2 = 10,000$ ksi, and the z axis is the NA. (a) In which material does the maximum normal stress occur when the beam is subjected to a bending moment M_z? (b) Determine the value of the minimum elastic section modulus, where $S_{min} \equiv M_z / \sigma_{max}$.

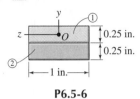

P6.5-6

Prob. 6.5-7. A bimetallic strip, whose cross section is depicted in Fig. P6.5-7, is used as the sensing element in a temperature-activated switch. The two metals are copper and nickel, whose elastic moduli are $E_c = 120$ GPa and $E_n = 210$ GPa, respectively. The z axis is the NA. (a) If the strip is subjected to a bending moment $M_z = 2$ N·m, what are the maximum stresses $(\sigma_c)_{max}$ and $(\sigma_n)_{max}$ in the copper and nickel, respectively? (b) What radius of curvature, ρ, is produced by this bending moment?

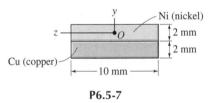

P6.5-7

Prob. 6.5-8. A simply supported beam spans a length of $L = 4$ m and carries a uniformly distributed load of magnitude w. The

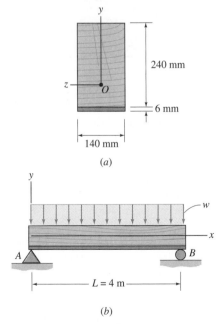

P6.5-8 and P6.5-9

cross section of the beam consists of a timber beam ($E_w = 10$ GPa) to which is attached a steel bottom-plate ($E_s = 210$ GPa). The z axis is the NA. (a) If the distributed load is $w = 4$ kN/m (including beam weight), find the maximum stresses in the wood and in the steel, $(\sigma_{max})_w$ and $(\sigma_{max})_s$. (b) Determine the radius of curvature, ρ, at the midspan.

Prob. 6.5-9. If the allowable stresses for the beam in Prob. 6.5-8 are $(\sigma_{allow})_w = 6$ MPa and $(\sigma_{allow})_s = 120$ MPa in the wood and steel, respectively, what is the allowable value of the distributed load w (including the weight of the beam)?

Prob. 6.5-10. A simply supported beam spans a length of 20 ft. and supports equal downward loads, P, as shown in Fig. P6.5-10. To strengthen the 6×12 timber beam (finished dimensions are 5.5 in. $\times$ 11.5 in., as shown) a 4 in. $\times \frac{3}{8}$ in. steel plate is bonded to the bottom of the wood beam. The elastic moduli are $E_w = 1.2 \times 10^6$ psi and $E_s = 30 \times 10^6$ psi for the wood and steel, respectively. The specific weights of the wood and steel are $\gamma_w = 40$ lb/ft³ and $\gamma_s = 490$ lb/ft³. (a) If the loads on the beam are $P = 2$ kips each, find the maximum stresses in the wood and in the steel, $(\sigma_{max})_w$ and $(\sigma_{max})_s$. Include the weight of the beam in your calculations. (b) Determine the radius of curvature in the section of the beam between the two loads.

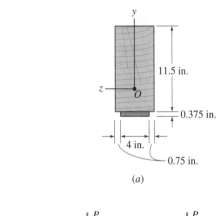

(a)

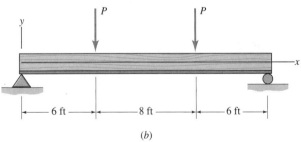

(b)

P6.5-10 and P6.5-11

Prob. 6.5-11. If the allowable stresses for the beam in Prob. 6.5-10 are $(\sigma_{allow})_w = 1,500$ psi and $(\sigma_{allow})_s = 15,000$ psi for the wood and steel, respectively, what is the allowable value of the load P that may be carried by the beam? (You may neglect the weight of the beam in your calculations.)

Prob. 6.5-12. A wood beam (E_w = 1,500 ksi) is to be reinforced by the addition of a steel plate (E_s = 30,000 ksi). The allowable stresses in the wood and steel are $(\sigma_{allow})_w$ = 1 ksi and $(\sigma_{allow})_s$ = 12 ksi, respectively. The dimensions of the wood beam are given (5.5 in. × 7.5 in), as is the thickness of the steel plate (t = 0.25 in). You are to determine the width of the steel plate such that the allowable stresses in the steel and wood are reached simultaneously (i.e., for the same value bending moment). Let the z axis be the NA.

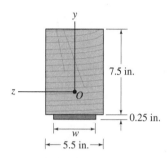

P6.5-12

Prob. 6.5-13. The cross section of a steel-reinforced concrete beam of width b is shown in Fig. P6.5-13a. The allowable compressive stress in the concrete is σ_c, and the allowable tensile stress in the steel is σ_s. The diameter of each of the three steel bars is d, and the ratio of the moduli of elasticity is E_s/E_c. Assume that the area of the steel bars is concentrated along a horizontal line at a distance of h from the top of the beam, and assume that the concrete is only effective in compression. That is, tension is carried solely by the steel bars. Calculate the maximum allowable bending moment for the beam. Let b = 12 in., h = 16 in., d = 0.875 in., σ_c = 1.5 ksi, σ_s = 18 ksi, and E_s/E_c = 10.

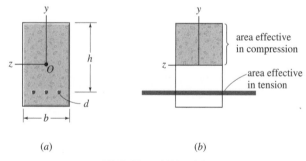

(a)   (b)

P6.5-13 and P6.5-14

Prob. 6.5-14. Repeat Prob. 6.5-13, using the following data: b = 200 mm, h = 300 mm, d = 16 mm, σ_c = 10 MPa, σ_s = 125 mPa, and E_s/E_c = 10.

Prob. 6.6-1. A 4×4 timber beam (see Table D.8 for finish dimension b) is subjected to a resultant bending moment of mag-

342

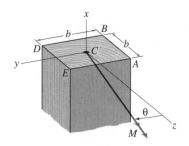

P6.6-1 and P6.6-2

nitude M = 5 kip·in. oriented at angle θ = 30°, where θ is the angle shown in Fig. P6.6-1. Determine the flexural stress at each corner of the cross section, and sketch the stress distribution (as in Fig. 6.23c).

Prob. 6.6-2. A square timber beam whose cross-sectional dimension is b = 150 mm is subjected to a resultant bending moment of magnitude M = 2 kN·m at an angle θ = 45°, where θ is the angle indicated in Fig. P6.6-2. Determine the flexural stress at each corner of the cross section, and sketch the stress distribution (as in Fig. 6.23c).

Prob. 6.6-3. A beam with rectangular cross section of dimensions b = 4 in. and h = 2 in. is subjected to a resultant bending moment of magnitude M = 2 kip·ft oriented at an angle θ = 10°, where θ is the angle shown in Fig. P6.6-3. (a) Determine the orientation of the neutral axis for this situation and sketch its location with respect to the cross section. (b) Determine the maximum tensile flexural stress and indicate where it occurs.

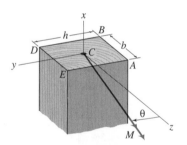

P6.6-3 and P6.6-4

Prob. 6.6-4. A rectangular beam with cross-sectional dimensions b = 50 mm and h = 25 mm is subjected to a resultant bending moment M oriented at an angle θ = 50°, where θ is the angle indicated in Fig. P6.6-4. (a) Locate the neutral axis for the given cross section and the given orientation of moment. (b) Determine the maximum moment that can be applied if the allowable flexural stress (tension or compression) is σ_{allow} = 120 MPa.

Prob. 6.6-5. Solve Example Problem 6.9, substituting a **W**12×50 wide-flange beam for the **S**12×50 I-beam in the present text example. Compare your answers with those of Example 6.9, and briefly discuss the similarities and differences in the stresses in the **S**-shape and the **W**-shape.

Prob. 6.6-6. A **W**250×45 wide-flange beam is subjected to a moment M that is intended to lie along the z axis (due to loading in the xy plane). (See Table D.2 for section properties.) (a) De-

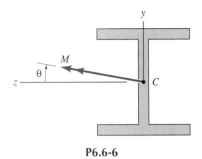

P6.6-6

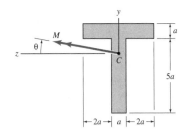

P6.6-8

termine the effect a "load misalignment" of $\theta = 1.5°$ has on the orientation of the neutral axis. Show the neutral axis on a sketch of the cross section. (b) Determine the maximum tensile flexural stress acting on the cross section and compare this stress with the maximum stress in the beam if the load were properly aligned (i.e., $\theta = 0°$). Let $M = 30$ kN·m.

Prob. 6.6-7. [This problem examines the effect of misalignment of the load on the T-beam in Example Problem 6.2 (Section 6.3).] The 4 kip·ft moment, M, acts through the centroid of the T-section at an inclination of $\theta = 2.5°$, as shown in Fig. P6.6-7. (a) Determine the orientation of the neutral axis, and show the neutral axis on a sketch of the cross section. (b) Determine the maximum tensile flexural stress acting on the cross section, and compare this stress with the maximum tensile stress of 5.76 ksi that was obtained in Example Problem 6.2 for a moment perfectly aligned with the z axis (i.e., $\theta = 0°$).

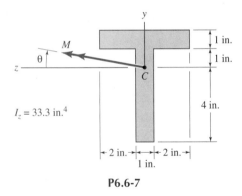

P6.6-7

Prob. 6.6-8. A moment M acts through the centroid of the T-section in Fig. P6.6-8 at an inclination angle $\theta = 4°$. (a) Determine the orientation of the neutral axis, and show the neutral axis on a sketch of the cross section. (b) Determine the maximum tensile flexural stress acting on the cross section, and compare this with the maximum tensile flexural stress the beam would experience if $\theta = 0°$. Express your answers in terms of M and the dimension a.

Prob. 6.6-9. Due to load misalignment, the bending moment acting on the channel sections in Fig. P6.6-9 is inclined at an angle of 3° with respect to the y axis. If the allowable flexural stress for this beam is $\sigma_{\text{allow}} = 16$ ksi, what is the maximum moment, M_{max}, that may be applied?

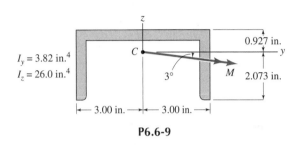

P6.6-9

Prob. 6.6-10. The structural tee whose cross section is shown in Fig. P6.6-10 *b* is used as a cantilever beam to support a concentrated load P that passes through the centroid of the end cross section, as shown in Fig. P6.6-10a. The load acts in the $x'z'$-plane, but the beam is rotated through a 2° angle about the $x(x')$ axis as indicated. Calculate the increase (or decrease, if that is the case) in the maximum compressive flexural stress that is caused by a 2° rotation of the beam. Let $I_y = 7.33 \times 10^5$ mm⁴, and $I_z = 2.77 \times 10^6$ mm⁴.

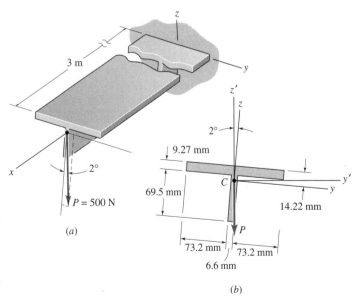

P6.6-10

343

Prob. 6.6-11. A $4 \times 4 \times \frac{1}{2}$ equal-leg angle (see Table D.5) is subjected to loading in the $x'z'$-plane that produces a moment $M_{y'} = 12$ kip·in. at the section shown in Fig. P6.6-11. (a) Determine the orientation of the neutral axis, and show the neutral axis on a sketch of the cross section. (b) Determine the maximum tensile flexural stress acting on the cross section, and indicate the location of this maximum stress.

(Note: From Eq. C-28, $I_y + I_z = I_{y'} + I_{z'}$.)

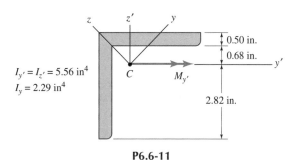

$I_{y'} = I_{z'} = 5.56$ in^4
$I_y = 2.29$ in^4

0.50 in.
0.68 in.
2.82 in.

P6.6-11

*Prob. 6.6-12. As shown in Fig. P6.6-12, the vector representing the bending moment on the $5 \times 3 \times \frac{1}{2}$ structural-angle cross section is along the y' direction. (a) From Table D.6, determine the location of the centroid of the cross section, the orientation of the principal axes, and the values of the centroidal principal moments of inertia. (b) Determine the orientation of the neutral axis at this section, and indicate the orientation of the neutral axis on a sketch of the cross section. (c) If the magnitude of the maximum allowable flexural stress (tensile or compressive) is $\sigma_{allow} = 16$ ksi, what is the maximum allowable value of the moment, $(M_{y'})_{max}$?

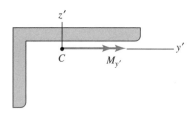

P6.6-12

*Prob. 6.6-13. As shown in Fig. P6.6-13, the bending moment M acts at an angle of 20° counterclockwise from the y' axis of an unequal-leg angle section. (a) Locate the centroid of the cross section, determine the orientation of the centroidal principal axes, and determine the principal moments of inertia. You may use the results of Example C-1 and C-4 of Appendix C. (b) Determine the orientation of the neutral axis at this section, and indicate the orientation of the neutral axis on a sketch of the cross section. (c) Calculate the maximum tensile flexural stress and the maximum compressive flexural stress at this cross section.

344

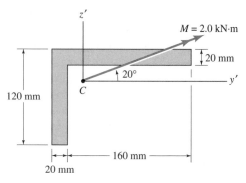

$M = 2.0$ kN·m
20 mm
20°
120 mm
160 mm
20 mm

P6.6-13

*Prob. 6.6-14. An aluminum **Z** section has the dimensions shown in Fig. P6.6-14. The yield strength (tension or compression) of the aluminum is $\sigma_Y = 40$ ksi, and a factor of safety of $FS = 2.0$ is to be maintained. (a) Determine $I_{y'}$, $I_{z'}$, and $I_{y'z'}$. (See Appendices C-2 and C-3.) (b) Determine the orientation of the centroidal principal axes for this cross section. (See Appendix C-4.) (c) Determine the orientation of the neutral axis for the given orientation of bending moment. (d) Determine the maximum allowable magnitude of bending moment $M_{y'}$.

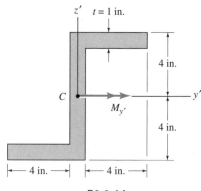

$t = 1$ in.
4 in.
4 in.
4 in.
4 in.

P6.6-14

Prob. 6.6-15. Consider a linearly elastic prismatic beam of arbitrary cross section subjected to bending-moment components M_y and M_z, as shown in Fig. P6.6-15. Let the x axis pass through

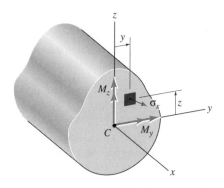

P6.6-15

the centroid of the cross section (i.e., $\bar{y} = \bar{z} = 0$), but let the y and z axes be arbitrarily oriented with respect to the cross section. With $F(x) = 0$, use Eqs. 6.32 through 6.34 to show that the flexural stress σ_x is given by

$$\sigma_x = \frac{-y(M_z I_y + M_y I_{yz}) + z(M_y I_z + M_z I_{yz})}{I_y I_z - I_{yz}^2}$$

For **Problems 6.7-1 through 6.7-20**, *assume that the material is elastic-plastic (i.e., linearly elastic, perfectly plastic), and has the same stress-strain properties for both tension and compression.*

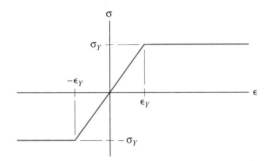

Prob. 6.7-1. A beam with rectangular cross section ($b = 1$ in., $h = 2$ in.) is made of structural steel with $\sigma_Y = 36$ ksi and $E = 30 \times 10^3$ ksi. (a) Calculate numerical values for the yield moment M_Y, the plastic moment M_P, and the shape factor f. (b) Plot the moment-curvature diagram for this particular beam for values of curvature up to $\kappa = 3\kappa_Y$.

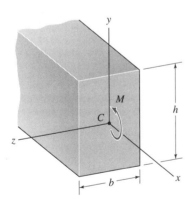

P6.7-1, P6.7-2, P6.7-17, and P6.7-18

Prob. 6.7-2. Repeat Problem 6.7-1 for a rectangular beam with $b = 50$ mm, $h = 150$ mm, $E = 210$ GPa, and $\sigma_Y = 250$ MPa.

Prob. 6.7-3. A rectangular box beam with height $h = 12$ in. and width $b = 8$ in. has a constant wall thickness $t = 0.50$ in. It is made of structural steel with $\sigma_Y = 36$ ksi and $E = 30 \times 10^3$ ksi. Calculate numerical values for the yield moment M_Y, the plastic moment M_P, and the shape factor f.

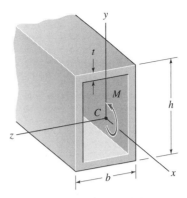

P6.7-3, P6.7-4, and P6.7-18

Prob. 6.7-4. Repeat Prob. 6.7-3 for a rectangular box beam with $b = 300$ mm, $h = 400$ mm and $t = 12.7$ mm, made of steel with $\sigma_Y = 250$ MPa and $E = 210$ GPa.

Prob. 6.7-5. A wide-flange beam with dimensions $b_f = 8.060$ in., $t_f = 0.660$ in., $h = 13.92$ in., and $t_w = 0.370$ in. is made of structural steel with $\sigma_Y = 36$ ksi and $E = 30 \times 10^3$ ksi. Calculate numerical values for the yield moment M_Y, the plastic moment M_P, and the shape factor f.

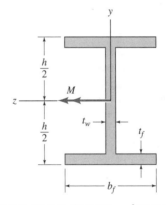

P6.7-5, P6.7-6, P6.7-13, and P6.7-14

Prob. 6.7-6. Repeat Prob. 6.7-5 for a wide-flange beam with dimensions $b_f = 400$ mm, $t_f = 30$ mm, $h = 850$ mm, and $t_w = 18$ mm. The beam is made of structural steel with $\sigma_Y = 250$ MPa and $E = 210$ GPa. Calculate numerical values for the yield moment M_Y, the plastic moment M_P, and the shape factor f.

Prob. 6.7-7. A **W**10×45 wide-flange beam is made of structural steel with $\sigma_Y = 36$ ksi and $E = 30 \times 10^3$ ksi. Calculate numerical values for the yield moment M_Y, the plastic moment M_P, and the shape factor f. (See Table D.1 for the section properties.)

Prob. 6.7-8. A **W**200×71 wide-flange beam is made of structural steel with $\sigma_Y = 250$ MPa and $E = 210$ GPa. Calculate numerical values for the yield moment M_Y, the plastic moment M_P, and the shape factor f. (See Table D.2 for the section properties.)

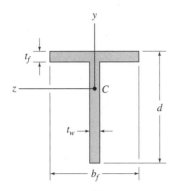

P6.7-9 and P6.7-10

Prob. 6.7-9. The structural tee section shown in Fig. P6.7-9 has the dimensions $d = 5.05$ in., $b_f = 8.02$ in., $t_w = 0.350$ in., $t_f = 0.620$ in., and it is made of steel with $\sigma_Y = 36$ ksi and $E = 30 \times 10^3$ ksi. Calculate numerical values for the yield moment M_Y, the plastic moment M_P, and the shape factor f.

Prob. 6.7-10. Repeat Prob. 6.7-9 for a structural tee beam with dimensions $d = 430$ mm, $b_f = 400$ mm, $t_w = 18$ mm, $t_f = 30$ mm that is made of structural steel with $\sigma_Y = 250$ MPa and $E = 210$ GPa.

Prob. 6.7-11. A beam with solid circular cross section of diameter d is made of elastic-plastic steel with yield point σ_Y and modulus of elasticity E. Determine expressions for the yield moment M_Y, the plastic moment M_P, and the shape factor f.

Prob. 6.7-12. A circular tube of outer diameter d_o and inner diameter d_i is made of elastic-plastic steel with yield point σ_Y and modulus of elasticity E. Determine expressions for the yield moment M_Y, the plastic moment M_P, and the shape factor f.

Prob. 6.7-13. Determine the value of the bending moment, call it M_F, that would cause the flange of the wide-flange beam in Prob. 6.7-5 to be fully plastic while the web remains linearly elastic.

Prob. 6.7-14. Repeat Prob. 6.7-13 for the wide flange beam in Prob. 6.7-6.

Prob. 6.7-15. Repeat Prob. 6.7-13 for the wide flange beam in Prob. 6.7-7.

Prob. 6.7-16. Repeat Prob. 6.7-13 for the wide flange beam in Prob. 6.7-8.

Prob. 6.7-17. A beam with solid circular cross section of diameter d is subjected to the plastic moment M_P; then the load is completely removed. The beam is made of elastic-plastic steel with yield point σ_Y. Determine an expression for the maximum residual stress, and sketch the residual-stress state (as is done for a rectangular beam in Fig. 6.36d).

Prob. 6.7-18. Repeat Prob. 6.7-17 for the rectangular box beam in Prob. 6.7-3.

Prob. 6.7-19. Repeat Prob. 6.7-17 for the **W**10×45 wide-flange beam in Prob. 6.7-7.

Prob. 6.7-20. Repeat Prob. 6.7-17 for the **W**200×71 wide-flange beam in Prob. 6.7-8.

*Prob. 6.7-21. A beam with rectangular cross section of width b and height h, as shown in Fig. P6.7-21a, is made of material whose stress-strain curve (in tension or compression) is approximately the form shown in Fig. P6.7-21b. Let $\sigma_{Y2} = \sigma_Y$, $\sigma_{Y1} = \frac{3}{4}\sigma_Y$, $\epsilon_{Y2} = \epsilon_Y$, and $\epsilon_{Y1} = \frac{1}{2}\epsilon_Y$. (a) Determine an expression for the moment that causes first yielding, M_{Y1}. (b) Determine an expression for the moment M_{Y2} that causes the outer fibers to reach the second yield stress, $\sigma_{Y2} = \sigma_Y$. (c) Determine an expression for the fully plastic moment, M_P, the moment that causes all fibers (except an infinitesimally small core near the neutral axis) to reach the second yield stress.

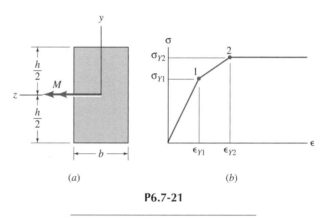

(a)   (b)

P6.7-21

Prob. 6.8-1. If the allowable shear stress (for shear parallel to the grain of the wood) for a 6 in. $\times$ 10 in. timber beam (see Appendix D.8 for actual finish dimensions b and h) is $(\tau_{\text{allow}})_w = 400$ psi, what is the maximum value of transverse shear force, V_{max}, that the beam can sustain, based on this shear-stress criterion?

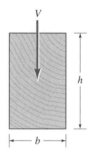

P6.8-1 and P6.8-2

Prob. 6.8-2. A timber beam is to be selected to sustain a maximum transverse shear of $V_{\text{max}} = 60$ kN without exceeding an allowable shear stress of 2 MPa (for shear parallel to the grain of the wood). If wood beams are available with cross sections having $h = 2b$, what is the minimum value of b (to the nearest even mm) that satisfies the shear stress criterion?

Prob. 6.8-3. A timber company wishes to increase its sales of timber by supplying log-cabin kits that can be assembled and used

346

as vacation cabins. As the structural engineer for the timber company, you must (a) determine an expression that relates the maximum shear stress in a log (circular cross section) to the transverse shear force V and the diameter d, and (b) determine the maximum span, L_{max}, of a simply supported log beam that must support a uniformly distributed load w (including beam weight) without exceeding the allowable shear stress τ_{allow}.

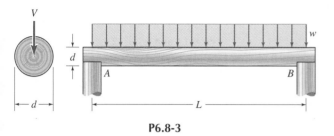

P6.8-3

Prob. 6.8-4. A simply supported beam of length L with rectangular cross section of width b and depth h is required to support a midspan concentrated load P, as shown in Fig. P6.8-4. (a) Determine an expression for P_σ, the maximum allowable load based on an allowable flexural stress σ_{allow}. (b) Determine an expression for P_τ, the maximum allowable load based on an allowable shear stress τ_{allow}. [Neglect the weight of the beam in (a) and (b).]

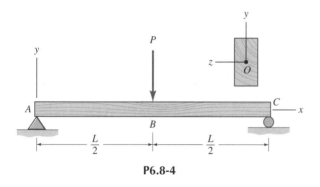

P6.8-4

Prob. 6.8-5. A beam with rectangular cross section supports two equal-magnitude loads $P = 16$ kN, as shown in Fig. P6.8-5.

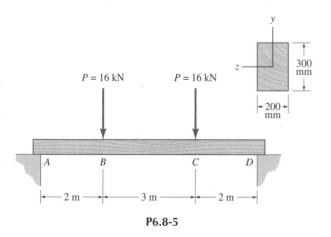

P6.8-5

(a) Determine the maximum flexural stress, σ_{max}, in the beam, and (b) determine the maximum shear stress, τ_{max}, in the beam. (Neglect the weight of the beam.)

Prob. 6.8-6. A simply supported beam AC of length $L = 10$ ft. supports a concentrated load $P = 15$ kips at B, as shown in Fig. P6.8-6. The cross section of the timber beam has nominal dimensions 8 in. $\times$ 8 in. (See Appendix D.8 for the actual cross-sectional dimensions.) Determine the flexural stress $\sigma(x, y)$ and shear stress $\tau(x, y)$ at three levels—$y = 0$ in., $y = 1.25$ in., and $y = 2.50$ in.—on the cross section just to the left of the load P.

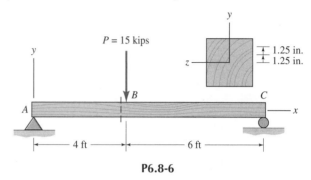

P6.8-6

Prob. 6.8-7. A 6-m-long cantilever beam AC supports a distributed "lift" load, as shown in Fig. P6.8-7. Determine the flexural stress $\sigma(0, y)$ and transverse shear stress $\tau(0, y)$ at three levels— $y = 0$ mm, $y = 30$ mm, and $y = 60$ mm—at the cantilever support, that is, at $x = 0$.

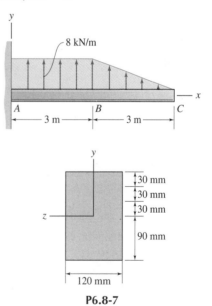

P6.8-7

Prob. 6.8-8. A simply supported timber beam, whose depth is $h = 9.5$ in., supports a concentrated load P at its midspan. The allowable flexural stress for the wood is $\sigma_{allow} = 1.2$ ksi, and the allowable shear stress is $\tau_{allow} = 150$ psi. Determine the length (call it L_{cr}) below which the shear-stress criterion governs the allowable load and above which the flexural stress criterion governs.

347

Prob. 6.8-9. A 40 mm×300 mm (finish dimensions) timber plank is used as a diving board. The diving board is held down at end A by a steel strap that is secured by anchor bolts, and it rests on a roller at section B. Calculate the maximum permissible load, P_{max}, that can be exerted on the beam by a diver if the allowable flexural stress in the wood is $\sigma_{allow} = 9$ MPa and allowable shear stress is $\tau_{allow} = 1$ MPa. Neglect rotational restraint at A, and neglect dynamic effects caused by the diver "springing" on the end of the diving board, that is, consider P to be a static load. (Assume that Eq. 6.65 is valid. Section 6.9 discusses shear in "wide beams" like this one.)

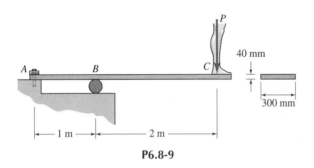

P6.8-9

Prob. 6.8-10. A timber beam, with nominal dimensions 4 in.×8 in., is supported and loaded as shown in Fig. 6.8-10. If the allowable shear stress for the wood is $(\tau_{allow})_w = 120$ psi, and if the load at C is always twice the load at A, that is, $2P$ and P, respectively, calculate the maximum load P that may be applied to this beam. Include the weight of the beam in your calculations, using $\gamma = 36$ lb/ft^3 for the specific weight of the wood. (See Appendix D.8 for the finish dimensions of lumber.)

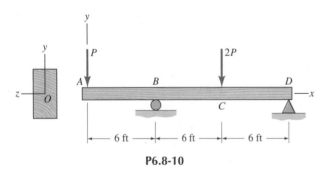

P6.8-10

Prob. 6.8-11. A 4×8 timber beam ($b = 3.5$ in., $d = 7.5$ in.) is strengthened and stiffened by the addition of two 0.25 in.×3.5 in. aluminum cover plates, which are glued to the top and bottom surfaces of the beam, as shown in Fig. P.6.8-11. Let $E_a = 10 \times 10^3$ ksi and $E_w = 2 \times 10^3$ ksi. (Note: This is a nonhomogeneous beam, so you will need to incorporate the bending analysis of Section 6.5 in solving this problem.) (a) If the "sandwich" beam is to support a vertical shear force $V = 1400$ lb, what is the shear stress that must be transferred between the cover plates and the wood core by the layer of glue? (b) What is the maximum shear stress in the wood for the given shear force V?

348

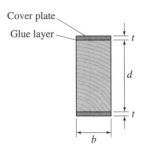

P6.8-11 and P6.8-12

Prob. 6.8-12. Repeat Prob. 6.8-11 for a beam that has a timber core with dimensions $b = 90$ mm, $d = 250$ mm ($E_w = 14$ GPa), to which are glued steel cover plates with dimensions $b = 90$ mm, $t = 6$ mm ($E_s = 210$ GPa). The beam is required to handle a vertical shear force $V = 10$ kN.

*Prob. 6.8-13. Consider the timber beam with single aluminum cover plate as described in Example Problem 6.8. The beam is subjected to a vertical shear force $V = 4$ kN. (a) What is the maximum shear stress in the wood? (b) What is the average shear stress in the glue layer between the timber beam and the cover plate?

Prob. 6.8-14. Using the shear stress formula, Eq. 6.65, determine an expression for the shear stress at the neutral axis level on the circular cross section shown in Fig. P6.8-14. Compare your results with the information given in Fig. 6.43b.

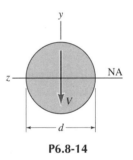

P6.8-14

Prob. 6.8-15. Using the shear stress formula, Eq. 6.65, determine an expression for the shear stress at the neutral axis level on the thick-wall circular tubular cross section shown in Fig. P6.8-15.

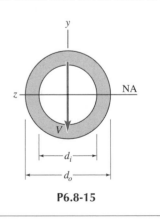

P6.8-15

Prob. 6.10-1. A channel section is subjected to a vertical shear force $V = 2$ kips at the section shown in Fig. P.6.10-1. Determine the values of the horizontal shear stress τ_A at the point designated A, and the vertical shear stress τ_B at the point labeled B. Assume the thickness t_f to be constant.

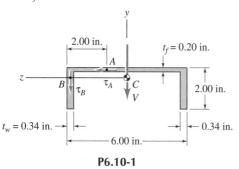

P6.10-1

Prob. 6.10-2. Repeat Prob. 6.10-1 for the channel section in Fig. P6.10-2, which is subjected to a vertical shear $V = 10$ kN.

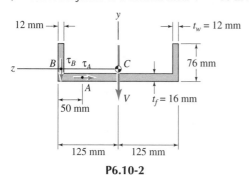

P6.10-2

Prob. 6.10-3. A **W**12×50 wide-flange beam and a **W**16×50 wide-flange beam are each subjected to the same magnitude of vertical shear force, V. Determine the ratio of the maximum shear stresses $(\tau_{max})_a/(\tau_{max})_b$. (See Table D.1 for the section properties of these beams.)

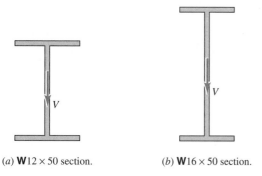

(a) **W**12 × 50 section.   (b) **W**16 × 50 section.

P6.10-3

Prob. 6.10-4. A structural tee beam has the dimensions shown in Fig. P6.10-4. (a) Determine an expression for $\tau_w(y)$, the (vertical) shear stress in the web, and (b) determine the percent of the shear force that is carried by the web (i.e., $\dfrac{V_w}{V} \times 100\%$).

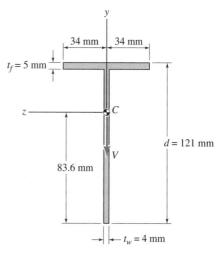

P6.10-4

Prob. 6.10-5. A tee beam has the relative cross-sectional dimensions shown in Fig. P6.10-5, and it is subjected to a vertical shear force V. (a) Determine an expression for $\tau_w(y)$, the (vertical) shear stress in the web, and (b) sketch the function $\tau_w(y)$ obtained in Part (a).

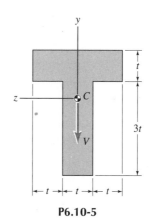

P6.10-5

Prob. 6.10-6. A tubular beam with circular cross section has an outer diameter $d_o = 12.75$ in. and wall thickness $t = 0.375$ in., as indicated in Fig. P6.10-6. If the beam is subjected to a vertical shear force $V = 40$ kips at this section, what is the maximum shear stress on the cross section?

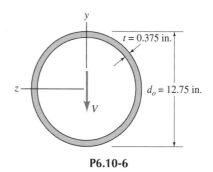

P6.10-6

Prob. 6.10-7. A rectangular box beam has the dimensions shown in Fig. P6.10-7 and is subjected to a vertical shear force $V = 60$ kN. Determine the web shear stresses τ_A and τ_B at the locations A and B indicated in the figure.

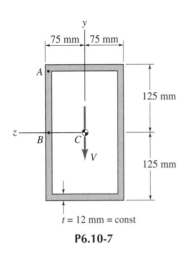

$t = 12$ mm = const

P6.10-7

Prob. 6.10-8. A rectangular box beam has the relative dimensions shown in Fig. P6.10-8. If the beam is to be subjected to a transverse shear force V, what is the ratio $(\tau_{max})_a/(\tau_{max})_b$ for the beam in orientations (a) and (b)? In each case, indicate where τ_{max} occurs.

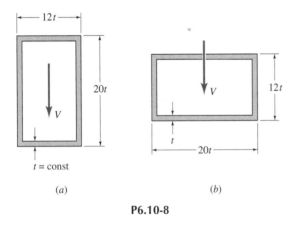

(a)

(b)

P6.10-8

Prob. 6.10-9. A wide-flange beam has the relative dimensions shown in Figs. P6.10-9. If the beam is to be subjected to a transverse shear force V, what is the ratio $(\tau_{max})_a/(\tau_{max})_b$ for the beam in orientations (a) and (b)? In each case, indicate where τ_{max} occurs.

Prob. 6.10-10. The beam with overhang, as shown in Fig. P6.10-10a, supports a uniformly distributed downward load of intensity $w = 2$ kips/ft. If the beam is a **W**12×35 wide-flange beam, determine the web shear stresses τ_C and τ_D at locations C and D in the cross section just to the right of the support at B.

350

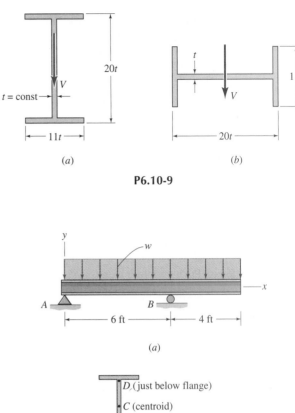

(a)

(b)

P6.10-9

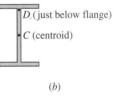

(a)

D (just below flange)

C (centroid)

(b)

P6.10-10 and P6.10-11

Prob. 6.10-11. For the **W**12×35 wide-flange beam loaded as shown in Fig. P6.10-11, let $w = 2$ kips/ft. (a) Determine the maximum flexural stress, σ_{max}, and (b) determine the maximum transverse shear stress, τ_{max}.

Prob. 6.10-12. The beam in Fig. P6.10-12 supports a concentrated downward load $P = 40$ kN at its left end and a uniformly distributed downward load $w = 40$ kN/m over the span BC. If the beam is a **W**310×52 wide-flange beam, (a) determine the maximum flexural stress, σ_{max}, and (b) determine the maximum transverse shear stress, τ_{max}.

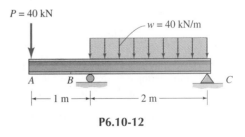

P6.10-12

Prob. 6.10-13. The simply supported tee beam, whose cross-sectional dimensions are given in Fig. P6.10-13b, supports a triangular distributed downward load whose maximum intensity, $w_0 = 1$ kip/ft, occurs at the midspan. (a) Determine the maximum flexural stress, σ_{max}, in this beam. (b) Determine the maximum transverse shear stress, τ_{max}.

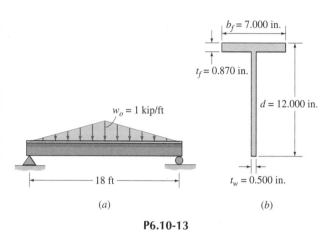

(a)   (b)

P6.10-13

Prob. 6.10-14. The tongue, AB, of a small utility trailer is the square box section shown in Fig. P6.10-14b. Assume that the total weight of the trailer and its load is W, and that its line of action is 1 ft forward of the axle of the trailer, as shown in Fig. P6.10-14a. Furthermore, assume that the tongue is effectively cantilevered from the trailer body at B. Neglect any axial towing loads on the tongue, and consider only the static loads when the trailer hitch at A is resting on a stationary trailer ball. Determine the maximum weight W (trailer plus load) if the allowable flexural stress in the tongue is $\sigma_{allow} = 12$ ksi and the allowable transverse shear stress in the tongue is $\tau_{allow} = 8$ ksi.

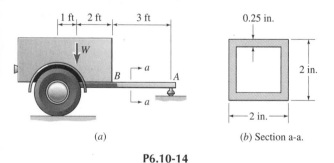

(a)   (b) Section a-a.

P6.10-14

Prob. 6.10-15. Two 2 in.×6 in. (nominal dimensions) boards are nailed and glued together to form a tee beam, as shown in Fig. P6.10-15b. Assume that the nails and glue are sufficient to cause the two planks to function together as a single beam. The allowable stress in horizontal shear (i.e., shear parallel to the grain of the wood) is $\tau_{allow} = 80$ psi. If this beam is to be used as a cantilever to support a triangularly distributed load, as shown in Fig. P6.10-15a, what is the maximum load intensity w_0 that can

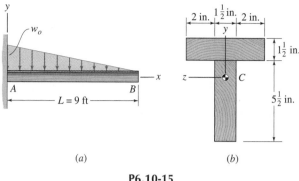

(a)   (b)

P6.10-15

be supported by this beam? In your calculations neglect the weight of the beam.

Prob. 6.10-16. Repeat Prob. 6.10-15, replacing the tee beam with the hollow box beam shown in Fig. P6.10-16 and increasing the beam's length to $L = 12$ ft.

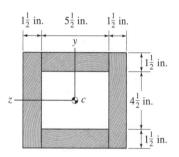

P6.10-16

Prob. 6.10-17. A wood beam is made by nailing and gluing together three planks to form the tee section shown in Fig. P6.10-17b. The beam spans 4 m and supports a concentrated load P at its center. The allowable stress in horizontal shear (i.e., shear parallel to the grain of the wood) is $\tau_{allow} = 600$ kPa. Assume that the planks are securely nailed and glued together so that they function as a single beam. What is the maximum load P that can be placed on this beam based on the horizontal-shear allowable? Neglect the weight of the beam.

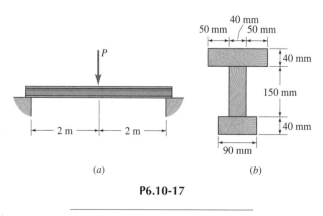

(a)   (b)

P6.10-17

Prob. 6.11-1. A glued-laminated (glulam) timber beam is made from three 2×4 ($1\frac{1}{2} \times 3\frac{1}{2}$ in. finish dimensions) boards, as shown in Fig. P6.11.1. The strength of the wood in horizontal shear is $\tau_{\text{allow}} = 80$ psi (which takes into account a factor of safety). What is the required shear strength of the glue joints (psi), if there is to be a factor of safety against failure of the glue joints of $FS = 3.0$? That is, how strong must the glue joint be so that the joints do not fail before the wood itself does? Neglect the thickness of the glue joint.

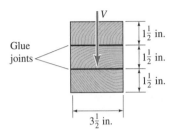

P6.11-1

Prob. 6.11-2. Lumber is available in three sizes for use in making glued-laminated (glulam) beams — 25 mm $\times$ 100 mm, 50 mm $\times$ 100 mm, and 100 mm square (all are finished-lumber dimensions). "Company A" fabricates the 150 mm $\times$ 100 mm beam shown in Fig. P6.11-2a, and "Company B" fabricates beams like the one in Fig. P6.11-2b. Both beams sell for the same price and are made of the same type and grade of lumber. On the basis of shear strength of the beams (i.e., the allowable transverse shear force V), which one would you choose? How much stronger (in shear) is it than its competitor?

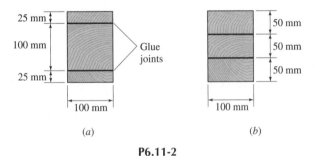

(a)   (b)

P6.11-2

Prob. 6.11-3. A steel plate girder is fabricated by the welding of two 16 in. $\times$ 1 in. flange plates to a 70 in. $\times \frac{3}{8}$ in. web plate using fillet welds whose allowable shear strength is $q_{\text{allow}} = 2$ kips/in. Determine the allowable vertical shear force, V, for this plate girder if the four fillet welds run continuously along the length of the girder.

Prob. 6.11-4. A steel plate girder is fabricated by the welding of two 300 mm $\times$ 25 mm flange plates to a 1 m $\times$ 10 mm web plate using fillet welds whose allowable shear strength is $q_{\text{allow}} = 500$ kN/m. If the four fillet welds run continuously along the length of the girder, what is the allowable vertical shear force, V, for this plate girder?

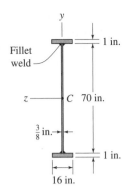

P6.11-3

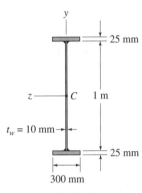

P6.11-4

Prob. 6.11-5. A $\mathbf{W}24 \times 62$ steel wide-flange beam is strengthened and stiffened by the addition of 6 in. $\times \frac{3}{8}$ in. steel plates that are welded to the flanges of the wide-flange beam by continuous fillet welds, as shown in Fig. P6.11-5. (a) Determine the moment of inertia I and the section modulus S for the modified beam. (Neglect the contribution of the weld area to these section properties.) (b) If the allowable shear flow for the weld beads is $q_{\text{allow}} = 2$ kips/in., determine the maximum shear force, V, allowed for this modified section.

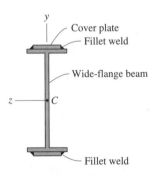

P6.11-5 and P6.11-6

Prob. 6.11-6. Repeat Prob. 6.11-5 for a $\mathbf{W}410 \times 60$ wide-flange beam to which two 152 mm $\times$ 10 mm cover plates are welded with continuous welds whose shear strength is $q_{\text{allow}} = 400$ kN/m.

*Prob. 6.11-7. A $\mathbf{W}18\times60$ steel wide-flange beam is strengthened and stiffened by capping the flanges with $\mathbf{C}10\times20$ steel channels that are bolted to the flanges as shown in Fig. P6.11-7. If each bolt has an allowable (direct) shear strength of 2.0 kips and the bolts are spaced at 12 in. longitudinally, what is the maximum allowable vertical shear force V for this modified-section beam? (Hint: Use Table D.4 to obtain the area and centroidal location for the channel for use in calculating its Q.)

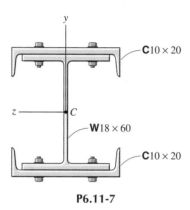

P6.11-7

Prob. 6.11-8. Four 2×6 boards (nominal dimensions) are nailed together to form the box beam shown in Fig. P6.11-8. If the nails are spaced at regular intervals of $\Delta x = 6$ in. along the beam, and if each nail has an allowable force in shear of $V_{nail} = 300$ lb, what is the maximum vertical shear force, V, for this built-up beam. ($d = 5\frac{1}{2}$ in., and $t = 1\frac{1}{2}$ in.)

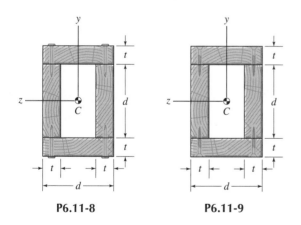

P6.11-8   P6.11-9

Prob. 6.11-9. Four 30 mm $\times$ 180 mm (actual dimensions) boards are attached together by wood screws to form a box beam (Fig. P6.11-9). If each screw has an allowable-shear-force capacity of $V_s = 1$ kN and the beam is to be subjected to a vertical shear force $V = 5$ kN, what is the maximum permissible longitudinal spacing, Δx, of the screws? ($t = 30$ mm and $d = 180$ mm.)

Prob. 6.11-10. A wood box beam is fabricated by placing wood screws at regular intervals, Δx, in the four locations indicated in Fig. P6.11-10. The vertical boards are 2×10's (actual dimensions: $t_1 = 1\frac{1}{2}$ in., $d = 9\frac{1}{4}$ in.) and the horizontal boards are 2×6's

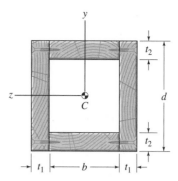

P6.11-10, P6.11-11, and P6.11-13a

(actual dimensions: $t_2 = 1\frac{1}{2}$ in., $b = 5\frac{1}{2}$ in.), and the screws have an allowable shear-force capacity of $V_s = 360$ lb. Determine the maximum spacing, Δx, if the beam is to be designed for a maximum vertical shear force $V = 24$ kips.

Prob. 6.11-11. Repeat Prob. 6.11-10 for a box beam constructed of four boards ($t_1 = 20$ mm, $d = 250$ mm; $t_2 = 40$ mm, $b = 150$ mm) in the configuration shown in Fig. P6.11-11. Let $V_s = 1.2$ kN and $V = 10$ kN, and determine the maximum longitudinal screw spacing, Δx.

Prob. 6.11-12. A box beam is constructed of four 2×6 planks that are nailed together in the configuration shown in Fig. P6.11-12b. The beam supports three equal loads $P = 500$ lb equally spaced on a total span of $L = 16$ ft, as shown in Fig. P6.11-12a. Consider the end conditions at A and E to be "pinned." (a) Determine the maximum horizontal shear stress in the beam. (b) If the nails used to assemble the beam have an allowable shear-force capacity of $V_{nail} = 500$ lb, what is the maximum allowable spacing of the nails in each segment of the beam, that is, Δx_{AB}, Δx_{BC}, and so on?

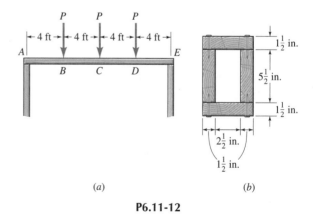

(a)   (b)

P6.11-12

Prob. 6.11-13. A wood box beam is fabricated by attaching four boards together in the configuration shown in Fig. P6.11-13(a) (see Prob. 6.11-10). The dimensions of the boards are: $t_1 = 25$ mm, $d = 250$ mm; $t_2 = 40$ mm, $b = 150$ mm. Each screw has an allowable shear capacity of 1.5 kN. If the box beam is to be

353

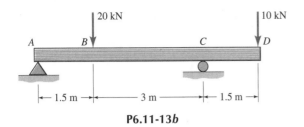

P6.11-13b

loaded and supported as indicated in Fig. P6.11-13(b), calculate the minimum required longitudinal spacing of the screws for each interval along the beam; that is, determine Δx_{AB}, Δx_{BC}, and Δx_{CD}.

Prob. 6.11-14. Two 2×8 boards are attached together by screws that are spaced at regular intervals $\Delta x = 12$ in. along the length of the beam to form the inverted tee beam shown in Fig. P6.11-14b. Each screw has an allowable shear-force capacity of $V_s = 480$ lb. If the beam supports a single concentrated load P at the center and is simply supported, what is the maximum allowable load P?

support a vertical shear force $V = 1400$ lb, with a factor of safety against shear failure of the screws of $FS = 3.0$, what is the minimum required shear capacity of each screw? (Note: This is a nonhomogeneous beam, so you will have to incorporate the bending analysis of Section 6.5 in solving this problem.)

For **Problems 6.12-1 through 6.12-15**, *assume that the thickness* t *is much smaller than other cross-sectional dimensions, so that* t^2, t^3, *and so on may be neglected in comparison with* t *(e.g., in the moment of inertia* I*). Use centerline dimensions for all derivations and calculations.*

Probs. 6.12-1 through 6.12-3. Determine the eccentricity, e, of the shear center, S, of each channel section. The eccentricity is measured, in each problem, from the centerline of the web, which is taken as the y axis for these problems.

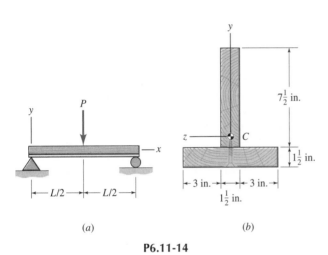

(a)   (b)

P6.11-14

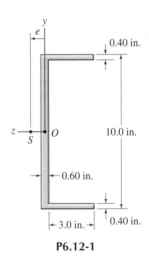

P6.12-1

*Prob. 6.11-15. A 4×8 timber beam is strengthened and stiffened by the attachment of two $\frac{1}{4}$ in. $\times 3\frac{1}{2}$ in. aluminum cover plates, using two rows of screws spaced at a regular interval $\Delta x = 6$ in. along the length of the beam. If the "sandwich" beam is to

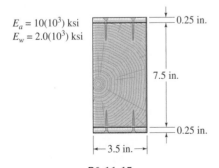

P6.11-15

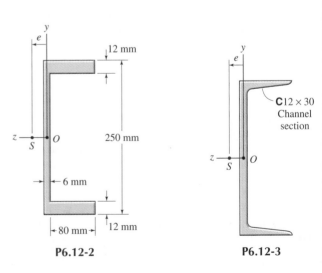

P6.12-2   P6.12-3

Prob. 6.12-4. The "hat" section in Fig. P6.12-4 has a constant wall thickness t. (a) Determine expressions for the shear flows $q_1(y)$ and $q_2(z)$. Express your answers in terms of the total shear force V and the dimensions of the cross section. (b) Determine expressions for V_1 and H_2, the resultants of $q_1(y)$ and $q_2(z)$, respectively. (c) Determine an expression for the eccentricity, e, of the shear center, S.

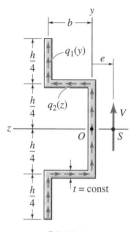

P6.12-4

Prob. 6.12-5. (a) Locate the shear center S of the "hat" section in Fig. P6.12-5 by determining the eccentricity, e. (b) If a vertical shear force $V = 10$ kips acts through the shear center of this hat section, what are the values of the shear stresses τ_A and τ_B at the locations and directions indicated in Fig. P6.12-5?

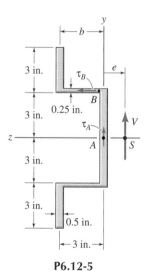

P6.12-5

Prob. 6.12-6. Consider the cross section shown in Fig. P6.12-6. (a) Determine expressions for the flange shear flows $q_1(z)$ and $q_2(z)$, as indicated on the figure. (b) Determine expressions for the resultant shear forces H_1 and H_2 corresponding to the shear flows q_1 and q_2. (c) Determine an expression for the eccentricity, e, of the shear center, S.

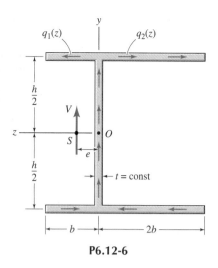

P6.12-6

Prob. 6.12-7. Determine the shear-center location (i.e., the eccentricity e) for the open-box section shown in Fig. P6.12-7.

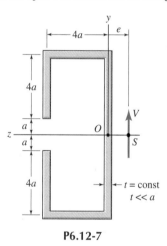

P6.12-7

Prob. 6.12-8. Determine the eccentricity, e, of the shear center, S, for the thin-wall box beam having a longitudinal slit along one side, as shown in Fig. P6.12-8.

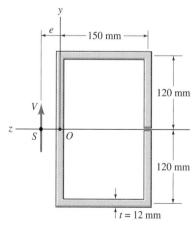

P6.12-8

355

*Prob. 6.12-9. A channel beam is to be fabricated from a web plate and two flange plates as shown in Fig. P6.12-9. The total depth is to be $d = 18.0$ in., the web thickness $t_w = 0.50$ in., and the flange width $b_f = 6.00$ in. If, for structural reasons, it is necessary for the shear center to lie 0.90 in. from the web plate, what is the required thickness, t_f, of the flange plates?

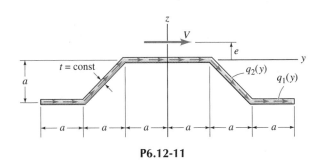

P6.12-11

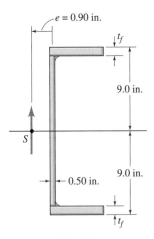

P6.12-9

Prob. 6.12-10. Consider the constant-thickness ($t \ll a$) channel section shown in Fig. P6.12-10. (a) Determine an expression for the shear flow $q_f(s)$ in the sloping flange, and (b) locate the shear center by determining an expression for the eccentricity, e. Express your answer in terms of the dimension a.

*Prob. 6.12-12. Vertical extension plates with cross-sectional dimensions 0.400 in. $\times$ 2.00 in. are welded to the flanges of a **W**8 $\times$ 21 wide-flange beam to form the cross section illustrated in Fig. P6.12-12a. (a) Determine an expression for the shear flows $q_1(z)$ and $q_2(z)$ in the two sections of flange, and determine an expression $q_3(y)$ for the shear flow in the vertical flange extensions, as indicated in Fig. P6.12-12b. (b) Determine the eccentricity, e, of the shear center of this modified wide-flange beam.

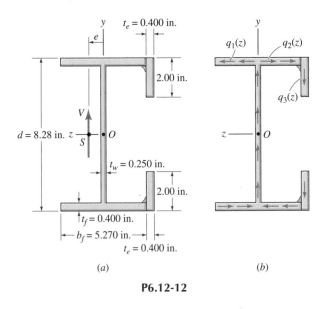

P6.12-12

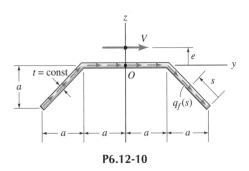

P6.12-10

Prob. 6.12-11. Consider the modified hat section shown in Fig. P6.12-11. (a) Determine an expression for the shear flows $q_1(y)$ and $q_2(y)$, and (b) determine an expression relating the shear-center eccentricity, e, to the dimension a.

*Prob. 6.12-13. A pipe-beam has a longitudinal slit along one side that makes it an open, thin-wall beam. The thickness t is much less than the mean radius r. (a) Determine an expression for the shear flow $q(\theta)$. (b) Determine the location of the shear center, S, by determining its eccentricity, e, from the center of the circular cross section.

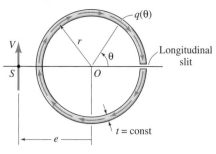

P6.12-13

*Prob. 6.12-14.** A segment of a thin-wall pipe, having an excluded angle 2α, acts as a beam, with loading parallel to the xy plane and passing through the shear center, S. (a) Determine an expression for the shear flow $q(\theta)$. (b) Determine the location of the shear center, S, by determining its eccentricity, e, from the center of the circular cross section.

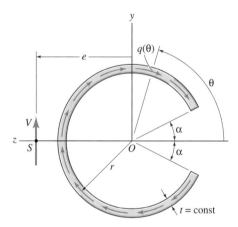

P6.12-14

Prob. 6.12-15. The special thin-wall channel section shown in Fig. P6.12-15 is formed by bending a 6-mm-thick steel plate.

(a) Determine an expression for $q_f(s)$, the shear flow in the flanges. (b) Determine the location of the shear center, S.

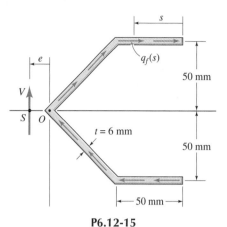

P6.12-15

Prob. 6.12-16. An equal-leg, thin-wall angle section is subjected to a transverse shear force V at an angle of 45° to the legs of the angle, as shown in Fig. P6.12-16. (a) Determine an expression for $q(s)$, the shear flow in the legs of the angle. (b) By integration, show that the resultant of this shear flow is the shear force V passing through the point S shown in the figure. (c) Determine an expression for the maximum shear stress.

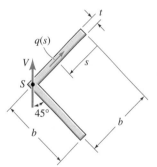

P6.12-16

7 DEFLECTION OF BEAMS

7.1 INTRODUCTION

Couples and transverse forces applied to beams cause them to deflect laterally, as illustrated in Fig. 6.1 and in Fig. 7.1. In Chapter 6 we determined a relationship between the curvature of the deflection curve of a beam and the bending moment at a cross section. In this chapter we will relate the deflection and slope of beams to their loading and support. As indicated in Fig. 7.1*a*, the *deflection curve* is characterized by a function $v(x)$ that gives the *transverse displacement* (i.e., displacement in the y direction) of the points that lie along the axis of the beam. The *slope* of the deflection curve is labeled $\theta(x)$.

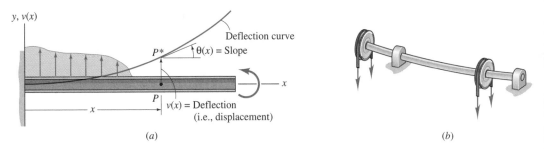

FIGURE 7.1 Examples of the deflection of beams and shafts (deflection is exaggerated).

FIGURE 7.2 Points of maximum deflection.

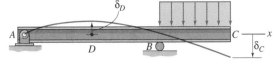

There are several reasons for considering the deflection of beams. For example, we may need to know the maximum deflection of a given beam under a given set of loads. As illustrated in Fig. 7.2, the maximum deflection might occur at an unsupported end of the beam, (δ_C), or it might occur at an interior section where the slope vanishes, (δ_D). For example, the beams of the equipment trailer in Fig. 7.3*a* must not deflect so much under load that the clearance between the trailer and the ground becomes unacceptably small. Also, the beams that support a bridge deck should be designed to have just the right initial upward camber (deflection) (Fig. 7.3*b*) so that they become straight when they are loaded (Fig. 7.3*c*).

Statically indeterminate beams provide another important reason for studying beam deflection. If a beam is statically indeterminate, its reactions and its internal stresses cannot be determined without considering deflections. Finally, we will need to relate the *transverse* deflection $v(x)$ to *axial* loads when we study the buckling of columns (Chapter 10).

Starting with the moment-curvature equation that was derived in Chapter 6, Eq. 6.12, we will develop *differential equations* that relate the deflection $v(x)$ to the bending moment $M(x)$, the transverse shear force $V(x)$, and the transverse load $p(x)$. Then we will discuss the *boundary conditions* and *continuity conditions* that must accompany these differential equations. Next, we will solve for the deflection and slope of statically determinate and statically indeterminate beams by *integration of the differential equations,* incorporating the given loads and the given boundary and continuity conditions (Section 7.3 and 7.4). *Discontinuity functions* are introduced in Section 7.5 to facilitate the integration of the differential equations for beams having several loads and/or supports. Finally, we will use *superposition of known solutions* to solve beam deflection problems, first using a *force-method* approach (Section 7.6) and then a *displacement-method* approach (Section 7.7).

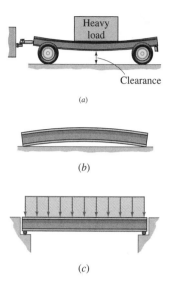

Clearance

(a)

(b)

(c)

FIGURE 7.3 Illustrations of beam deflection.

7.2 DIFFERENTIAL EQUATIONS OF THE DEFLECTION CURVE

Moment-Curvature Equation. The moment-curvature equation that we derived in Section 6.3, Eq. 6.12, is the starting point for an analysis of the deflection of linearly elastic beams.

$$M(x) = \frac{E(x)I(x)}{\rho(x)} \tag{7.1}$$

where $\rho(x)$ is the radius of curvature of the deflection curve at section x.[1] The sign conventions for $M(x)$, $v(x)$, and $\rho(x)$ are reviewed in Fig. 7.4. Since the moment-curvature equation relates a force-type quantity, $M(x)$, to a displacement-type quantity, $\rho(x)$, it is very important for these sign conventions to be observed at all times.

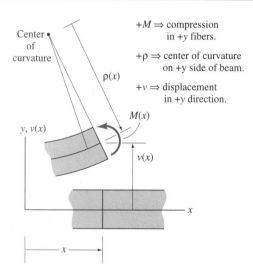

Center of curvature

$\rho(x)$

$M(x)$

$y, v(x)$

$v(x)$

x

x

$+M \Rightarrow$ compression in $+y$ fibers.

$+\rho \Rightarrow$ center of curvature on $+y$ side of beam.

$+v \Rightarrow$ displacement in $+y$ direction.

FIGURE 7.4 The sign convention for beam deflection analysis.

[1]For nonhomogeneous, linearly elastic beams, Eq. 6.22 or Eq. 6.27 must be used.

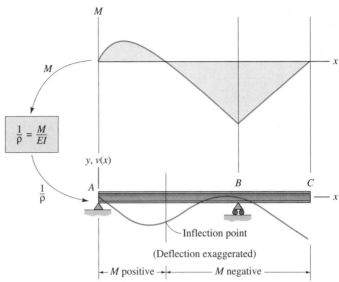

FIGURE 7.5 Use of the moment-curvature equation in sketching the deflection curve.

Before actually calculating the slope and deflection of a beam, it is helpful to sketch the *anticipated deflection curve* of the beam. This can be done with the help of Eq. 7.1 and a moment diagram. In fact, we did this in Chapter 5 as a means of checking moment diagrams. Figure 7.5 is taken from Example 5.6. First, there can be no deflection of the beam at supports A and B, but the slope of the beam is unrestricted at the supports and everywhere else along the beam. At any section where the moment is negative, the center of the curvature is on the $-y$ side of the beam, that is, the beam is concave downward. There is an *inflection point* where the moment changes from positive to negative.

The inclination angle, $\theta(x)$, is related to the derivative of the deflection by

$$\theta = \arctan\left(\frac{dv}{dx}\right) \qquad (7.2)$$

If the slope is small (i.e., $\dfrac{dv}{dx} \ll 1$), then the approximation

$$\theta \doteq \frac{dv}{dx} \qquad (7.3)$$

may be treated as an equality. Deflection $v(x)$, slope $\theta(x)$, and radius of curvature $\rho(x)$ are illustrated in Fig. 7.6. From calculus, the radius of curvature is related to the derivatives of the deflection by

$$\frac{1}{\rho} = \frac{\dfrac{d^2v}{dx^2}}{\left[1 + \left(\dfrac{dv}{dx}\right)^2\right]^{3/2}} \qquad (7.4)$$

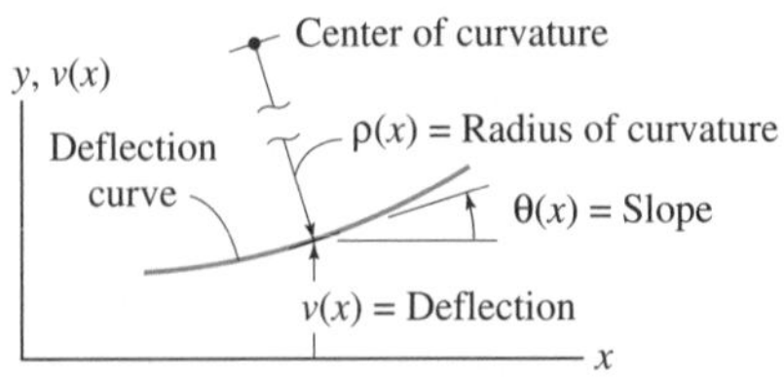

FIGURE 7.6 Geometric relationships of the deflection curve.

Again, where the slope is small compared with unity, the approximation

$$\frac{1}{\rho} \doteq \frac{d^2v}{dx^2} \tag{7.5}$$

can be treated as an equality. Then the moment-curvature equation (Eq. 7.1) becomes

$$\frac{d^2v(x)}{dx^2} = \frac{M(x)}{E(x)I(x)} \tag{7.6}$$

To simplify the equations in this chapter, we will adopt the notation of using primes to denote differentiation with respect to x. Thus,

$$v' \equiv \frac{dv}{dx}, \; v'' \equiv \frac{d^2v}{dx^2}, \; M' \equiv \frac{dM}{dx}, \text{ etc.} \tag{7.7}$$

With the above prime notation, Eq. 7.6 may be written as

$$\boxed{EIv'' = M} \qquad \text{\textbf{Moment-curvature equation}} \tag{7.8}$$

Although the curvature, $\kappa \equiv 1/\rho$, does not appear explicitly in it, we will refer to this equation as the *moment-curvature equation*. It is a *second-order, ordinary differential equation* that we can integrate to determine the slope and deflection, given an expression for $M(x)$ and appropriate boundary conditions and continuity conditions.

Load-Deflection Equation.
Equations relating load, transverse shear, and bending moment were derived in Chapter 5. Equations 5.2 and 5.1, repeated here, are

$$M'(x) = V(x) \tag{7.9}$$

$$V'(x) = p(x) \tag{7.10}$$

The sign conventions for p, V, and M are reviewed in Fig. 7.7.

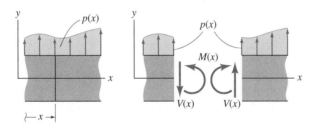

FIGURE 7.7 The sign conventions for positive load, shear, and moment.

Combining Eqs. 7.8, 7.9, and 7.10 we get

$$\boxed{(EIv'')' = V} \qquad \text{\textbf{Shear-deflection equation}} \tag{7.11}$$

and

$$\boxed{(EIv'')'' = p} \qquad \text{\textbf{Load-deflection equation}} \tag{7.12}$$

The load-deflection equation, which is a *fourth-order, ordinary differential equation,* is especially useful.

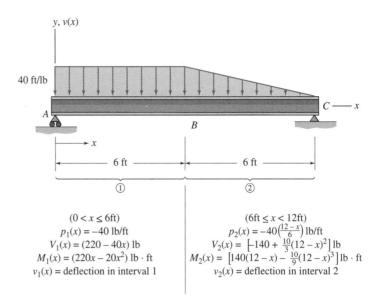

FIGURE 7.8 A two-interval beam.

$(0 < x \le 6\text{ft})$	$(6\text{ft} \le x < 12\text{ft})$
$p_1(x) = -40$ lb/ft	$p_2(x) = -40\left(\frac{12-x}{6}\right)$ lb/ft
$V_1(x) = (220 - 40x)$ lb	$V_2(x) = \left[-140 + \frac{10}{3}(12-x)^2\right]$ lb
$M_1(x) = (220x - 20x^2)$ lb·ft	$M_2(x) = \left[140(12-x) - \frac{10}{9}(12-x)^3\right]$ lb·ft
$v_1(x)$ = deflection in interval 1	$v_2(x)$ = deflection in interval 2

If the flexural rigidity, EI, is constant, then Eqs. 7.11 and 7.12 become the following third-order and fourth-order differential equations, respectively.

$$EIv''' = V \tag{7.13}$$

$$EIv'''' = p \tag{7.14}$$

We will refer to Eq. 7.11 (or 7.13) as the *shear-deflection equation*, and to Eq. 7.12 (or 7.14) as the *load-deflection equation*.

A beam that requires more than one expression for bending moment because of the presence of concentrated loads, discontinuities in distributed loads, or interior supports will be referred to as a *multi-interval beam*. Consider the two-interval beam in Fig. 7.8 (repeated from Example Problem 5.7). Each interval has its own differential equations, and each interval will have its own expression for the deflection curve. As indicated in Fig. 7.8, we will use numerical subscripts to indicate the range of validity of each variable.[2]

Boundary Conditions and Continuity Conditions. As noted in Table 7.1, when the above differential equations (Eqs. 7.8, 7.11, and 7.12) are integrated, there will

TABLE 7.1   Number of Constants of Integration to be Evaluated

Type	Differential Equation	Order of DE	Number of BC/CC per Interval
(1)	$EIv'' = M$	2	2
(2)	$(EIv'')' = V$	3	3
(3)	$(EIv'')'' = p$	4	4

---

[2]Some texts employ shifting of the origin and/or reflection of the coordinate reference frame. A shift of origin necessitates the use of more than one independent variable (e.g., x_1, x_2). In the case of a reflected x axis (i.e., right-to-left instead of left-to-right), ''tricky'' changes of signs are required. To simplify matters, in this text the same xy reference frame is used for the entire beam; so no change of sign convention is required, and no subscript is needed on the independent variable x. (See Fig. 7.8 and Example Problem 7.3.)

TABLE 7.2a Boundary Conditions and Continuity Conditions for the Second-Order Deflection Equation

	Type	Symbol	BC/CC		Type	Symbol	BC/CC
	$EIv'' = M$ (second-order)				**$EIv'' = M$ (second-order)**		
BC	Fixed end		$v = 0,$ $v' = 0$	**CC**	Pin		$v_1 = v_2$
	Simple support		$v = 0$		Concentrated force		$v_1 = v_2,$ $v'_1 = v'_2$
	Free end		No BC		Concentrated couple		$v_1 = v_2,$ $v'_1 = v'_2$
CC	Roller		$v_1 = v_2 = 0,$ $v'_1 = v'_2$		Discontinuity in load function		$v_1 = v_2,$ $v'_1 = v'_2$

TABLE 7.2b Boundary Conditions and Continuity Conditions for the Fourth-Order Deflection Equation

	Type	Symbol	BC/CC		Type	Symbol	BC/CC
	$(EIv'')'' = p$ (fourth-order)				**$(EIv'')'' = p$ (fourth-order)**		
BC	Fixed end		$v = 0,$ $v' = 0$	**CC**	Roller		$v_1 = v_2 = 0,$ $v'_1 = v'_2,$ $M_1 = M_2$
	Simple support		$v = 0,$ $M = 0$		Pin, with force		$v_1 = v_2,$ $V_2 - V_1 = P_0,$ $M_1 = M_2 = 0$
	Free end		$V = 0,$ $M = 0$		Concentrated force		$v_1 = v_2,\ v'_1 = v'_2,$ $V_2 - V_1 = P_0,$ $M_1 = M_2$
	Concentrated force		$V = P_0,$ $M = 0$		Concentrated couple		$v_1 = v_2,\ v'_1 = v'_2,$ $V_1 = V_2,$ $M_2 - M_1 = -M_0$
	Concentrated couple		$V = 0,$ $M = -M_0$		Discontinuity in load function		$v_1 = v_2,$ $v'_1 = v'_2,$ $V_1 = V_2,$ $M_1 = M_2$

be *constants of integration* that must be evaluated by use of *boundary conditions (BC) and continuity conditions (CC)*; these are listed in Table 7.2.

We will use only the second-order equation and the fourth-order equation, so let us consider the specific boundary and continuity conditions that apply to each. Recall from your study of differential equations that the constants of integration of a certain order ODE can only be evaluated using known values, or relationships of values, of the lower derivatives, including the function itself. Hence, only values of v and v' can be used in boundary conditions and continuity conditions for evaluating constants of integration for a second-order equation. Values of v, v', M, and V are used to evaluate constants of integration when the load-deflection equation (a fourth-order ODE) is used. Table 7.2 lists the various boundary conditions and support conditions that may arise.

In the next two sections we integrate differential equations to solve for the deflection and slope of statically determinate beams (Section 7.3) and of statically indeterminate beams (Section 7.4). We will use the same basic procedure to solve all beam-deflection problems, with only a minor modification required in the treatment of statically indeterminate problems.

7.3 SLOPE AND DEFLECTION BY INTEGRATION—STATICALLY DETERMINATE BEAMS

Let us now illustrate the solution of slope and deflection problems by integration of either the second-order differential equation, Eq. 7.8, or the fourth-order differential equation, Eq. 7.12. Since all solutions follow a straightforward procedure, we will outline that procedure here and will dispense with the *Plan the Solution* discussion of each problem.

The solution of each problem involves application of the three fundamentals of deformable-body mechanics—*Equilibrium, Force-Deformation Behavior (Material Behavior), and Geometry of Deformation.* The following solution procedure is recommended:

PROCEDURES FOR DETERMINING SLOPE AND DEFLECTION OF STATICALLY DETERMINATE BEAMS BY INTEGRATION

1. Sketch the anticipated **deflection curve**. In some cases this can be done by reference to the original load diagram. In other situations, a moment diagram is very helpful (e.g., Fig. 7.5).

2. *Second-order method*—Sketch free-body diagrams and develop a **moment function** $M_i(x)$ for each interval of the beam. *Fourth-order method*—From the load diagram, develop a **load function** $p_i(x)$ for each interval of the beam. (This includes $p_i(x) = 0$ for intervals that have no distributed loading.)

   In Step 2 the *sign conventions* in Fig. 7.7 must be rigorously adhered to.

3. *Second-order method*—Using Eq. 7.8 and the expressions for $M_i(x)$ developed in Step 2, write down a **moment-curvature equation** for each interval.

$$\boxed{(EIv'')_i = M_i(x)} \qquad \begin{matrix}(7.8)\\ \text{repeated}\end{matrix}$$

Fourth-order method—Using Eq. 7.12 and the expressions for $p_i(x)$ developed in Step 2, write down a **load-deflection equation** for each interval.

$$\boxed{(EIv'')_i'' = p_i(x)} \qquad \begin{matrix}(7.12)\\ \text{repeated}\end{matrix}$$

4. **Integrate the differential equations** developed in Step 3. Include appropriate constants of integration.

5. Write down the appropriate **boundary conditions** and **continuity conditions** using Table 7.2a or 7.2b as a guide.

Second-order method—Use Table 7.2a as a guide in setting up boundary conditions and continuity conditions. Note that all boundary conditions involve only the slope, v', or the deflection, v, since Eq. 7.8 is a second-order differential equation.

Fourth-order method—Use Table 7.2b as a guide. In addition to conditions on v' and v, boundary and continuity conditions on the shear, V, and the bending moment, M, may also apply.

6. Use the boundary and continuity conditions of Step 5 to **evaluate the constants of integration** introduced in Step 4.

   The first integral of Eq. 7.12 leads to an equation of the form

$$(EIv'')_i' = \int p_i(x)dx + C_1$$

Since $(EIv'')' = V$ (Eq. 7.11), boundary and continuity conditions on the shear, V, are appropriate. A second integration leads to an equation of the form

$$(EIv'')_i = \int \left[\int p_i(x)dx + C_1 \right] dx + C_2$$

From Eq. 7.8, EIv'' is just the bending moment, M, so boundary and continuity conditions on M apply. Table 7.2b lists these conditions on V and M.

7. Complete the solution by evaluating the slope and deflection at points where they are required, by evaluating deflection maxima, etc.

Example Problem 7.1 is solved by both the second-order method and the fourth-order method. Others are solved by only one of the two methods.

■■■■■■■■■■■■■■□ E X A M P L E 7 . 1 □□□□□□□□□□□□□□■■

A uniform cantilever beam is subjected to a transverse load P_B and a couple M_B at its "free" end, as shown in Fig. 1. Using the second-order method and the fourth-order method: (a) Determine expressions for the slope, $v'(x)$, and the deflection, $v(x)$. (b) Determine expressions for the deflection and slope at B.

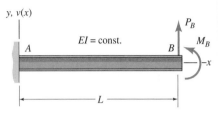

Fig. 1

Solution A (2nd-Order Method)
(A.a) Determine formulas for slope and deflection.

Sketch the Deflection Curve: Both loads in Fig. 1 will bend the beam upward. Therefore, the deflection should resemble the deflection curve in Fig. 2.

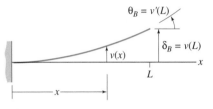

Fig. 2 The anticipated deflection curve.

Determine M(x): The free-body diagram in Fig. 3 can be used to determine an expression for M(x).

$$+\circlearrowleft \left(\sum M\right)_a = 0: \quad M(x) - P_B(L - x) - M_B = 0 \tag{1}$$

Write the Moment-Curvature Equation: From Eqs. 7.8 and (1),

$$EIv'' = M(x) = M_B + P_B(L - x) \tag{2}$$

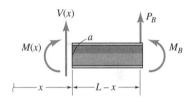

Fig. 3 A free-body diagram.

Integrate the Differential Equation: We need to integrate Eq. (2) twice.

$$EIv' = M_B x + P_B L x - P_B\left(\frac{x^2}{2}\right) + C_1 \tag{3a}$$

$$EIv = M_B\left(\frac{x^2}{2}\right) + P_B L\left(\frac{x^2}{2}\right) - P_B\left(\frac{x^3}{6}\right) + C_1 x + C_2 \tag{3b}$$

Identify the Boundary Conditions: End A is "fixed." That is, slope and deflection must be zero at $x = 0$. So,

$$v'(0) = v(0) = 0 \tag{4a,b}$$

Evaluate the Constants of Integration: From Eqs. (3a) and (4a),

$$EIv'|_{x=0} = C_1 = 0 \tag{5a}$$

and, from Eqs. (3b), (4b), and (5a),

$$EIv|_{x=0} = C_2 = 0 \tag{5b}$$

Therefore,

$$v'(x) = \frac{1}{EI}\left[M_B x + P_B L x - P_B\left(\frac{x^2}{2}\right) \right]$$

$$v(x) = \frac{1}{EI}\left[M_B\left(\frac{x^2}{2}\right) + P_B L\left(\frac{x^2}{2}\right) - P_B\left(\frac{x^3}{6}\right) \right]$$

$$\left.\right\} \quad \textbf{Ans. (A.a)} \quad (6)$$

(A.b) Determine the tip deflection and slope. We can now evaluate the tip deflection and slope using Eqs. (6).

$$\delta_B \equiv v(L) = \frac{1}{EI}\left[M_B\left(\frac{L^2}{2}\right) + P_B\left(\frac{L^3}{3}\right) \right]$$

$$\theta_B \equiv v'(L) = \frac{1}{EI}\left[M_B(L) + P_B\left(\frac{L^2}{2}\right) \right]$$

$$\left.\right\} \quad \textbf{Ans. (A.b)} \quad (7)$$

For example, due to P_B alone, we get

$$\delta_B = \frac{P_B L^3}{3EI}, \qquad \theta_B = \frac{P_B L^2}{2EI}$$

These results, together with the results that are obtained when M_B is applied alone, are listed in Table E.1 of Appendix E.

Solution B (4th-Order Method)

Sketch the Deflection Curve: See Fig. 2.

Determine p(x): From the load diagram, Fig. 1,

$$p(x) = 0 \tag{8}$$

Write the Load-Deflection Equation: From Eqs. 7.12 and (8),

$$(EIv'')'' = 0 \tag{9}$$

Integrate the Differential Equation: We need to integrate Eq. (9) four times. We can also make use of Eqs. 7.11 and 7.8.

$$(EIv'')' = V(x) = D_1 \tag{10a}$$

$$EIv'' = M(x) = D_1 x + D_2 \tag{10b}$$

$$EIv' = D_1\left(\frac{x^2}{2}\right) + D_2 x + D_3 \tag{10c}$$

$$EIv = D_1\left(\frac{x^3}{6}\right) + D_2\left(\frac{x^2}{2}\right) + D_3 x + D_4 \tag{10d}$$

Identify the Boundary Conditions: We can refer to Table 7.2b. We know the displacement and slope at $x = 0$ and the moment and shear at $x = L$. Therefore, observing all sign conventions, we get

$$V(L) = -P_B, \qquad M(L) = M_B \qquad \text{(11a,b)}$$

$$v'(0) = 0, \qquad v(0) = 0 \qquad \text{(11c,d)}$$

Evaluate the Constants of Integration: Combining Eqs. (10) and (11), we get

$$D_1 = -P_B \qquad \text{(12a)}$$

$$D_1 L + D_2 = M_B \rightarrow D_2 = M_B + P_B L \qquad \text{(12b)}$$

$$D_3 = 0, \qquad D_4 = 0 \qquad \text{(12c,d)}$$

Finally, combining Eqs. (10) and (12), we get

$$\left.\begin{aligned}
v'(x) &= \frac{1}{EI}\left[M_B x + P_B L x - P_B\left(\frac{x^2}{2}\right) \right] \\[2mm]
v(x) &= \frac{1}{EI}\left[M_B\left(\frac{x^2}{2}\right) + P_B L\left(\frac{x^2}{2}\right) - P_B\left(\frac{x^3}{6}\right) \right]
\end{aligned}\right\} \qquad \text{Ans. (B.a)}$$

as we did in the second-order solutions, Eq. (6).

Review the Solution The same results were obtained by two methods, so they are probably correct. A check of the dimensionality of each term in the slope and deflection expressions shows that each term has the proper dimensions.

The results that were obtained in Example Problem 7.1 can be quite useful in solving beam deflection problems by superposition (Sections 7.6 and 7.7). For the cantilever beam of Fig. 7.9, Eqs. (7) of Example Problem 7.1 may be written in the form:

$$\begin{aligned}
\delta &= \left(\frac{L^3}{3EI}\right)P + \left(\frac{L^2}{2EI}\right)M \\[2mm]
\theta &= \left(\frac{L^2}{2EI}\right)P + \left(\frac{L}{EI}\right)M
\end{aligned} \qquad \text{(7.15)}$$

On the other hand, we can solve these equations for P and M in terms of δ and θ and get

$$\begin{aligned}
P &= \left(\frac{12EI}{L^3}\right)\delta - \left(\frac{6EI}{L^2}\right)\theta \\[2mm]
M &= -\left(\frac{6EI}{L^2}\right)\delta + \left(\frac{4EI}{L}\right)\theta
\end{aligned} \qquad \text{(7.16)}$$

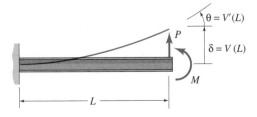

FIGURE 7.9 Load-deflection relationships for a uniform cantilever beam.

367

The next example problem illustrates the usefulness of the fourth-order method when a beam is subjected to a distributed loading that is not simply constant or linear.

■■■■■■■■■■■■■■■□□ E X A M P L E 7 . 2 □□□□□□□□□□□□□■■■■

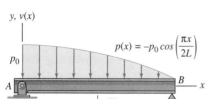

Fig. 1 A simply supported beam with cosine load.

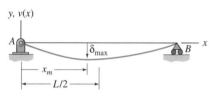

Fig. 2 The expected deflection curve.

Use the fourth-order method to analyze the uniform, simply supported beam in Fig. 1. (a) Determine expressions for the slope and deflection of the beam. (b) Determine the maximum deflection of the beam.

Solution

(a) Determine formulas for the slope and deflection of the beam.

Sketch the Deflection Curve: The beam will obviously deflect downward over its entire length, as sketched in Fig. 2. The maximum deflection should occur to the left of the center of the beam.

Determine p(x): The load distribution, $p(x)$, is given in Fig. 1. There is a negative sign because the load acts in the $-y$ direction.

$$p(x) = -p_0 \cos\left(\frac{\pi x}{2L}\right) \tag{1}$$

Write the Load-Deflection Equation: From Eqs. 7.12 and (1),

$$(EIv'')'' = p(x) = -p_0 \cos\left(\frac{\pi x}{2L}\right) \tag{2}$$

Integrate the Differential Equation: We need to integrate Eq. (2) four times, making use of Eqs. 7.11 and 7.8 and introducing constants of integration.

$$V(x) = (EIv'')' = -p_0\left(\frac{2L}{\pi}\right)\sin\left(\frac{\pi x}{2L}\right) + C_1 \tag{3a}$$

$$M(x) = EIv'' = p_0\left(\frac{2L}{\pi}\right)^2 \cos\left(\frac{\pi x}{2L}\right) + C_1 x + C_2 \tag{3b}$$

$$EIv' = p_0\left(\frac{2L}{\pi}\right)^3 \sin\left(\frac{\pi x}{2L}\right) + C_1\left(\frac{x^2}{2}\right) + C_2 x + C_3 \tag{3c}$$

$$EIv = -p_0\left(\frac{2L}{\pi}\right)^4 \cos\left(\frac{\pi x}{2L}\right) + C_1\left(\frac{x^3}{6}\right) + C_2\left(\frac{x^2}{2}\right) + C_3 x + C_4 \tag{3d}$$

Identify the Boundary Conditions: The moment vanishes at both ends, and the displacement is also zero at both ends (see Table 7.2b). Therefore,

$$M(0) = M(L) = 0 \tag{4a,b}$$

$$v(0) = v(L) = 0 \tag{4c,d}$$

Evaluate the Constants of Integration: Combining Eqs. (3) and (4) we get (the source equations are cited in square brackets at the left margin):

[4a, 3b] $$M(0) = p_0\left(\frac{2L}{\pi}\right)^2 + C_2 = 0$$

so

$$C_2 = -p_0\left(\frac{2L}{\pi}\right)^2 \tag{5a}$$

[4b, 3b]

$$M(L) = C_1L + C_2 = 0$$

so

$$C_1 = \frac{p_0}{L}\left(\frac{2L}{\pi}\right)^2 \tag{5b}$$

[4c, 3d]

$$EIv(0) = -p_0\left(\frac{2L}{\pi}\right)^4 + C_4 = 0$$

so

$$C_4 = p_0\left(\frac{2L}{\pi}\right)^4 \tag{5c}$$

[4d, 3d]

$$EIv(L) = C_1\left(\frac{L^3}{6}\right) + C_2\left(\frac{L^2}{2}\right) + C_3L + C_4 = 0$$

$$C_3 = \frac{p_0 L^3}{3\pi^4}(4\pi^2 - 48) \tag{5d}$$

Finally, from Eqs. (3) and (5),

$$v' = \frac{p_0 L^3}{\pi^4 EI}\left[8\pi \sin\left(\frac{\pi x}{2L}\right) + 2\pi^2\left(\frac{x}{L}\right)^2 - 4\pi^2\left(\frac{x}{L}\right) + \frac{4}{3}\pi^2 - 16 \right] \tag{6a}$$

$$v = \frac{p_0 L^4}{\pi^4 EI}\left(-16\cos\left(\frac{\pi x}{2L}\right) + \frac{2\pi^2}{3}\left(\frac{x}{L}\right)^3 - 2\pi^2\left(\frac{x}{L}\right)^2 \right.$$

$$\left. + \left(\frac{4\pi^2}{3} - 16\right)\left(\frac{x}{L}\right) + 16 \right] \tag{6b}$$

Ans. (a)

(b) Determine the maximum deflection of the beam. The maximum deflection occurs when the slope is zero. Figure 2 suggests that this should occur in the interval $0 < \frac{x_m}{L} < \frac{1}{2}$. This means that we must set the expression in square brackets in Eq. (6a) equal to zero and find a root x_m for which $v'(x_m) = 0$. We can do this by trial and error or by using some root-finder, and we get

$$x_m = 0.485L$$

Using Eq. (6b) to evaluate $v(x_m)$, we get

$$\delta_{\max} = |v(0.485L)| = 8.70(10^{-3})\frac{p_0 L^4}{EI} \quad \text{**Ans. (b)**} \tag{7}$$

Review the Solution The dimensionality of all terms is correct. We get the correct signs by spot-checking $v'(0)$, $v'(L)$ and $v(L/2)$. We can use Table E.2 to check the magnitude of $\delta_{\max}$ in Eq. (7), since the cosine load is between a triangular load (Fig. 3a) and a uniform load (Fig. 3b). Therefore, the answer seems to be correct.

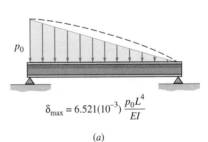

p_0

$\delta_{\max} = 6.521(10^{-3})\dfrac{p_0 L^4}{EI}$

(a)

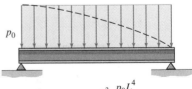

p_0

$\delta_{\max} = 13.02(10^{-3})\dfrac{p_0 L^4}{EI}$

(b)

Fig. 3 Comparison deflections.

Observe that by using the fourth-order method in Example Problem 7.2, we avoid having to determine an expression for $M(x)$ at the outset. Example Problems 7.1 and 7.2 are single-span problems involving only one loading interval. Let us now consider a couple of two-interval problems. Example Problem 7.3 is solved by the second-order method; Example Problem 7.4 by the fourth-order method.

■■■■■■■■■■■■■■■ EXAMPLE 7.3 ■■■■■■■■■■■■■■■

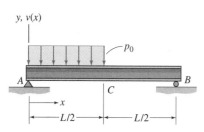

Fig. 1 The load diagram of a simply supported beam.

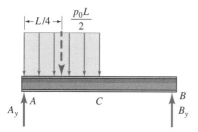

(a) Overall free-body diagram.

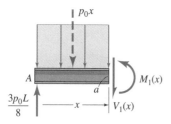

(b) Free-body diagram for load interval 1 – $(0 < x < L/2)$.

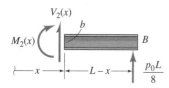

(c) Free-body diagram for load interval 2 – $(L/2 < x < L)$.

Fig. 3 Free-body diagrams.

A uniform, simply supported beam has a uniform load over half of its length, as shown in Fig. 1. (a) Using the second-order integration method, determine expressions for the slope and deflection in load intervals AC and CB. (b) Determine the maximum deflection.

Solution

(a) Determine formulas for the slope and deflection of the beam.

Sketch the Deflection Curve: The deflection will be downward over the entire length, with the maximum deflection occurring between A and C, as illustrated in the deflection diagram, Fig. 2.

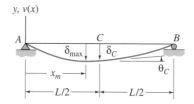

Fig. 2 The expected deflection curve.

Equilibrium—Determine $M_1(x)$ and $M_2(x)$: Since this beam has two load intervals, we need two moment equations. First, however, we need an overall free-body diagram in order to determine the reactions at A and B. From the free-body diagram in Fig. 3a, we get the equilibrium equations

$$\left(\sum M\right)_A = 0: \qquad B_y = \frac{p_0 L}{8}$$

$$\left(\sum M\right)_B = 0: \qquad A_y = \frac{3p_0 L}{8}$$

From the free-body diagrams in Figs. 3b and 3c, respectively, we get[3]

$$\left(\sum M\right)_a = 0: \qquad M_1(x) = \frac{3p_0 L x}{8} - \frac{p_0 x^2}{2} \qquad (1a)$$

$$\left(\sum M\right)_b = 0: \qquad M_2(x) = \frac{p_0 L}{8}(L - x) \qquad (1b)$$

Write the Moment-Curvature Equations: There is a moment-curvature equation (Eq. 7.8) for each interval.

$$EIv_1'' = M_1(x) = \frac{3p_0 L x}{8} - \frac{p_0 x^2}{2} \qquad (2a)$$

$$EIv_2'' = M_2(x) = \frac{p_0 L}{8}(L - x) \qquad (2b)$$

---
[3]See Footnote 2.

Integrate the Differential Equations:

$$EIv_1' = \frac{3p_0Lx^2}{16} - \frac{p_0x^3}{6} + C_1 \tag{3a}$$

$$EIv_2' = -\frac{p_0L}{16}(L-x)^2 + C_2 \tag{3b}$$

$$EIv_1 = \frac{p_0Lx^3}{16} - \frac{p_0x^4}{24} + C_1x + D_1 \tag{4a}$$

$$EIv_2 = \frac{p_0L}{48}(L-x)^3 + C_2x + D_2 \tag{4b}$$

Identify the Boundary Conditions and Continuity Conditions: The deflection is zero at the two ends, *A* and *B*.

$$v_1(0) = 0, \qquad v_2(L) = 0 \tag{5a,b}$$

The displacement and slope are continuous at *C*, since the beam is not broken there.

$$v_1'\left(\frac{L}{2}\right) = v_2'\left(\frac{L}{2}\right), \qquad v_1\left(\frac{L}{2}\right) = v_2\left(\frac{L}{2}\right) \tag{5c,d}$$

Evaluate the Constants of Integration: Using the source equations indicated in brackets, we get:

[4a, 5a] $$\qquad EIv_1(0) = D_1 = 0 \tag{6a}$$

[4b, 5b] $$\qquad EIv_2(L) = C_2L + D_2 = 0 \tag{6b}$$

[3a, 3b, 5c] $$\quad \frac{3p_0L}{16}\left(\frac{L}{2}\right)^2 - \frac{p_0}{6}\left(\frac{L}{2}\right)^3 + C_1 = -\frac{p_0L}{16}\left(\frac{L}{2}\right)^2 + C_2 \tag{6c}$$

[4a, 4b, 5d, 6a] $$\quad \frac{p_0L}{16}\left(\frac{L}{2}\right)^3 - \frac{p_0}{24}\left(\frac{L}{2}\right)^4 + C_1\left(\frac{L}{2}\right)$$

$$= \frac{p_0L}{48}\left(\frac{L}{2}\right)^3 + C_2\left(\frac{L}{2}\right) + D_2 \tag{6d}$$

Combining Eqs. (6b), (6c), and (6d), we get

$$C_1 = -\frac{9}{384}\left(p_0L^3\right) \tag{7a}$$

$$C_2 = \frac{7}{384}\left(p_0L^3\right) \tag{7b}$$

$$D_2 = -\frac{7}{384}\left(p_0L^4\right) \tag{7c}$$

Finally, inserting the constants of integration into Eqs. (3) and (4), we get the following expressions for the slope and deflection in load intervals 1 and 2:

$$v_1' = \frac{p_0 L^3}{384EI}\left[-9 + 72\left(\frac{x}{L}\right)^2 - 64\left(\frac{x}{L}\right)^3\right] \tag{8a}$$

$$v_2' = \frac{p_0 L^3}{384EI}\left[-17 + 48\left(\frac{x}{L}\right) - 24\left(\frac{x}{L}\right)^2\right] \tag{8b}$$

$$v_1 = \frac{p_0 L^4}{384EI}\left[-9\left(\frac{x}{L}\right) + 24\left(\frac{x}{L}\right)^3 - 16\left(\frac{x}{L}\right)^4\right] \tag{8c}$$

$$v_2 = \frac{p_0 L^4}{384EI}\left[1 - 17\left(\frac{x}{L}\right) + 24\left(\frac{x}{L}\right)^2 - 8\left(\frac{x}{L}\right)^3\right] \tag{8d}$$

Ans. (a)

(b) Determine the maximum deflection of the beam. As observed in Fig. 2, the maximum deflection occurs in the loaded interval, AC. Therefore, we let $v_1'(x_m) = 0$ to determine the location where the slope is zero. This gives the cubic equation

$$-64\left(\frac{x_m}{L}\right)^3 + 72\left(\frac{x_m}{L}\right)^2 - 9 = 0$$

The value of x_m can be found by trial and error or by some other root-finding method. It is

$$x_m = 0.460L$$

Then, the maximum deflection is

$$\delta_m = |v_1(0.460L)| = 6.56(10^{-3})\frac{p_0 L^4}{EI} \qquad \text{Ans. (b)}$$

Review the Solution The answers have been written in terms of a dimensionless length (x/L) so that all dimensional quantities can be collected into one coefficient term. The coefficient terms:

$$v' \sim \left(\frac{p_0 L^3}{EI}\right) \quad \text{and} \quad v \sim \left(\frac{p_0 L^4}{EI}\right)$$

are dimensionally correct. As with many beam deflection problems, the answers to this problem can be found in Appendix E. If they were not, we could check the value of δ_m by comparing it with the midpoint deflection of a uniformly loaded beam. From Table E.2, that value is

$$|v(L/2)| = \frac{5}{384}\left(\frac{p_0 L^4}{EI}\right)$$

The value of δ_m should be slightly greater than half of this, which it is.

Use the fourth-order integration method to determine expressions for slope and deflection in intervals *AD* and *DB* of the uniform cantilever beam in Fig. 1. Also, determine specific expressions for the slope and deflection at $x = L$.

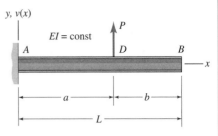

Solution

Sketch the Deflection Curve: As sketched in Fig. 2, the deflection will be upward along the entire length of the beam. Since there is no load to the right of point *D*, the beam will remain straight from *D* to *B*.

Fig. 1 A concentrated load on a cantilever beam.

Determine $p_1(x)$ and $p_2(x)$: There are no distributed loads on this beam, so, for Interval 1 and Interval 2, respectively,

$$p_1(x) = p_2(x) = 0 \qquad (1a,b)$$

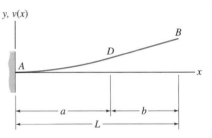

Write the Load-Deflection Equations: Combine Eqs. (1) with Eq. 7.12 to give

$$(EIv_1'')'' = 0 \qquad (2a)$$

$$(EIv_2'')'' = 0 \qquad (2b)$$

Fig. 2 The expected deflection curve.

Integrate the Differential Equations: For interval 1, the four integrals of Eq. (2a) are:

$$V_1(x) = (EIv_1'')' = C_1 \qquad (3a)$$

$$M_1(x) = EIv_1'' = C_1x + D_1 \qquad (3b)$$

$$EIv_1' = C_1\left(\frac{x^2}{2}\right) + D_1x + E_1 \qquad (3c)$$

$$EIv_1 = C_1\left(\frac{x^3}{6}\right) + D_1\left(\frac{x^2}{2}\right) + E_1x + F_1 \qquad (3d)$$

For interval 2, the integrals of Eq. (2b) are:

$$V_2(x) = (EIv_2'')' = C_2 \qquad (4a)$$

$$M_2(x) = EIv_2'' = C_2x + D_2 \qquad (4b)$$

$$EIv_2' = C_2\left(\frac{x^2}{2}\right) + D_2x + E_2 \qquad (4c)$$

$$EIv_2 = C_2\left(\frac{x^3}{6}\right) + D_2\left(\frac{x^2}{2}\right) + E_2x + F_2 \qquad (4d)$$

Identify the Boundary Conditions and Continuity Conditions: Since there are eight constants of integration to be evaluated, we need a total of eight boundary and/or continuity conditions. The deflection and slope vanish at $x = 0$ and the moment and shear vanish at $x = L$. Therefore,

$$v_1(0) = v_1'(0) = 0 \qquad (5a,b)$$

$$V_2(L) = M_2(L) = 0 \qquad (6a,b)$$

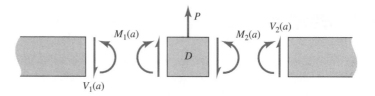

Fig. 3 Equilibrium of the node at D.

There is slope continuity and displacement continuity at $x = a$, so

$$v_1'(a) = v_2'(a) \tag{7a}$$

$$v_1(a) = v_2(a) \tag{7b}$$

The final two conditions result from equilibrium of the "joint" at D. (They are also listed in Table 7.2b.) From the free-body diagram in Fig. 3,

$$V_2(a) - V_1(a) = P \tag{8a}$$

$$M_2(a) = M_1(a) \tag{8b}$$

Evaluate the Constants of Integration: Combining Eqs. (3) through (8) we get the following constants of integration:

$$C_1 = -P, \quad D_1 = Pa, \quad E_1 = F_1 = 0 \tag{9a–d}$$

$$C_2 = D_2 = 0, \quad E_2 = \frac{Pa^2}{2}, \quad F_2 = -\frac{Pa^3}{6} \tag{10a–d}$$

Finally, the above constants can be inserted into Eqs. (3c,d) and (4c,d) to give the following slope and deflection equations:

$$v_1'(x) = \frac{Pa^2}{2EI}\left[-\left(\frac{x}{a}\right)^2 + 2\left(\frac{x}{a}\right) \right] \tag{10a}$$

$$v_1(x) = \frac{Pa^3}{6EI}\left[-\left(\frac{x}{a}\right)^3 + 3\left(\frac{x}{a}\right)^2 \right] \tag{10b}$$

$$v_2'(x) = \frac{Pa^2}{2EI} \tag{10c}$$

$$v_2(x) = \frac{Pa^3}{6EI}\left[3\left(\frac{x}{a}\right) - 1 \right] \tag{10d}$$

Ans.

Determine the slope and deflection at the tip of the beam. Evaluating Eqs. (10c) and (10d) at $x = L$, we get

$$v_2'(L) = \frac{Pa^2}{2EI} \tag{11a}$$

$$v_2(L) = \frac{Pa^3}{6EI}\left(\frac{3L}{a} - 1 \right) \tag{11b}$$

Ans.

Review the Solution The above results can be checked by consulting Table. E.1.

Example Problem 7.4 could have been solved easily by the second-order integration method. Note that the only free-body diagram used in Example Problem 7.4 was a very "local" free-body diagram of the joint used to relate V_1 to V_2 and M_1 to M_2 at the joint. **No overall free-body diagrams are needed for a fourth-order solution**. The boundary conditions and continuity (jump) conditions for fourth-order solutions are obtained from (local) joint free-body diagrams, and the results are listed in Table 7.2b.

Next, we will use the second-order method and the fourth-order method to solve slope and deflection problems for statically indeterminate beams.

7.4 SLOPE AND DEFLECTION BY INTEGRATION— STATICALLY INDETERMINATE BEAMS

In Section 7.2 the second-order and fourth-order ordinary differential equations that govern the deflection of linearly elastic beams were derived, and in Section 7.3 they were used to solve for the slope and deflection of statically determinate beams. In this section we will extend the solution procedure to treat statically indeterminate beam deflection problems.

For example, compare the beam in Example Problem 7.3 with the beam in Fig. 7.10. Since the simply supported beam in Example Problem 7.3 is statically determinate, it is possible to determine all reactions using equilibrium equations alone. In Fig. 7.10 there are four reactions, but only three independent equations of equilibrium can be written, so this beam is *statically indeterminate*. The addition of a rotation *constraint* at A gives rise to a *redundant reaction*, M_A.

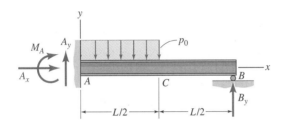

FIGURE 7.10 A statically indeterminate beam.

Each additional constraint, beyond those needed to prevent collapse of a beam, gives rise to an additional redundant reaction at that constraint, whose value can only be determined by considering the deflection of the entire beam (see Example Problem 5.3). In the example problems that follow, it will be demonstrated that **each extra constraint that is imposed on a beam (or system of beams) leads to a redundant reaction** that, along with the constants of integration, can be evaluated by enforcing boundary conditions and continuity conditions. Thus, there is a slight modification of Steps 2 and 5 of the procedure outlined in Section 7.3 in order to account for the redundant reactions.

EXAMPLE 7.5

A couple, M_B, is applied to a uniform, propped cantilever beam, as shown in Fig. 1. Use the second-order integration method: (a) to determine expressions for the slope and deflection of the beam; and (b) to determine the reactions at A and B.

Solution
(a) Determine expressions for the slope and deflection of the beam.

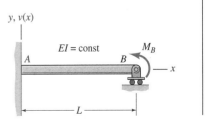

Fig. 1 A uniform, propped-cantilever beam with specified end couple.

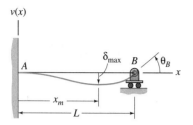

$v(x)$

δ_{max} B θ_B

A

x_m

L

Fig. 2 The expected deflection curve.

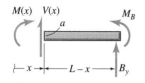

$M(x)$ $V(x)$ M_B

a

x $L - x$ B_y

Fig. 3 A free-body diagram.

Sketch the Deflection Curve: The couple M_B will tend to rotate end B counterclockwise. Therefore, the beam should take the shape indicated in Fig. 2.

Equilibrium—Determine an Expression for M(x): If we take the free-body diagram shown in Fig. 3, we will get an expression for $M(x)$ that includes the given end couple (moment) and the unknown reaction at B. We can let this reaction at B, which arises due to the added constraint at B, be the unknown *redundant*.

$$\left(\sum M \right)_a = 0: \qquad M(x) = M_B + B_y(L - x) \qquad (1)$$

Write the Moment-Curvature Equation: From Eq. 7.8 and Eq. (1),

$$EIv'' = M(x) = M_B + B_y(L - x) \qquad (2)$$

Integrate the Differential Equation: The two integrations of Eq. (2) give

$$EIv' = M_Bx + B_yLx - B_y\left(\frac{x^2}{2}\right) + C_1 \qquad (3a)$$

$$EIv = M_B\left(\frac{x^2}{2}\right) + B_yL\left(\frac{x^2}{2}\right) - B_y\left(\frac{x^3}{6}\right) + C_1x + C_2 \qquad (3b)$$

Identify the Boundary Conditions: Since we have three unknowns in Eqs. (3), two constants of integration plus the unknown redundant reaction force B_y, we need a total of three boundary conditions. They are:

$$v'(0) = v(0) = v(L) = 0 \qquad (4a,b,c)$$

Evaluate the Constants: The source equations are identified in brackets.

[4a, 3a] $\qquad\qquad\qquad EIv'(0) = C_1 = 0 \qquad (5a)$

[4b, 3b, 5a] $\qquad\qquad EIv(0) = C_2 = 0 \qquad (5b)$

[4c, 3b, 5a, 5b] $\quad EIv(L) = M_B\left(\frac{L^2}{2}\right) + B_y\left(\frac{L^3}{3}\right) = 0$

or $\qquad\qquad\qquad\qquad B_y = -\frac{3}{2}\frac{M_B}{L} \qquad (5c)$

Combining Eqs. (5) with Eqs. (3), we get the following expressions for the slope and deflection of this propped-cantilever beam.

$$v'(x) = \frac{M_BL}{4EI}\left[3\left(\frac{x}{L}\right)^2 - 2\left(\frac{x}{L}\right) \right] \qquad (6a)$$

$$v(x) = \frac{M_BL^2}{4EI}\left[\left(\frac{x}{L}\right)^3 - \left(\frac{x}{L}\right)^2 \right] \qquad (6b)$$

Ans. (a)

(b) Determine the reactions at A and B. We found the redundant reaction in Eq. (5c). To determine the reactions at $x = 0$, we can use a free-body diagram of the entire beam (Fig. 4).

Equilibrium:

$$\sum F_y = 0: \qquad A_y = \frac{3}{2}\frac{M_b}{L} \qquad\qquad (7a)$$

$$\left(\sum M\right)_A = 0: \qquad M_A = M_B - \frac{3}{2}\left(\frac{M_B}{L}\right)L = -\frac{M_B}{2} \qquad (7b)$$

In summary,

$$A_y = \frac{3}{2}\frac{M_B}{L}, \qquad M_A = -\frac{M_B}{2} \qquad\qquad \textbf{Ans. (b)}$$

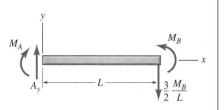
Fig. 4 A free-body diagram.

Review the Solution The deflection and slope expressions in Eqs. (6) have the proper dimensions. From Eq. (6a), the slope at $x = L$ is positive, which is correct. The couple M_B would lift end B up, were it not for the constraint. Therefore, B_y has the correct sign. Finally, M_A has the correct sign according to our preliminary sketch in Fig. 2.

The above solution of Example Problem 7.5 differs little from the second-order solution of statically determinate problems in Section 7.3. The presence of one constraint (boundary condition) beyond the minimum required to prevent collapse of the beam led to an additional unknown constant to be evaluated. To every additional degree of redundancy there will be a corresponding additional boundary condition to use in evaluating the unknowns.

Let us now use the fourth-order method to solve a problem that is very similar to Example Problem 7.5. Note how statical indeterminancy enters into this solution.

■■■■■■■■■■■■■■■■■□ E X A M P L E   7 . 6 □■■■■■■■■■■■■■■■■■

End B of a cantilever beam is guided so that the slope at B remains zero as the beam deflects upward (or downward) due to a transverse load at B. The beam in Fig. 1 is linearly elastic and uniform. Assume that the rollers at B permit enough horizontal displacement so that the beam remains free of axial force as it deflects laterally. Determine: (a) expressions for the slope and deflection of the beam; (b) the displacement at B; and (c) the reactions at A and B.

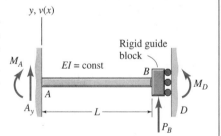
Fig. 1 A cantilever beam with constrained slope at its "free" end.

Solution

(a) Determine the slope and deflection functions for the beam.

Sketch the Deflected Beam: The end B is pushed upward by the force P_B, but because of the constraint provided by the guide block, it cannot rotate, as illustrated in Fig. 2.

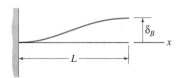
Fig. 2 The expected deflection curve.

377

Determine an Expression for p(x): From Fig. 1 there is no distributed load on the beam, so

$$p(x) = 0 \tag{1}$$

Write the Load-Deflection Equation: From Eq. 7.12 and Eq. (1), we get the homogeneous, fourth-order equation

$$(EIv'')'' = 0 \tag{2}$$

Integrate the Differential Equation: We integrate Eq. (2) four times, incorporating Eqs. 7.11 and 7.8. Thus,

$$V(x) = (EIv'')' = C_1 \tag{3a}$$

$$M(x) = EIv'' = C_1x + C_2 \tag{3b}$$

$$EIv' = C_1\left(\frac{x^2}{2}\right) + C_2x + C_3 \tag{3c}$$

$$EIv = C_1\left(\frac{x^3}{6}\right) + C_2\left(\frac{x^2}{2}\right) + C_3x + C_4 \tag{3d}$$

Identify the Boundary Conditions: There are four constants of integration to be determined.[4] There are three slope/displacement constraints:

$$v'(0) = v(0) = v'(L) = 0 \tag{4a,b,c}$$

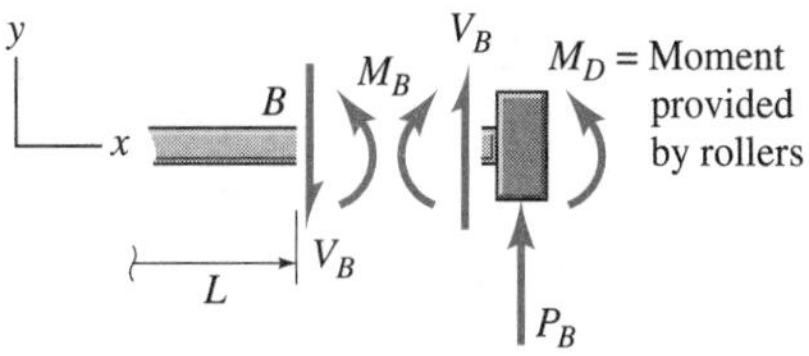

And, there is one force-type constraint at end B. Taking a free-body diagram of the rigid end-block at B, we get

$$\sum F_y = 0: \qquad\qquad V_B \equiv V(L) = -P_B \tag{4d}$$

[Equation (4d) could also be "extrapolated" from the concentrated force example in Table 7.2b, but it is best to rely on joint free-body diagrams and equilibrium equations. Adherence to the sign convention is essential!]

Evaluate the Constants: The source equations are indicated in brackets.

[4a, 3c]] $\qquad\qquad EIv'(0) = C_3 = 0 \tag{5a}$

[4b, 3d, 5a] $\qquad\qquad EIv(0) = C_4 = 0 \tag{5b}$

[4d, 3a] $\qquad\qquad V(L) = C_1 = -P_B \tag{5c}$

[4c, 3c, 5a, 5c] $\qquad EIv'(L) = -P_B\left(\frac{L^2}{2}\right) + C_2L = 0$

or $\qquad\qquad\qquad\qquad C_2 = \dfrac{P_BL}{2} \tag{5d}$

Fig. 3 A free-body diagram of the joint at *B*.

In the figure: y, x, B, M_B, V_B, M_D = Moment provided by rollers, V_B, L, P_B.

---

[4]In a fourth-order solution, *redundant reactions* do not appear explicitly. For example, the moment supplied by the rollers at *B* does not appear explicitly, as B_y did in Example Problem 7.5.

378

Combining the constants in Eqs. (5) with the slope and deflection expressions in Eqs. (3c, d), we get

$$v'(x) = \frac{P_B L^2}{2EI}\left[-\left(\frac{x}{L}\right)^2 + \left(\frac{x}{L}\right) \right] \quad\quad (6a)$$

$$v(x) = \frac{P_B L^3}{12EI}\left[-2\left(\frac{x}{L}\right)^3 + 3\left(\frac{x}{L}\right)^2 \right] \quad\quad (6b)$$

Ans. (a)

(b) Determine the displacement at end B. By evaluating Eq. (6b) at $x = L$, we get

$$\delta_B \equiv v(L) = \frac{P_B L^3}{12EI} \quad\quad \text{Ans. (b)} \quad (7)$$

(c) Determine the reactions at A and B. Equations (3a) and (3b) are evaluated at ends $x = 0$ and $x = L$ to find the reactions at A and B. From Eq. (4d) we already know the value of V_B. But there are still three unknown reactions. We could use the free-body diagram in Fig. 4 to get two of these if we knew the third. We will use this approach later as a check. Now, however, we will combine Eqs. (3a, b) with the constants from Eq. (5) to get

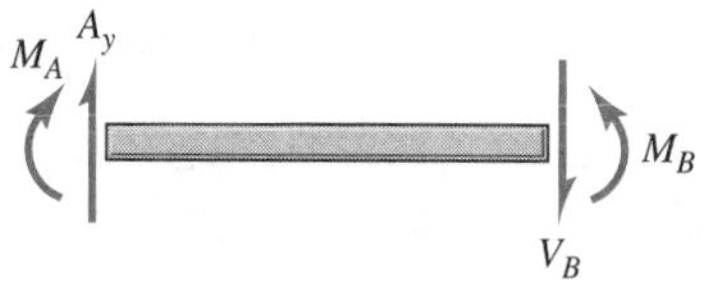

Fig. 4 A free-body diagram showing the reactions at A and B.

$$V(x) = C_1 = -P_B$$

$$M(x) = C_1 x + C_2 = -P_B x + \frac{P_B L}{2}$$

So,

$$A_y \equiv V(0) = -P_B \quad\quad (8a)$$

$$V_B \equiv V(L) = -P_B \quad\quad (8b)$$

$$M_A \equiv M(0) = \frac{P_B L}{2} \quad\quad \text{Ans. (c)} \quad (8c)$$

$$M_B \equiv M(L) = -\frac{P_B L}{2} \quad\quad (8d)$$

Review the Solution   The answers in Eqs. (6) have the proper dimensions, and they can easily be shown to satisfy the slope and displacement boundary conditions at A and B [Eqs. (4)]. From Fig. 1, we see that there must be a downward reaction of magnitude of P_B at A [Eq. (8a)], and from the sketch in Fig. 2 we expect from symmetry considerations that M_A and M_B will have the same magnitude but opposite signs [Eqs. (8c, d)]. Finally, the answers in Eqs. (8) satisfy equilibrium for the free-body diagram in Fig. 4. Therefore, our answers seem to be correct.

The next example problem illustrates the solution of statically indeterminate multi-interval problems using the second-order method.

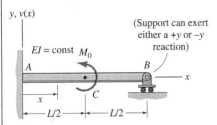

Fig. 1 A two-interval uniform, propped cantilever beam.

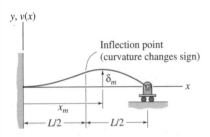

Fig. 2 The expected deflection curve.

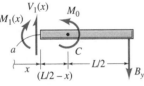

(a) Interval 1 ($0 < x < L/2$).

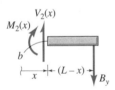

(b) Interval 2 ($L/2 < x < L$).

Fig. 3 Free-body diagrams.

For the uniform linearly elastic beam in Fig. 1 (see Example Problem 5.3), use the second-order integration method: (a) to determine expressions for the slope and deflection in the intervals AC and CB; and (b) to determine the reactions at A and B.

Solution

(a) Determine the slope and deflection functions for this beam.

Sketch the Deflection Curve: If end B were not restrained against vertical displacement, the moment M_0 would cause the beam to deflect upward all along its entire length. Therefore, we can expect a deflected shape *similar* to the curve sketched in Fig. 2. We need a moment diagram to tell just how the curvature changes in going from A to B, but the deflection curve in Fig. 2 should be *approximately* correct.

Equilibrium—Determine Expressions for $M_1(x)$ and $M_2(x)$: Since the beam has one excess support, there is one redundant reaction. Since there is only one unknown reaction at B, let that be the redundant reaction. Therefore, it is best to draw free-body diagrams that include end B rather than end A.[5]

From Fig. 3a,

$$\left(\sum M\right)_a = 0: \qquad M_1(x) = M_0 - B_y(L - x) \tag{1a}$$

and, from Fig. 3b,

$$\left(\sum M\right)_b = 0: \qquad M_2(x) = -B_y(L - x) \tag{1b}$$

Write the Moment-Curvature Equations:

$$EIv_1'' = M_1(x) = M_0 - B_y(L - x) \tag{2a}$$

$$EIv_2'' = M_2(x) = -B_y(L - x) \tag{2b}$$

Integrate the Differential Equations:

$$EIv_1' = M_0x - B_yLx + B_y\left(\frac{x^2}{2}\right) + C_1 \tag{3a}$$

$$EIv_2' = -B_yLx + B_y\left(\frac{x^2}{2}\right) + C_2 \tag{3b}$$

$$EIv_1 = M_0\left(\frac{x^2}{2}\right) - B_yL\left(\frac{x^2}{2}\right) + B_y\left(\frac{x^3}{6}\right) + C_1x + D_1 \tag{4a}$$

$$EIv_2 = -B_yL\left(\frac{x^2}{2}\right) + B_y\left(\frac{x^3}{6}\right) + C_2x + D_2 \tag{4b}$$

---

[5]If we were to use free-body diagrams that include end A, we would have to write an overall equilibrium equation to relate the two reactions at A to each other.

Identify the Boundary Conditions and Continuity Conditions: From Table 7.2a and Fig. 2, we get the following boundary conditions at A and B:

$$v_1'(0) = v_1(0) = v_2(L) = 0 \qquad\qquad (5a,b,c)$$

and at C we have the following continuity conditions:

$$v_1'\left(\frac{L}{2}\right) = v_2'\left(\frac{L}{2}\right) \qquad\qquad (5d)$$

$$v_1\left(\frac{L}{2}\right) = v_2\left(\frac{L}{2}\right) \qquad\qquad (5e)$$

Thus, we have five equations to use to determine the four constants of integration and the unknown reaction B_y.

Evaluate the Unknown Constants: Using the source equations listed in brackets, we get the following constants:

[5a, 3a]	$C_1 = 0$	(6a)
[5b, 4a, 6a]	$D_1 = 0$	(6b)
[5d, 3a, 3b, 6a]	$C_2 = \dfrac{M_0 L}{2}$	(6c)
[5e, 4a, 4b, 6a, 6b, 6c]	$D_2 = \dfrac{-M_0 L^2}{8}$	(6d)
[5c, 4b, 6c, 6d]	$B_y = \dfrac{9M_0}{8L}$	(6e)

Substituting the above constants into Eqs. (3) and (4), we obtain the following expressions for slope and deflection in intervals 1 and 2:

$$v_1'(x) = \frac{M_0 L}{16EI}\left[9\left(\frac{x}{L}\right)^2 - 2\left(\frac{x}{L}\right) \right] \qquad\qquad (7a)$$

$$v_2'(x) = \frac{M_0 L}{16EI}\left[9\left(\frac{x}{L}\right)^2 - 18\left(\frac{x}{L}\right) + 8 \right] \qquad\qquad (7b)$$

$$v_1(x) = \frac{M_0 L^2}{16EI}\left[3\left(\frac{x}{L}\right)^3 - \left(\frac{x}{L}\right)^2 \right] \qquad\qquad (7c)$$

$$v_2(x) = \frac{M_0 L^2}{16EI}\left[3\left(\frac{x}{L}\right)^3 - 9\left(\frac{x}{L}\right)^2 + 8\left(\frac{x}{L}\right) - 2 \right] \qquad\qquad (7d)$$

Ans. (a)

381

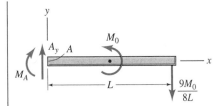

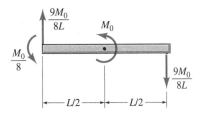

Fig. 4 A free-body diagram.

Fig. 5 The reactions to a couple M_0 at $x = L/2$.

(b) Determine the reactions at A and B. We determined the value of the redundant reaction, B_y, in Eq. (6e). The remaining reactions can be obtained from the overall free-body diagram in Fig. 4.

$$\sum F_y = 0: \qquad\qquad A_y = \frac{9M_0}{8L} \tag{8a}$$

$$\left(\sum M\right)_A = 0: \qquad\qquad M_A = -\frac{M_0}{8} \tag{8b}$$

In summary, the reactions are shown on Fig. 5.

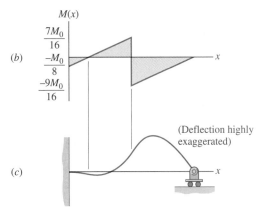

Fig. 6 (*a*) Shear diagram, (*b*) moment diagram, and (*c*) revised deflection curve.

Review the Solution Obviously curvature of the beam near $x = 0$ in our "expected deflection" sketch in Fig. 2 does not agree with the negative value that we got for M_A in Eq. (8b). Either the sketch is not entirely correct, or our solution is not correct, or both! So, let us sketch V and M diagrams based on Fig. 5 to see if the moment diagram corresponds to a deflection curve that is similar to Fig. 2. Of course, we could also just plot Eqs. (7c) and (7d). The revised deflection curve in Fig. 6*c*, which has been drawn with the aid of the moment diagram in Fig. 6*b*, $\left(M = \dfrac{EI}{\rho}\right)$, is *similar* to our original estimate in Fig. 2, so our solution is probably correct. But, we should recheck our solution just to be sure that no error has been made.

The next example problem illustrates how to use the fourth-order integration method to solve a statically indeterminate problem of a beam with a distributed load. The beam is clamped at both ends, leading to a *fixed-end solution*. Fixed-end solutions like this enter into the superposition solutions discussed in Section 7.7; several are tabulated in Table E.3 in Appendix E.

The uniform, linearly elastic beam in Fig. 1 supports a triangularly distributed load. Use the fourth-order integration method: (a) to determine expressions for the shear, the bending moment, the slope, and the deflection of the beam; and (b) to determine the reactions at A and B.

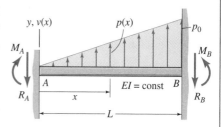

Fig. 1 A fixed-fixed beam with triangular load.

Solution

(a) Determine expressions for the shear, the bending moment, the slope, and the deflection of the beam.

Sketch the Deflection Curve: The beam will deflect upward, as sketched in Fig. 2.

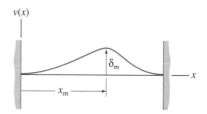

Fig. 2 The expected deflection curve.

Determine an Expression for p(x): From similar triangles, we get

$$\frac{p(x)}{x} = \frac{p_0}{L}$$

or

$$p(x) = p_0\left(\frac{x}{L}\right) \tag{1}$$

Write the Load-Deflection Equation: From Eq. 7.12 and Eq. (1), we get

$$(EIv'')'' = p(x) = p_0\left(\frac{x}{L}\right) \tag{2}$$

Integrate the Differential Equation: We need to integrate Eq. (2) four times, incorporating Eqs. 7.11 and 7.8.

$$(EIv'')' = V(x) = p_0\left(\frac{x^2}{2L}\right) + C_1 \tag{3a}$$

$$EIv'' = M(x) = p_0\left(\frac{x^3}{6L}\right) + C_1x + C_2 \tag{3b}$$

$$EIv' = p_0\left(\frac{x^4}{24L}\right) + C_1\left(\frac{x^2}{2}\right) + C_2x + C_3 \tag{3c}$$

$$EIv = p_0\left(\frac{x^5}{120L}\right) + C_1\left(\frac{x^3}{6}\right) + C_2\left(\frac{x^2}{2}\right) + C_3x + C_4 \tag{3d}$$

Identify the Boundary Conditions: The fixed-fixed boundary conditions are:

$$v'(0) = v(0) = v'(L) = v(L) = 0 \tag{4a–d}$$

383

Evaluate the Constants: The source equations are in brackets.

[4a, 3c] $$C_3 = 0 \qquad (5a)$$

[4b, 3d, 5a] $$C_4 = 0 \qquad (5b)$$

[4c, 3c, 5a] $$12C_1L + 24C_2 = -p_0L^2$$

[4d, 3d, 5a, 5b] $$20C_1L + 60C_2 = -p_0L^2$$

From these last two equations we get

$$C_1 = -\frac{3}{20}p_0L \qquad (5c)$$

$$C_2 = \frac{1}{30}p_0L^2 \qquad (5d)$$

We can combine the constants in Eqs. (5) with the expressions in Eqs. (3) to obtain the answers to Part (a).

$$V(x) = \frac{p_0L}{20}\left[10\left(\frac{x}{L}\right)^2 - 3\right] \qquad (6a)$$

$$M(x) = \frac{p_0L^2}{60}\left[10\left(\frac{x}{L}\right)^3 - 9\left(\frac{x}{L}\right) + 2\right] \qquad (6b)$$

$$v'(x) = \frac{p_0L^3}{120EI}\left[5\left(\frac{x}{L}\right)^4 - 9\left(\frac{x}{L}\right)^2 + 4\left(\frac{x}{L}\right)\right] \qquad (7a)$$

$$v(x) = \frac{p_0L^4}{120EI}\left[\left(\frac{x}{L}\right)^5 - 3\left(\frac{x}{L}\right)^3 + 2\left(\frac{x}{L}\right)^2\right] \qquad (7b)$$

Ans. (a)

(b) Determine the reactions at A and B. Taking note the relationship between $V(x)$ and $M(x)$ and the symbols used in Fig. 1, and using Eqs. (6a) and (6b), we get the following reactions:

$$R_A \equiv -V(0) = \frac{3}{20}p_0L \qquad (8a)$$

$$M_A \equiv M(0) = \frac{1}{30}p_0L^2 \qquad (8b)$$

Ans. (b)

$$R_B \equiv V(L) = \frac{7}{20}p_0L \qquad (8c)$$

$$M_B \equiv M(L) = \frac{1}{20}p_0L^2 \qquad (8d)$$

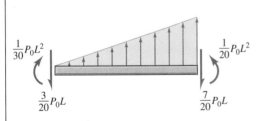

$\frac{1}{30}P_0L^2$ $\frac{1}{20}P_0L^2$

$\frac{3}{20}P_0L$ $\frac{7}{20}P_0L$

Fig. 3 The reactions.

Review the Solution Let us show these reactions on a sketch like Fig. 1 to see if they look reasonable.

The reactions satisfy overall equilibrium, with the larger moment and shear at the more heavily loaded end, B. Therefore, our solution seems to be correct.

*7.5 USE OF DISCONTINUITY FUNCTIONS TO DETERMINE BEAM DEFLECTIONS

In Section 7.2 the *moment-curvature differential equation,* Eq. 7.8, and the *load-deflection differential equation,* Eq. 7.12, were derived, and in Sections 7.3 and 7.4 they were integrated to determine the slope and deflection of beams.

$$EIv'' = M(x)$$

(7.8)
repeated

$$(EIv'')' = p(x)$$

(7.12)
repeated

In Example Problems 7.3 and 7.7 you discovered that it is a very lengthy process to solve deflection problems when there are discontinuities in loading and/or support along the length of the beam. Such cases required the writing of an expression for the bending moment for each interval of the beam and the use of compatibility conditions at points of discontinuity due to loading or supports. *Discontinuity functions* were introduced in Section 5.5 and were used there to obtain concise expressions for load, shear, and moment for beams. In this section we will illustrate how the use of discontinuity functions greatly simplifies the solution of multi-interval beam-deflection problems.[6] The procedure outlined below applies both to statically determinate beams and to statically indeterminate beams.

One question that arises is, Are constants of integration needed when expressions involving discontinuity functions are integrated? The answer is, Yes, constants of integration are needed to represent the shear at $x = 0$, $V(0)$; the moment at $x = 0$, $M(0)$; the slope of the beam at $x = 0$, $v'(0)$; and the displacement at $x = 0$, $v(0)$. When these constants of integration are introduced, they are introduced with the unit step function in the form $C\langle x \rangle^0$, as indicated in Eqs. 7.17.[7] In subsequent integrals, the Macaulay bracket is then integrated in the usual manner.

$$V(x) = EI(v'')' = \int_{0^-}^{x} p(\xi)d\xi + V(0)\langle x \rangle^0$$

$$M(x) = EIv'' = \int_{0^-}^{x} V(\xi)d\xi + M(0)\langle x \rangle^0$$

$$v'(x) = \int_{0^-}^{x} \left(\frac{M(\xi)}{EI(\xi)}\right) d\xi + v'(0)\langle x \rangle^0$$

(7.17)

$$v(x) = \int_{0^-}^{x} v'(\xi)d\xi + v(0)\langle x \rangle^0$$

---

[6]This method of solving beam deflection problems is called *Clebsch's Method* after its developer, the German engineer and mathematician A. Clebsch (1833–1872) who proposed the method. An excellent review of the method and its applications may be found in ''Clebsch's Method for Beam Deflection,'' by W. D. Pilkey, [Ref. 7-1].

[7]It is not essential to include the unit step $\langle x \rangle^0$ when introducing the constants of integration, but it is done here in order to maintain uniformity of notation among the various discontinuity terms being integrated.

The following straightforward procedure may be used to solve slope/deflection problems for either statically determinate beams or statically indeterminate beams. Therefore, it is a very good method to use either for pencil-and-paper solutions or for computer solutions. However, the displacement-method procedure described in Section 7.7 leads to the more powerful and more generally applicable finite element method.

PROCEDURE FOR USING DISCONTINUITY FUNCTIONS TO DETERMINE BEAM DEFLECTION

1. Sketch the expected deflection curve. Draw a free-body diagram of the entire beam, labeling all reactions appropriately.

2. Using information from the *Load* column of Table 5.2, write a discontinuity-function expression for the load $p(x)$. Introduce all reactions as unknowns.

3. Integrate the load expression four times, using information from Table 5.1 or Table 5.2 or using Eqs. 5.12b, 5.13, and 5.14. Introduce a constant of integration in the form $C\langle x \rangle^0$ with each integral, as indicated in Eq. 7.17.

4. Identify the force-type (shear and moment) and displacement-type (displacement and slope) boundary conditions. Use these, together with the *shear and moment closure equations*

$$V(L^+) = 0, \qquad M(L^+) = 0 \qquad (7.18)$$

to evaluate all constants of integration.

As the following examples will illustrate, **the shear and moment closure equations automatically enforce overall equilibrium of the beam**. Nevertheless, a free-body diagram of the entire beam should be drawn so that unknown reactions can be labeled clearly.

■■■■■■■■■■■■■■■■■ E X A M P L E 7 . 9 ■■■■■■■■■■■■■■■■■

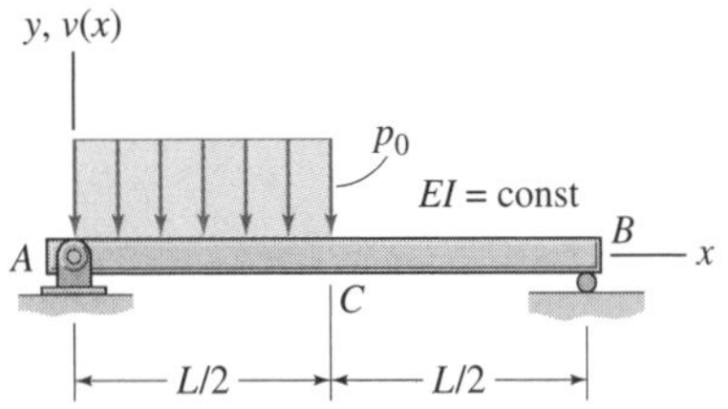

Fig. 1

Use discontinuity functions to solve for the deflection of the beam in Example Problem 7.3 (see Fig. 1).

Solution We will follow the procedure outlined above. The sketch of the expected deflection curve may be found as Fig. 2 of Example Problem 7.3.

Free-body Diagram and Load Equation: This is a statically determinate problem, so we could immediately use equilibrium to solve for the reactions. However, these will be provided "automatically" by the shear and moment closure equations as we carry out the steps outlined above in the procedure. Using the discontinuity functions listed under Cases 2 and 3 in Table 5.2 together with the free-body diagram in Fig. 2, we can write the following discontinuity-function expression for $p(x)$.

$$p(x) = A_y\langle x \rangle^{-1} - p_0[\langle x \rangle^0 - \langle x - L/2 \rangle^0] + B_y\langle x - L \rangle^{-1} \qquad (1)$$

Integration of the Discontinuity Equations: From Eq. 7.12,

$$(EIv'')'' = p(x) \qquad (2)$$

Therefore, we can combine Eqs. (1) and (2) and integrate once to get

$$V(x) = (EIv'')' = A_y\langle x \rangle^0 - p_0[\langle x \rangle^1 - \langle x - L/2 \rangle^1] \qquad (3a)$$
$$+ B_y\langle x - L \rangle^0$$

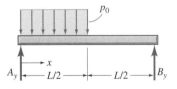

Fig. 2 A free-body diagram.

In this case there is an unknown reaction at $x = 0$, giving a term $A_y\langle x \rangle^0$, so we do not need an additional constant of integration $V(0)\langle x \rangle^0$. Continuing the integrations,

$$M(x) = EIv'' = A_y\langle x \rangle^1 - \frac{p_0}{2}[\langle x \rangle^2 - \langle x - L/2 \rangle^2]$$
$$+ B_y\langle x - L \rangle^1 + M(0)\langle x \rangle^0 \tag{3b}$$

Since $M(0) = 0$, we can immediately eliminate the last term in Eq. (3b).

$$EIv' = \frac{A_y}{2}\langle x \rangle^2 - \frac{p_0}{6}[\langle x \rangle^3 - \langle x - L/2 \rangle^3]$$
$$+ \frac{B_y}{2}\langle x - L \rangle^2 + EIv'(0)\langle x \rangle^0 \tag{3c}$$

$$EIv = \frac{A_y}{6}\langle x \rangle^3 - \frac{p_0}{24}[\langle x \rangle^4 - \langle x - L/2 \rangle^4]$$
$$+ \frac{B_y}{6}\langle x - L \rangle^3 + EIv'(0)\langle x \rangle^1 + EIv(0)\langle x \rangle^0 \tag{3d}$$

Since $v(0) = 0$, we can immediately eliminate the last term in Eq. (3d).

Boundary Conditions and Closure Conditions: We have one remaining force-type boundary condition and one remaining displacement-type boundary condition. They are:

$$M(L) = 0, \qquad v(L) = 0 \tag{4a,b}$$

In addition, we have the shear closure equation

$$V(L^+) = 0 \tag{4c}$$

(We do not need to use $M(L^+) = 0$ since Eq. (4a) takes care of moment closure.)

Let us first apply Eq. (4c), the shear closure equation. Combining Eqs. (3a) and (4c), we get

$$V(L^+) = A_y - p_0(L/2) + B_y = 0 \tag{5a}$$

Note that this is just the equilibrium equation $\Sigma F_y = 0$. Next, let us satisfy Eq. (4a) using $M(x)$ from Eq. (3b). We just get the equilibrium equation $(\Sigma M)_B = 0$. That is,

$$M(L) = A_y L - \frac{p_0}{2}\left[L^2 - \left(\frac{L}{2}\right)^2\right] = 0$$

from which

$$A_y = \frac{3p_0 L}{8} \tag{5b}$$

Combining Eqs. (5a) and (5b), we get

$$B_y = \frac{p_0 L}{8} \tag{5c}$$

(Note that only force-type boundary conditions plus the shear and moment closure conditions are required in order for us to solve for all reactions, even though this is a two-interval beam.)

Equation (4b) can now be evaluated to obtain the constant $v'(0)$. Thus,

$$EIv(L) = \frac{p_0 L}{16} L^3 - \frac{p_0}{24}\left[L^4 - \left(\frac{L}{2}\right)^4 \right] + EIv'(0)L = 0$$

so,

$$EIv'(0) = -\frac{9}{384} p_0 L^3 \tag{5d}$$

Inserting Eqs. (5b), (5c), and (5d) into Eq. (3d), we get

$$EIv(x) = \frac{p_0 L}{16}\langle x \rangle^3 - \frac{p_0}{24}\left(\langle x \rangle^4 - \langle x - L/2 \rangle^4 \right) \quad \text{Ans.} \tag{6}$$
$$+ \frac{p_0 L}{48}\langle x - L \rangle^3 - \frac{3 p_0 L^3}{128}\langle x \rangle^1$$

The $\langle x - L \rangle$ term is, of course, zero throughout the length of the beam, so it could be dropped.

Review the Solution By evaluating the Macaulay brackets in Eq. (6) for interval 1 ($0 < x < L/2$) and for interval 2 ($L/2 < x < L$) we see that the resulting expressions are the same as the expressions obtained in Example Problem 7.3, namely

$$v_1 = \frac{p_0 L^4}{384EI}\left[-9\left(\frac{x}{L}\right) + 24\left(\frac{x}{L}\right)^3 - 16\left(\frac{x}{L}\right)^4 \right], \quad (0 \le x \le L/2)$$

$$v_2 = \frac{p_0 L^4}{384EI}\left[1 - 17\left(\frac{x}{L}\right) + 24\left(\frac{x}{L}\right)^2 - 8\left(\frac{x}{L}\right)^3 \right], \quad (L/2 \le x \le L)$$

Let us now use discontinuity functions to re-solve the statically indeterminate, multi-interval deflection problem in Example 7.7. Note how the discontinuity-function approach simplifies the solution, in comparison with the solution in Example 7.7.

Use discontinuity functions to obtain expressions for the reactions and for the deflection of the propped cantilever beam in Fig. 1 (repeated from Example Problem 7.7).

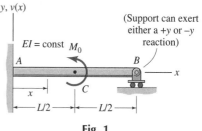

Fig. 1

Solution   We will follow the procedure outlined on page 386 for solving this type of problem. The expected deflection curve is sketched in Example Problem 7.7, Fig. 2.

Free-body Diagram and Load Equation: Utilizing Cases 1 and 2 of Table 5.2, and referring to the free-body diagram in Fig. 2 we can write the following discontinuity-function expression for the load $p(x)$, including the unknown reactions.

$$p(x) = M_A\langle x\rangle^{-2} + A_y\langle x\rangle^{-1} - M_0\langle x - L/2\rangle^{-2} + B_y\langle x - L\rangle^{-1} \quad (1)$$

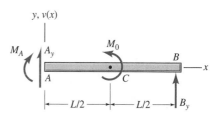

Fig. 2   A free-body diagram.

Integration of Discontinuity Equations: Integrating Eq. (1), with the aid of Eq. 7.17a, we get

$$V(x) = (EIv'')' = \underline{M_A\langle x\rangle^{-1}} + A_y\langle x\rangle^0 - \underline{M_0\langle x - L/2\rangle^{-1}} \quad (2a)$$
$$+ B_y\langle x - L\rangle^0$$

where the A_y term represents the total shear reaction at A and serves as the constant of integration. The moment terms in Eq. (2a) are underlined because they are singularity terms, not true shear forces. They are nonzero only at the locations where the couples M_A and M_0 are applied.

Continuing with the integrations (as in Eqs. 7.17),

$$M(x) = EIv'' = M_A\langle x\rangle^0 + A_y\langle x\rangle^1 - M_0\langle x - L/2\rangle^0 \quad (2b)$$
$$+ B_y\langle x - L\rangle^1$$

$$EIv' = M_A\langle x\rangle^1 + \frac{A_y}{2}\langle x\rangle^2 - M_0\langle x - L/2\rangle^1 \quad (2c)$$
$$+ \frac{B_y}{2}\langle x - L\rangle^2 + EIv'(0)\langle x\rangle^0$$

$$EIv = \frac{M_A}{2}\langle x\rangle^2 + \frac{A_y}{6}\langle x\rangle^3 - \frac{M_0}{2}\langle x - L/2\rangle^2 \quad (2d)$$
$$+ \frac{B_y}{6}\langle x - L\rangle^3 + EIv'(0)\langle x\rangle^1 + EIv(0)\langle x\rangle^0$$

Boundary Conditions and Closure Conditions: We have five unknowns—A_y, B_y, M_A, $v'(0)$, and $v(0)$—and we have five equations—three displacement-type boundary conditions and two closure equations. The displacement-type boundary conditions are:

$$v'(0) = 0, \quad v(0) = 0, \quad v(L) = 0 \qquad \text{(3a,b,c)}$$

and the closure equations are:

$$V(L^+) = 0, \qquad M(L^+) = 0 \qquad \text{(3c,d)}$$

The $v'(0)$ and $v(0)$ terms in Eqs. (2) are zero because of the fixed boundary at A (Eqs. (3a,b)). Since $\langle x - a \rangle^{-1} = 0$ except at $x = a$, Eq. (2a) can be combined with Eq. (3d) to give

$$V(L^+) = +A_y + B_y = 0 \qquad \text{(4a)}$$

which is just the force-equilibrium equation, $\Sigma F_y = 0$. From Eqs. (2b) and the moment closure equation, Eq. (3d),

$$M(L^+) = M_A + A_y L - M_0 = 0 \qquad \text{(4b)}$$

which is just the moment-equilibrium equation, $(\Sigma M)_B = 0$. From Eqs. (2d) and (3a) through (3c),

$$EIv(L) = \frac{M_A}{2}L^2 + \frac{A_y}{6}L^3 - \frac{M_0}{2}(L/2)^2 = 0 \qquad \text{(4c)}$$

Finally, combining Eqs. (4a) through (4c) we get

$$A_y = \frac{9}{8}\frac{M_0}{L}, \quad B_y = -\frac{9}{8}\frac{M_0}{L}, \quad M_A = -\frac{1}{8}M_0 \qquad \text{Ans.} \quad \text{(5)}$$

These are the same results that we obtained in Example Problem 7.7.

The deflection equation is obtained by substituting Eqs. (5) into Eq. (2d). Thus,

$$v = \frac{M_0}{EI}\left[-\frac{1}{16}\langle x \rangle^2 + \frac{3}{16L}\langle x \rangle^3 - \frac{1}{2}\langle x - L/2 \rangle^2 \right.$$
$$\left. -\frac{3}{16L}\langle x - L \rangle^3 \right] \qquad \text{Ans.} \quad \text{(6)}$$

This single discontinuity-function expression combines the results expressed in Eqs. (7c) and (7d) of Example Problem 7.7.

In conclusion, the preceding examples illustrate the simplifications that are introduced by using the discontinuity-function method to solve deflection problems that involve load discontinuities or interior supports. These simplifications are: only a single load equation need be written and integrated; the slope and displacement compatibility equations at interval boundaries do not have to be written explicitly; and there are generally fewer constants of integration to be evaluated.

7.6 SLOPE AND DEFLECTION OF BEAMS BY SUPERPOSITION— FORCE METHOD

In Section 7.2 we derived the differential equations governing the deflection of linearly elastic beams, and in Sections 7.3 through 7.5 we solved for the slope and deflection of beams using Eq. 7.8, the second-order moment-curvature equation, and Eq. 7.12, the fourth-order load-deflection equation. These equations are *linear differential equations,* since the deflection function $v(x)$ and its derivatives appear linearly, that is, only to the first power. Therefore, the slope and deflection of a beam that simultaneously supports multiple loads can be obtained by linear superposition (i.e., by addition) of the effects of the loads acting separately. For example, the simply supported beam in Fig. 7.11*a* may be analyzed by adding the solutions for the two separate loads shown in Figs. 7.11*b* and 7.11*c*. These solutions are available in Table E.2. Superposition holds for *all* quantities: reactions, internal shear and bending moment, slope, deflection, and so on. Example Problems 7.11 and 7.12 illustrate this type of superposition problem.

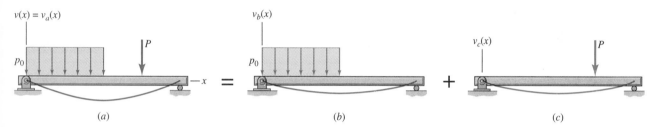

FIGURE 7.11 Superposition of two load cases: $v(x) \equiv v_a(x) = v_b(x) + v_c(x)$.

Superposition can also be applied to another useful way, as illustrated by Fig. 7.12. This may be referred to as *differential-load superposition.* In Example Problem 7.4 we obtained expressions for the slope and deflection of a cantilever beam with a single concentrated load applied at an arbitrary location along the beam. In Fig. 7.12 we can say that a differential load $dP(\xi)$ at location ξ produces a corresponding differential displacement $dv(x, \xi)$ at every location x along the beam, and we can determine the total displacement $v(x)$ by summing up the effect of the entire distributed load by integrating (with respect to ξ) over the loaded portion of the beam.[8] Example Problem 7.13 illustrates this procedure.

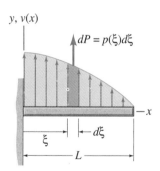

FIGURE 7.12
Superposition of differential loads.

Finally, the superposition of solutions can be applied to statically indeterminate problems as well as statically determinate problems. In this case there will be constraints on slope and/or displacement (boundary conditions and/or continuity conditions) that must be satisfied. The equations of constraint are used to solve for the unknown redundant reactions. Example Problems 7.14 and 7.15 illustrate the use of this force-method version of superposition to solve statically indeterminate beam problems. A displacement-method version of superposition is presented in Section 7.7.

Procedure for Solving Superposition Problems (Force Method). We

refer to the superposition procedure described here as a force-method procedure because the expressions that are used in the superposition process are displacements and slopes expressed in terms of force-type quantities (distributed forces, concentrated forces, and couples). The following solution procedure is suggested.

---

[8]The differential displacement $dv(x, \xi)$ depends on the magnitude and point of application of the differential load, $dP(\xi)$, and also on the point x where dv is being evaluated.

1. Carefully study the boundary conditions and the loading given in the problem statements, and sketch the expected deflection curve of the beam. Explore various ways to break the given problem down into *statically determinate subproblems* for which the solution is given in Table E.1 (Cantilever Beams) or in Table E.2 (Simply Supported Beams). (In some cases there may be several alternative ways to choose the constituent solutions from Tables E.1 and E.2.) Sketch the deflection curves of these subproblems.

2. Write *superposition equations* for any quantities that are required by the problem statement, for example, slope, deflec-

tion, etc., using information from the tables to express slope and displacement in terms of force-type quantities.

3. If the problem is statically indeterminate, write down the *boundary conditions and/or continuity conditions* that are not automatically satisfied by the constituent sub problems that you selected in Step 1. Corresponding to each such equation there will be a redundant force (or couple). Use these equations of constant to solve for the redundants.

4. Complete the solution (e.g., determine maximum deflection if requested, determine other reactions).

Statically Determinate Beams. Example Problems 7.11 through 7.13 illustrate the linear superposition of known solutions to solve statically determinate beam-deflection problems.

EXAMPLE 7.11

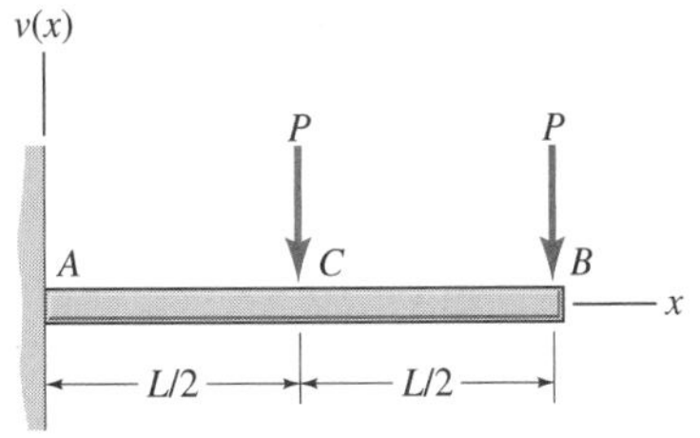

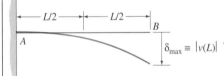

Fig. 1 A cantilever beam with two concentrated loads.

Determine the maximum deflection of the uniform, linearly elastic cantilever beam in Fig. 1.

Solution

Sketch the Deflection Curve: Since both loads push the beam downward, the maximum deflection occurs at $x = L$, as indicated in the deflection diagram, Fig. 2.

Select the Subproblems: We only need to examine the cantilever beam candidate solutions in Table E.1, specifically E.1(3) and E.1(4).

Write the Superposition Equation: We only need an equation for the total deflection at B. Thus, from Fig. 3,

$$\delta_{max} \equiv |v(L)| = \delta_b + \delta_c = |v_b(L)| + |v_c(L)| \tag{1}$$

Fig. 2 The expected deflection curve.

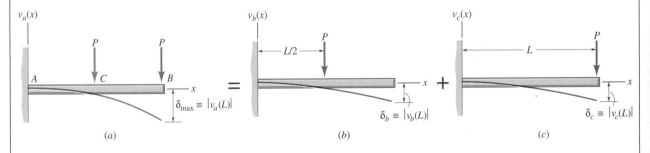

Fig. 3 The superposition of subproblems.

392

where, from Table E.1(4)

$$\delta_b = \frac{5PL^3}{48EI} \qquad (2a)$$

and, from Table E.1(3)

$$\delta_c = \frac{PL^3}{3EI} \qquad (2b)$$

Complete the Solution: From Eqs. (1) and (2),

$$\delta_{max} = \frac{21PL^3}{48EI} = 0.438\frac{PL^3}{EI} \qquad \text{Ans.} \quad (3)$$

Review the Solution The deflection at the tip of a cantilever beam due to a concentrated tip load is $\delta = \frac{PL^3}{3EI}$, as given in Eq. (2b). This is a frequently used expression, and it is useful in estimating the accuracy of answers of many problems, including this one. For example, the deflection of the beam in Fig. 1 should be greater than the deflection of the same beam without the load at $x = L/2$, but less than the deflection would be with both loads at $x = L$. Equation (3) satisfies this inequality:

$$\frac{PL^3}{3EI} < \frac{21PL^3}{48EI} < \frac{2PL^3}{3EI}$$

■■■■■■■■■■■■■■■□□ **E X A M P L E 7 . 1 2** □□□■■■■■■■■■■■■■■

A uniform simply supported beam similar to the one in Fig. 7.11 is subjected to a uniform distributed load and a concentrated load as shown in Fig. 1. Consider the particular case: $a = L/2$, $b = 3L/4$, $P = p_0L/2$. Use super-position of solutions from Table E.2 to solve for the maximum deflection of this beam.

Solution

Sketch the Deflection Curve: The beam will deflect downward throughout its entire length, as shown in Fig. 2. The loads in Fig. 1 divide the beam into three intervals: *AC, CD,* and *DB.* Certainly, the maximum deflection will not occur in interval *DB*. If *P* were zero, x_m would definitely fall in interval *AC*, but, if there were no distributed load on the beam, x_m would fall between *C* and *D*. Therefore, as a part of our solution, we will have to examine an expression for the slope at *C* to determine whether the beam slopes upward at *C*, as shown in Fig. 2, or whether it actually slopes downward at *C*.

Select the Subproblems: Obviously, we only need to examine solutions for simply supported beams. We will use the letter subscripts, like *a* and *b*, to denote the constituent subproblems, and we will use number subscripts, progressing from left to right, to denote the interval of validity. For example,

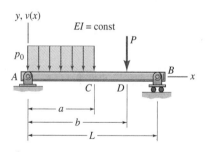

Fig. 1 A simply supported beam with two applied loads.

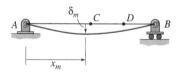

Fig. 2 The expected deflection curve.

393

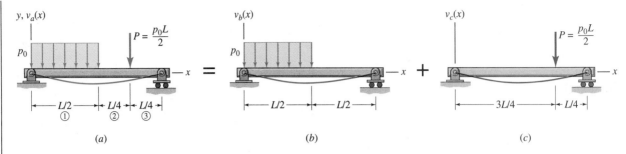

Fig. 3 Superposition of deflection subproblems.

$v_{b1}(x)$ denotes the deflection in interval 1 of the beam loaded as shown in Fig. 3b, and so on.

Write the Superposition Equations: Since we are to solve for the maximum deflection, we must first locate the point $x = x_m$ where the <u>total</u> slope v_a' vanishes, that is where $v_a'(x_m) = 0.$[9] From Table E.2(6) and E.2(4), respectively, we can write down the following superposition equations, taking proper account of the signs.

$$v_{a1}'(x) = v_{b1}'(x) + v_{c1}'(x) \tag{1}$$

$$= \frac{-p_0 L^3}{384EI}\left[64\left(\frac{x}{L}\right)^3 - 72\left(\frac{x}{L}\right)^2 + 9\right]$$

$$- \frac{PL^2}{384EI}\left[-48\left(\frac{x}{L}\right)^2 + 15\right]$$

Setting $P = \dfrac{p_0 L}{2}$, we get

$$v_{a1}'(x) = \frac{-p_0 L^3}{768EI}\left[128\left(\frac{x}{L}\right)^3 - 192\left(\frac{x}{L}\right)^2 + 33\right] \tag{2}$$

If $v_{a1}'(L/2) > 0$, we know that the point of zero slope lies in interval 1. Therefore, before writing other superposition equations, let us check the value of $v_{a1}'\left(\dfrac{L}{2}\right)$. Evaluating Eq. (2) at $x = \dfrac{L}{2}$, we obtain

$$v_{a1}'\left(\frac{L}{2}\right) = \frac{-p_0 L^3}{768EI}\left[128\left(\frac{1}{2}\right)^3 - 192\left(\frac{1}{2}\right)^2 + 33\right]$$

$$v_{a1}'\left(\frac{L}{2}\right) = \frac{-p_0 L^3}{768EI} \tag{3}$$

Since $v_{a1}'\left(\dfrac{L}{2}\right) < 0$, the beam must become horizontal in interval 2. So, we use information from Table E.2(4) and E.2(6) to write expressions for the slope

---

[9]It is the total deflection that is to be maximized. The maximum deflections for the subproblems cannot be superimposed because, in general, they do not occur at the same beam location.

and deflection in interval 2. The slope is given by

$$v'_{a2}(x) = v'_{b2}(x) + v'_{c2}(x) \tag{4a}$$

$$= \frac{-p_0 L^3}{384EI}\left[24\left(\frac{x}{L}\right)^2 - 48\left(\frac{x}{L}\right) + 17\right]$$

$$+ \frac{PL^2}{384EI}\left[48\left(\frac{x}{L}\right)^2 - 15\right]$$

The deflection in interval 2 is given by

$$v_{a2}(x) = v_{b2}(x) + v_{c2}(x) \tag{4b}$$

$$= \frac{-p_0 L^4}{384EI}\left[8\left(\frac{x}{L}\right)^3 - 24\left(\frac{x}{L}\right)^2 + 17\left(\frac{x}{L}\right) - 1\right]$$

$$+ \frac{PL^3}{384EI}\left[16\left(\frac{x}{L}\right)^3 - 15\left(\frac{x}{L}\right)\right]$$

Setting $P = \dfrac{p_0 L}{2}$ in Eqs. (4), we get

$$v'_{a2}(x) = \frac{p_0 L^3}{768EI}\left[96\left(\frac{x}{L}\right) - 49\right] \tag{5a}$$

$$v_{a2}(x) = \frac{p_0 L^4}{768EI}\left[48\left(\frac{x}{L}\right)^2 - 49\left(\frac{x}{L}\right) + 2\right] \tag{5b}$$

Complete the Solution: We can now evaluate Eq. (5a) to determine the value of x_m at which $v'_{a2}(x_m) = 0$. The maximum deflection occurs at this point.

$$v'_{a2}(x_m) = 0 \rightarrow x_m = \frac{49}{96}L \tag{6a}$$

Then, since x_m falls in interval 2,

$$\delta_m \equiv |v_{a2}(x_m)| = \frac{p_0 L^4}{768EI}\left|48\left(\frac{49}{96}\right)^2 - 49\left(\frac{49}{96}\right) + 2\right|$$

or

$$\delta_m = \frac{p_0 L^4}{768EI}(10.51) = 1.37(10^{-2})\frac{p_0 L^4}{EI} \qquad \textbf{Ans.} \tag{6b}$$

Review the Solution   Since we have a total load of $p_0 L$ on the beam, half of which is distributed, a good check on our answer in Eq. (6b) would be to determine the midspan deflection of a simply supported beam with a uniform distributed load over its total length. From Table E.2(5) we get

$$\delta_C = \delta_{\max} = \frac{5p_0 L^4}{384EI} = 1.30(10^{-2})\frac{p_0 L^4}{EI}$$

Thus, our answer in Eq. (6b) appears to be correct, and we will, therefore, assume that we have made no "error" [except that we originally misguessed the location of the point of maximum deflection (Fig. 2)].

Next, let us illustrate the *differential-load superposition* approach introduced earlier in Fig. 7.12.

Procedure for Differential-Load Superposition.
As indicated earlier in the discussion of Fig. 7.12, we can superpose differential-load solutions if we have a solution for the desired quantity due to a concentrated load at an arbitrary position on the beam. The steps that may be used are as follows.

PROCEDURE FOR DIFFERENTIAL-LOAD SUPERPOSITION

1. Sketch the load diagram and the deflection curve.
2. In the Tables of Slopes and Deflection of uniform beams, Table E.1 or Table E.2, identify the solution that will provide the desired quantity due to a concentrated force at an arbitrary location on the beam.
3. Form an expression for the distributed load as a function of position along the beam, that is, form an expression for $p(x)$.
4. Use an integral to sum up the effect of the differential load dP, as indicated in Fig. 7.12.

Example Problem 7.13 illustrates the above procedure.

■■■■■■■■■■■■■■■■ **E X A M P L E   7 . 1 3** ■■■■■■■■■■■■■■■■■

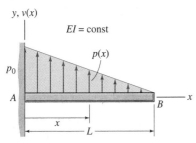

Fig. 1 A cantilever beam with triangular loading.

Use the differential-load superposition approach to determine expressions for the deflection and slope at the tip of a uniform, linearly elastic cantilever beam with triangular load, as illustrated in Fig. 1.

Solution

Sketch the Deflection Curve: The deflection curve is shown in Fig. 2.

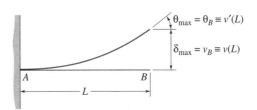

Fig. 2 The expected deflection curve.

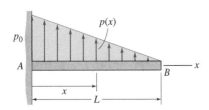

Fig. 3 Load diagram.

Form the Load Expression: (Since we only need $v'(L)$ and $v(L)$, and not functions of position like $v'(x)$ and $v(x)$, we do not need to use a dummy variable, like the variable ξ (Greek xi) shown in Fig. 7.12.) From similar triangles in Fig. 3, we get

$$\frac{p(x)}{L - x} = \frac{p_0}{L}$$

or

$$p(x) = p_0 \left(1 - \frac{x}{L} \right) \tag{1}$$

Form the Deflection Expression and Integrate: We can use the results in Eqs. (10c) and (10d) of Example Problem 7.4. These are also given in Table E.1(4).

$$dv'(L,x) = \frac{dPx^2}{2EI} \qquad (2a)$$

$$dv(L, x) = \frac{dP}{6EI}\left(3Lx^2 - x^3\right) \qquad (2b)$$

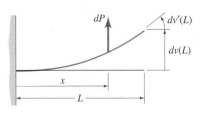

Fig. 4 Deflection due to differential load dP.

In Fig. 4 the increment of load is $dP = p(x)dx$, and since the load extends from $x = 0$ to $x = L$, we have the integrals

$$v'(L) = \frac{1}{2EI}\int_0^L x^2[p(x)dx] \qquad (3a)$$

$$v(L) = \frac{1}{6EI}\int_0^L (3Lx^2 - x^3)[p(x)dx] \qquad (3b)$$

or, combining Eqs. (1) and (3),

$$v'(L) = \frac{p_0}{2EIL}\int_0^L (x^2L - x^3)\,dx \qquad (4a)$$

$$v(L) = \frac{p_0}{6EIL}\int_0^L (3L^2x^2 - 4Lx^3 + x^4)\,dx \qquad (4b)$$

Finally,

$$v'(L) = \frac{p_0L^3}{24EI}, \qquad v(L) = \frac{p_0L^4}{30EI} \qquad \textbf{Ans.} \quad (5a,b)$$

Review the Solution We can obviously check the above answer by referring to Table E.1(8). But, if the table did not have this particular triangular load, it would undoubtedly have the solution for a uniformly distributed load on a cantilever beam [Table E.1(5)]. We could expect the answer in Eq. (5b) to be somewhat less than half the deflection at the tip of a uniformly loaded cantilever beam, namely $\delta = \frac{p_0L^4}{8EI}$. In fact, our answer is about one-fourth of this value, which seems reasonable.

Statically Indeterminate Beams. As noted earlier in this section, if a beam, or a system of beams, is statically indeterminate, there are more boundary conditions or other constraints than the minimum that is required to prevent collapse of the beam or beam system. For each additional *constraint,* there will be a *redundant force or moment* that can be determined by satisfying the constraint equation(s).

In many cases there are alternative ways to construct a superposition solution for a statically indeterminate problem. That is, there may be alternative ways to select the redundant(s) to be used, so long as the required solutions are available (e.g., in Appendix E). The next two example problems illustrate alternative ways to select the redundant and solve a statically indeterminate propped-cantilever-beam problem. In planning a superposition solution, you should mentally explore alternatives and try to solve the problem in the most efficient manner (i.e., Plan the Solution).

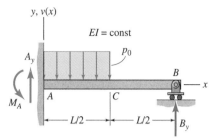

Fig. 1 A propped-cantilever beam.

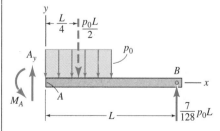

Fig. 2 The expected deflection curve.

Solve for the reactions on the beam shown in Fig. 1. Use information on cantilever beams from Table E.1.

Solution

Sketch the Deflection Curve: The expected deflection curve is shown in Fig. 2.

Select the Subproblems: We are to select subproblems from the cantilever-beam table, Table E.1. This means that B_y is to be considered the *redundant reaction*. The constituent subproblems are shown in Fig. 3.

Write the Superposition Equation: The reaction B_y is obtained by writing a *constraint equation* for the support B.

$$v(L) = -(\delta_B)_b + (\delta_B)_c = 0 \qquad (1)$$

From Table E.1(6) and E.1(3), we get

$$-\frac{7p_0L^4}{384EI} + \frac{B_yL^3}{3EI} = 0$$

so

$$B_y = \left(\frac{7}{128}\right)p_0L \qquad \textbf{Ans.} \quad (2)$$

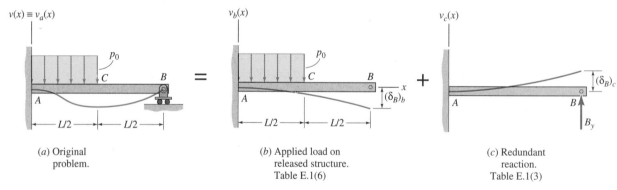

(a) Original problem.	(b) Applied load on released structure. Table E.1(6)	(c) Redundant reaction. Table E.1(3)

Fig. 3 The superposition of cantilever-beam subproblems.

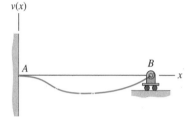

Fig. 4 A free-body diagram.

Complete the Problem: To solve for the reactions at A we need the free-body diagram (Fig. 4) and equilibrium equations.

$$\sum F_y = 0: \qquad A_y = \frac{p_0L}{2} - \frac{7p_0L}{128} = \left(\frac{57}{128}\right)p_0L$$

$$\left(\sum M\right)_A = 0: \qquad M_A = -\frac{7p_0L^2}{128} + \frac{p_0L^2}{8} = \left(\frac{9}{128}\right)p_0L^2$$

Therefore, the shear-force and bending-moment reactions at A are

$$A_y = \left(\frac{57}{128}\right)p_0 L, \qquad M_A = \left(\frac{9}{128}\right)p_0 L^2 \qquad \textbf{Ans.} \quad (3)$$

Review the Solution If the support at B were to be removed, the reactions at A would be $A_y = p_0 L/2$, $M_A = p_0 L^2/8$. The presence of a "prop" at B should reduce these values. The amount of reduction exhibited by our answers in Eqs. (3) seem reasonable.

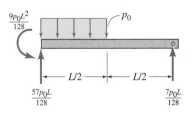

Fig. 5 Summary of reactions on the propped-cantilever beam.

■■■■■■■■■■■■■■■□ **E X A M P L E   7 . 1 5** □■■■■■■■■■■■■■■■

Solve for the moment at A on the propped-cantilever beam in Example Problem 7.14. This time, use information for simply supported beams from Table E.2.

Solution Figures 1 and 2 from Example Problem 7.14 will not be repeated here.

Select the Subproblems: If we select subproblems from the table of simply-supported-beam displacement functions, the moment M_A becomes the redundant reaction. The constituent subproblems are shown in Fig. 1.

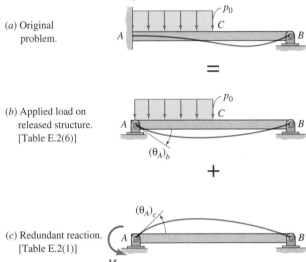

(a) Original problem.

=

(b) Applied load on released structure. [Table E.2(6)]

+

(c) Redundant reaction. [Table E.2(1)]

Fig. 1 The superposition of simply-supported-beam subproblems.

Write the Superposition Equation: The reaction M_A is obtained by writing a *constraint equation* on the rotation at A.

$$v'(0) = -(\theta_A)_b + (\theta_A)_c = 0 \qquad (1)$$

From Table E.2(6) and E.2(1), we get

$$-\frac{3p_0 L^3}{128EI} + \frac{M_A L}{3EI} = 0 \qquad (2)$$

399

so

$$M_A = \frac{9p_0L^2}{128}$$

Ans.

Review the Solution This is the same expression for M_A that we got in Example Problem 7.14. Other reactions could be obtained from a free-body diagram like Fig. 4 in Example Problem 7.14.

The next example problem illustrates how to solve statically indeterminate beam problems that have more than one redundant.

■■■■■■■■■■■■■■■ **EXAMPLE 7.16** ■■■■■■■■■■■■■■■

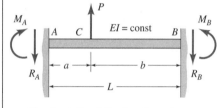

Fig. 1 A fixed-fixed uniform beam.

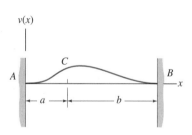

Fig. 2 The expected deflection curve (for $a < b$).

A concentrated load P is applied to a uniform fixed-fixed beam, as illustrated in Fig. 1. Determine the reactions R_B and M_B.

Solution

Sketch the Deflection Curve: The deflection will be upward, and the slope and deflection are zero at A and B, as indicated in the deflection diagram (Fig. 2).

Select the Subproblems: The results of Example Problem 7.4 [see also Table E.1(4)] can be used, along with the deflection and slope information of Example Problem 7.1 [see also Table E.1(3) and E.1(1)]. These subproblems are shown in Fig. 3.

Write the Superposition Equations: The displacement constraint and the slope constraint must be enforced at end B.

$$v_B \equiv v(L) = (\delta_B)_b - (\delta_B)_c + (\delta_B)_d = 0 \tag{1a}$$

$$v_B' \equiv v'(L) = (\theta_B)_b - (\theta_B)_c + (\theta_B)_d = 0 \tag{1b}$$

With information from Table E.1, these two constraint equations become

$$\frac{Pa^3}{6EI}\left(\frac{3L}{a} - 1\right) - \frac{R_BL^3}{3EI} + \frac{M_BL^2}{2EI} = 0 \tag{2a}$$

$$\frac{Pa^2}{2EI} - \frac{R_BL^2}{2EI} + \frac{M_BL}{EI} = 0 \tag{2b}$$

Solving Eqs. (2) simultaneously, we get

$$R_B = 3P\left(\frac{a}{L}\right)^2 - 2P\left(\frac{a}{L}\right)^3 \tag{3a}$$

$$M_B = P\left(\frac{a}{L}\right)^2(L - a) \tag{3b}$$

Ans.

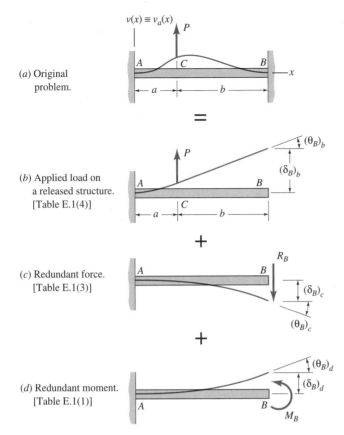

(a) Original problem.

(b) Applied load on a released structure. [Table E.1(4)]

$(\theta_B)_b$

$(\delta_B)_b$

+

(c) Redundant force. [Table E.1(3)]

R_B

$(\delta_B)_c$

$(\theta_B)_c$

+

(d) Redundant moment. [Table E.1(1)]

$(\theta_B)_d$

$(\delta_B)_d$

M_B

Fig. 3 Statically indeterminate superposition involving two redundant reactions.

Review the Solution The above expressions for R_B and M_B have the proper dimensions, F and $F \cdot L$, respectively. If Eqs. (3) are evaluated for $a = 0$, we get $R_B = M_B = 0$, which makes sense. If we evaluate Eqs. (3) at $a = L$, we get $R_B = P$, $M_B = 0$, which also makes sense. Finally, if we evaluate Eq. (3a) at $a = L/2$, we get $R_B = P/2$, which is the correct answer. It appears that our solution is correct.

A similar problem, but for a nonuniform beam, is solved by the *displacement method* in Example Problem 7.18.

The statically indeterminate problems in Example Problems 7.14 through 7.16 had displacement and slope constraints of the type $v(L) = 0$, $v'(0) = 0$, and so on. That is, the displacement and slope were zero at certain points along the beam. Figure 7.13 illustrates two situations where the constraint is provided by another flexible structure. The next problem illustrates the solution of statically indeterminate problems of this type.

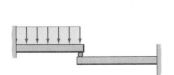

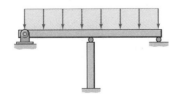

(a) The redundant support supplied by a second beam.

(b) The redundant support supplied by an axial-deformation member.

FIGURE 7.13 Beams with flexible supports.

401

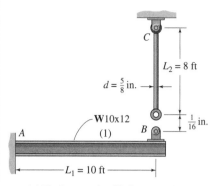

(a) The beam and rod before assembly.

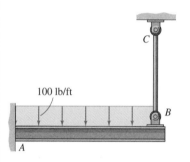

(b) The beam-rod system with uniform load.

Fig. 1 A beam-rod system.

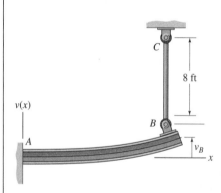

Fig. 2 The deflection curve for Part (a).

As shown in Fig. 1a, a steel beam, AB, is designed to be cantilevered from a rigid wall at A and supported by a steel hanger rod, BC, that is pinned to a rigid support at C. The beam is a **W**10×12, and the rod diameter is $d = \frac{5}{8}$ in. Use $E_{\text{steel}} = 30 \times 10^6$ psi. (a) If the rod is manufactured $\frac{1}{16}$ in. too short, how much stress will be induced in the rod by stretching it, inserting the pin at B, and then releasing the external forces required to mate the parts? (b) How much additional stress is induced in the rod by a uniformly distributed load of 100 lb/ft subsequently applied to the beam, as shown in Fig. 1b? What is the final displacement of B?

Plan the Solution   The rod BC is an axial-deflection member that can be treated as the members in Section 3.7 were treated. In Part (a) the beam will be loaded only by the force of the rod at B. This is a "misfit" problem similar to the ones in Section 3.7. For Part (b) we will have to add the distributed load by superposing another subproblem for the beam. We can follow the same steps that were used in Example 7.14, except for adding the rod at B. We should find that the rod is in tension in Part (a), and that the tension becomes greater in Part (b).

Solution

(a) Determine the initial stress induced in the rod when the rod-beam system is assembled. Let us follow the same steps used previously in solving statically indeterminate problems by superposition (force method).

Sketch the Deflection Curve: As illustrated in Fig. 2, the rod will pull the beam upward, and the beam will stretch the rod. Note: The short distance between B and the end of the beam is neglected, and the dimension V_B is treated as the deflection at the end of the beam.

Select the Subproblems: Here we can choose a cantilever beam subproblem from Table E.1(3), and we also have an axial-deformation (rod) subproblem. Let T_0 be the "initial force" in the rod, that is, the force in the rod in Fig. 3. Subscript 1 refers to the beam AB, and subscript 2 refers to the rod BC.

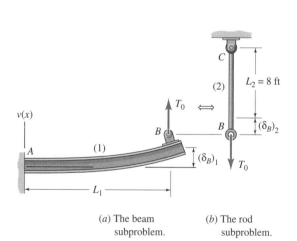

(a) The beam subproblem.

(b) The rod subproblem.

Fig. 3 The subproblems for Part (a).

Write the Superposition Equation: For the rod and beam to be connected together by a pin at B, we have the following *deformation-compatibility equation.* (The terminology is defined in Figs. 2 and 3.)

$$v_B \equiv v(L_1) = (\delta_B)_1 = \frac{1}{16} \text{ in.} - (\delta_B)_2 \tag{1}$$

From Table E.1(3) we get

$$(\delta_B)_1 = \left(\frac{L^3}{3EI}\right)_1 T_0 \tag{2a}$$

and, from Eq. 3.12, the elongation of the rod is

$$(\delta_B)_2 \equiv e_2 = \left(\frac{L}{AE}\right)_2 T_0 \tag{2b}$$

Complete Part (a): Combining Eqs. 1 and 2 we get

$$\frac{T_0 L_1^3}{3E_1 I_1} + \frac{T_0 L_2}{A_2 E_2} = \frac{1}{16} \text{ in.} \tag{3a}$$

or

$$T_0 \left[\frac{(120 \text{ in.})^3}{3(30 \times 10^6 \text{ psi})(53.8 \text{ in}^4)} + \frac{96 \text{ in.}}{\pi(\frac{5}{16} \text{ in.})^2 (30 \times 10^6 \text{ psi})} \right]$$

$$= \frac{1}{16} \text{ in.} \tag{3b}$$

Therefore, the "initial force" in the rod is

$$T_0 = 170 \text{ lb} \tag{4}$$

so the "initial stress" in the rod will be

$$\sigma_0 = \frac{T_0}{A_2} = \frac{170 \text{ lb}}{\pi(\frac{5}{16} \text{ in.})^2} = 555 \text{ psi} \qquad \text{Ans. (a)} \tag{5}$$

(b) Determine the effects of adding the distributed load. We can follow the same steps used in Part (a), adding a distributed-load subproblem for the beam.

Sketch the Deflection Curve: At this point, we do not know whether the beam will still be deflected upward, as it was in Part (a), or whether the distributed load will cause the final deflection at B to be downward. In Fig. 4 we have assumed that the beam deflects upward.

Select the Subproblems: To the two subproblems in Fig. 3 we need to add a beam with distributed load. Let T (without subscript) be the tension in the rod in Part (b). The subproblems are shown in Fig. 5.

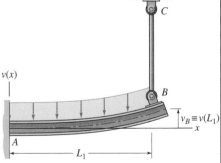

Fig. 4 The deflection curve for Part (b).

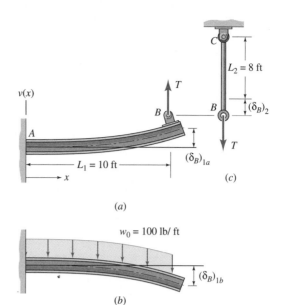

$v(x)$

A

T

B

$L_1 = 10\ \text{ft}$

x

$(\delta_B)_{1a}$

(a)

C

$L_2 = 8\ \text{ft}$

B

$(\delta_B)_2$

T

(c)

$w_0 = 100\ \text{lb/ft}$

$(\delta_B)_{1b}$

(b)

Fig. 5 The beam subproblems (a,b) and the rod subproblem (c).

Write the Superposition Equations: Again, for the rod and the beam to be connected by a pin at B, we must satisfy the compatibility equation

$$v_B \equiv v(L_1) = (\delta_B)_{1a} - (\delta_B)_{1b} = \frac{1}{16} \text{ in.} - (\delta_B)_2 \qquad (6)$$

From Table E.1(3) and E.1(5) we get

$$(\delta_B)_{1a} = \frac{TL_1^3}{3E_1I_1} \qquad (7a)$$

$$(\delta_B)_{1b} = \frac{w_0L_1^4}{8E_1I_1} \qquad (7b)$$

and from Eq. 3.12 we get

$$(\delta_B)_2 \equiv e_2 = \frac{TL_2}{A_2E_2} \qquad (7c)$$

Complete Part (b):

$$\frac{TL_1^3}{3E_1I_1} - \frac{w_0L_1^4}{8E_1I_1} = \frac{1}{16} \text{ in.} - \frac{TL_2}{A_2E_2} \qquad (8a)$$

or

$$T\left[\frac{(120 \text{ in.})^3}{3(30 \times 10^6 \text{ psi})(53.8 \text{ in}^4)} + \frac{96 \text{ in.}}{\pi(\frac{5}{16} \text{ in.})^2(30 \times 10^6 \text{ psi})}\right]$$

$$= \frac{1}{16} \text{ in.} + \left[\frac{(100 \text{ lb/ft})(10 \text{ ft})(120 \text{ in.})^3}{8(30 \times 10^6 \text{ psi})(53.8 \text{ in}^4)}\right] \quad (8b)$$

Therefore,

$$T = 535 \text{ lb} \quad (9)$$

so the stress in rod BC is

$$\sigma = \frac{T}{A_2} = \frac{535 \text{ lb}}{\pi(\frac{5}{16} \text{ in.})^2} = 1742 \text{ psi}$$

$$\sigma = 1740 \text{ psi} \qquad \text{Ans. (b)} \quad (10)$$

We can use Eq. (6) to evaluate the tip deflection.

$$v_B = \frac{1}{16} \text{ in.} - (\delta_B)_2 = \frac{1}{16} \text{ in.} - \frac{TL_2}{A_2E_2} \quad (11)$$

Therefore,

$$v_B = \frac{1}{16} \text{ in.} - \frac{(535 \text{ lb})(96 \text{ in.})}{\pi(\frac{5}{16} \text{ in.})^2(30 \times 10^6 \text{ psi})} = 0.057 \text{ in.} \quad \text{Ans. (b)}$$

Review the Solution   If all of the distributed load (i.e., 1000 lb) were to be applied directly to the rod, it would elongate

$$e = \left(\frac{PL}{AE}\right)_2 = \frac{(1000 \text{ lb})(96 \text{ in.})}{\pi(\frac{5}{16} \text{ in.})^2(30 \times 10^6 \text{ psi})} = 0.0104 \text{ in.}$$

which is much less than $\frac{1}{16}$ in. ($= 0.0625$ in.). Therefore, the rod is so stiff that it acts almost like a rigid support. By comparison, if the full 1000 lb were to be hung from the cantilever beam at B, the deflection of the tip of the beam would be

$$\delta = \frac{PL^3}{3EI} = \frac{(1000 \text{ lb})(120 \text{ in.})^3}{3(30 \times 10^6 \text{ psi})(53.8 \text{ in}^4)} = 0.357 \text{ in.}$$

Thus, we can see that this beam is very flexible in comparison with the rod. As we can see by examining the terms on the left-hand side of Eq. (3b), most of the $\frac{1}{16}$-in. gap is closed by deflection of the beam, not by stretching of the rod.

As beam-deflection problems get more difficult, superposition of solutions by the *force method,* as described in Section 7.6, may get quite lengthy and tedious, and a solution by the *displacement method* is preferable. Stress-analysis and structural-analysis computer programs are generally based on the displacement method, as illustrated in Appendix G.6.

In the force-method solutions of the previous section, as with force-method solutions in Chapters 3 and 4, *deflections (and slopes) were expressed in terms of forces,* and the deflection (and slope) expressions were superposed. In the case of statically indeterminate structures, displacement (and slope) constraint equations were solved to obtain the redundants. In this section you will discover that, when the displacement method is applied to beams, *nodal equilibrium equations are expressed in terms of displacements,* just as was done in Chapters 3 and 4 for axial-deformation and torsion problems, respectively.

Force-Deformation Equations for a Uniform Bernoulli-Euler Beam Element with Linearly Varying Distributed Load.

The first step in formulating a displacement-method solution procedure for transversely loaded beams is to establish the *force-deformation relations* for a single-beam element (Fig. 7.14).[10]

Unlike the force-deformation equation for a uniform axial-deformation element, with its single axial force F (Eqs. 3.13 and 3.14), or the torque-twist equation for a uniform torsion element with its single torque T (Eqs. 4.18 and 4.19), the shear forces V_i and V_j acting at the ends of the beam element in Fig. 7.14 are not necessarily equal, and neither are the moments M_i and M_j. Therefore, it is first necessary to give distinct "names" to the two ends. We will use the names "i" and "j" to distinguish, respectively, the left end of the beam element and its right end, as indicated in Fig. 7.14a. The *sign convention* adopted for V_i, M_i, V_j, and M_j is that positive stress resultants act in the same directions as the corresponding displacements δ_i, θ_i, δ_j, and θ_j.[11]

Combining the force-deformation results in Eq. 7.16 and the fixed-end reactions given in Eqs. (8) of Example 7.8, we can write the following *force-deformation relations* for the

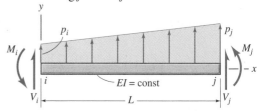

(a) The distributed load and the stress resultants V and M.

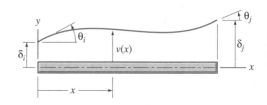

FIGURE 7.14 A uniform beam element with transverse loading.

(b) The deflection curve $v(x)$.

---

[10]This parallels Section 3.4 for the uniform axial-deformation element and Section 4.3 for a uniform element subjected to pure torsion.

[11]The sign conventions for V_j and M_i are opposite to the shear and moment sign conventions in Chapter 6. That is, $V(L) = -V_j$, and $M(0) = -M_i$. For small displacements it is permissible to equate the slope $v'(x)$ with the rotation angle θ. Thus, $\theta_i = v'(0)$ and $\theta_j = v'(L)$.

$$V_i = \frac{EI}{L^3}(12\delta_i + 6L\theta_i - 12\delta_j + 6L\theta_j) - \frac{7p_iL}{20} - \frac{3p_jL}{20}$$

$$M_i = \frac{EI}{L^2}(6\delta_i + 4L\theta_i - 6\delta_j + 2L\theta_j) - \frac{p_iL^2}{20} - \frac{p_jL^2}{30}$$

$$V_j = \frac{EI}{L^3}(-12\delta_i - 6L\theta_i + 12\delta_j - 6L\theta_j) - \frac{3p_iL}{20} - \frac{7p_jL}{20}$$

$$M_j = \frac{EI}{L^2}(6\delta_i + 2L\theta_i - 6\delta_j + 4L\theta_j) + \frac{p_iL^2}{30} + \frac{p_jL^2}{20}$$

Force-deformation equations with fixed-end forces   (7.19)

The δ and θ terms in Eqs. 7.19 form the *force-deformation relations* of a uniform beam with only nodal loads, that is, with no distributed loading.[12] The p_i and p_j terms in Eqs. 7.19 are called *fixed-end forces*. As in Example 7.8, these are the shear and moment reactions of a uniform beam whose ends are fixed (i.e., $v(0) = v'(0) = v(L) = v'(L) = 0$).

The deflection curve $v(x)$ along the beam can be expressed in terms of the end displacements (δ_i and δ_j), the end slopes (θ_i and θ_j), and the distributed-load amplitudes (p_i and p_j) by the following equation:[13]

$$
\begin{aligned}
v(x) = {} & \delta_i\left[1 - 3\left(\frac{x}{L}\right)^2 + 2\left(\frac{x}{L}\right)^3\right] + L\theta_i\left[\left(\frac{x}{L}\right) - 2\left(\frac{x}{L}\right)^2 + \left(\frac{x}{L}\right)^3\right] \\
& + \delta_j\left[3\left(\frac{x}{L}\right)^2 - 2\left(\frac{x}{L}\right)^3\right] + L\theta_j\left[-\left(\frac{x}{L}\right)^2 + \left(\frac{x}{L}\right)^3\right] \\
& + \frac{p_iL^4}{120EI}\left[2\left(1 - \frac{x}{L}\right)^2 - 3\left(1 - \frac{x}{L}\right)^3 + \left(1 - \frac{x}{L}\right)^5\right] \\
& + \frac{p_jL^4}{120EI}\left[2\left(\frac{x}{L}\right)^2 - 3\left(\frac{x}{L}\right)^3 + \left(\frac{x}{L}\right)^5\right]
\end{aligned}
\tag{7.20}
$$

---

[12]See Homework Problems 7.7-16 and 7.7-17.

[13]See Homework Problem 7.7-17 for a derivation of the δ_i term in Eq. 7.20.

SUPERPOSITION PROCEDURE—DISPLACEMENT METHOD

1. Sketch the expected deflection curve.
2. Consider the beam to consist of uniform *elements* connected together by *nodes,* or joints. Sketch each element and each node, and label the forces and couples that act on each.
3. *Element Force-Deformation Relations:* Express the shear force and bending moment at the element-to-node interface in terms of the nodal transverse displacement and slope. These are given by Eqs. 7.19.
4. *Fixed-End Forces:* If there are distributed loads on an element, include the appropriate *fixed-end force* terms from Eqs. 7.19. (Additional fixed-end forces are given in Table E.3.)
5. *Nodal Equilibrium:* Write force and moment equilibrium equations for each node. Include the external concentrated force and/or couple (if either acts on the node) and the shear and moment from the adjoining beam elements (from Steps 3 and 4).
6. *Geometry of Deformation; Compatibility:* Where an element attaches to a node, express all element end-displacement and slope quantities in terms of the transverse displacement and slope of the node itself.
7. Solve the equilibrium equations for the unknown nodal displacements and slopes.
8. Solve for any other required quantities (reactions, deflection-curve formula, etc.).

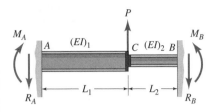

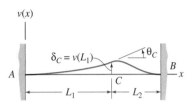

Fig. 1 A stepped, fixed-fixed beam.

Fig. 2 The expected deflection curve.

A fixed-fixed beam has a stepped cross section as indicated in Fig. 1. A concentrated load P is applied at the point where the cross-sectional properties change from $(EI)_1$ to $(EI)_2$. Using the *displacement method,* (a) determine the displacement and slope at node C, and (b) determine the reactions R_B and M_B.

Solution

(a) Determine the displacement and slope at node C.

Sketch the Deflection Curve: The beam will obviously deflect upward from end A to end B. The point of maximum deflection may be to the left of C or to the right of C, depending on the EI values and the L values. In Fig. 2 we assume that the maximum deflection occurs to the right of C, so $\theta_C > 0$.

Separate the Nodes and Elements: Since the key step in a displacement-method solution is writing nodal equilibrium equations in terms of the displacement and slope at each node and the external loads applied to the node, let us treat the beam as two elements, (1) and (2), connected together by a node at C, as indicated in Fig. 3.

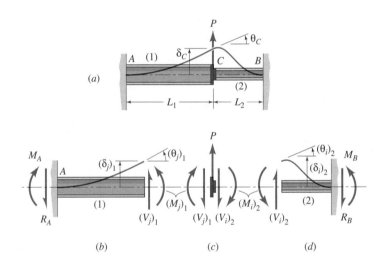

Fig. 3 The elements and connecting node.

Next we directly apply the three fundamentals of deformable-body mechanics—*equilibrium, force-deformation behavior, and geometry of deformation.*

Nodal Equilibrium: Write force and moment equilibrium equations for the node at C, where the transverse displacement and slope are unknown. The free-body diagram of node C is shown in Fig. 3c.

$$\sum F_y = 0: \qquad (V_j)_1 + (V_i)_2 = P \qquad\qquad (1a)$$

$$\left(\sum M\right)_C = 0: \qquad (M_j)_1 + (M_i)_2 = 0 \qquad\qquad (1b)$$

Equilibrium

Beam Element Force-Deformation Behavior: Here we want the force-type quantities, that is V and M, to be expressed in terms of displacement-type quantities, that is δ and θ. For this, we can use Eq. 7.19 (or see Table E.3). From Eqs. 7.19 we get

$$(V_j)_1 = \left(\frac{EI}{L^3}\right)_1 [12(\delta_j)_1 - 6L_1(\theta_j)_1] \tag{2a}$$

$$(M_j)_1 = \left(\frac{EI}{L^2}\right)_1 [-6(\delta_j)_1 + 4L_1(\theta_j)_1] \tag{2b}$$

Force-Deformation Behavior

$$(V_i)_2 = \left(\frac{EI}{L^3}\right)_2 [12(\delta_i)_2 + 6L_2(\theta_i)_2] \tag{2c}$$

$$(M_i)_2 = \left(\frac{EI}{L^2}\right)_2 [6(\delta_i)_2 + 4L_2(\theta_i)_2] \tag{2d}$$

Geometry of Deformation: The δ and θ quantities in Eqs. (2) and in Figs. 3*b* and 3*d* can be related to the displacement and slope at node C, shown in Fig. 3*a*. Thus,

$$(\theta_j)_1 = (\theta_i)_2 = \theta_C \tag{3a}$$

Geometry of Deformation

$$(\delta_j)_1 = (\delta_i)_2 = \delta_C \tag{3b}$$

Combining Eqs. (1) through (3) in the order (3) $\rightarrow$ (2) $\rightarrow$ (1), we get the *equilibrium equations expressed in terms of displacement and slope at node C.*

$$12\left[\left(\frac{EI}{L_3}\right)_1 + \left(\frac{EI}{L^3}\right)_2\right]\delta_C - 6\left[\left(\frac{EI}{L^2}\right)_1 - \left(\frac{EI}{L^2}\right)_2\right]\theta_C = P \tag{4a}$$

$$-6\left[\left(\frac{EI}{L^2}\right)_1 - \left(\frac{EI}{L^2}\right)_2\right]\delta_C + 4\left[\left(\frac{EI}{L}\right)_1 + \left(\frac{EI}{L}\right)_2\right]\theta_C = 0 \tag{4b}$$

Complete Part (a): Equations (4) are two simultaneous, algebraic equations for the unknown joint displacement δ_C and slope θ_C. This is called a *displacement-method solution* because the unknowns in Eqs. (4) are displacements. Given numerical values of P, $(EI)_i$, and L_i, we could easily solve for numerical values of these joint displacements.[14]

(b) Solve for the reactions R_B and M_B. We can use Eqs. 7.19(c) and 7.19(d) to solve for R_B and M_B using the nodal displacements obtained from the solution of Eqs. (4). Thus,

$$R_B \equiv -(V_j)_2 = -\left(\frac{EI}{L^3}\right)_2 (-12\delta_C - 6L_2\theta_C) \tag{5a}$$

$$M_B \equiv (M_j)_2 = \left(\frac{EI}{L^2}\right)_2 (6\delta_C + 2L_2\theta_C) \tag{5b}$$

---

[14]The computer program BEAMDEF, which is described in Appendix G.6, may be used to carry out a displacement-method solution for this type of beam-deflection problem.

Review the Solution   As one check to verify that the above solutions will lead to correct answers for δ_C, θ_C, R_B, and M_B, let us take the special case of the symmetrical problem, with $EI = $ const, and $L_1 = L_2 = L/2$. This should lead to the result $\theta_C = 0$. Substituting the above values into Eq. (4b) we get

$$16\left(\frac{EI}{L}\right)\theta_C = 0 \rightarrow \theta_C = 0$$

which is the result that we would expect.

If all of the loads on a beam appear as concentrated loads (forces and/or couples) at nodes (joints), the problem is called a *nodal-load problem*. The preceding example problem is a typical nodal-load beam problem solved by the displacement method.

The next example illustrates how distributed loads are treated in a displacement-method solution. A key ingredient of a distributed-load solution is the inclusion of appropriate *fixed-end forces* for each individual element (Step 4 in the PROCEDURE on p. 407).

■■■■■■■■■■■■■■■■■■■■ E X A M P L E   7 . 1 9 ■■■■■■■■■■■■■■■■■■■■■■

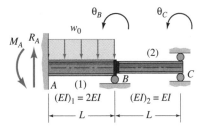

Fig. 1 A two-span continuous beam with uniform distributed loading over span *AB*.

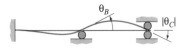

Fig. 2 The expected deflection curve.

The two-span continuous beam in Fig. 1 has a uniform distributed load over span *AB*. Use the displacement method: (a) to determine the unknown slopes θ_B and θ_C, and (b) to determine the reactions R_A and M_A at the fixed end *A*.

Plan the Solution   We need to write moment equilibrium equations for the joints at *B* and *C*, where the beam is free to rotate. Equations 7.19 can be used to relate element end moments to the corresponding element end rotations and to the fixed-end moment due to the distributed load on element (1).

Solution

(a) Determine the nodal slopes (rotations) θ_B and θ_C.

Sketch the Deflection Curve: The angles θ_B and θ_C are taken positive counterclockwise, as illustrated in Fig. 1. Therefore, the magnitude of the expected clockwise angle at node *C* is labeled $|\theta_C|$ in Fig. 2.

Show the Elements and Nodes Separately: All of the reactions, shear forces, and bending moments are shown on the nodes and elements in Fig. 3. As in Example Problem 7.18, we will apply equilibrium, force-deformation behavior, and geometry of deformation to complete our solution.

Nodal Equilibrium: Referring to the free-body diagrams of nodes *B* and *C*, we can write the following two moment-equilibrium equations; these correspond to the unknown rotations at *B* and *C*.

$$\left(\sum M\right)_B = 0: \qquad (M_j)_1 + (M_i)_2 = 0 \qquad\qquad \text{(1a)}$$

Equilibrium

$$\left(\sum M\right)_C = 0: \qquad\qquad (M_j)_2 = 0 \qquad\qquad \text{(1b)}$$

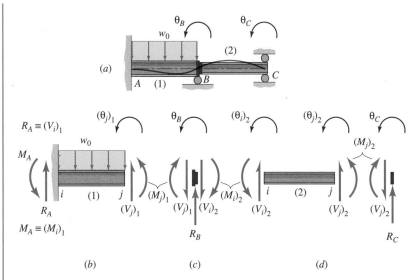

Fig. 3 The continuous beam separated into elements and nodes.

Element Force-Deformation Equations with Fixed-End Forces: Equation 7.19(d) is used to express $(M_j)_1$ in terms of the unknown slope $(\theta_j)_1$ and the fixed-end moment due to the uniform distributed load $(p_i)_1 = (p_j)_1 = -w_0$. Then Eqs. 7.19(b) and (d) are used to write expressions for the moments $(M_i)_2$ and $(M_j)_2$, respectively. Observe that $(EI)_1 = 2EI$ and that $L_1 = L_2 = L$.

$$(M_j)_1 = \frac{8EI}{L}(\theta_j)_1 - \frac{w_0 L^2}{12} \tag{2a}$$

$$(M_i)_2 = \frac{2EI}{L}[2(\theta_i)_2 + (\theta_j)_2] \tag{2b}$$

$$(M_j)_2 = \frac{2EI}{L}[(\theta_i)_2 + 2(\theta_j)_2] \tag{2c}$$

Force-Deformation Equations with Fixed-End Forces

Geometry of Deformation; Compatibility: In Fig. 1 the rotation angles θ_B and θ_C are taken positive counterclockwise in order to be consistent with the sign convention for θ_i and θ_j in Fig. 7.14*b*. Therefore,

$$(\theta_j)_1 = (\theta_i)_2 = \theta_B \tag{3a}$$

$$(\theta_j)_2 = \theta_C \tag{3b}$$

Geometry of Deformation

Equation (3a) is a *compatibility equation* that expresses the continuity of slope at *B*.

Complete Part (a): Finally, we substitute Eqs. (3) → (2) → (1), in displacement-method fashion, and get the *equilibrium equations in terms of unknown displacements*:

$$\left[\frac{8EI}{L} + \frac{4EI}{L}\right]\theta_B + \frac{2EI}{L}\theta_C = \frac{w_0 L^2}{12} \tag{4a}$$

$$\frac{2EI}{L}\theta_B + \frac{4EI}{L}\theta_C = 0 \tag{4b}$$

411

Finally, solving Eqs. (4) simultaneously, we get

$$\theta_B = \frac{w_0 L^3}{132 EI}, \qquad \theta_C = -\frac{w_0 L^3}{264 EI} \qquad \textbf{Ans. (a)} \quad (5)$$

(b) Determine the reactions R_A and M_A. As noted in Fig. 3, $R_A \equiv (V_i)_1$ and $M_A \equiv (M_i)_1$. Also, $(p_i)_1 = (p_j)_1 = -w_0$. Using Eqs. 7.19(a) and 7.19(b), together with Eqs. (3) above, we get

$$R_A \equiv (V_i)_1 = \frac{2EI}{L^2}(6\theta_B) + \frac{1}{2}w_0 L \qquad (6a)$$

$$M_A \equiv (M_i)_1 = \frac{2EI}{L}(2\theta_B) + \frac{1}{12}w_0 L^2 \qquad (6b)$$

Substitution of Eq. (5a) into Eqs. (6) gives

$$R_A = \frac{1}{11}w_0 L + \frac{1}{2}w_0 L \qquad (7a)$$

$$M_A = \frac{1}{33}w_0 L^2 + \frac{1}{12}w_0 L^2 \qquad (7b)$$

or

$$R_A = \frac{13}{22}w_0 L, \qquad M_A = \frac{15}{132}w_0 L^2 \qquad \textbf{Ans. (b)} \quad (8)$$

Review the Solution   If we consider element (1), we can see that the right end of element (1) is somewhere between fully clamped (no rotation) and propped (no moment). The solutions to the reactions for these two statically indeterminate problems are shown in Fig. 4. It can be seen that the reactions R_A and M_A obtained in Eq. (8) do, indeed, lie between the values of the reactions for these two beams.

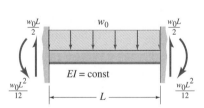   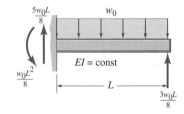

(a) A uniform fixed-fixed beam.   (b) A uniform propped-cantilever beam.

Fig. 4   Two beams related to element (1) of this problem.

Appendix G.6 is a continuation of this section. The computer program BEAMDEF, which is introduced in Appendix G.6, is based on the displacement-method solution procedure presented here in Section 7.7.

412

7.8 PROBLEMS ■■■

Problems 7.3-1 through 7.3-15. *These problems are to be solved by* **integrating the second-order differential equation**, *Eq. 7.8. The flexural rigidity, EI, is constant for each beam.*

Prob. 7.3-1. For the simply supported beam shown in Fig. P7.3-1, (a) determine the equation of the deflection curve, $v(x)$, (b) determine the angle of rotation θ_A at the left end, and (c) determine the maximum deflection, $\delta \equiv \max |v(x)|$.

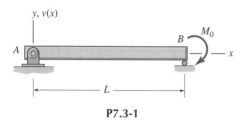

P7.3-1

Prob. 7.3-2. For the simply supported beam shown in Fig. P7.3-2, (a) determine the equation of the deflection curve, $v(x)$, (b) determine the angle of rotation θ_A at the left end, and (c) determine the maximum deflection, $\delta \equiv \max |v(x)|$.

P7.3-2 and P7.3-25

Prob. 7.3-3. The structural steel ($E = 29 \times 10^3$ ksi) simply supported **W**12×50 wide-flange beam in Fig. P7.3-3 is subjected to a couple M_A that produces a maximum deflection of 0.10 in. (a) What is the value of M_A? (b) What is the maximum flexural stress in the beam under this loading? See Table D.1 for the cross-sectional properties of the beam.

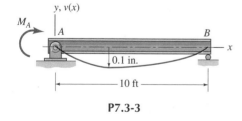

P7.3-3

Prob. 7.3-4. For the simply supported beam in Fig. P7.3-4, (a) determine the equation of the deflection curve, $v(x)$; (b) determine the angle of rotation θ_A at the left end; and (c) determine the maximum deflection, $\delta \equiv \max |v(x)|$. (Note:

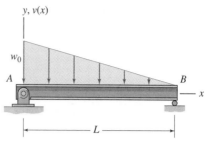

P7.3-4 and P7.3-26

You may have to use a trial and error solution to determine where the maximum deflection occurs.)

Prob. 7.3-5. Repeat Prob. 7.3-4 for the beam in Fig. P7.3-5.

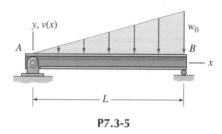

P7.3-5

Prob. 7.3-6. For the (two-interval) simply supported beam in Fig. P7.3-6, (a) determine the equations of the deflection curve, $v_a(x)$ and $v_b(x)$, where $0 \le x \le \dfrac{L}{2}$ for $v_a(x)$ and $\dfrac{L}{2} \le x \le L$ for $v_b(x)$; and (b) determine the vertical displacement at the center of the beam, δ_B.

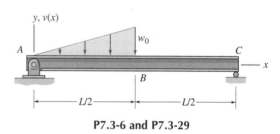

P7.3-6 and P7.3-29

Prob. 7.3-7. Repeat Prob. 7.3-6 for the beam in Fig. P7.3-7.

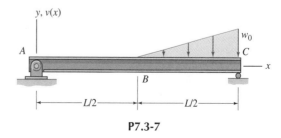

P7.3-7

Prob. 7.3-8. For the (two-interval) simply supported beam in Fig. P7.3-8, determine the equations of the deflection curve, $v_a(x)$ and $v_b(x)$, where $0 \leq x \leq a$ for $v_a(x)$ and $a \leq x \leq (a + b)$ for $v_b(x)$. Express your answers in terms of M_0, E, I, a, b, and the coordinate x.

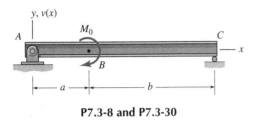

P7.3-8 and P7.3-30

Prob. 7.3-9. A uniform shaft is supported by bearings at A and C and is subjected to a downward load P through the pulley at B. Neglect the width of the pulley and bearings, and assume that the bearings provide only vertical support to the shaft. (a) Determine expressions for the deflection curve $v_a(x)$ in section AB and the deflection curve $v_b(x)$ in the section BC. (b) The maximum deflection of the shaft between bearings A and C will occur between B and C, as shown in Fig. P7.3-9. Determine an expression for $\delta \equiv \max |v_b(x)|$.

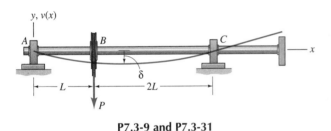

P7.3-9 and P7.3-31

Prob. 7.3-10. A uniform load of intensity w_0 per unit length is applied to the overhanging segment, AB, of the beam in Fig. P7.3-10. (a) Determine expressions for the deflection curve $v_a(x)$ in segment AB and the deflection curve $v_b(x)$ in the section BC. (b) Determine the deflection δ_A at the left end of the beam. (c) Determine the maximum (upward) deflection, $\delta \equiv \max[v_b(x)]$, in the section BC.

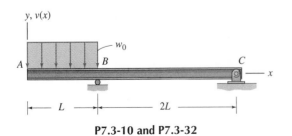

P7.3-10 and P7.3-32

Prob. 7.3-11. A person standing at end C of a diving board exerts a downward force W, as shown in Fig. P7.3-11. (a) Determine expressions for the deflection curve $v_a(x)$ in segment AB and the deflection curve $v_b(x)$ in the section BC. (b) Determine the deflection δ_C where the diver is standing. (c) Determine the maximum (upward) deflection, $\delta \equiv \max[v_a(x)]$, in segment AB.

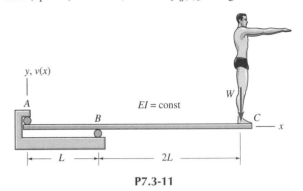

P7.3-11

Prob. 7.3-12. A 10-ft-long **W**8×15 wide-flange beam AB is supported by two 6-ft-long rods AC and BD of cross-sectional area $A_r = 1.0$ in^2. The beam supports a uniformly distributed load $w_0 = 1$ kip/ft, as shown in Fig. P7.3-12. Determine the maximum deflection of the beam, taking into account the elongation of the two rods but neglecting the weight of the beam and the rods. Use $E = 30 \times 10^3$ ksi for the rods and the beam. (Hint: Solve first for the elongation of the rods, and then use these elongations as the boundary conditions $v(0)$ and $v(L)$ for the beam.)

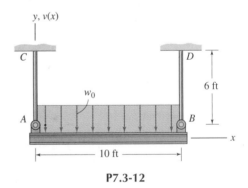

P7.3-12

Prob. 7.3-13. The cantilever beam AB in Fig. P7.3-13 supports a triangularly distributed load of maximum intensity w_0. (a) Determine an expression for the deflection curve, $v(x)$, for this beam, and (b) determine expressions for the slope θ_B and the deflection δ_B at end B.

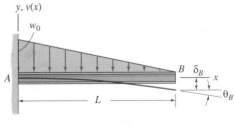

P7.3-13 and P7.3-33

Prob. 7.3-14. Repeat Prob. 7.3-13 for the cantilever beam AB loaded as shown in Fig. P7.3-14.

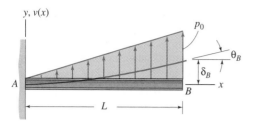

P7.3-14 and P7.3-34

Prob. 7.3-15. Loads P_A and P_B are applied to a uniform cantilever beam AC, as shown in Fig. P7.3-15, causing deflections δ_A and δ_B at A and B, respectively. Using the *second-order integration method*, determine expressions for P_A and P_B in terms of E, I, L, δ_A, and δ_B.

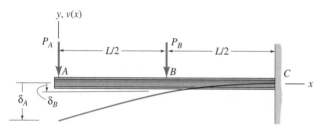

P7.3-15

Problems 7.3-16 through 7.3-24. *In solving these problems, you may use deflection formulas given in Tables E.1 and E.2 of Appendix E. The flexural rigidity, EI, is constant for each beam in this group of problems.*

Prob. 7.3-16. The simply supported beam AB in Fig. P7.3-16 carries a uniformly distributed load of intensity w_0 on a span of $L = 8$ ft. Determine the maximum deflection, δ, if the depth of the beam is $h = 6$ in., the maximum flexural stress in the beam is $\sigma_{max} = 12$ ksi, and the beam is made of aluminum with $E = 10 \times 10^3$ ksi.

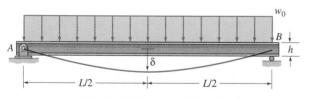

P7.3-16, P7.3-17, and P7.3-18

Prob. 7.3-17. A $\mathbf{W}200 \times 59$ wide-flange steel beam is simply supported and carries a uniformly distributed load of $w_0 = 10$ kN/m on a span of $L = 5$ m (see Fig. P7.3-17). For $E = 200$ GPa, (a) determine the maximum deflection of the beam, and (b) determine the maximum flexural stress in the beam. (See Table D.2 of Appendix D for the cross-sectional properties of the beam.)

Prob. 7.3-18. What is the depth h of a uniformly loaded, simply supported beam (see Fig. 7.3-18) if the maximum bending stress is $\sigma_{max} = 8$ ksi, the maximum deflection is $\delta = 0.20$ in., the span is $L = 10$ ft, and the modulus of elasticity is $E = 30 \times 10^3$ ksi?

Prob. 7.3-19. A simply supported beam AC carries a concentrated load P at its midspan, B (Fig. P7.3-19). Determine the maximum deflection, δ, if the span is $L = 8$ ft, the depth of the beam is $h = 6$ in., the maximum flexural stress is $\sigma_{max} = 12$ ksi, and the beam is made of aluminum with $E = 10 \times 10^3$ ksi.

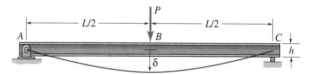

P7.3-19, P7.3-20, and P7.3-21

Prob. 7.3-20. A simply supported beam AC carries a concentrated load P at its midspan point B (see Fig. P7.3-20). Determine the depth of the beam h in mm if the maximum bending stress is $\sigma_{max} = 50$ MPa, the maximum deflection is $\delta = 5$ mm, the span is $L = 3$ m, and the modulus of elasticity is $E = 200$ GPa.

Prob. 7.3-21. A $\mathbf{W}8 \times 40$ wide-flange steel beam is simply supported and carries a concentrated load of $P = 10$ kips at its midspan, B (see Fig. P7.3-21). The span of the beam is $L = 12$ ft and its modulus of elasticity is $E = 30 \times 10^3$ ksi. (a) Determine the maximum deflection of the beam, and (b) determine the maximum flexural stress in the beam. (See Table D.1 of Appendix D for the cross-sectional properties of the beam.)

Prob. 7.3-22. A cantilever beam AB carries a uniformly distributed load of intensity w_0 on a span of $L = 8$ ft (Fig. P7.3-22). Determine the maximum deflection, δ, if the depth of the beam is $h = 6$ in., the maximum flexural stress in the beam is $\sigma_{max} = 12$ ksi, and the beam is made of steel with $E = 30 \times 10^3$ ksi.

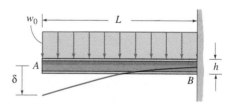

P7.3-22, P7.3-23, and P7.3-24

Prob. 7.3-23. A $\mathbf{W}16 \times 40$ wide-flange steel cantilever beam AB carries a uniformly distributed load of $w_0 = 2.7$ kips/ft on a span of $L = 8$ ft (see Fig. P7.3-23). $E = 30 \times 10^3$ ksi. (a) Determine the maximum deflection, δ, and (b) determine the maximum flexural stress in the beam. (See Table D.1 of Appendix D for the cross-sectional properties of the beam.)

Prob. 7.3-24. What is the maximum bending stress in the uniformly loaded cantilever beam in Fig. P7.3-24 if the depth of the beam is $h = 250$ mm, the maximum deflection is $\delta = 5$ mm, the span is $L = 3$ m, and the modulus of elasticity is $E = 70$ GPa?

415

Problems 7.3-25 through 7.3-34. *These problems are to be solved by* **integrating the fourth-order differential equation**, *Eq. 7.12. The flexural rigidity, EI, is constant for each beam.*

Prob. 7.3-25. Use the fourth-order method to solve the problem as stated in Prob. 7.3-2.

Prob. 7.3-26. Use the fourth-order method to solve the problem as stated in Prob. 7.3-4.

Prob. 7.3-27. The simply supported beam in Fig. P7.3-27 is subjected to a distributed load of intensity $p(x) = -w_0 \sin \dfrac{\pi x}{L}$. [Note: This load acts downward, so there is a minus sign in the expression for $p(x)$.] (a) Use the fourth-order method to determine an expression for the deflection curve of this beam, and (b) determine the slope, θ_A, of the beam at end A.

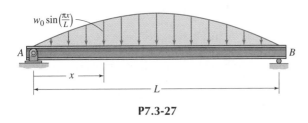

P7.3-27

Prob. 7.3-28. The simply supported beam in Fig. P7.3-28 is subjected to a distributed load of intensity $p(x) = -w_0 \left(\dfrac{x}{L}\right)^2$. [Note: This load acts downward, so there is a minus sign in the expression for $p(x)$.] (a) Use the fourth-order method to determine the equation of the deflection curve, $v(x)$, and (b) determine the maximum deflection, $\delta \equiv \max |v_b(x)|$, of this beam.

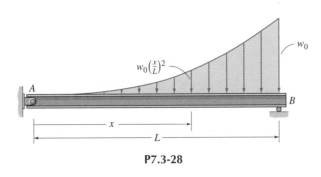

P7.3-28

Prob. 7.3-29. Use the fourth-order method to solve the problem as stated in Prob. 7.3-6.

Prob. 7.3-30. Use the fourth-order method to solve the problem as stated in Prob. 7.3-8.

Prob. 7.3-31. Use the fourth-order method to solve the problem as stated in Prob. 7.3-9.

Prob. 7.3-32. Use the fourth-order method to solve the problem as stated in Prob. 7.3-10.

Prob. 7.3-33. Use the fourth-order method to solve the problem as stated in Prob. 7.3-13.

Prob. 7.3-34. For the beam in Prob. 7.3-14, (a) use the fourth-order integration method to determine an expression for the tip deflection, δ_B, and then (b) determine an expression for the length L that would make the maximum tensile stress in the beam equal to σ_{allow} and the corresponding maximum deflection equal to δ_{allow}. Express your answer in terms of E, δ_{allow}, σ_{allow}, and the depth of the beam, h.

Problems 7.4-1 through 7.4-11. *These problems are to be solved by* **integrating the second-order differential equation**, *Eq. 7.8. The flexural rigidity, EI, is constant for the beams in Probs. 7.4-1 through 7.4-10.*

Prob.7.4-1. For the uniformly loaded propped cantilever beam in Fig. P7.4-1, use the second-order integration method (a) to solve for the reactions at A and B, and (b) to determine an expression for the deflection curve $v(x)$.

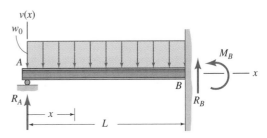

P7.4-1, P7.4-12, P7.6-33, P7.6-34, and P11.3-40

Prob.7.4-2. The propped cantilever beam in Fig. P7.4-2 has a **W**8 × 40 cross section and supports a uniformly distributed load of $w_0 = 5$ kips/ft on a span of $L = 8$ ft. (a) Use the second-order integration method to solve for the reactions at A and B. (Solve the problem using symbols, w_0, EI, etc., and then substitute numerical values for these quantities.) (b) Sketch the shear and moment diagrams for this beam.

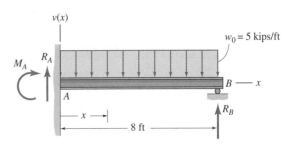

P7.4-2, P7.6-35, P7.6-36, P11.8-18, and P11.8-42

Prob. 7.4-3. The propped cantilever beam AB in Fig. P7.4-3 supports a linearly varying load of maximum intensity w_0. (a) Use the second-order integration method to solve for the reactions at

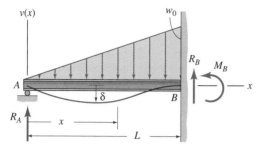

P7.4-3, P7.4-13, P7.6-37, P7.6-38, P11.8-17, and P11.8-41

A and B and for the deflection curve $v(x)$. (b) Determine an expression for the maximum deflection, $\delta \equiv \max |v(x)|$. (c) Sketch the shear diagram, $V(x)$.

Prob. 7.4-4. The uniformly loaded beam in Fig. P7.4-4 is completely fixed at ends A and B. (a) Use the second-order integration method to determine the reactions R_A and M_A and to determine an expression for the deflection curve $v(x)$. (b) Sketch the shear diagram, $V(x)$, and the moment diagram, $M(x)$.

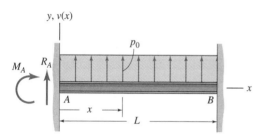

P7.4-4, P7.6-39, and P7.6-40

Prob. 7.4-5. The beam AB in Fig. P7.4-5 supports a linearly varying load of maximum intensity w_0. (a) Use the second-order integration method to determine the reactions at A and B and to determine an expression for the deflection curve $v(x)$. (b) Determine an expression for the maximum deflection, $\delta \equiv \max |v(x)|$. (c) Sketch the shear diagram, $V(x)$.

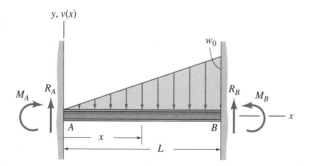

P7.4-5, P7.4-14, P7.6-41, P7.6-42, P11.3-41, and P11.8-46

*Prob. 7.4-6. At end B, the cantilever beam in Fig. P7.4-6 is pinned to a uniform rod whose cross-sectional area is A_2, whose length is L_2, and whose modulus of elasticity is E_2. The beam

supports a uniformly distributed load of intensity w_0; its flexural rigidity is $E_1 I_1$, and its length is L_1. (a) Use the second-order integration method to determine the reactions R_A and M_A at A, and the tension, F_2, in the rod. (b) Determine an expression for the deflection curve, $v(x)$, of the beam.

Prob. 7.4-7. The propped cantilever beam in Fig. P7.4-7 is subjected to a concentrated load P at distance a from end A. (a) Use the second-order integration method to determine the reactions at A and C and the deflection curves $v_a(x)$ and $v_b(x)$ for the segments of the beam to the left of load P and to the right of load P, respectively. (b) Letting $a = L/3$, sketch the shear diagram, $V(x)$, and the moment diagram, $M(x)$.

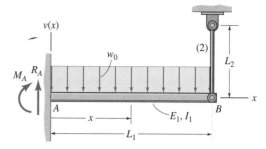

P7.4-6, P7.6-43, and P11.8-47

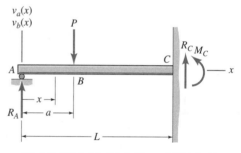

P7.4-7, P7.6-44, P7.6-45, and P11.8-19

*Prob. 7.4-8. The fixed-fixed beam in Fig. P7.4-8 is subjected to a concentrated load P at distance a from end A. (a) Use the second-order integration method to determine the reactions at A and C and the deflection curves $v_a(x)$ and $v_b(x)$ for the segments of the beam to the left of load P and to the right of load P, respectively. (b) Letting $a = L/3$, sketch the shear diagram, $V(x)$, and the moment diagram, $M(x)$.

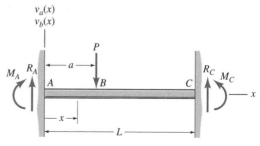

P7.4-8, P7.4-15, P7.6-46, and P11.8-20

417

*Prob. 7.4-9. The fixed-fixed beam in Fig. P7.4-9 is subjected to a uniformly distributed load of intensity w_0 over the interval AB (i. e., $0 \le x \le a$). (a) Use the second-order integration method to determine the reactions at A and C and the deflection-curve expressions $v_a(x)$ (for $0 \le x \le a$) and $v_b(x)$ (for $a \le x \le L$) (b) Letting $a = L/2$, sketch the complete shear diagram, $V(x)$, for this beam.

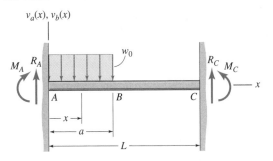

P7.4-9, P7.4-16, P7.6-47, P7.6-48, P11.8-23, and P11.8-44

Prob. 7.4-10. The continuous beam in Fig. P7.4-10 is simply supported at ends A and C and is propped at its midpoint B. It supports a uniformly distributed load of intensity w_0 on the span AB. (a) Use the second-order integration method to determine the reactions at A, B, and C. (b) Determine expressions for the deflection curves $v_a(x)$ in span AB and $v_b(x)$ in span BC. (c) Sketch the shear diagram, $V(x)$, and the moment diagram, $M(x)$.

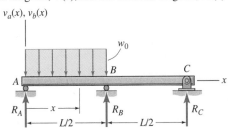

P7.4-10, P7.6-49, P11.8-21, and P11.8-43

*Prob. 7.4-11. The beam AC in Fig. P7.4-11 is fixed to a rigid wall at A and is supported by props at B and C. In span AB the flexural rigidity is EI, but in span BC the flexural rigidity is $2EI$. The beam supports a uniformly distributed load over span BC. (a) Use the second-order integration method to determine the reactions at A, B, and C. (b) Determine expressions for the deflection curves $v_a(x)$ in span AB and $v_b(x)$ in span BC. (c) Sketch the shear diagram, $V(x)$, and the moment diagram, $M(x)$.

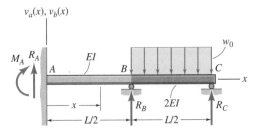

P7.4-11, P7.6-50, and P11.8-22

Problems 7.4-12 through 7.4-20. *These problems are to be solved by* **integrating the fourth-order differential equation**, *Eq. 7.12. The flexural rigidity, EI, is constant for all beams in this group of problems.*

Prob. 7.4-12. Re-solve Prob. 7.4-1 using the fourth-order integration method.

Prob. 7.4-13. Re-solve Prob. 7.4-3 using the fourth-order integration method.

Prob. 7.4-14. Re-solve Prob. 7.4-5 using the fourth-order integration method.

*Prob. 7.4-15. Re-solve Prob. 7.4-8 using the fourth-order integration method. Let $a = L/3$.

*Prob. 7.4-16. Re-solve Prob. 7.4-9 using the fourth-order integration method. Let $a = L/2$.

Problems 7.4-17 through 7.4-20. *For each of these problems, use the* **fourth-order integration method** *to determine the reactions at ends A and B and to determine expressions for the deflection curve $v(x)$.*

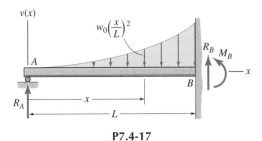

P7.4-17

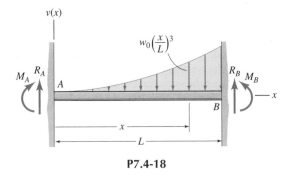

P7.4-18

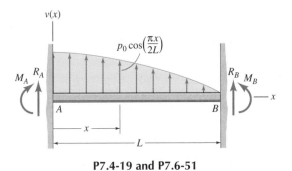

P7.4-19 and P7.6-51

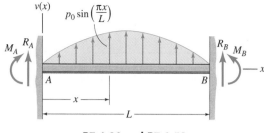

P7.4-20 and P7.6-52

Use the discontinuity-function method *of Section 7.5 to solve each of the following problems. Let EI = const for each beam.*

Probs. 7.5-1 through 7.5-4. For the beams shown in Figs. P7.5-1 through 7.5-4, (a) determine discontinuity-function expressions for the slope $\theta(x)$ and the deflection $v(x)$ for $0 \leq x \leq L$, and (b) evaluate the displacement at points B and C.

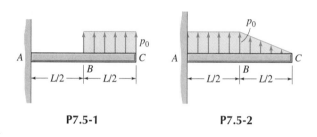

P7.5-1   **P7.5-2**

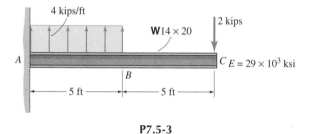

P7.5-3

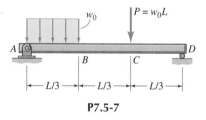

P7.5-4

Prob. 7.5-5. For the simply supported beam shown, (a) determine discontinuity-function expressions for the slope $\theta(x)$ and the deflection $v(x)$ for $0 \leq x \leq L$, and (b) evaluate the deflection $v(\frac{L}{4})$.

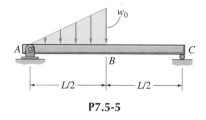

P7.5-5

Prob. 7.5-6. For the overhanging beam shown in Fig. P7.5-6, (a) determine discontinuity-function expressions for the slope $\theta(x)$ and the deflection $v(x)$ for $0 \leq x \leq 3L$; (b) determine expressions for the slope and deflection at $x = 0$, that is, for $\theta(0)$ and $v(0)$; and (c) determine an expression for the maximum deflection between the supports B and C.

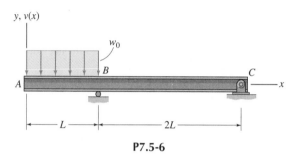

P7.5-6

Prob. 7.5-7. For the simply supported beam shown, (a) determine expressions for the end slopes $\theta_A \equiv \theta(0)$ and $\theta_D \equiv \theta(L)$, (b) determine discontinuity-function expressions for $\theta(x)$ and $v(x)$ for $0 \leq x \leq L$, and (c) determine an expression for the midspan deflection $v(\frac{L}{2})$.

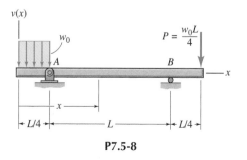

P7.5-7

Prob. 7.5-8. For the three-segment overhanging beam shown in Fig. P7.5-8, (a) determine discontinuity-function expressions for the slope $\theta(x)$ and the deflection $v(x)$ for $0 \leq x \leq \frac{3L}{2}$; and (b) determine the maximum upward deflection of the beam.

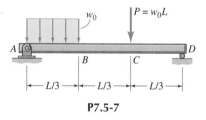

P7.5-8

419

Prob. 7.5-9. For the beam shown in Fig. P7.5-9, use the discontinuity-function method to determine an expression for the end couple M_0, in terms of P and L, such that the deflection at end D is zero.

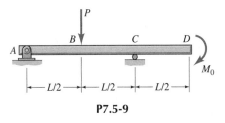

P7.5-9

Probs.7.5-10 through 7.5-12. For the propped cantilever beams shown in Figs. P7.5-10 through 7.5-12, use the discontinuity-function method (a) to determine expressions for the reactions R_A, R_B, M_B; and (b) to determine expressions for the slope $\theta(x)$ and the deflection $v(x)$ for $0 \le x \le L$.

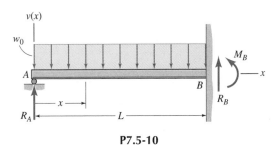

P7.5-10

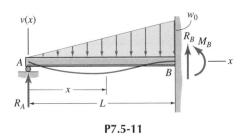

P7.5-11

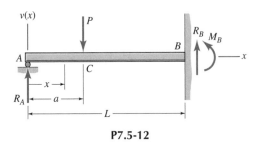

P7.5-12

Prob. 7.5-13. For the propped cantilever beam shown, use the discontinuity-function method (a) to determine expressions for

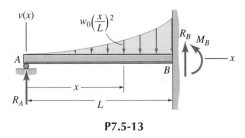

P7.5-13

the reactions R_A, R_B, and M_B, and (b) to determine expressions for the slope $\theta(x)$ and the deflection $v(x)$ for $0 \le x \le L$.

Probs.7.5-14 and 7.5-15. For the clamped-clamped beam shown, use the discontinuity-function method (a) to determine expressions for the reactions at A and B, that is, R_A, M_A, R_B, and M_B, and (b) to determine expressions for the slope $\theta(x)$ and the deflection $v(x)$ of the beam.

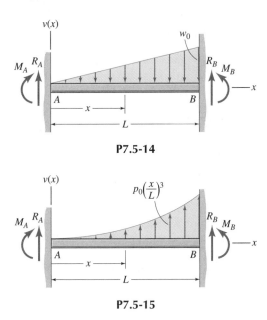

P7.5-14

P7.5-15

Probs.7.5-16 through 7.5-18. For the clamped-clamped beam shown, use the discontinuity-function method (a) to determine expressions for the reactions at A and C, that is, R_A, M_A, R_C, and M_C; (b) to determine discontinuity-function expressions for the slope $\theta(x)$ and the deflection $v(x)$ for $0 \le x \le L$; and (c) to determine an expression for the deflection at point B.

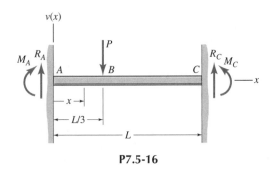

P7.5-16

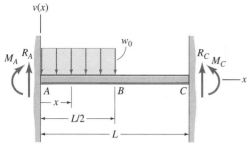

P7.5-17

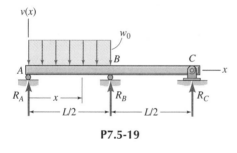

P7.5-18

*Prob. 7.5-19 For the two-span continuous beam shown, use the discontinuity-function method (a) to determine the reactions R_A, R_B, and R_C; (b) to determine discontinuity-function expressions for the slope $\theta(x)$ and the deflection $v(x)$ for $0 \leq x \leq L$; and (c) to evaluate the slope expression $\theta(x)$ at support A.

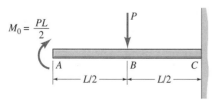

P7.5-19

Problems 7.6-1 through 7.6-56. *Using Tables E.1 and E.2, De-*
flections and Slopes of Beams, **apply the force method of su-**
perposition *to solve each of the problems of this section. Let the*
flexural rigidity, EI, be constant for each beam.
Probs. 7.6-1 and 7.6-2. Determine expressions for the slope θ_A
and the deflection δ_A at end A.

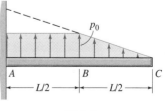

P7.6-1

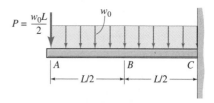

P7.6-2

Probs. 7.6-3 and 7.6-4. Determine expressions for the slope θ_C
and the deflection δ_C at end C.

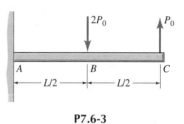

P7.6-3

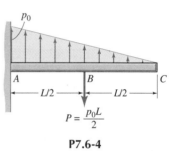

P7.6-4

Probs. 7.6-5 and 7.6-6. Determine expressions for the slope θ_C
and the deflection δ_C at end C. (Hint: You can subtract the effect
of loading indicated by the dashed lines.)

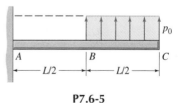

P7.6-5

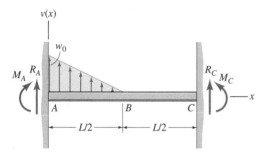

P7.6-6

Probs. 7.6-7 and 7.6-8. Determine expressions for the slope θ_A at end A and the deflection δ_C at section C.

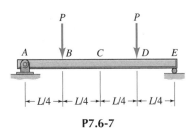

P7.6-7

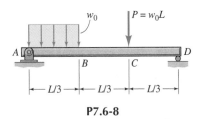

P7.6-8

Prob. 7.6-9. Determine expressions for the slope θ_A at end A and the midspan deflection, δ_B.

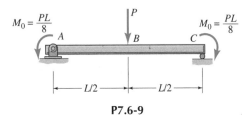

P7.6-9

Prob. 7.6-10. Determine an expression for the value of the end couple M_0, in terms of P and L, such that the deflection at end D is $\delta_D = 0$.

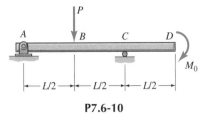

P7.6-10

Probs. 7.6-11 and 7.6-12. (a) Determine expressions for the slope $v'(x)$ and the deflection $v(x)$ in the segment AB, and (b) determine the maximum upward deflection of the beam.

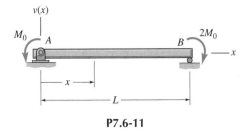

P7.6-11

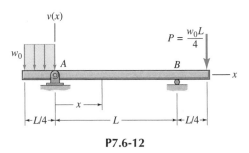

P7.6-12

*Prob. 7.6-13. For the antisymmetric loading in Fig. P7.6-13, (a) determine expressions for the slope $v'(x)$ and the deflection $v(x)$ in the segment AB, and (b) determine the maximum upward deflection of the beam, which occurs in segment AB.

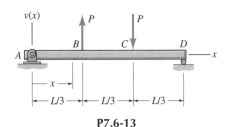

P7.6-13

*Prob.7.6-14. For the beam shown in Fig. P7.6-14, (a) determine expressions for the slope $v'(x)$ and the deflection $v(x)$ in the segment BC, and (b) determine the maximum (downward) deflection of the beam.

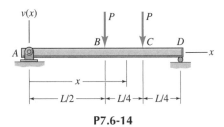

P7.6-14

Probs. 7.6-15 and 7.6-16. Determine expressions for the slope θ_C and the deflection δ_C at end C.

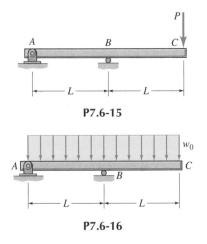

P7.6-15

P7.6-16

Problems 7.6-17 through 7.6-24. *Use the method of Example Problem 7.13 to solve Probs. P7.6-17 through P7.6-24.*

Probs. 7.6-17 and 7.6-18. Determine expressions for the slope θ_B and deflection δ_B at end B.

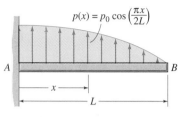

$$p(x) = p_0 \cos\left(\frac{\pi x}{2L}\right)$$

P7.6-17

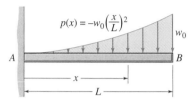

$$p(x) = -w_0\left(\frac{x}{L}\right)^2$$

P7.6-18

Prob.7.6-19. Determine expressions for the slope θ_D and deflection δ_D at end D.

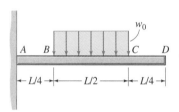

P7.6-19

Prob. 7.6-20. Determine expressions for the slope θ_C and deflection δ_C at end C.

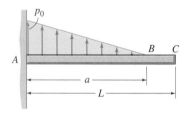

P7.6-20

Problems 7.6-21 through 7.6-24. *Determine expressions for the slope θ_A at end A and deflection δ_B at section B.*

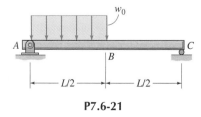

P7.6-21

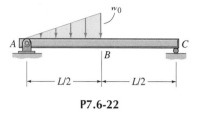

P7.6-22

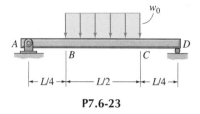

P7.6-23

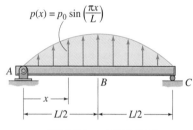

$$p(x) = p_0 \sin\left(\frac{\pi x}{L}\right)$$

P7.6-24 and P11.8-30

Prob. 7.6-25. Determine the slope θ_C and deflection δ_C at end C of the $\mathbf{W}14 \times 120$ wide-flange beam in Fig. P7.6-25. (See Table D.1 for the cross-sectional properties of the beam.) Let $E_{\text{steel}} = 29 \times 10^3$ ksi.

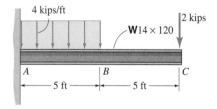

P7.6-25, PC7.7-1, P11.8-7, and P11.8-32

Prob. 7.6-26. Determine the slope θ_C and deflection δ_C at end C of the timber beam in Fig. P7.6-26. Let $E_w = 11$ GPa.

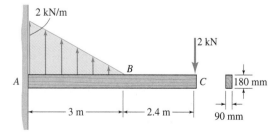

P7.6-26, PC7.7-2, P11.8-8, and P11.8-33

423

Prob. 7.6-27. An 8-in. (nominal) standard steel pipe acts as a cantilever beam that supports two 7-kip loads from hanger rods as shown in Fig. P7.6-27. (See Table D.7 for the cross-sectional properties of the pipe.) Determine the deflections δ_B at section B and δ_C at the right end. Let $E_{steel} = 29 \times 10^3$ ksi.

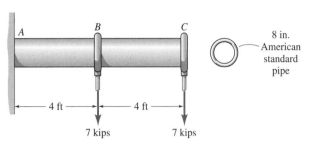

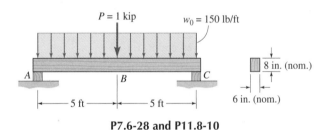

P7.6-27, PC7.7-3, and P11.8-9

Prob. 7.6-28. A 6 in. $\times$ 8 in. (nominal) simply-supported timber beam supports a uniformly distributed load of $w_0 = 150$ lb/ft and a concentrated midspan load of 1 kip. (See Table D.8 for the cross-sectional properties of this beam.) Calculate the slope θ_A at end A and the deflection δ_B at midspan. Let $E_w = 1,600$ ksi.

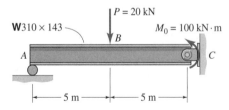

P7.6-28 and P11.8-10

Prob. 7.6-29. Determine the slope θ_A at end A and the deflection δ_B at the midspan section B of the $\mathbf{W}310 \times 143$ steel beam in Fig. P7.6-29. (See Table D.2 for the cross-sectional properties of the beam.) Let $E_{steel} = 200$ GPa.

P7.6-29, P7.6-30, PC7.7-4, and P11.8-11

Prob. 7.6-30. Determine the maximum deflection of the wide-flange beam AC in Fig. P7.6-30.

*Prob. 7.6-31. A wide-flange beam supports a uniform load of 2 kips/ft on a 14-ft span. The beam has a moment of inertia of $I = 300$ in^4, and at end B it is supported by a steel rod with a cross-sectional area of 0.50 in^2. Let $E_{steel} = 29 \times 10^3$ ksi. (a) Determine the deflection of the beam at B. (b) Determine

expressions for the slope $v'(x)$ and deflection $v(x)$ of the beam by superposing the deflection of a simply supported flexible beam and the deflection that a "rigid" beam would experience due to the stretching of rod BC. (c) Determine the maximum deflection of the beam.

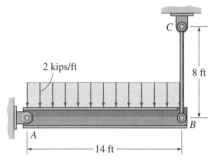

P7.6-31 and P11.8-34

*Prob. 7.6-32. An aluminum-alloy beam has a moment of inertia $I = 50 \times 10^6$ mm^4 and is supported by a pin at A and a 3-m-long aluminum rod that is attached to the beam at C. The cross-sectional area of the rod CD is 200 mm^2. Let $E_a = 70$ GPa. (a) Determine the displacement of the beam at C. (b) By superposing the deflection of a simply-supported flexible beam and the deflection that a "rigid" beam would experience due to the stretching of rod CD, determine the total deflection at the load point, B.

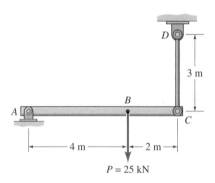

P7.6-32 and P11.8-35

Prob. 7.6-33. Use superposition of beam-deflection solutions from Table E.1 to solve Prob. 7.4-1.

Prob. 7.6-34. Use superposition of beam-deflection solutions from Table E.2 to solve Prob. 7.4-1.

Prob. 7.6-35. Use superposition of beam-deflection solutions from Table E.1 to solve Prob. 7.4-2.

Prob. 7.6-36. Use superposition of beam-deflection solutions from Table E.2 to solve Prob. 7.4-2.

Prob. 7.6-37. Use superposition of beam-deflection solutions from Table E.1 to solve Prob. 7.4-3.

Prob. 7.6-38. Use superposition of beam-deflection solutions from Table E.2 to solve Prob. 7.4-3.

Prob. 7.6-39. Use superposition of beam-deflection solutions from Table E.1 to solve Prob. 7.4-4.

Prob. 7.6-40. Use superposition of beam-deflection solutions from Table E.2 to solve Prob. 7.4-4.

Prob. 7.6-41. Use superposition of beam-deflection solutions from Table E.1 to solve Prob. 7.4-5.

Prob. 7.6-42. Use superposition of beam-deflection solutions from Table E.2 to solve Prob. 7.4-5.

Prob. 7.6-43. Use superposition of beam-deflection solutions from Table E.1 to solve Prob. 7.4-6.

Prob 7.6-44. Use superposition of beam-deflection from Table E.1 to solve Prob. 7.4-7.

Prob. 7.6-45. Use superposition of beam-deflection solutions from Table E.2 to solve Prob. 7.4-7.

Prob. 7.6-46. Use superposition of beam-deflection solutions from Table E.2 to solve Prob. 7.4-8.

Prob. 7.6-47. Use superposition of beam-deflection solutions from Table E.1 to solve Prob. 7.4-9.

Prob. 7.6-48. Use superposition of beam-deflection solutions from Table E.2 to solve Prob. 7.4-9.

Prob. 7.6-49. Use superposition of beam-deflection solutions from Table E.2 to solve Prob. 7.4-10.

Prob. 7.6-50. Use superposition of beam-deflection solutions from Table E.2 to solve Prob. 7.4-11.

Prob. 7.6-51. Use superposition of beam-deflection solutions from Table E.1 to solve Prob. 7.4-19.

Prob. 7.6-52. Use superposition of beam-deflection solutions from Table E.2 to solve Prob. 7.4-20.

Prob. 7.6-53. The cantilever beam AC in Fig. P7.6-53 has additional support from cable CD. The beam supports a uniform load of intensity w_0 over half of its length, as shown. Before the load is applied, the cable is taut, but force-free. Determine an expression that relates the cable tension to the load intensity, w_0; the flexural rigidity of the beam, EI; the axial rigidity of the cable, EA; and the lengths of the beam and the cable.

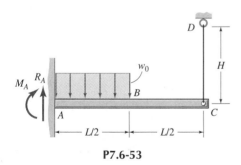

P7.6-53

Prob. 7.6-54. The cantilever beam AC in Fig. P7.6-54 has a moment of inertia $I = 50 \times 10^6$ mm^4 and is supported by rod CD, whose cross-sectional area is $A = 200$ mm^2. Let $E_{\text{beam}} = E_{\text{rod}}$. A concentrated load $P = 25$ kN is applied to the beam at B. The rod CD is force-free prior to application of the load P. Determine the tension induced in rod CD.

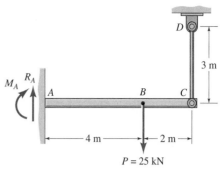

P7.6-54

Prob. 7.6-55. The simply supported beam AC in Fig. P7.6-55 has additional support from a cable BD that is attached to the beam at its midspan, B. The beam supports a uniform load of intensity w_0 over its entire length, as shown. Before the load is applied, the cable is taut but force-free. Determine an expression that relates the cable tension to the load intensity, w_0; the flexural rigidity of the beam, EI; the axial rigidity of the cable, EA; and the lengths of the beam and the cable.

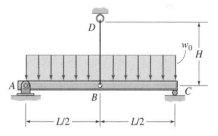

P7.6-55 and P7.6-56

*Prob. 7.6-56. (a) Let T be the tension in cable BD, and draw shear and moment diagrams of the beam AC in Fig. P7.6-56. (b) Determine the relationship between T and the total load w_0L that minimizes the maximum magnitude of the moment in the beam AC. (c) If all values except H are specified (i.e., EI, AE, and L are given), what value of H would give the condition described in Part (b)?

The problems in this section are to be solved by using displacement method superposition. *The key equations relating forces to displacements are Eqs. 7.19 and 7.20 (p. 407). For distributed-load problems, Table E.3 provides the basic* fixed-end forces. *The bending stiffness, EI, is constant unless otherwise noted.*

Prob. 7.7-1. A couple M_0 is applied at the propped end of the uniform propped-cantilever beam in Fig. P7.7-1. Determine the reaction R_A.

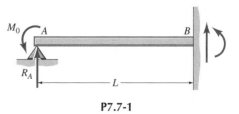

P7.7-1

Probs. 7.7-2 and 7.7-3. For the fixed-fixed beams shown, (a) determine the transverse displacement δ_B and the slope θ_B at node B, and (b) determine the reactions R_A and M_A at end A. (Note, the lengths are $L_1 = L$ and $L_2 = 2L$.)

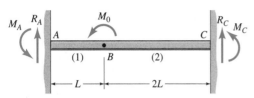

P7.7-2, PC7.7-5, and P11.8-24

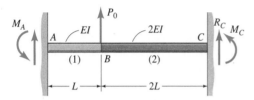

P7.7-3, PC7.7-7, P11.8-25, and P11.8-45

Prob. 7.7-4. An external couple M_0 is applied at node B of the two-span beam shown in Fig. P7.7-4. (a) Determine θ_B, the angle of rotation at B, and (b) determine the reactions R_A, M_A, and R_B.

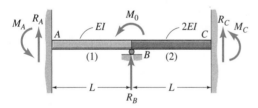

P7.7-4 and P11.8-52

Probs. 7.7-5 through 7.7-7. For the two-span beams with distributed loading as shown, (a) determine θ_B, the angle of rotation at B, and (b) determine the reactions R_A, M_A, and R_B.

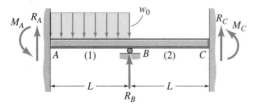

P7.7-5 and PC7.7-8

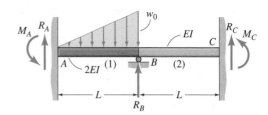

P7.7-6 and PC7.7-9

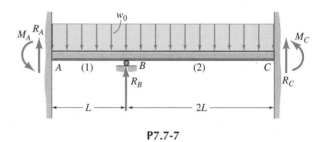

P7.7-7

Probs. 7.7-8 and 7.7-9. For the uniform fixed-fixed beams shown, (a) determine the transverse displacement δ_B and the slope θ_B at node B, and (b) determine the reactions R_A and M_A at end A.

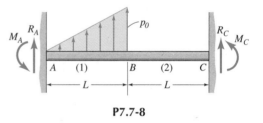

P7.7-8

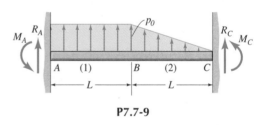

P7.7-9

Probs. 7.7-10 and 7.7-11. For the two-span beams with distributed loading as shown, (a) determine θ_A and θ_B, the angles of rotation at nodes A and B, respectively, and (b) determine the reactions R_A and R_B at the respective nodes.

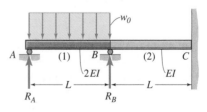

P7.7-10 and P11.8-48

426

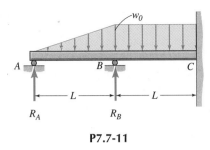

P7.7-11

Probs. 7.7-12 through 7.7-15. For the two-span continuous beams shown, (a) determine the rotation angles at the three supports: θ_A, θ_B, and θ_C; and (b) determine the reaction at A, R_A.

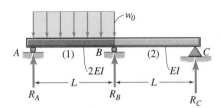

P7.7-12, PC7.7-6, and P11.8-49

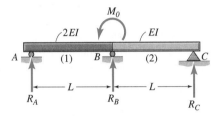

P7.7-13 and P11.8-50

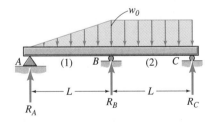

P7.7-14

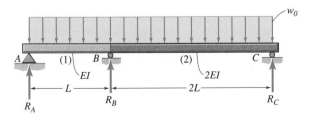

P7.7-15, PC7.7-10, and P11.8-51

Prob. 7.7-16. Starting with Eqs. 7.16 and using equilibrium and superposition, show that, for a uniform beam with *nodal* (end) *displacements* δ_A, θ_A, δ_B, and θ_B, the corresponding *nodal forces* P_A, M_A, P_B, and M_B, are given by the equations:

$$P_A = \left(\frac{EI}{L^3}\right)(12\delta_A + 6L\theta_A - 12\delta_B + 6L\theta_B)$$

$$M_A = \left(\frac{EI}{L^3}\right)(6L\delta_A + 4L^2\theta_A - 6L\delta_B + 2L^2\theta_B)$$

$$P_B = \left(\frac{EI}{L^3}\right)(-12\delta_A - 6L\theta_A + 12\delta_B - 6L\theta_B)$$

$$M_B = \left(\frac{EI}{L^3}\right)(6L\delta_A + 2L^2\theta_A - 6L\delta_B + 4L^2\theta_B)$$

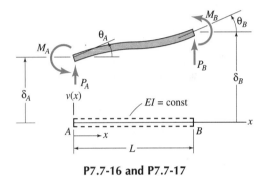

P7.7-16 and P7.7-17

Prob. 7.7-17. Starting with the second-order deflection equation $EIv''(x) = M(x)$, derive the four equations given in Prob. 7.7-16 for the special case when only δ_A is nonzero, that is, when $\theta_A = \delta_B = \theta_B = 0$, but $\delta_A \neq 0$. (Hint: The boundary conditions are: $v(0) = \delta_A$, $v'(0) = v'(L) = v(L) = 0$.) What is the expression for the deflection curve $v(x)$ due to δ_A?

COMPUTER EXERCISES—Section 7.7.
*Use the **MechSOLID** computer program BEAMDEF to solve the homework exercises C7.7-1 through C7.7-10. Refer to Appendix G.6 for a description of this computer program and for an example of its use. Hand in computer printouts of the following **MechSOLID** "screens": BEAMDEF: Model, BEAMDEF: Boundary Conditions, BEAMDEF: Results, BEAMDEF: Deflection Diagram, BEAMDEF: Shear Diagram, and BEAMDEF: Moment Diagram. Samples of these screens are illustrated in Example G-7.*

Prob. C7.7-1. Use BEAMDEF to solve Prob. 7.6-25. Let node A in Fig. P7.6-25 (Fig. PC7.7-1) be relabeled as node ⊡, node B be relabeled as node ②, and node C be relabeled as node ③.

Prob. C7.7-2. Use BEAMDEF to solve Prob. 7.6-26. Let node A in Fig. P7.6-26 (Fig. PC7.7-2) be relabeled as node ⊡, node B be relabeled as node ②, and node C be relabeled as node ③.

Prob. C7.7-3. Use BEAMDEF to solve 7.6-27. Let node A in Fig. P7.6-27 (Fig. PC7.7-3) be relabeled as node ⊡, node B be relabeled as node ②, and node C be relabeled as node ③.

Prob. C7.7-4. Use BEAMDEF to solve Prob. 7.6-29. Let node A in Fig. P7.6-29 (Fig. PC7.7-4) be relabeled as node $\boxed{1}$, node B be relabeled as node $\boxed{2}$, and node C be relabeled as node $\boxed{3}$.

*Prob. C7.7-5. The BEAMDEF program can be used to solve nonnumeric problems like Prob. 7.7-2 if you remember that nodal rotations will have the form $\theta_i = \hat{\theta}_i \left(\dfrac{M_0 L}{EI} \right)$, where $\hat{\theta}_i$ is a nondimensional factor that is obtained by BEAMDEF if you input the following data: $M_0 = L = E = I = 1$. Likewise, transverse displacements will have the form $v = \hat{v} \left(\dfrac{M_0 L^2}{EI} \right)$. Also, element shear forces will have the form $V_i = \hat{V}_i \left(\dfrac{M_0}{L} \right)$, and element bending moments will have the form $M_i = \hat{M}_i M_0$, where $\hat{V}_i$ and $\hat{M}_i$ are the shear force and bending moment at end i as output by BEAMDEF, and similarly for V_j and M_j at end j. (a) Use BEAMDEF to solve Prob. 7.7-2. Let the nodes in Fig. P7.7-2 (Fig. PC7.7-5) be relabeled as follows: $A \rightarrow \boxed{1}$, $B \rightarrow \boxed{2}$, and node C be relabeled as node $\boxed{3}$. (b) Using the BEAMDEF results for the displacement and rotation at node $\boxed{2}$ (i. e., node B), and using Eq. G.6-4 of Appendix G, determine an expression for $\delta_{\max}$, the maximum transverse displacement of this beam.

*Prob. C7.7-6. The BEAMDEF program can be used to solve nonnumeric problems like Prob. 7.7-12 if you remember that nodal rotations will have the form $\theta_i = \hat{\theta}_i \left(\dfrac{M_0 L}{EI} \right)$, where $\hat{\theta}_i$ is a nondimensional factor that is obtained by BEAMDEF if you input the following data: $M_0 = L = E = I = 1$. Likewise, element shear forces will have the form $V_i = \hat{V}_i \left(\dfrac{M_0}{L} \right)$, and element bending moments will have the form $M_i = \hat{M}_i M_0$, where $\hat{V}_i$ and $\hat{M}_i$ are the shear force and bending moment at end i as output by BEAMDEF, and similarly for V_j and M_j at end j. (a) Use BEAMDEF to solve Prob. 7.7-12. Let the nodes in Fig. P7.7-12 (Fig. PC7.7-6) be relabeled as follows: $A \rightarrow \boxed{1}$, $B \rightarrow \boxed{2}$, and node C be relabeled as node $\boxed{3}$. (b) Re-solve this problem making the beam more flexible by letting $(EI)_1 = EI$, rather than $2EI$. How does this change of flexural rigidity of element (1) affect the values of maximum shear force and maximum bending moment in these two-span beams?

*Prob. C7.7-7. The BEAMDEF program can be used to solve nonnumeric problems like Prob. 7.7-3 if you remember that nodal rotations will have the form $\theta_i = \hat{\theta}_i \left(\dfrac{P_0 L^2}{EI} \right)$, where $\hat{\theta}_i$ is a nondimensional factor that is obtained by BEAMDEF if you input the following data: $P_0 = L = E = I = 1$. Likewise, transverse displacements will have the form $v = \hat{v} \left(\dfrac{P_0 L^3}{EI} \right)$. Also, element shear forces will have the form $V_i = \hat{V}_i P_0$, and element bending moments will have the form $M_i = \hat{M}_i (P_0 L)$, where $\hat{V}_i$ and $\hat{M}_i$ are the shear force and bending moment at end i as output by BEAMDEF, and similarly for v_j and M_j at end j. (a) Use BEAMDEF to solve Prob. 7.7-3. Let the nodes in Fig. P7.7-3 (Fig. PC7.7-7) be relabeled as follows: $A \rightarrow \boxed{1}$, $B \rightarrow \boxed{2}$, and node C be relabeled

as node $\boxed{3}$. (b) Using the BEAMDEF results for the displacement and rotation at node $\boxed{2}$ (i.e., node B), and using Eq. G.6-4 of Appendix G, determine an expression for $\delta_{\max}$, the maximum transverse displacement of this beam.

*Prob. C7.7-8. The BEAMDEF program can be used to solve nonnumeric problems like Prob. 7.7-5 if you remember that nodal rotations will have the form $\theta_i = \hat{\theta}_i \left(\dfrac{w_0 L^3}{EI} \right)$, where $\hat{\theta}_i$, is a nondimensional factor that is obtained by BEAMDEF if you input the following data: $w_0 = L = E = I = 1$. Likewise, element shear forces will have the form $V_i = \hat{V}_i (w_0 L)$, and element bending moments will have the form $M_i = \hat{M}_i (w_0 L^2)$, where $\hat{V}_i$ and $\hat{M}_i$ are the shear force and bending moment at end i as output by BEAMDEF, and similarly for V_j and M_j at end j. (a) Use BEAMDEF to solve Prob. 7.7-5. Let the nodes in Fig. P7.7-5 (Fig. PC7.7-8) be relabeled as follows: $A \rightarrow \boxed{1}$, $B \rightarrow \boxed{2}$, and node C be relabeled as node $\boxed{3}$. (b) Plot shear-force and bending-moment diagrams for this two-span beam.

*Prob. C7.7-9. The BEAMDEF program can be used to solve nonnumeric problems like Prob. 7.7-6 if you remember that nodal rotations will have the form $\theta_i = \hat{\theta}_i \left(\dfrac{w_0 L^3}{EI} \right)$, where $\hat{\theta}_i$ is a nondimensional factor that is obtained by BEAMDEF if you input the following data: $w_0 = L = E = I = 1$. Likewise, element shear forces will have the form $V_i = \hat{V}_i (w_0 L)$, and element bending moments will have the form $M_i = \hat{M}_i (w_0 L^2)$, where $\hat{V}_i$ and $\hat{M}_i$ are the shear force and bending moment at end i as output by BEAMDEF, and similarly for V_j and M_j at end j. (a) Use BEAMDEF to solve Prob. 7.7-6. Let the nodes in Fig. P7.7-6 (Fig. PC7.7-9) be relabeled as follows: $A \rightarrow \boxed{1}$, $B \rightarrow \boxed{2}$, and node C be relabeled as node $\boxed{3}$. (b) Sketch shear-force and bending-moment diagrams for this two-span beam. (c) Change the flexural rigidity for element (1) to $EI_1 = EI$ and, again using a BEAMDEF solution, plot shear-force and bending-moment diagrams for the modified beam. Discuss any changes that you observe in the shear and moment distributions.

*Prob. C7.7-10. The BEAMDEF program can be used to solve nonnumeric problems like Prob. 7.7-15 if you remember that nodal rotations will have the form $\theta_i = \hat{\theta}_i \left(\dfrac{w_0 L^3}{EI} \right)$, where $\hat{\theta}_i$ is a nondimensional factor that is obtained by BEAMDEF if you input the following data: $w_0 = L = E = I = 1$. Likewise, element shear forces will have the form $v_i = \hat{V}_i (w_0 L)$, and element bending moments will have the form $M_i = \hat{M}_i (w_0 L^2)$, where $\hat{V}_i$ and $\hat{M}_i$ are the shear force and bending moment at end i as output by BEAMDEF, and similarly for V_j and M_j at end j. (a) Use BEAMDEF to solve Prob. 7.7-15. Let the nodes in Fig. P7.7-15 (Fig. PC7.7-10) be relabeled as follows: $A \rightarrow \boxed{1}$, $B \rightarrow \boxed{2}$, and node C be relabeled as node $\boxed{3}$. (b) Sketch shear-force and bending-moment diagrams for this two-span beam. (c) Change the flexural rigidity for element (2) to $(EI)_2 = EI$ and, again using a BEAMDEF solution, plot shear-force and bending-moment diagrams for the modified beam. Discuss any changes that you observe in the shear and moment distributions.

TRANSFORMATION OF STRESS AND STRAIN; MOHR'S CIRCLE

8

8.1 INTRODUCTION

In earlier chapters you were introduced to *stress analysis,* and, in particular, you learned about the stress distribution on cross sections of long, slender members loaded axially ($\sigma = F/A$), in torsion ($\tau = Tp/J$), or in bending ($\sigma = -My/I$, $\tau = VQ/It$). In Section 2.8 you learned that, even though only axial loading is applied to a member, the stress distribution on an inclined sectioning-plane consists of shear stress as well as normal stress (Fig. 8.1). You also learned that the normal stress and shear stress on an oblique plane are directly related to the axial stress through stress-transformation equations. In Chapter 9 we will consider *combined loading* of slender members, that is, loads that produce various combinations of axial force, torque, bending moment, and shear force. But first, in order to completely characterize the *state of stress* produced by a single type of load or by a combination of loads, we must develop *stress transformation* equations and procedures.

We begin, in Section 8.2, by introducing a two-dimensional state of stress called *plane stress.* The stress-transformation equations that are derived in Section 8.3 are a generalization of the stress-transformation equations developed in Section 2.8. In Section 8.4 we locate maxima and minima of normal stress and shear stress, and in Section 8.5 a very important graphical representation of this stress transformation, called *Mohr's circle,* is introduced.

In Section 2.10 you were introduced to the stresses at a point acting on three mutually orthogonal planes (Fig. 2.38), and you learned that six independent stress quantities (e.g., σ_x, σ_y, σ_z, τ_{xy}, τ_{xz}, and τ_{yz}) are required to completely specify the stress on the three mutually orthogonal planes. In Sections 8.6 and 8.7 we will briefly examine stress transformations for three-dimensional states of stress.

As you will discover in Section 8.3, the *stress transformation equations* are based solely on *equilibrium;* they do not depend on material properties (linearly elastic behavior, etc.) or on the geometry of deformation. However, there are similar *strain transformation equations*; these are based solely on the *geometry of deformation.* The transformation of strain is discussed in Sections 8.8 through 8.12.

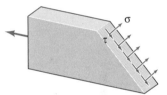

FIGURE 8.1 Normal stress and shear stress on an inclined section of an axial-deformation member.

429

In general, if we were to pass three mutually orthogonal planes through any point in a deformable body under load (Fig. 8.2a), we would find normal and/or shear stresses on all three planes (Fig. 8.2b). This is called a *three-dimensional state of stress*. Fortunately, there are many important instances where some of these stresses vanish, so that there is then a *two-dimensional state of stress*. For example, consider the stresses acting on a small element taken from the web of a plate girder of a bridge (Fig. 8.3). Because the web is thin, and because there are no loads applied directly to the surface of the web, the three stresses associated with the z axis (σ_z, $\tau_{zx} = \tau_{xz}$, and $\tau_{zy} = \tau_{yz}$) are all zero. The nonzero stresses lie in the xy plane.[1]

If the stresses σ_z, τ_{xz}, and τ_{yz} vanish everywhere, that is, if

$$\sigma_z = \tau_{xz} = \tau_{yz} = 0 \qquad (8.1)$$

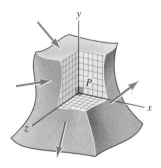

(a) A set of three mutually orthogonal planes through an arbitrary point P.

(b) A three-dimensional state of stress referred to rectangular cartesian axes. (Stresses are shown only on visible faces.)

Stresses are shown only on the positive faces; oppositely-directed stresses act on the negative faces.

$\tau_{yx} = \tau_{xy}$
$\tau_{zx} = \tau_{xz}$
$\tau_{zy} = \tau_{yz}$

FIGURE 8.2 A general three-dimensional state of stress.

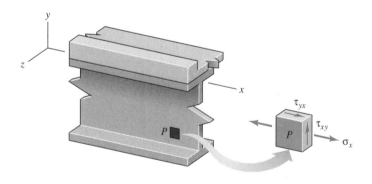

(a) A portion of a bridge girder.

(b) A web element, with stresses shown.

FIGURE 8.3 An example of plane stress.

---

[1] Strictly speaking, the nonzero stresses act on planes whose normal vectors lie in the xy plane. Hence, the arrows representing these stresses all lie in the xy plane. The nonzero stresses implied by Eqs. 8.1 (σ_x, τ_{xy}, τ_{xz}) lie in the xy plane. A state of plane stress would also exist if the stresses were in the xz plane only or the yz plane only.

the state of stress in a body is said to be *plane stress*. Figure 8.4 shows three-dimensional and two-dimensional views of an element in *general plane stress*. From moment equilibrium it was found (Eqs. 2.31) that

$$\tau_{yx} = \tau_{xy} \tag{8.2}$$

Since a two-dimensional view can depict the relevent (i.e., nonzero) stress information, it is not really necessary to use a three-dimensional view when discussing plane stress. Special cases of plane stress are *uniaxial stress* (Fig. 8.5*a*), *pure shear* (Fig. 8.5*b*), and *biaxial stress* (Fig. 8.5*c*).

Deformable bodies that are in a state of plane stress are usually thin, plate-like members, like the rectangular bar in Fig. 8.1 or the web of the plate girder in Fig. 8.3.[2] The stress-transformation equations that are derived in Section 8.3 are applicable to such plane-stress problems, but they are also applicable to a much wider class of two-dimensional stress problems, where Eqs. 8.1 are valid locally, but not everywhere in the body. For example, some portions of the surface of a deformable body are subjected to distributed loading or concentrated loading, but at those parts of the surface where no loading is directly applied, the stress state may be classified as *locally two-dimensional,* or *locally plane stress.* A circular rod subjected to an axial load and a torque, as illustrated in Fig. 8.6, is an example of this type of local plane stress. Since there are no stresses on the surface of the body at the point in question, Eq. 8.1 is satisfied locally, and we can treat the stress at this point of the surface as plane stress.

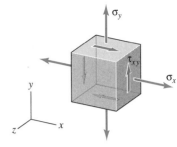

(*a*) Three-dimensional view.

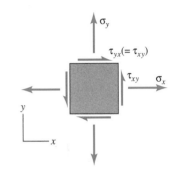

(*b*) Two-dimensional view.

FIGURE 8.4 A state of plane stress depicted in 3-D and in 2-D.

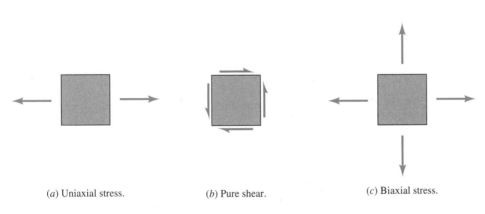

(*a*) Uniaxial stress. (*b*) Pure shear. (*c*) Biaxial stress.

FIGURE 8.5 Special cases of plane stress.

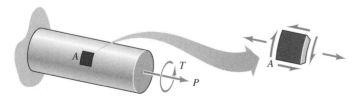

(*a*) A rod with axial load and torque. (*b*) A two-dimensional stress state.

FIGURE 8.6 An example of two-dimensional stress at the surface of a deformable body.

---

[2]Chapter 2 of *Theory of Elasticity,* by S. P. Timoshenko and J. N. Goodier, [Ref. 8-1], discusses plane stress and plane strain problems.

431

State of Stress at a Point. If the state of plane stress at a point is known with reference to particular coordinates, say the stresses (σ_x, σ_y, and τ_{xy}) in the xy reference coordinates, what are the values of the stresses at the same point in the body if we rotate the frame of reference? That is, what will be the values of ($\sigma_{x'}$, $\sigma_{y'}$, and $\tau_{x'y'}$) on planes oriented along $x'y'$ axes? In Chapter 6 you learned how to use the flexure formula and the shear formula to determine the stresses on the x and y faces of an element located at an arbitrary point (x, y) in a beam, as illustrated in Fig. 8.7a. Since we have no assurance that the stresses on those particular faces are the critical (i.e., maximum) ones, however, it is very important for us to be able to determine the stresses on inclined planes, like the n face and the t face in Fig. 8.7b.

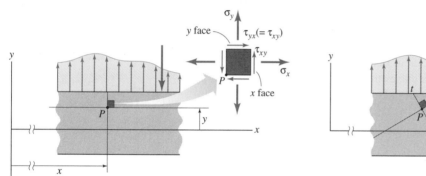

(a) The stress state at point P referred to x and y axes.

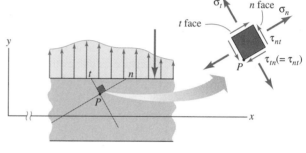

(b) The stress state at point P, referred to n and t axes.

FIGURE 8.7 The state of stress at a point, as represented in two reference frames.

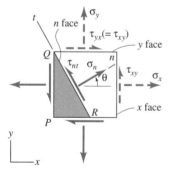

FIGURE 8.8 The relationship of an arbitrarily-oriented n face to the reference (xy) axes.

Notation and Sign Convention. Let us suppose that the stresses σ_x, σ_y, and τ_{xy} are known at some point (e.g., point P in Fig. 8.7a). To determine the normal stress and the shear stress on other faces, like the n and t faces in Fig. 8.7b, we need only to consider one arbitrarily oriented face and to relate the normal stress and shear stress on that face to the xy stresses. As illustrated in Fig. 8.8, we orient the arbitrary face, which we call the n *face,* by rotating its outward normal n through the counterclockwise angle $\theta \equiv \theta_{xn}$, starting with the positive x axis.

The t axis is oriented so that the ntz axes, like the xyz axes, form a right-handed coordinate frame. As always, the normal stress σ_n is *positive in tension,* and a *positive shear stress on the $+n$ face,* τ_{nt}, *acts in the $+t$ direction.*

Stress Transformation for Plane Stress.

Let us assume that the stresses σ_x, σ_y, and τ_{xy} are known, and that the stresses σ_n and τ_{nt} are to be determined. As was pointed out earlier, the stress tranformation equations are based solely on *equilibrium.* To set up equations of equilibrium, we need a *free-body diagram.* The triangular element PQR in Fig. 8.8, which is repeated as Fig. 8.9a, shows the stresses acting on the three faces of the element. The areas of the various faces are given in Fig. 8.9b and, by multiplying each stress by the area of the face on which that stress acts, we obtain the corresponding free-body diagram in Fig. 8.9c.

By writing the equilibrium equations for the free body in Fig. 8.9c and substituting the areas from Fig. 8.9b, and by using the fact that $\tau_{yx} = \tau_{xy}$ (Eq. 8.2), we obtain

$$+\nearrow \sum F_n = 0:$$
$$\sigma_n \Delta A - \sigma_x (\Delta A \cos \theta) \cos \theta - \sigma_y (\Delta A \sin \theta) \sin \theta$$
$$- \tau_{xy}(\Delta A \cos \theta) \sin \theta - \tau_{xy}(\Delta A \sin \theta) \cos \theta = 0$$

$$+\nwarrow \sum F_t = 0:$$
$$\tau_{nt} \Delta A + \sigma_x (\Delta A \cos \theta) \sin \theta - \sigma_y (\Delta A \sin \theta) \cos \theta$$
$$- \tau_{xy}(\Delta A \cos \theta) \cos \theta + \tau_{xy}(\Delta A \sin \theta) \sin \theta = 0$$

Dividing these equations by ΔA and collecting terms, we get

$$\sigma_n = \sigma_x \cos^2 \theta + \sigma_y \sin^2 \theta + \tau_{xy}(2 \sin \theta \cos \theta) \tag{8.3}$$
$$\tau_{nt} = -(\sigma_x - \sigma_y) \cos \theta \sin \theta + \tau_{xy}(\cos^2 \theta - \sin^2 \theta)$$

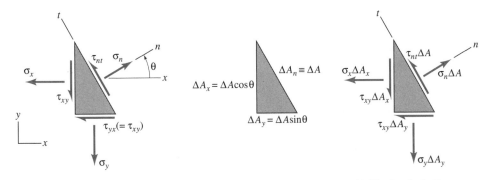

(a) A triangular stress element.　　(b) The areas of the faces.　　(c) The free-body diagram.

FIGURE 8.9 The free-body diagram based on a plane-stress element.

Equations 8.3 can be expressed in a more convenient form by incorporating the following trigonometric identities:

$$2 \sin \theta \cos \theta = \sin 2\theta$$

$$\sin^2\theta = \frac{1}{2}(1 - \cos 2\theta)$$

$$\cos^2\theta = \frac{1}{2}(1 + \cos 2\theta)$$

(8.4)

Then, from Eqs. 8.3, the stress-transformation equations for plane stress become

$$\sigma_n = \left(\frac{\sigma_x + \sigma_y}{2}\right) + \left(\frac{\sigma_x - \sigma_y}{2}\right) \cos 2\theta + \tau_{xy} \sin 2\theta$$

$$\tau_{nt} = -\left(\frac{\sigma_x - \sigma_y}{2}\right) \sin 2\theta + \tau_{xy} \cos 2\theta$$

(8.5)

It should be emphasized again that, to obtain this stress transformation, we had to multiply stresses times areas to get the *forces* acting on the free body in Fig. 8.9c. It is <u>not</u> correct to just sum the stresses on a stress element, like the one in Fig. 8.9a!

To emphasize the fact that Eqs. 8.5 enable us to compute the normal stress and the shear stress on any face, that is, on a face at any orientation, let us generalize the stress plot of Fig. 2.30 and obtain a plot of σ_n and τ_{nt} for the particular state of stress indicated in Fig. 8.10a. The stresses are all referred to a common stress magnitude, σ_0, and their senses are indicated by the arrows in Fig. 8.10a. A range of θ from $-90°$ to $+90°$ is sufficient to represent all possible planes that can be passed through a point, so the plot in Fig. 8.10b extends from $-90°$ to $+90°$.

The following example illustrates how equilibrium equations may be used to calculate the normal stress and shear stress on an oblique face.

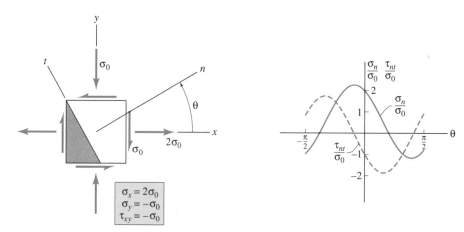

(a) A sample state of plane stress.

(b) Normal stress and shear stress as functions of the orientation of the stress face.

FIGURE 8.10 The normal stress and shear stress corresponding to a sample state of plane stress.

The state of plane stress at a point is indicated in Fig. 1a. Determine the normal stress $\sigma_{x'}$ and the shear stress $\tau_{x'y'}$ on the x' face, which is rotated 30° counterclockwise from the x face, as illustrated in Fig. 1b.

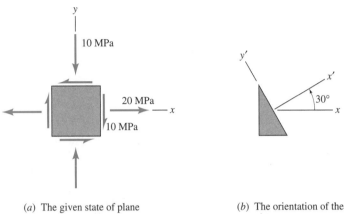

(a) The given state of plane stress at a point.

(b) The orientation of the x' face.

Fig. 1   Stress-transformation data.

Plan the Solution   We can follow the procedure used to derive Eqs. 8.3, using the specific angle of 30° rather than the general angle θ.

Solution

Equilibrium: The free-body diagram is shown in Fig. 2.

$+\nearrow \sum F_{x'} = 0$:

$$\sigma_{x'}\Delta A - 20\Delta A_x \cos(30°) + 10\Delta A_x \sin(30°)$$
$$+ 10\Delta A_y \sin(30°) + 10\Delta A_y \cos(30°) = 0$$

So,

$$\sigma_{x'} = 3.84 \text{ MPa} \qquad \text{Ans.} \quad (1a)$$

$+\nwarrow \sum F_{y'} = 0$:

$$\tau_{x'y'}\Delta A + 20\Delta A_x \sin(30°) + 10\Delta A_x \cos(30°)$$
$$- 10\Delta A_y \sin(30°) + 10\Delta A_y \cos(30°) = 0$$

or

$$\tau_{x'y'} = -17.99 \text{ MPa} \qquad \text{Ans.} \quad (1b)$$

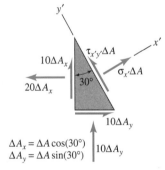

$\Delta A_x = \Delta A \cos(30°)$
$\Delta A_y = \Delta A \sin(30°)$

Fig. 2   The free-body diagram.

Stresses on Orthogonal Faces.   The two stress-transformation equations for plane stress, Eqs. 8.5, relate the normal stress σ_n and the shear stress τ_{nt} on an arbitrary face to a given set of xy stresses. However, we frequently need to make reference to the stresses

435

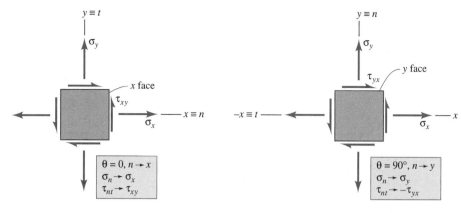

(a) The x face is the n face.

(b) The y face is the n face.

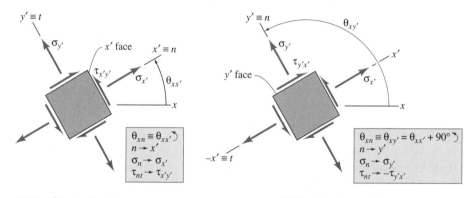

FIGURE 8.11 The stresses on orthogonal faces.

(c) The x' face is the n face.

(d) The y' face is the n face.

on two orthogonal faces—for example, the x face and the y face. Figure 8.11a illustrates the case of $\theta = 0$. Then, as indicated on Fig. 8.11a, $n \rightarrow x$, $t \rightarrow y$, $\sigma_n \rightarrow \sigma_x$, and $\tau_{nt} \rightarrow \tau_{xy}$. When we consider the y face (Fig. 8.11b), we must remember that the ntz axes form a right-handed coordinate system. Therefore, when $\theta = 90°$, $n \rightarrow y$, $t \rightarrow -x$, $\sigma_n \rightarrow \sigma_y$, and $\tau_{nt} \rightarrow -\tau_{yx}(= -\tau_{xy})$. The last of these equivalencies, $\tau_{nt} \rightarrow -\tau_{xy}$, results from the fact that the t axis corresponds to the $-x$ axis, not the $+x$ axis.

Now let us consider a pair of arbitrarily oriented orthogonal faces, for example, the x' face and y' face in Fig. 8.11c and Fig. 8.11d. The orientation of the faces x' and y' is specified by the one angle $\theta_{xx'}$. Figure 8.11c relates σ_n and τ_{nt} to the stresses on the x' face, and Fig. 8.11d relates σ_n and τ_{nt} to the stresses on the y' face. We can calculate these stresses using Eqs. 8.5, noting that $2\theta_{xy'} = 2(\theta_{xx'} + 90°) = 2\theta_{xx'} + 180°$.

$$
\begin{aligned}
\sigma_{x'} &\equiv \sigma_n(\theta_{xx'}) = \left(\frac{\sigma_x + \sigma_y}{2}\right) + \left(\frac{\sigma_x - \sigma_y}{2}\right)\cos 2\theta_{xx'} + \tau_{xy}\sin 2\theta_{xx'} \\[2mm]
\tau_{x'y'} &\equiv \tau_{nt}(\theta_{xx'}) = -\left(\frac{\sigma_x - \sigma_y}{2}\right)\sin 2\theta_{xx'} + \tau_{xy}\cos 2\theta_{xx'} \\[2mm]
\sigma_{y'} &\equiv \sigma_n(\theta_{xx'} + 90°) = \left(\frac{\sigma_x + \sigma_y}{2}\right) - \left(\frac{\sigma_x - \sigma_y}{2}\right)\cos 2\theta_{xx'} - \tau_{xy}\sin 2\theta_{xx'} \\[2mm]
\tau_{y'x'} &\equiv -\tau_{nt}(\theta_{xx'} + 90°) = -\left(\frac{\sigma_x - \sigma_y}{2}\right)\sin 2\theta_{xx'} + \tau_{xy}\cos 2\theta_{xx'} = \tau_{x'y'}
\end{aligned}
\tag{8.6}
$$

Equations 8.5 are primarily useful if one needs (1) to calculate the normal stress and the shear stress on a face at a specified angle, (2) to generate a plot like Fig. 8.10b, or (3) to derive expressions for the maximum normal stress and the maximum shear stress at a point. These latter expressions will be derived in Section 8.4, but let us first illustrate the use of these formulas to determine stresses on the faces of a rotated element, like the $x'y'$ element in Fig. 8.11.

■■■■■■■■■■■■■■□□ E X A M P L E 8 . 2 □□■■■■■■■■■■■■■■

The state of plane stress at a point is indicated in Fig. 1a. Use Eqs. 8.5 to determine the stresses on faces that are rotated 30° counterclockwise from the orientation of the element in Fig. 1a, as illustrated in Fig. 1b. Show the stresses on a rotated stress element.

Plan the Solution For the x' face $\theta = 30°$, while for the y' face $\theta = 30° + 90° = 120°$ (or we could use $\theta = -60°$). This is a straightforward "plug-in" type problem employing Eqs. 8.5 (or 8.6). From Fig. 1a we have:

$$\sigma_x = 20 \text{ MPa}, \quad \sigma_y = -10 \text{ MPa}, \quad \tau_{xy} = -10 \text{ MPa}$$

Solution For the x' face, $\theta = 30°$, so $2\theta = 60°$, and, referring to Fig. 8.11c, $n \to x'$, $t \to y'$. Then, Eq. 8.5a gives

$$\sigma_{x'} = \left(\frac{\sigma_x + \sigma_y}{2}\right) + \left(\frac{\sigma_x - \sigma_y}{2}\right) \cos 2\theta + \tau_{xy} \sin 2\theta$$

$$\sigma_{x'} = \left(\frac{20 \text{ MPa} - 10 \text{ MPa}}{2}\right) + \left(\frac{20 \text{ MPa} + 10 \text{ MPa}}{2}\right) \cos(60°)$$

$$+ (-10 \text{ MPa}) \sin(60°)$$

or

$$\sigma_{x'} = 3.84 \text{ MPa} \qquad \text{Ans.} \quad (1a)$$

Similarly, Eq. 8.5b gives

$$\tau_{x'y'} = -\left(\frac{\sigma_x - \sigma_y}{2}\right) \sin 2\theta + \tau_{xy} \cos 2\theta$$

$$\tau_{x'y'} = -\left(\frac{20 \text{ MPa} + 10 \text{ MPa}}{2}\right) \sin(60°) + (-10 \text{ MPa}) \cos(60°)$$

or

$$\tau_{x'y'} = -17.99 \text{ MPa} \qquad \text{Ans.} \quad (1b)$$

For the y' face, $\theta = 120°$, so $2\theta = 240°$, and, referring to Fig. 8.11d, $n \to y'$, $t \to -x'$. Then, from Eqs. 8.5,

$$\left.\begin{array}{ll} \sigma_{y'} \equiv \sigma_n(120°), & \sigma_{y'} = 6.16 \text{ MPa} \\ \tau_{y'x'} \equiv -\tau_{nt}(120°), & \tau_{y'x'} = -17.99 \text{ MPa} \end{array}\right\} \quad \text{Ans.} \quad (2)$$

Equations (1) determine the stresses on the x' face, and Eqs. (2) determine the stresses on the y' face. These results are illustrated in Fig. 2.

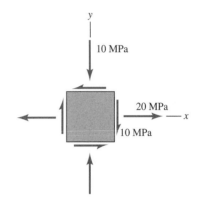

(a) The given state of plane stress at a point.

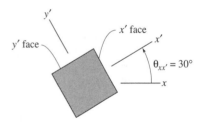

(b) An element rotated 30° counterclockwise.

Fig. 1 Stress-transformation data.

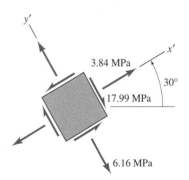

Fig. 2 Stresses on an $x'y'$ element at $\theta = 30°$.

Review the Solution This is a very short problem whose results should be verified by first seeing if the magnitude of each answer is reasonable. Then the calculations should be spot-checked for accuracy. In the present case, the stresses given on Fig. 1a are in the same proportions as the stresses in Fig. 8.10a. Hence, we can look at the values of σ_n/σ_0 and τ_{nt}/σ_0 in Fig. 8.10b in order to confirm our answers for this problem.

It is not a coincidence that in Example Problem 8.2, $(\sigma_x + \sigma_y) = (\sigma_{x'} + \sigma_{y'})$. By adding Eqs. 8.6a and 8.6c, we get

$$\boxed{\sigma_{x'} + \sigma_{y'} = \sigma_x + \sigma_y} \tag{8.7}$$

Therefore, the sum of the normal stresses on orthogonal faces is a constant, which is called a *stress invariant*. That is, the sum of normal stresses on orthogonal faces does not vary with the angle θ.

It is apparent from Fig. 8.10b that the maxima and minima of normal stress and shear stress do not necessarily act on the x and y faces, that is, on the faces whose stresses are "given." In the next section we will derive equations that enable us to locate the planes on which these maxima (and minima) act.

8.4 PRINCIPAL STRESSES AND MAXIMUM SHEAR STRESS

Since a structural member or machine component may fail because of excessive normal stress or excessive shear stress, it is important to be able to determine the maximum normal stress and the maximum shear stress at a point. Figure 8.10b illustrates the fact that, for a plane-stress situation, there are certain planes on which the maximum and minimum normal stresses act and planes on which the maximum and minimum shear stresses act. In this section we will determine how to locate these planes and how to calculate these special stresses for the plane-stress case. Three-dimensional stress states are discussed in Section 8.6.

Principal Stresses. The maximum and minimum normal stresses are called *principal stresses*.[3] As is evident from Fig. 8.10b, the principal stresses occur on planes that satisfy the equation

$$\frac{d\sigma_n}{d\theta} = 0 \tag{8.8}$$

where $\sigma_n(\theta)$ is given by Eq. 8.5a. Then,

$$\frac{d\sigma_n}{d\theta} = -(\sigma_x - \sigma_y) \sin 2\theta_p + 2\tau_{xy} \cos 2\theta_p = 0$$

The angles θ_p determine the orientation of the *principal planes,* the planes on which the principal stresses act. They are obtained by solving for the two values of θ_p that satisfy

---

[3]The word princip<u>al</u> is often confused with the word princip<u>le</u>. Principal stresses are the <u>al</u> type, meaning *most important* stresses.

the equation

$$\tan 2\theta_p = \frac{\tau_{xy}}{\left(\dfrac{\sigma_x - \sigma_y}{2}\right)} \qquad (8.9)$$

Figure 8.12a illustrates how to use the tangent value given by Eq. 8.9 to determine the angles θ_p. There are two angles between 0° and 360° that satisfy Eq. 8.9. As illustrated by Fig. 8.12b, these two values of $2\theta_p$, labeled $2\theta_{p1}$ and $2\theta_{p2}$, differ by 180°, so the principal planes are oriented at 90° to each other. That is,

$$\theta_{p2} = \theta_{p1} \pm 90° \qquad (8.10)$$

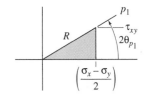

(a) Determination
of angle $2\theta_p$.

To determine the actual values of the principal stresses, that is, the normal stresses on the principal planes, we must substitute θ_{p1} and θ_{p2} into Eq. 8.5a. From Fig. 8.12a we can see that

$$R = \sqrt{\left(\frac{\sigma_x - \sigma_y}{2}\right)^2 + \tau_{xy}^2} \qquad (8.11)$$

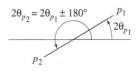

(b) Two angles satisfy Eq. 8.9.

FIGURE 8.12
Determination of the angles that locate principal planes.

The quantity R is positive, and it has the units of stress. Also, from Fig. 8.12,

$$\sin 2\theta_{p1} = \frac{\tau_{xy}}{R}, \qquad \cos 2\theta_{p1} = \frac{\left(\dfrac{\sigma_x - \sigma_y}{2}\right)}{R} \qquad (8.12a)$$

and

$$\sin 2\theta_{p2} = \frac{-\tau_{xy}}{R}, \qquad \cos 2\theta_{p2} = \frac{-\left(\dfrac{\sigma_x - \sigma_y}{2}\right)}{R} \qquad (8.12b)$$

Combining Eq. 8.5a with Eqs. 8.11 and 8.12, we get the two expressions for the two *principal stresses*:

$$\begin{aligned} \sigma_1 &\equiv \sigma_n(\theta_{p1}) = \sigma_{\text{avg}} + R \\ \sigma_2 &\equiv \sigma_n(\theta_{p2}) = \sigma_{\text{avg}} - R \end{aligned} \qquad (8.13)$$

where

$$\sigma_{\text{avg}} = \frac{\sigma_x + \sigma_y}{2} \qquad (8.14)$$

and where R is given by Eq. 8.11. The designation $\sigma_1 \equiv \sigma_n(\theta_{p1})$ denotes the *maximum normal stress* at the point, while $\sigma_2 \equiv \sigma_n(\theta_{p2})$ denotes the *minimum normal stress* at the point. It is easy to see that σ_1 and σ_2 satisfy Eq. 8.7.

With angles θ_{p1} and θ_{p2} determined by Eqs. 8.12, and σ_1 and σ_2 determined by Eqs. 8.13 (with 8.11 and 8.14), we can show the two principal stresses on a properly oriented

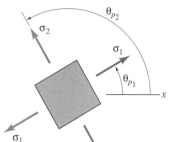

FIGURE 8.13 A principal-stress element.

element. First, however, let us determine the values of the shear stress on the principal planes by substituting Eqs. 8.12 into Eq. 8.5b. We get

$$\tau_{nt}(\theta_{p1}) = 0, \qquad \tau_{nt}(\theta_{p2}) = 0 \tag{8.15}$$

This illustrates the important fact that **there is no shear stress on principal planes**. Therefore, Fig. 8.13 represents the stresses on a properly oriented *principal-stress element*.

In Section 8.5 you will learn a convenient graphical procedure, called *Mohr's circle,* that is very useful for locating principal planes and calculating principal stresses. Both the equations of this section and the Mohr's circle method introduced in Section 8.5 determine *in-plane stresses,* that is, stresses in what has been designated as the *xy* plane. Section 8.6 extends the notion of principal stresses to three dimensions, and Section 8.7 treats maximum shear stress in three dimensions.

Maximum In-Plane Shear Stress. We can locate the planes of maximum in-plane shear stress and calculate the value of the maximum in-plane shear stress by following the same procedure that we used to determine the principal planes and the principal stresses. By differentiating Eq. 8.5b we obtain

$$\frac{d\tau_{nt}}{d\theta} = -(\sigma_x - \sigma_y)\cos 2\theta_s - 2\tau_{xy}\sin 2\theta_s = 0$$

where θ_s designates a plane on which the shear is a maximum.[4] Therefore,

$$\tan 2\theta_s = -\frac{\left(\dfrac{\sigma_x - \sigma_y}{2}\right)}{\tau_{xy}} \tag{8.16}$$

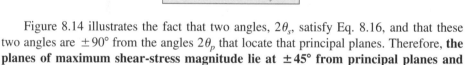

FIGURE 8.14
Determination of the angles that locate planes of maximum shear stress.

Figure 8.14 illustrates the fact that two angles, $2\theta_s$, satisfy Eq. 8.16, and that these two angles are $\pm 90°$ from the angles $2\theta_p$ that locate that principal planes. Therefore, **the planes of maximum shear-stress magnitude lie at $\pm 45°$ from principal planes and are oriented at 90° to each other**. From Fig. 8.14 we get

$$\sin 2\theta_{s1} = \frac{-\left(\dfrac{\sigma_x - \sigma_y}{2}\right)}{R}, \qquad \cos 2\theta_{s1} = \frac{\tau_{xy}}{R} \tag{8.17a}$$

and

$$\sin 2\theta_{s2} = \frac{\left(\dfrac{\sigma_x - \sigma_y}{2}\right)}{R}, \qquad \cos 2\theta_{s2} = \frac{-\tau_{xy}}{R} \tag{8.17b}$$

---

[4]Since the maximum (i.e., algebraically largest) shear stress and the minimum (i.e., algebraically least) shear stress have the same magnitude, we will just refer to the ''maximum'' (see Eq. 8.18).

Substituting these expressions into Eq. 8.5b, we get the following expressions for *maximum* (and minimum) *in-plane shear stress*:

$$\tau_{s1} \equiv \tau_{nt}(\theta_{s1}) = R = \tau_{max}$$

$$\tau_{s2} \equiv \tau_{nt}(\theta_{s2}) = -R = -\tau_{max}$$

(8.18)

Thus, we find that the planes of maximum in-plane shear stress magnitude satisfy the equation

$$\theta_s = \theta_p \pm 45°$$

(8.19)

and that the shear stresses on these planes are equal in magnitude.

Before sketching a maximum-shear-stress element, we need to determine the values of normal stress on each of the planes of maximum shear stress. Substituting Eqs. 8.17 into Eq. 8.5a, we get

$$\sigma_{s1} \equiv \sigma_n(\theta_{s1}) = \frac{\sigma_x + \sigma_y}{2} = \sigma_{avg}$$

$$\sigma_{s2} \equiv \sigma_n(\theta_{s2}) = \frac{\sigma_x + \sigma_y}{2} = \sigma_{avg}$$

(8.20)

Thus, **the planes of maximum shear are not free of normal stress** (unless $\sigma_x = -\sigma_y$). On the contrary, they both have the same normal stress, σ_{avg}. Figure 8.15b depicts the *maximum-shear-stress element* that corresponds to the principal stress element of Fig. 8.13 (repeated as Fig. 8.15a). Note how the planes of maximum shear are oriented at 45° to the principal planes, and note that the *maximum-shear diagonal* (in Fig. 8.15b the shear arrows all point to this diagonal) lies along the p_1 principal direction. Mohr's circle, which is discussed in the next section, will provide more insight into the relationships illustrated in Fig. 8.15.

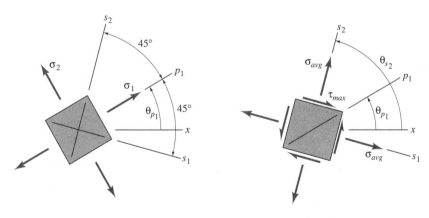

(*a*) Principal-stress element.

(*b*) Maximum shear-stress element.

FIGURE 8.15 The relationship of the planes of maximum shear stress to the principal planes.

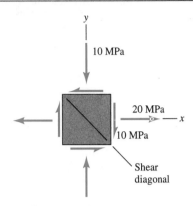

Fig. 1 A state of plane stress.

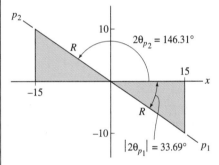

Fig. 2 A sketch representing principal directions.

For the plane-stress state depicted in Fig. 1 (same as Fig. 1a of Example 8.1), (a) determine the orientation of the principal planes; determine the principal stresses; and illustrate the principal stresses on a properly oriented principal-stress element. (b) Determine the orientation of the planes of maximum in-plane shear stress; determine the value of the maximum shear stress; and illustrate the maximun in-plane shear stress on a properly oriented element.

Plan the Solution From Fig. 1, we have

$$\sigma_x = 20 \text{ MPa}, \quad \sigma_y = -10 \text{ MPa}, \quad \tau_{xy} = -10 \text{ MPa}$$

For Part (a), we could determine the two values of $2\theta_p$ from Eq. 8.9, but it will be more instructive to use Eqs. 8.12 to construct a sketch like Fig. 8.12. The values of σ_1 and σ_2 are easily calculated by using Eqs. 8.13.

For Part (b), we could determine the two values of $2\theta_s$ from Eq. 8.16, but, again, it will be more instructive if we use Eqs. 8.17 to construct a sketch like Fig. 8.14. The stresses on the planes of maximum shear are given by Eqs. 8.18.

Solution

(a) Principal Stresses and Principal Planes: To determine the orientation of the principal planes, we will first sketch a figure like Fig. 8.12. To do this, we use Eqs. 8.12a and 8.12b to sketch Fig. 2.

$$\frac{\sigma_x - \sigma_y}{2} = \frac{20 \text{ MPa} + 10 \text{ MPa}}{2} = 15 \text{ MPa}$$

and

$$\tau_{xy} = -10 \text{ MPa}$$

The value of R can be computed by referring to Fig. 2 or by using Eq. 8.11. This gives

$$R = \sqrt{\left(\frac{\sigma_x - \sigma_y}{2}\right)^2 + (\tau_{xy})^2} = \sqrt{(15 \text{ MPa})^2 + (-10 \text{ MPa})^2} \quad (1a)$$

$$R = \sqrt{325} \text{ MPa} = 18.03 \text{ MPa} \quad (1b)$$

From Eq. 8.12a, as illustrated in Fig. 2,

$$2\theta_{p1} = \left[\sin^{-1}\left(\frac{-10}{\sqrt{325}}\right); \cos^{-1}\left(\frac{15}{\sqrt{325}}\right)\right] = -33.69° \quad (2a)$$

$$\theta_{p1} = -16.8° \qquad \text{Ans. (a)} \quad (2b)$$

Equation (2a) is to be read, ''$2\theta_{p1}$ is the angle whose sine is s and whose cosine is c, as given in the square bracket $[\sin^{-1}(s); \cos^{-1}(c)]$.'' Likewise, from Eq. 8.12b, as illustrated in Fig. 2,

$$2\theta_{p2} = \left[\sin^{-1}\left(\frac{10}{\sqrt{325}}\right); \cos^{-1}\left(\frac{-15}{\sqrt{325}}\right)\right] = 146.31° \quad (3a)$$

$$\theta_{p2} = 73.2° \qquad \text{Ans. (a)} \quad (3b)$$

From Eq. 8.14,

$$\sigma_{\mathrm{avg}} = \frac{\sigma_x + \sigma_y}{2} = \frac{(20 \text{ MPa} - 10 \text{ MPa})}{2} = 5 \text{ MPa} \qquad (4)$$

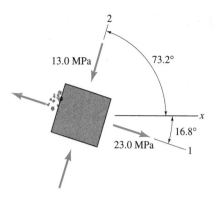

and, from Eqs. 8.13,

$$\sigma_1 = \sigma_{\mathrm{avg}} + R = 23.0 \text{ MPa} \qquad \text{Ans. (a)} \quad (5a)$$

$$\sigma_2 = \sigma_{\mathrm{avg}} - R = -13.0 \text{ MPa} \qquad \text{Ans. (a)} \quad (5b)$$

These stresses are shown on the principal-stress element in Fig. 3.

Fig. 3 The principal-stress element.

(b) Maximum In-Plane Shear Stress: To determine the orientation of the planes of maximum in-plane shear stress, we will sketch a figure like Fig. 8.14. (We could also use the fact that the maximum shear stress element is rotated 45° from the principal stress element.) Equations 8.17 enable us to determine the cosine and sine legs of angles $2\theta_{s1}$ and $2\theta_{s2}$ and, from them, to sketch Fig. 4.

From Eq. 8.17a, as illustrated in Fig. 4,

$$2\theta_{s1} = \left[\sin^{-1}\left(\frac{-15}{\sqrt{325}}\right); \cos^{-1}\left(\frac{-10}{\sqrt{325}}\right) \right] = -123.69° \qquad (6a)$$

$$\theta_{s1} = -61.8° \qquad \text{Ans. (b)} \quad (6b)$$

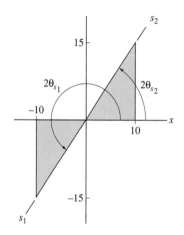

Similarly, from Eq. 8.17b, as illustrated in Fig. 4,

$$2\theta_{s2} = \left[\sin^{-1}\left(\frac{15}{\sqrt{325}}\right); \cos^{-1}\left(\frac{10}{\sqrt{325}}\right) \right] = 56.31° \qquad (7a)$$

$$\theta_{s2} = 28.2° \qquad \text{Ans. (b)} \quad (7b)$$

From Eqs. 8.18,

Fig. 4 A sketch representing maximum shear stress directions.

$$\tau_{s1} \equiv \tau_{nt}(\theta_{s1}) = R = 18.0 \text{ MPa} \qquad \text{Ans. (b)} \quad (8a)$$

$$\tau_{s2} \equiv \tau_{nt}(\theta_{s2}) = -R = -18.0 \text{ MPa} \qquad \text{Ans. (b)} \quad (8b)$$

The maximum in-plane shear stress element is sketched in Fig. 5 using the data in Eq. (4) and in Eqs. (6) through (8).

Review the Solution The principal stress σ_1 should be $\geq$ the greater normal stress on Fig. 1, which it is; and principal stress σ_2 should be $\leq$ the lesser normal stress on Fig. 1, which it is. The p_1 principal direction should lie between the axis of the greater normal stress, here the x axis, and the shear diagonal. The p_1 direction of $-16.8°$ satisfies this requirement. Therefore, the principal stresses and principal directions are probably correct.

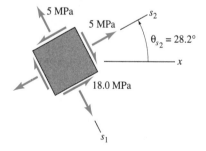

Fig. 5 The maximum in-plane shear stress element.

The maximum in-plane shear stress must be greater than the given shear on Fig. 1, and the directions of maximum shear must bisect the principal directions. These requirements are satisfied by Fig. 5, so our maximum shear stresses are also probably correct.

Problems like this one will generally be solved by using Mohr's Circle (Section 8.5), which provides a helpful visualization of the entire problem.

Mohr's circle is an ingenious graphical representation of the *plane stress transformation equations,* Eqs. 8.5. It permits an easy visualization of the normal stress and shear stress on arbitrary planes, and it greatly facilitates the solution of plane-stress problems like Example Problems 8.1 through 8.3.[5]

Derivation of Mohr's Circle. Let us begin our derivation of Mohr's circle for plane stress by rewriting Eqs. 8.5 in the form

$$\sigma_n - \sigma_{\text{avg}} = \left(\frac{\sigma_x - \sigma_y}{2}\right) \cos 2\theta + \tau_{xy} \sin 2\theta$$

$$\tau_{nt} = -\left(\frac{\sigma_x - \sigma_y}{2}\right) \sin 2\theta + \tau_{xy} \cos 2\theta$$

(8.21)

Squaring both sides of each of these equations, and adding the resulting squares, we get

$$(\sigma_n - \sigma_{\text{avg}})^2 + \tau_{nt}^2 = \left(\frac{\sigma_x - \sigma_y}{2}\right)^2 + \tau_{xy}^2$$

or

$$(\sigma_n - \sigma_{\text{avg}})^2 + \tau_{nt}^2 = R^2 \qquad (8.22)$$

This is the equation of a circle in (σ, τ) coordinates, with center at $(\sigma_{\text{avg}}, 0)$ and radius R. The plane-stress transformation equations 8.5 or 8.21, are just parametric equations of a circle, with the parameter being θ, and with the coordinates of point N on the circle representing the normal stress σ_n and shear stress τ_{nt} on the n plane at orientation $\theta \equiv \theta_{xn}$.

To determine more of the properties of Mohr's circle, consider now the circle shown in Fig. 8.16. After discussing many properties of Mohr's circle, using Fig. 8.16, we will

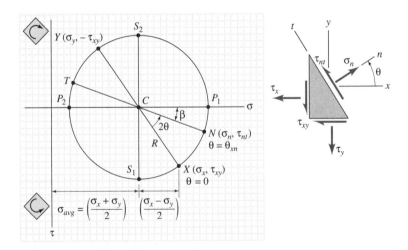

FIGURE 8.16 Properties of Mohr's circle for plane stress.

[5]Otto Mohr (1835–1918), a German structural engineer and professor of engineering mechanics, introduced the graphical representation of stress at a point, Mohr's circle; developed a failure criterion (see Sect. 12.3); and made many other important contributions to mechanics of structures and materials. [Ref. 8-2] Mohr's circle is also useful in the computation of strain transformations (Section 8.10) and the computation of moments and products of inertia (Appendix C.3).

suggest a procedure for you to use in constructing a Mohr's circle from given stress data, and we will illustrate how to solve problems like Example Problems 8.1 through 8.3 using Mohr's circle.

Mohr's circle is drawn on a set of rectilinear axes with the horizontal axis (axis of abscissas) representing the normal stress σ, and with the vertical axis (axis of ordinates) representing the shear stress τ. Note that the **positive τ axis is downward**. The sign convention for θ is the same one that was introduced in Section 8.3. The angle $\theta \equiv \theta_{xn}$ is, as previously, measured **counterclockwise** from the x axis to the n axis **on the body** undergoing plane stress. Correspondingly, an angle of $2\theta \equiv 2\theta_{xn}$ is measured **counterclockwise on Mohr's circle**.[6] *Every point on Mohr's circle corresponds to the stresses σ and τ on a particular face*; for the generic point N, the stresses are (σ_n, τ_{nt}). To emphasize this, we will **label points on the circle with the same label as the face represented by that point**, except that a capital letter will designate the point on Mohr's circle. The x face is represented by point X on the circle; the n face is represented by point N on the circle, and so on.

To reinforce the sign convention for plotting shear stresses on Mohr's circle, the small shear stress icons in Fig. 8.16 indicate that **the shear stress on a face plots as positive shear (i.e., plots downward) if the shear stress on the face would tend to rotate a stress element counterclockwise**. Conversely, the shear stress on a face plots as negative shear (i.e., plots upward) if the shear stress on that face would tend to rotate a stress element clockwise. Note that this sign convention for plotting τ causes the y face to be represented by the point Y at $(\sigma_y, -\tau_{xy})$ just as was explained in Fig. 8.11b.

Before we look at examples of how Mohr's circle is used, let us establish the fact that Mohr's circle does, indeed, provide a graphical representation of the *stress transformation equations,* Eqs. 8.5. Equations 8.23 and 8.24 come directly from the trigonometry of the circle in Fig. 8.16.

$$\sigma_n = \sigma_{\text{avg}} + R \cos \beta, \qquad \tau_{nt} = R \sin \beta \tag{8.23}$$

and

$$\frac{\sigma_x - \sigma_y}{2} = R \cos(2\theta + \beta), \qquad \tau_{xy} = R \sin(2\theta + \beta) \tag{8.24}$$

The angle β on Fig. 8.16 is introduced just to facilitate the derivation that follows. The trigonometric identities for the cosine and sine of the sum of two angles are

$$\cos(\alpha + \beta) = \cos \alpha \cos \beta - \sin \alpha \sin \beta \tag{8.25}$$

$$\sin(\alpha + \beta) = \sin \alpha \cos \beta + \cos \alpha \sin \beta$$

Letting $\alpha = 2\theta$, we can convert Eqs. 8.24 to the form

$$\frac{\sigma_x - \sigma_y}{2} = R(\cos 2\theta \cos \beta - \sin 2\theta \sin \beta) \tag{8.26a}$$

$$\tau_{xy} = R(\sin 2\theta \cos \beta + \cos 2\theta \sin \beta) \tag{8.26b}$$

If we multiply Eq. 8.26a by $\cos 2\theta$ and Eq. 8.26b by $\sin 2\theta$ and add the resulting equations, we get

$$\left(\frac{\sigma_x - \sigma_y}{2}\right) \cos 2\theta + \tau_{xy} \sin 2\theta = R \cos \beta \tag{8.27}$$

---

[6]The sign convention for Mohr's circle is sometimes a cause of confusion, leading to many different treatments in texts on the mechanics of deformable bodies. Just remember, in this text, τ **is positive downward**, and **angles are turned in the same direction on both the body (angle θ) and on Mohr's circle (angle 2θ)**.

but, combining this with Eq. 8.23a leads to

$$\sigma_n = \sigma_{\text{avg}} + \left(\frac{\sigma_x - \sigma_y}{2} \right) \cos 2\theta + \tau_{xy} \sin 2\theta$$

(8.5a) repeated

which, of course, is just Eq. 8.5a. Similarly, multiplying Eq. 8.26a by $\sin 2\theta$ and Eq. 8.26b by $\cos 2\theta$ and subtracting the latter from the former we get

$$- \left(\frac{\sigma_x - \sigma_y}{2} \right) \sin 2\theta + \tau_{xy} \cos 2\theta = R \sin \beta$$

(8.28)

but, Eq. 8.23b reveals that this can be written

$$\tau_{nt} = - \left(\frac{\sigma_x - \sigma_y}{2} \right) \sin 2\theta + \tau_{xy} \cos 2\theta$$

(8.5b) repeated

which is just Eq. 8.5b. Again, we have shown that Eqs. 8.5 are just the parametric equations of a circle in (σ, τ) coordinates.

Having established that **Mohr's circle of stress is a graphical representation of the transformation equations for plane stress**, let us examine other properties that can be easily deduced from Mohr's circle.

Properties of Mohr's Circle. Referring to Fig. 8.16, we can conclude the following:

- The center of Mohr's circle lies on the σ axis at $(\sigma_{\text{avg}}, 0)$.
- Points on the circle that lie above the σ axis (i.e., τ negative) correspond to faces that have a clockwise-acting shear; points that lie below the σ axis (i.e., τ positive) correspond to faces that have a counterclockwise-acting shear, as illustrated by Fig. 8.17.
- The radius of the circle is determined by applying the Pythagorean theorem to the triangle with sides τ_{xy} and $\left(\dfrac{\sigma_x - \sigma_y}{2} \right)$, giving

$$R = \sqrt{ \left(\frac{\sigma_x - \sigma_y}{2} \right)^2 + \tau_{xy}^2 }$$

(8.11) repeated

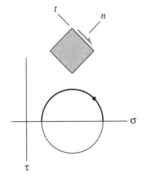

FIGURE 8.17 The Mohr's circle shear-stress sign convention.

(a) Clockwise shear stress.

(b) Counterclockwise shear stress.

- Two planes that are 90° apart on the physical body are represented by the two points at the extremities of a diameter, such as points X and Y or P_1 and P_2 in Fig. 8.16.

- If we rotate counterclockwise by an angle θ_{ab} to go from face a to face b of the physical body, we must rotate in that same direction through the angle $2\theta_{ab}$ to get from point A on Mohr's circle to point B. Figure 8.18 illustrates this property of the Mohr's circle sign convention. In equations, a positive angle is always counterclockwise.

- The *principal planes* are represented by points P_1 and P_2 at the intersections of Mohr's circle with the σ axis (Fig. 8.16). The corresponding *principal stresses* are $\sigma_1 = \sigma_{avg} + R$ and $\sigma_2 = \sigma_{avg} - R$.

- The planes of maximum shear stress are represented by points S_1 and S_2 that lie directly below and above the center of the Mohr's circle (Fig. 8.16). The corresponding stresses are: (σ_{avg}, R) on face s_1, and $(\sigma_{avg}, -R)$ on face s_2.

- Since the stresses on orthogonal planes n and t are represented by the points at each end of a diameter of Mohr's circle,

$$\sigma_n + \sigma_t = \sigma_x + \sigma_y$$

(8.7)
repeated

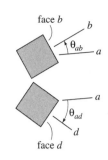

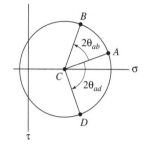

FIGURE 8.18 Consistent angles.

PROCEDURE FOR CONSTRUCTING AND USING MOHR'S CIRCLE OF STRESS

To solve plane-stress problems, such as determining the stresses on a particular face (e.g., Example Problems 8.1 and 8.2) or determining principal stresses and maximum in-plane shear stresses (Example Problem 8.3), the following procedure is suggested:

Draw Mohr's Circle[7]

1. Establish a set of (σ, τ) axes, with the same scale on both axes. It is good idea to use paper that has a grid, like graph paper or "engineering paper." Use a scale that will result in a circle of reasonable size.

2. Assuming that σ_x, σ_y, and τ_{xy} are given (or can be determined from a given stress element), locate point X at (σ_x, τ_{xy}) and point Y at $(\sigma_y, -\tau_{xy})$.

3. Connect points X and Y with a straight line, and locate the center of the circle where this line crosses the σ axis at $(\sigma_{avg}, 0)$.

4. Draw a circle with center at $(\sigma_{avg}, 0)$ and passing through points X and Y. It is best to use a compass to draw the circle.

Compute the Required Information

5. Form the triangle with sides τ_{xy} and $\left(\dfrac{\sigma_x - \sigma_y}{2}\right)$, and compute

$$R = \sqrt{\left(\frac{\sigma_x - \sigma_y}{2}\right)^2 + \tau_{xy}^2}$$

6. If the stresses on a particular face, call it face n, are required, locate point N on the circle by turning an angle 2θ counterclockwise (or clockwise) on the circle, corresponding to rotating an angle θ counterclockwise (clockwise) from some reference face on the stress element. Using trigonometry, calculate σ_n and τ_{nt}.[8]

7. If the principal stresses and the orientation of the principal planes are required, use

$$\sigma_1 = \sigma_{avg} + R, \qquad \sigma_2 = \sigma_{avg} - R$$

to calculate the principal stresses, and use trigonometry to determine some angle, such as $2\theta_{xp1}$ that can be used to locate a principal plane, say p_1, with respect to some known face, say the x face.

8. Use a procedure similar to Step 7 if the maximum in-plane shear stress and the planes of maximum shear stress are required.

---

[7]After you become proficient with Mohr's circle, a simple sketch will suffice. However, at first it is helpful if you draw stress magnitudes (at least roughly) to scale and draw angles the correct size.

[8]Although it was suggested that the circle be accurately drawn to some scale, you should use trigonometry to compute the required answers and only scale magnitudes and angles off of the Mohr's circle as a check of your calculations.

EXAMPLE 8.4

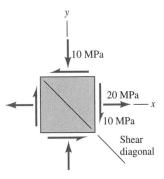

Fig. 1 A state of plane stress.

For the plane-stress state in Example Problems 8.1 and 8.2 (Fig. 1), do the following: (a) Draw Mohr's circle. (b) Determine the stresses on all faces of an element that is rotated 30° counterclockwise from the orientation of the stress element in Fig. 1. (c) Determine the orientation of the principal planes; determine the principal stresses. (d) Determine the orientation of the planes of maximum shear stress; determine the value of the maximum shear stress.

Solution We can just follow the procedure outlined above.

(a) Mohr's Circle: On grid paper (Fig. 2), plot point X at (20 MPa, -10 MPa) and plot point Y, at (-10 MPa, 10 MPa). The center of the circle is obtained by connecting X and Y. The diameter crosses the σ axis at C: $(\sigma_{avg}, 0)$, where

$$\sigma_{avg} = \frac{20 \text{ MPa} - 10 \text{ MPa}}{2} = 5 \text{ MPa} \tag{1}$$

The circle is drawn with center at C and passing through points X and Y. The radius R is calculated from the shaded triangle XCB in Fig. 2.

$$R = \sqrt{(15 \text{ MPa})^2 + (10 \text{ MPa})^2} = \sqrt{325} \text{ MPa} = 18.03 \text{ MPa} \tag{2}$$

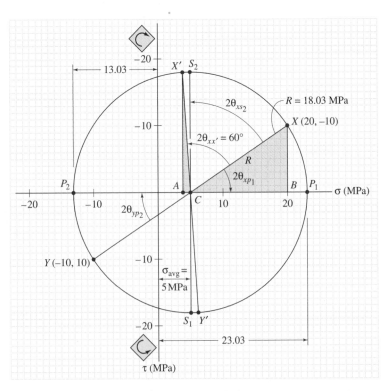

Fig. 2 Mohr's circle.

448

(b) *Stresses on* x′ *and* y′ *Faces:* Locate the points on Mohr's circle that correspond to rotating the stress element by 30°. This means rotating 60° counterclockwise from the XY diameter on Mohr's circle. We label these two points X′ and Y′.

To determine the stresses at points X′ and Y′, we need to establish the geometry (trigonometry) of the triangle X′CA. To determine $\angle X'CA$ we need to first determine the (clockwise) angle $2\theta_{xp1}$, in Fig. 2. Using the triangle, XCB, we get

$$2\theta_{xp1} = \tan^{-1}\left(\frac{10}{15}\right) = 33.69° \tag{3a}$$

$$\theta_{xp1} = 16.8° \ \circlearrowright \tag{3b}$$

Therefore,

$$\angle X'CA = 180° - 60° - 2\theta_{xp1} = 86.31° \tag{4}$$

From the triangle X′CA,

$$\overline{AC} = R\cos(\angle X'CA) = 18.03\cos(86.31°) \tag{5a}$$

or

$$\overline{AC} = 1.16 \text{ MPa} \tag{5b}$$

Therefore,

$$\left.\begin{array}{l} \sigma_{x'} = \sigma_{\text{avg}} - \overline{AC} = 3.84 \text{ MPa} \\ \sigma_{y'} = \sigma_{\text{avg}} + \overline{AC} = 6.16 \text{ MPa} \end{array}\right\} \quad \text{Ans. (b)} \tag{6}$$

Also, from triangle X′CA we get

$$\tau_{x'y'} = -R\sin(\angle X'CA) = -18.03\sin(86.31°)$$

or

$$\tau_{x'y'} = -18.0 \text{ MPa} \qquad \text{Ans. (b)} \tag{7}$$

Equations (6) and (7) are the same answers that we obtained in Example Problem 8.1 by using formulas directly.

(c) *Principal Planes and Principal Stresses:* We have already calculated θ_{xp1} in Eq. (3). From Fig. 2,

$$2\theta_{yp2} = 2\theta_{xp1} = 33.69° \tag{8a}$$

$$\theta_{yp2} = 16.8° \ \circlearrowright \tag{8b}$$

Also, from Fig. 2,

$$\sigma_1 = \sigma_{\text{avg}} + R = 5 \text{ MPa} + 18.03 \text{ MPa} = 23.0 \text{ MPa}$$

$$\sigma_2 = \sigma_{\text{avg}} - R = 5 \text{ MPa} - 18.03 \text{ MPa} = -13.0 \text{ MPa}$$

or

$$\sigma_1 = 23.0 \text{ MPa}, \qquad \sigma_2 = -13.0 \text{ MPa} \qquad \text{Ans. (c)} \quad (9)$$

(d) Maximum In-Plane Shear Stress: The planes of maximum in-plane shear stress are represented by the points S_1 and S_2 on Mohr's circle (Fig. 2). From Fig. 2,

$$2\theta_{xs1} = 90° + 2\theta_{xp1} = 90° + 33.69° = 123.69°$$

so

$$\theta_{xs1} = 61.8° \quad \circlearrowright$$

Also, by referring to Fig. 2, we see that

$$2\theta_{xs2} = 90° - 2\theta_{xp1} = 90° - 33.69° = 56.31°$$
$$\theta_{xs2} = 28.2°$$

The maximum in-plane shear stress occurs on plane s_1 and on plane s_2 and is given by

$$\tau_{s1s2} = R = 18.0 \text{ MPa} \qquad \text{Ans. (d)}$$

On the planes of maximum shear stress, the normal stress is

$$\sigma_{s1} = \sigma_{s2} = \sigma_{avg} = 5 \text{ MPa} \qquad \text{Ans. (d)}$$

Review the Solution It is very important to construct the Mohr's circle properly. Once that is done, subsequent results can be checked visually by (roughly) estimating the values of normal stresses and shear stresses that are required and the angles that are required.

The comments in the *Review the Solution* section of Example Problem 8.3 apply to the results that we obtained above by using Mohr's circle.

MOHR, a computer program for constructing Mohr's circles of stress and strain, is one of the *MechSOLID* collection of computer programs. It is described in Appendix G.7. Computer homework exercises for Section 8.5 follow the regular Section 8.5 homework problems at the end of the chapter.

*8.6 THREE-DIMENSIONAL STRESS STATES— PRINCIPAL STRESSES

In Sections 8.2 through 8.5 we were dealing with plane stress—formulating stress transformation equations, determining expressions for principal stresses and maximum in-plane shear stresses, and establishing a graphical representation of the plane-stress transformation equations, called Mohr's circle. Since plane stress is only a special case of the more-general three-dimensional state of stress, however, we need to look further at three-dimensional stress states. In particular, we need to be able to determine principal stresses and absolute maximum shear stresses. In this section we will examine principal stresses for a general state of stress; the next section treats absolute maximum shear stresses.

Principal Stresses—Plane-Stress Case. The procedure employed in Section 8.4 to obtain principal stresses and principal directions for states of plane stress is restricted to two-dimensional analysis. However, there is an alternative procedure that may be applied to the plane-stress case, and also be readily extended to three-dimensional stress states. Before considering principal stresses in three dimensions, let us apply this alternative procedure to the plane-stress case.

Let Fig. 8.19 represent a free body whose oblique face is one of the principal faces (hence, the label p), and let the direction cosines of the normal to this face be $l(= \cos \theta_p)$ and $m(= \sin \theta_p)$. By summing forces in the x and y directions and noting, as before, that $\tau_{yx} = \tau_{xy}$, we get

$$\overset{+}{\rightarrow} \sum F_x = 0: \qquad (\sigma_p \Delta A)l - \sigma_x(l\Delta A) - \tau_{xy}(m\Delta A) = 0$$

$$+\uparrow \sum F_y = 0: \qquad (\sigma_p \Delta A)m - \sigma_y(m\Delta A) - \tau_{xy}(l\Delta A) = 0$$

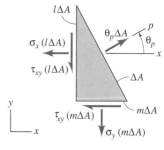

$l \equiv \cos \theta_p, \; m \equiv \sin \theta_p, \; l^2 + m^2 = 1$

FIGURE 8.19 A free-body diagram for calculating principal stress.

Dividing these equations by ΔA and rearranging, we obtain a set of two simultaneous equations in l and m, which we can write as[9]

$$(\sigma_x - \sigma_p)l + \tau_{xy}m = 0 \qquad (8.29a)$$

$$\tau_{xy}l + (\sigma_y - \sigma_p)m = 0 \qquad (8.29b)$$

Since this is a set of homogeneous algebraic equations, and since we cannot solve them by merely setting $l = m = 0$ (since $l^2 + m^2 = 1$), the determinant of the coefficients must vanish, giving

$$\begin{vmatrix} (\sigma_x - \sigma_p) & \tau_{xy} \\ \tau_{xy} & (\sigma_y - \sigma_p) \end{vmatrix} = 0 \qquad (8.30)$$

Expanding this determinate leads to the quadratic equation

$$\sigma_p^2 - \sigma_p(\sigma_x + \sigma_y) + (\sigma_x\sigma_y - \tau_{xy}^2) = 0 \qquad (8.31)$$

The roots of this quadratic equation are the two principal stresses

$$\sigma_1 \equiv \sigma_{p1} = \sigma_{avg} + R$$
$$\sigma_2 \equiv \sigma_{p2} = \sigma_{avg} - R$$

$\qquad\qquad (8.13)$
$\qquad\qquad$ repeated

just as we obtained by a different procedure in Section 8.4.[10]

---

[9]This is called an *algebraic eigenvalue problem*. There are many different physical situations, in addition to stress analysis, that lead to this type of problem; for example, dynamics, vibrations, stability, control, etc. The solution of algebraic eigenvalue problems is discussed in books on linear algebra, like *Elementary Linear Algebra*, by Howard Anton, [Ref. 8-3].

[10]The principal directions corresponding to σ_1 and σ_2 could be obtained from Eqs. 8.29, but here we will only consider principal stresses.

Principal Stresses—General State of Stress. Let us now apply the above approach to determine the principal stresses for a general three-dimensional state of stress, like the one illustrated in Fig. 8.2b. In Fig. 8.20a we take a tetrahedral (i.e., four-faced) block, whose oblique face is a *principal plane*, that is, *a plane on which there is no shear stress.* If the area of the triangular oblique face ABC is ΔA, the areas of the other three faces are given by

$$\Delta A_x \equiv \text{area } (BOC) = l\Delta A$$
$$\Delta A_y \equiv \text{area } (AOC) = m\Delta A \qquad (8.32)$$
$$\Delta A_z \equiv \text{area } (AOB) = n\Delta A$$

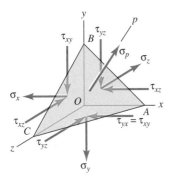

(a) Tetrahedral stress element.

where (l, m, n) are the direction cosines, as illustrated in Fig. 8.20b. Each stress shown in Fig. 8.20a must be multiplied by the area of the face on which it acts to obtain the corresponding force acting on the tetrahedron. The free-body diagram will not be drawn here. Equilibrium of the tetrahedron leads to the three-dimensional version of Eqs. 8.29, namely to

$$\sum F_x = 0: \qquad (\sigma_p\Delta A)l - \sigma_x(l\Delta A) - \tau_{xy}(m\Delta A) - \tau_{xz}(n\Delta A) = 0$$
$$\sum F_y = 0: \qquad (\sigma_p\Delta A)m - \sigma_y(m\Delta A) - \tau_{yz}(n\Delta A) - \tau_{xy}(l\Delta A) = 0$$
$$\sum F_z = 0: \qquad (\sigma_p\Delta A)n - \sigma_z(n\Delta A) - \tau_{xz}(l\Delta A) - \tau_{yz}(m\Delta A) = 0$$

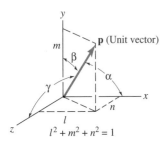

(b) Direction cosines.

FIGURE 8.20 A tetrahedral element for deriving the principal stress σ_p, which acts on a plane whose normal is **p**.

where the shear-stress equalities $\tau_{yx} = \tau_{xy}$, and so on, are incorporated to reduce the number of shear stresses. Dividing by ΔA and collecting terms, we get

$$(\sigma_x - \sigma_p)l + \tau_{xy}m + \tau_{xz}n = 0$$
$$\tau_{xy}l + (\sigma_y - \sigma_p)m + \tau_{yz}n = 0 \qquad (8.33)$$
$$\tau_{xz}l + \tau_{yz}m + (\sigma_z - \sigma_p)n = 0$$

As before, since we cannot satisfy these equations by taking $l = m = n = 0$ (since $l^2 + m^2 + n^2 = 1$ for the direction cosines), and since these are homogeneous equations, we must find values of σ_p that make the determinant of the coefficients in Eq. 8.33 equal to zero.

$$\begin{vmatrix} (\sigma_x - \sigma_p) & \tau_{xy} & \tau_{xz} \\ \tau_{xy} & (\sigma_y - \sigma_p) & \tau_{yz} \\ \tau_{xz} & \tau_{yz} & (\sigma_z - \sigma_p) \end{vmatrix} = 0 \qquad (8.34)$$

When this determinant is expanded, we get the following cubic equation:

$$\sigma_p^3 - A\sigma_p^2 + B\sigma_p - C = 0 \qquad (8.35)$$

where

$$A = \sigma_x + \sigma_y + \sigma_z$$
$$B = \sigma_x\sigma_y + \sigma_x\sigma_z + \sigma_y\sigma_z - \tau_{xy}^2 - \tau_{xz}^2 - \tau_{yz}^2 \qquad (8.36)$$
$$C = \sigma_x\sigma_y\sigma_z + 2\tau_{xy}\tau_{xz}\tau_{yz} - \sigma_x\tau_{yz}^2 - \sigma_y\tau_{xz}^2 - \tau_z\tau_{xy}^2$$

The quantities *A*, *B*, and *C* are called *stress invariants* since they do not depend on the particular orientation of the *x*, *y*, and *z* axes.[11]

The three roots of the cubic polynomial in Eq. 8.35 are the three *principal stresses*. Let us call them

$$\sigma_1 \equiv \sigma_{\text{maximum}}, \quad \sigma_2 \equiv \sigma_{\text{intermediate}}, \quad \sigma_3 \equiv \sigma_{\text{minimum}}$$

that is, $\sigma_1 \geq \sigma_2 \geq \sigma_3$.[12] To each principal stress σ_i there is a unit normal vector (l_i, m_i, n_i) that defines the corresponding *principal direction*.[13] If we draw the stress element whose faces are all principal planes, we get Fig. 8.21. Since all faces of this element are free of shear stress, this element is said to be in a *triaxial state of stress*.

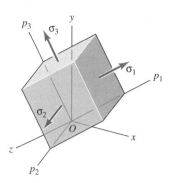

FIGURE 8.21 Principal stresses acting on a three-dimensional element.

8.7 ABSOLUTE MAXIMUM SHEAR STRESS

In Section 8.4 we examined **maximum in-plane** *shear stress* for the case of plane stress. For a general state of stress at a point, including the case of plane stress, we need to look for the **absolute maximum** *shear stress*. To do so, it is convenient to assume that we already know the principal directions and the principal stresses at the point.[14] Figure 8.22*a* represents the element on which the principal stresses act at the point. The principal directions are labeled p_1, p_2, and p_3, with $\sigma_1 \geq \sigma_2 \geq \sigma_3$.

Absolute Maximum Shear Stress—General Stress State. The *absolute maximum shear stress at a point* is the greatest (i.e., largest magnitude) shear stress acting in **any direction** on **any plane** passing through the point. Let us begin our search for the absolute maximum shear stress at a point by examining the stresses on a plane whose normal *n* is perpendicular to the p_2 direction, that is, a plane like the oblique plane shown in Fig. 8.22*b*. Since there is no shear stress at all on the p_2 face, the shear stress on the *n*-face has no p_2-component, as is indicated in Fig. 8.22*b*. Figure 8.23*a* is a two-dimensional view of the wedge in Fig. 8.22*b* showing the forces obtained by multiplying each stress

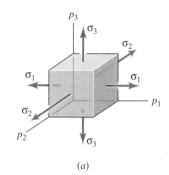

(a)

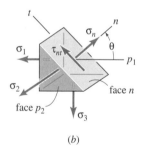

(b)

FIGURE 8.22 (*a*) An element in triaxial stress, and (*b*) a wedge for stress transformation based on triaxial stresses.

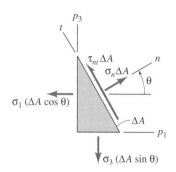

(*a*) A free-body diagram for determining stresses in the $p_1 p_3$ plane. (σ_2 is omitted because it is normal to the $p_1 p_3$ plane.)

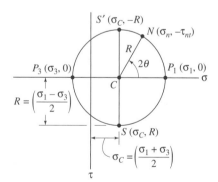

(*b*) Mohr's circle for stresses in the $p_1 p_3$ plane.

FIGURE 8.23 Transformation of stresses in the $p_1 p_3$ plane.

---

[11]See Section 78 of *Theory of Elasticity*, by S. P. Timoshenko and J. N. Goodier, [Ref. 8-1], for further discussion of the stress invariants.

[12]In the terminology of linear algebra, the principal stresses are called *eigenvalues*.

[13]In the terminology of linear algebra, the principal directions are called *eigenvectors*. See Section 77 of *Theory of Elasticity*, by S. P. Timoshenko and J. N. Goodier, [Ref. 8-1], for further discussion of the principal directions.

[14]Equations 8.33 and 8.35 of the preceding section may be used to determine these.

by the area of the face on which it acts. By summing forces on the free-body diagram of Fig. 8.23a we obtain a special case of Eqs. 8.5, namely,

$$\sigma_n = \frac{\sigma_1 + \sigma_3}{2} + \frac{\sigma_1 - \sigma_3}{2} \cos 2\theta = \sigma_{\text{avg}} + R \cos 2\theta$$

$$-\tau_{nt} = \frac{\sigma_1 - \sigma_3}{2} \sin 2\theta = R \sin 2\theta$$

(8.37)

It is clear that Eqs. 8.37 locate the point N on the Mohr's circle in Fig. 8.23b.

The maximum shear stress in the $p_1 p_3$ plane is given by the radius of the circle in Fig. 8.23b, that is

$$(\tau_{\text{max}})_{p_1 p_3} \equiv \tau_s = \frac{\sigma_1 - \sigma_3}{2}$$

(8.38)

where the designation $(\tau_{\text{max}})_{p_1 p_3}$ refers to the maximum in-plane shear stress in the $p_1 p_3$ plane. It can be shown[15] that this shear stress is also the **absolute maximum** *shear stress* at the point. Therefore,

$$\boxed{\tau_{\substack{\text{abs} \\ \text{max}}} \equiv \tau_s = \frac{\sigma_1 - \sigma_3}{2}}$$

(8.39)

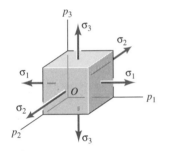

(a) Stress element with principal planes as faces.

and this absolute maximum shear stress acts on planes whose normal s bisects the 90° angle between the p_1 and p_3 directions, as illustrated in Fig. 8.24b. The normal stress acting on the planes of maximum shear stress is

$$\sigma_s = \frac{\sigma_1 + \sigma_3}{2}$$

(8.40)

Note that the τ_s arrows and the σ_s arrows in Fig. 8.24b both have components that point in the p_1 direction.

Stress transformations, like Eqs. 8.37, and Mohr's circles, like Fig. 8.23b, can be developed for faces whose normal n lies in the $p_1 p_2$ plane or the $p_2 p_3$ plane. However, since the absolute maximum shear occurs on the planes whose normal lies in the $p_1 p_3$ plane (Fig. 8.24b), we will not examine the other cases here.

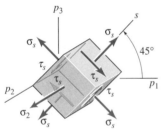

$$\sigma_s = \left(\frac{\sigma_1 + \sigma_3}{2}\right), \ \tau_s = \left(\frac{\sigma_1 - \sigma_3}{2}\right)$$

(b) The element on which the absolute maximum shear stress acts.

FIGURE 8.24 Planes of absolute maximum shear stress.

Absolute Maximum Shear Stress—Plane-Stress State.

In Section 8.4 we examined the principal stresses and the maximum in-plane shear stresses for the plane-stress states, where the conditions for the plane stress are $\sigma_z = \tau_{xz} = \tau_{yz} = 0$. Since there is no shear stress on the z faces, the z axis is one of the three principal directions at the point. In determining the absolute maximum shear stress, the question, then, is whether the stress $\sigma_z = 0$ is the maximum principal stress (σ_1), the intermediate principal stress (σ_2), or the minimum principal stress (σ_3). Figure 8.25 illustrates these three options.[16] In each case a solid Mohr's circle is drawn for the in-plane (i.e., xy plane) stress transformation discussed in Sections 8.4 and 8.5. The dashed circles are for the $p_1 p_3$ stress trans-

---

[15]See Section 79 of *Theory of Elasticity*, by S. P. Timoshenko and J. N. Goodier, [Ref. 8-1].

[16]The use of Mohr's circle is extended to three-dimensional stress states by drawing a separate circle through each pair of principal stresses.

(a) Case I,
$(\sigma_3 \le \sigma_2 \le 0)$

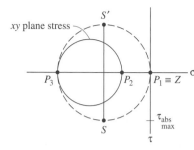

(a1) Mohr's circle.

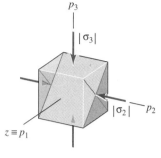

(a2) A maximum shear plane.

(b) Case II,
$(\sigma_3 \le 0 \le \sigma_1)$

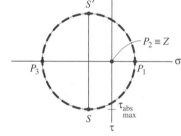

(b1) Mohr's circle.

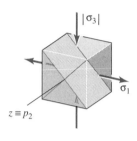

(b2) A maximum shear plane.

(c) Case III,
$(0 \le \sigma_2 \le \sigma_1)$

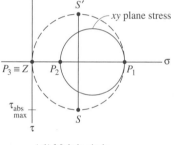

(c1) Mohr's circle.

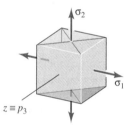

(c2) A maximum shear plane.

FIGURE 8.25 The absolute maximum shear stress for various plane stress states; planes of absolute maximum shear stress.

formation, which leads to the absolute maximum shear stress (see Fig. 8.23b). The value of the absolute maximum shear stress is always given by Eq. 8.39, and the corresponding normal stress is always given by Eq. 8.40.

$$
\begin{aligned}
&\text{Case I } (\sigma_3 \le \sigma_2 \le 0){:} && \tau_{\substack{\text{abs} \\ \text{max}}} \equiv \tau_s = \frac{-\sigma_3}{2}; \; \sigma_s = \frac{\sigma_3}{2} \\[2mm]
&\text{Case II } (\sigma_3 \le 0 \le \sigma_1){:} && \tau_{\substack{\text{abs} \\ \text{max}}} \equiv \tau_s = \frac{\sigma_1 - \sigma_3}{2}; \; \sigma_s = \frac{\sigma_1 + \sigma_3}{2} \quad (8.41) \\[2mm]
&\text{Case III } (0 \le \sigma_2 \le \sigma_1){:} && \tau_{\substack{\text{abs} \\ \text{max}}} \equiv \tau_s = \frac{\sigma_1}{2}; \; \sigma_s = \frac{\sigma_1}{2}
\end{aligned}
$$

It should be clear from Fig. 8.25 that, **only when the in-plane principal stresses have opposite signs is the maximum in-plane shear stress also the absolute maximum** shear stress. When both in-plane principal stresses are negative (Fig. 8.25*a*) or when both are positive (Fig. 8.25*c*), the absolute maximum shear stress acts on planes at 45° to the free surface, and the maximum in-plane shear stress is not the absolute maximum shear stress. **Even though the *z* faces are stress free, they must be taken into account in determining the absolute maximum shear stress!** But, in every case, from Eqs. 8.39 and 8.40 we have[17]

$$\tau_{\substack{abs\\max}} \equiv \tau_s = \frac{\sigma_{max} - \sigma_{min}}{2}$$

$$\sigma_s = \frac{\sigma_{max} + \sigma_{min}}{2}$$

(8.42)

[17]The stresses σ_{max} and σ_{min} are signed quantities (i.e., tension positive, compression negative); they are not just magnitudes.

■■■■■■■■■■■■■ EXAMPLE 8.5 □□□□□□□□□□□□■■

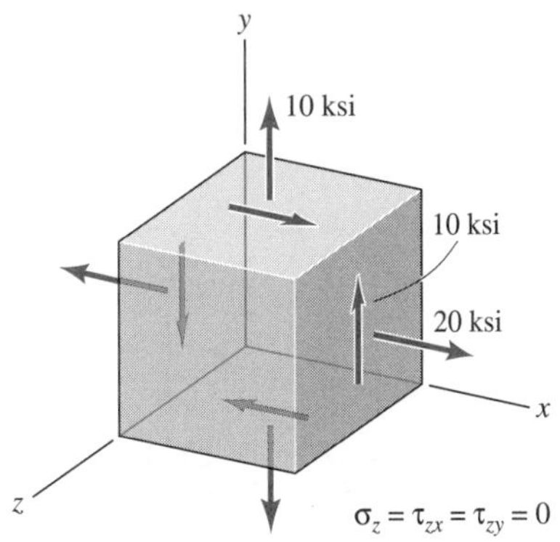

Fig. 1 An element in plane stress.

$\sigma_z = \tau_{zx} = \tau_{zy} = 0$

An element in plane stress has the stresses shown in Fig. 1. (a) Determine the three principal stresses. Use a Mohr's circle to determine in-plane stresses. (b) Determine the maximum in-plane shear stress. (c) Determine the orientation of the principal planes, and sketch the principal-stress element. (d) Determine the absolute maximum shear stress. Show an element oriented so that the absolute maximum shear stress acts on the element.

Plan the Solution We need to determine the principal directions and in-plane principal stresses for the *xy* plane using the Mohr's circle technique of Section 8.5. From Mohr's circle we can also get the maximum in-plane shear stress. In Part (c) we will have to order the principal stresses $\sigma_1 \geq \sigma_2 \geq \sigma_3$, and compare the three principal stresses in this problem (the two in-plane principal stresses plus $\sigma_z = 0$) with the three cases depicted in Fig. 8.25. The maximum absolute shear stress is calculated using Eq. 8.39.

Solution

(a) Principal Stresses: One of the principal stresses is $\sigma_z = 0$, since $\tau_{zx} = \tau_{zy} = 0$. The other two principal stresses are obtained from the Mohr's circle in Fig. 2.

From triangle *XCA* we get

$$R = \sqrt{(\overline{CA})^2 + (\overline{XA})^2} = \sqrt{(5 \text{ ksi})^2 + (10 \text{ ksi})^2}$$

So,

$$R = \sqrt{125} \text{ ksi} = 11.18 \text{ ksi} \tag{1}$$

Since all points on the Mohr's circle in Fig. 2 have $\sigma > 0$, $\sigma_z = 0$ is the minimum principal stress. Therefore, the intersections of Mohr's circle with

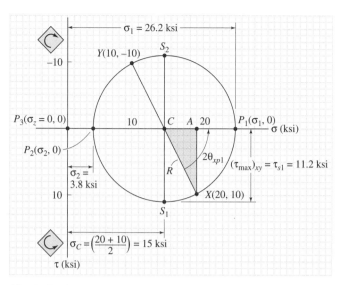

Fig. 2 Mohr's circle for the xy plane, with P_3 shown for reference.

the σ axis are labeled P_1 and P_2. From the circle in Fig. 2,

$$\sigma_1 = \sigma_C + R = 15 \text{ ksi} + 11.2 \text{ ksi} = 26.2 \text{ ksi} \qquad (2)$$
$$\sigma_2 = \sigma_C - R = 15 \text{ ksi} - 11.2 \text{ ksi} = 3.8 \text{ ksi}$$

Therefore, the three principal stresses are

$$\sigma_1 = 26.2 \text{ ksi}, \quad \sigma_2 = 3.8 \text{ ksi}, \quad \sigma_3 = 0 \qquad \textbf{Ans. (a)} \quad (3)$$

(b) Maximum In-Plane Shear Stress: The maximum shear stress in the xy plane is the shear stress at point S_1 in Fig. 2, or

$$(\tau_{\max})_{xy} = R = 11.2 \text{ ksi} \qquad \textbf{Ans. (b)} \quad (4)$$

(c) Principal-Stress Element: To orient the principal-stress element ($p_1 p_2 p_3$ axes) relative to the xyz axes we only need to relate p_1 and p_2 to x and y, since we already know that $p_3 \equiv z$ (since $\sigma_z = 0$). From Fig. 2 we can determine the angle $2\theta_{xp1}$. From triangle XCA we get

$$2\theta_{xp1} = \tan^{-1}\left(\frac{10}{5}\right) = 63.43° \qquad (5a)$$

$$\theta_{xp1} = 31.7° \qquad \textbf{Ans. (c)} \quad (5b)$$

A properly oriented principal-stress element is shown in Fig. 3.

(d) Absolute Maximum Shear Stress: The plane-stress Mohr's circle in Fig. 2 corresponds to Case III (Fig. 8.25c). Therefore, we need to construct a $p_1 p_3$ Mohr's circle. For clarity, we will draw another figure, Fig. 4, repeating part

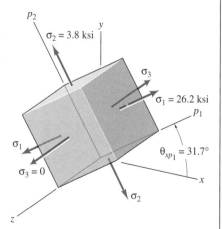

Fig. 3 The principal-stress element.

457

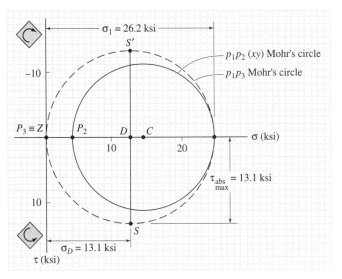

Fig. 4 Mohr's circle for determining $\tau_{abs\atop max}$.

of Fig. 2. From the dashed p_1p_3 Mohr's circle in Fig. 4 we get

$$\tau_{abs\atop max} = \frac{\sigma_1}{2} = 13.1 \text{ ksi} \qquad \textbf{Ans. (d)} \quad (6)$$

Figures 5a through 5c depict the planes of absolute maximum shear stress. First, in Figs. 5a and 5b the orientations of the planes of maximum shear stress at 45° to the p_1 and p_3 axes (faces) are illustrated. Finally, in Fig. 5c a two-dimensional view of the p_1p_3 plane is shown, looking "down" the p_2 axis.

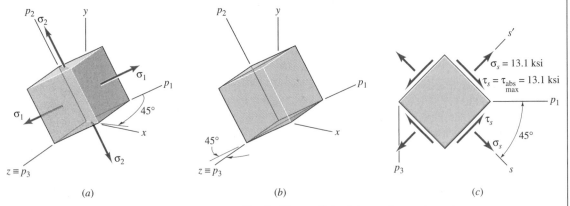

Fig. 5 Planes of absolute maximum shear stress.

Review the Solution We should first check to make sure the points X and Y in Fig. 2 correctly represent the stresses on the x and y faces in Fig. 1, especially making sure that the sign of the shear stress is correct at X and Y. Since the answers in Eqs. (3), (4), (5), and (6) came directly from the Mohr's circles in Figs. 2 and 3, we can visually check to see if they are reasonable.

458

8.8 PLANE STRAIN

Definitions of extensional strain ϵ and shear strain γ were given in Section 2.9. These strains vary with position in a body and with the orientation of the reference directions. For example, at point P in (or on the surface of) a deformable body, the extensional strains ϵ_x, ϵ_y, and ϵ_z are determined by examining the change in length of short, mutually orthogonal line segments Δx, Δy, and Δz; and the shear strains γ_{xy}, γ_{xz}, and γ_{yz} are determined by the changes in the right angles that originally exist between these lines. Figure 8.26 illustrates these incremental line segments. There are situations when it is necessary to determine the extensional strain ϵ_n or the shear strain γ_{nt}, given the xyz-referenced strains and the n and t directions. In Sections 8.9 and 8.10 we consider only two-dimensional strain analysis, but in Section 8.11 we will return to the topic of three-dimensional strain analysis.

One example of two-dimensional strain is called plane strain. If the strains satisfy the equations

$$\epsilon_z = \gamma_{xz} = \gamma_{yz} = 0 \tag{8.43}$$

everywhere in a deformable body, the body is said to be in a *state of plane strain*. The vanishing of extensional strain in the z direction requires that the body somehow be restrained from expanding or contracting in the z direction, and the shear-strain equations further require that all planes that are originally parallel to the xy plane remain plane. One example of plane strain is a very long, cylindrical object, like the dam illustrated in Fig. 8.27. Gravitational loads and upstream water-pressure loads on the dam are all parallel to the xy plane, and because of the (assumed) rigid abutments at its ends, the dam is not free to expand or contract in the z direction. Therefore, it can be assumed that $\epsilon_z = \gamma_{xz} = \gamma_{yz} = 0$ everywhere in the dam.

Plane stress and plane strain are very different and should not be confused. As discussed in Section 8.2, *plane stress* is defined by the equations

$$\sigma_z = \tau_{xz} = \tau_{yz} = 0 \tag{8.1 repeated}$$

and it most frequently occurs in thin, plate-like members, like the web of the plate girder in Fig. 8.3. If we let σ_z (and ΔT) be zero in the last of Eqs. 2.32, we get that, for linearly elastic, isotropic materials,

$$\sigma_z = 0 \rightarrow \epsilon_z = \frac{-\nu}{E}(\sigma_x + \sigma_y) \tag{8.44}$$

On the other hand, if we let ϵ_z (and ΔT) be zero in the last of Eqs. 2.34, we get, for linearly elastic, isotropic materials,

$$\epsilon_z = 0 \rightarrow \sigma_z = \frac{E\nu}{(1 + \nu)(1 - 2\nu)}(\epsilon_x + \epsilon_y) \tag{8.45}$$

Consequently, plane stress generally does not lead to $\epsilon_z = 0$, and plane strain normally requires a nonzero value of σ_z.

One very important example of two-dimensional strain analysis is the experimental determination of strains on the surface of a deformable body. Since it is the *strains at a point*, like point A on the surface of the rod in Fig. 8.6a, that are of interest, and since the surface is *locally plane*, we can concentrate on the analysis of strains in a plane.

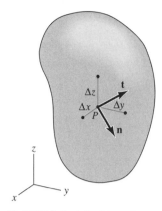

FIGURE 8.26 Mutually orthogonal line segments used in defining extensional strains and shear strains.

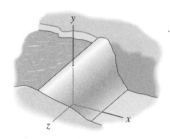

FIGURE 8.27 An example of plane strain.

459

Before we derive the strain-transformation equations, let us examine a specific example of how strains vary with the orientation of the reference axes. Figure 8.28 shows the "before-deformation" and "after-deformation" pictures of a plane membrane. Before deformation, a grid of horizontal and vertical lines is marked on the sheet, with a uniform grid spacing of length a. Points P_1, P_2, and P_3 are the origins of axis systems x_1y_1, x_2y_2, and x_3y_3 in the orientations shown. The sheet is deformed by uniformly stretching it to twice its original length in the horizontal direction. At the same time, the sheet is prevented from expanding or contracting in the vertical direction. Hence, each $a \times a$ square before deformation becomes a $2a \times a$ rectangle after deformation. (This would be considered large deformation.) Clearly, the *deformation* is the same at P_1, P_2, and P_3, but we will get different values of ϵ_n and γ_{nt} depending on the orientation of the reference axes. For example, the shear strain $\gamma_{x_1y_1} = 0$, but $\gamma_{x_2y_2}$ and $\gamma_{x_3y_3}$ are clearly not zero, since there are changes in the right angles at P_2 and P_3. Likewise, $\epsilon_{y_1} = 0$, but ϵ_{y_2} and ϵ_{y_3} are not zero, since the lengths of P_2R_2 and P_3R_3 are changed by the deformation.

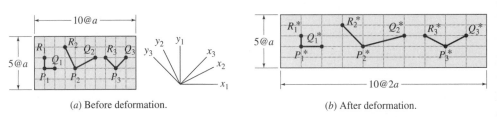

(a) Before deformation. (b) After deformation.

FIGURE 8.28 An example of how strain quantities depend on the orientation of the reference axes.

Whereas the stress-transformation equations, Eqs. 8.3 (or Eqs. 8.5) are based on equilibrium only, the **strain-transformation equations are based solely on the geometry of deformation** (including some small-angle approximations). Figure 8.29a, 8.29b, and 8.29c depict the separate effects of ϵ_x, ϵ_y, and γ_{xy}, respectively. These effects can be added together to get the total expressions for ϵ_n and γ_{nt} as functions of the angle $\theta \equiv \theta_{xn} \equiv \theta_{yt}$ and the three strains ϵ_x, ϵ_y, and γ_{xy}.

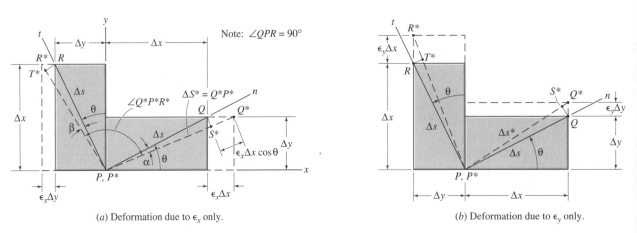

(a) Deformation due to ϵ_x only. (b) Deformation due to ϵ_y only.

FIGURE 8.29 The geometry of deformation used to derive strain transformation equations.

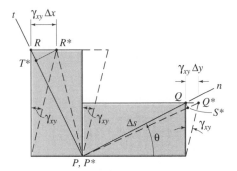

(c) Deformation due to γ_{xy} only.

FIGURE 8.29 (continued).

Contribution of ϵ_x, ϵ_y, and γ_{xy} to ϵ_n and γ_{nt}.

Using Fig. 8.29a, we will derive expressions that relate ϵ_n and γ_{nt} to ϵ_x. The contributions of ϵ_y to ϵ_n and γ_{nt}, and of γ_{xy} to ϵ_n and γ_{nt} can be determined in an analogous manner, so these derivations are left as homework problems. (See Homework Problems 8.9-1 and 8.9-2.) Therefore, assuming that all strains are small, we can write the following superposition expressions for ϵ_n and γ_{nt}:

$$\epsilon_n = \epsilon_n' + \epsilon_n'' + \epsilon_n''' \tag{8.46a}$$

$$\gamma_{nt} = \gamma_{nt}' + \gamma_{nt}'' + \gamma_{nt}''' \tag{8.46b}$$

where ϵ_n and γ_{nt} depend on ϵ_x, ϵ_y, and γ_{xy}, and also on the angle θ. That is,

$$\epsilon_n \equiv \epsilon_n(\epsilon_x, \epsilon_y, \gamma_{xy}; \theta)$$

$$\epsilon_n' \equiv \epsilon_n(\epsilon_x; \theta)$$

$$\epsilon_n'' \equiv \epsilon_n(\epsilon_y; \theta)$$

$$\epsilon_n''' \equiv \epsilon_n(\gamma_{xy}; \theta)$$

and analogously for γ_{nt}.

To construct Fig. 8.29a, we first construct a rectangle whose diagonal, PQ, is oriented at the angle θ counterclockwise from the x axis and whose length is Δs. The sides of this rectangle have lengths

$$\Delta x = \Delta s \cos \theta, \qquad \Delta y = \Delta s \sin \theta$$

Next, in order to make the angle $\angle QPR$ a right angle, we form a rectangle whose diagonal, PR, is oriented at the angle θ counterclockwise from the y axis and whose length is also Δs. Having constructed the shaded figure that is composed of two rectangles and that represents part of the *undeformed body*, we hold point P fixed and draw dashed lines to show the shape of that part of the deformed body when ϵ_x is positive and $\epsilon_y = \gamma_{xy} = 0$.

The extensional strain in the direction n is defined by

$$\epsilon_n(P) = \lim_{Q \to P \text{ along } n} \left(\frac{\Delta s^* - \Delta s}{\Delta s} \right) \tag{2.29}$$
$$\text{repeated}$$

and, the shear strain with respect to axes n and t by

$$\gamma_{nt}(P) = \lim_{\substack{Q \to P \text{ along } n \\ R \to P \text{ along } t}} \left(\frac{\pi}{2} - \angle Q^*P^*R^* \right)$$

(2.30)
repeated

In the present case we can dispense with the limit operation and just write

$$\epsilon_n(P) = \left(\frac{\Delta s^* - \Delta s}{\Delta s} \right)$$

(8.47)

$$\gamma_{nt} = \left(\frac{\pi}{2} - \angle Q^*P^*R^* \right)$$

(8.48)

At the same time, we will make several small-angle approximations.

Contribution of ϵ_x to ϵ_n: We will use the geometry of Fig. 8.29a to develop expressions for $\epsilon_n'(\epsilon_x; \theta)$ and $\gamma_{nt}'(\epsilon_x; \theta)$. From Fig. 8.29a and Eq. 8.47,

$$\epsilon_n' = \frac{\Delta s^* - \Delta s}{\Delta s} = \frac{\overline{Q^*S^*}}{\overline{QP}}$$

(8.49)

To determine the elongation $\overline{Q^*S^*}$, we drop a perpendicular from point Q to the line P^*Q^*. In the process, we make the small-angle approximation that $\overline{P^*S^*} \doteq \overline{PQ}$, that is, that the perpendicular QS^* is (approximately) the arc of a circle of radius Δs with center at P. Since $\overline{QQ^*} = \epsilon_x \Delta x$, the elongation of the rectangle due to the strain ϵ_x, and since angle $\angle QQ^*S^*$ is (approximately) equal to θ, we have

$$\overline{Q^*S^*} = \epsilon_x \Delta x \cos \theta = \epsilon_x (\Delta s \cos \theta) \cos \theta$$

Therefore,

$$\epsilon_n' = \frac{\epsilon_x \Delta s \cos^2 \theta}{\Delta s} = \epsilon_x \cos^2 \theta$$

(8.50)

The contributions ϵ_n'' and ϵ_n''' due to ϵ_y and γ_{xy} can be derived using Figs. 8.29b and 8.29c, respectively. Summing the three contributions (Eq. 8.46a), we get the following *extensional-strain-transformation formula*:

$$\epsilon_n = \epsilon_x \cos^2 \theta + \epsilon_y \sin^2 \theta + \gamma_{xy} \sin \theta \cos \theta$$

(8.51)

Contribution of ϵ_x to γ_{nt}: From Fig. 8.29a and Eq. 8.48 we have

$$\gamma_{nt}' = \frac{\pi}{2} - (\angle Q^*P^*R^*)' = -\alpha - \beta$$

(8.52)

where α and β are angles that are defined in Fig. 8.29a. Again using a small angle approximation, we have

$$\overline{QS^*} \doteq \alpha \Delta s, \qquad \overline{RT^*} \doteq \beta \Delta s$$

so

$$\gamma_{nt}' \doteq -\left(\frac{\overline{QS^*}}{\Delta s} \right) - \left(\frac{\overline{RT^*}}{\Delta s} \right)$$

But,

$$\overline{QS^*} \doteq \epsilon_x \Delta x \sin \theta = \epsilon_x (\Delta s \cos \theta) \sin \theta$$

$$\overline{RT^*} \doteq \epsilon_x \Delta y \cos \theta = \epsilon_x (\Delta s \sin \theta) \cos \theta$$

Finally, combining these equations, we get

$$\gamma'_{nt} = -2\epsilon_x \sin \theta \cos \theta$$

Combining this expression with the contributions of ϵ_y and γ_{xy} to γ_{nt}, as in Eq. 8.46b, we get the following *shear-strain-transformation formula*:

$$\gamma_{nt} = -2(\epsilon_x - \epsilon_y) \sin \theta \cos \theta + \gamma_{xy}(\cos^2 \theta - \sin^2 \theta) \qquad (8.53)$$

Equations 8.51 and 8.53 are the strain analogs of Eqs. 8.3a and 8.3b, with the one exception that *there is a difference of a factor of two in the shear-strain terms* in Eqs. 8.51 and 8.53 compared with corresponding terms in the stress-transformation equations 8.3a and 8.3b. For example, Eq. 8.53 has the same form as Eq. 8.3b if we divide Eq. 8.53 by two and write it in the form

$$\frac{\gamma_{nt}}{2} = -(\epsilon_x - \epsilon_y) \sin \theta \cos \theta + \frac{\gamma_{xy}}{2}(\cos^2 \theta - \sin^2 \theta)$$

Like the stress-transformation equations, the *strain-transformation equations* can be simplified by incorporating the double-angle trigonometric identities, giving

$$\epsilon_n = \left(\frac{\epsilon_x + \epsilon_y}{2}\right) + \left(\frac{\epsilon_x - \epsilon_y}{2}\right) \cos 2\theta + \left(\frac{\gamma_{xy}}{2}\right) \sin 2\theta \qquad (8.54a)$$

$$\frac{\gamma_{nt}}{2} = -\left(\frac{\epsilon_x - \epsilon_y}{2}\right) \sin 2\theta + \left(\frac{\gamma_{xy}}{2}\right) \cos 2\theta \qquad (8.54b)$$

Note that, unlike the shear-stress terms in Eqs. 8.5, all shear strains in Eq. 8.54 are divided by two.

Since the strain-transformation equations, Eqs. 8.54, are completely analogous to the transformation equations for plane stress, Eqs. 8.5, formulas for determining principal directions and principal stresses, and other formulas that are based on Eqs. 8.5 can be converted to formulas for comparable strain-related quantities. In the next section we will use Mohr's circle to solve strain-transformation problems.

8.10 MOHR'S CIRCLE FOR TWO-DIMENSIONAL STRAIN ■■■■■■■■■■■

The strain-transformation equations, Eqs. 8.54, are completely analogous to the transformation equations for plane stress, on which the derivation of Mohr's circle of stress in Section 8.5 was based, namely, Eqs. 8.5. By following the same procedure that was used in Section 8.5, we obtain the following equations that characterize *Mohr's circle of strain*:

$$(\epsilon_n - \epsilon_{avg})^2 + \left(\frac{\gamma_{nt}}{2}\right)^2 = R^2 \qquad (8.55a)$$

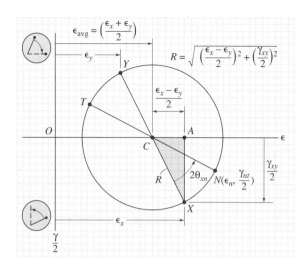

FIGURE 8.30 Mohr's circle of strain.

where

$$R = \sqrt{\left(\frac{\epsilon_x - \epsilon_y}{2}\right)^2 + \left(\frac{\gamma_{xy}}{2}\right)^2} \tag{8.55b}$$

$$\epsilon_{avg} = \frac{\epsilon_x + \epsilon_y}{2} \tag{8.55c}$$

Equation 8.55a represents the equation of a circle in the $\left(\epsilon, \frac{\gamma}{2}\right)$ plane with center at $(\epsilon_{avg}, 0)$ and radius R (Fig. 8.30). (All strain quantities are dimensionless.) Thus, Eqs. 8.54 are just the parametric equations of this Mohr's circle of strain, with the angle 2θ being the parameter.

To clarify the sign convention of Mohr's circle of strain, particularly the sign convention for shear strain γ, let us recall the sign convention for Mohr's circle of stress. There are two equivalent ways to establish the sign of the τ-coordinate of a point in the (σ, τ) plane:

Method 1: If the ntz axes form a right-handed coordinate system, point N has the coordinates $(\sigma_n, +\tau_{nt})$, while point T has the coordinates $(\sigma_t, -\tau_{nt})$.

Method 2: If the point N is plotted with positive shear stress (i.e., downward), then the shear stress τ on the N face would tend to rotate the corresponding nt stress element counterclockwise. (See Fig. 8.17 and also the icons on the τ axis of Fig. 8.16.)

There are two analogous methods for Mohr's circle of strain:

Method 1: If the ntz axes form a right-handed coordinate system, point N has the coordinates $\left(\epsilon_n, +\frac{\gamma_{nt}}{2}\right)$, while point T has the coordinates $\left(\epsilon_t, -\frac{\gamma_{nt}}{2}\right)$.

Method 2: If the point N is plotted with positive shear strain (i.e., downward), then a line element in the n direction would tend to rotate counterclockwise. (See the icons on the $\frac{\gamma}{2}$ axis in Fig. 8.30).

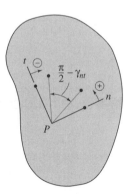

FIGURE 8.31 An explanation of the shear-strain sign convention for Mohr's circle of strain.

Figure 8.31 illustrates the fact that a positive shear strain at point P represents a reduction in the angle between the n and t axes by an amount γ_{nt}, and, in the process, the incremental line element in the n direction rotates counterclockwise. The icons at the ends of the $\dfrac{\gamma}{2}$ axis in Fig. 8.30 indicate the direction of rotation of a generic line element n, depending on the sign associated with the shear-strain coordinate of point N.

PROCEDURE FOR CONSTRUCTING AND USING MOHR'S CIRCLE OF STRAIN

The procedure for constructing Mohr's circle of strain is virtually identical to the procedure for constructing Mohr's circle of stress.

Draw Mohr's Circle: Figure 8.30 illustrates Steps 1 through 6.

1. Establish a set of $\left(\epsilon, \dfrac{\gamma}{2}\right)$ axes, with the same scale on both axes. Use paper that has a grid, and use a scale that results in a circle of reasonable size. The **positive** $\dfrac{\gamma}{2}$ axis points **downward**.

2. Assuming that ϵ_x, ϵ_y, and γ_{xy} are given, locate the point X at $\left(\epsilon_x, \dfrac{\gamma_{xy}}{2}\right)$ and the point Y at $\left(\epsilon_y, -\dfrac{\gamma_{xy}}{2}\right)$.

3. Connect points X and Y with a straight line, and locate the center of the circle, C, where the line crosses the ϵ axis at $(\epsilon_{\text{avg}}, 0)$.

4. Draw a circle that has its center at $(\epsilon_{\text{avg}}, 0)$ and that passes through points X and Y. (It is a good idea to use a compass in drawing the circle.)

Compute the Required Information:

5. Form the shaded triangle XCA with sides $\left(\dfrac{\gamma_{xy}}{2}\right)$ and $\left(\dfrac{\epsilon_x - \epsilon_y}{2}\right)$, and compute the radius of the circle.

$$R = \sqrt{\left(\dfrac{\epsilon_x - \epsilon_y}{2}\right)^2 + \left(\dfrac{\gamma_{xy}}{2}\right)^2} \qquad \begin{array}{l}(8.55b)\\ \text{repeated}\end{array}$$

6. If the extensional strain in a particular direction, say ϵ_n, is required, locate point N on Mohr's circle by turning an angle 2θ counterclockwise (or clockwise) on the circle, corresponding to rotating an angle θ counterclockwise (or clockwise) from some reference direction such as the x direction. Construct a diameter through point N and the center of the circle, and use trigonometry to calculate the value of ϵ_n (and ϵ_t and γ_{nt} if required).

7. If the *principal strains* and the orientations of the *principal directions of strain* are required, use

$$\epsilon_1 = \epsilon_{\text{avg}} + R \qquad (8.56)$$
$$\epsilon_2 = \epsilon_{\text{avg}} - R$$

to calculate the principal strains, where $\epsilon_{\text{avg}} = \epsilon_C$ is given by Eq. 8.55c. Use trigonometry to determine some angle, like $2\theta_{xp_1}$, that can be used to locate a principal direction of strain, say p_1, with respect to some known reference direction. There is no shear strain between the principal directions of strain.

8. Use a procedure similar to Step 7 if the maximum in-plane shear strain and maximum shear directions are required. The directions of maximum shear strain are at $\pm 45°$ to the principal directions of strain.

Figure 8.32 illustrates Steps 7 and 8.

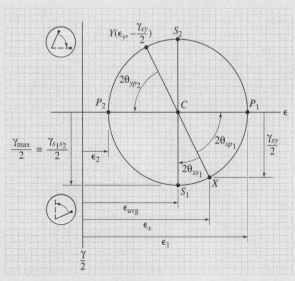

FIGURE 8.32 A Mohr's circle of strain with the principal strains and maximum in-plane shear strains identified.

EXAMPLE 8.6

At a certain point on a deformable body, the in-plane strains referred to a set of xy axes are:

$$\epsilon_x = 120\mu, \quad \epsilon_y = -40\mu, \quad \text{and} \quad \gamma_{xy} = -120\mu$$

where μ is the unit of *microstrain* (i.e., 10^{-6} in./in., or microinches per inch). (a) Using a square of unit length in the x and y directions to represent the undeformed body at this point, draw a sketch of this elemental square before and after deformation. (Exaggerate the deformation.) (b) Sketch a Mohr's circle of strain for these in-plane strains. (c) Determine the extensional strain in the direction n that is 30° *clockwise* from the x axis. (d) Determine the principal strains and the principal directions of strain, and sketch a deformed element that is oriented in the principal directions. (e) Finally, compute the maximum in-plane shear strain and the associated extensional strains in the directions of the axes of maximum in-plane shear strain. Sketch a deformed element that is oriented in the directions of maximum in-plane shear strain.

Solution This example problem can be solved using the eight steps suggested under *Procedure for Constructing and Using of Mohr's Circle of Strain.*

(a) State of Strain: The shaded square in Fig. 1 is the undeformed element. The dashed lines indicated the shape of the deformed element. Since the shear strain, γ_{xy}, is negative, the right angle at P increases by a total angle of 120μ. This shear strain angle is apportioned as $\left|\dfrac{\gamma_{xy}}{2}\right|$ clockwise to the x edge PQ and $\left|\dfrac{\gamma_{xy}}{2}\right|$ counterclockwise to the y edge PR.

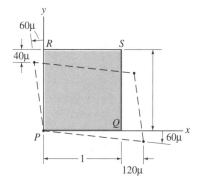

Fig. 1 An example of in-plane strains.

(b) Mohr's Circle of Strain: On grid paper (Fig. 2), the point X is plotted at $\left(\epsilon_x, +\dfrac{\gamma_{xy}}{2}\right)$, which is the point $(120\mu, -60\mu)$. Point Y is then plotted at $\left(\epsilon_y, -\dfrac{\gamma_{xy}}{2}\right)$. Note that the x edge PQ rotates clockwise in Fig. 1, and that this agrees with the icons on the negative (upper) end of the $\dfrac{\gamma}{2}$ axis.

The center of the circle, C, lies at $(\epsilon_{\text{avg}}, 0)$, where

$$\epsilon_{\text{avg}} = \frac{\epsilon_x + \epsilon_y}{2} = \frac{(120\mu - 40\mu)}{2} = 40\mu \tag{1}$$

The radius of the circle is length $\overline{XC}$, which is given by

$$R = \sqrt{\overline{CA}^2 + \overline{XA}^2} = \sqrt{(80\mu)^2 + (60\mu)^2} = 100\mu \tag{2}$$

(c) Extensional Strain ϵ_n: To determine the extensional strain in the n direction shown on Fig. 3, we locate the radial line CN on Mohr's circle in Fig. 2 at an angle $2\theta_{xn} = 60°$ *clockwise* from the radial line CX. From Fig. 2, we get

$$\epsilon_n = \overline{OC} + \overline{CB} = \epsilon_{\text{avg}} + R\cos 2\theta_{p1n} \tag{3}$$

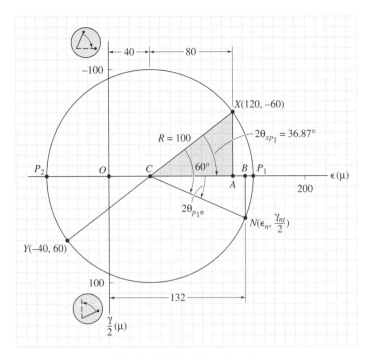

Fig. 2 Mohr's circle of strain.

To determine the angle $2\theta_{p_1n}$, we can use triangle XCA to first determine the angle $2\theta_{xp_1}$.

$$2\theta_{xp_1} = \tan^{-1}\left(\frac{\overline{XA}}{\overline{CA}}\right) = \tan^{-1}\left(\frac{60\mu}{80\mu}\right) = 36.87° \qquad (4)$$

Then,

$$2\theta_{p_1n} = 60° - 2\theta_{xp_1} = 23.13° \quad \circlearrowright \qquad (5)$$

So, combining Eqs. (1) through (5), we get

$$\epsilon_n = \overline{OC} + \overline{CB} = 40\mu + 100\mu \cos(23.13°)$$

or

$$\epsilon_n = 132\mu \qquad \textbf{Ans. (c)} \quad (6)$$

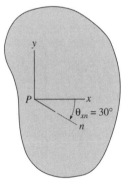

Fig. 3 The orientation of the n axis.

(d) Principal Strains and Principal Directions: To avoid cluttering up Fig. 2, we will repeat the basic Mohr's circle as Fig. 4. The principal directions are represented on Fig. 4 by points P_1 and P_2 at $(\epsilon_1, 0)$ and $(\epsilon_2, 0)$, respectively. From Fig. 4,

$$\epsilon_1 = \epsilon_{\text{avg}} + R = 40\mu + 100\mu = 140\mu \qquad (7)$$

$$\epsilon_2 = \epsilon_{\text{avg}} - R = 40\mu - 100\mu = -60\mu$$

467

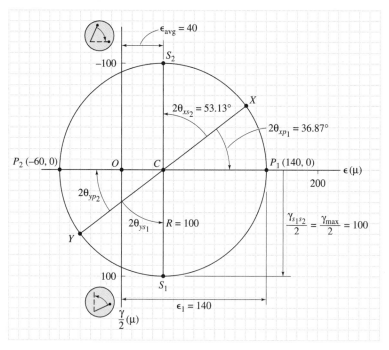

Fig. 4 Mohr's circle showing principal-strain points P_1 and P_2 and maximum-shear-strain points S_1 and S_2.

Points P_1 and P_2 can be located relative to points X and Y, respectively, by *clockwise* angles

$$2\theta_{xp_1} = 2\theta_{yp_2} = 36.87° \quad \circlearrowright \tag{8}$$

as calculated in Eq. (4). Therefore, the principal strains and corresponding principal directions are:

$$\left.\begin{array}{ll} \epsilon_1 = 140\mu, & \theta_{xp_1} = 18.4° \\ \epsilon_2 = -60\mu, & \theta_{yp_2} = 18.4° \end{array}\right\} \quad \textbf{Ans. (d)} \tag{10}$$

To sketch the undeformed and deformed principal-strain elements, let us begin with a unit square oriented at

$$\theta_{xp_1} = \theta_{yp_2} = 18.4° \quad \circlearrowright$$

with respect to the xy axes (Fig. 5). The strains ϵ_1 and ϵ_2 are then used in sketching the deformed principal element (dashed lines).

(e) Maximum In-Plane Shear Strain: The points corresponding to maximum in-plane shear strain, points S_1 and S_2, are located at the lowest and highest points on the Mohr's circle in Fig. 4. From Fig. 4,

$$\frac{\gamma_{s_1s_2}}{2} = R = 100\mu \tag{11}$$

The directions s_1 and s_2 are determined by the angles $2\theta_{xs_2} = 2\theta_{ys_1}$ on Fig. 4.

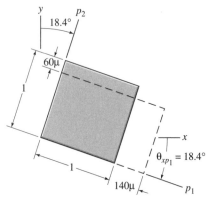

Fig. 5 An element illustrating the principal strains.

468

Thus,

$$2\theta_{xs_2} = 2\theta_{ys_1} = 90 - 2\theta_{xp_1} = 90 - 36.87° = 53.13° \qquad (12)$$

so,

$$\theta_{xs_2} = \theta_{ys_1} = 26.6° \;) \qquad (13)$$

Therefore, the maximum in-plane shear strain is given by

$$\gamma_{s_1s_2} = 200\mu, \quad \theta_{xs_2} = \theta_{ys_1} = 26.6° \qquad \text{Ans. (e)} \quad (14)$$

The extensional strain in the s_1 and s_2 directions is

$$\epsilon_{s_1} = \epsilon_{s_2} = \epsilon_{\text{avg}} = 40\mu \qquad \text{Ans. (e)} \quad (15)$$

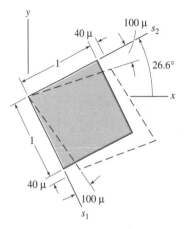

Fig. 6 An element depicting the maximum in-plane shear strain.

To sketch the maximum shear strain element, let us begin with a unit square oriented at the angle given in Eq. (14). We draw the element so that s_1s_2z forms a right-handed coordinate system. (Note: The $+y$ direction and the $-y$ direction are 180° apart on the deformable body, so they are 360° "apart" on Mohr's circle, that is, the point Y represents both the $+y$ axis and the $-y$ axis.) To get a right-handed coordinate system s_1s_2z in Fig. 6 we should actually rewrite Eq. (14b) as

$$\theta_{xs_2} = \theta_{-ys_1} = 26.6° \;) \qquad (16)$$

Review the Solution   The key to correct solution of in-plane strain problems using Mohr's circle is accurate drawing of the circle from information about the strain state, that is, correctly plotting points X and Y. We have to remember to divide all γ values by two before plotting! The Mohr's circle in Fig. 2 is accurately drawn.

It is also useful to sketch the deformed element using information about the state of strain. This has been done in Fig. 1. By comparing Fig. 5 and Fig. 6 with Fig. 1, we can see that all three figures are in agreement; that is, all exhibit stretching in the direction of the QR diagonal in Fig. 1. Therefore, our solutions are probably correct.

MOHR, a computer program for constructing Mohr's circles of stress and strain, is one of the **MechSOLID** collection of computer programs. It is described in Appendix G.7. Computer homework exercises for Section 8.10 follow the regular Section 8.10 homework problems at the end of the chapter.

*8.11   ANALYSIS OF THREE-DIMENSIONAL STRAIN ■■■■■■■■■■■

An analysis of the three-dimensional geometry of deformation of a body establishes the fact that, at any point in the body, there are three directions, called *principal strain directions*, that are mutually perpendicular both before and after deformation.[18] That is, there

---

[18]*Theory of Elasticity*, by S. P. Timoshenko and J. N. Goodier, [Ref. 8-1], Sections 81 and 82.

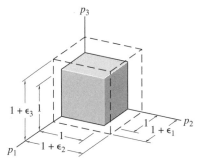

FIGURE 8.33 The principal strains at a point.

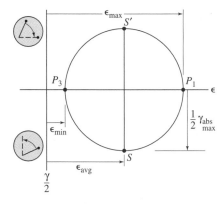

FIGURE 8.34 Mohr's circle for the plane of principal strains ϵ_1 and ϵ_3.

is no shear strain between the principal strain directions, as is illustrated in Fig. 8.33. (For convenience, the undeformed element in Fig. 8.33 is taken to be a unit cube.) The principal strains are labeled in the order

$$\epsilon_1 \equiv \epsilon_{maximum}, \quad \epsilon_2 \equiv \epsilon_{intermediate}, \quad \epsilon_3 \equiv \epsilon_{minimum} \tag{8.57}$$

and their values can be determined by using geometrical equations that are analogous to the equilibrium equations used in Section 8.6 to determine principal stresses.

The *absolute maximum shear strain* occurs between two perpendicular axes that lie in the plane of p_1 and p_3, just as in the case of absolute maximum shear stress.[19] Hence, a Mohr's circle of strain can be constructed for the $p_1 p_3$ strain plane, and from this Mohr's circle (Fig. 8.34), the absolute maximum shear strain is found to be

$$\gamma_{\substack{abs \\ max}} = \epsilon_{max} - \epsilon_{min} \tag{8.58}$$

Plane Strain. For the special case of *plane strain*,

$$\epsilon_z = \gamma_{xz} = \gamma_{yz} = 0 \tag{8.43}$$
repeated

The determination of the absolute maximum shear strain for this case is competely analogous to the analysis of absolute maximum shear stress for the case of plane stress.[20] For example, if Example Problem 8.6 is a plane-strain problem, then $\epsilon_1 = 140\mu$, $\epsilon_2 = \epsilon_z = 0$, $\epsilon_3 = -60\mu$, and the xy plane corresponds to the $p_1 p_3$ plane when the three-dimensional strain is considered.

Plane Stress. Where a state of plane stress exists in a body, for example at a free surface of the body, the stresses satisfy

$$\sigma_z = \tau_{xz} = \tau_{yz} = 0 \tag{8.1}$$
repeated

---

[19]See Section 8.7.

[20]See Section 8.7, particularly Fig. 8.25.

If the body is linearly elastic and isotropic, its material behavior satisfies Eqs. 2.32 and 2.34. Therefore, we get

$$\epsilon_z = \frac{-\nu}{E}(\sigma_x + \sigma_y)$$

(8.44)
repeated

and

$$\gamma_{xz} = \gamma_{yz} = 0$$

(8.59)

Therefore, the z direction is a principal-strain direction, but the principal strain ϵ_z is not zero. Therefore, it is necessary to determine the two principal strains in the xy plane (e.g., using a Mohr's circle as in Section 8.10) and then to order all three principal strains as in Eq. 8.57 before using Eq. 8.58 to compute the absolute maximum shear strain.

8.12 MEASUREMENT OF STRAIN; STRAIN ROSETTES

There are several experimental techniques that may be used to measure strain. These are described in texts on *experimental stress analysis*.[21] The most straightforward technique employs wire or metal foil *electrical-resistance strain gages,* like the ones pictured in Fig. 8.35. Figure 8.35a shows a *metal foil strain gage,* Fig. 8.35b depicts two *strain rosettes,* and Fig. 8.35c illustrates a typical strain-gage installation. The gage consists of a wire or

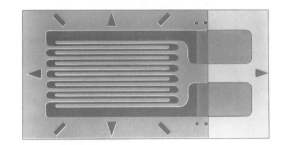

(a) A metal foil strain gage.

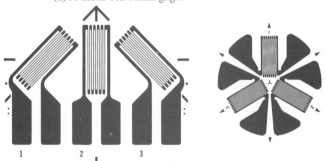

(b) A 45° strain rosette and an equiangular rosette.

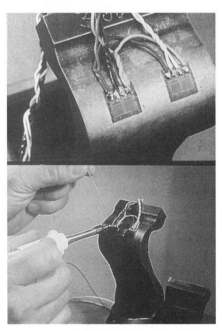

(c) A typical strain gage installation.

FIGURE 8.35 Electrical-resistance strain gages. (Courtesy of Micro-Measurements Division of Measurements Group, Inc.)

[21]*Mechanical Measurements*, by T. G. Beckwith, R. D. Marangoni, and J. H. Lienhard V, [Ref. 8-4].

foil "grid" on a thin paper or plastic backing that can be bonded (e.g., glued) directly to the surface whose strain is to be measured. Extensional strain along the axis of the grid stretches (or contracts) the metal grid element, causing a change in electrical resistance of the grid. This change in resistance can be converted directly to the extensional strain ϵ_n along the direction n of the axis of the gage.

Since electrical-resistance strain gages can directly measure only extensional strain, it is not possible with a single gage to measure shear strain or, for example, to determine the directions and magnitudes of the principal strains at a point. Therefore, it is common practice to employ a *strain rosette* consisting of three electrical-resistance strain gages mounted on a common backing sheet, like the rosettes depicted in Fig. 8.35b. Let the three gages of a rosette be oriented along axes that are labeled *a, b,* and *c,* as shown in Fig. 8.36. The extensional strain along each of these axes can be related to the three strain quantities (ϵ_x, ϵ_y, γ_{xy}), where the xy reference frame may be established in any convenient orientation. For example, the axis of the *a* gage may be taken as the x axis by setting $\theta_a \equiv 0$. From Eq. 8.51 we get, for arbitrary angles,

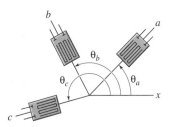

FIGURE 8.36 Notation for a strain rosette with arbitrary angles.

$$
\begin{aligned}
\epsilon_a &= \epsilon_x \cos^2 \theta_a + \epsilon_y \sin^2 \theta_a + \gamma_{xy} \sin \theta_a \cos \theta_a \\
\epsilon_b &= \epsilon_x \cos^2 \theta_b + \epsilon_y \sin^2 \theta_b + \gamma_{xy} \sin \theta_b \cos \theta_b \\
\epsilon_c &= \epsilon_x \cos^2 \theta_c + \epsilon_y \sin^2 \theta_c + \gamma_{xy} \sin \theta_c \cos \theta_c
\end{aligned}
\tag{8.60}
$$

The two common rosette arrangements are the 45° *rectangular rosette* and the 60° *equiangular rosette* shown in Fig. 8.35b. Substituting $\theta_a = 0°$, $\theta_b = 45°$, and $\theta_c = 90°$ into Eqs. 8.60, we get the following equations for the 45° rosette:

$$
\begin{aligned}
\epsilon_x &= \epsilon_a \\
\epsilon_y &= \epsilon_c \\
\gamma_{xy} &= 2\epsilon_b - \epsilon_a - \epsilon_c
\end{aligned}
\tag{8.61}
$$

For the equiangular rosette, we can let $\theta_a = 0°$, $\theta_b = 60°$, and $\theta_c = 120°$. Then, Eqs. 8.60 give the following:

$$
\begin{aligned}
\epsilon_x &= \epsilon_a \\
\epsilon_y &= \frac{1}{3}(2\epsilon_b + 2\epsilon_c - \epsilon_a) \\
\gamma_{xy} &= \frac{2}{\sqrt{3}}(\epsilon_b - \epsilon_c)
\end{aligned}
\tag{8.62}
$$

Once strains ϵ_x, ϵ_y, and γ_{xy} have been calculated using Eq. 8.60 (or 8.61 or 8.62), a Mohr's circle of strain can be drawn for the surface strains at the rosette location.

In most instances where strain rosettes are used, there is a need to determine the principal stresses and, perhaps, the absolute maximum shear stress. The rosette is affixed to a surface, which is usually stress-free. In that case, one of the principal stresses is $\sigma_z = 0$. Therefore, if the material constants E and ν are known, we can use Eqs. 2.32 to determine the in-plane principal stresses and also the extensional strain ϵ_z. Then,

$$\epsilon_1 = \frac{1}{E}(\sigma_1 - \nu\sigma_2)$$

$$\epsilon_2 = \frac{1}{E}(\sigma_2 - \nu\sigma_1)$$

which can be solved for σ_1 and σ_2 to give

$$\boxed{\begin{aligned} \sigma_1 &= \frac{E}{1 - \nu^2}(\epsilon_1 + \nu\epsilon_2) \\ \sigma_2 &= \frac{E}{1 - \nu^2}(\epsilon_2 + \nu\epsilon_1) \end{aligned}}$$

(8.63)

In terms of stresses, ϵ_z is given by Eq. 8.44,

$$\epsilon_z = \frac{-\nu}{E}(\sigma_x + \sigma_y) = \frac{-\nu}{E}(\sigma_1 + \sigma_2)$$

And, in terms of the in-plane strains, ϵ_z is given by

$$\epsilon_z = \frac{-\nu(\epsilon_x + \epsilon_y)}{1 - \nu} = \frac{-\nu(\epsilon_1 + \epsilon_2)}{1 - \nu}$$

(8.64)

■■■■■■■■■■■■■■■■ EXAMPLE 8.7 ■■■■■■■■■■■■■■■■■■■■

The landing-gear strut of an airplane is instrumented with a 45° strain rosette to measure the strains in the strut during landing. The a-gage is oriented along the axial direction of the strut, as shown in Fig. 1. At one instant during a landing, the measured strains are:

$$\epsilon_a = -700\mu, \qquad \epsilon_b = 0, \qquad \epsilon_c = -100\mu$$

(a) Letting the x axis be oriented along the a-gage and the y axis be oriented along the c-gage, determine the strain values ϵ_x, ϵ_y, and γ_{xy}. (b) Sketch a Mohr's circle of strain. (c) Determine the principal (surface) strains at the rosette location. (d) Determine the principal stresses at the rosette location. Let $E = 15 \times 10^3$ ksi and $\nu = 0.3$.

Plan the Solution Since this is a standard 45° rosette, we can use Eqs. 8.61 to compute the required xy strains. Then we can use the points $X\left(\epsilon_x, +\dfrac{\gamma_{xy}}{2}\right)$ and $Y\left(\epsilon_y, -\dfrac{\gamma_{xy}}{2}\right)$ to plot Mohr's circle. The state of stress at the surface is plane stress, so the in-plane principal stresses are given by Eqs. 8.63.

Fig. 1 A strain rosette installation.

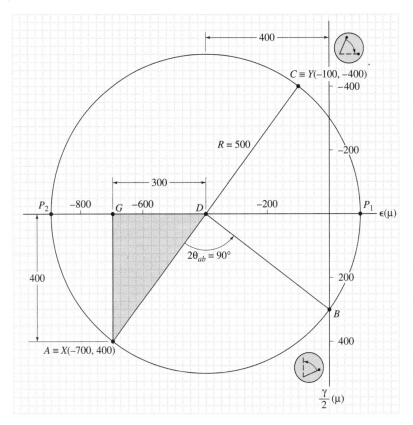

Fig. 2 Mohr's circle of strain.

Solution

(a) Rosette Equations: For the 45° rosette, Eqs. 8.61 give

$$\epsilon_x = \epsilon_a = -700\mu$$

$$\epsilon_y = \epsilon_c = -100\mu$$

$$\gamma_{xy} = 2\epsilon_b - \epsilon_a - \epsilon_c = 0 + 700\mu + 100\mu = 800\mu$$

$$\epsilon_x = -700\mu, \quad \epsilon_y = -100\mu, \quad \gamma_{xy} = 800\mu \quad \textbf{Ans. (a)} \quad (1)$$

(b) Mohr's Circle of Strain: We can use the strain values in Eqs. (1) to plot points $X(\equiv A)$ and $Y(\equiv C)$, and then draw the Mohr's circle of strain, as shown in Fig. 2.

(c) Principal Strains: The values of principal strains in the xy plane are given by

$$(\epsilon_1)_{xy} = \epsilon_{\text{avg}} + R, \qquad (\epsilon_2)_{xy} = \epsilon_{\text{avg}} - R \qquad (2)$$

where

$$\epsilon_{\text{avg}} = \frac{\epsilon_x + \epsilon_y}{2} - \frac{-700\mu - 100\mu}{2} = -400\mu \qquad (3)$$

and

$$R = \sqrt{(\overline{GD})^2 + (\overline{GA})^2} = \sqrt{(300\mu)^2 + (400\mu)^2} = 500\mu \qquad (4)$$

Therefore, combining Eqs.(2) through (4), we get the following principal strains in the xy plane:

$$(\epsilon_1)_{xy} = 100\mu, \qquad (\epsilon_2)_{xy} = -900\mu \qquad \text{Ans. (c)} \quad (5)$$

(d) Principal Stresses: The principal in-plane strains in Eqs. (5) can be substituted into Eqs. 8.63 to give the in-plane (i.e., in the xy plane) principal stresses.

$$(\sigma_1)_{xy} = \frac{E}{1 - \nu^2}(\epsilon_1 + \nu\epsilon_2)_{xy}$$

$$(\sigma_2)_{xy} = \frac{E}{1 - \nu^2}(\epsilon_2 + \nu\epsilon_1)_{xy}$$

$$(6)$$

Then,

$$(\sigma_1)_{xy} = \frac{(15 \times 10^3 \text{ ksi})}{1 - (0.3)^2}[100 + (0.3)(-900)](10^{-6}) \qquad (7a)$$

$$= -2.80 \text{ ksi}$$

$$(\sigma_2)_{xy} = \frac{(15 \times 10^3 \text{ ksi})}{1 - (0.3)^2}[-900 + (0.3)(100)](10^{-6}) \qquad (7b)$$

$$= -14.3 \text{ ksi}$$

If we order the principal stresses, including $\sigma_z = 0$, in the order $\sigma_3 \leq \sigma_2 \leq \sigma_1$, we get

$$\sigma_1 = 0 \text{ ksi}, \quad \sigma_2 = -2.8 \text{ ksi}, \quad \sigma_3 = -14.3 \text{ ksi} \qquad \text{Ans. (d)}$$

Review the Solution We have a good check on Parts (a) and (b) by observing that point B, which was not used directly in plotting Mohr's circle, has the correct extensional strain, $\epsilon_b = 0$. We can scale off points P_1 and P_2 on Fig. 2 and see that the values of ϵ_1 and ϵ_2 in Eqs. (5) are correct. It seems strange that $(\sigma_1)_{xy}$ in Eq. (7a) is negative, while ϵ_1 in Eq. (5a) is positive. But, by double-checking Eqs. 8 we see that this result is correct. Hence, we observe that **principal stresses need not always have the same sign as the corresponding principal strains**.

MOHR, a computer program for constructing Mohr's circles of stress and strain, is one of the *MechSOLID* collection of computer programs. It is described in Appendix G.7. Computer homework exercises for Section 8.12 follow the regular Section 8.12 homework problems at the end of the chapter.

8.13 PROBLEMS ■■

Problems 8.3-1 through 8.3-4. *For each of these problems, you are to sketch a free-body diagram like the one in Fig. 8.9c, and use equilibrium equations to solve for the normal stress and shear stress on the indicated inclined plane NN.*

Prob. 8.3-1. The state of plane stress at a point is given by the stresses $\sigma_x = 0$ ksi, $\sigma_y = 35.6$ ksi, and $\tau_{xy} = 10.0$ ksi, as indicated on the figure below.

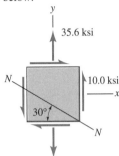

P8.3-1 and P8.5-1

Prob. 8.3-2. The state of plane stress at a point is given by the stresses $\sigma_x = 48$ MPa, $\sigma_y = -32$ MPa, and $\tau_{xy} = 16$ MPa, as illustrated on the figure below.

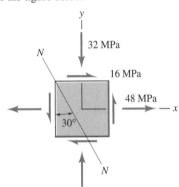

P8.3-2 and P8.5-2

Prob. 8.3-3. The state of plane stress at a point is given by the stresses $\sigma_x = 3000$ psi, $\sigma_y = -5000$ psi, and $\tau_{xy} = -5000$ psi, as illustrated on the figure below.

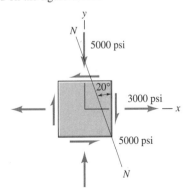

P8.3-3 and P8.5-3

Prob. 8.3-4. The state of plane stress at a point is given by the stresses $\sigma_x = -48$ MPa, $\sigma_y = -24$ MPa, and $\tau_{xy} = -60$ MPa, as illustrated on the figure below.

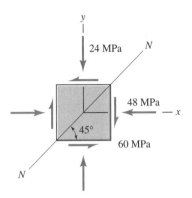

P8.3-4 and P8.5-4

Problems 8.3-5 through 8.3-10. *For each of these problems, an element in plane stress is subjected to the stresses σ_x, σ_y, and τ_{xy}, as indicated. Use the plane-stress-transformation equations, Eqs. 8.6, to determine the stresses $\sigma_{x'}$, $\sigma_{x'}$, and $\tau_{x'y'}$ on an element that is rotated by the angle $\theta \equiv \theta_{xx'}$. Show the calculated stresses on an element oriented at this angle.*

Prob. 8.3-5. The given stresses are: $\sigma_x = 0$ ksi, $\sigma_y = 35.6$ ksi, and $\tau_{xy} = 10.0$ ksi. The angle is $\theta = 60°$.

Prob. 8.3-6. Repeat Prob. 8.3-5 for $\theta = -60°$.

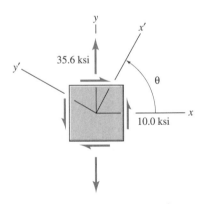

P8.3-5, P8.3-6, P8.5-5, P8.5-6, and PC8.5-1

Prob. 8.3-7. The given stresses are $\sigma_x = 48$ MPa, $\sigma_y = -32$ MPa, and $\tau_{xy} = 16$ MPa. The angle is $\theta = 30°$.

Prob. 8.3-8. Repeat Prob. 8.3-7 for $\theta = -30°$.

476

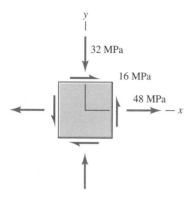

P8.3-7, P8.3-8, P8.5-7, P8.5-8, and PC8.5-2

Prob. 8.3-9. The given stresses are $\sigma_x = 3000$ psi, $\sigma_y = -5000$ psi, and $\tau_{xy} = -5000$ psi. The angle is $\theta = 20°$.

Prob. 8.3-10. Repeat Prob. 8.3-9 for $\theta = -20°$.

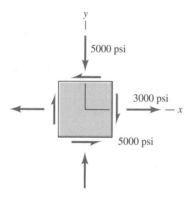

P8.3-9, P8.3-10, P8.5-9, P8.5-10, and PC8.5-3

Prob. 8.3-11. On a thin bracket, the state of plane stress referred to the $x'y'$ axes has been found to be $\sigma_{x'}$, $\sigma_{y'}$, and $\tau_{x'y'}$, as shown on the figure below. Use the plane-stress-transformation equations to determine the stresses σ_x, σ_y, and τ_{xy}, the stresses referred to the xy axes. Show these stresses on a properly oriented stress element.

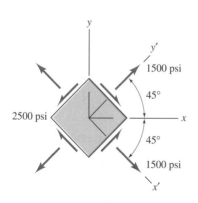

P8.3-11 and P8.5-22

Prob. 8.3-12. Repeat Prob. 8.3-11 for the state of plane stress shown on the figure below.

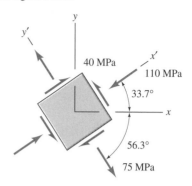

P8.3-12

Prob. 8.3-13. Point A on the surface of a machine component is subjected to plane stress. The stresses σ_x, σ_y, τ_{xy}, $\sigma_{x'}$, and $\tau_{x'y'}$ are known, but the angle $\theta \equiv \theta_{xx'}$ and the normal stress $\sigma_{y'}$ are unknown. Use the information shown on the two stress elements (both are at point A, but are rotated relative to each other) to determine the angle θ and the value of $\sigma_{y'}$.

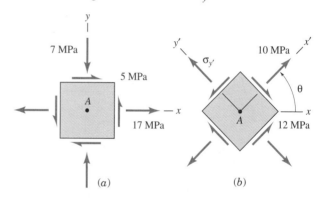

P8.3-13, P8.5-23, and PC8.5-4a

Prob. 8.3-14. Repeat Prob. 8.3-13 using the data in the figure below.

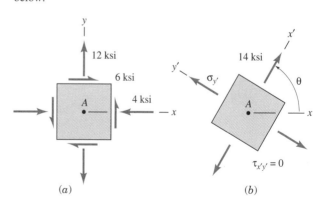

P8.3-14 and PC8.5-4b

Prob. 8.3-15. The state of plane stress at point A on the surface of an axially loaded bar is shown in Fig. P8.3-15b. Determine the axial load P (in Newtons) if the cross-sectional dimensions of the bar are as indicated in Fig. P8.3-15a.

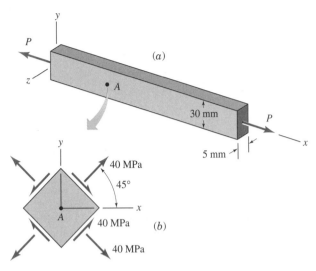

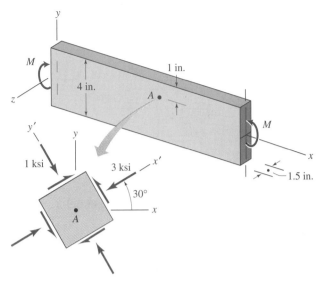

P8.3-17 and P8.5-26

P8.3-15 and P8.5-24

Prob. 8.3-16. The state of plane stress at point A on the surface of an axially loaded bar shown in Fig. P8.3-16. Determine the compressive axial load P (in kips) if the cross-sectional dimensions of the bar are as indicated in Fig. P8.3-16.

Prob. 8.3-18. The rectangular bar in Fig. P8.3-18 is subjected to bending in the xy plane and, simultaneously, to an axial tensile force P. The state of plane stress at point A at the top edge of the bar is shown in Fig. P8.3-18. If the bending moment is $M = 2$ kN $\cdot$ m, what is the magnitude of the axial force P? (Hint: Consider Eq. 8.7.)

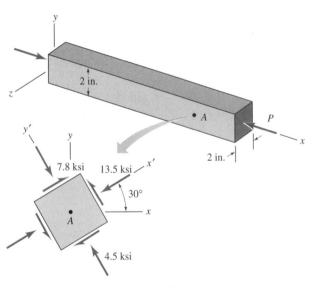

P8.3-16 and P8.5-25

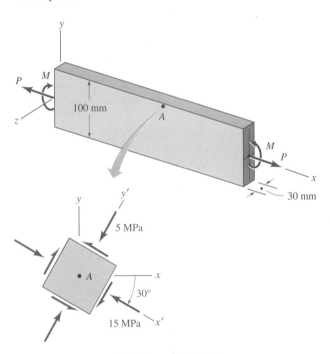

P8.3-18 and P8.5-27

Prob. 8.3-17. The state of plane stress at point A on the surface of a rectangular bar subjected to pure bending in the xy plane is shown in Fig. P8.3-17. Determine the bending moment M (in kip $\cdot$ in.) if the location of A and the cross-sectional dimensions of the bar are as indicated in Fig. P8.3-17. (Hint: Consider Eq. 8.7.)

Prob. 8.3-19. A block of wood is subjected to a vertical compressive stress of magnitude $3\sigma_0$ and, simultaneously, to a horizontal compressive stress of magnitude σ_0, as indicated in Fig. P8.3-19. The wood block will fail if either (a) the compressive stress perpendicular to the grain exceeds 260 psi (C) or (b) the shear stress parallel to the grain exceeds 130 psi. Determine the maximum allowable value of σ_0.

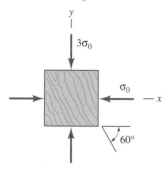

P8.3-19, P8.5-28, and Prob. PC8.5-5

Problems 8.4-1 through 8.4-6. *At a certain point in a member subjected to plane stress, the stresses σ_x, σ_y, and τ_{xy} have the values listed below. (a) Determine the principal stresses and show them on a sketch of a properly oriented stress element, and (b) determine the shear stress and normal stress on the planes of maximum shear stress, and show these on a sketch of a properly oriented stress element.*

Prob. 8.4-1. $\sigma_x = 3000$ psi, $\sigma_y = -5000$ psi, and $\tau_{xy} = -5000$ psi.

Prob. 8.4-2. $\sigma_x = 48$ MPa, $\sigma_y = -32$ MPa, and $\tau_{xy} = 16$ MPa.

Prob. 8.4-3. $\sigma_x = 3$ ksi, $\sigma_y = 21$ ksi, and $\tau_{xy} = -12$ ksi.

Prob. 8.4-4. $\sigma_x = 12$ MPa, $\sigma_y = -8$ MPa, and $\tau_{xy} = -4$ MPa.

Prob. 8.4-5. $\sigma_x = -4$ ksi, $\sigma_y = 12$ ksi, and $\tau_{xy} = 6$ ksi.

Prob. 8.4-6. $\sigma_x = 2,400$ psi, $\sigma_y = 14,400$ psi, and $\tau_{xy} = 14,400$ psi.

Problems 8.4-7 through 8.4-12. *At a certain point in a member subjected to plane stress, the stresses σ_x, σ_y, and τ_{xy} have the values shown on Figs. P8.4-7 through P8.4-12. (a) Determine the principal stresses and show them on a sketch of a properly oriented stress element, and (b) determine the shear stress and the normal stress on the planes of maximum shear stress, and show these on a sketch of a properly oriented stress element.*

Prob. 8.4-7. The state of plane stress for this problem is shown in Fig. P8.4-7.

Prob. 8.4-8. The state of plane stress for this problem is shown in Fig. P8.4-8.

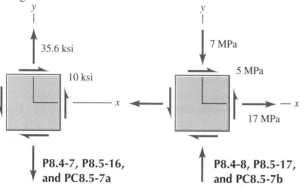

P8.4-7, P8.5-16, and PC8.5-7a

P8.4-8, P8.5-17, and PC8.5-7b

Prob. 8.4-9. The state of plane stress for this problem is shown in Fig. P8.4-9.

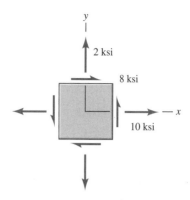

P8.4-9, P8.5-18, and PC8.5-8a

Prob. 8.4-10. The state of plane stress for this problem is shown in Fig. P8.4-10.

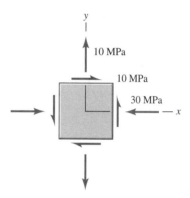

P8.4-10, P8.5-19, and PC8.5-8b

Prob. 8.4-11. The state of plane stress for this problem is shown in Fig. P8.4-11.

Prob. 8.4-12. The state of plane stress for this problem is shown in Fig. P8.4-12.

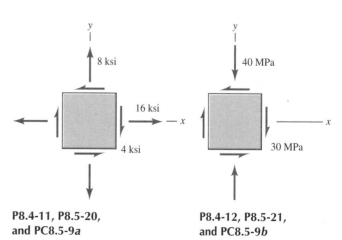

P8.4-11, P8.5-20, and PC8.5-9a

P8.4-12, P8.5-21, and PC8.5-9b

Prob. 8.4-13. At a certain point on the surface of a machine part, the normal stresses on two mutually perpendicular faces (the x and y faces) are 40 MPa (C) and 10 MPa (T), as shown in Fig. P8.4-13. The maximum in-plane shear stress at this point is $\tau_{\max} = 50$ MPa. Determine the magnitude of the shear stress, τ, on the x and y faces, and determine the in-plane principal stresses at this point.

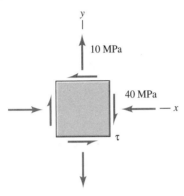

P8.4-13, P8.5-29, and PC8.5-10

Prob. 8.4-14. At a certain point on the surface of an airplane wing, the state of plane stress can be described by Fig. P8.4-14. The maximum in-plane shear stress at this point is $\tau_{\max} = 13$ ksi. (a) Determine the magnitude of the shear stress, τ, on the x and y faces. (b) Determine the two in-plane principal stresses at this point and show them on a properly oriented principal-stress element.

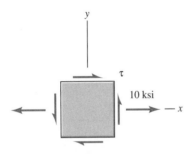

P8.4-14, P8.5-30, and PC8.5-11

Prob. 8.4-15. The state of plane stress at a point can be described by a known tensile stress $\sigma_x = 70$ MPa, an unknown tensile stress

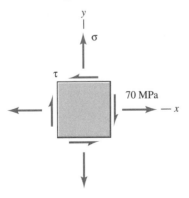

P8.4-15 and P8.5-31

σ, and an unknown shear stress τ, as indicated in Fig. P8.4-15. At this point the maximum in-plane shear stress is 78 MPa, and one of the two in-plane principal stresses is 22 MPa (T). Determine the values of the two unknown stresses, labeled σ and τ on the figure, and determine the second in-plane principal stress.

The problems in Section 8.5 are to be solved using Mohr's circle. Consider only the in-plane stresses.

Problems 8.5-1 through 8.5-4. *Use Mohr's circle to solve for the normal stress and the shear stress on the indicated inclined plane NN.*

Prob. 8.5-1. Use Mohr's circle to solve Prob. 8.3-1.

Prob. 8.5-2. Use Mohr's circle to solve Prob. 8.3-2.

Prob. 8.5-3. Use Mohr's circle to solve Prob. 8.3-3.

Prob. 8.5-4. Use Mohr's circle to solve Prob. 8.3-4.

Problems 8.5-5 through 8.5-10. *Use Mohr's circle to solve for the stresses $\sigma_{x'}$, $\sigma_{y'}$, and $\tau_{x'y'}$ on an element that is rotated by the angle $\theta \equiv \theta_{xx'}$. Show the calculated stresses on a stress element oriented at this angle.*

Prob. 8.5-5. Use Mohr's circle to solve Prob. 8.3-5.

Prob. 8.5-6. Use Mohr's circle to solve Prob. 8.3-6.

Prob. 8.5-7. Use Mohr's circle to solve Prob. 8.3-7.

Prob. 8.5-8. Use Mohr's circle to solve Prob. 8.3-8.

Prob. 8.5-9. Use Mohr's circle to solve Prob. 8.3-9.

Prob. 8.5-10. Use Mohr's circle to solve Prob. 8.3-10.

Prob. 8.5-11. (a) Construct Mohr's circle for an element in a uniaxial state of stress (Fig. P8.5-11). (b) Use this Mohr's circle to derive the following equations for the normal stress σ_n and shear stress τ_{nt} on the n-face (see Eqs. 2.19).

$$\sigma = \frac{\sigma_x}{2}(1 + \cos 2\theta), \qquad \tau_{nt} = -\left(\frac{\sigma_x}{2}\right)\sin 2\theta$$

(c) Use the Mohr's circle to determine the planes on which the maximum shear stress acts. Sketch a properly oriented maximum-shear-stress element and indicate the normal and shear stresses acting on its faces.

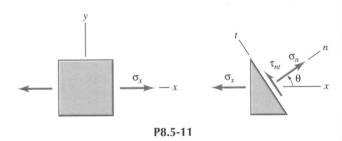

P8.5-11

Prob. 8.5-12. (a) Construct Mohr's circle for an element undergoing pure shear (Fig. P8.5-12). (b) Use this Mohr's circle to derive the following equations for the normal stress σ_n and shear stress τ_{nt} on the n-face.

$$\sigma_t = \tau_{xy} \sin 2\theta, \qquad \tau_{nt} = \tau_{xy} \cos 2\theta$$

(c) Use the Mohr's circle to determine the principal planes. Sketch a properly oriented principal-stress element and show the principal stresses acting on its faces.

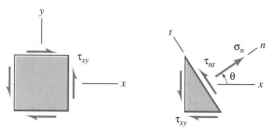

P8.5-12

Problems 8.5-13 through 8.5-21. *For each of these problems, (a) construct a Mohr's circle of stress. Using this Mohr's circle, (b) determine the principal stresses and show them on a properly oriented stress element, and (c) determine the maximum shear stress and the normal stress on the planes of maximum shear, and show these on a sketch of a properly oriented stress element.*

Prob. 8.5-13. Use the stresses given in Prob. 8.4-1.

Prob. 8.5-14. Use the stresses given in Prob. 8.4-2.

Prob. 8.5-15. Use the stresses given in Prob. 8.4-3.

Prob. 8.5-16. Use the stresses given in Prob. 8.4-7.

Prob. 8.5-17. Use the stresses given in Prob. 8.4-8.

Prob. 8.5-18. Use the stresses given in Prob. 8.4-9.

Prob. 8.5-19. Use the stresses given in Prob. 8.4-10.

Prob. 8.5-20. Use the stresses given in Prob. 8.4-11.

Prob. 8.5-21. Use the stresses given in Prob. 8.4-12.

Problems 8.5-22 through 8.5-31. *Use Mohr's circle to solve the named problems from Sections 8.3 and 8.4.*

Prob. 8.5-22. Use Mohr's Circle to solve Prob. 8.3-11.

Prob. 8.5-23. Use Mohr's Circle to solve Prob. 8.3-13.

Prob. 8.5-24. Use Mohr's Circle to solve Prob. 8.3-15.

Prob. 8.5-25. Use Mohr's Circle to solve Prob. 8.3-16.

Prob. 8.5-26. Use Mohr's Circle to solve Prob. 8.3-17.

Prob. 8.5-27. Use Mohr's Circle to solve Prob. 8.3-18.

Prob. 8.5-28. Use Mohr's Circle to solve Prob. 8.3-19.

Prob. 8.5-29. Use Mohr's Circle to solve Prob. 8.4-13.

Prob. 8.5-30. Use Mohr's Circle to solve Prob. 8.4-14.

Prob. 8.5-31. Use Mohr's Circle to solve Prob. 8.4-15.

Computer Exercises—Sect. 8.5. *Use the MechSOLID computer program MOHR to solve the homework exercises C8.5-1 through C8.5-11. Refer to Appendix G.7 for a description of this computer program and for examples of its use. Hand in computer printouts of the following MechSOLID "screens": MOHR: Results, and Mohr's Circle. Samples of these screens are illustrated in Example G-8.*

Prob. C8.5-1. Use the computer program MOHR (a) to solve Prob. 8.3-5, and (b) to solve Prob. 8.3-6.

Prob. C8.5-2. Use the computer program MOHR (a) to solve Prob. 8.3-7, and (b) to solve Prob. 8.3-8.

Prob. C8.5-3. Use the computer program MOHR (a) to solve Prob. 8.3-9, and (b) to solve Prob. 8.3-10.

Prob. C8.5-4. Use the computer program MOHR (a) to solve Prob. 8.3-13, and (b) to solve Prob. 8.3-14.

Prob. C8.5-5. Use the computer program MOHR to solve Prob. 8.3-19.

Prob. C8.5-6. Use the computer program MOHR (a) to solve Prob. 8.5-13, and (b) to solve Prob. 8.5-14.

Prob. C8.5-7. Use the computer program MOHR (a) to solve Prob. 8.5-16, and (b) to solve Prob. 8.5-17.

Prob. C8.5-8. Use the computer program MOHR (a) to solve Prob. 8.5-18, and (b) to solve Prob. 8.5-19.

Prob. C8.5-9. Use the computer program MOHR (a) to solve Prob. 8.5-20, and (b) to solve Prob. 8.5-21.

*Prob. C8.5-10. Use the computer program MOHR to solve Prob. 8.5-29.

*Prob. C8.5-11. Use the computer program MOHR to solve Prob. 8.5-30.

Prob. 8.6-1. (a) Using the free-body diagram of Fig. 8.19, and using equilibrium equations for this free body, derive Eqs. 8.29. (b) Show that the two roots of the quadratic equation in Eq. 8.31 are given by the formulas for σ_1 and σ_2 in Eq. 8.13.

*Prob. 8.6-2. From Fig. 8.19, note that $\tan \theta_p = \dfrac{m}{l}$. (a) Use this formula, together with Eq. 8.29a, to show that Eq. 8.9 is the formula for $\tan 2\theta_p$. (Hint: Use the double-angle trigonometric identity.) (b) Show that Eq. 8.9 can also be derived starting with Eq. 8.29b.

Prob. 8.6-3. A state of *generalized plane stress* exists if one of the coordinate faces, say the z face, is free of shear stress but not free of normal stress; that is, if $\tau_{xz} = \tau_{yz} = 0$, but the remaining stresses—σ_x, σ_y, σ_z, τ_{xy}—are not zero. (a) Show that, for generalized plane stress, Eq. 8.34 reduces to

$$[\sigma_p^2 - \sigma_p(\sigma_x + \sigma_y) + (\sigma_x\sigma_y - \tau_{xy}^2)] \, (\sigma_p - \sigma_z) = 0 \quad (1)$$

(b) Show that the three roots of Eq.(1) are

$$\sigma_{p1} = \sigma_{avg} + R, \; \sigma_{p2} = \sigma_{avg} - R, \; \sigma_{p3} = \sigma_z \quad (2)$$

where

$$\sigma_{avg} = \frac{\sigma_x + \sigma_y}{2} \quad, \quad R = \sqrt{\left(\frac{\sigma_x - \sigma_y}{2}\right)^2 + \tau_{xy}^2}$$

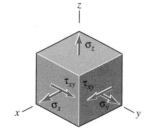

P8.6-3. Generalized plane stress acting on an element.

Prob. 8.6-4. Use the results of Prob. 8.6-3, that is, Eqs. (1) and (2) in the problem statement, to solve for the three principal stresses—σ_{p1}, σ_{p2}, and σ_{p3}—for the state of generalized plane stress depicted in Fig. P8.6-4.

Prob. 8.6-5. Repeat Prob. 8.6-4 for the state of generalized plane stress shown in Fig. P8.6-5.

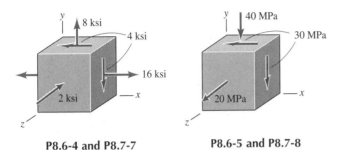

P8.6-4 and P8.7-7   **P8.6-5 and P8.7-8**

Prob. 8.6-6. Repeat Prob. 8.6-4 for the state of generalized plane stress shown in Fig. P8.6-6.

Prob. 8.6-7. Repeat Prob. 8.6-4 for the state of generalized plane stress shown in Fig. P8.6-7.

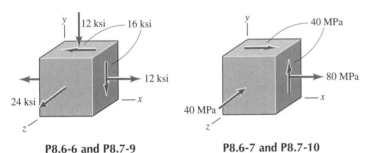

P8.6-6 and P8.7-9   **P8.6-7 and P8.7-10**

Prob. 8.6-8. Use Eqs. (1) and (2) in the problem statement of Prob. 8.6-3 to solve for the three principal stresses—σ_{p1}, σ_{p2}, and σ_{p3}—for the following state of generalized plane stress: $\sigma_x = \sigma_y = 10$ ksi, $\tau_{xy} = 6$ ksi, $\sigma_z = -4$ ksi, $\tau_{xz} = \tau_{yz} = 0$.

Problems 8.7-1 through 8.7-6. *For the states of plane stress shown, (a) Sketch Mohr's circle for the stresses in the* xy *plane, and (b) determine the absolute maximum shear stress,* $\tau_{\text{abs}\atop\text{max}}$.

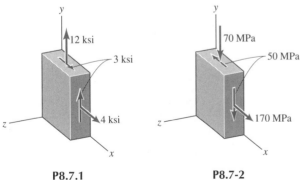

P8.7.1   **P8.7-2**

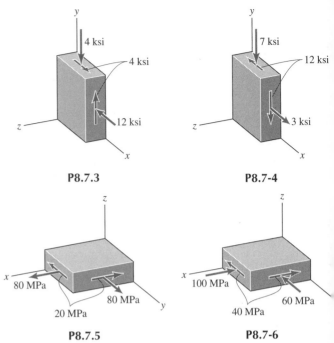

P8.7.3   **P8.7-4**

P8.7.5   **P8.7-6**

Problems 8.7-7 through 8.7-10. *For the states of generalized plane stress indicated, (a) plot Mohr's circle for the stresses in the* xy *plane, (b) determine the three principal stress—σ_{p1}, σ_{p2}, and σ_{p3}—that correspond to the given state of stress, and (c) determine the absolute maximum shear stress,* $\tau_{\text{abs}\atop\text{max}}$.

Prob. 8.7-7. Use the stress state described in Prob. 8.6-4.

Prob. 8.7-8. Use the stress state described in Prob. 8.6-5.

Prob. 8.7-9. Use the stress state described in Prob. 8.6-6.

Prob. 8.7-10. Use the stress state described in Prob. 8.6-7.

Prob. 8.9-1. Using Fig. 8.29b, derive expressions that relate ϵ_y to ϵ_n and γ_{nt}. (These expressions appear in Eqs. 8.51 and 8.53, respectively.)

Prob. 8.9-2. Using Fig. 8.29c, derive expressions that relate γ_{xy} to ϵ_n and γ_{nt}. (These expressions appear in Eqs. 8.51 and 8.53, respectively.)

*Prob. 8.9-3. At point P on the surface of a flat plate, the strains are given by ϵ_x, ϵ_y, and γ_{xy}. Thus, the small rectangle in Fig. P8.9-3(a), whose diagonal PQ defines the direction n, is deformed into the parallelogram shape in Fig. P8.9-3(b). Use trigonometry to derive Eq. 8.51, which relates ϵ_n to the given strains. (Hint: Use the law of cosines.)

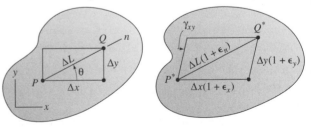

P8.9-3

Prob. 8.9-4. Following the procedure employed in Section 8.4 to derive expressions for directions of principal stress and the corresponding principal stresses, and starting with Eq. 8.54a, (a) derive expressions for the *directions of principal strain,* that is, the directions θ_{p1} and θ_{p2} along which ϵ_n assumes its maximum and minimum values, respectively. (b) Obtain expressions for the *principal strains,* ϵ_1 and ϵ_2, in terms of ϵ_x, ϵ_y, and γ_{xy}.

Prob. 8.9-5. Following the procedure employed in Section 8.4 to derive expressions for the maximum in-plane shear stress, and starting with Eq. 8.54b, obtain an expression for the maximum in-plane shear strain $\gamma_{max} \equiv \gamma_{s_1 s_2}$ in terms of ϵ_x, ϵ_y, and γ_{xy}.

Prob. 8.9-6. A thin rectangular bar is subjected to an axial load P, as shown in Fig. P8.9-6. Near the center of the bar, line segments AB and AC at 45° to the axis of the bar define the directions n and t. (a) Determine the extensional strains ϵ_x and ϵ_y. (Note: $\gamma_{xy} = 0$ for axial loading.) (b) Determine the extensional strains ϵ_n and ϵ_t. (c) Determine the shear strain γ_{nt}.

$$E = 30(10^3)\text{ksi}, \quad \nu = 0.3, \quad t \equiv \text{thickness} = 0.25 \text{ in.}$$

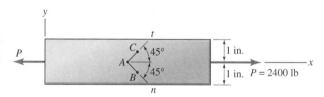

P8.9-6, P8.10.1, and PC8.10-1

Prob. 8.9-7. Repeat Prob. 8.9-6 for the rectangular bar in Fig. P8.9-7.

$$E = 200\text{GPa}, \quad \nu = 0.3, \quad t \equiv \text{thickness} = 10 \text{ mm}$$

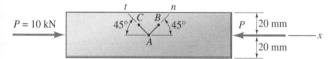

P8.9-7, P8-10.2, and PC8.10-2

Prob. 8.9-8. A rectangular bar whose cross-sectional dimensions are 5 in. ×1 in. is subjected to an axial load $P = 48$ kips that produces extensional strains $\epsilon_n = 640\mu$, $\epsilon_t = 0$ along the n and t directions indicated in Fig. P8.9-8. Determine the values of Young's modulus, E, and Poisson's ratio, ν, for this bar. Assume linearly elastic, isotropic material behavior.

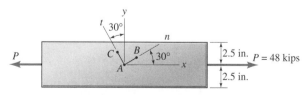

P8.9-8, P8.10-3, and PC8.10-3

Prob. 8.9-9. On the outer surface of a pressure vessel, strains $\epsilon_n = 200\mu$ and $\epsilon_t = 400\mu$ are measured along the n and t directions indicated in Fig. P8.9-9. (a) Determine the extensional strains ϵ_x and ϵ_y. (Note: $\gamma_{xy} = 0$.) (b) Determine the shear strain γ_{nt}. (c) For pressure vessels, $\sigma_y = 2\sigma_x$ (see Section 9.2). Therefore, what is the value of Poisson's ratio, ν, for this tank? Assume linearly elastic, isotropic material behavior.

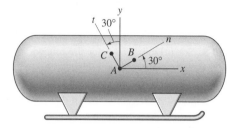

P8.9-9, P8.10-4, and PC8.10-4

Prob. 8.9-10. At a certain point on the surface of a pressure vessel (like the one shown in Fig. P8.9-9) the strains are given by: $\epsilon_x = 100\mu$, $\epsilon_y = 500\mu$, and $\gamma_{xy} = 0$. (a) Determine the state of strain for nt axes rotated counterclockwise by 30° from the xy axes, as shown in Fig. P8.9-10; that is, determine ϵ_n, ϵ_t, and γ_{nt}. (b) Sketch the shape of the elemental square $\overline{ABDC}$ after deformation. Let the original lengths of the sides be $\overline{AB} = \overline{AC} = 1$, but exaggerate the deformation.

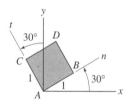

P8.9-10, P8.10-5, and PC8.10-5

Prob. 8.10-1. Use Mohr's strain circle to solve Prob. 8.9-6.

Prob. 8.10-2. Use Mohr's strain circle to solve Prob. 8.9-7.

Prob. 8.10-3. Use Mohr's strain circle to solve Prob. 8.9-8.

Prob. 8.10-4. Use Mohr's strain circle to solve Prob. 8.9-9.

Prob. 8.10-5. Use Mohr's strain circle to solve Prob. 8.9-10.

Problems 8.10-6 through 8.10-10. *Use the generic figures below in solving Probs. 8.10-6 through 8.10-10.*

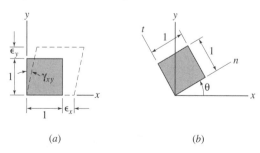

(*a*)   (*b*)

P8.10-6 through P8.10-10, and PC8.10-6 through PC8.10-10

Prob. 8.10-6. An element of material is subjected to the following state of plane strain: $\epsilon_x = 200\mu$, $\epsilon_y = -100\mu$, and $\gamma_{xy} = -200\mu$. (a) Use Mohr's circle to calculate the strains ϵ_n, ϵ_t, and γ_{nt} for an element rotated (counterclockwise) by an angle $\theta = 20°$. (b) Use the given strains to produce a sketch of the deformed xy element, similar to the sketch in Fig. P8.10-6a. (c) Use the calculated strains to produce a sketch of the deformed nt element, starting with the undeformed unit element in Fig. P8.10-6b.

Prob. 8.10-7. Solve Prob. 8.10-6 for the following strains and angle: $\epsilon_x = 300\mu$, $\epsilon_y = 750\mu$, $\gamma_{xy} = 450\mu$, $\theta = 30°$.

Prob. 8.10-8. Solve Prob. 8.10-6 for the following strains and angle: $\epsilon_x = 0$, $\epsilon_y = 400\mu$, $\gamma_{xy} = -300\mu$, $\theta = -30°$.

Prob. 8.10-9. Solve Prob. 8.10-6 for the following strains and angle: $\epsilon_x = 50\mu$, $\epsilon_y = 0$, $\gamma_{xy} = 120\mu$, $\theta = 45°$.

Prob. 8.10-10. Solve Prob. 8.10-6 for the following strains and angle: $\epsilon_x = 150\mu$, $\epsilon_y = 300\mu$, $\gamma_{xy} = 200\mu$, $\theta = -22.5°$.

Problems 8.10-11 through 8.10-16. *For each of the listed states of plane strain, (a) use Mohr's circle to determine the in-plane principal strains and principal-strain directions, and show how a unit square oriented in the principal-strain directions deforms; and (b) determine the maximum in-plane shear strain.*

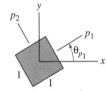

P8.10-11 through P8.10-16, and PC8.10-11 through PC8.10-16

Prob. 8.10-11. $\epsilon_x = 400\mu$, $\epsilon_y = -200\mu$, $\gamma_{xy} = -400\mu$.
Prob. 8.10-12. $\epsilon_x = 200\mu$, $\epsilon_y = 500\mu$, $\gamma_{xy} = -300\mu$.
Prob. 8.10-13. $\epsilon_x = 0$, $\epsilon_y = -400\mu$, $\gamma_{xy} = 300\mu$.
Prob. 8.10-14. $\epsilon_x = 100\mu$, $\epsilon_y = 0$, $\gamma_{xy} = 240\mu$.
Prob. 8.10-15. $\epsilon_x = -20\mu$, $\epsilon_y = 220\mu$, $\gamma_{xy} = 100\mu$.
Prob. 8.10-16. $\epsilon_x = 150\mu$, $\epsilon_y = -300\mu$, $\gamma_{xy} = 200\mu$.

*Prob. 8.10-17. Let the x and y axes at point P on the surface of a flat plate undergoing plane stress (see Section 8.2) be oriented in the directions of the *principal stresses*, σ_1 and σ_2, respectively, and let the x' and y' axes be oriented in the directions of the *principal strains*, ϵ_1 and ϵ_2. Assume that the plate is linearly elas-

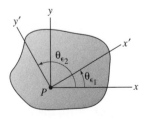

P8.10-17

tic. Do the *axes of principal strain* (the $x'y'$ axes) coincide with the *axes of principal stress* (the xy axes)? Either prove that they coincide, or, if they do not coincide, derive a formula for the angle θ_{ϵ_1} that orients the ϵ_1 axis with respect to the σ_1 axis.

Prob. 8.10-18. A thin plate is subjected to a state of plane stress with the following stresses: $\sigma_x = 8.4$ ksi, $\sigma_y = 2.4$ ksi, $\tau_{xy} = $ ($\sigma_z = \tau_{xz} = \tau_{yz} = 0$ also.) The plate satisfies Hooke's Law, w $E = 10(10^3)$ ksi and $\nu = \frac{1}{3}$. Sketch a Mohr's stress circle also a Mohr's strain circle, and comment on the similarities a differences between these two circles.

Prob. 8.10-19. Solve Prob. 8.10-18 for the following state plane stress: $\sigma_x = 450$ MPa, $\sigma_y = 150$ MPa, $\tau_{xy} = 0$. Let E 100 GPa and $\nu = \frac{1}{3}$.

Prob. 8.10-20. Solve Prob. 8.10-18 for the following state plane stress: $\sigma_x = 18$ ksi, $\sigma_y = 0$ ksi, $\tau_{xy} = 6$ ksi. Let E $10(10^3)$ ksi, $\nu = 0.30$.

*Prob. 8.10-21. Draw Mohr's strain circle for the follow plane strain state: $\epsilon_x = 90\mu$, $\epsilon_y = 270\mu$, $\epsilon_{p1} = 330\mu$. Determ the following strains: γ_{xy}, ϵ_{p2}, θ_{p1}.

*Prob. 8.10-22. Draw Mohr's strain circle for the follow plane strain state: $\epsilon_x = -240\mu$, $\epsilon_y = 0$, $\epsilon_{p2} = -250\mu$. De mine the following strains: γ_{xy}, ϵ_{p1}, θ_{p2}.

Computer Exercises—Sect. 8.10. *Use the MechSOLID c puter program MOHR to solve the homework exercises C8.1 through C8.10-16. Refer to Appendix G.7 for a description of computer program and for examples of its use. Hand in comp printouts of the following MechSOLID "screens": MOHR: sults, and Mohr's Circle. Samples of these screens are illustra in Example G-8.*

Prob. C8.10-1. Use the computer program MOHR to solve Pr 8.10-1.

Prob. C8.10-2. Use the computer program MOHR to solve Pr 8.10-2.

Prob. C8.10-3. Use the computer program MOHR to solve Pr 8.10-3.

Prob. C8.10-4. Use the computer program MOHR to solve P 8.10-4.

Prob. C8.10-5. Use the computer program MOHR to solve Pr 8.10-5.

Prob. C8.10-6. Use the computer program MOHR to solve Pr 8.10-6.

Prob. C8.10-7. Use the computer program MOHR to solve Pr 8.10-7.

Prob. C8.10-8. Use the computer program MOHR to solve Pr 8.10-8.

Prob. C8.10-9. Use the computer program MOHR to solve P 8.10-9.

Prob. C8.10-10. Use the computer program MOHR to so Prob. 8.10-10.

Prob. C8.10-11. Use the computer program MOHR to so Prob. 8.10-11.

Prob. C8.10-12. Use the computer program MOHR to solve Prob. 8.10-12.

Prob. C8.10-13. Use the computer program MOHR to solve Prob. 8.10-13.

Prob. C8.10-14. Use the computer program MOHR to solve Prob. 8.10-14.

Prob. C8.10-15. Use the computer program MOHR to solve Prob. 8.10-15.

Prob. C8.10-16. Use the computer program MOHR to solve Prob. 8.10-16.

Prob. 8.12-1. Starting with Eqs. 8.60, derive Eq. 8.61 for a 45° strain-gage rosette and Eqs. 8.62 for an equiangular strain-gage rosette.

Prob. 8.12-2. Use Mohr's circle for strain in Fig. P8.12-2 to derive the third equation of Eqs. 8.61, that is,

$$\gamma_{xy} = 2\epsilon_b - \epsilon_a - \epsilon_c$$

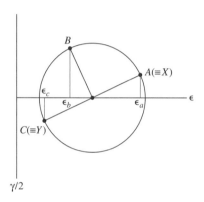

P8.12-2

Prob. 8.12-3. The 45° rectangular rosette in Fig. P8.12-3 was used to obtain the following strains—$\epsilon_a = 175\mu$, $\epsilon_b = 140\mu$, and $\epsilon_c = -105\mu$—at a point on the free surface of a machine component being tested. (a) Determine the in-plane strains ϵ_x, ϵ_y, and γ_{xy}. (b) Use a Mohr's strain circle to determine the in-plane principal strains ϵ_1 and ϵ_2 at the rosette's location. (c) Determine the orientations of the principal-strain axes relative to the orientation of the "a" gage.

Prob. 8.12-4. Solve Prob. 8.12-3 for the following measured strains: $\epsilon_a = 440\mu$, $\epsilon_b = 510\mu$, and $\epsilon_c = 340\mu$.

Prob. 8.12-5. During stress testing of a new design for a titanium-alloy helicopter transmission case, the following strains were measured on the outer surface of the transmission case by use of a 45° equiangular rosette (see Prob. 8.12-3 for the rosette configuration): $\epsilon_a = -270\mu$, $\epsilon_b = 371\mu$, and $\epsilon_c = 670\mu$. (a) Determine the strain components ϵ_x, ϵ_y, and γ_{xy}. (b) Sketch a Mohr's circle for this strain state, and determine the principal strains and the maximum shear strain at the rosette's location. (c) Letting $E = 100$ GPa and $\nu = 0.33$, determine the principal stresses and the maximum in-plane (i.e., in the plane of the gage) shear stress. Sketch a stress element oriented in the principal directions.

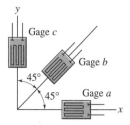

P8.12-3 through P8.12-5 and PC8.12-1 through PC8.12-3

Prob. 8.12-6. The 60° equiangular rosette shown in Fig. P8.12-6 was used to obtain the following strains—$\epsilon_a = 160\mu$, $\epsilon_b = 520\mu$, and $\epsilon_c = 160\mu$—at a point on the free surface of a machine component being tested. (a) Determine the in-plane strains ϵ_x, ϵ_y, and γ_{xy}. (b) Use a Mohr's circle to determine the in-plane principal strains ϵ_1 and ϵ_2 at the rosette's location. (c) Determine the orientations of the principal-strain axes relative to the orientation of the "a" gage.

Prob. 8.12-7. Solve Prob. 8.12-6 for the following measured strains: $\epsilon_a = -200\mu$, $\epsilon_b = 450\mu$, and $\epsilon_c = 125\mu$.

Prob. 8.12-8. At a point on the outer surface of a gas-turbine engine, the strains $\epsilon_a = 935\mu$, $\epsilon_b = 167\mu$, and $\epsilon_c = 668\mu$ were measured using a 60° equiangular rosette (see figure with Prob. 8.12-6). (a) Determine the strain components ϵ_x, ϵ_y, and γ_{xy}. (b) Sketch a Mohr's circle for this state of strain, and determine the principal strains and the maximum shear strain at the rosette's location. (c) Letting $E = 30 \times 10^3$ ksi and $\nu = 0.33$, determine the principal stresses and the maximum in-plane (i. e., in the plane of the gage) shear stress. Sketch a stress element oriented in the principal directions.

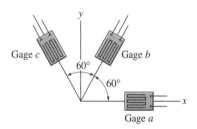

P8.12-6 through P8.12-8 and PC8.12-4 through PC8.12-6

Prob. 8.12-9. Starting with Eqs. 8.60, derive expressions for ϵ_x, ϵ_y, and γ_{xy} in terms of measured extensional strains ϵ_a, ϵ_b, and ϵ_c for each of the two rosette orientations shown in Fig. P8.12-9a,b.

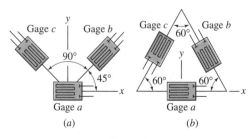

P8.12-9

485

Prob. 8.12-10. Repeat Prob. 8.12-9 for the two rosette configurations shown in Fig. P8.12-10a,b. How do your results compare with Eqs. 8.61 and 8.62, respectively?

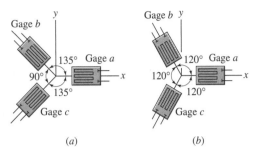

(a)   (b)

P8.12-10

Prob. 8.12-11. At a point on the surface of a steel machine component, the strain rosette shown in Fig. P8.12-11 measured the following extensional strains: $\epsilon_a = 700\mu$, $\epsilon_b = 560\mu$, and $\epsilon_c = -280\mu$. (a) Determine the strain components ϵ_x, ϵ_y, and γ_{xy} at the rosette location. (b) Determine the stress components σ_x, σ_y, and τ_{xy} at the rosette location. (c) Using a Mohr's circle for strain, determine the principal strains and the maximum shear strain at the point. (d) Letting $E = 30 \times 10^3$ ksi and $\nu = 0.30$, determine the principal stresses and the absolute maximum shear stress at the point. (Recall Section 8.7.)

Prob. 8.12-12. Use Mohr's circle to determine the extensional strains ϵ_a, ϵ_b, and ϵ_c that would be indicated by the rosette shown in Fig. P8.12-12 (see Prob. 8.12-11) if the two in-plane principal strains at the rosette location are $\epsilon_1 = -184\mu$ and $\epsilon_2 = 736\mu$. The direction of principal strain ϵ_1 is 30° clockwise from the x axis.

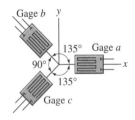

P8.12-11, P8.12-12, and PC8.12-7

Prob. 8.12-13. At a point on an aluminum-alloy bracket, the strain rosette shown in Fig. P8.12-13 measured the following extensional strains: $\epsilon_a = 592\mu$, $\epsilon_b = -444\mu$, and $\epsilon_c = 740\mu$. (a) Determine the strain components ϵ_x, ϵ_y, and γ_{xy} at the rosette

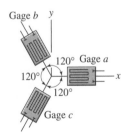

P8.12-13, P8.12-14, and PC8.12-8

486

location. (b) Letting $E = 70$ GPa and $\nu = 0.33$, determine the stress components σ_x, σ_y, and τ_{xy} at the rosette location. (c) Using a Mohr's circle for strain, determine the principal strains and the maximum in-plane shear strain at the point. (d) Determine the principal stresses and the absolute maximum shear stress at the point. (Recall Section 8.7.)

Prob. 8.12-14. Use Mohr's circle to determine the extensional strains ϵ_a, ϵ_b, and ϵ_c that would be indicated by the rosette shown in Fig. P8.12-14 (see Prob. 8.12-13) if the two in-plane principal strains at the rosette location are $\epsilon_1 = 580\mu$ and $\epsilon_2 = -260\mu$. The direction of the principal strain ϵ_1 is oriented 15° clockwise from the x axis (i.e., from the orientation of the "a" gage).

***Prob. 8.12-15.** A *torsion load cell* (to measure torque T) is constructed by mounting two strain gages on a tubular shaft, with the gages oriented at $\pm 45°$ to the axis of the tube, as indicated in Fig. P8.12-15. The gages are wired so that the measurement circuit gives an output $\epsilon_t \equiv \epsilon_b - \epsilon_a$. Determine the relationship between the applied torque T and the measured strain difference, ϵ_t, if the tube has the following properties: r_o = outer radius, r_i = inner radius, E = modulus of elasticity, and ν = Poisson's ratio.

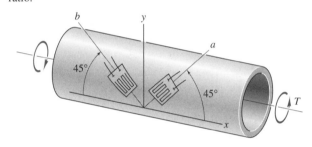

P8.12-15

Computer Exercises—Sect. 8.12. *Use the **MechSOLID** computer program MOHR to solve the homework exercises C8.12-1 through C8.112-8. Refer to Appendix G.7 for a description of this computer program and for examples of its use. Hand in computer printouts of the following **MechSOLID** "screens": MOHR: Results, and Mohr's Circle. Samples of these screens are illustrated in Example G-8.*

Prob. C8.12-1. Use the computer program MOHR to solve Prob. 8.12-3.

Prob. C8.12-2. Use the computer program MOHR to solve Prob. 8.12-4.

Prob. C8.12-3. Use the computer program MOHR to solve Prob. 8.12-5.

Prob. C8.12-4. Use the computer program MOHR to solve Prob. 8.12-6.

Prob. C8.12-5. Use the computer program MOHR to solve Prob. 8.12-7.

Prob. C8.12-6. Use the computer program MOHR to solve Prob. 8.12-8.

Prob. C8.12-7. Use the computer program MOHR to solve Prob. 8.12-11.

Prob. C8.12-8. Use the computer program MOHR to solve Prob. 8.12-13.

STRESSES DUE TO COMBINED LOADING; PRESSURE VESSELS

9.1 INTRODUCTION

In previous chapters, specifically Chapters 2, 3, 4, and 6, formulas were derived that relate the normal stress and the shear stress on any cross section of a slender member to the stress resultants on the cross section. Some typical formulas that were derived are listed in Table 9.1. In many practical situations, two or more stress resultants occur on a cross section, so we need to determine the *state of stress at a point* due to various combined loads.

The combined effect of normal and shear stresses can be conveniently analyzed (for plane stress) by the use of *Mohr's Circle for stress* (Section 8.5). In this chapter we will investigate the stress distribution in slender members under several combinations of loading. In addition, we will examine the state of stress in thin-wall pressure vessels. The biaxial state of stress in thin-wall pressure vessels is also conveniently analyzed by using Mohr's circle.

TABLE 9.1.   Formulas for Stresses

Stress Resultant	Symbol	Formula	References
Normal force	F	$\sigma = \dfrac{F}{A}$	Sections 2.2, 3.2
Torsional moment	T	$\tau = \dfrac{T\rho}{I_p}$	Section 4.2
Bending moment	M	$\sigma = \dfrac{-My}{I}$	Section 6.3
Transverse shear force	V	$\tau = \dfrac{VQ}{It}$	Section 6.8

The following three-step procedure will be useful in solving for stresses due to combined loading:

1. *Determine the internal resultants*: This, of course, involves drawing free-body diagrams and writing equilibrium equations. For statically indeterminate problems, material behavior and geometry of deformation must also be considered.

2. *Calculate the individual stresses*: Formulas like those listed in Table 9.1 are used to compute the stress distributions that result from the various stress resultants.

3. *Combine the individual stresses*: This step involves algebraically summing like stresses (e.g., two σ's on the same face), or using Mohr's circle when the stresses are dissimilar (e.g., σ_x and σ_y). In most cases, the principal stresses and the maximum shear stress are required, and these can be obtained from Mohr's circle for stress.

9.2 · THIN-WALL PRESSURE VESSELS

(*a*) A scuba diver with cylindrical air tank.

(*b*) Spherical oxygen and propellant tanks on a spacecraft.

FIGURE 9.1 Examples of thin-wall pressure vessels.

One form of "combined loading" occurs in thin-wall pressure vessels like the cylindrical and spherical tanks shown in Fig. 9.1. Thin-wall pressure vessels vary in size, for example, from 2-in.-diameter hair-spray, shaving cream, or spray-paint cans to gas-storage tanks that are sixty feet or more in diameter. In this section we will examine cylindrical and spherical tanks under internal (gas) pressure. These are special cases of thin-wall structures called *thin shells*.[1]

Although a metallic or plastic thin-wall pressure vessel does not expand as much as a balloon does when it is pressurized, in both cases the pressurization stretches the "skin" or walls of the pressure vessel, enlarging the vessel. In general, the walls of a vessel are considered to be thin if the radius-to-wall-thickness ratio is ten or more (i.e., $\frac{r}{t} \geq 10$). In that case, the tensile stress in the vessel wall varies insignificantly (<5%) from the inside of the vessel wall to the outside.

Cylindrical Pressure Vessels. The diver's air tank in Fig. 9.1*a* is an example of a thin-wall cylindrical pressure vessel. Figure 9.2*a* shows a circular-cylinder pressure vessel with end closures. In the cylindrical section the normal (tensile) stresses in the longitudinal, or axial, direction and in the circumferential, or hoop, direction are called the *axial stress* σ_a and the *hoop stress* σ_h, respectively. We will assume that the vessel contains a pressurized gas whose weight can be neglected, and we will use appropriate free-body diagrams in relating σ_a and σ_h to the internal pressure p.[2] It is also assumed that σ_a and σ_h are constant through the thickness of the wall of the pressure vessel.

Axial (Longitudinal) Stress: To determine the axial stress, we can employ the free-body diagram in Fig.9.2*c*, where a "cut" has been made through the shell wall and the gas at an arbitrary cross section x at some distance from the end closure.[3] Summing the forces

---

[1]*Thin Elastic Shells* by Harry Kraus, [Ref. 9-1].

[2]This is the *gage pressure,* that is, the internal pressure minus the external (or atmospheric) pressure. If the pressure is larger on the outside than on the inside, a thin-wall shell may be subject to collapse by buckling. We will consider only internal pressurization.

[3]The effect of the end closure is discussed later in this section.

in the x direction, we get

$$\sum F_x = 0: \qquad\qquad \sigma_a A_a - pA_{pa} = 0 \qquad\qquad (9.1)$$

The longitudinal stress acts on the cut section of the vessel wall. The gas pressure, however, acts on the gas still occupying the vessel below the cutting plane at x. Thus, Eq. 9.1, together with areas A_a and A_{pa} based on Fig. 9.2b, becomes

$$\sigma_a[\pi(r + t)^2 - \pi r^2] - p\pi r^2 = 0$$

But, since $t \ll r$, this can be approximated by

$$\sigma_a(2\pi rt) - p\pi r^2 = 0$$

so, the longitudinal stress is given by

$$\boxed{\sigma_a = \frac{pr}{2t}} \qquad\qquad (9.2)$$

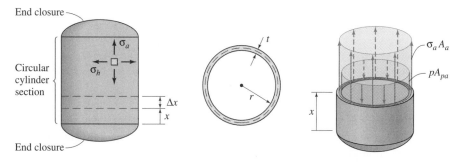

End closure

Circular cylinder section

End closure

(a) A circular cylinder with end closures.   (b) Shell dimensions.   (c) A free-body diagram for determining the axial stress σ_a.

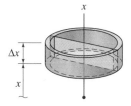

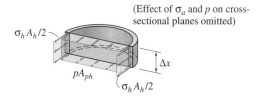

(Effect of σ_a and p on cross-sectional planes omitted)

(d) The diametral cutting plane for a hoop-stress free-body diagram.   (e) A free-body diagram for determining the hoop stress σ_h.

FIGURE 9.2 Figures used in the analysis of stresses in circular cylinders under internal pressure.

Hoop Stress: The free-body diagram in Fig. 9.2e, which is based on the cross-sectional and longitudinal cutting planes illustrated in Fig. 9.2d, may be used in determining the hoop stress, σ_h. By taking an arbitrary longitudinal cutting plane that cuts the vessel in half (i.e., it contains the longitudinal axis), we obtain two surfaces on which the stress is σ_h. Also, the two σ_h-forces in Fig. 9.2e are parallel. Thus, summing forces in the hoop (i.e., tangential to the circumference) direction, we get

$$\sum F_h = 0: \qquad\qquad \sigma_h A_h - pA_{ph} = 0 \qquad\qquad (9.3)$$

The areas are $A_h = 2t\Delta x$ and $A_{ph} = 2r\Delta x$, so Eq. 9.3 gives the following expression for hoop stress:

$$\sigma_h = \frac{pr}{t} \tag{9.4}$$

Spherical Pressure Vessels. Consider now a thin-wall spherical pressure vessel of inner radius r and wall thickness t ($r/t \geq 10$). Because of the spherical symmetry, the normal stress will be the same in any direction in the shell. This stress, which is labeled σ_s to identify it with a spherical pressure vessel, can be determined by taking a free-body diagram that consists of half of the spherical vessel plus the pressurized gas occupying the half-sphere. By summing the forces on the free-body diagram in Fig. 9.3b, we get

$$\sigma_s A_s - pA_{ps} = 0 \tag{9.5}$$

The approximations that led from Eq. 9.1 to Eq. 9.2 hold here as well, so Eq. 9.5 gives

$$\sigma_s = \frac{pr}{2t} \tag{9.6}$$

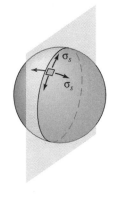

(a) A spherical pressure vessel with diametral cutting plane.

The stress in a spherical pressure vessel is equal to the longitudinal stress in a cylindrical pressure vessel having the same r/t ratio; and it is just half the value of the hoop stress in the cylinder. Thus, for a given pressure, a spherical pressure vessel can have a thinner wall than a cylindrical vessel can.

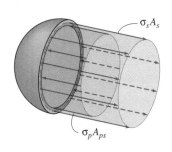

State of Stress in Pressure-Vessel Walls. The previous analyses give us the *"in-plane"* stresses in the walls of cylindrical and spherical pressure vessels. Since there is no shear stress on the cutting planes in Fig. 9.2c, Fig. 9.2e, or Fig. 9.3b, the stresses σ_a, σ_h, and σ_s are all in-plane principal stresses. The radial normal stress is the third principal stress in both cases. At the inner surface of the pressure vessel wall, the radial normal stress is $\sigma_r = -p$, since the pressure pushes on the inside surface. On the outer surface the (gage) pressure is zero, so $\sigma_r = 0$. Since r/t is assumed to be equal to, or greater than, ten, the radial stress is much smaller than the in-plane stresses, so it is usually ignored.[4] To determine the state of stress in the walls of cylindrical and spherical pressure vessels, we can use Eqs. 9.2 and 9.4 for cylinders and Eq. 9.6 for spheres to plot a Mohr's circle of stress for each. Since σ_s is the normal stress in any in-plane direction in a spherical vessel, the Mohr's circle for in-plane stresses in Fig. 9.4b degenerates to a single point on the σ axis. From the analysis in Section 8.7 we know that

(b) A free-body diagram for determining the normal stress in a spherical shell.

FIGURE 9.3 A spherical shell under internal pressure.

$$\tau_{\substack{abs \\ max}} = \frac{\sigma_1 - \sigma_3}{2} \tag{8.39}$$
repeated

From Figs. 9.4a and 9.4b this gives

$$\left(\tau_{\substack{abs \\ max}}\right)_{cyl} = \frac{pr}{2t}$$
$$\left(\tau_{\substack{abs \\ max}}\right)_{sph} = \frac{pr}{4t} \tag{9.7}$$

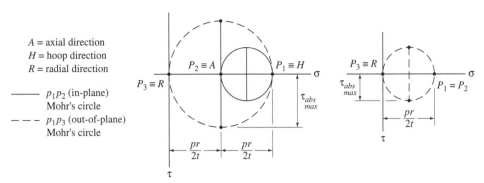

(a) Mohr's circles for a cylinder.

(b) Mohr's circles for a sphere.

FIGURE 9.4 Mohr's circles for in-plane and out-of-plane stresses.

Effects of End Closures and Other Discontinuities. Of course, to fill a pressure vessel with gas or liquid, or take the gas or liquid out, there must be a hole in the pressure-vessel wall and some sort of "connector," as illustrated in Fig. 9.5a. The stress formulas developed above do not apply to stresses in the immediate vicinity of such discontinuities in the pressure-vessel wall.

The vicinity of the joint between the cylindrical section and the end closure of a pressure vessel (Fig. 9.2a) is also a location where the previous analysis does not apply. This is illustrated in Figs. 9.5b and 9.5d. If the end closure and the cylindrical section of the pressure vessel were permitted to expand freely under the effect of internal pressure, the cylinder would expand radially more than the end closure would, as illustrated in Fig. 9.5b. However, since the closure must be welded, or otherwise attached, to the cylinder, both the cylinder and the end closure undergo significant local deformation that involves localized bending stresses. Analysis of the stress distribution at such discontinuities is beyond the scope of this text.[5] Flat-plate end closures, like the ones illustrated in Fig. 9.5c, are especially undesirable and should be avoided if possible.

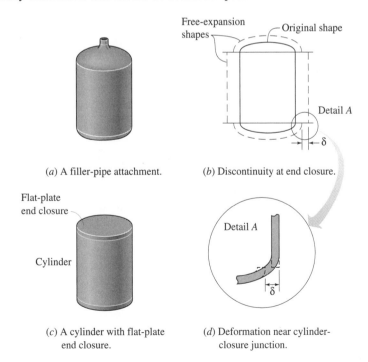

(a) A filler-pipe attachment.

(b) Discontinuity at end closure.

(c) A cylinder with flat-plate end closure.

(d) Deformation near cylinder-closure junction.

FIGURE 9.5 Two types of discontinuities in thin-wall pressure vessels.

[5]This topic is treated in textbooks on thin shells; for example, see Section 116 in *Theory of Plates and Shells* by S. Timoshenko and S. Woinowsky-Krieger, [Ref. 9-2].

A cylindrical pressure vessel 2.50 m in diameter is fabricated by shaping two 10-mm-thick steel plates and butt-welding the plates along helical arcs, as shown in Fig. 1. The maximum internal pressure in the pressure vessel is 1200 kPa. For this pressure level, calculate the following quantities: (a) the axial stress and the hoop stress; (b) the absolute maximum shear stress; and (c) the normal stress, σ_n, perpendicular to the weld line, and the shear stress, τ_{nt}, tangent to the weld line.

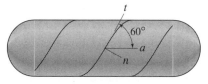

Fig. 1

Solution

(a) In-Plane Stresses: The axial stress and the hoop stress in the cylindrical section of the pressure vessel are given by Eqs. 9.2 and 9.4, respectively. For the given pressure vessel and internal-pressure loading, the axial stress and hoop stress are:

$$\sigma_a = \frac{pr}{2t} = \frac{(1200 \text{ kPa})(1.25 \text{ m})}{2(10 \text{ mm})} = 75 \text{ MPa}$$

$$\sigma_h = \frac{pr}{t} = \frac{(1200 \text{ kPa})(1.25 \text{ m})}{(10 \text{ mm})} = 150 \text{ MPa}$$

Since there is no shear on longitudinal or circumferential cutting surfaces, these stresses are principal in-plane stresses. Therefore,

$$\sigma_1 = \sigma_h = 150 \text{ MPa}, \qquad \sigma_2 = \sigma_a = 75 \text{ MPa} \qquad \text{Ans. (a)} \quad (1)$$

These stresses are shown on the biaxial stress element in Fig. 2a.

(b) Absolute Maximum Shear Stress: The in-plane Mohr's circle for stress is shown as the solid-line circle in Fig. 2b. Since both the axial stress and the hoop stress are tension, the absolute maximum shear stress is not the maximum in-plane shear stress. Rather, this is a stress state with $\sigma_1 \geq \sigma_2 \geq 0$, as described in Fig. 8.25c. Since $r/t = 125$, we can neglect the internal pressure, p, in comparison with the in-plane stresses and draw a Mohr's circle passing through P_1 and the origin, P_3. Thus,

$$\tau_{\text{abs} \atop \text{max}} = \frac{\sigma_1}{2} = \frac{150 \text{ MPa}}{2}$$

or

$$\tau_{\text{abs} \atop \text{max}} = 75 \text{ MPa} \qquad \text{Ans. (b)} \quad (2)$$

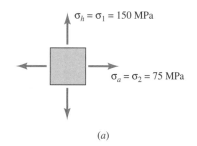

$\sigma_h = \sigma_1 = 150$ MPa

$\sigma_a = \sigma_2 = 75$ MPa

(a)

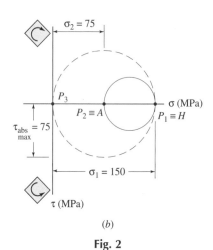

$\sigma_2 = 75$

P_3

$P_2 \equiv A$

σ (MPa)

$P_1 \equiv H$

$\tau_{\text{abs}} = 75$
max

$\sigma_1 = 150$

τ (MPa)

(b)

Fig. 2

(c) *Weld-line Stresses:* The in-plane Mohr's circle is redrawn in Fig. 3a. The point N represents the stresses on a face parallel to the weld, since the direction n in Fig. 1 is perpendicular to the weld line. From the Mohr's circle in Fig. 3, we get

$$\sigma_n = \overline{OB} = 112.5 \text{ MPa} - (37.5 \text{ MPa}) \cos 60° = 93.75 \text{ MPa}$$

$$\tau_{nt} = -\overline{NB} = -(37.5 \text{ MPa}) \sin 60° = -32.48 \text{ MPa}$$

or

$$\sigma_n = 93.8 \text{ MPa}, \qquad \tau_{nt} = -32.5 \text{ MPa} \qquad \text{Ans. (c)} \quad (3)$$

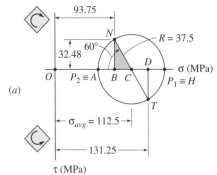

(a)

The normal stress σ_t is given by

$$\sigma_t = \overline{OD} = 112.5 \text{ MPa} + (37.5 \text{ MPa}) \cos 60° = 131.25 \text{ MPa} \qquad (b)$$

These stresses are shown on the rotated element in Fig. 3b.

The normal stress and shear stress on the weld can be converted to normal-force-per-unit-length and shear-force-per-unit-length (shear flow) as was done in Section 6.8.

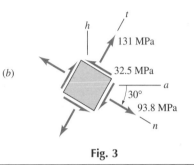

Fig. 3

9.3 STRESS DISTRIBUTION IN BEAMS

To illustrate the distribution of stress in a transversely loaded beam, let us consider the simply supported rectangular beam in Fig. 9.6. There are localized stress concentrations at the loading point and at the two support points, so the stresses in the immediate vicinity of these points cannot be computed by using the flexure formula ($\sigma = -My/I$) and the shear formula ($\tau = VQ/It$) developed in Chapter 6. However, according to St. Venant's Principle, which was introduced in Section 3.3, the state of stress away from these stress concentrations can be based on the formulas of *elementary beam theory*. The normal stress

FIGURE 9.6 A simply supported beam used for examining the state of stress in a beam.

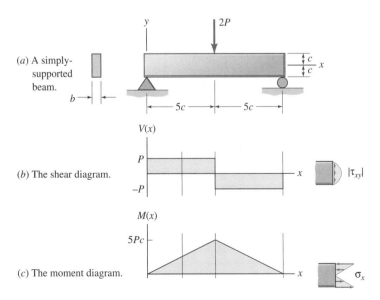

(a) A simply-supported beam.

(b) The shear diagram.

(c) The moment diagram.

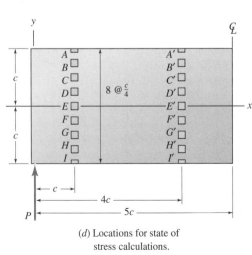

(d) Locations for state of stress calculations.

and the shear stress may be combined by using the stress-transformation equations of Section 8.3, Eqs. 8.3 or 8.5, or by using Mohr's circle for stress to represent the state of (plane) stress at any point in the beam that is sufficiently distant from the load and support points.

For the simply supported beam in Fig. 9.6, we will determine the state of stress at nine equally spaced points at $x = c$, where the bending moment is small, and at nine other equally spaced points at $x = 4c$, where the moment is near its maximum value (Fig. 9.6d). This will permit us to see how the principal stresses σ_1 and σ_2 vary from point to point in this particular beam. We can determine whether, for example, the maximum normal stress is always equal to the maximum flexural stress, or whether some combination of flexural stress and transverse shear stress may lead to a larger normal stress at some point in the beam other than the point(s) where the maximum flexural stress occurs.

For the half-span $0 < x < 5c$ the flexural stress and transverse shear stress are given by

$$\sigma_x = \frac{-Pxy}{I} = -3\sigma_o\left(\frac{xy}{c^2}\right) \tag{9.8}$$

$$\tau_{xy} = -\frac{3\sigma_o}{2}\left(1 - \frac{y^2}{c^2}\right) \tag{9.9}$$

where $\sigma_o = P/A$. These are illustrated in Figs. 9.6c and 9.6b, respectively. The maximum normal stress occurs at midspan and is $(\sigma_x)_{\text{max.}} = 15\sigma_o$. Sample Mohr's-circle calculations for point C are illustrated in Fig. 9.7.

The principal stresses at the two cross sections at $x = c$ and $x = 4c$ are tabulated graphically in Fig. 9.8. Note how the orientation of the principal stresses varies from top to bottom of the beam at each cross section, with the principal stresses being parallel to the x and y axes at the top and bottom, and at $\pm 45°$ to the x axis at the neutral axis ($y = 0$). Also note that, even at $x = c$ where the moment is relatively small, the maximum tension (compression) occurs at the bottom (top) of the beam and not at some intermediate height where there is a nonzero transverse shear.

This example is typical of rectangular-beam problems in that the maximum tensile and compressive stresses are simply flexural stresses $\left(\sigma = \dfrac{-My}{I}\right)$ at sections where

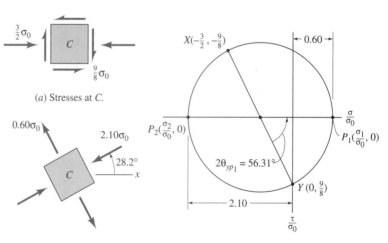

FIGURE 9.7 Determination of the principal stresses at point C in Fig. 9.6d.

(a) Stresses at C.

(c) Principal stresses at C.

(b) Mohr's circle for point C.

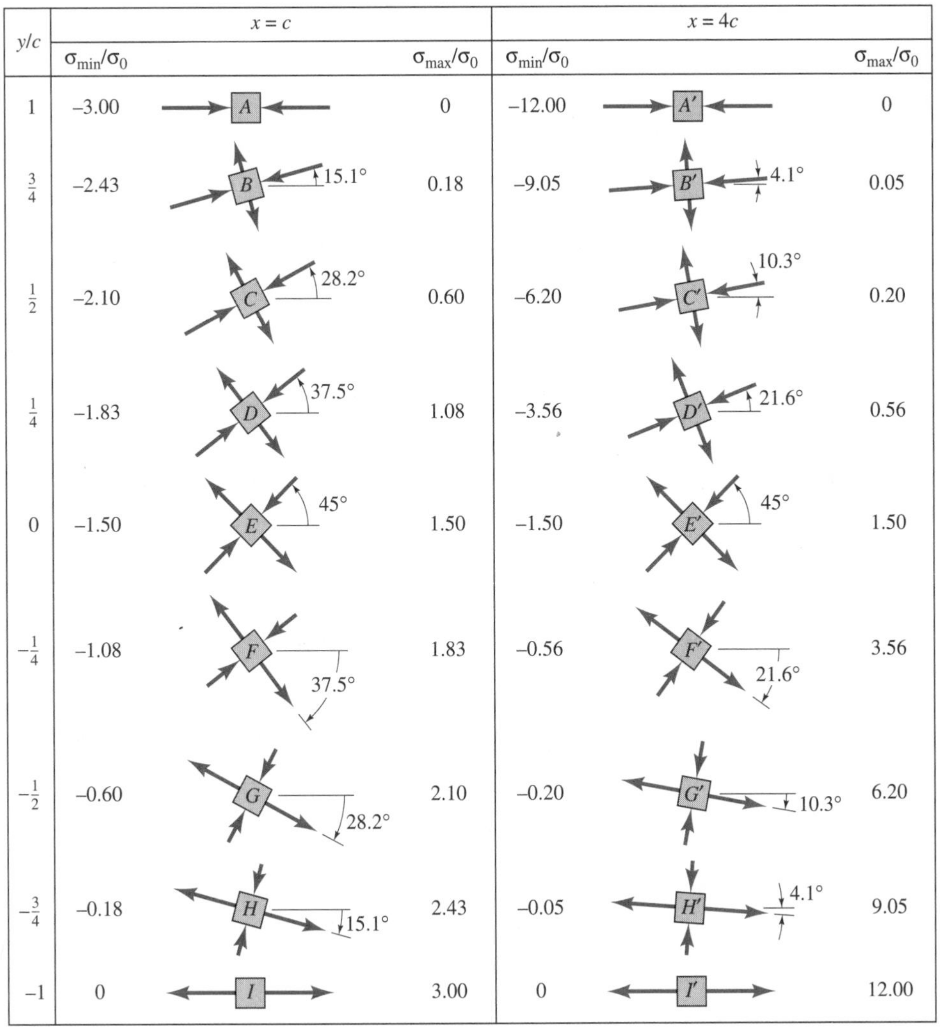

y/c	$x = c$			$x = 4c$		
	σ_{min}/σ_0	(element)	σ_{max}/σ_0	σ_{min}/σ_0	(element)	σ_{max}/σ_0
1	-3.00	A	0	-12.00	A'	0
$\frac{3}{4}$	-2.43	B ($15.1°$)	0.18	-9.05	B' ($4.1°$)	0.05
$\frac{1}{2}$	-2.10	C ($28.2°$)	0.60	-6.20	C' ($10.3°$)	0.20
$\frac{1}{4}$	-1.83	D ($37.5°$)	1.08	-3.56	D' ($21.6°$)	0.56
0	-1.50	E ($45°$)	1.50	-1.50	E' ($45°$)	1.50
$-\frac{1}{4}$	-1.08	F ($37.5°$)	1.83	-0.56	F' ($21.6°$)	3.56
$-\frac{1}{2}$	-0.60	G ($28.2°$)	2.10	-0.20	G' ($10.3°$)	6.20
$-\frac{3}{4}$	-0.18	H ($15.1°$)	2.43	-0.05	H' ($4.1°$)	9.05
-1	0	I	3.00	0	I'	12.00

FIGURE 9.8 The principal stresses at two cross sections of the simply supported beam in Fig. 9.6.

$|M(x)|$ has its maximum value. Therefore, the transverse shear has no effect in determining the maximum normal stresses in the beam. In wide-flange beams, where the shear stress τ_{xy} may be large near the outer fibers where σ_x is also large, a complete investigation (i.e., using Mohr's circle) should be conducted to determine the principal stresses at cross sections where V and M are both significant. Such an analysis is carried out in Example Problem 9.2.

Figure 9.8 gives the entire principal-stress picture at a total of 18 points in the beam in Fig. 9.6. It is not possible to provide complete principal-stress information like this (i.e., values of σ_1 and σ_2 and orientation of principal directions) at every point in a beam. However, it is possible to provide two diagrams that permit a visualization of the principal stresses at every point in a beam. One diagram shows the orientation of the two principal directions at every point. Curves, called *stress trajectories,* are drawn so that they are tangent to the principal directions at every point. Since the two principal directions at any point are orthogonal, two stress trajectories pass through every point, and they are perpendicular to each other. Figure 9.9 shows the stress trajectories for a simply-supported beam. (Stress concentrations are not accounted for in this figure.) A typical use of stress trajectories is to determine the direction of the principal tensile stress in beams made of brittle material (e.g., concrete) so that reinforcement can be provided to carry the tensile

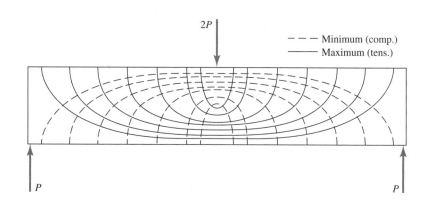

FIGURE 9.9 Stress trajectories for a simply-supported beam with a single midspan load.

stresses. The stress trajectories show the directions of principal stress, but they provide no information about the magnitudes of the principal stresses. A plot of *stress contours* contains curves that connect points of equal principal stress.

Using the *finite-element method* to perform the stress calculations, and using colors to represent different stress magnitudes, it is possible to produce color images of the stress distribution not only in beams, but also in more complex members. The color insert near the beginning of this textbook illustrates stress plots that were produced from finite element solutions.

In Section 6.4, *Design of Beams for Strength*, the sizing of the cross section of a beam to carry specified loads and to have specified supports was based on the maximum flexural stress in the beam. Hence, for example, the maximum moment for the beam in Example Problem 6.5 occurs at $x = 5$ ft, and this moment, together with a given value of allowable stress, was used to determine the most efficient (least weight) cross section of beam to be used for this application. However, since a wide-flange beam was selected for this application, it is possible that the maximum principal stress at a section where both M and V are large may exceed the maximum flexural stress on which the design in Example Problem 6.5 was based. The following example problem explores this possibility.

■■■■■■■■■■■■■■■□ **EXAMPLE 9.2** ■■■■■■■■■■■■■■■

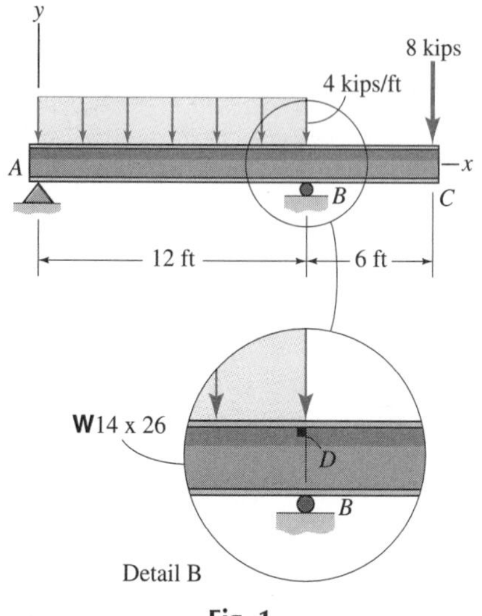

Fig. 1

Just to the left of point B in Fig. 1*a* (same as Fig. 1 of Example Problem 6.5) the transverse shear force is $V = -28$ kips, and the bending moment is $M = -48$ kip·ft. Ignoring any stress concentration due to the support at B, determine the principal stresses at point D in Fig. 1*b*. Compare the maximum tensile stress at D with the maximum flexural stress in the beam, which occurs at $x = 5$ ft, where $M(5 \text{ ft}) = 50$ kip·ft and $V(5 \text{ ft}) = 0$. The beam is a **W**14×26. (See Example Problems 6.5 and 6.15.)

Plan the Solution We can use the flexure formula to determine the normal stress on the cross section at B, and the shear stress distribution on this cross section was determined in Example Problem 6.15. We can use Mohr's circle to combine these and to determine the principal stress magnitudes and directions at point D.

Solution The shear force and bending moment just to the left of section B are shown in Fig. 2*a*, and the essential cross-sectional dimensions of the **W**14×26 beam are shown in Fig. 2*b*.

The normal stress at D is

$$(\sigma_x)_D = -\frac{My}{I} = -\frac{(-48 \text{ kip} \cdot \text{ft})(12 \text{ in./ft})(6.535 \text{ in.})}{245 \text{ in}^4}$$

or

$$(\sigma_x)_D = 15.364 \text{ ksi } (T)$$

The shear stress at point D has a magnitude

$$\tau_D \equiv |\tau_{xy}|_D = \frac{|V|Q}{It} = \frac{28 \text{ kips}(5.025 \text{ in.})(0.420 \text{ in.})(6.745 \text{ in.})}{(245 \text{ in}^4)(0.255 \text{ in.})}$$

or

$$\tau_D = 6.380 \text{ ksi } \uparrow$$

We can construct a Mohr's circle for the stresses in the plane of the web at point D.

From the Mohr's circle in Fig. 3,

$$R = \sqrt{(7.682)^2 + (6.380)^2} = 9.986 \text{ ksi} \tag{1}$$

$$\sigma_{1_D} = 7.682 \text{ ksi} + 9.986 \text{ ksi} = 17.67 \text{ ksi} \qquad \textbf{Ans.} \tag{2}$$

$$\sigma_{2_D} = 7.68 \text{ ksi} - 9.98 \text{ ksi} = -2.30 \text{ ksi}$$

$$2\theta_{xp_1} = \tan^{-1}\left(\frac{6.380}{7.682}\right) = 39.71° \tag{3}$$

$$\theta_{xp_1} = 19.9°$$

At $x = 5$ ft, the maximum tensile stress is just the flexural stress at the bottom of the beam (since $V = 0$). Therefore,

$$\sigma_x(5 \text{ ft}, -6.955 \text{ in.}) = \frac{-My}{I} = \frac{-(50 \text{ kip} \cdot \text{ft})(12 \text{ in./ft})(-6.955 \text{ in.})}{245 \text{ in}^4}$$

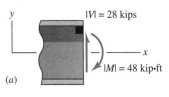

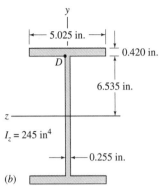

Fig. 2

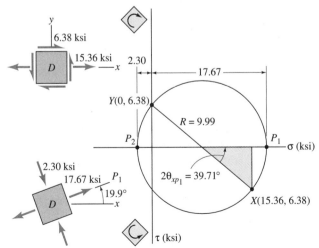

Fig. 3 Mohr's circle for point D.

or

$$\sigma_x(5 \text{ ft}, -6.955 \text{ in.}) = 17.03 \text{ ksi}$$

Therefore, the tensile principal stress at D, 17.67 ksi, is slightly larger than the maximum flexural stress in the beam, 17.03 ksi. However, since an allowable stress of 19 ksi was used in Example Problem 6.5 in selecting the $\mathbf{W}14 \times 26$ beam cross section, the beam has enough stress margin to be safe, even though the principal stress at D is slightly larger than the maximum flexural stress on which the beam design was originally based.

Review the Solution The calculations in this problem are quite simple, so they can just be rechecked for accuracy. Since there is a significant value of transverse shear force at a cross section where the moment is nearly its maximum value, and since the cross section has heavy flanges and a thin web, we should not be surprised to find principal stresses, like the σ_1 stress at point D, that exceed the maximum flexural stress in the beam.

9.4 STRESSES DUE TO COMBINED LOADS

The analysis Procedure outlined in Section 9.1 will now be applied to solve several stress analysis problems that involve various combinations of load types—axial, torsional, and bending.

EXAMPLE 9.3

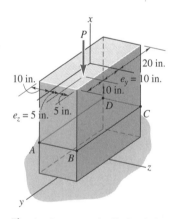

Fig. 1 An eccentrically loaded short compression member.

An axial compressive load of 800 kips is applied eccentrically to a short rectangular compression member, as shown in Fig. 1. (The effect of eccentric compressive loading on longer members is treated in Section 10.4.) Determine the distribution of normal stress on a cross section, say *ABCD*, that is far enough from the point of load application that stress concentration effects may be neglected. Sketch the stress distribution and identify the location of the neutral axis in cross section *ABCD*.

Plan the Solution The eccentric load P produces axial deformation plus bending about the y and z axes. Therefore, this problem involves a superposition of the stresses due to F, M_y, and M_z. Equation 6.36 treats bending about two principal axes.

Solution

Stress Resultants: Figure 2 shows the stress resultants on cross section *ABCD*. The sign conventions for stresses and stress resultants is the same as the sign conventions adopted previously (e.g., Chapters 2, 4, and 6).

Applying equilibrium to a free-body diagram of the member above section *ABCD* we get the following expressions for the stress resultants:

$$
\begin{aligned}
F &= -P \\
M_y &= -Pe_z \\
M_z &= Pe_y
\end{aligned}
\tag{1}
$$

Individual Normal Stresses: Combining Eq. 3.8, for the normal stress due to the axial force *F*, with Eq. 6.36, for the normal stress due to the bending-moment components, we get

$$\sigma_x = \frac{F}{A} + \frac{M_y z}{I_y} - \frac{M_z y}{I_z} \tag{2}$$

Taking each stress contribution separately, and combining Eqs. (1) and (2), we obtain the following:

$$(\sigma_x)_F = \frac{-P}{A} = \frac{-800 \text{ kips}}{(40 \text{ in.})(20 \text{ in.})} = -1000 \text{ psi}$$

$$(\sigma_x)_{M_y} = \frac{(-Pe_z)z}{I_y} = \frac{(-800 \text{ kips})(5 \text{ in.})z}{[\frac{1}{12}(40 \text{ in.})(20 \text{ in.})^3]} = -150z \text{ psi} \tag{3}$$

$$(\sigma_x)_{M_z} = \frac{-(Pe_y)y}{I_z} = \frac{-(800 \text{ kips})(10 \text{ in.})y}{[\frac{1}{12}(20 \text{ in.})(40 \text{ in.})^3]} = -75y \text{ psi}$$

These stress contributions are sketched in Fig. 3. The maximum values of the bending stresses are

$$\text{Max}|(\sigma_x)_{M_y}| = 150z|_{z=10\text{in.}} = 150(10 \text{ in.}) = 1500 \text{ psi}$$

$$\text{Max}|(\sigma_x)_{M_z}| = 75y|_{y=20 \text{ in.}} = 75(20 \text{ in.}) = 1500 \text{ psi}$$

Superposition of Stresses: Using Figs. 3*a* through 3*c*, we can combine, algebraically, the individual stress contributions at four corners to get

$$(\sigma_x)_A = -1000 + 1500 - 1500 = -1000 \text{ psi}$$

$$(\sigma_x)_B = -1000 - 1500 - 1500 = -4000 \text{ psi}$$

$$(\sigma_x)_C = -1000 - 1500 + 1500 = -1000 \text{ psi} \tag{4}$$

$$(\sigma_x)_D = -1000 + 1500 + 1500 = 2000 \text{ psi}$$

With the aid of these corner stresses, we can sketch the combined stress distribution. Since Eq. (2) is linear in *y* and *z*, the *stress surface* will be a plane (which has been "folded" to show tension and compression as they are shown in Fig. 3). In Fig. 4 the member has been rotated about the *x* axis to provide

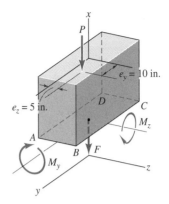

Fig. 2 Stress resultants.

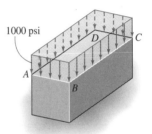

(*a*) Stress distribution due to F.

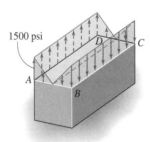

(*b*) Stress distribution due to M_y.

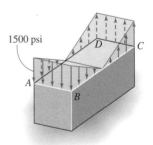

(*c*) Stress distribution due to M_z.

Fig. 3 Individual stress contributions.

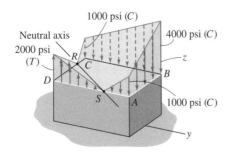

Fig. 4 The combined stresses on section *ABCD*.

a better perspective for viewing the stress surface and the *neutral axis,* which is the intersection of the stress surface with the *ABCD* plane.

The equation of the neutral axis line is given by setting $\sigma_x(y^*, z^*) = 0$, where (y^*, z^*) are coordinates of points on the neutral axis. Combining Eqs. (2) and (3) we get

$$\sigma_x(y^*, z^*) = -1000 - 75y^* - 150z^* = 0 \qquad (5)$$

The intersection, R, of the neutral axis with edge CD is given by setting $y^* = -20$ in. This gives $(y^*, z^*)_R = (-20 \text{ in.}, 3.33 \text{ in.})$. Similarly, $(y^*, z^*)_S = (6.67 \text{ in.}, -10 \text{ in.})$.

Review the Solution   It is obvious from the location of the compressive force P in Fig. 1 that corner B will have the highest compressive stress of any point in the cross section. This is confirmed by Eq. (4b) and by Fig. 4. The load P is far enough from the axis of the member that it actually causes tension at D. This means that between B and D the stress changes from compression to tension. Hence, there is a neutral axis ($\sigma_x = 0$) that passes between A and D. Therefore, Fig. 4 appears to represent accurately the stress distribution due to the eccentric load P.

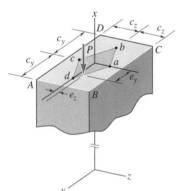

FIGURE 9.10   The kern of a rectangular cross section.

If a compression member, like the one in Example Problem 9.3, is made of a brittle material like concrete, it is not desirable to permit tensile stresses to occur at any point in the cross section. This requires that the compressive load P be located within a small region near the axis of the member. This region is called the *kern* of the cross section, or *core* of the cross section. For a rectangular cross section, like the one in Fig. 9.10, the kern is bounded by four straight lines. The equations of the four lines that bound the kern can be derived by combining Eqs. (1) and (2) of Example Problem 9.3 to give the normal stress on a cross section like section *ABCD* in Fig. 9.10. Thus,

$$\sigma_x(y, z) = \frac{-P}{A} + \frac{-Pe_z z}{I_y} + \frac{-Pe_y y}{I_z} \qquad (9.10)$$

This normal stress will be zero at corner D $(-c_y, -c_z)$ if P is applied anywhere along the line "d," whose equation is

$$\left(\frac{c_y}{I_z}\right)e_{yd} + \left(\frac{c_z}{I_y}\right)e_{zd} = \frac{1}{A} \qquad (9.11)$$

Similar equations can be derived for the kern boundary lines a, b, and c. (See Homework Problem 9.4-5.)

The next example problem illustrates the superposition of stresses due to combined axial and torsional loading.

During the drilling of an oil well, the section of the drill pipe at A (above ground level) is under combined loading due to a tensile force $P = 70$ kips and a torque $T = 6$ kip·ft, as illustrated in Fig. 1. The drill pipe has an outside diameter of 4.0 in. and an inside diameter of 3.640 in. Determine the maximum shear stress at point A on the outer surface of the drill pipe. The radial stress at this point is zero. The yield strength in tension of this drill pipe is 95 ksi.

Plan the Solution We can use Mohr's circle to combine the normal stress σ due to force P and the shear stress τ due to T, but we may also need to consider the three-dimensional aspect of the stress at A to determine the absolute maximum shear stress.

Solution

Stress Resultants: The stress resultants are given in the problem statement:

$$F = P = 70 \text{ kips}, \qquad T = 6 \text{ kip·ft} \tag{1}$$

Individual Stresses: From Eq. 3.8, we get the normal stress

$$\sigma = \frac{F}{A} = \frac{70 \text{ kips}}{\pi[(2 \text{ in.})^2 - (1.820 \text{ in.})^2]} = 32.41 \text{ ksi} \tag{2}$$

From Eq. 4.13, we get the torsional shear stress

$$\tau = \frac{Tr_o}{I_p} = \frac{(6 \text{ kip·ft})(12 \text{ in./ft})(2 \text{ in.})}{\frac{\pi}{2}[(2 \text{ in.})^4 - (1.820 \text{ in.})^4]} = 18.23 \text{ ksi} \tag{3}$$

Figure 2 summarizes the "in-plane" stresses on the surface of the drill pipe at point A. The radial stress, normal to the surface, is zero.

Superposition of Stresses: Mohr's circle may be used to combine the in-plane stresses in Fig. 2. From the Mohr's circle in Fig. 3.

$$R = \sqrt{(16.20 \text{ ksi})^2 + (18.23 \text{ ksi})^2} = 24.39 \text{ ksi} \tag{4}$$

Then,

$$\sigma_1 = \sigma_{avg} + R = 16.20 + 24.39 = 40.6 \text{ ksi} \tag{5}$$
$$\sigma_3 = \sigma_{avg} - R = 16.20 - 24.39 = -8.2 \text{ ksi}$$

The in-plane principal stresses are labeled σ_1 and σ_3, since the out-of-plane principal stress, $\sigma_r = 0$, is the intermediate principal stress. Then, from Eq. 8.38,

$$\tau_{\substack{abs \\ max}} = \frac{\sigma_1 - \sigma_3}{2} = \frac{40.6 \text{ ksi} - (-8.2 \text{ ksi})}{2} \tag{6a}$$

or

$$\tau_{\substack{abs \\ max}} = 24.4 \text{ ksi} \qquad \textbf{Ans.} \tag{6b}$$

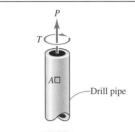

Fig. 1 Portions of an oilwell drill string.

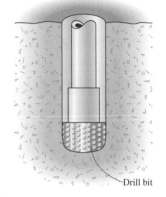

Fig. 2 In-plane stresses at point A.

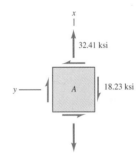

Fig. 3 Mohr's circle.

501

Review the Solution   The calculations in Eqs. (2) and (3) should be re-checked. Points X and Y are plotted correctly, so σ_1 and σ_3 appear to be correct. Finally, since the working stresses in this example should not produce yielding of the drill pipe, the absolute maximum shear stress should be much less than half the tensile yield strength. Therefore, the answer in Eq. (6b) seems reasonable.

Many interesting applications of deformable-body mechanics in the field of oilwell drilling engineering are presented in *Oilwell Drilling Engineering—Principles and Practice,* by H. Rabia, [Ref. 9-3].

In the final example problem on stresses due to combined loading, we consider a problem that involves all types of stress resultants: *F, T, M,* and *V.*

■■■■■■■■■■■■■■■ E X A M P L E   9 . 5 ■■■■■■■■■■■■■■■

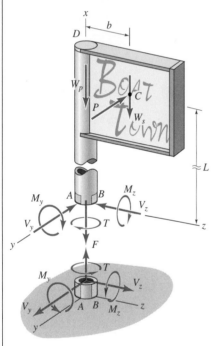

Fig. 1   A cantilevered sign.

Wind blowing on a sign produces a pressure whose resultant, P, acts in the $-y$ direction at point C, as shown in Fig. 1. The weight of the sign, W_s, acts vertically through point C, and the thin-wall pipe that supports the sign has a weight W_p.

Following the procedure outlined in Section 9.1, determine the principal stresses at points A and B, where the pipe column is attached to its base. Use the following numerical data.[6]

> Pipe OD = 3.50 in., A = 2.23 in^2, $I_y = I_z$ = 3.02 in^4, I_p = 6.03 in^4, W_s = 125 lb, W_p = 160 lb, P = 75 lb, b = 40 in., L = 220 in.

Plan the Solution   It will be a good idea to tabulate the stress resultants, stress formulas, and so forth, so that no stress contribution will be missed. The weight W_s contributes to the axial force, and it also produces a moment about the y axis. The wind force P produces a transverse shear force in the y direction, and it also causes a torque about the x axis and a moment about the z axis. A correct free-body diagram is essential.

Solution

Stress Resultants: All six stress resultants on the cross section at the base of the pipe are shown in Fig. 1. The upper portion of Fig. 1 can serve as a free-body diagram for determining these six stress resultants. The sign convention is the one introduced in Fig. 2.34. Let us tabulate the equilibrium equations and indicate what stress is produced by each stress resultant and label each individual stress.

Individual Stresses: Using the formulas from Table 9.1, we can compute the numerical value of each of the nonzero stresses listed in Table 1.

$$\sigma_{A1} = \sigma_{B1} = \frac{F}{A} = \frac{-(125 \text{ lb}) - (160 \text{ lb})}{2.23 \text{ in}^2} = -128 \text{ psi} \qquad (7)$$

The shear stress τ_{B2} is due to the transverse shear force V_y. The basic shear

---

[6]Cross-sectional properties of the 3.50-in.-OD pipe are from Table D.7.

502

TABLE 1. A Table of Stress Resultants and the Stresses Produced

Eq. No.	Equilibrium Equation		Stress at A	Stress at B
(1)	$\sum F_x = 0$	$F = -W_s - W_p$	σ_{A1}	σ_{B1}
(2)	$\sum F_y = 0$	$V_y = -P$	—	τ_{B2}
(3)	$\sum F_z = 0$	$V_z = 0$	—	—
(4)	$\sum M_x = 0$	$T = Pb$	τ_{A4}	τ_{B4}
(5)	$\sum M_y = 0$	$M_y = -W_s b$	—	σ_{B5}
(6)	$\sum M_z = 0$	$M_z = -PL$	σ_{A6}	—

stress formula is

$$\tau_{B2} = \frac{V_y Q}{I_z t} \qquad (8)$$

where Q has to be calculated for the shaded area in Fig. 2. In Example Problem 6.13, it was shown that the shear stress in this case (stress at the neutral axis of a thin-wall pipe) is given by

$$\tau = \frac{2V}{A} \qquad (9a)$$

Therefore,

$$\tau_{B2} = \frac{2(75 \text{ lb})}{2.23 \text{ in}^2} = 67 \text{ psi} \qquad (9b)$$

$$\tau_{A4} = \tau_{B4} = \frac{T r_o}{I_p} = \frac{(Pb) r_o}{I_p} \qquad (10a)$$

so

$$\tau_{A4} = \tau_{B4} = \frac{(75 \text{ lb})(40 \text{ in.})(1.75 \text{ in.})}{6.03 \text{ in}^4} = 871 \text{ psi} \qquad (10b)$$

The flexural stresses due to M_y and M_z are given by Eq. 6.36.

$$\sigma_{B5} = \frac{M_y r_o}{I_y} = \frac{(-W_s b) r_o}{I_y} \qquad (11a)$$

$$\sigma_{B5} = \frac{-(125 \text{ lb})(40 \text{ in.})(1.75 \text{ in})}{3.02 \text{ in}^4} = -2897 \text{ psi} \qquad (11b)$$

$$\sigma_{A6} = \frac{-M_z r_o}{I_z} = \frac{-(-PL) r_o}{I_z} \qquad (12a)$$

$$\sigma_{A6} = \frac{(75 \text{ lb})(220 \text{ in.})(1.75 \text{ in.})}{3.02 \text{ in}^4} = 9561 \text{ psi} \qquad (12b)$$

Superposition of Stresses: Using the above values, and taking proper note of the physical significance of the sign of each term by referring to Fig. 1, we get the stresses shown in Fig. 3.

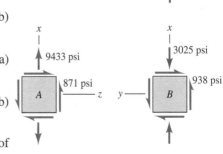

Fig. 2

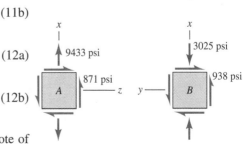

Fig. 3 The states of stress at points A and B.

503

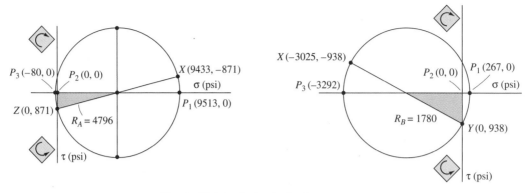

Fig. 4 Mohr's circles for in-plane stresses.

Using the stresses shown in Fig. 3, we can construct a Mohr's circle for the state of plane stress points A and B on the pipe surface. The radial normal stress is $\sigma_r = 0$ at both places. From Fig. 4a,

$$R_A = \sqrt{(9433/2)^2 + (871)^2} = 4796 \text{ psi} \tag{13}$$

$$(\sigma_1)_A = (9433/2) + 4796 = 9513 \text{ psi}$$
$$(\sigma_3)_A = 9433/2) - 4796 = -80 \text{ psi} \tag{14}$$

and, from Fig. 4b,

$$R_B = \sqrt{(-3025/2)^2 + (938)^2} = 1780 \text{ psi} \tag{15}$$

$$(\sigma_1)_B = (-3025/2) + 1780 = 267 \text{ psi}$$
$$(\sigma_3)_B = (-3025/2) - 1780 = -3292 \text{ psi} \tag{16}$$

In summary, the principal stresses at points A and B, rounded to three significant figures, are:

$$(\sigma_1)_A = 9510 \text{ psi}, \quad (\sigma_2)_A = 0, \quad (\sigma_3)_A = -80 \text{ psi} \qquad \textbf{Ans.}$$
$$(\sigma_1)_B = 267 \text{ psi}, \quad (\sigma_2)_B = 0, \quad (\sigma_3)_B = -3290 \text{ psi}$$

Review the Solution By showing all six possible internal resultants at the cross section where stresses are to be calculated, by writing down and solving all six possible equilibrium equations, and by carefully considering what stress(es) is (are) produced by each stress resultant, we have accounted for the effects of all loads on the structure. As noted earlier, we have been careful to make sure that each stress component acts in the direction that "makes sense." For example, the force P bends the pipe in the direction that produces tension at point A, and so forth.

The maximum flexural stress at the base occurs at neither A nor B. Equation 6.36 could be used to combine the flexural stresses due to M_y and M_z, and we would also have to consider the effect of shear stress. (See Homework Problem 9.4-17.)

For all pressure-vessel problems for Section 9.2, the pressure p is the gage pressure, that is, the absolute internal pressure minus the absolute external pressure. All cylinders are right circular cylinders.

Prob. 9.2-1. A steel oxygen cylinder used by a welder has an inner radius of $r_i = 4$ in. and a wall thickness of 0.5 in. The cylinder is pressurized to $p = 2000$ psi. (a) Determine the axial stress σ_a and the hoop stress σ_h in the cylindrical body of the tank. (b) Determine the tensile force per inch length of the weld between the hemispherical head and the cylindrical body of the tank.

Weld

P9.2-1

Prob. 9.2-2. A steel propane tank for a barbecue grill has a 12-in. inside diameter and a wall thickness of $\frac{1}{8}$ in. The tank is pressurized to 200 psi. (a) Determine the axial stress σ_a and the hoop stress σ_h in the cylindrical body of the tank. (b) Determine the tensile force per inch length of the weld between the upper and lower sections of the tank. (c) Determine the absolute maximum shear stress in the cylindrical portion of the tank.

Weld

Prob. 9.2-3. Solve Prob. 9.2-2 for a tank with 300-mm inside diameter and a wall thickness of 4 mm, if the tank is pressurized to 1.5 MPa.

P9.2-2 and P9.2-3

Prob. 9.2-4. A scuba diver's aluminum air tank has an outer diameter of $d_o = 7.0$ in. and a wall thickness of $t = 0.5$ in., and it is pressurized to a service pressure of $p = 3000$ psi. (a) Determine the principal stresses and the maximum in-plane shear stress in the cylindrical portion of the tank. (b) Determine the absolute maximum shear stress, τ_{abs} .
                                                            max

Prob. 9.2-5. Repeat Prob. 9.2-4 for an air tank with an outer diameter of $d_o = 180$ mm and a wall thickness of $t = 13$ mm that it is pressurized to a service pressure of $p = 20$ MPa.

P9.2-4 and P9.2-5

Prob. 9.2-6. The cylindrical portion of a compressed-air tank is fabricated of steel plate that is welded along a helix that makes an angle of $\alpha = 70°$ with respect to the longitudinal axis of the tank. The inside diameter of the cylinder is 48 in., the wall thickness is 0.5 in., and the internal pressure is 240 psi. Determine the following quantities for the cylindrical portion of the tank: (a) the axial stress σ_a and the hoop stress σ_h, (b) the normal stress and

shear stress on planes parallel and perpendicular to the weld, (c) the maximum in-plane shear stress, and (d) the absolute maximum shear stress.

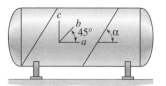

P9.2-6 through P9.2-8

Prob. 9.2-7. Solve Prob. 9.2-6 for a tank with $\alpha = 65°$, $d_i = 1$ m, $t = 20$ mm, and $p = 2$ MPa.

Prob. 9.2-8. A 45° strain gage rosette is placed on the compressed air tank with gage "a" oriented parallel to the axis of the tank. (See Fig. P9.2-8.) The tank is made of steel with Young's modulus $E = 29(10^6)$ psi and Poisson's ratio $\nu = 0.3$. If the inside diameter of the tank is 40 in., the wall thickness is $\frac{3}{8}$ in., and the internal pressure is 180 psi, what would be the readings for ϵ_a, ϵ_b, and ϵ_c?

Problems 9.2-9 and 9.2-10 treat cylindrical tanks used as vertical fluid-storage reservoirs, or standpipes. Problem 9.2-11 treats a water-filled vertical pipe. The hoop stress, σ_h, is given by Eq. 9.4, just as in the case of uniform internal pressure in a circular cylinder. According to Pascal's Law, the fluid develops hydrostatic pressure $p = \gamma h$, where γ is the specific weight of the fluid and h is the depth below the fluid surface of the point where the pressure is being calculated. For water, $\gamma = 62.4$ $lb/ft^3 = 9.81$ kN/m^3.

Prob. 9.2-9. (See previous Note.) A vertical standpipe has an inside diameter of $d_i = 3$ m and is filled with water to a depth of $h = 5$ m. If the allowable hoop stress is 80 MPa, what is the minimum wall thickness of the tank to the nearest millimeter? (Neglect the restraint that the base exerts on the cylindrical tank.)

Prob. 9.2-10. Solve Prob. 9.2-9 for a standpipe with inside diameter of $d_i = 10$ ft that is filled to a depth of $h = 40$ ft. The allowable hoop stress is 12 ksi. (Neglect the restraint that the base exerts on the cylindrical tank.)

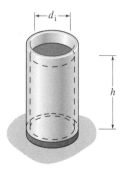

P9.2-9 and P9.2-10

505

Prob. 9.2-11. A pump at the base of a vertical steel pipe creates a pressure at the base of the pipe that is sufficient to lift the water in the pipe to a height of 24 ft, where the water is discharged at atmospheric pressure into a cooling tower. If the pipe has an inside diameter of 2 ft and a wall thickness of $\frac{1}{4}$ in., determine (a) the maximum hoop stress in the pipe, and (b) the absolute maximum shear stress in the pipe. For Part (b), assume that only the lower 18 ft of the steel pipe is supported by the base, with the upper 6 ft of pipe supported by the elbow attached to the cooling tower. The specific weight of the steel is $\gamma = 490$ lb/ft^3.

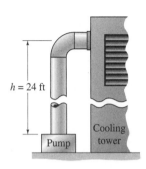

P9.2-11

Prob. 9.2-12. Hemispherical end caps are welded to a cylindrical main body to form a propane tank. The tank has an inside diameter of $d_i = 40$ in. and is to be subjected to a maximum internal pressure of $p = 120$ psi. The allowable tensile stress in the wall of the tank is 12 ksi, and the allowable tensile stress in the weld is 8 ksi. (a) Determine the minimum wall thickness t_c of the cylindrical part of the tank. (b) Determine the minimum wall thickness t_s of the hemispherical end caps. (c) What is the minimum thickness of the weld?

P9.2-12 through P9.2-14

Prob. 9.2-13. Solve Prob. 9.2-12 for a tank with an inside diameter of $d_i = 750$ mm that is subjected to a maximum internal pressure of $p = 750$ kPa. The allowable tensile stress in the wall of the tank is 80 MPa, and the allowable tensile stress in the weld is 50 MPa.

Prob. 9.2-14. A cylindrical tank with closed ends contains compressed propane gas at a gage pressure of 250 psi. The inside diameter of the tank is 5 ft, and the wall thickness is $\frac{3}{4}$ in. (a) Determine the axial stress σ_a and the hoop stress σ_h. (b) Draw a Mohr's circle for the in-plane stresses in the cylinder wall, and determine the maximum in-plane shear stress. Show a properly oriented maximum-shear-stress element. (c) Calculate the absolute maximum shear stress, $\tau_{\text{abs max}}$, in the wall of the cylindrical tank.

Prob. 9.2-15. Hydraulic pressure p acts on a piston at A, which in turn exerts a force P on the object at B. The allowable tensile stress in the wall of the hydraulic cylinder is 100 MPa. If the inside diameter of the cylinder is $d_i = 125$ mm, the wall thickness of the cylinder is $t = 6$ mm, and the diameter of the piston rod is $d_r = 20$ mm, determine the maximum force P that can be exerted by the piston rod. (Assume that the stress in the cylinder wall is the only factor that limits the value of P.)

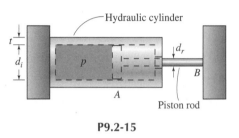

P9.2-15

Problems C9.3-1 through C9.3-3 *involve computer solutions. For Problems C9.3-1 and C9.3-2, unless your instructor indicates otherwise, you may use any software "tool" with which you are familiar. You may use a programming language such as BASIC or FORTRAN, or you may use a spreadsheet program or some other mathematical application software. Where results are to be plotted, you should choose appropriate scales for the axes and should properly label the axes. Appendix G.7 describes the use of the MechSOLID program MOHR, which is to be used in solving Problem C9.3-3.*

Prob. C9.3-1. (a) Write a computer program to evaluate the flexural stress σ_x (Eq. 9.8) and the shear stress τ_{xy} (Eq. 9.9) for the simply supported beam in Fig. 9.6. As input use the normalized x and y locations defined by $\hat{x} \equiv x/c$ and $\hat{y} \equiv y/c$ where $0 < \hat{x} < 5$ and $-1 \le \hat{y} \le 1$. Express the output as normalized stresses defined by $\hat{\sigma}_x \equiv (\sigma_x/\sigma_0)$ and $\hat{\tau} \equiv (\tau_{xy}/\sigma_0)$. (b) Using the appropriate formulas from Section 8.4, extend the computer program of Part (a) to determine the normalized principal stresses $(\hat{\sigma}_1/\sigma_0)$ and $(\hat{\sigma}_2/\sigma_0)$ and the principal directions θ_{p1} and θ_{p2}. (c) Use your computer program to solve for $\hat{\sigma}_x$ and $\hat{\tau}_{xy}$ at the following locations $(\hat{x},\hat{y})$: $A''(2,1)$, $B''(2,0.75)$, $C''(2,0.50)$, $D''(2,0.25)$, $E''(2,0)$, $F''(2,-0.25)$, $G''(2,-0.50)$, $H''(2,-0.75)$, $I''(2,-1)$. (d) Use your computer program to determine the principal directions and normalized principal stresses for the points A'' through I'' listed in Part (c). (e) For points A'' through I'', sketch principal stress elements like the elements shown in Fig. 9.8.

Prob. C9.3-2. (a) Write a computer program to determine the flexural stress, σ_x, and the shear stress, τ_{xy}, in the web of the **W**14 × 26 wide-flange beam in Example Problem 9.2 for $0 < x < 12$ ft and -6.535 in. $\le y \le 6.535$ in., that is, for the web between the supports at A and B. (b) Extend your program to include calculation of the two in-plane principal stresses, σ_1 and σ_2, for the web between the supports at A and B. (c) Determine

506

the principal stresses, σ_1 and σ_2, just to the left of the support at B and at the following distances from the neutral axis: $y_D = 6.535$ in., $y_E = 4.0$ in., $y_F = 2.0$ in., $y_G = 0.0$ in., as indicated in Fig. P9.3-2.

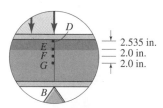

PC9.3-2

Prob. C9.3-3. Using the MOHR computer program described in Appendix G.7, plot a Mohr's circle for each of the following points in the simply-supported beam in Fig. 9.6: F, G, F', and G'. Plot nondimensional stresses $\hat\sigma \equiv (\sigma/\sigma_0)$ and $\hat\tau \equiv (\tau/\sigma_0)$. Hand in a listing of your input for each case and the resulting Mohr's circle of stress plotted by the MOHR program.

Prob. 9.4-1. The rectangular beam in Fig. P9.4-1 is subjected to a load at the centroid of end, C. The components of the load are $P_x = 10$ kips, $P_y = 10$ kips, and $P_z = 4$ kips. The dimensions of the beam are $b = 4$ in., $h = 6$ in., and $L = 100$ in. Use Mohr's circle to determine the principal stresses at point A and at point B.

Prob. 9.4-2. Repeat Prob. 9.4-1 for $L = 3$ m, $b = 100$ mm, and $h = 150$ mm for loads of $P_x = 40$ kN, $P_y = 25$ kN, and $P_z = -20$ kN.

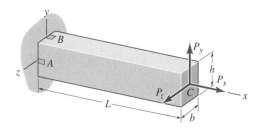

P9.4-1 and P9.4-2

Prob. 9.4-3. A section of oilfield drill pipe is being lifted by a crane, as shown in Fig. P9.4-3. The pipe has an outside diameter

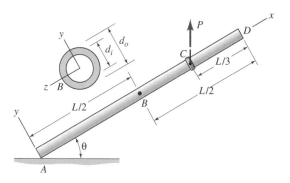

P9.4-3 and P9.4-4

of $d_o = 4.50$ in., an inside diameter of $d_i = 3.64$ in., and a total length of $L = 30$ ft. It weighs $w = 20$ lb/ft. Use Mohr's circle to determine the principal stresses at point B when the inclination of the pipe is $\theta = 30°$.

Prob. 9.4-4. Repeat Prob. 9.4-3 for $L = 10$ m, $d_o = 114$ mm, $d_i = 92$ mm, $w = 300$ N/m, and $\theta = 45°$.

Prob. 9.4-5. Show that the boundary line a of the kern of the rectangular cross section in Fig. 9.10 is given by the expression

$$\left(\frac{-c_y}{I_z}\right)e_{ya} + \left(\frac{c_z}{I_y}\right)e_{za} = \frac{1}{A}$$

Note: The formula for the boundary line d is given in Eq. 9.11.

Prob. 9.4-6. A portion of a structural steel frame consists of a $\mathbf{W}16 \times 100$ member. The stress resultants in the frame at section C are shown in Fig. P9.4-6. (a) Using Mohr's circle, determine the principal stresses at point A, a point on the web of the beam just next to one flange. Show the orientation of the principal stresses on a sketch. (b) From Mohr's circle, determine the maximum in-plane shear stress at point A. Show the orientation of the maximum shear stresses on a sketch.

See Table D.1 for the cross-sectional dimensions of the $\mathbf{W}16 \times 100$ member. Neglect the weight of the frame in calculating the stresses at point A.

Prob. 9.4-7. A chair on a ski lift is supported by a steel pipe whose outer diameter is $d_o = 60$ mm and whose inner diameter is $d_i = 52$ mm. The weight of the pipe may be neglected in comparison with the weight of the chair and its occupants, which is $W = 2$ kN. (a) Determine the stresses σ_x, σ_y, and τ_{xy} at point C, which is on the front of the pipe at the indicated cross section. The x axis is parallel to the 45° section of pipe, AB. (b) Using a Mohr's circle, determine the principal stresses and the maximum in-plane shear stress at point C. (c) Determine the maximum tensile stress in the straight section of the pipe, DE.

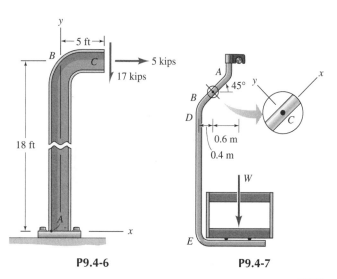

P9.4-6　　　　　　　　**P9.4-7**

Prob. 9.4-8. A wide-flange beam is subjected to axial and transverse loads as shown. Determine the principal stresses in the beam at points A and B. The dimensions of the cross section are: $h = 4$ in., $b = 3$ in., $t_f = 0.23$ in., and $t_w = 0.15$ in.

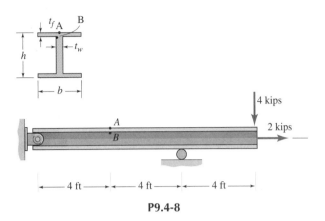

P9.4-8

Prob. 9.4-9. The boom of a crane has a rectangular box section with the dimensions shown. Determine the principal stresses in the boom at points A and B. Neglect the weight of the boom, pulley, and cable.

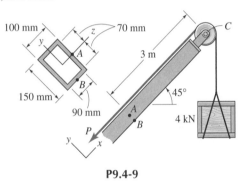

P9.4-9

Prob. 9.4-10. A vertical force of 40 lb is applied to a pipe wrench, whose handle is parallel to the z axis, as shown in Fig. P9.4-10. Using Mohr's circles, determine the principal stresses at points A and B in the cross section where the pipe threads begin. The pipe has a nominal diameter of 1 in. See Table D.7 of Appendix D for cross-sectional dimensions of the pipe.

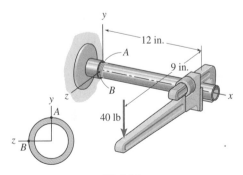

P9.4-10

Prob. 9.4-11. A force of 150 N acts at point B on an L-shaped lug wrench, as shown in Fig. P9.4-11a. The force acts vertically downward, perpendicular to the plane of the wrench. The handle of the lug wrench is a steel rod with a diameter of 12.5 mm, and its planform is shown in Fig. P9.4-11b. Determine the principal stresses and the maximum shear stress at point A, which is on the top of the wrench handle.

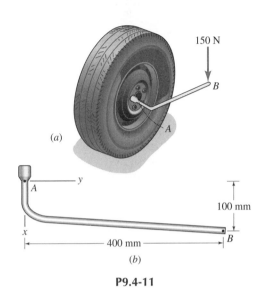

P9.4-11

Prob. 9.4-12. At a particular cross section of the drive axle of a race car, the stress resultants are as shown in Fig. P9.4-12. Use Mohr's circle to determine the principal stresses and the maximum shear stress at the following points in this cross section: (a) Point A, the top point in the cross section, and (b) Point B, the point ($y = 0$, $z = 0.5$ in.) in the cross section.

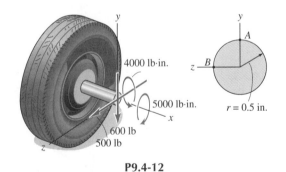

P9.4-12

Prob. 9.4-13. A post-hole digger is mounted on a tractor (not shown). The power unit of the machine applies a torque of 3000 lb · in. to the auger, and it also exerts a downward force of 200 lb on the auger. If the shaft of the auger is a solid circular rod with a diameter of 2.0 in., determine the principal stresses and the maximum shear stress at a typical point A on the surface of the shaft of the auger near the power unit.

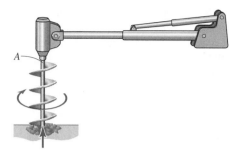

P9.4-13

Prob. 9.4-14. A solid circular shaft whose diameter is 50 mm is subjected to a torque of 500 N · m and two components of transverse shear force as shown in Fig. P9.4-14. Use a Mohr's circle to determine the principal stresses and the maximum shear stress on the surface of the shaft at point A.

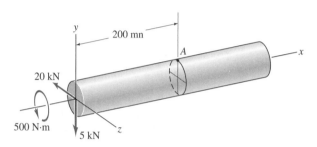

P9.4-14

Prob. 9.4-15. A solid circular shaft of radius $r = 1$ in. is subjected to a bending moment $M = 4.71$ kip · in. and a torque T as shown. At point A in the cross section, the maximum shear stress is $\tau_{max} = 6$ ksi. Using a Mohr's circle, determine the value of the torque T.

Prob. 9.4-16. A solid circular shaft of radius $r = 25$ mm is subjected to a torque $T = 98.2$ N · m and a bending moment M as shown. At point A in the cross section, the maximum compressive stress is $\sigma_2 = -8$ MPa. Using a Mohr's circle, determine the value of the bending moment M.

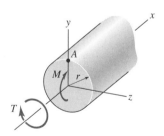

P9.4-15 and P9.4-16

*__Prob. 9.4-17.__ At the base of the cantilevered sign in Fig. 1 of Example 9.5, the stress resultants were found to be the following:

$$F = -285 \text{ lb}, \ V_y = -75 \text{ lb}, \ V_z = 0,$$

$$T = 3000 \text{ lb} \cdot \text{in.}, \ M_y = -5000 \text{ lb} \cdot \text{in.}, \ M_z = 16{,}500 \text{ lb} \cdot \text{in.}$$

(a) Determine the angle θ that the resultant moment vector M in Fig. P9.4-17 makes with the z axis, and locate the neutral axis of the base cross section. (b) Referring to Section 6.6, determine the maximum normal stress, σ_x, on the cross section at the base of the signpost in Example 9.5. (c) Neglecting the shear stress on the base cross section due to the transverse shear force V_y, determine the maximum normal stress at the base of the signpost. Indicate the location of the point in the base cross section where this maximum normal stress occurs.

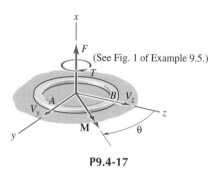

(See Fig. 1 of Example 9.5.)

P9.4-17

Problems **C9.4-1 and C9.4-2** *involve computer solutions. Unless your instructor indicates otherwise, for Problem C9.4-1 you may use any software "tool" with which you are familiar. You may use a programming language such as BASIC or FORTRAN, or you may use a spreadsheet program or some other mathematical application software. Where results are to be plotted, you should choose appropriate scales for the axes and should properly label the axes. Appendix G.7 describes the use of the **MechSOLID** program MOHR, which is to be used in solving Problem C9.4-2.*

Prob. C9.4-1. (a) Determine an expression for the normal stress σ_x at the base of the sign post in Fig. 1 of Example Problem 9.5 as a function of the angle θ measured from the y axis, as shown in Fig. PC9.4-1. That is, determine an expression for $\sigma_x(\theta)$ at the cross section $x = 0$. Note: You will need to use multi-axis bending theory to determine the bending component of σ_x. (See Sect. 6.6) (b) Using the computer, plot the expression for $\sigma_x(\theta)$ you derived in Part (a). At what values of θ does σ_x attain its maximum and minimum values on this cross section?

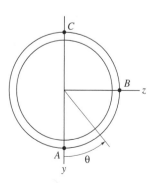

PC9.4-1 and PC9.4-2

Prob. C9.4-2. Figure PC9.4-2 shows the cross section at $x = 0$ for sign in Fig. 1 of Example Prob. 9.5. (a) Using the MOHR computer program described in Appendix G.7, verify the two Mohr's circles shown in Fig. 4 of Example Problem 9.5. Hand in a listing of your input for each case, together with the Mohr's circle of stress plotted by the MOHR software. (b) Using a table, like Table 1 in Example Problem 9.5, determine the normal stress σ_C and shear stress τ_C at point C, which is on the opposite side of the column from point A in Fig. 1 of Example Problem 9.5. Use the MOHR software, as in Part (a), to draw a Mohr's circle of stress for point C.

Computer-Aided Design Exercise—Section 9.4. *Develop a computer program using, for example, a mathematical programming language or a spreadsheet program, to perform a design trade-off study to determine suitable design parameters for the design described in Prob. C9.4-3.*

Prob. C9.4-3. A strength trainer has decided that there is money to be made in fabricating low-cost exercise machines, such as the *leg-press exercise machine* shown in Fig. PC9.4-3. The leg-press exercise consists of pushing on a footpedal that exerts a horizontal resisting force. The footpedal is attached to a rod AE that slides through a frictionless guide hole in the vertical post of the frame. Welded to the post at D is a vertical pipe column CD. Attached by a pin to the footpedal at A is a rod AB of length L, which is pinned at B to a block that can slide up and down on the pipe column CD. Assume that the coefficient of friction (static or sliding) between the slider block B and the pipe column is $\mu = 0.1$. As the pedal is moved horizontally, rod AB makes an angle θ with the horizontal. A spring surrounds the pipe column, and exerts a force against the stop at C and the slider block at B; the force varies linearly with the compression of the spring. The minimum value of θ is θ_0, at which position the spring is fully extended (i.e., the spring force is zero) and the footpedal is at its greatest distance from the post. Let x be the distance the pedal is moved, that is, $x = L(\cos\theta_0 - \cos\theta)$.

The *design objective* is to develop a low-cost leg-press machine that exerts a resisting pedal force within the range 150 lb to 300 lb for 2 in. $\leq x \leq$ 15 in.

The *design parameters* for this leg-press machine are: (1) L = the length of rod AB, (2) k = the spring constant of the resistance spring, (3) θ_0 = the minimum angle between rod AB and the horizontal (the angle at which the spring force is zero), and (4) the size of the (steel) pipe column.

The *design constraints* are: (1) The maximum pedal force must not exceed 300 lb. (2) The normal stress σ on the pipe column cross section at D must not exceed 16 ksi. (3) The angle θ is limited by a minimum value of 15° and by a maximum value of 70°. (4) To help minimize the cost of the leg-press machine, the pipe column must be a standard-weight steel pipe selected from Table D7; it must be the smallest-diameter pipe that satisfies the stress constraint (constraint 2).

Assignment: (a) For the following design-parameter cases, develop graphs of (1) pedal force versus x, (2) maximum normal stress at D versus x, and (3) angle θ versus x, for $0 \leq x \leq$ 15 in.:

Case A: $k = 30$ lb/in., $L = 24$ in., $\theta_0 = 20°$, $1\frac{1}{2}$-in. steel pipe.

Case B: $k = 26$ lb/in., $L = 24$ in., $\theta_0 = 15°$, $1\frac{1}{2}$-in. steel pipe.

Case C: $k = 18$ lb/in., $L = 30$ in., $\theta_0 = 20°$, $1\frac{1}{2}$-in. steel pipe.

Case D: $k = 15$ lb/in., $L = 30$ in., $\theta_0 = 15°$, $1\frac{1}{4}$-in. steel pipe.

Note: The (compressive) spring force is given by

$$F_s = kL(\sin\theta - \sin\theta_0)$$

In addition to the plots, hand in complete derivations (including free-body diagrams, displacement diagrams, etc.) of the equations that you have plotted. (b) Select the design that you consider to be the "best" of the above four designs. Write a paragraph discussing why you consider this design to be the best one. If any of the four designs fail to satisfy all of the constraints, discuss which designs fail and how they fail. In what respects do you think this spring-type leg-press machine is better (or worse) than a machine that involves pulleys, cables, and weights?

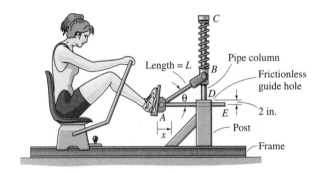

PC9.4-3

BUCKLING OF COLUMNS
10

10.1 INTRODUCTION

What constitutes *failure* of a structure or a machine part? As an engineer you must consider several possible *modes of failure* when designing a structure or a machine component. For example, the stress must be kept small enough so that the component will not fail by yielding or by tensile fracture. It might also be important to limit the *deflection* of the component. Further, if the part will be subjected to repeated cycles of loading, the stress must be limited in order to prevent failure by progressive fracture (called *fatigue failure*). To prevent the above types of failures, design criteria based on *strength* (stress) and *stiffness* (deflection) must be taken into consideration. Therefore, the preceding nine chapters were devoted to methods for calculating the stress distribution in, and the deflection of, members subjected to various types of loading. In this chapter we consider another important mode of failure, *buckling*.[1] Figure 10.1 shows the impressive shape into which a thin cylindrical shell deforms when subjected to its axial buckling load. Beams, plates, shells and other structural members may buckle under a variety of loading conditions.[2] The discussion in this chapter, however, is limited to compressive axial loading of slender members.

You can easily perform a buckling "experiment" by applying an axial load to a thin ruler or yard stick (meter stick). Figure 10.2a shows buckling of a thin rod (column), and Fig. 10.2b shows buckling of a truss member under compression. Weights are added until the *critical load* in the compression member, P_{cr}, is reached, and the member suddenly deflects laterally. In the analysis of axial deformation in Chapters 2, 3, and 9, we implicitly assumed that, even under compressive loading, the member undergoing axial deformation remained straight, and that the only deformation was a shortening or lengthening of the member. However, as the ruler demonstration shows, at some value of compressive axial load the ruler no longer remains straight, but suddenly deflects laterally, bending like a beam. Buckling failures are often sudden and catastrophic, which makes it all the more important for you to know how they can be prevented.

---

[1]The term *buckling* will be defined later in this section; it is illustrated in Figs. 10.1 and 10.2.

[2]*Theory of Elastic Stability,* by S. P. Timoshenko and J. M. Gere, [Ref. 10-1]; and *Buckling of Bars, Plates, and Shells,* by D. O. Brush and B. O. Almroth, [Ref. 10-2], treat the buckling of several types of members under various types of loading. *Structural Stability—Theory and Implementation,* by W. F. Chen and E. M. Lui, [Ref. 10-3], presents both theory and design curves.

FIGURE 10.1 Buckling failure of an axially compressed thin cylindrical shell. (Photo by W. H. Horton; from *Computerized Buckling Analysis of Shells*, by D. Bushnell, 1985. Reprinted by permission of Kluwer Academic Publishers.)

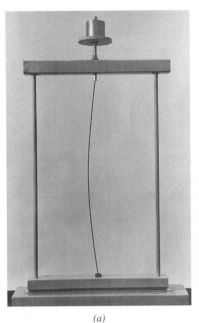

(a)

(b)

FIGURE 10.2 Buckling demonstrations.

Stability of Equilibrium. In this section we will examine the basic phenomenon of buckling; then we will examine the buckling behavior of slender columns. In Chapter 1 we noted that static deformable-body-mechanics problems always involve equilibrium; now, however, we must look more closely at the topic of equilibrium and consider the *stability of equilibrium.* This concept can be demonstrated very easily by considering the equilibrium of a ball on three different surfaces, as illustrated in Fig. 10.3. In all three situations the solid-colored ball is in an equilibrium position, that is, it satisfies $\Sigma F_x = 0$, $\Sigma F_y = 0$, and $\Sigma M = 0$. In Fig. 10.3a the ball is said to be in *stable equilibrium* because, if it is slightly displaced to one side and then released, it will move back toward the equilibrium position at the bottom of the "valley." The ball on the top of the "hill" in Fig. 10.3c is also in an equilibrium configuration, because it also satisfies $\Sigma F_x = 0$, $\Sigma F_y = 0$, and $\Sigma M = 0$. In this case, however, the ball is in an *unstable-equilibrium* configuration; if slightly displaced to either side, the ball will tend to move farther from the equilibrium position at the top of the "hill." Finally, if the ball is on a perfectly flat, level surface, as in Fig. 10.3b, it is said to be in a *neutral-equilibrium* configuration. If slightly displaced to either side, it has no tendency to move either away from or toward the original position, since it is in equilibrium in the displaced position as well as the original position.

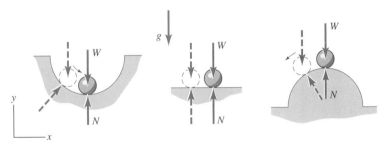

FIGURE 10.3 Stability of equilibrium.

(a) Stable equilibrium.   *(b)* Neutral equilibrium.   *(c)* Unstable equilibrium.

512

Buckling. Now, let us see how stability of equilibrium applies to compression members where some form of elastic deformation is possible. In Fig. 10.4 the member AB is assumed to be perfectly straight and perfectly rigid, and the supporting pin at A is assumed to be frictionless. A force P is applied vertically downward at B, which is on the axis of the member. The torsional spring at A has a spring constant k_θ, so it produces a *restoring moment, M_{Ar},* at A that is directly proportional to the angle of deflection of member AB from the vertical. That is,

$$M_{Ar} = k_\theta \theta \tag{10.1}$$

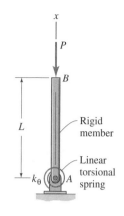

(a) The vertical equilibrium configuration.

The vertical configuration of member AB, shown in Fig. 10.4a, is certainly an *equilibrium configuration.* The question is: For what values of P is the vertical equilibrium configuration stable? neutral? unstable? To answer this question, we explore what happens when the member AB is rotated to one side through a small angle θ, as shown in Fig. 10.4b. Let M_{Ad} be the *disturbing moment,* that is, the sum of all moments that tend to make the angle θ increase; and let M_{Ar} be the *restoring moment,* that is the sum of all moments that tend to make θ decrease. The restoring moment, M_{Ar}, in Fig. 10.4c is given by Eq. 10.1; the disturbing moment is given by

$$M_{Ad} = PL \sin \theta \tag{10.2}$$

In analogy with the balls in Fig. 10.3, the system has the following requirements for stable equilibrium, neutral equilibrium, and unstable equilibrium, respectively:

Stable Equilibrium: $\qquad\qquad M_{Ad} < M_{Ar} \rightarrow PL \sin \theta < k_\theta \theta$

Neutral Equilibrium: $\qquad\qquad M_{Ad} = M_{Ar} \rightarrow PL \sin \theta = k_\theta \theta \qquad (10.3)$

Unstable Equilibrium: $\qquad\qquad M_{Ad} > M_{Ar} \rightarrow PL \sin \theta > k_\theta \theta$

Since we are interested in behavior of the system at, and very near, the vertical configuration, we let $\theta \rightarrow 0$ in Eqs. 10.3 and use the approximation $\sin \theta \doteq \theta$. Then, Eqs. 10.3 give

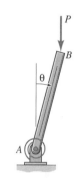

(b) A displaced configuration.

Stable Equilibrium: $\qquad\qquad\qquad P < P_{cr}$

Neutral Equilibrium: $\qquad\qquad\qquad P = P_{cr} \qquad (10.4)$

Unstable Equilibrium: $\qquad\qquad\qquad P > P_{cr}$

where

$$P_{cr} = \frac{k_\theta}{L} \tag{10.5}$$

The value of the load at which the transition from stable equilibrium to unstable equilibrium occurs is called the *critical load, P_{cr}.* This loss of stability of equilibrium is called *buckling,* so we also call P_{cr} the *buckling load.*

Consider now how the buckling load of the compression member AB in Fig. 10.4 could be obtained by performing an experiment. We could grasp member AB (with compressive force P applied), rotate it to the left (or to the right) through a small angle, and then release it. If $P < P_{cr}$, member AB would return to the vertical ($\theta = 0$) equilibrium position. If, on the other hand, $P > P_{cr}$, the column would tend to move farther from the

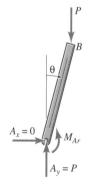

(c) A free-body diagram.

FIGURE 10.4 A simplified model of column buckling.

513

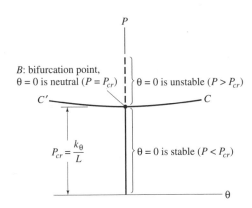

FIGURE 10.5 An equilibrium diagram for an idealized compression member.

$\theta = 0$ position once it is slightly disturbed. The critical load would be the value of P for which there was no visible tendency for the member either to return to the $\theta = 0$ position or to move farther from it.

A useful way to illustrate the relationship between applied load and stability is the *equilibrium diagram* shown in Fig. 10.5. It is a plot of load P versus deflection angle θ. The point labeled B in Fig. 10.5, where the equilibrium diagram branches, is called the *bifurcation point*.[3] Above the bifurcation point the vertical (i.e., $\theta = 0$) configuration (shown dashed) is an unstable-equilibrium configuration; but there are alternative stable-equilibrium configurations along curves BC and BC', with $\theta \neq 0$. Right at point B, where $P = P_{cr}$, the equilibrium is neutral.

In the next section we will determine the critical load for a straight, uniform, axially loaded, linearly elastic, pin-ended column.

10.2 THE IDEAL PIN-ENDED COLUMN

To investigate the stabilty of real columns with distributed flexibility, as contrasted with the rigid member with torsional spring that was used in the previous section to model column behavior, we begin by considering the *ideal pin-ended column,* as illustrated in Fig. 10.6a.[4] We make the following simplifying assumptions:

- The column is initially perfectly straight, and it is made of linearly elastic material.
- The column is free to rotate, at its ends, about frictionless pins; that is, it is restrained like a simply supported beam. Each pin passes through the centroid of the cross section.
- The column is symmetric about the xy plane, and any lateral deflection of the column takes place in the xy plane.
- The column is loaded by an axial compressive force P applied by the pins.

---

[3]The word *bifurcate* means to divide into two branches, or to fork.

[4]It is customary to refer to the pin-ended compression member being analyzed here as a column, and to illustrate it by a vertical member subjected to a downward load. This reflects the very important application to columns in buildings and other structures. However, the analysis applies equally to any pin-ended member, such as a member in a truss (Fig. 10.2b), so long as the member is in compression.

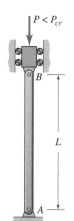

(a) Ideal column.

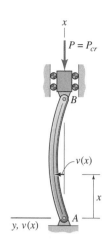

(b) Buckled configuration.

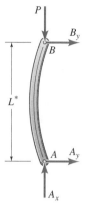

(c) FBD of entire column.

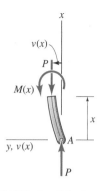

(d) FBD of partial column.

FIGURE 10.6 Buckling of a pin-ended column.

Buckled Configuration: If the axial load P is less than the critical load, P_{cr}, the column will remain straight, and it will shorten under a uniform (compressive) axial stress $\sigma = P/A$, as illustrated in Fig. 10.6a.[5] In the straight configuration with $P < P_{cr}$, the column is in *stable equilibrium*. However, if a load $P = P_{cr}$ is applied to the column, the straight configuration becomes a *neutral-equilibrium* configuration, and neighboring configurations, like the buckled shape in Fig. 10.6b, also satisfy equilibrium requirements. Therefore, to determine the value of the critical load, P_{cr}, and the shape of the buckled column, we will determine the value of the load P such that the (slightly) bent shape of the column in Fig. 10.6b is an equilibrium configuration.[6]

Equilibrium of the Buckled Column: First, using the free-body diagram of Fig. 10.6c, we get $A_x = P$ (from $\Sigma F_x = 0$), $A_y = 0$ [from $(\Sigma M)_B = 0$], and $B_y = 0$. Therefore, on the free-body diagram in Fig. 10.6d, we show only a vertical force P acting on the pin at A, and we show no horizontal force $V(x)$ at section x. The sign convention adopted for the moment $M(x)$ in Fig. 10.6d is the sign convention that was used in Chapter 6, namely, that a positive bending moment produces compression in the $+y$ fibers (which are on the left in Fig. 10.6d as a result of our choice of xy axes). From Fig. 10.6d we get

$$\left(\sum M\right)_A = 0: \qquad\qquad M(x) = -Pv(x) \qquad\qquad (10.6)$$

Differential Equation of Equilibrium, and End Conditions: Substituting $M(x)$ into the moment-curvature equation, Eq. 7.8, we obtain

$$EIv''(x) = M(x) = -Pv(x)$$

---

[5]Since column buckling always occurs under compressive normal stress, in the remainder of this chapter we will let σ denote <u>compressive</u> normal stress, digressing from the normal-stress sign convention that has been followed in previous chapters.

[6]This approach is called the *equilibrium method* or *bifurcation method*. There are several alternative ways to determine the critical load. See, for example, *Principles of Structural Stability,* by H. Ziegler, [Ref. 10-4], where four different methods are reviewed.

or

$$EIv''(x) + Pv(x) = 0 \tag{10.7}$$

This is the *differential equation* that governs the deflected shape of a pin-ended column. It is a homogeneous, linear, second-order, ordinary differential equation. The *boundary conditions* (see Table 7.2*a*) for the pin-ended member are

$$v(0) = 0, \qquad v(L) = 0 \tag{10.8}$$

Solution of the Differential Equation: The presence of the $v(x)$ term in Eq. 10.7 means that we cannot simply integrate twice to get the solution, as was done in Chapter 7. In fact, only when $EI =$ const is there a simple solution to Eq. 10.7. Therefore, for the remainder of this chapter we will consider only uniform columns.[7] Then Eq. 10.7 is an ordinary differential equation with constant coefficients. Let

$$\lambda^2 = \frac{P}{EI} \tag{10.9}$$

For the uniform column, therefore, Eq. 10.7 becomes

$$v'' + \lambda^2 v = 0 \tag{10.10}$$

The general solution to this homogeneous equation is

$$v(x) = C_1 \sin \lambda x + C_2 \cos \lambda x \tag{10.11}$$

We seek a value of λ and constants of integration C_1 and C_2 such that the two boundary conditions of Eqs. 10.8 are satisfied. Thus,

$$v(0) = 0 \rightarrow C_2 = 0 \tag{10.12}$$
$$v(L) = 0 \rightarrow C_1 \sin (\lambda L) = 0$$

Obviously, if we make both C_1 and C_2 equal to zero, the deflection $v(x)$ is zero everywhere (Eq. 10.11), and we just have the original straight configuration. If we want an alternative equilibrium configuration, like the one in Fig. 10.6*b*, we must pick a value of λ that satisfies Eq. 10.12b with $C_1 \neq 0$, that is

$$\sin(\lambda_n L) = 0 \rightarrow \lambda_n = \left(\frac{n\pi}{L}\right), \qquad n = 1, 2, \dots \tag{10.13}$$

Combining Eqs. 10.9 and 10.13 gives the following formula for the possible buckling loads:

$$P_n = \frac{n^2 \pi^2 EI}{L^2} \tag{10.14}$$

---

[7]Tapered columns are treated in Chapter 9 of *Guide to Stability Design Criteria for Metal Structures,* ed. by T. V. Galambos, [Ref. 10-5].

The deflection curve that corresponds to each load P_n is obtained by combining Eqs. 10.11 through 10.13 to get

$$v(x) = C \sin\left(\frac{n\pi x}{L}\right) \tag{10.15}$$

The function that represents the shape of the deflected column is called a *mode shape,* or *buckling mode.* The constant C, which determines the direction (sign) and amplitude of the deflection, is arbitrary, but it must be small. The existence of neighboring equilibrium configurations is analogous to the fact that the ball in Fig. 10.3b can be placed at neighboring locations on the flat, horizontal surface and still be in equilibrium.

The value of P at which buckling will actually occur is obviously the smallest value given by Eq. 10.14 (i.e., $n = 1$). Thus, the *critical load* is

$$\boxed{P_{\text{cr}} = \frac{\pi^2 EI}{L^2}} \tag{10.16}$$

and the corresponding *fundamental buckling mode* is

$$\boxed{v(x) = C \sin\left(\frac{\pi x}{L}\right)} \tag{10.17}$$

as illustrated in Fig. 10.7b. The critical load for an ideal column is known as the *Euler buckling load,* after the famous Swiss mathematician Leonhard Euler (1707–1783), who was first to establish a theory of buckling of columns.[8]

Although the column could theoretically buckle in the *second buckling mode,* illustrated in Fig. 10.7c, if a load $P_{\text{cr}_2} = 4\pi^2 EI/L^2$ were to be applied, this could only happen if there were some lateral bracing at $x = L/2$ to prevent the column from buckling in the first mode (Fig. 10.7b) at the much smaller Euler load of $P_{\text{cr}} = \pi^2 EI/L^2$.

Let us examine some important implications of the Euler buckling-load formula, Eq. 10.16. We can express this in terms of *critical* (buckling) *stress*

$$\sigma_{\text{cr}} \equiv \frac{P_{\text{cr}}}{A} = \frac{\pi^2 E(Ar^2)}{AL^2}$$

or

$$\boxed{\sigma_{\text{cr}} = \frac{\pi^2 E}{(L/r)^2}} \tag{10.18}$$

where

σ_{cr} = the critical (elastic buckling) stress.

E = the modulus of elasticity.

$r = \sqrt{I/A}$ = the radius of gyration.

L = the length of the member between supports.

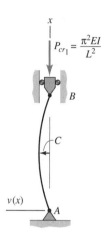

(a) Undeflected column.

(b) First buckling mode ($n = 1$).

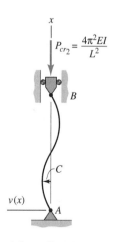

(c) Second buckling mode ($n = 2$).

FIGURE 10.7 Two examples of buckling modes.

[8]See Footnote 2 of Chapter 6.

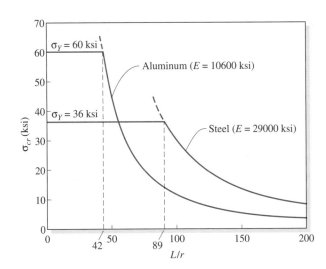

FIGURE 10.8 Graphs of
Euler's formula for structural
steel and an aluminum alloy.

The quantity L/r is called the *slenderness ratio* of the column. Curves of σ_{cr} versus L/r for structural steel and for an aluminum alloy are plotted in Fig. 10.8.

From Eq. 10.16 and from Fig. 10.18 we observe the following characteristics of *elastic buckling* of ideal columns:

(a)

(b)

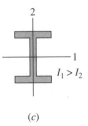

(c)

FIGURE 10.9 Column
cross sections.

- The only material property that enters directly into either the elastic buckling load or the buckling stress is the modulus of elasticity, E, which represents the stiffness of the material. Therefore, one way to increase the elastic buckling load of a member would be to use a member that is made of material with a higher E value.

- The elastic buckling load is inversely proportional to the square of the length of the column. Figure 10.8 illustrates this length effect.

- Euler's formula is valid only for ''long'' columns, that is, columns whose L/r ratio leads to a critical stress below the compressive proportional limit, σ_{PL}. (Since σ_{PL} is generally not available, the compressive yield stress σ_Y is usually substituted for it.) The values of L/r marking the limit of validity of Euler's formula for steel and for an aluminum alloy are illustrated in Fig. 10.8.

- The buckling load can be increased by increasing the value of the cross-sectional moment of inertia, I. This can be done, without increasing the cross-sectional area, by using thin-wall tubular members (e.g., Figs. 10.9a, b). However, if the column wall is too thin, *local buckling* can occur. (Figure 10.1 illustrates buckling of a short, thin-wall member under compression.)

- If the principal moments of inertia of the column cross section are unequal, as illustrated in Fig. 10.9c, the column will buckle about the axis of the cross section that has the least moment of inertia, unless boundary restraints or intermediate bracing force it to do otherwise (see Section 10.3).

- If the slenderness ratio is very large, say $L/r > 200$, the stress at buckling will be very small. Therefore, the strength of the material is under-utilized. The design should be modified by, for example, adding lateral bracing or changing the boundary conditions (see Section 10.3).

In the sections that follow, we will explore the following exceptions to the buckling behavior treated in this section: other boundary conditions (Section 10.3), eccentric loading (Section 10.4), imperfections (Section 10.5), and inelastic buckling (Section 10.6).

What is the maximum compressive load that can be applied to an aluminum-alloy compression member (Fig. 1) of length $L = 4$ m if the member is loaded in a manner that permits free rotation at its ends and if a factor of safety of 1.5 against buckling failure is to be applied? Recall, from Eq. 2.36, that

$E = 70$ GPa
$\sigma_Y = 270$ MPa
$r_o = 45$ mm
$r_i = 40$ mm

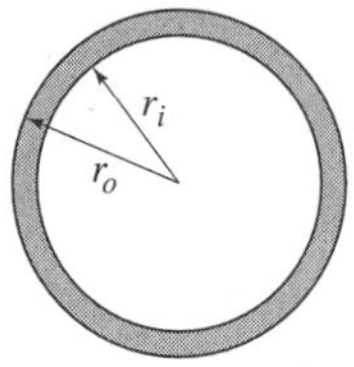

$$FS = \frac{\text{failure load}}{\text{allowable load}}$$

Fig. 1 Cross section.

Plan the Solution The allowable load is the critical load divided by the factor of safety. Since the ends of the member are free to rotate (pin-ended), the critical load is determined by Euler's formula, Eq. 10.16, provided that the corresponding stress is less than the yield stress of the material.

Solution From Eq. 10.16,

$$P_{cr} = \frac{\pi^2 EI}{L^2} \qquad (1)$$

$$= \frac{\pi^2 (70 \times 10^9 \text{ N/m}^2)(\frac{\pi}{4})[(0.045)^4 - (0.040)^4] \text{ m}^4}{(4 \text{ m})^2}$$

$$P_{cr} = 52.2 \text{ kN} \qquad \textbf{Ans.}$$

The corresponding average compressive stress in the member is

$$\sigma_{cr} = \frac{P_{cr}}{A} = \frac{52.2 \text{ kN}}{\pi[(0.045)^2 - (0.040)^2] \text{ m}^2}$$

$$\sigma_{cr} = 39.1 \text{ MPa} < \sigma_Y$$

so this member will undergo elastic buckling. (For this column, $L/r = 133$.) Finally, the allowable load is

$$P_{allow} = \frac{P_{cr}}{FS} = \frac{52.2 \text{ kN}}{1.5} = 34.8 \text{ kN} \qquad \textbf{Ans.}$$

10.3 THE EFFECT OF END CONDITIONS ON COLUMN BUCKLING

■■■■■■■■■■■■

Rarely, if ever, is a compression load actually applied to a member through frictionless pins. For example, the column might be bolted to a heavy base at the bottom and framed into other members at the top, as illustrated in Fig. 10.10a. However, an understanding of the effect of *idealized support conditions,* like those illustrated in Figs. 10.10b through 10.10d, enables an engineer to estimate the effect that actual end conditions, like those in Fig. 10.10a, would have on the buckling load of a real column.

We will begin by deriving an expression for the elastic buckling load of the fixed-pinned column in Fig. 10.10d. Then, we will indicate how the concept of the *effective length* of a column can be used to obtain the buckling load of columns with various end conditions.

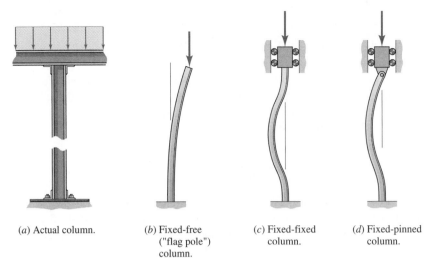

(*a*) Actual column.   (*b*) Fixed-free   (*c*) Fixed-fixed   (*d*) Fixed-pinned
                              ("flag pole")   column.   column.
                               column.

FIGURE 10.10   Various
column end conditions.

Buckling Load of an Ideal Fixed-Pinned Column.
We will assume the same ideal conditions that were listed at the beginning of Section 10.2, except that the pin at end A of the column is replaced by complete restraint of that end.

Buckled Configuration: To assist us in drawing free-body diagrams of the column in a buckled equilibrium configuration near the straight equilibrium configuration (Fig. 10.11*a*), we sketch a feasible shape of the column that satisfies the end conditions (Fig. 10.11*b*).

Equilibrium of the Buckled Column: By examining Fig. 10.11*b*, we can see that the curvature at A corresponds to a moment M_A in the sense shown in Fig. 10.11*c*. From $\Sigma F_x = 0$, we get $A_x = P$, and from $(\Sigma M)_A = 0$, we see that the pin at B must exert a horizontal force H_B, as indicated on Fig. 10.11*c*. Now we can construct a free-body diagram of the column below section x or above section x. We have done the latter in Fig. 10.11*d*, from which we get

$$\left(\sum M\right)_O = 0: \qquad\qquad M(x) = (L - x)H_B - Pv(x) \qquad\qquad (10.19)$$

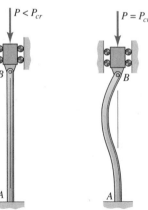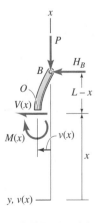

FIGURE 10.11   A fixed-
pinned column.

(*a*) Unbuckled   (*b*) Buckled shape   (*c*) FBD of entire   (*d*) FBD of partial
     column.   of column.   column.   column.

Differential Equation and End Conditions: Substituting $M(x)$ from Eq. 10.19 into the moment-curvature equation, Eq. 7.8, we get

$$EIv''(x) + Pv(x) = H_B L - H_B x \qquad (10.20)$$

As in Section 10.2, let us consider only uniform columns, and let us employ the definition of λ in Eq. 10.9. Then, Eq. 10.20 can be written in the form

$$v'' + \lambda^2 v = \frac{H_B L}{EI} - \frac{H_B x}{EI} \qquad (10.21)$$

Instead of the homogeneous differential equation that we got for the pinned-pinned column (Eq. 10.10), we have obtained a nonhomogeneous, linear, second-order, ordinary differential equation with constant coefficients.

The boundary conditions of the fixed-pinned column are

$$v(0) = 0, \quad v'(0) = 0, \quad v(L) = 0 \qquad (10.22)$$

Solution of the Differential Equation: The solution of Eq. 10.21 under the end conditions of Eq. 10.22 consists of a *complementary solution* plus a *particular solution.* To get the complementary solution, we set the right-hand side of Eq. 10.21 to zero. But this just gives us Eq. 10.10, so we can use Eq. 10.11 as the complementary solution. Since the right-hand side of Eq. 10.21 consists of a constant term and a term that is linear in x, let us try the following particular solution:

$$v_p(x) = C_3 + C_4 x \qquad (10.23)$$

Substituting this into Eq. 10.21, noting that $v_p''(x) = 0$, and recalling that $\lambda^2 = P/EI$, we get

$$\frac{P}{EI}(C_3 + C_4 x) = \frac{H_B L}{EI} - \frac{H_B x}{EI} \qquad (10.24)$$

Therefore, the particular solution is

$$v_p(x) = \frac{H_B L}{P} - \frac{H_B x}{P} \qquad (10.25)$$

Finally, the complete *general solution* of Eq. 10.21 is

$$v(x) = \frac{H_B L}{P} - \frac{H_B x}{P} + C_1 \sin \lambda x + C_2 \cos \lambda x \qquad (10.26)$$

We have three end conditions, Eqs. 10.22, to be used to determine the four constants—λ, H_B, C_1, and C_2.

$$v(0) = 0 \rightarrow C_2 = -\frac{H_B L}{P}$$

$$v'(0) = 0 \rightarrow C_1 = \frac{H_B}{\lambda P} \qquad (10.27)$$

$$v(L) = 0 \rightarrow C_1 \sin \lambda L + C_2 \cos \lambda L = 0$$

Equations 10.27 can be combined to give

$$C_1[\sin \lambda L - \lambda L \cos \lambda L] = 0 \tag{10.28}$$

This replaces the much simpler condition, Eq. 10.12b, that we got for the pinned-pinned column. Again, there are two solutions, but the solution $C_1 = 0$ leads to $H_B = C_2 = 0$, so we get the "trivial" solution of the straight equilibrium configuration, $v(x) \equiv 0$.

Alternate equilibrium configurations are possible, however, if λ satisfies the following equation:

$$\sin(\lambda_n L) - (\lambda_n L) \cos(\lambda_n L) = 0$$

or

$$\tan(\lambda_n L) = \lambda_n L, \qquad n = 1, 2, \ldots \tag{10.29}$$

This equation is called the *characteristic equation*. It has an infinite number of solutions, but, as in the case of the pinned-pinned column, we are only interested in the smallest value of λL that satisfies Eq. 10.29. One way to solve Eq. 10.29 is to plot $f(\lambda L) \equiv \tan(\lambda L)$ versus λL, and $g(\lambda L) \equiv \lambda L$ versus λL. The smallest value of λL where the curves $f(\lambda L)$ and $g(\lambda L)$ intersect is

$$\lambda_1 L = 4.4934 \tag{10.30}$$

Combining this with Eq. 10.9 we get

$$P_{\text{cr}} = (20.19)\frac{EI}{L^2} \tag{10.31}$$

Thus, replacing the pin at one end of a column by complete restraint raises the buckling load significantly—from the Euler load of Eq. 10.16 to the value given in Eq. 10.31. That is, the buckling load is increased by

$$\left(\frac{20.19 - \pi^2}{\pi^2}\right)(100\%) = 105\%$$

By comparing Eqs. 10.16 and 10.31, we can see that the elastic buckling load of any column can be expressed as some constant times the factor (EI/L^2). Thus, all of the comments regarding the effects of the parameters E, I, and L on the buckling of pinned-pinned columns also hold true for columns with other end conditions.

To get the fundamental buckling-mode shape of the fixed-pinned column, we can combine Eqs. 10.26, 10.27, and 10.30 to get

$$v(x) = C\left\{\sin\left(\frac{4.493x}{L}\right) + 4.493\left[1 - \left(\frac{x}{L}\right) - \cos\left(\frac{4.493x}{L}\right)\right]\right\} \tag{10.32}$$

This mode-shape expression, which is plotted in Fig. 10.12, is not nearly as simple as the $\sin(\pi x/L)$ buckling mode of the pinned-pinned column (Eq. 10.17). However, we again find the *shape* is specified, but the amplitude coefficient is arbitrary.

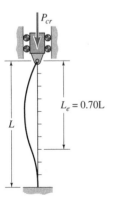

FIGURE 10.12 The fundamental buckling mode of a fixed-pinned column.

Effective Length of Columns. The Euler buckling load, given by Eq. 10.16, was developed for the ideal pinned-pinned column. We have just found, however, that a change of end conditions leads to an expression for the buckling load that differs from that in the Euler formula only in the value of a multiplicative constant. Therefore, the Euler formula can be extended to apply to the elastic buckling of columns with arbitrary end conditions if we write it as

$$P_{cr} = \frac{\pi^2 EI}{L_e^2} \qquad (10.33)$$

where L_e is the *effective length* of the column. That is, L_e is the length of a pin-ended column having the same buckling load as the actual column. For example, for the fixed-pinned column, we equate the expression for P_{cr} in Eq. 10.31, to the expression in Eq. 10.33 and get

$$\frac{\pi^2}{L_e^2} = \frac{20.19}{L^2}$$

or

$$L_e = 0.70L$$

This effective length of the fixed-pinned column is indicated in Fig. 10.12.

Physically, the effective length of a column is the distance between points of zero moment when the column is deflected in its fundamental elastic buckling mode. Figure 10.13 illustrates the effective lengths of columns with several types of end conditions. (Homework Problem 10.3-13 is a derivation based on Fig. 10.13e; Problem 10.14 is a similar derivation.) Some design codes employ a dimensionless coefficient K, called the *effective-length factor,* where

$$L_e \equiv KL \qquad (10.34)$$

Then the critical load is given by

$$P_{cr} = \frac{\pi^2 EI}{(KL)^2} \qquad (10.35)$$

FIGURE 10.13 The effective length of various columns.

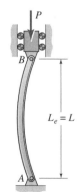

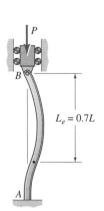

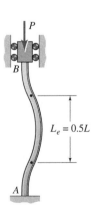

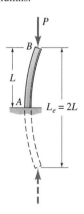

(a) Pinned-pinned column, $K = 1$.

(b) Fixed-pinned column, $K = 0.7$.

(c) Fixed-fixed column, $K = 0.5$.

(d) Partially-restrained column. $0.5 < K < 1$.

(e) Fixed-free column, $K = 2$.

The appropriate value of K is listed for each column shown in Fig. 10.13. From Eq. 10.35 we obtain the following expression for the critical stress:

$$\sigma_{cr} = \frac{\pi^2 E}{\left(\dfrac{KL}{r}\right)^2} \qquad (10.36)$$

where (KL/r) is the *effective slenderness ratio* of the compression member.

■■■■■■■■■■■■■□□□ EXAMPLE 10.2 □□□□□□□□□□□□■■■■

A stiff beam BC (assumed to be rigid) is supported by two identical columns whose flexural rigidity is EI (for bending in the xz plane). Assuming that the columns are prevented from rotating at either end by this arrangement and that sidesway is permitted, as illustrated in Fig. 1b, *estimate* the elastic buckling load, P_{cr}, by estimating the effective length of the columns and using Eq. 10.35.

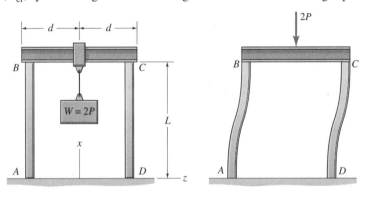

(a) Unbuckled configuration. (b) Buckled configuration.

Fig. 1

Plan the Solution   The problem statement asks us to estimate the effective length and then use Eq. 10.35. We can compare the columns in this problem to the effective-length column samples in Fig. 10.13.

Solution   In terms of effective length, the elastic buckling load is given by Eq. 10.33:

$$P_{cr} = \frac{\pi^2 EI}{L_e^2} \qquad (1)$$

Figure 1b bears some similarity to Fig. 10.13c, the fixed-fixed column. If we reflect the columns about their tops, as shown in Fig. 2, we have the equivalent of the fixed-fixed column situation of Fig. 10.13c. Therefore,

$$L_e = 0.5(2L) = L \qquad (2)$$

so, the critical load for each column is just the Euler buckling load,

$$P_{cr} = \frac{\pi^2 EI}{L^2} \qquad \text{Ans.} \quad (3)$$

and the effective-length factor is $K = 1$.

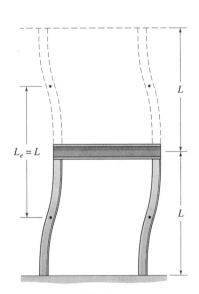

Fig. 2

EXAMPLE 10.3

In Example 10.2, buckling of columns AB and CD in the xz plane was considered. However, there is nothing to prevent the columns from buckling in the y direction. For the frame in Fig. 1a, determine whether the **W**6×20 columns AB and CD will buckle in the xz plane (y axis buckling), or whether they will buckle in the y direction (z axis buckling), and determine the buckling load. Assume that the joints at B and C are rigidly welded joints, that the beam BC is rigid, and that it applies a vertical load P at the centroid of the top of each column.

Let $E = 29(10^3)$ksi, $\sigma_Y = 36$ ksi, $I_y = 13.3$ in^4, $I_z = 41.4$ in^4, $A = 5.87$ in^2, and $L = 16$ ft.

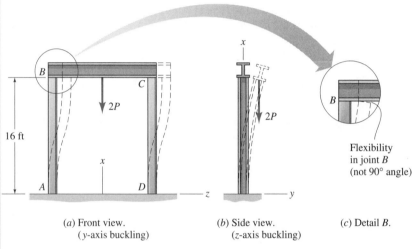

(a) Front view.
(y-axis buckling)

(b) Side view.
(z-axis buckling)

(c) Detail B.

Flexibility in joint B (not 90° angle)

Fig. 1 Two possible buckling modes.

Solution From Eq. 10.35,

$$P_{cr} = \frac{\pi^2 EI}{(KL)^2} \qquad (1)$$

Let P_{cr_y} be the critical load for buckling in the xz plane as illustrated in Fig. (1a), and let P_{cr_z} be the critical load for the buckling mode illustrated in Fig. (1b). Then,

$$P_{cr_y} = \frac{\pi^2 EI_y}{(K_y L)^2}, \qquad P_{cr_z} = \frac{\pi^2 EI_z}{(K_z L)^2} \qquad (2)$$

where, from Example 10.2, $K_y = 1$, and from Fig. 10.13e, $K_z = 2$. Then,

$$P_{cr_y} = \frac{\pi^2 [29(10^3) \text{ ksi}](13.3 \text{ in}^4)}{[1(16 \text{ ft})(12 \text{ in./ft})]^2} = 103.3 \text{ kips} \qquad (3a)$$

$$P_{cr_z} = \frac{\pi^2 [29(10^3) \text{ ksi}](41.4 \text{ in}^4)}{[2(16 \text{ ft})(12 \text{ in./ft})]^2} = 80.4 \text{ kips} \qquad (3b)$$

Since $P_{cr_z} < P_{cr_y}$, the frame will buckle in the out-of-plane mode indicated in Fig. (1b) at a buckling load of

$$P_{cr} = (P_{cr_z})_{cant} = 80.4 \text{ kips} \qquad \text{Ans.} \quad (4)$$

Review the Solution In the problem statement you were told to assume that beam BC is rigid and that the joints at B and C are rigidly welded (i.e., that the angles between the columns and the beam BC remain $90°$). Before accepting the value of P_{cr} in Eq. (4) as the true buckling load, let us reconsider these assumptions. Since a real beam BC would not be perfectly rigid, and since the joints at B and C may actually be flexible enough to allow the respective columns to rotate some at B and C relative to the beam BC, as indicated in Fig. (1c), we should take these possibilities into account. A ''worst-case'' assumption would be that the columns are pinned, not fixed, to the beam BC. Since the beam BC is free to translate horizontally, the value of K_y for a cantilever (fixed-free) column would apply; that is, $K_y = 2$. Then, Eq. (3a) would become

$$(P_{cr_y})_{cant} = \frac{\pi^2 [29(10^3) \text{ ksi}](13.3 \text{ in}^4)}{[2(16 \text{ ft})(12 \text{ in./ft})]^2} = 25.8 \text{ kips} \qquad (5)$$

By comparing Eqs. (4) and (5) we see the importance of correctly characterizing the end conditions of a column and applying an appropriate factor of safety to account for uncertainty in the end conditions. The most conservative value of buckling load for columns AB and CD is given by Eq. (5), that is

$$P_{cr} = (P_{cr_y})_{cant} = 25.8 \text{ kips}$$

The average compressive stress at this load is

$$\sigma_{cr} = \frac{P_{cr}}{A} = \frac{25.8 \text{ kips}}{5.87 \text{ in}^2} = 4.40 \text{ ksi} < 36 \text{ ksi}$$

so the assumption of elastic (Euler) buckling is valid.

■■■■■■■■■■■ ***10.4 ECCENTRIC LOADING; THE SECANT FORMULA**

In Sections 10.2 and 10.3 we considered ideal columns, that is, columns that are initially perfectly straight and whose compressive load is applied through the centroid of the cross section of the member. Such ideal conditions never exist in reality, since perfectly straight structural members cannot be fabricated, and since the point of application of the load seldom, if ever, lies exactly at the centroid of the cross section.

Beam-Column Behavior.[9] Figure 10.14a shows a column with eccentric load applied through a bracket. We will analyze the ''pinned-pinned'' column with eccentric loading shown in Fig. 10.14b.[10] When the eccentricity, e, is zero, we get the Euler column.

---

[9] A straight member that is simultaneously subjected to axial compression and lateral bending is called a *beam-column*.

[10] The pin supports at A and B are omitted to simplify this figure.

When $e \neq 0$, we use the free-body diagram in Fig. 10.14c to get

$$\left(\sum M\right)_A = 0: \qquad\qquad M(x) = -P[e + v(x)] \qquad\qquad (10.37)$$

which is substituted into the moment-curvature equation, Eq. (7.8), to give

$$EIv''(x) + Pv(x) = -Pe \qquad\qquad (10.38)$$

As in Eq. 10.9, we consider a uniform member and let $\lambda^2 = P/EI$. Then, Eq. 10.38 may be written as

$$v'' + \lambda^2 v = -\lambda^2 e \qquad\qquad (10.39)$$

The particular solution of this differential equation is $v_p(x) = -e = \text{const}$, so the general solution is

$$v(x) = C_1 \sin \lambda x + C_2 \cos \lambda x - e \qquad\qquad (10.40)$$

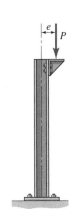

(a) Cantilever column.

The boundary conditions $v(0) = v(L) = 0$ are used to evaluate the constants in Eq. 10.40.

$$v(0) = 0 \rightarrow C_2 = e$$
$$v(L) = 0 \rightarrow C_1 \sin(\lambda L) + e(\cos(\lambda L) - 1) = 0$$

or, since $1 - \cos \phi = 2 \sin^2(\phi/2)$ and $\sin \phi = 2 \sin(\phi/2)\cos(\phi/2)$,

$$C_1 = e \tan\left(\frac{\lambda L}{2}\right)$$

Then,

$$\boxed{v(x) = e\left[\tan\left(\frac{\lambda L}{2}\right) \sin(\lambda x) + \cos(\lambda x) - 1 \right]} \qquad (10.41)$$

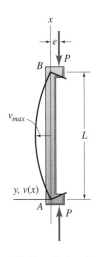

(b) Pinned-pinned column.

As indicated in Fig. 10.14b, the maximum deflection of the beam-column occurs at $x = L/2$. Its value is

$$\boxed{v_{max} = v(L/2) = e\left[\sec(\lambda L/2) - 1 \right]} \qquad\qquad (10.42)$$

Unlike an Euler column, which deflects laterally only if P equals or exceeds the Euler buckling load, P_{cr}, lateral deflection of an eccentrically loaded member occurs for any value of load P. It is convenient to illustrate this beam-column deflection by plotting Eq. 10.42, written as

$$\boxed{v_{max} = e\left[\sec\left(\frac{\pi}{2}\sqrt{\frac{P}{P_{cr}}}\right) - 1 \right]} \qquad\qquad (10.43)$$

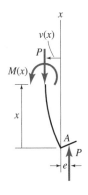

(c) Free-body diagram.

FIGURE 10.14
Eccentrically loaded columns.

527

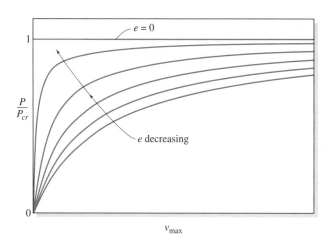

FIGURE 10.15 Load-deflection diagram for an eccentrically loaded column.

for several values of e (Fig. 10.15). As P approaches the Euler load, P_{cr}, the lateral deflection of the beam-column increases without bound. In the limit as $e \to 0$, the curve becomes two straight lines that represent the straight configuration ($P < P_{cr}$) and the buckled configuration ($P = P_{cr}$).

The above beam-column analysis is valid only as long as the (compressive) stress does not exceed the compressive proportional limit. It is important to note that the transverse deflection v is a *nonlinear function* of the load P, even though it depends linearly on the eccentricity e. Thus,

$$v_{max}(P_1 + P_2) \neq v_{max}(P_1) + v_{max}(P_2)$$

as can be seen by examining Eq. 10.43 or Fig. 10.15. This nonlinear load-deflection relationship results from the fact that the bending moment $M(x)$ of the beam-column depends on the deflection $v(x)$, as given by Eq. 10.37. By contrast, the beam deflection produced by lateral loads or by applied couples varies linearly with the value of each load (Chapter 7), since there is no v-term in the bending-moment expression for those types of loading.

Secant Formula. A member under beam-column action is subjected to a combination of compressive axial force P and bending moment $M(x)$, as indicated by the free-body diagram of Fig. 10.14c. The maximum (magnitude) moment occurs at $x = L/2$ and is obtained by combining Eqs. 10.37 and 10.42 in the manner indicated in Fig. 10.16. Thus,

$$M_{max} = |M(L/2)| = Pe \sec\left(\frac{\lambda L}{2}\right) \tag{10.44}$$

FIGURE 10.16
Superposition of beam-column stresses.

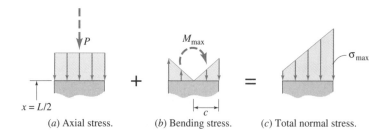

(*a*) Axial stress. (*b*) Bending stress. (*c*) Total normal stress.

The maximum compressive stress is therefore

$$\sigma_{max} = \frac{P}{A} + \frac{M_{max}c}{I} \tag{10.45}$$

Combining Eqs. 10.44 and 10.45, and recalling the definition of λ in Eq. 10.9, we get

$$\sigma_{max} = \frac{P}{A}\left[1 + \left(\frac{ec}{r^2}\right)\sec\left(\frac{L_e}{2r}\sqrt{\frac{P}{AE}}\right)\right] \tag{10.46}$$

where

σ_{max} = the maximum compressive beam-column stress.

P = the (eccentric) axial compressive load.

A = the cross-sectional area of the compression member.

e = the eccentricity of the load.

c = the distance from the centroid to the extreme fiber where σ_{max} acts (Fig. 10.16b).

E = the modulus of elasticity.

I = the area moment of inertia about the centroidal bending axis.

$r = \sqrt{I/A}$ = the radius of gyration.

$L_e = KL$ = the effective length of the member.

Equation 10.46 is called the *secant formula*. For example, to determine the maximum compressive load that can be applied at a given eccentricity to a column of given length and given material, without causing yielding of the material, we can set $\sigma_{max} = \sigma_Y$, the yield point in compression, and solve Eq. 10.46 numerically for P/A, the average stress.[11] Figure 10.17 is a plot of the average stress P/A versus the *slenderness ratio* L_e/r for structural steel for several values of the *eccentricity ratio* ec/r^2. For all curves on Fig. 10.17 except the curve labeled *Euler's formula*, $\sigma_{max} = \sigma_Y$, and therefore $P = P_Y$, the

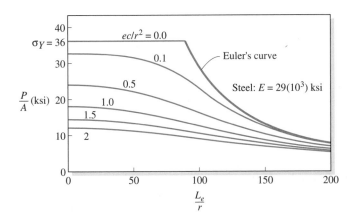

FIGURE 10.17 The average stress, P/A, corresponding to $\sigma_{max} = \sigma_Y$, based on the secant formula (Eq. 10.46).

[11]We assume that the material remains linearly elastic up to the stress σ_Y, that is, we assume that $\sigma_{PL} = \sigma_Y$.

load at which yielding first occurs.

If $e = 0$ and $\sigma_{max} = \sigma_Y$, Eq. 10.46 simply gives $\dfrac{P}{A} = \sigma_Y$, and this plots as a horizontal line at stress σ_Y. However, long columns with zero eccentricity buckle elastically at the Euler critical load, and their maximum stress is less than σ_Y. Thus, the Euler load is an upper bound on the value of P for long columns, and therefore the Euler curve is shown in Fig. 10.17.[12]

Clearly, the secant formula expresses a nonlinear relationship between the applied compressive load P and the maximum compressive stress σ_{max}. Therefore, the principal of superposition does not apply to the calculation of the stress in beam-columns.[13]

---

[12]The behavior of short and medium-length ideal columns is examined further in Section 10.6, *Inelastic Buckling*.

[13]The secant term in Eq. 10.46 approaches unity as $L_e/r \to 0$. Therefore, for very short members, superposition can be applied, as in Example 9.3.

PROCEDURE FOR DETERMINING THE ALLOWABLE LOAD FOR AN ECCENTRICALLY LOADED COLUMN

The *allowable load* for a given eccentrically loaded column may be determined by the following procedure:

1. Obtain, or estimate, the value of the eccentricity e.
2. Substitute the value of e into the secant formula, along with the geometric parameters r, c, A, and L, and the material properties E and σ_Y (i.e., let $\sigma_{max} = \sigma_Y$), and solve for the load P_Y.

3. Divide the load P_Y by the appropriate factor of safety to determine the allowable load. (Note that, because of the nonlinear relationship between load and stress, the factor of safety must be applied directly to the load. It is not applied to the stress to determine an allowable stress from which an allowable load is then calculated!)

The following example problem illustrates the use of the secant formula.

◼◼◼◼◼◼◼◼◼◼◼◼◻◻◻◻ **EXAMPLE 10.4** ◻◻◻◻◻◻◻◻◻◻◻◻◼◼◼

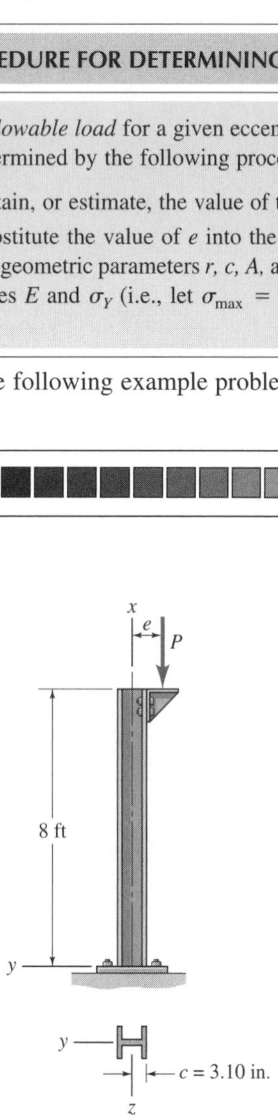

Fig. 1 Eccentrically loaded column.

A **W**6×20 structural steel ($E = 29(10^3)$ ksi, $\sigma_Y = 36$ ksi) column is eccentrically loaded as shown in Fig. 1. Assume that the load is applied directly at the top cross section even though it is applied through a bracket that is attached to a flange. Also assume that the column is supported in a manner that prevents out-of-plane buckling, that is, buckling in the z direction.

$$A = 5.87 \text{ in}^2, \quad r = 2.66 \text{ in.}, \quad c = 3.10 \text{ in.}$$

(a) If a compressive load $P = 20$ kips is applied at an eccentricity $e = 4.0$ in., what is the maximum compressive stress in the column? Where does it occur? (b) What is the factor of safety against initial yielding of the column under the above loading?

Solution

(a) Maximum Compressive Stress: The maximum compressive stress (see Fig. 2) can be calculated directly by using the secant formula, Eq. 10.46.

$$\sigma_{max} = \frac{P}{A}\left[1 + \left(\frac{ec}{r^2}\right) \sec\left(\frac{KL}{2r}\sqrt{\frac{P}{AE}}\right)\right] \quad (1)$$

The column is fixed at its base and free to translate sideways at the top. Therefore, it is a fixed-free, or cantilever, column, and from Fig. 10.13e we get $K = 2$. Thus,

$$\sigma_{max} = \frac{20}{5.87}\left[1 + \left(\frac{4.0(3.10)}{(2.66)^2}\right)\sec\left(\frac{2(96)}{2(2.66)}\sqrt{\frac{20}{5.87(29,000)}}\right)\right]$$

$$= 9.866 \text{ ksi}$$

$$\sigma_{max} = 9.87 \text{ ksi} \qquad\qquad \text{Ans.}$$

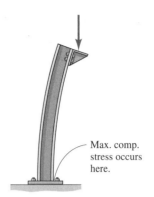

Max. comp. stress occurs here.

Fig. 2

(b) Factor of Safety: Since the secant formula is nonlinear, we must determine the value of load P_Y that make $\sigma_{max} = \sigma_Y$, the compressive yield stress. We may be able to get an estimate from Fig. 10.17. For this column

$$\frac{KL}{r} = \frac{2(96 \text{ in.})}{(2.66 \text{ in.})} = 72.2$$

$$\frac{ec}{r^2} = \frac{(4.0 \text{ in.})(3.10 \text{ in.})}{(2.66 \text{ in.})^2} = 1.753$$

By interpolating between the curves for $\left(\frac{ec}{r^2}\right) = 1.0$ and $\left(\frac{ec}{r^2}\right) = 2.0$ at $\frac{L_e}{r} \doteq 75$ on Fig. 10.17, we get $\frac{P_Y}{A} \doteq 12$ ksi. Therefore, since $A = 5.87 \text{ in}^2$, we can expect a solution of $P_Y \doteq 70$ kips.

We need to solve for the value of P_Y that satisfies Eq. (1) with $\sigma_{max} = \sigma_Y = 36$ ksi, that is, the equation

$$36 \text{ ksi} = \frac{P_Y}{(5.87 \text{ in}^2)}\left[1 + 1.753 \sec\left(36.1\sqrt{\frac{P_Y}{(5.87 \text{ in}^2)(29,000 \text{ ksi})}}\right)\right]$$

or

$$211 = P_Y\left[1 + 1.753 \sec\left(0.0875\sqrt{P_Y}\right)\right] \qquad\qquad (2)$$

Therefore, the load that will cause first yielding is

$$P_Y = 64.2 \text{ kips}$$

and, since the actual load is 20 kips, that factor of safety with respect to yielding is

$$n_Y = \frac{P_Y}{P} = \frac{64.2 \text{ kips}}{20 \text{ kips}} = 3.21$$

$$n_Y = 3.2 \qquad\qquad \text{Ans.}$$

Note that the factor of safety is based on loads (Eq. 2.36), and is not given by σ_Y/σ_{max}.

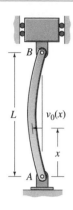

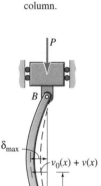

(a) Intially crooked column.

The previous section indicated how the behavior of a straight column is affected if the load is applied eccentrically rather than at the centroid of the cross section of the member. The behavior of a column is also affected by any initial lack of straightness of the axis of the column, or *initial imperfection* of the column. The ideal pin-ended column of Fig. 10.6a is replaced, in Fig. 10.18a, by the column that has an initial crookedness $v_0(x)$.

Although $v_0(x)$ is usually small, its exact functional form differs from column to column and is unknown. However, we can get an idea of the effect of initial crookedness by assuming that $v_0(x)$ can be represented by

$$v_0(x) = \delta_0 \sin\left(\frac{\pi x}{L}\right) \tag{10.47}$$

which has the same shape as the fundamental buckling mode of an ideal column (Eq. 10.17). From the free-body diagram of Fig. 10.18c we get

$$M(x) = -P[v(x) + v_0(x)] \tag{10.48}$$

where $v(x)$ is the deflection caused by the load P. Combining Eqs. 10.47 and 10.48 with the moment-curvature equation, Eq. 7.8, gives the differential equation

$$EIv''(x) + Pv(x) = -P\delta_0 \sin\left(\frac{\pi x}{L}\right) \tag{10.49}$$

The solution to Eq. 10.49 with the boundary conditions $v(0) = v(L) = 0$ is

$$v(x) = \left(\frac{\alpha\delta_0}{1 - \alpha}\right) \sin\left(\frac{\pi x}{L}\right) \tag{10.50}$$

where

$$\alpha \equiv \frac{P}{P_{cr}} = \frac{PL^2}{\pi^2 EI} \tag{10.51}$$

From Eq. 10.50 we can determine the maximum deflection, δ_{max}, and the maximum bending moment, M_{max}, as follows:

$$\delta_{max} = \delta_0 + v(L/2) = \frac{\delta_0}{1 - \alpha} \tag{10.52}$$

$$M_{max} = P\delta_{max} = \frac{P\delta_0}{1 - \alpha} \tag{10.53}$$

Then,

$$\sigma_{max} = \frac{P}{A} + \frac{M_{max}c}{I} = \frac{P}{A}\left[1 + \frac{\delta_0 c}{r^2(1 - \alpha)}\right] \tag{10.54}$$

(b) Crooked column under compressive loading.

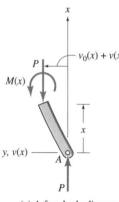

(c) A free-body diagram.

FIGURE 10.18 A column with initial crookedness $v_0(x)$.

where $r^2 = I/A$.

532

Since $\alpha = P/P_{cr}$, Eqs. 10.52 through 10.54 are all nonlinear in the load P. Equation 10.54 for the maximum compressive stress is similar to the secant formula for eccentric loading of columns, Eq. 10.46. If the *imperfection ratio* $\delta_0 c/r^2$ is used to determine a family of curves of P_Y/A versus L_e/r for a given compressive yield stress $\sigma_{max} = \sigma_Y$, the result is a set of curves very similar to those in Fig. 10.17. Therefore, the analysis of columns with initial crookedness is very similar to the analysis of columns with eccentric loading.

*10.6 INELASTIC BUCKLING OF IDEAL COLUMNS

Let us reconsider the buckling of ideal (i.e., initially straight, axially loaded) pin-ended columns. If an ideal column is sufficiently long, it will buckle elastically at the critical stress given by Euler's formula, Eq. 10.18, with $\sigma_{cr} \leq \sigma_{PL}$. Therefore, we can determine the value of the slenderness ratio, L_e/r, above which elastic buckling occurs, by setting $\sigma_{cr} = \sigma_{PL}$ in Eq. 10.18. This gives

$$\frac{L_{ec}}{r} = \sqrt{\frac{\pi^2 E}{\sigma_{PL}}} \qquad (10.55)$$

Columns for which $L_e > L_{ec}$ are called *long columns*. Long columns fail at their *elastic buckling limit*, namely, the Euler buckling stress.

If an ideal column is very short, it will not buckle at all, but it will simply fail by crushing of the material at the ultimate compressive stress σ_U. Columns that fail in this manner are called *short columns*, and they fail at their *compressive-strength limit*, as indicated on Fig. 10.19. Between short columns and long columns lies a range of columns, called *intermediate columns*, whose failure mode is referred to as *inelastic buckling*. They fail when the average stress P/A reaches the *inelastic-buckling limit*, indicated by the curve BC in Fig. 10.19.

Let us now examine the phenomenon of inelastic buckling to determine the inelastic buckling limit of an ideal column whose compressive stress-strain diagram has the form

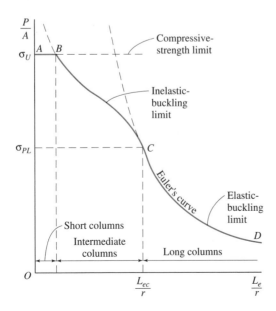

FIGURE 10.19 Buckling behavior of an ideal column.

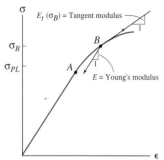

FIGURE 10.20
Compressive stress-strain
diagram.

shown in Fig. 10.20.[14] Assume that the column fails in the range of intermediate columns, so buckling does not occur at a stress below the proportional limit σ_{PL}. And assume that when its average compressive stress P/A reaches the value σ_B (point B on Fig. 10.20), the column is just on the verge of inelastic buckling. There are two principal theories that predict the value of σ_B at which a column will buckle inelastically. The simplest theory is the *tangent-modulus theory,* developed in 1889 by F. Engesser, a German engineer. The tangent-modulus formula is obtained by replacing Young's modulus, E, in Euler's buckling formula by the *tangent modulus* E_t, which is the slope of the compressive stress-strain curve. At stress σ_B the tangent modulus is given by

$$E_t(\sigma_B) = \left.\frac{d\sigma}{d\epsilon}\right|_{\sigma = \sigma_B} \tag{10.56}$$

The meaning of E_t is illustrated on Fig. 10.20. Thus, the *tangent modulus buckling stress* is given by

$$\sigma_t = \frac{\pi^2 E_t}{(L_e/r)^2} \tag{10.57}$$

The tangent-modulus curve for inelastic buckling is labeled σ_t in Fig. 10.21.

A slightly higher inelastic buckling stress σ_r is predicted by the *reduced-modulus theory.* In the reduced-modulus formula

$$\sigma_r = \frac{\pi^2 E_r}{(L_e/r)^2} \tag{10.58}$$

the *reduced-modulus* E_r ($E_t < E_r < E$) depends on E, E_t, and also the shape of the cross section. Since actual buckling tests of intermediate-length columns result in buckling loads that tend to be close to the tangent-modulus load, and also since the tangent-modulus load is far easier to calculate than the reduced-modulus load, the tangent-modulus theory is the generally preferred theory for inelastic buckling of columns [Ref. 10-6].

The following example illustrates the use of the tangent-modulus theory to determine the inelastic buckling stress.

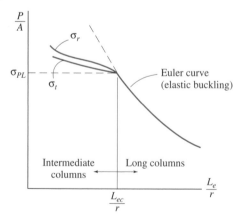

FIGURE 10.21 Critical
stress versus slenderness ratio.

---

[14]Because actual columns are not perfectly straight and are not loaded exactly along the centroidal axis, and because stresses (called *residual stresses*) are induced into columns when they are manufactured, it is not possible to determine an analytical expression for the inelastic buckling limit of real columns. However, use of the analytical expressions presented here, together with the results of many column buckling experiments, makes it possible to obtain empirical column-design formulas, several of which are discussed in Section 10.7.

A $4 \times 4 \times 1/2$ tubular column, whose cross section is shown in Fig. 1, has an effective length of 100 in. ($A = 6.36$ in^2, $r = 1.39$ in.). The column is made of a material whose compressive stress-strain curve is given by the curve ABC in Fig. 2. The corresponding curve of tangent modulus E_t versus compressive stress σ is the curve $DEFG$ in Fig. 2, with values of the tangent modulus on the upper scale. Compute the buckling load for this column.

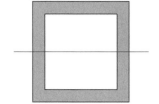

Fig. 1

Solution Let us first determine whether elastic or inelastic buckling governs, by using Eq. 10.55 to determine L_{ec}/r. From the tangent-modulus curve of Fig. 2 we can estimate that $E = 30(10^3)$ ksi, and $\sigma_{PL} \approx 29$ ksi. Then,

$$\frac{L_{ec}}{r} = \sqrt{\frac{\pi^2 E}{\sigma_{PL}}} = \sqrt{\frac{\pi^2 (30 \times 10^3 \text{ ksi})}{30 \text{ ksi}}} = 101.0$$

The actual value of L_e/r is

$$\frac{L_e}{r} = \frac{100 \text{ in.}}{1.39 \text{ in.}} = 71.9$$

Since $L_e/r < L_{ec}/r$, inelastic buckling occurs.

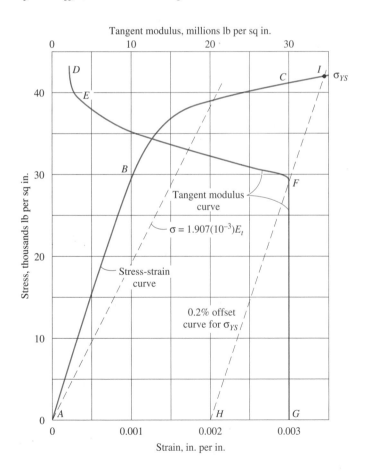

Fig. 2 The compressive stress-strain curve and the tangent modulus curve for a structural material.

From the tangent-modulus formula, Eq. 10.57, we get

$$\sigma_t = \frac{\pi^2 E_t}{(L_e/r)^2} = \frac{\pi^2 E_t}{(71.9)^2} = 0.001907\, E_t \qquad (1)$$

This straight line is plotted in Fig. 2. Its intersection with the tangent-modulus curve gives the solution for the tangent-modulus stress. First, try $\sigma_1 = 33$ ksi. From Fig. 2, we estimate that the corresponding value of the tangent modulus is $E_1 \equiv E_t(\sigma_1) \approx 17(10^3)$ ksi. From Eq. (1),

$$\sigma_{t1} = 0.001907\,(17(10^3)\ \text{ksi}) = 32.4\ \text{ksi}$$

which is lower than our first-guess value of 33 ksi. We need a slightly higher value of E_t, so try $E_t(\sigma_2) \approx 17.2(10^3)$ ksi. Then Eq. (1) gives

$$\sigma_{t1} = 32.8\ \text{ksi}$$

which agrees with the value taken from the tangent-modulus curve. Then,

$$P_t = \sigma_t A = (32.8\ \text{ksi})(6.36\ \text{in}^2) = 209\ \text{kips}$$

or, rounded to two significant figures,

$$P_t = 210\ \text{kips} \qquad \textbf{Ans.}$$

10.7 DESIGN OF CENTRALLY LOADED COLUMNS

In Sections 10.2 through 10.6 we examined the behavior of columns of known geometry (perfectly straight or with a specified form of imperfection), known material behavior (free of any residual stress and having a known compressive stress-strain diagram), known end conditions (pinned, fixed, or free), and known line of action of load. For real columns, all of these factors are subjected to variability (e.g., see Ref. 10-5), and this variability must be properly accounted for when a column is designed. Therefore, design codes typically specify the use of *column design formulas* that are obtained by curve-fitting data from laboratory tests of many real columns and that incorporate appropriate factors of safety. In this section we will consider only design formulas that apply to columns that are nominally straight and whose load is applied along the axis that passes through the centroid of the cross section at each end.

In Section 10.3 it was shown that the *effective slenderness ratio, KL/r*, provides a convenient way to relate the behavior of columns with various end conditions to the behavior of pin-ended columns. In Section 10.6 it was indicated that the behavior of columns can be classified according to KL/r range as short columns, intermediate-length columns, and long columns. For long columns the *Euler formula* for elastic buckling, Eq. 10.36, is used as the basis for determining the critical buckling stress σ_{cr}. That is, for long columns

$$\sigma_{\text{cr}} = \frac{\pi^2 E}{\left(\dfrac{KL}{r}\right)^2} \qquad (10.59)$$

Extensive series of laboratory tests have been performed on real columns in the short and intermediate ranges of KL/r, and empirical design formulas that fit these test data have been developed. A factor of safety, which may depend on KL/r, is specified in order to convert the critical stress σ_{cr} to allowable stress σ_{allow}. That is,

$$\sigma_{allow} = \frac{\sigma_{cr}}{FS} \tag{10.60}$$

In this section we will consider several representative column-design formulas for centrally loaded columns of steel, aluminum alloy, and wood. Each design formula applies only to columns made of a specific material and having an effective slenderness ratio in a specific range. The purpose of this section is just to illustrate the basic aspects of the column-design process. The reader should consult the appropriate design code or specification before designing a column for actual fabrication and use.

Structural Steel Columns. The design of columns of structural steel is based on formulas proposed by the Structural Stability Research Council (SSRC), formerly called the Column Research Council (CRC). The American Institute of Steel Construction (AISC), an organization that publishes specifications for use in the design of steel structures, has adopted the SSRC formulas and prescribed the safety factors to be used in the design of steel columns [Ref. 10-7].

For long columns, the Euler formula, Eq. 10.59, is used. This formula is theoretically valid as long as σ_{cr} does not exceed the proportional limit of the material in compression. (The yield stress σ_Y may be substituted for σ_{PL}.) However, it has been observed that the process of rolling that is used to produce structural-steel shapes such as wide-flange sections produces large residual compressive stresses so that yielding may occur at a compressive stress as low as half the nominal compressive yield stress. Therefore, the AISC limits the range of validity of the Euler formula to those values of KL/r for which $\sigma_{cr} \leq 0.5\sigma_Y$, that is, for values of $KL/r \geq (KL/r)_c$, with

$$\left(\frac{KL}{r}\right)_c = \left(\frac{\pi^2 E}{0.5\sigma_Y}\right)^{\frac{1}{2}} \tag{10.61}$$

The value of E given by AISC is $29(10^3)$ ksi.[15]

Combining Eqs. 10.59 and 10.61, we can write Euler's formula in the form

$$\frac{\sigma_{cr}}{\sigma_Y} = \frac{\pi^2 E}{\sigma_Y \left(\frac{KL}{r}\right)^2} = \frac{\left(\frac{KL}{r}\right)_c^2}{2\left(\frac{KL}{r}\right)^2}, \qquad \left(\frac{KL}{r}\right) \geq \left(\frac{KL}{r}\right)_c \tag{10.62}$$

This equation is plotted in Fig. 10.22, where its range of validity is indicated by the solid-line portion of the curve labeled *Euler's formula.*

For shorter columns, that is, for values of KL/r that do not exceed $(KL/r)_c$, the Column Reseach Council proposed use of the parabolic curve[16]

$$\frac{\sigma_{cr}}{\sigma_Y} = 1 - \frac{\left(\frac{KL}{r}\right)^2}{2\left(\frac{KL}{r}\right)_c^2}, \qquad \left(\frac{KL}{r}\right) \leq \left(\frac{KL}{r}\right)_c \tag{10.63}$$

---

[15]This value is for room-temperature conditions (70°F); the assumed value of E decreases linearly to a value of $25(10^3)$ ksi at 900°F.

[16]This is sometimes called the *Johnson column formula.*

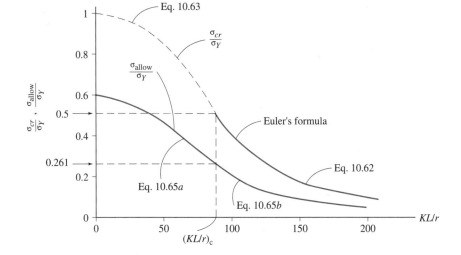

FIGURE 10.22 Design curves for buckling of structural steel columns.

This curve has a horizontal tangent at $KL/r = 0$, and it merges smoothly with the Euler curve at $(KL/r)_c$, as shown in Fig. 10.22.

The factor of safety corresponding to Eq. 10.63 is given by the AISC as

$$FS_1 = \frac{5}{3} + \frac{3\left(\dfrac{KL}{r}\right)}{8\left(\dfrac{KL}{r}\right)_c} - \frac{\left(\dfrac{KL}{r}\right)^3}{8\left(\dfrac{KL}{r}\right)_c^3}, \qquad \left(\frac{KL}{r}\right) \le \left(\frac{KL}{r}\right)_c \tag{10.64a}$$

and, corresponding to Eq. 10.62 the factor of safety is the constant

$$FS_2 = \frac{23}{12}, \qquad \left(\frac{KL}{r}\right) \ge \left(\frac{KL}{r}\right)_c \tag{10.64b}$$

The AISC formulas for σ_{allow} are obtained by combining Eqs. 10.60 and 10.62 through 10.64 to give

$$\frac{\sigma_{\text{allow}}}{\sigma_Y} = \frac{1}{FS_1}\left[1 - \frac{\left(\dfrac{KL}{r}\right)^2}{2\left(\dfrac{KL}{r}\right)_c^2}\right], \qquad \left(\frac{KL}{r}\right) \le \left(\frac{KL}{r}\right)_c \tag{10.65a}$$

$$\frac{\sigma_{\text{allow}}}{\sigma_Y} = \frac{1}{FS_2}\left[\frac{\left(\dfrac{KL}{r}\right)^2}{2\left(\dfrac{KL}{r}\right)^2}\right], \qquad \left(\frac{KL}{r}\right)_c \le \left(\frac{KL}{r}\right) \le 200 \tag{10.65b}$$

These formulas are also plotted in Fig. 10.22.

Finally, it is noted in Eq. 10.65b and on Fig. 10.22 that the AISC restricts columns to values of $KL/r \le 200$.

Aluminum-Alloy Columns. The Aluminum Association provides specifications for the design of aluminum-alloy structures. Euler's formula is used for long columns, and

FIGURE 10.23 The typical design curve for buckling of aluminum-alloy columns.

straight lines are prescribed for short and intermediate columns. The general form of these curves is indicated in Fig. 10.23. Specific design formulas, which depend on alloy, temper, and usage, are provided by the Aluminum Association [Ref. 10-8]. For the alloy 2014-T6, an alloy used in building structures, the column design formulas are:

$$\sigma_{\text{allow}} = 28 \text{ ksi}, \qquad 0 \leq \frac{KL}{r} \leq 12 \tag{10.66a}$$

$$\sigma_{\text{allow}} = \left[30.7 - 0.23\left(\frac{KL}{r}\right) \right] \text{ ksi}, \qquad 12 < \frac{KL}{r} \leq 55 \tag{10.66b}$$

$$\sigma_{\text{allow}} = \frac{54{,}000 \text{ ksi}}{\left(\dfrac{KL}{r}\right)^2}, \qquad 55 < \frac{KL}{r} \tag{10.66c}$$

These formulas incorporate a factor of safety.

Timber Columns. The design of wood structural members is governed by the specifications published by the American Forest and Paper Association [Ref. 10-9]. For example, the three formulas for allowable stresses for *rectangular columns* are

$$\sigma_{\text{allow}} = F_c, \qquad 0 \leq \frac{KL}{d} \leq 11 \tag{10.67a}$$

$$\sigma_{\text{allow}} = F_c\left[1 - \frac{1}{3}\frac{\left(\dfrac{KL}{d}\right)^4}{\left(\dfrac{KL}{d}\right)_c^4} \right], \qquad 11 < \frac{KL}{d} \leq \left(\frac{KL}{d}\right)_c \tag{10.67b}$$

$$\sigma_{\text{allow}} = \frac{0.3E}{\left(\dfrac{KL}{d}\right)^2}, \qquad \left(\frac{KL}{d}\right)_c < \left(\frac{KL}{d}\right) \leq 50 \tag{10.67c}$$

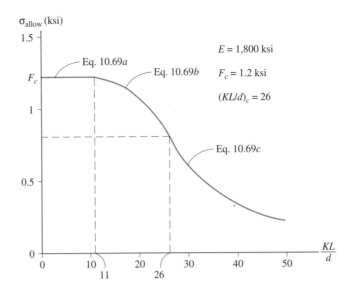

FIGURE 10.24 A (three-segment) design curve for buckling of wood columns.

In the above formulas, F_c is the value of allowable stress for compression parallel to the grain of the wood, and E is the modulus of elasticity. A factor of safety is included in the value of F_c. The values of F_c and E vary greatly with type of wood, moisture content, duration of load, and other factors, with typical values of F_c being in the range of 400 psi to 2000 psi, and typical values of E ranging from $1.2(10^6)$ psi to about $1.8(10^6)$ psi. The effective slenderness ratio, KL/d, is taken as the larger of the two values $(KL/d)_1$ and $(KL/d)_2$ for buckling in the direction of the principal axes of the cross section, with d_1 and d_2 being the finished dimensions of the rectangular cross section. The value of the effective slenderness ratio that separates intermediate columns from long columns is determined by the formula

$$\left(\frac{KL}{d}\right)_c = \left(\frac{0.45E}{F_c}\right)^{\frac{1}{2}} \tag{10.68}$$

The three-segment design curve, based on Eqs. 10.67 and 10.68 with $E = 1.8(10^6)$ psi and $F_c = 1200$ psi, is plotted in Fig. 10.24.

The column-design formulas presented in Eqs. 10.65, 10.66, and 10.67 can be used in a very straightforward manner to determine the allowable load for a column of a given material with a given effective length and given cross section. To design a column—that is, to determine a suitable cross section for a column of given material and given effective length, to support a given load—an iterative procedure is necessary.

COLUMN-DESIGN PROCEDURE

The basic steps of this design procedure are:

Step 1: Select a trial cross section.

Step 2: Determine the value of the effective slenderness ratio by comparing the values $(KL/r)_1$ and $(KL/r)_2$ for buckling in the directions of the principal axes of the cross section. (For a wood column with rectangular cross section, use KL/d instead of KL/r.)

Step 3: Using the column-design formula that is appropriate to the selected material and to the effective slenderness ratio calculated in Step 2, determine the allowable stress σ_{allow}. Multiply σ_{allow} by the cross-sectional area of the trial cross section to determine the allowable load.

Step 4: If the allowable load calculated in Step 3 is equal to or slightly greater than the given design load, accept the trial section. If the allowable load is less than the given design load, repeat Steps 1 through 3 for another trial section. If the allowable load is much greater than required, try a smaller cross section.

Determine the size of a square S4S (surfaced on four sides) structural-timber column of 10-ft unbraced length to carry an axial load of 10 kips (see Fig. 1). Assume that the square ends of the column give an effective length factor $K = 1$, that $E = 1.8(10^6)$ psi, and that $F_c = 1200$ psi.

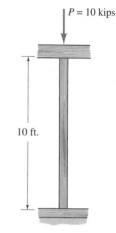

$P = 10$ kips

10 ft.

Fig. 1

Solution

Step 1: Try a nominal 6×6-in. section (5.5 in. $\times$ 5.5 in. finished dimensions).

Step 2: Since this is a square cross section,

$$\left(\frac{KL}{d}\right) = \frac{1(10 \text{ ft})(12 \text{ in./ft})}{5.5 \text{ in.}} = 21.82$$

for both principal directions. From Eq. 10.68 $(KL/d)_c = 26.0$, so this column falls in the intermediate-length range, for which Eq. 10.67b is valid.

Step 3: From Eq. 10.67b

$$\sigma_{\text{allow}} = 1.20\left[1 - \frac{1}{3}\left(\frac{KL/d}{26.0}\right)^4\right] \text{ ksi}$$

$$= 1.20\left[1 - \frac{1}{3}\left(\frac{21.82}{26.0}\right)^4\right] = 1.002 \text{ ksi}$$

$$P_{\text{allow}} = \sigma_{\text{allow}}A = (1.002 \text{ ksi})(5.5 \text{ in.})(5.5 \text{ in.})$$
$$= 30.3 \text{ kips} > 10 \text{ kips}$$

The 6×6-in. section is obviously adequate, but let us see if a 4×4-in. section would be sufficient.

Step 1: Try a nominal 4×4-in. section (3.5 in. $\times$ 3.5 in. finished dimensions).

Step 2:

$$\left(\frac{KL}{d}\right) = \frac{1(10 \text{ ft})(12 \text{ in./ft})}{3.5 \text{ in.}} = 34.3 > 26$$

This column falls in the long-column range, for which Eq. 10.67c is valid.

$$\sigma_{\text{allow}} = \frac{540 \text{ ksi}}{(34.3)^2} = 0.459 \text{ ksi}$$

$$P_{\text{allow}} = \sigma_{\text{allow}}A = (0.459 \text{ ksi})(3.5 \text{ in.})(3.5 \text{ in.})$$
$$= 5.63 \text{ kips} < 10 \text{ kips}$$

Since we were asked to select S4S structural timber with square cross section, we must use the 6×6-in. section. **Ans.**

The design formulas presented in this section are only applicable to centrally loaded columns. The design of eccentrically loaded columns is beyond the scope of this book.

Problems 10.1-1 through 10.1-4. *In solving the problems for Section 10.1, assume that the bars are rigid, that the springs are linearly elastic, and that the displacements and angles of rotation are small. Also, neglect gravity, and assume that buckling takes place in the plane of the figure.*

Prob. 10.1-1. Determine the critical load P_{cr} for the bar-spring system shown in Fig. P10.1-1. The load P remains vertical as the bar rotates about the pin at A. The force in spring BD is proportional to the elongation of the spring, with spring constant k. That is, $F_s = ke$, where e is the elongation of the spring. The spring is unstretched when $\theta = 0$.

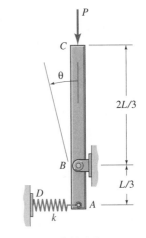

P10.1-3

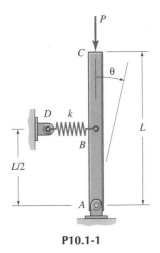

P10.1-1

Prob. 10.1-2. Determine the critical load P_{cr} for the bar-spring system shown in Fig. P10.1-2. The load P remains horizontal. End A is free to move horizontally, and point B moves in a circular arc about C. A frictionless pin connects bars AB and BC at point B.

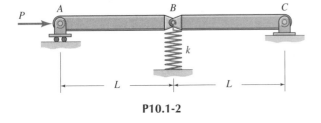

P10.1-2

Prob. 10.1-3. Determine the critical load P_{cr} for the bar-spring system shown in Fig. P10.1-3. The load P remains vertical.

Prob. 10.1-4. Determine the critical load P_{cr} for the bar-spring system shown in Fig. P10.1-4. The load P remains horizontal. End A is free to move horizontally, and point B moves in a circular arc about C. A frictionless pin connects bars AB and BC at point B. Like the rotational spring in Fig. 10.4, the rotational spring at C exerts a restoring moment $(M_C)_r = k_\theta \theta$ when bar BC rotates away from the horizontal position.

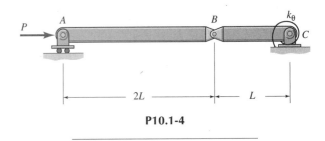

P10.1-4

In solving the problems for Section 10.2, assume that the compression member in question is an ideal slender, prismatic, pin-ended, elastic column.

Prob. 10.2-1. Determine the critical load P_{cr} of a square tubular steel column that is 20 ft long and that has the cross section shown in Fig. P10.2-1. Let $E_{st} = 29(10^3)$ ksi and $\sigma_Y = 46$ ksi.

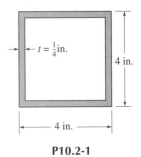

P10.2-1

Prob. 10.2-2. Determine the critical load P_{cr} of a steel pipe column that is 20 ft long; that has a $\frac{1}{4}$-in. wall thickness, as shown in Fig. P10.2-2; and that weighs the same amount per foot as the square tubular column in Fig. P10.2-1. Let $E_{st} = 29(10^3)$ ksi and $\sigma_Y = 46$ ksi.

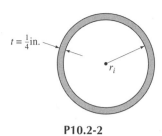

$$t = \tfrac{1}{4}\text{in.}$$

r_i

P10.2-2

Prob. 10.2-3. A builder plans to nail together two boards to form a column. Which of the configurations in Fig. P10.2-3 would give the greater buckling load, configuration A, or configuration B? Assume that the ends of the columns are "pinned" in a manner that permits buckling in any direction. Show calculations to support your answer.

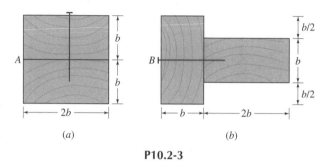

(a)

(b)

P10.2-3

Prob. 10.2-4. (a) Calculate the critical load P_{cr} for a **W**8×35 steel column having a length of 24 ft. Let $E_{st} = 29(10^3)$ ksi and $\sigma_Y = 36$ ksi, and assume that the column is free to buckle in any direction. See Table D.1 of Appendix D for wide-flange section properties. (b) Would a **W**12×35 column be better (i.e., have a higher critical load) than the 8-in. section? (c) If $\sigma_Y = 50$ ksi rather than 36 ksi, would there be any effect on the critical load?

Prob. 10.2-5. A pin-ended column that is 16 ft long must support a compressive axial load of $P = 300$ kips with a factor of safety with respect to buckling of 2.0. Assume that the column has "pinned" ends that allow buckling in any direction. From Table D.1 of Appendix D, select the lightest **W** shape that will support the load. Let $E_{st} = 29(10^3)$ ksi and $\sigma_Y = 36$ ksi.

Prob. 10.2-6. A solid-waste compactor, which is illustrated in Fig. P10.2-6, can be purchased in a "standard" configuration or a "heavy-duty" configuration. One difference between the designs is the size of the push rod that is used. The push rod may be considered to be a pin-ended column of length 1.5 m. The push rod on the standard model is a rectangular bar with dimensions $b = 15$ mm, $h = 30$ mm. It is made of steel with $E_{st} = 200$ GPa, $\sigma_Y = 250$ MPa. (a) Determine the compaction-force capacity of the standard model if the push rod is designed with a factor of safety of 3.0 with respect to elastic buckling. (b) If the push rod of the heavy-duty model is made of the same steel, and if it also has a rectangular cross section with $h = 2b$, what

dimension b is required (to the nearest mm) if the heavy-duty model is to have a capacity of 50 kN?

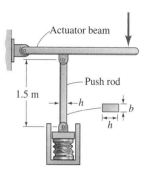

P10.2-6

Prob. 10.2-7. Three pin-ended columns have the same length and the same cross-sectional area. They are made of the same material, and they are free to buckle in any direction. The cross sections of the columns are: (1) a circle, (2) a square, and (3) an equilateral triangle. Determine expressions for the elastic buckling loads—P_1, P_2, and P_3—of the respective columns.

Prob. 10.2-8. An equal-leg angle section of length $L = 12$ ft is to be used to support a compressive load of $P = 20$ kips with a factor of safety of 2.5 against elastic buckling.[17] Assume that the ends of the column are pinned in a manner that permits it to buckle in any direction. From Table D.5 of Appendix D, select the lightest steel section that meets the design requirements. Let $E_{st} = 29(10^3)$ ksi and $\sigma_Y = 36$ ksi.

Prob. 10.2-9. Solve Prob. 10.2-8 if the load is $P = 10$ kips and the length is $L = 10$ ft.

Prob. 10.2-10. From Table D.7 of Appendix D, select a standard steel pipe that would meet the design requirements of Prob. 10.2-8.

Prob. 10.2-11. It is suspected that a small control rod on the spacecraft depicted in Fig. P10.2-11 "failed" as a result of elastic buckling when the rod became exposed to direct sunlight that increased its temperature uniformly by an amount ΔT. Assume that the ends of the pushrod are pinned to structure that is "rigid," and assume that the control rod is stress-free when $\Delta T = 0$. Determine an expression for the value of ΔT that would be required to cause the buckling failure. Express your answer in terms of cross-sectional and material properties of the bar, that is, in terms of I, A, α, etc., and the length L.

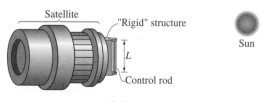

P10.2-11

[17]Because of the difficulty of applying axial loads exactly passing through the centroid of an angle section, such members are not recommended for use as columns.

*Prob. 10.2-12. A "rigid" horizontal beam AD is supported by two pin-ended columns as shown in Fig. P10-2.12. The columns have the same cross-sectional dimensions and are made of the same material (i.e., same I and same E). (a) Determine an expression for the load W_a that will cause elastic buckling of one of the columns, and indicate which column will buckle first. (b) Determine an expression for load W_b that would cause the second column to buckle also. Assume that the force in the column that buckles first remains constant after that column buckles.

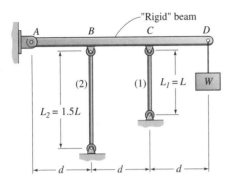

P10.2-12 and P10.2-13

*Prob. 10.2-13. Solve Prob. 10.2-12 with $L_1 = L_2 = L$.

Prob. 10.2-14. An aluminum-alloy column is 15 ft long and is pinned at both ends. What is the allowable compressive load P_{allow} if the factor of safety against elastic buckling is 2.75 and the column is a 6.00×4.00 standard I-beam section weighing 4.692 lb/ft? (See Table D.9 of Appendix D.) Let $E_{al} = 10.0(10^3)$ ksi and $\sigma_{PL} = 38$ ksi.

Prob. 10.2-15. The truss shown in Fig. 10.2-15 is used as part of a rig to pull pilings out of the ground. If the two truss members are steel pipes with outer diameter of $d_o = 2.875$ in. and wall thickness of $t = 0.275$ in., determine the largest tension that can be exerted on the piling without causing elastic buckling of a truss member. The cable that exerts tension T on the piling passes over a pulley that is free to rotate about the same pin that connects the two truss members together at B. Let $E_{st} = 29(10^3)$ ksi and $\sigma_Y = 36$ ksi.

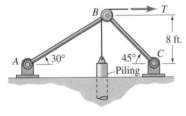

P10.2-15

Prob. 10.2-16. The truss shown in Fig. P10.2-16 is made of steel rods, each of which has a solid, circular cross section with a diameter of 1 in. The ends of the members are pinned. Determine the maximum load W that can be supported at D without causing elastic buckling of any member of the truss. Let $E_{st} = 29(10^3)$ ksi and $\sigma_Y = 36$ ksi.

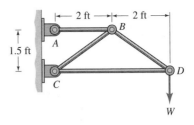

P10.2-16

Prob. 10.2-17. The compression member BC of the truss ABC in Fig. P10.2-17 is a square solid steel bar with 1.75 in. $\times$ 1.75 in. cross section. The tension member AB is a solid steel rod that has a circular cross section with diameter d. Let $E_{st} = 29(10^3)$ ksi and $\sigma_Y = 36$ ksi. (a) Determine the maximum load W that can be supported at B without causing elastic buckling of member BC. (b) Determine the diameter of member AB if it is to yield in tension just when the load W reaches the value determined in Part (a). The yield stress in tension is $\sigma_Y = 36$ ksi.

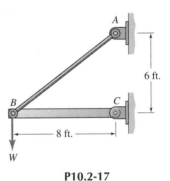

P10.2-17

Prob. 10.2-18. Member AB of the truss in Fig. P10.2-18 has a fixed length L_1, is pinned to a fixed support at A, and remains horizontal. Determine an expression for the load W that would produce elastic buckling of member BC. Express your answer in terms of the distance H, which determines the (variable) location of support point C and also determines the length of member BC.

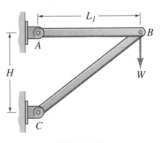

P10.2-18

544

Prob. 10.2-19. The steel compression strut BC of the frame ABC in Fig. P10.2-19 is a steel tube with an outer diameter of $d_o = 48$ mm and a wall thickness of $t = 5$ mm. Determine the factor of safety against elastic buckling if a distributed load of 10 kN/m is applied to the horizontal frame member AB as shown. Let $E_{st} = 210$ GPa and $\sigma_Y = 340$ MPa.

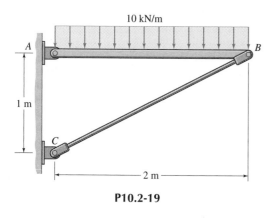

10 kN/m

A

B

1 m

C

2 m

P10.2-19

Problems 10.3-1 through 10.3-14. *In solving the problems of Section 10.3, assume that the compression member in question is an ideal slender, prismatic, elastic column with the stated end conditions. Also, assume that deflections and slopes are small.*

Prob. 10.3-1. A $\mathbf{W}14 \times 26$ wide-flange steel column has a length of $L = 18$ ft and is supported at its ends so that it is free to buckle in any direction. Using the effective-length method, calculate the elastic buckling load P_{cr} for columns with the following end conditions: (a) pinned-pinned, (b) fixed-pinned, and (c) fixed-fixed. See Figs. 10.13a, b, and c, respectively, for columns with these end conditions, and see Table D.1 of Appendix D for the properties of the $\mathbf{W}14 \times 26$ section. Let $E_{st} = 29(10^3)$ ksi and $\sigma_Y = 50$ ksi.

Prob. 10.3-2. Solve Prob. 10.3-1 for a standard steel pipe column with nominal diameter of $d = 3$ in. and a length of $L = 20$ ft. See Table D.7 of Appendix D for the cross-sectional properties of the pipe column. Let $E_{st} = 29(10^3)$ ksi and $\sigma_Y = 50$ ksi.

Prob. 10.3-3. Solve Prob. 10.3-1 for a steel pipe column with an outer diameter of $d_o = 60$ mm, a wall thickness of $t = 4$ mm, and a length of $L = 4$ m. Let $E_{st} = 210$ GPa and $\sigma_Y = 340$ MPa.

Prob. 10.3-4. Two $4 \times 3 \times \frac{1}{2}$ steel angles are attached together as shown in Fig. P10.3-4 to form a column with a tee cross section. (a) Determine the radii of gyration of the cross section about the y axis and about the z axis using the cross sectional properties given in Table. D.6 of Appendix D. (b) If the column has fixed-free end conditions, has a length of $L = 8$ ft, and is free to buckle in any direction, what is the factor of safety of the column against elastic buckling if the compressive axial load on the column is $P = 5$ kips? Let $E_{st} = 29(10^3)$ ksi and $\sigma_Y = 36$ ksi.

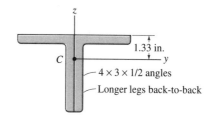

z

1.33 in.

C

y

$4 \times 3 \times 1/2$ angles

Longer legs back-to-back

P10.3-4

Prob. 10-3.5. A square, tubular steel column has the dimensions shown in Fig. P10.3.5, has a length of $L = 3$ m, and has an axial compression load of $P = 50$ kN. If the column has fixed-free end conditions and is free to buckle in any direction, what is the factor of safety with respect to elastic buckling? Let $E_{st} = 210$ GPa and $\sigma_Y = 250$ MPa.

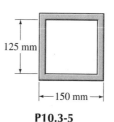

125 mm

150 mm

P10.3-5

Prob. 10.3-6. The $6 \times 6 \times \frac{1}{2}$ equal-leg angle section in Fig. P10.3-6 has a cross-sectional area $A = 5.75$ in^2 and radii of gyration $r_y = r_z = 1.86$ in. The minimum radius of gyration for this section is $r_m = 1.18$ in. If this section is to be used as a fixed-fixed column of length $L = 22$ ft, what is the largest compressive load that can be applied through the centroid of the cross section without causing elastic buckling?[18] Let $E_{st} = 29(10^3)$ ksi and $\sigma_Y = 36$ ksi.

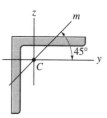

z

m

45°

C

y

P10.3-6 and P10.3-7

Prob. 10.3-7. Solve Prob. 10.3-6 for an equal-leg steel column with cross-sectional area $A = 9700$ mm^2, and with radii of gyration $r_y = r_z = 62$ mm and $r_m = 40$ mm. The section is to be used as a fixed-free column of length $L = 4$ m. Let $E_{st} = 210$ GPa and $\sigma_Y = 340$ MPa.

---

[18]Because of the difficulty of applying axial loads exactly passing through the centroid of an angle section, such members are not recommended for use as columns.

Prob. 10.3-8. Determine the elastic buckling load of a 12-ft-long 6×6 timber column that can be considered to be fixed at its base and free at the top. Let the modulus of elasticity (in compression parallel to the grain) be $E_w = 1.6(10^3)$ ksi, and let $\sigma_Y = 6$ ksi. (Recall that the finished dimensions of the wood column are smaller than the nominal dimensions stated above. See Table D.8 of Appendix D.)

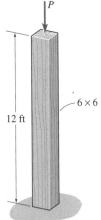

P10.3-8

Prob. 10.3-9. Rigid walls at each end of a solid slender rod AB provide fixed-fixed boundary conditions for the member in Fig. P10.3-9. At the reference temperature, T_0, the rod is perfectly stress-free. (a) Derive a formula that expresses the uniform increase in temperature, ΔT_{cr}, required to cause elastic buckling of the compression member. Express your answer in terms of the following material and geometric parameters: the coefficient of thermal expansion, α, and the slenderness ratio, L/r. (b) Determine the value of ΔT_{cr} in $°F$ required to cause elastic buckling of a stainless steel rod with a diameter of $d = 1$ in. and a length of $L = 4$ ft. The coefficient of thermal expansion for stainless steel is $\alpha = 9.6 \times 10^{-6}/°F$.

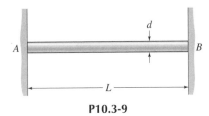

P10.3-9

Prob. 10.3-10. A straight, slender rod is fixed to a rigid support at end A and pinned to a rigid support at end B, as shown in Fig. P10.3-10. At the reference temperature, T_0, the rod is perfectly stress-free. (a) Derive a formula that expresses the uniform increase in temperature, ΔT_{cr}, required to cause elastic buckling of the compression member. Express your answer in terms of the following material and geometric parameters: the coefficient of thermal expansion, α, and the slenderness ratio, r/L. (b) Determine the value of ΔT_{cr} in $°C$ required to cause elastic buckling of an aluminum rod with a diameter of $d = 20$ mm and a length of $L = 1$ m. The coefficient of thermal expansion for aluminum is $\alpha = 23 \times 10^{-6}/°C$.

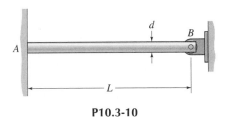

P10.3-10

Prob. 10.3-11. A rigid beam AB that supports a uniformly distributed load of intensity w (force per unit length) is pinned to a rigid support at end A and is supported by a square timber column at end B. Neglecting the weight of the beam AB, derive a formula for the value of the distributed load, w_{cr}, that would cause elastic buckling of the support column BC. Since the exact end conditions of the column are not known, express your answer in terms of the following material and geometric parameters: the modulus of elasticity of the wood column, E_w; the (finished) cross-sectional dimension of the square column, b; the length a of the horizontal beam AB; and the effective length of the column, KL.

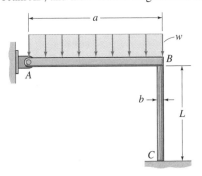

P10.3-11

*Prob. 10.3-12. A beam AB that supports a uniformly distributed load of intensity w (force per unit length) is simply supported at both ends, and it rests on a pipe column at its center C. Neglecting the weight of the beam AB, derive a formula for the value of the distributed load, w_{cr}, that would cause elastic buckling of the support column CD. Assume that the column is fixed to its base at D and pinned to the beam at C. Express your answer in terms of the following material and geometric parameters: the modulus of elasticity, E (same for the beam AB and the column CD); the length of the beam, a, and the length of the column, L; the moment of inertia of the beam, I_b, and of the column, I_c; and the

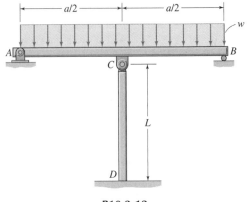

P10.3-12

cross-sectional area of the column, A_c. (Note: This must be solved as a statically indeterminate problem. Use the fact that the deflection of the beam at C and the shortening of the column just prior to buckling are equal.)

*Prob. 10.3-13. Following a procedure similar to the one used in Fig. 10.11 and in Eqs. 10.20 through 10.32 to determine an expression for the elastic buckling load of a fixed-pinned column, derive an expression for the buckling load of a fixed-free column, as shown in Fig. 10.13e and in Fig. P10.3-13. (Hint: Take a free-body diagram of length $(L - x)$ from the top. The maximum deflection, δ, can be eliminated by setting $v(L) \equiv \delta$.)

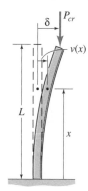

P10.3-13

*Prob. 10.3-14. Following a procedure similar to the one used in Fig. 10.11 and in Eqs. 10.20 through 10.32 to determine an expression for the elastic buckling load of a fixed-pinned column, derive an expression for the buckling load of the fixed-free column shown in Fig. P10.3-14. The portion of the column from A to B is flexible, with modulus of elasticity E and moment of inertia I, but the portion of the column from B to the point of application of the load P can be considered to be perfectly rigid. (Hint: Take a free-body diagram from section x to the top of the column, where the force P is applied. Use the end conditions that $v(L) \equiv \delta$ and $v'(L) \equiv \theta$.)

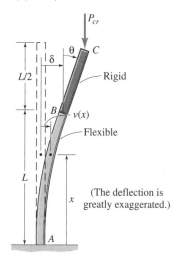

P10.3-14

Problems 10.4-1 through 10.4-12. *In solving these problems, assume that the compression member in question is an ideal slender, prismatic, elastic column; and assume that bending occurs in the xy plane.*

Prob. 10.4-1. The cantilever column in Fig. P10.4-1 has a vertical load P that is applied at an eccentricity e with respect to the axis of the column. Starting with a free-body diagram, formulate the differential equation that governs the beam-column action of this member. Then, solving the differential equation, determine expressions for (a) the maximum transverse deflection, v_{max}, and (b) the maximum bending moment, M_{max}.

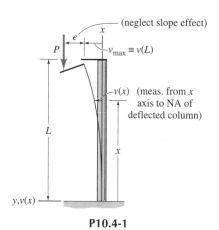

P10.4-1

Prob. 10.4-2. A compressive load $P = 2$ kips is applied parallel to the axis of the column in Fig. P10.4-2 at an eccentricity $e = 0.2$ in. from the axis. The steel column ($E = 30 \times 10^3$ ksi) has a circular cross section with diameter $d = 1$ in.; its length is $L = 5$ ft. (a) Determine the maximum deflection, v_{max}, of this eccentrically loaded column, and (b) determine the maximum bending moment, M_{max}.

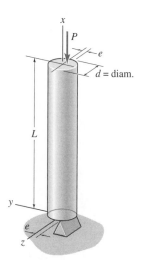

P10.4-2, P10.4-7, P10.4-9, and PC10.4-1

Prob. 10.4-3. A compressive load $P = 100$ kN is applied parallel to the axis of the column in Fig. P10.4-3 at an eccentricity $e = 30$ mm. from the axis. The aluminum-alloy column ($E = 70$ GPa) has a thin-wall, square, box cross section with outer dimension $b = 130$ mm and wall thickness $t = 10$ mm. Its length is $L = 4$ m. (a) Determine the maximum deflection, v_{max}, of this eccentrically loaded column, and (b) determine the maximum bending moment, M_{max}.

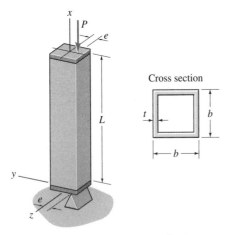

P10.4-3, P10.4-8, P10.4-10, and PC10.4-2

Prob. 10.4-4. A load $P = 120$ kips is applied at an eccentricity $e = 8$ in. in the plane of the web of the **W**12×50 wide-flange steel column shown in Fig. P10.4-4. Let $E = 30 \times 10^3$ ksi., and see Table D.1 for the cross-sectional properties of this beam. What is the maximum length L_{allow} that this column can have if the transverse deflection at the top of the column must not exceed $v_{max} = 1$ in.? (Note: Since this is a cantilever column, you will need to consider its "effective length.")

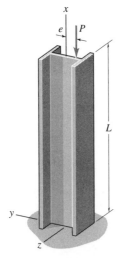

P10.4-4 and P10.4-5

Prob. 10.4-5. Repeat Prob. 10.4-4 for a **W**360×79 wide-flange steel column as follows: $E = 70$ GPa, $P = 300$ kN, $e = 240$ mm, and $v_{max} = 25$ mm. (See Table D.2 for the cross-sectional properties of this column.)

Prob. 10.4-6. A compressive load P is applied in the mid-plane of a square pinned-end column at an eccentricity e from the axis of the column, as shown in Fig. P10.4-6. (a) If $P = P_{cr}/4$ and $e = b/4$, determine the maximum deflection of the column, v_{max}. (P_{cr} is the Euler buckling load.) (b) For the same conditions as in (a), determine an expression for the maximum bending moment M_{max}.

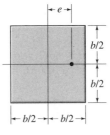

P10.4-6

Prob. 10.4-7. Consider the steel column in Fig. P10.4-7 (see Prob. 10.4-2). Let $d = 1$ in., $e = 0.2$ in., $L = 5$ ft, and $E = 30 \times 10^3$ ksi. (a) If the applied compressive load is $P = 2.5$ kips, what is the magnitude of the maximum normal stress, σ_{max}? (b) What is the allowable compressive load, P_{allow}, if the steel has a yield strength of $\sigma_Y = 36$ ksi and the factor of safety with respect to yielding of the material is $FS = 3$?

Prob. 10.4-8. Consider the aluminum-alloy column in Fig. P10.4-8 (see Prob. 10.4-3). Let $b = 130$ mm, $t = 10$ mm, $e = 30$ mm, $L = 4$ m, and $E = 70$ GPa. (a) If the applied compressive load is $P = 200$ kN, what is the magnitude of the maximum normal stress, σ_{max}? (b) What is the allowable compressive load, P_{allow}, if the steel has a yield strength of $\sigma_Y = 250$ MPa and the factor of safety with respect to yielding of the material is $FS = 2.5$?

*Prob. 10.4-9. Consider the steel column in Fig. P10.4-9 (see Prob. 10.4-2). Let $E = 30 \times 10^3$ ksi, $L = 8$ ft, $\sigma_Y = 36$ ksi, and the factor of safety with respect to yielding be $FS = 2.5$. (a) If the maximum compressive load to be applied to the column is $P = 2.0$ kips, and it is to be applied at an eccentricity $e = 0.25$ in., what is the required (i.e., minimum) column diameter, d, to the nearest 0.10 in.? (b) If the diameter of the column is $d = 2.0$ in., and the maximum compressive load to be applied to the column is $P = 5.0$ kips, what (to the nearest 0.10 in.) is the maximum eccentricity, e_{max}, at which this load may be applied?

*Prob. 10.4-10. Consider the aluminum-alloy column in Fig. P10.4-10 (see Prob. 10.4-3). Let $E = 70$ GPa, $L = 6$ m, $t = 15$ mm, $\sigma_Y = 250$ GPa, and the factor of safety with respect to yielding be $FS = 2$. (a) If the maximum compressive load to be applied to the column is $P = 250$ kN, and it is to be applied at an eccentricity $e = 50$ mm, what is the required (i.e., minimum) column outer cross-sectional dimension, b, to the nearest mm? (b) If the outer cross-sectional dimension of the column is $b = 150$ mm, and the maximum compressive load to be applied to the column is $P = 250$ kN, what, to the nearest 1 mm, is the maximum eccentricity, e_{max}, at which this load may be applied?

*Prob. 10.4-11. A compressive load P is applied parallel to the axis of the steel pin-ended pipe column (d_o = outer diameter; t = wall thickness) in Fig. P10.4-11 at an eccentricity e from the axis. Bending is restricted to the plane containing the axis of the column and the line of action of the loading. Let $E = 30 \times 10^3$ ksi, $L = 10$ ft, $t = 0.25$ in., and $e = 6$ in. (a) If the outer diameter of the column is $d_o = 6$ in., and the compressive load is $P = 25$ kips, what is the maximum normal stress in the column? (b) If the outer diameter of the column is $d_o = 6$ in., the yield strength of the material is $\sigma_Y = 50$ ksi, and the factor of safety with respect to yielding is $FS = 2.5$, what is the allowable load, P_{allow}? (c) If the allowable load is $P_{allow} = 10$ kips, the yield strength of the steel is $\sigma_Y = 50$ ksi, and the factor of safety with respect to yielding is $FS = 2.5$, what is the required (minimum) outer diameter, d_o, to the nearest 0.1 in.?

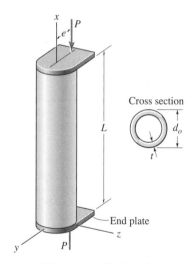

P10.4-11 and P10.4-12

*Prob. 10.4-12. For the steel pin-ended pipe column in Fig. P10.4-12, let $E = 210$ GPa, $L = 4$ m, $t = 10$ mm, and $e = 50$ mm. (a) If the outer diameter of the column is $d_o = 220$ mm, and the compressive load is $P = 500$ kN, what is the maximum normal stress in the column? (b) If the outer diameter of the column is $d_o = 220$ mm, the yield strength of the material is $\sigma_Y = 250$ MPa, and the factor of safety with respect to yielding is $FS = 2.5$, what is the allowable load P_{allow}? (c) If the allowable load is $P_{allow} = 200$ kN, the yield strength of the steel is $\sigma_Y = 250$ MPa, and the factor of safety with respect to yielding is $FS = 2.5$, what is the required (minimum) outer diameter, d_o, to the nearest mm?

*Prob. 10.4-13. A compressive load P is applied parallel to the x axis of the slender, rectangular aluminum-alloy member in Fig. P10.4-13. The load acts in the xy plane at an eccentricity e from the z axis. The member is supported at its ends in a manner that permits bending in any direction. Let $E = 70$ GPa, $L = 1$ m, $b = 10$ mm, $h = 40$ mm, and $e = 10$ mm. (a) Determine the load $(P_{cr})_y$ for elastic buckling of the member for bending in the xz plane. (b) If the yield strength of the material is $\sigma_Y = 270$ MPa, what load $(P_Y)_z$ would cause yielding due to bending in the xy plane? (c) From your answers to Parts (a) and (b), what can you conclude about the most likely failure mode of this member?

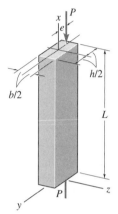

P10.4-13 and P10.4-14

*Prob. 10.4-14. A compressive load P is applied parallel to the x axis of the slender, rectangular titanium-alloy member in Fig. P10.4-14. The load acts in the xy plane at an eccentricity e from the z axis. The member is supported at its ends in a manner that permits bending in any direction. Let $E = 16 \times 10^3$ ksi, $L = 20$ in., $b = 0.25$ in., $h = 0.50$ in., and $e = 0.25$ in. (a) Determine the load $(P_{cr})_y$ for elastic buckling of the member for bending in the xz plane. (b) If the yield strength of the material is $\sigma_Y = 120$ ksi, what load $(P_Y)_z$ would cause yielding due to bending in the xy plane? (c) From your answers to parts (a) and (b), what can you conclude about the most likely failure mode of this member?

Computer Exercises—Section 10.4. *Develop a computer program (e.g., using a mathematical programming language or a spreadsheet program) to generate the plots required in Probs. C10.4-1 and C10.4-2.*

Prob. C10.4-1. Consider the pin-ended column in Prob. 10.4-2, with $E = 30 \times 10^3$ ksi, $L = 8$ ft, $d = 2$ in., and $e = 0.5$ in. (a) Generate a plot of the maximum normal stress, σ_{max}, versus the compressive load, P, for $0 \leq P \leq 15$ kips. Assume that the maximum normal stress remains less than σ_Y for this range of loading. (b) Plot the maximum transverse displacement, v_{max}, for values of the compressive load $0 \leq P \leq 15$ kips. (c) Based on the plots you have generated in (a) and (b), write a short paragraph discussing the difference between using Eq. 2.36 to define "factor of safety" and using Eq. 2.37 to define "factor of safety." (Note that the *Procedure for Determining the Allowable Load for an Eccentrically Loaded Column* specifies the use of Eq. 2.36.)

Prob. C10.4-2. Consider the pin-ended column in Prob. 10.4-3, with $E = 70$ GPa, $L = 4$ m, $b = 130$ mm, $t = 10$ mm, and $e = 30$ mm. (a) Generate a plot of the maximum normal stress, σ_{max}, versus the compressive load, P, for the $0 \leq P \leq 350$ kN. Assume that the maximum normal stress remains less than σ_Y for this range of loading. (b) Plot the maximum transverse displacement, v_{max}, for values of the compressive load $0 \leq P \leq 350$ kN. (c) Based on the plots you have generated in (a) and (b), write a short paragraph discussing the difference between using Eq. 2.36 to define "factor of safety" and using Eq. 2.37 to define "factor of safety." (Note that the *Procedure for Determining the Allowable Load for an Eccentrically Loaded Column,* given in Section 10.4, specifies the use of Eq. 2.36.)

Problems 10.5-1 through 10.5-5. *In solving these problems, assume that the compression member in question is a slender, prismatic, elastic column with an initial deflection in the form of a half-sine curve (Eq. 10.47) with amplitude δ_0. Also assume that the compressive load P acts at the centroid of the cross section.*

Prob. 10.5-1. A compressive load $P = 2$ kips is applied parallel to the axis of an imperfect pin-ended column, like the one shown in Fig. 10.18. The amplitude of the initial imperfection is $\delta_0 \equiv v_0(L/2) = 0.25$ in. The steel column ($E = 30 \times 10^3$ ksi) has a circular cross section with diameter $d = 1.25$ in., and its length is $L = 6$ ft. (a) Determine the maximum total deflection, δ_{max}, of this eccentrically loaded column, and (b) determine the maximum normal stress, σ_{max}.

Prob. 10.5-2. A compressive load $P = 200$ kN is applied parallel to the axis of an imperfect pin-ended column, like the one shown in Fig. 10.18. The amplitude of the initial imperfection is $\delta_0 \equiv v_0(L/2) = 30$ mm. The aluminum-alloy column ($E = 70$ GPa) has a thin-wall, square, box cross section with outer dimension $b = 200$ mm and wall thickness $t = 10$ mm. Its length is $L = 5$ m. (a) Determine the maximum total deflection, δ_{max}, of this eccentrically loaded column, and (b) determine the maximum normal stress, σ_{max}.

Prob. 10.5-3. A compressive load P is applied parallel to the axis of a pin-ended, imperfect elastic (modulus E) column, like the one shown in Fig. 10.18. The cross section of the column is a square with edge dimension b, and its length is L. (a) If $P = P_{cr}/4$ and $\delta_0 = b/4$, determine an expression for the maximum total deflection of the column, δ_{max}. (P_{cr} is the Euler buckling load.) Assume that $\sigma_{max} < \sigma_Y$. (b) For the same conditions as in (a), determine an expression for the maximum normal stress σ_{max}.

*Prob. 10.5-4. A compressive load P is applied to an imperfect, steel, pin-ended pipe column, like the one shown in Fig. 10.18 (d_o = outer diameter, t = wall thickness). Let $E = 30 \times 10^3$ ksi, $L = 10$ ft, $t = 0.25$ in., and $\delta_0 = 0.5$ in. (a) If the outer diameter of the column is $d_o = 6$ in., and the compressive load is $P = 25$ kips, what is the maximum normal stress in the col-

umn? (b) If the outer diameter of the column is $d_o = 6$ in., the yield strength of the material is $\sigma_Y = 50$ ksi, and the factor of safety with respect to yielding is $FS = 2.5$, what is the allowable load P_{allow}? (c) If the allowable load is $P_{allow} = 80$ kips, the yield strength of the steel is $\sigma_Y = 50$ ksi, and the factor of safety with respect to yielding is $FS = 2.5$, what is the required (minimum) outer diameter, d_o, to the nearest 0.1 in.?

*Prob. 10.5-5. A compressive load P is applied to an imperfect, steel, pin-ended pipe column, like the one shown in Fig. 10.18. Let $E = 210$ GPa, $L = 3$ m, $t = 20$ mm, and $\delta_0 = 10$ mm. (a) If the outer diameter of the column is $d_o = 100$ mm, and the compressive load is $P = 300$ kN, what is the maximum normal stress in the column? (b) If the outer diameter of the column is $d_o = 100$ mm, the yield strength of the material is $\sigma_Y = 270$ MPa, and the factor of safety with respect to yielding is $FS = 3.0$, what is the allowable load P_{allow}? (c) If the allowable load is $P_{allow} = 300$ kN, the yield strength of the steel is $\sigma_Y = 270$ MPa, and the factor of safety with respect to yielding is $FS = 3$, what is the required (minimum) outer diameter, d_o, to the nearest mm?

Computer Exercises—Section 10.5. *Develop a computer program (e.g., using a mathematical programming language or a spreadsheet program) to generate the plots required in Probs. C10.5-1 and C10.5-2. In solving these problems, assume that the compression member in question is a slender, prismatic, elastic column with an initial deflection in the form of a half-sine curve (Eq. 10.47) with amplitude δ_0. Also assume that the compressive load P acts at the centroid of the cross section.*

*Prob. C10.5-1. A compressive load P is applied to an imperfect pin-ended column, like the one shown in Fig. 10.18. The steel column ($E = 30 \times 10^3$ ksi, $\sigma_Y = 36$ ksi) has a circular cross section with diameter $d = 2$ in. (a) Let the compressive load be $P = P_Y$, the load that makes $\sigma_{max} = \sigma_Y$. For an initial imperfection $\delta_0 = 0.25$ in., generate a plot of the "average normal stress," P/A, versus the normalized length, L/r, for $10 \leq L/r \leq 200$. (b) Change δ_0 from 0.25 in. to 0.5 in., and generate a second plot. (c) Write a short paragraph discussing the effect of initial imperfection amplitude and column length on the ability of columns to carry compressive loads.

*Prob. C10.5-2. A compressive load P is applied to an imperfect pin-ended column, like the one shown in Fig. 10.18. The steel column ($E = 200$ GPa, $\sigma_Y = 340$ MPa) has a square, box cross section with cross-sectional dimensions $b = 100$ mm and $t = 12.5$ mm. (a) Let the compressive load be $P = P_Y$, the load that makes $\sigma_{max} = \sigma_Y$. For an initial imperfection $\delta_0 = 10$ mm, generate a plot of the "average normal stress," P/A, versus the normalized length, L/r, for $10 \leq L/r \leq 200$. (b) Change δ_0 from 10 mm to 20 mm, and generate a second plot. (c) Write a short paragraph discussing the effect of initial imperfection amplitude and column length on the ability of columns to carry compressive loads.

Problems 10.6-1 through 10.6-3. *In solving these problems, assume that the compression member in question is a slender, prismatic ideal column made of the material whose compressive stress-strain properties are plotted.*

Prob. 10.6-1. A 6-in. (nominal diameter) standard-weight steel pipe column has an effective length of 120 in. (See Table D.7 for the cross-sectional properties of this pipe column.) The steel of which the column is made has a compressive stress-strain curve that is given by the curve *ABC* in Fig. 2 of Example 10.5. The compressive tangent modulus curve for this material is the curve *DEFG* in this same figure. (a) Calculate the buckling load for this column. (Assume that $\sigma_{EL} = \sigma_{PL}$, and estimate σ_{PL} from the σ vs E_t curve.) (b) How long would the column have to be for it to buckle elastically (i.e., according to the Euler formula for elastic buckling)?

Prob. 10.6-2. A 4 in. $\times$ 4 in. $\times \frac{3}{8}$ in. (actual dimensions; see Fig. 10.6-2a) aluminum-alloy box column has an effective length of 72 in. The material of which the column is made has a compressive stress-strain curve that is given by the curve *ABC* in Fig. P10.6-2b. The compressive tangent modulus curve (i.e., the σ vs E_t curve) for this material is the curve *DEFG* in this same figure. (a) Calculate the buckling load for this column. (Assume that $\sigma_{EL} = \sigma_{PL}$, and estimate σ_{PL} from the σ vs E_t curve.) (b) How long would the column have to be for it to buckle elastically (i.e., according to the Euler formula for elastic buckling)?

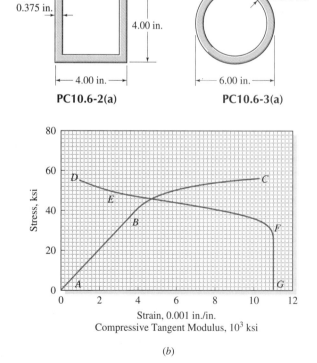

PC10.6-2(a) PC10.6-3(a)

(b)

P10.6-2 and P10.6-3

Prob. 10.6-3. An aluminum-alloy pipe column has the material properties shown in Fig. 10.6-3b. The (actual) dimensions of the pipe column, shown in Fig. 10.6-3a, are: outer diameter $d_o = 6.000$ in., wall thickness $t = 0.375$ in., and length $L = 8$ ft. (a) Calculate the buckling load for this column. (Assume that $\sigma_{EL} = \sigma_{PL}$, and estimate σ_{PL} from the σ vs E_t curve.) (b) How long would the column have to be for it to buckle elastically (i.e., according to the Euler formula for elastic buckling)?

Problems for Section 10.7. *In solving these problems, assume that the compressive axial load is centrally applied at the ends of the column and, unless indicated otherwise, that the column is supported in a manner that permits buckling in any direction.*

Prob. 10.7-1. Determine the allowable compressive axial loads $P (= P_{\text{allow.}})$ for pin-supported **W**8$\times$40 wide-flange steel columns having the following lengths: $L = 12$ ft, $L = 16$ ft, and $L = 24$ ft. Let $E = 29 \times 10^3$ ksi and $\sigma_Y = 36$ ksi.

Prob. 10.7-2. Determine the allowable compressive axial loads $P (= P_{\text{allow.}})$ for pin-supported **W**12$\times$65 wide-flange steel columns having the following lengths: $L = 10$ ft, $L = 20$ ft, and $L = 30$ ft. Let $E = 29 \times 10^3$ ksi and $\sigma_Y = 50$ ksi.

Prob. 10.7-3. Determine the allowable compressive axial loads $P (= P_{\text{allow.}})$ for pin-supported **W**14$\times$53 wide-flange steel columns having the following lengths: $L = 18$ ft, $L = 22$ ft, and $L = 26$ ft. Let $E = 29 \times 10^3$ ksi and $\sigma_Y = 36$ ksi.

Prob. 10.7-4. Determine the allowable compressive axial loads $P (= P_{\text{allow.}})$ for pin-supported **W**200$\times$59 wide-flange steel columns having the following lengths: $L = 4$ m, $L = 6$ m, and $L = 8$ m. Let $E = 200$ GPa and $\sigma_Y = 250$ MPa.

Prob. 10.7-5. Determine the allowable compressive axial loads $P (= P_{\text{allow.}})$ for pin-supported **W**250$\times$45 wide-flange steel columns having the following lengths: $L = 4$ m, $L = 5$ m, and $L = 6$ m. Let $E = 200$ GPa and $\sigma_Y = 250$ MPa.

Prob. 10.7-6. Determine the allowable compressive axial loads $P (= P_{\text{allow.}})$ for pin-supported **W**310$\times$97 wide-flange steel columns having the following lengths: $L = 6$ m, $L = 8$ m, and $L = 10$ m. Let $E = 200$ GPa and $\sigma_Y = 340$ MPa.

Prob. 10.7-7. Determine the maximum allowable length for a pin-supported **W**10$\times$60 wide-flange steel column if the compressive axial load is (a) 150 kips; (b) 300 kips. Let $E = 29 \times 10^3$ ksi and $\sigma_Y = 36$ ksi.

Prob. 10.7-8. Determine the maximum allowable length for a pin-supported **W**12$\times$50 wide-flange steel column if the compressive axial load is (a) 100 kips; (b) 200 kips. Let $E = 29 \times 10^3$ ksi and $\sigma_Y = 36$ ksi.

Prob. 10.7-9. Determine the maximum allowable length for a pin-supported **W**16$\times$100 wide-flange steel column if the compressive axial load is (a) 300 kips; (b) 450 kips. Let $E = 29 \times 10^3$ ksi and $\sigma_Y = 50$ ksi.

Prob. 10.7-10. The 27-ft-long **W**12×50 wide-flange steel column shown in Fig. P10.7-10 has a ''fixed'' base and receives equal downward (compressive) loading from the beams that frame into it at the top. Consider only buckling in the *xy* plane. Because of the flexibility of the adjoining beams, the end conditions of the column fall somewhere between ''fixed-fixed'' and ''pinned-pinned.'' It is estimated that the effective length factor, *K*, for the beam falls in the range $0.6 \le K \le 0.8$. (Recall that $K = 0.5$ for a fixed-ended column and $K = 1.0$ for a pin-ended column.) Using the AISC formulas, determine the allowable loads for the following effective-length factors: $K = 0.6$, $K = 0.7$, and $K = 0.8$. Let $E = 29 \times 10^3$ ksi and $\sigma_Y = 36$ ksi.

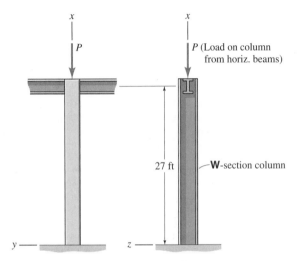

P10.7-10, P10.7-11, and P10.7-12

Prob. 10.7-11. Repeat Prob. 10.7-10 for a 24-ft-long **W**16×40 wide-flange steel column.

Prob. 10.7-12. Repeat Prob. 10.7-10 for a 28-ft-long **W**8×40 wide-flange steel column.

Prob. 10.7-13. Using the Aluminum Association column design formulas, determine the allowable compressive axial loads *P*

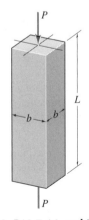

P10.7-13, P10.7-14, and P10.7-15

552

($= P_{\text{allow}}$) for 2.5 in.×2.5 in. (i.e., $b = 2.5$ in. in Fig. P10.7-13) pin-supported 2014-T6 aluminum-alloy columns having the following lengths: $L = 3$ ft, $L = 4$ ft, and $L = 5$ ft.

Prob. 10.7-14. Using the Aluminum Association column design formulas, determine the maximum allowable length for the 3 in.×3 in. pin-supported 2014-T6 aluminum-alloy column in Fig. P10.7-14 if the compressive axial load is (a) 150 kips; (b) 175 kips.

Prob. 10.7-15. To the nearest $\frac{1}{16}$ in., determine the minimum cross-sectional dimension *b* for a square cross-section, pin-supported 4-ft-long 2014-T6 aluminum-alloy column (Fig. P10.7-15) if the compressive axial load is (a) 150 kips; (b) 175 kips. Use the Aluminum Association column design formulas.

Prob. 10.7-16. Determine the maximum allowable length for the 2 in.×4 in. pin-supported 2014-T6 aluminum-alloy column in Fig. P10.7-16 if the compressive axial load is (a) 100 kips; (b) 150 kips. Use the Aluminum Association column design formulas, and assume that the column is free to buckle in any direction.

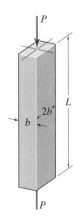

P10.7-16 and P10.7-17

Prob. 10.7-17. To the nearest $\frac{1}{16}$ in., determine the minimum cross-sectional dimension *b* for a $b \times 2b$ rectangular-cross-section pin-supported 2-ft-long 2014-T6 aluminum-alloy column (Fig. P10.7-17) if the compressive axial load is 100 kips. Use the Aluminum Association column design formulas, and assume that the column is free to buckle in any direction.

Prob. 10.7-18. Determine the maximum allowable length for the pin-supported 2014-T6 aluminum-alloy pipe column in Fig. P10.7-18 if $d_o = 5$ in., $t = 0.5$ in., and if the compressive axial load is (a) 75 kips; (b) 150 kips. Use the Aluminum Association column design formulas.

Prob. 10.7-19. The pin-supported 2014-T6 aluminum-alloy pipe column in Fig. P10.7-19 is $L = 10$ ft long. (a) If the outer diameter is $d_o = 6$ in. and the wall thickness is $t = 0.5$ in., what is the allowable compressive axial load? (b) To the nearest $\frac{1}{16}$ in., determine the minimum wall thickness *t* if the outer diameter of the column is $d_o = 8$ in. and the compressive axial load is 250 kips. Use the Aluminum Association column design formulas.

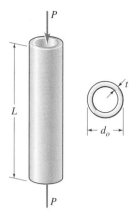

P10.7-18 and P10.7-19

Prob. 10.7-20. Determine the allowable compressive axial loads P ($= P_{\text{allow}}$) for 4×4 (nominal dimensions; see Table D.8 for actual finish dimensions) S4S timber columns (Fig. P10.7-20) having the following lengths: $L = 6$ ft and $L = 8$ ft. Use the American Wood Council design formulas with $E = 1800$ ksi and $F_c = 1.2$ ksi, and assume "pin supports" (i.e., $K = 1$).

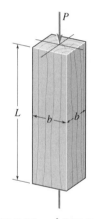

P10.7-20 and P10.7-21

Prob. 10.7-21. Determine the maximum allowable length for the 4×4 (nominal dimensions; see Table D.8) S4S timber columns in Fig. P10.7-21 if the compressive axial load is (a) 6 kips; (b) 9 kips; and (c) 12 kips. Use the American Wood Council design formulas with $E = 1800$ ksi and $F_c = 1.2$ ksi, and assume "pin supports" (i.e., $K = 1$).

Prob. 10.7-22. The timber column shown in Fig. P10.7-22 frames into heavy beams at its top and at its base, so that partial fixity with $K = 0.8$ can safely be assumed. The column is 12 ft long and must support an axial compressive load of $P = 30$ kips.

From Table D.8, select the minimum size of square-cross-section S4S timber that can be used for this column. Use the American Wood Council design formulas with $E = 1800$ ksi and $F_c = 1.2$ ksi.

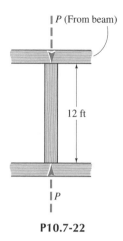

P10.7-22

Prob. 10.7-23. Determine the allowable compressive axial loads P ($= P_{\text{allow.}}$) for 4×6 (nominal dimensions; see Table D.8 for actual finish dimensions) S4S timber columns (Fig. P10.7-23) having the following lengths: $L = 6$ ft, $L = 8$ ft, and $L = 10$ ft. Use the American Wood Council design formulas with $E = 1800$ ksi and $F_c = 1.2$ ksi, and assume "pin supports" (i.e., $K = 1$).

Prob. 10.7-24. From Table D.8, select the minimum-size $6 \times h_{\text{nom}}$ (nominal dimensions) S4S timber column 8-ft long (Fig. P10.7-24) that will support an axial compressive load of up to $P = 40$ kips. Use the American Wood Council design formulas with $E = 1800$ ksi and $F_c = 1.2$ ksi, and assume "pin supports" (i.e., $K = 1$).

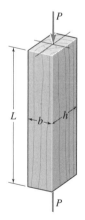

P10.7-23 and P10.7-24

11 ENERGY METHODS

11.1 INTRODUCTION

In Chapters 1 through 10 we employed the three fundamental concepts of deformable-body mechanics (equilibrium, geometry of deformation, and constitutive behavior of materials) to examine the response of several types of structural members to applied loads and/or temperature changes. We determined the distribution of normal stress and shear stress in members and the deformation of the members. We also examined the stability of members undergoing axial compression. We turn now to the important topic of *energy methods* in deformable-body mechanics. Before the advent of the digital computer, energy methods were the most powerful tools available for solving deflection problems, especially statically indeterminate problems. And now, energy methods form a basis of the finite-element method, the most popular and most powerful current method for analyzing deformable bodies (machines, structures, etc.). You will see that the energy methods presented in this chapter again incorporate the three essentials of deformable-body mechanics—equilibrium, geometry of deformation, and constitutive behavior of materials.

In mechanics the term *work* refers to a quantity that is basically (*force* × *distance*). When work is done on a deformable body, some or all of the work done on the body goes into *strain energy* stored in the body. For example, when you stretch a rubber band by pulling on it, the work that you do on the rubber band is stored as strain energy. When you release the applied force, the rubber band releases this energy as it returns to its undeformed shape.

In this chapter we will define a number of work-energy terms: work of external forces, complementary work, strain energy, complementary strain energy, virtual displacements, virtual forces, and virtual work. The *work-energy principle* will be employed to calculate the deflection of members subjected to single static loads. The more powerful virtual-work principles—the *principal of virtual displacements* and the *principle of virtual forces*—will also be introduced and will be used to solve more complex problems, such as statically indeterminate problems for structures with several loads acting simultaneously. The relationship of the energy principles to the *displacement method* and the *force method,* introduced in Chapter 3 and used in subsequent chapters, will be pointed out. Finally, energy methods will be used to solve several simple *impact-loading* problems.

11.2 WORK AND COMPLEMENTARY WORK; STRAIN ENERGY AND COMPLEMENTARY ENERGY

This section is devoted to providing the definitions of important work-energy terms that will be used in later sections of this chapter. There, several work-energy principles are introduced, and deformable-body mechanics problems are solved using these work-energy principles.

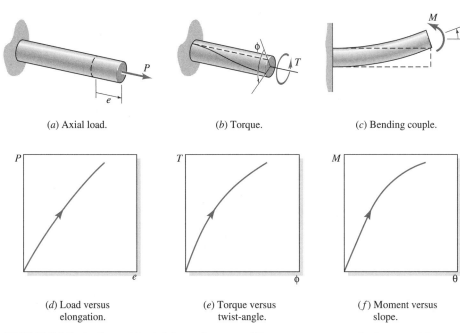

(a) Axial load.

(b) Torque.

(c) Bending couple.

(d) Load versus elongation.

(e) Torque versus twist-angle.

(f) Moment versus slope.

FIGURE 11.1 Several load-deformation cases. (To conserve space, only the part of the load-deformation curves corresponding to the positive loads are shown. Negative loading is, however, equally permissible.)

Figure 11.1 shows members with three types of applied loads—axial (Fig. 11.1a), torsional (Fig. 11.1b), and bending (Fig. 11.1c). Below each of these figures is a plot of the load-deformation curve that might typically be obtained in each case by slowly increasing the magnitude of the applied load. In the definitions that follow we will refer primarily to the axial-loading case, but the definitions can easily be extended to the other two cases in Fig. 11.1 as well as to more general loading by forces and couples.

Work and Complementary Work. Consider the case of axial loading of a slender rod by a single load P, as indicated in Fig. 11.1a. The *work* done by the force P to elongate the rod by an amount e_1 will be designated by $\mathcal{W}_P(e_1)$, and it is given by the integral of force times distance

$$\mathcal{W}_P(e_1) = \int_0^{e_1} P(e)\, de \qquad \textbf{Work} \quad (11.1)$$

On Fig. 11.2a this appears as the area (shaded red) <u>below</u> the $P - e$ curve up to $e = e_1$. Since the force varies with elongation (i.e., with ''distance''), the integral expression in Eq. 11.1 must be used to calculate the work done as the force is increased up to the value $P_1 \equiv P(e_1)$.

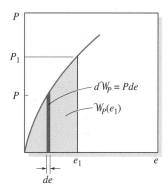

(a) The area representing the work $\mathcal{W}$.

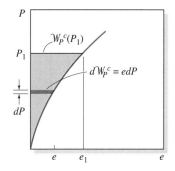

(b) The area representing the complementary work.

FIGURE 11.2 Load versus elongation curves for axial deformation.

555

In a similar manner, the work done by a torsional couple T and by a bending couple M, respectively, are

$$\mathcal{W}_T(\phi_1) = \int_0^{\phi_1} T(\phi) \, d\phi, \qquad \mathcal{W}_M(\theta_1) = \int_0^{\theta_1} M(\theta) \, d\theta \qquad (11.2)$$

If the load-displacement relationship is linear, the curves in Fig. 11.2 become straight lines, and the area under the curve is triangular. Therefore, for the linear case,

$$\mathcal{W}_P = \frac{1}{2} Pe, \quad \mathcal{W}_T = \frac{1}{2} T\phi, \quad \mathcal{W}_M = \frac{1}{2} M\theta \qquad (11.3)$$

■■■■■■■■■■■■■■■□□ E X A M P L E   1 1 . 1 ■□■□■□■□■□■□■□■■

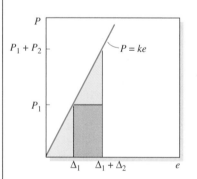

(a) Load P_1 applied.   (b) Load P_2 added.

(c) Load versus elongation curve.

Fig. 1 An illustration of the work due to multiple loads.

The axial deformation member in Figs. 1a and 1b has the linear load-elongation curve shown in Fig. 1c. Using the load-elongation curve, explain why $\mathcal{W}(\Delta_1 + \Delta_2) \neq \mathcal{W}(\Delta_1) + \mathcal{W}(\Delta_2)$.

Plan the Solution   The work done by an axial load acting on a linearly elastic member is given by Eq. 11.3a and is represented by the triangular area under the load-elongation curve.

Solution   From Eq. 11.3a and Fig. 1c, $\mathcal{W}(\Delta_1)$ is the lower-left triangular area

$$\mathcal{W}(\Delta_1) = \frac{1}{2} P_1 \Delta_1 \qquad (1)$$

The work done by force P over an elongation Δ_2, only, $\mathcal{W}(\Delta_2)$, is the upper-right triangular area in Fig. 1c, given by

$$\mathcal{W}(\Delta_2) = \frac{1}{2} P_2 \Delta_2 \qquad (2)$$

and $\mathcal{W}(\Delta_1 + \Delta_2)$ is the entire area under the load-elongation curve up to $\Delta = \Delta_1 + \Delta_2$.

$$\mathcal{W}(\Delta_1 + \Delta_2) = \frac{1}{2}(P_1 + P_2)(\Delta_1 + \Delta_2)$$

$$= \frac{1}{2} P_1 \Delta_1 + \left[\frac{1}{2} P_1 \Delta_2 + \frac{1}{2} P_2 \Delta_1 \right] + \frac{1}{2} P_2 \Delta_2 \qquad (3)$$

$$= \frac{1}{2} P_1 \Delta_1 + [P_1 \Delta_2] + \frac{1}{2} P_2 \Delta_2$$

where the equation $P = ke$ has been used to obtain the $P_1 \Delta_2$ term from the bracketed term on the preceding line. Therefore, due to the presence of the terms in square brackets in Eq. (3),

$$\mathcal{W}(\Delta_1 + \Delta_2) \neq \mathcal{W}(\Delta_1) + \mathcal{W}(\Delta_2) \qquad (4)$$

The term in square brackets in Eq. (3) corresponds to the darker-shaded rectangular area in Fig. 1c. It is the additional work done by the first load P_1 when the second load P_2 is applied, stretching the rod an additional amount Δ_2.

Review the Solution The graphical interpretation of the equations agrees with the terms of the equations.

The lesson that we learn from this example is that work is not a linear function of load (or deformation), so we must exercise care in the calculation of work, especially when multiple loads are involved.

A second work-type quantity that is useful in solid mechanics is called complementary work. The load-elongation curve in Fig. 11.2b is the same as that in Fig. 11.2a, but Fig. 11.2b illustrates the *complementary work*, $\mathcal{W}^c(P_1)$, which is the area (shaded blue) <u>above</u> the $P - e$ curve, given by

$$\mathcal{W}^c(P_1) = \int_0^{P_1} e(P)\, dP \qquad \textbf{Complementary Work} \qquad (11.4)$$

(Note that $\mathcal{W}^c$ is expressed as a function of force, whereas the work $\mathcal{W}$ is expressed as a function of elongation.)

Strain Energy and Complementary Strain Energy.
As indicated in Section 2.5, a material is said to be *elastic* (or, to behave elastically) if the load-deflection path traced on unloading is the same as the load-deflection path traced on loading. The work done on an elastic member during loading is stored in the member as strain energy, and that strain energy is recovered upon unloading. For an elastic body, the work, $\mathcal{W}$, done on the body by the applied force is stored as *elastic strain energy, $\mathcal{U}$.* That is

$$\mathcal{U}(e) = \mathcal{W}(e)\big|_{\text{done on elastic body}} \qquad (11.5)$$

Similarly, for an elastic body, the complementary work, $\mathcal{W}^c$, done on the body is stored as *complementary strain energy, $\mathcal{U}^c$.* Thus,

$$\mathcal{U}^c(P) = \mathcal{W}^c(P)\big|_{\text{done on elastic body}} \qquad (11.6)$$

The strain energy $\mathcal{U}$ (shaded red) and complementary strain energy $\mathcal{U}^c$ (shaded blue) are illustrated in Fig. 11.3. The arrows in Fig. 11.3 denote the fact that the same load-elongation curve is traced when the load is decreasing as when it is increasing.

Linearly Elastic Behavior.
When the material is linearly elastic, the load versus deflection curve has the form illustrated in Fig. 11.4. In this case[1]

$$P = ke, \qquad e = fP \qquad (11.7)$$

where k and f are the stiffness coefficient and flexibility coefficient, respectively, ($f = 1/k$). In this case the triangular areas that represent $\mathcal{U}$ and $\mathcal{U}^c$ are equal, and the value can

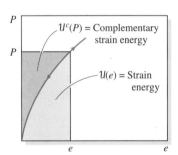

FIGURE 11.3 An illustration of strain energy and complementary strain energy.

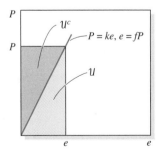

FIGURE 11.4 The load versus elongation diagram of a linearly elastic member.

---

[1]Recall Section 3.3.

be expressed in any of the following ways:

$$\mathcal{U} = \mathcal{U}^c = \frac{1}{2}Pe = \frac{1}{2}ke^2 = \frac{1}{2}fP^2 \qquad (11.8)$$

It is usually convenient to think of $\mathcal{U}$ as being a function of e, that is, $\mathcal{U} = \frac{1}{2}ke^2$, and

$\mathcal{U}^c$ as being a function of P, that is, $\mathcal{U}^c = \frac{1}{2}fP^2$. However, for linearly elastic members the alternative forms listed in Eqs. 11.8 may sometimes be useful (see Section 11.8), and it is sometimes convenient to think of the strain energy stored in a linearly elastic member as $\left(\frac{1}{2}P\right)e$, that is, the average force times the total elongation.

EXAMPLE 11.2

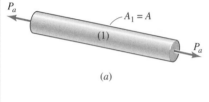

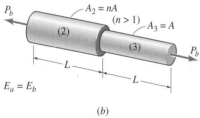

$E_a = E_b$

(b)

Fig. 1 Two rods undergoing axial deformation.

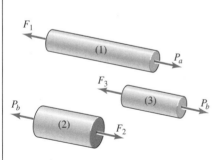

Fig. 2 Free-body diagrams.

The two linearly elastic rods in Fig. 1 have the same modulus E and are to be compared on the basis of the elastic strain energy that is stored in each: (a) when the maximum stress in each of the two rods is the same; and (b) when the total elongation of each of the two rods is the same.

Plan the Solution Since $\sigma_i = F_i/A_i$, the maximum axial stress in rod (b) occurs in section (3), and since $A_1 = A_3 = A$, the loads P_a and P_b will be equal in (a). In (b) we want $e_b = e_a$. Let us begin by plotting $P - e$ curves, like Fig. 11.4, for both rods. Then we can easily compare the values of $\mathcal{U}_a$ and $\mathcal{U}_b$ for the two cases. We can make use of Eqs. 11.7 (or 3.13 and 3.14) to relate P to e for each rod.

Solution Let subscripts 1 through 3 refer to the elements labeled in Fig. 1, and let subscripts a and b refer to the rods in Fig. 1a and 1b, respectively.

Equilibrium: The free-body diagrams for the three rods elements are shown in Fig. 2.

$$\left(\sum F\right)_1 = 0: \qquad F_1 = P_a$$

$$\left(\sum F\right)_2 = 0: \qquad F_2 = P_b \qquad (1)$$

$$\left(\sum F\right)_3 = 0: \qquad F_3 = P_b$$

Force-Elongation Relations: From Eqs. 11.7, the equations

$$e_i = f_iF_i = \left(\frac{L}{AE}\right)_i F_i, \qquad F_i = k_ie_i = \left(\frac{AE}{L}\right)_i e_i \qquad (2)$$

give the force-elongation relations of the individual elements. For compatibility of deformation we get

$$e_a = e_1, \qquad e_b = e_2 + e_3 \qquad (3)$$

Therefore,

$$e_a = \left(\frac{2L}{AE}\right) P_a \equiv f_a P_a$$

$$e_b = \left(\frac{L}{nAE}\right) P_b + \left(\frac{L}{AE}\right) P_b = \left(\frac{n+1}{n}\right)\left(\frac{L}{AE}\right) P_b \equiv f_b P_b$$

(4)

and, therefore,

$$P_a = \frac{1}{2}\left(\frac{AE}{L}\right) e_a \equiv k_a e_a$$

$$P_b = \left(\frac{n}{n+1}\right)\left(\frac{AE}{L}\right) e_b \equiv k_b e_b$$

(5)

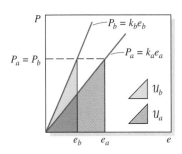

(a) Strain energy relationship when $P_b = P_a$.

Let us plot Eqs. (5) on the same $P - e$ graph (Fig. 3). Since $n > 1$, $k_b > k_a$, so the slope of the $P_b - e_b$ line is greater than the slope of the $P_a - e_a$ line in Fig. 3. That is, due to its larger cross section, rod (b) is stiffer.

Strain Energy: From Eqs. 11.8 and Fig. 11.4, strain energy is the triangular area under the load-deflection curve. The area is given by

$$\mathcal{U} = \frac{1}{2} k e^2 = \frac{1}{2} f P^2$$

(6)

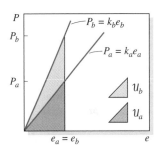

(b) Strain energy relationship when $e_b = e_a$.

Fig. 3 Strain energy comparisons.

Complete Part (a), the equal-stress case: The maximum stresses in the two rods are equal if $P_a = P_b$. From Fig. 3a it is clear that $\mathcal{U}_a > \mathcal{U}_b$. That is, **the rod with enlarged cross section stores less strain energy than the uniform rod does when the maximum stress in each rod is equal to that in the other rod**. In equation form,

$$\frac{\mathcal{U}_b}{\mathcal{U}_a} = \frac{\frac{1}{2} f_b P_b^2}{\frac{1}{2} f_a P_a^2} = \frac{f_b}{f_a} = \frac{n+1}{2n} < 1 \qquad \text{Ans (a)} \quad (7)$$

Complete Part (b), the equal-elongation case: If the elongation of the two rods is the same, there must be a greater force acting on the stiffer rod. That is, $P_b > P_a$, as seen in Fig. 3b. In this case, $\mathcal{U}_b > \mathcal{U}_a$.

$$\frac{\mathcal{U}_b}{\mathcal{U}_a} = \frac{\frac{1}{2} k_b e_b^2}{\frac{1}{2} k_a e_a^2} = \frac{k_b}{k_a} = \frac{2n}{n+1} > 1 \qquad \text{Ans. (b)} \quad (8)$$

Therefore, **if the two rods in Fig. 1 are made to elongate the same amount, the rod with increased cross section will store more strain energy than the uniform rod does**.

Review the Solution Since the strain energy discussion above is graphically based, the answers are correct if we did not make a mistake in calculating the f's in Eq. (7) and k's in Eq. (8). We know that rod b is stiffer because of the enlarged section. Therefore, $k_b > k_a$. Since this is true in Eqs. (8), our answers seem to be correct.

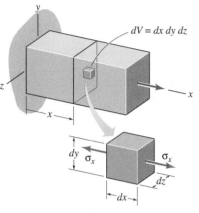

FIGURE 11.5 An element of volume subjected to uniaxial stress σ_x.

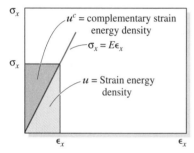

FIGURE 11.6 Strain energy density and complementary strain energy density for a uniaxial stress state.

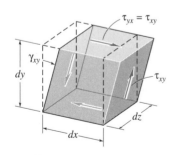

FIGURE 11.7 An elemental volume undergoing deformation due to shear stress $\tau_{xy} = \tau_{yx}$.

Strain Energy Density for Linearly Elastic Bodies. To introduce the concepts of strain energy and complementary strain energy, the simple case of axial deformation of an elastic member by a single load P was considered above. However, to treat more complex situations we need to consider how the stored strain energy is distributed throughout the deformed body.[2] This leads us to the topic of strain energy per unit volume, or, simply, *strain energy density*. Consider a small volume element dV in a linearly elastic rod undergoing axial deformation, as illustrated in Fig. 11.5.

Let $d\mathcal{U}$ be the strain energy stored in an elemental volume of a linearly elastic body, like dV in Fig. 11.5. Furthermore, let the strain energy stored in the elemental volume be expressed in the form

$$d\mathcal{U} = u \, dV$$

where u is the *strain energy density,* that is, the strain energy per unit volume at the location of the differential volume dV. Then, the *total strain energy* is obtained by summing the $d\mathcal{U}$s over the volume of the entire body, giving the integral expression

$$\mathcal{U} = \int_V u \, dV \tag{11.9}$$

For the uniaxial stress state depicted in Fig. 11.5 we have

$$d\mathcal{U} = \underbrace{(\tfrac{1}{2} \sigma_x dy \, dz) \, (\epsilon_x dx)}_{\text{avg. force} \times \text{displ.}} \equiv u_{\sigma_x} \, dV$$

where

$$u_{\sigma_x} = \frac{1}{2}\sigma_x\epsilon_x \tag{11.10}$$

is the strain energy density for linearly elastic deformation due to a uniaxial stress state σ_x. Since $\sigma_x = E\epsilon_x$, Eq. 11.10 can also be written in the following alternative uniaxial stress-state forms:

$$u_{\sigma_x} = \frac{1}{2}E\epsilon_x^2 = \frac{1}{2}\frac{\sigma_x^2}{E} \tag{11.11}$$

This strain energy density is depicted in the stress-strain diagram in Fig. 11.6. The strain energy density is the <u>area below the stress-strain curve</u>. (The complementary strain energy density u^c is also shown in Fig. 11.6, but it will not be discussed further until Section 11.7.)

Consider now an elemental volume dV subjected only to shear stress $\tau_{xy} = \tau_{yx}$, as shown in Fig. 11.7. For this case we have

$$d\mathcal{U} = \underbrace{(\tfrac{1}{2} \tau_{xy} dx \, dz) \, (\gamma_{xy} dy)}_{\text{avg. force} \times \text{displ.}} \equiv u_{\tau_{xy}} dV$$

---

[2]We will return to the discussion of complementary strain energy in Section 11.7.

Therefore, the strain energy density related to τ_{xy} is

$$u_{\tau_{xy}} = \frac{1}{2}\tau_{xy}\gamma_{xy} \qquad (11.12)$$

Since $\tau_{xy} = G\gamma_{xy}$, alternative forms of Eq. 11.12 are

$$u_{\tau_{xy}} = \frac{1}{2}G\gamma_{xy}^2 = \frac{1}{2}\frac{\tau_{xy}^2}{G} \qquad (11.13)$$

In a similar manner, we could derive expressions for the strain energy density associated with the remaining components of stress. The general expression for *strain energy density* in a linearly elastic body is

$$u = \frac{1}{2}(\sigma_x\epsilon_x + \sigma_y\epsilon_y + \sigma_z\epsilon_z + \tau_{xy}\gamma_{xy} + \tau_{xz}\gamma_{xz} + \tau_{yz}\gamma_{yz}) \qquad (11.14)$$

In the next section we will use this strain-energy-density expression as the basis for deriving expressions for the strain energy in slender members undergoing axial deformation, torsion, and bending.

11.3 ELASTIC STRAIN ENERGY FOR VARIOUS TYPES OF LOADING

In order to apply work-energy methods to solve problems, we must first obtain expressions for the elastic strain energy stored when various types of loads are applied—axial loads, torsional loads, and bending loads. We will restrict our attention in this section to linearly elastic behavior.

Axial Deformation. In Section 3.2 the basic equations for normal stress and extensional strain were developed. Recall that, for axial deformation of a member such as the one illustrated in Fig. 11.8, and for the special case of $E = E(x)$ (i.e., the material is homogeneous at any cross section, but its properties may vary with x),

$$\epsilon_x(x) = \frac{du(x)}{dx} \qquad (11.15)$$

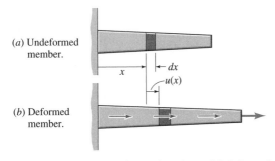

(a) Undeformed member.

(b) Deformed member.

FIGURE 11.8 A member undergoing axial deformation.

and

$$\sigma_x = \frac{F(x)}{A(x)} \tag{11.16}$$

where $u(x)$ is the displacement of the cross section at x and $F(x)$ is the (internal) force on the cross section at x. Therefore, since $dV = A(x)dx$ for the axial-deformation member, we can combine Eqs. 11.9, 11.11, 11.15, and 11.16 to get

$$\mathcal{U} = \frac{1}{2}\int_0^L EA(u')^2dx \tag{11.17}$$

and

$$\mathcal{U} = \frac{1}{2}\int_0^L \frac{F^2dx}{EA} \tag{11.18}$$

where $(\ \ )' \equiv \dfrac{d(\ \ )}{dx}$; and F, E, A, and u may all be functions of x. We will have occasion to use both of the above expressions for the strain energy of linearly elastic bars undergoing axial deformation.

For a uniform linearly elastic rod subjected to axial end loads, as illustrated in Fig. 11.9, Eq. 11.18 gives

$$\mathcal{U} = \frac{P^2L}{2EA} \tag{11.19}$$

FIGURE 11.9 An end-loaded uniform rod.

Torsion of Circular Members. Although it is possible to obtain expressions for strain energy stored in torsional members with various cross sections, here we will only consider the simplest case of linearly elastic circular rods that are homogeneous at each cross section [i.e., $G = G(x)$], as shown in Fig. 11.10. Recall from Eqs. 4.1 and 4.12 that, for this case, the shear strain γ is given by

$$\gamma(x, \rho) = \rho\frac{d\phi(x)}{dx} \tag{11.20}$$

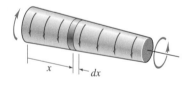

(a)

and the shear stress τ is related to the internal torque at a cross section by the equation

$$\tau(x, \rho) = \frac{T(x)\rho}{I_p(x)} \tag{11.21}$$

(b)

Combining Eqs. 11.9, 11.13, 11.20, and 11.21, we obtain the following expressions for the strain energy due to torsion:

FIGURE 11.10 A circular rod undergoing torsional deformation.

$$\mathcal{U} = \frac{1}{2}\int_0^L \int_A G(x)\left[\rho\frac{d\phi(x)}{dx}\right]^2 dAdx$$

or, since $I_p = \displaystyle\int_A \rho^2dA$,

$$\mathcal{U} = \frac{1}{2}\int_0^L GI_p(\phi')^2dx \tag{11.22}$$

An alternative expression for $\mathcal{U}$ in terms of the torque is

$$\mathcal{U} = \frac{1}{2}\int_0^L \int_A \frac{1}{G(x)}\left[\frac{T(x)\rho}{I_p(x)}\right]^2 dA\,dx$$

or

$$\mathcal{U} = \frac{1}{2}\int_0^L \frac{T^2 dx}{GI_p} \qquad (11.23)$$

For a uniform circular linearly elastic rod subjected to end torques T, as shown in Fig. 11.11, Eq. 11.23 gives

$$\mathcal{U} = \frac{T^2 L}{2GI_p} \qquad (11.24)$$

Bending of Beams—Flexural Strain Energy.

A beam, like the one in Fig. 11.12, has flexural stress σ_x and shear stress τ_{xy} that vary with position in the beam. From Eq. 11.14 it can be seen that the contributions of normal stress and shear stress to the strain energy can be treated separately. We begin by considering the contribution of the flexural stress σ_x (and strain ϵ_x), and we employ the Bernoulli-Euler beam theory of Sections 6.2, 6.3, and 7.2. Recall from Eqs. 6.3 and 7.5 that the extensional strain ϵ_x is related to the local radius of curvature, $\rho(x)$, of the deflection curve and to the deflection $v(x)$ by

FIGURE 11.11 A uniform, end-loaded torsion member.

$$\epsilon_x(x,\ y) = \frac{-y}{\rho(x)} \doteq -y\frac{d^2v(x)}{dx^2} \qquad (11.25)$$

where y is the distance from the neutral axis of the cross section. When Young's modulus is independent of position in the cross section, that is, $E = E(x)$, the flexural stress σ_x is related to the internal bending moment, $M(x)$, by the flexure formula, that is, by

$$\sigma_x(x,\ y) = \frac{-M(x)y}{I(x)} \qquad (11.26)$$

From Eqs. 11.9, 11.11, 11.25, and 11.26 we get

$$\mathcal{U} = \frac{1}{2}\int_0^L \int_A E(x)\left[\frac{-y}{\rho(x)}\right]^2 dA\,dx \doteq \frac{1}{2}\int_0^L \int_A E(x)\left(\frac{d^2v(x)}{dx^2}\right)^2 y^2\,dA\,dx$$

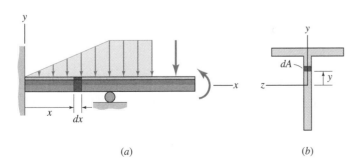

FIGURE 11.12 A beam undergoing flexural and shear deformation.

(a)   (b)

or

$$\mathcal{U}_\sigma = \frac{1}{2} \int_0^L EI(v'')^2 dx \qquad (11.27)$$

where the subscript σ emphasizes the fact that this is the *strain energy due to flexural stress*. In terms of the bending moment $M(x)$, the strain energy due to flexure is given by

$$\mathcal{U}_\sigma = \frac{1}{2} \int_0^L \int_A \frac{1}{E(x)} \left[\frac{-M(x)y}{I(x)} \right]^2 dA dx$$

or

$$\mathcal{U}_\sigma = \frac{1}{2} \int_0^L \frac{M^2 dx}{EI} \qquad (11.28)$$

Bending of Beams—Shear-Strain Energy. The shear stress in a beam also contributes to the strain energy that is stored in the beam. In Section 6.8 an equilibrium analysis led to the shear-stress formula of Eq. 6.65, namely

$$\tau_{xy} = \frac{V(x)Q(x, y)}{I(x)t(y)} \qquad (11.29)$$

where $V(x)$ is the transverse shear force. Combining Eqs. 11.9, 11.13, and 11.29, we get

$$\mathcal{U}_\tau = \frac{1}{2} \int_0^L \int_A \frac{1}{G(x)} \left[\frac{V(x)Q(x, y)}{I(x)t(y)} \right]^2 dA dx$$

or

$$\mathcal{U}_\tau = \frac{1}{2} \int_0^L \left[\frac{V^2(x)}{G(x)I^2(x)} \int_A \frac{Q^2(x, y)}{t^2(y)} dA \right] dx \qquad (11.30)$$

To simplify this expression for $\mathcal{U}_\tau$, let us define a new cross-sectional property f_s, called the *form factor for shear*. Let

$$f_s(x) \equiv \frac{A(x)}{I^2(x)} \int_A \frac{Q^2(x, y)}{t^2(y)} dA \qquad (11.31)$$

(The form factor is a dimensionless number that depends only on the shape of the cross section, so it rarely actually varies with x.) Combining Eqs. 11.30 and 11.31 we get the following expression for the *strain energy due to shear in bending*:

$$\mathcal{U}_\tau = \frac{1}{2} \int_0^L \frac{f_s V^2 dx}{GA} \qquad (11.32)$$

The form factor for shear must be evaluated for each shape of cross section. For example, for a rectangular cross section of width b and height h, the expression

$$Q = \frac{b}{2} \left(\frac{h^2}{4} - y^2 \right)$$

TABLE 11.1   Form Factor f_s for Shear

Section		f_s
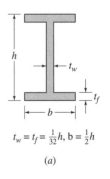 Rectangle		$\dfrac{6}{5}$
Circle		$\dfrac{10}{9}$
Thin tube		2
I-section or box section		$\approx \dfrac{A}{A_{\text{web}}}$

was obtained in Example Problem 6.14. Therefore, from Eq. 11.31 we get

$$f_s = \frac{bh}{(\frac{1}{12}bh^3)^2} \int_{-h/2}^{h/2} \frac{1}{b^2}\left[\frac{b}{2}\left(\frac{h^2}{4} - y^2\right)\right]^2 b\,dy = \frac{6}{5} \tag{11.33}$$

The form factor for other cross-sectional shapes is determined in a similar manner. Several of these are listed in Table 11.1. The approximation for an I-section or box section is based on assuming that the shear force is uniformly distributed over the depth of the web(s).

■■■■■■■■■■■■■■■■■■ E X A M P L E   1 1 . 3 ■■■■■■■■■■■■■■■■■■

Using the form factor approximation for I-sections in Table 11.1, determine the ratio of the shear strain energy $\mathcal{U}_\tau$ to the flexural strain energy $\mathcal{U}_\sigma$ for the wide-flange beam shown in Fig. 1.[3] Assume that $E = 2.6G$, and express your answer in terms of the depth-to-length ratio h/L.

Plan the Solution   We can use equilibrium to determine expressions for $M(x)$ and $V(x)$ that can be substituted into Eqs. 11.28 and 11.32, respectively, to calculate $\mathcal{U}_\sigma$ and $\mathcal{U}_\tau$.

Solution

Equilibrium: From the free-body diagram in Fig. 2,

$$\sum F_y = 0: \qquad\qquad V(x) = -P \tag{1}$$

$$\left(\sum M\right)_a = 0: \qquad\qquad M(x) = -Px$$

Strain Energy: From Eq. 11.28, the strain energy due to flexure is given by

$$\mathcal{U}_\sigma = \frac{1}{2}\int_0^L \frac{M^2 dx}{EI} \tag{2a}$$

$$t_w = t_f = \tfrac{1}{32}h, \ b = \tfrac{1}{2}h$$

(a)

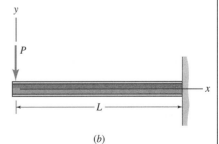

(b)

Fig. 1   A wide-flange beam.

[3]The shape of this cross section is approximately that of a **W**8 × 13 beam.

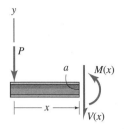

Fig. 2 Free-body diagram.

and from Eq. 11.32, the strain energy due to transverse shear is

$$\mathcal{U}_\tau = \frac{1}{2}\int_0^L \frac{f_s V^2 dx}{GA} \qquad (2b)$$

From Table 11.1, f_s can be approximated by

$$f_s = \frac{A}{A_{\text{web}}} \qquad (3)$$

For the cross section shown in Fig. 1a,

$$A = 2bt_f + (h - 2t_f)t_w = \frac{62}{1024}h^2 = 0.060547h^2$$

$$I = \frac{1}{12}[bh^3 - (b - t_w)(h - 2t_f)^3] = 0.009480h^4 \qquad (4)$$

$$f_s = \frac{A}{A_w} = \frac{62}{30} = 2.066667$$

Combining Eqs. (1) and (2), we get

$$\mathcal{U}_\tau = \frac{1}{2}\frac{f_s}{GA}\int_0^L (-P)^2 dx = \frac{f_s P^2 L}{2GA}$$
$$\mathcal{U}_\sigma = \frac{1}{2}\left(\frac{1}{EI}\right)\int_0^L (-Px)^2 dx = \frac{P^2 L^3}{6EI} \qquad (5)$$

Therefore,

$$\frac{\mathcal{U}_\tau}{\mathcal{U}_\sigma} = \frac{\dfrac{f_s P^2 L}{2GA}}{\dfrac{P^2 L^3}{6EI}} \qquad (6)$$

Using the given ratio of $E/G = 2.6$, together with other values from Eqs. (4), we get

$$\frac{\mathcal{U}_\tau}{\mathcal{U}_\sigma} = \frac{3EI f_s}{GAL^2} \cong 2.5\left(\frac{h}{L}\right)^2 \qquad \textbf{Ans.} \quad (7)$$

A value of $\mathcal{U}_\tau/\mathcal{U}_\sigma = 0.05$ (i.e., 5%) corresponds to an L/h ratio of approximately 7, which is a relatively short beam. Therefore, shear strain energy and, correspondingly, shear deformation, may be neglected except in the case of short, stubby beams.

Review the Solution   The answer in Eq. (7) is dimensionless, as it should be. Furthermore, the dependence on h/L seems reasonable. That is, shear becomes important when the length of the beam is too short to have large values of bending moment. Finally, we should check the answer by rechecking each formula and calculation.

11.4 WORK-ENERGY PRINCIPLE FOR CALCULATING DEFLECTIONS

In the remaining sections of this chapter, we will examine several energy principles, that is, several ways in which energy methods can be used to solve deformable-body problems. The simplest (but most limited) of these is the *principle of work and energy*. We will consider only the case of slowly applied loading, so that the kinetic energy (e.g., 1/2 mass × velocity-squared) can be ignored. We will not consider other forms of energy such as thermal energy, chemical energy, and electromagnetic energy. Therefore, **if the stresses in a body do not exceed the elastic limit, all of the work done on a body by external forces is stored in the body as elastic strain energy.** The formula

$$\boxed{\mathcal{W}_{\text{ext}} = \mathcal{U}}$$

Work-Energy Principle (11.34)

is a statement of this *Work-Energy Principle*. This principle is not restricted to linearly elastic behavior. The next two Example Problems illustrate the use of the above Work-Energy Principle to calculate the displacement due to a single load applied to a deformable body.

■■■■■■■■■■■■■■■■□ E X A M P L E   1 1 . 4 ■■■■■■■■■■■■■■■■

Use the Work-Energy Principle to calculate the transverse deflection of the beam in Fig. 1*a* and the end slope of the beam in Fig. 1*b*. Assume that both beams remain linearly elastic under the given loads, and let *EI* = constant. Neglect shear strain energy of the beams.

Plan the Solution We can use Eqs. 11.3a and 11.3c to give the work done by the external loads *P* and *M*, respectively. The strain energy in each case can be determined by the use of Eq. 11.28. Finally, the deflection Δ and slope θ will result from applying the Work-Energy Principle, Eq. 11.34.

Solution In the following solution, let subscript *a* apply to the beam in Fig. 1*a* and subscript *b* to the beam in Fig. 1*b*, and let subscript 1 apply to the span $0 < x < L$ and the subscript 2 apply to the span $L < x < 2L$.

Equilibrium: To obtain expressions for *M(x)* for each beam, we first draw the necessary free-body diagrams.

From Fig. 2*a*,

$$\left(\sum M\right)_C = 0: \qquad\qquad B_y = 2P \qquad\qquad (1a)$$

From Fig. 2*b*,

$$\left(\sum M\right)_{C_1} = 0: \qquad\qquad M_{a_1} = -Px \qquad\qquad (1b)$$

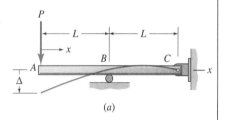

(a)

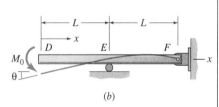

(b)

Fig. 1 Deflection of the beams.

567

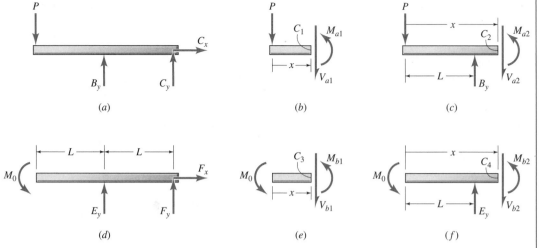

Fig. 2 Free-body diagrams.

From Fig. 2c,

$$\left(\sum M\right)_{C_2} = 0: \quad M_{a2} = -Px + 2P(x - L) = P(x - 2L) \tag{1c}$$

From Fig. 2d,

$$\left(\sum M\right)_F = 0: \quad E_y = \frac{M_0}{L} \tag{1d}$$

From Fig. 2e,

$$\left(\sum M\right)_{C_3} = 0: \quad M_{b_1} = -M_0 \tag{1e}$$

From Fig. 2f,

$$\left(\sum M\right)_{C_4} = 0: \quad M_{b2} = -M_0 + \frac{M_0}{L}(x - L) = \frac{M_0}{L}(x - 2L) \tag{1f}$$

Strain Energy: The strain energy stored in a beam due to flexure is given by Eq. 11.28 (modified for length $2L$).

$$\mathcal{U} = \frac{1}{2}\int_0^{2L} \frac{M^2 dx}{EI} = \frac{1}{2}\int_0^L \frac{M_1^2 dx}{EI} + \frac{1}{2}\int_L^{2L} \frac{M_2^2 dx}{EI} \tag{2}$$

Combining Eqs. (1) and (2) we get

$$\mathcal{U}_a = \frac{1}{2}\int_0^L \frac{(-Px)^2 dx}{EI} + \frac{1}{2}\int_L^{2L} \frac{[P(x - 2L)]^2 dx}{EI} = \frac{P^2 L^3}{3EI} \tag{3a}$$

$$\mathcal{U}_b = \frac{1}{2}\int_0^L \frac{(-M_0)^2 dx}{EI} + \frac{1}{2}\int_L^{2L} \frac{\left[\frac{M_0}{L}(x - 2L)\right]^2 dx}{EI} = \frac{2M_0^2 L}{3EI} \quad (3b)$$

Work of External Loads: Equations 11.3a and 11.3b give

$$\mathcal{W}_a = \frac{1}{2}P\Delta, \qquad \mathcal{W}_b = \frac{1}{2}M_0\theta \qquad (4a\text{-}b)$$

Work-Energy Principle: Applying the Work-Energy Principle, Eq. 11.34, we get

$\mathcal{W}_a = \mathcal{U}_a$: $\qquad\qquad \Delta = \dfrac{2PL^3}{3EI}$ $\qquad\qquad$ **Ans.** (5a)

$\mathcal{W}_b = \mathcal{U}_b$: $\qquad\qquad \theta = \dfrac{4M_0L}{3EI}$ $\qquad\qquad$ **Ans.** (5b)

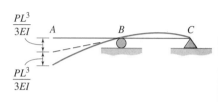

Review the Solution The preceding answers can be checked by applying the method of superposition of deflections discussed in Section 7.6 and using data from Table E.1 and E.2 of Appendix E. The results are shown in Fig. 3.

A "ballpark" estimate of each answer can be obtained by just ignoring the slope at B and E. From Table E.1 this would give estimates of

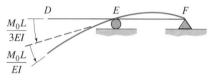

$$\Delta > \frac{PL^3}{3EI}, \ \theta > \frac{M_0L}{EI}$$

Fig. 3

Calculating the displacement of various points in a "complex" structure, like a planar truss, is not a simple matter. However, if the "right question" is asked, the Work-Energy Principle provides a simple, straightforward way to calculate displacements, as the next example problem illustrates.

■■■■■■■■■■■■■□ E X A M P L E 1 1 . 5 □■■■■■■■■■■■■■■

For the two-bar truss in Fig. 1, use the Work-Energy Principle to compute the displacement Δ_B of node B in the direction of the load P. Let $A_1 = A_2 = 1.2$ in^2, and $E_1 = E_2 = 10 \times 10^3$ ksi. The members remain linearly elastic under the given loading.

Plan the Solution For each member in the truss, we can use Eq. 11.19 to determine the strain energy, using the member axial forces obtained from equilibrium of the joint at B. Since strain energy is a scalar quantity, the total strain energy is the sum of the strain energies in all the members. Since the structure remains linearly elastic, and since Δ_B is in the same direction as P, the work of the external load is given by Eq. 11.3a, as in the previous example. Finally, the Work-Energy Principle, Eq. 11.34, will lead to an equation for the unknown displacement Δ_B.

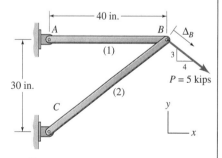

Fig. 1 A two-member planar truss.

Solution

Equilibrium: Since the truss in Fig. 1 is statically determinate, we can compute the member forces using only equilibrium equations and the free-body diagram

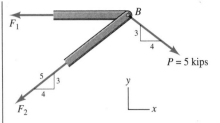

Fig. 2 The free-body diagram.

shown in Fig. 2.

$$\sum F_y = 0: \qquad -\frac{3}{5}F_2 - \frac{3}{5}P = 0 \rightarrow F_2 = -P \qquad (1a)$$

$$\sum F_x = 0: \qquad -F_1 - \frac{4}{5}(-P) + \frac{4}{5}P = 0 \rightarrow F_1 = \frac{8}{5}P \qquad (1b)$$

Strain Energy: Since strain energy is a scalar quantity,

$$\mathcal{U} = \mathcal{U}_1 + \mathcal{U}_2 \qquad (2)$$

For each member, the strain energy is given by Eq. 11.19. Therefore,

$$\mathcal{U} = \frac{F_1^2 L_1}{2A_1 E_1} + \frac{F_2^2 L_2}{2A_2 E_2}$$

$$\mathcal{U} = \frac{(8 \text{ kips})^2 (40 \text{ in.}) + (-5 \text{ kips})^2 (50 \text{ in.})}{2(1.2 \text{ in}^2)(10 \times 10^3 \text{ ksi})} \qquad (3)$$

Therefore,

$$\mathcal{U} = 0.1588 \text{ kip} \cdot \text{in.} \qquad (4)$$

Work of External Load: The work of the external load P is

$$\mathcal{W}_{\text{ext}} = \frac{1}{2}P\Delta_B = \frac{1}{2}(5 \text{ kips})(\Delta_B \text{ in.}) = 2.5\Delta_B \text{ kip} \cdot \text{in.} \qquad (5)$$

since, as noted above, the structure is linear, and Δ_B is in the same direction as P.

Work-Energy Principle: From Eq. 11.34,

$$\mathcal{W}_{\text{ext}} = \mathcal{U} \qquad (6)$$

Therefore,

$$2.5\Delta_B \text{ kip} \cdot \text{in.} = 0.1588 \text{ kip} \cdot \text{in.}$$

and, finally,

$$\Delta_B = 0.064 \text{ in.} \qquad \textbf{Ans.} \qquad (7)$$

Review the Solution We could solve this problem by the method discussed in Section 3.8. However, we just need to see if our answer is reasonable. We know that the elongation of a single axial member is given by $e_i = F_i L_i / A_i E_i$ (Eq. 3.13). Therefore,

$$e_1 = \frac{(8 \text{ kips.})(40 \text{ in.})}{(1.2 \text{ in.}^2)(10 \times 10^3 \text{ ksi})} = 0.0267 \text{ in.}$$

$$e_2 = \frac{(-5 \text{ kips})(50 \text{ in.})}{(1.2 \text{ in.}^2)(10 \times 10^3 \text{ ksi})} = -0.0208 \text{ in.}$$

Without carrying out the geometrical calculations necessary to relate e_1 and e_2 to Δ_B, we can see that our answer in Eq. (7) is reasonable.

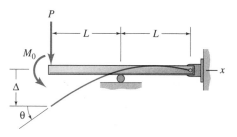

FIGURE 11.13 A beam with two loads.

The Work-Energy Principle provides a relatively simple, straightforward procedure for calculating the displacement of a body at the point of application of a single load and in the direction of the load, as illustrated in Example Problems 11.4 and 11.5. However, suppose the two loads, P and M_0, are applied slowly and simultaneously to a beam, as illustrated in Fig. 11.13. Following the procedure of Example Problem 11.4, we have

$$M_1(x) = -Px - M_0 \tag{11.35}$$

$$M_2(x) = \left(P + \frac{M_0}{L}\right)(x - 2L)$$

Then, substitution of these moments into Eq. 11.28 gives

$$\mathcal{U} = \frac{P^2L^3}{3EI} + \frac{5PM_0L^2}{6EI} + \frac{2M_0^2L}{3EI} \tag{11.36}$$

The work done by P and M_0 when they are applied simultaneously to the beam is

$$\mathcal{W}_{\text{ext}} = \frac{1}{2}P\Delta + \frac{1}{2}M_0\theta \tag{11.37}$$

It is clear that we cannot use the Work-Energy Principle, $\mathcal{W}_{\text{ext}} = \mathcal{U}$, to determine either Δ or θ since there are two unknowns and only one equation.

In conclusion, **the Work-Energy Principle is useful only for determining the displacement at the point of application of a single load and in the direction of that load**. And, of course, it applies only to elastic deformation. Therefore, we now turn to more "powerful" energy methods that are based on quantities called virtual work and complementary virtual work.

*11.5 VIRTUAL WORK   ■■■■■■■■■■■

Because of the limited applicability of the Work-Energy Principle discussed in Section 11.4, more widely applicable methods based on virtual work and complementary virtual work were developed.[4] In Sections 11.5 and 11.6 we will consider analytical methods based on virtual work, and, in Sections 11.7 and 11.8 we will examine complementary virtual work methods.

---

[4]In the present mechanics context, the relevant definition of the word *virtual* is "being in essence or effect, but not in fact." (Ref.: *Webster's New Collegiate Dictionary*.)

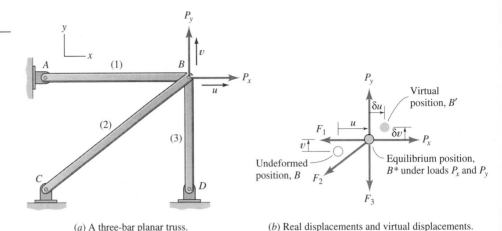

(a) A three-bar planar truss. (b) Real displacements and virtual displacements.

FIGURE 11.14 An example of real displacements and virtual displacements.

Virtual Displacements and Virtual Work. Consider the pin that connects the three two-force members of the planar truss in Fig. 11.14a to be a "particle" that is acted on by the forces shown in Fig. 11.14b. (A similar truss was analyzed in Example Problem 3.15.) When there are no loads on the truss, the truss is undeformed, and the pin is at its <u>initial position</u>, B. When loads P_x and P_y are applied, the pin moves a distance u in the x direction and a distance v in the y direction to its (true) <u>equilibrium position</u>, B*. Now, imagine that the pin moves infinitesimal distances δu in the x direction and δv in the y direction. These infinitesimal, imaginary displacements are called *virtual displacements.*[5]

> A **virtual displacement** *is an infinitesimal, imaginary displacement, denoted by the symbol $\delta(\cdot)$, that is consistent with all kinematic constraints. That is, virtual displacements must satisfy the same boundary conditions and compatibility conditions that the real displacements must satisfy. Real forces are not altered by virtual displacements.*

(Note carefully the distinction between the true displacements u and v caused by the loads, and the infinitesimal virtual displacements that "exist" only in our imagination.)

> **Virtual work**, *designated by δW, is the work done by real forces when virtual displacements occur. The real forces are assumed to remain constant.*

To be precise, virtual work is the work that would be done if the points of application of the forces were to actually move through the (imaginary) virtual displacements. Hence, for the example in Fig. 11.14b,

$$\delta W = \left(\sum F_x \right) \delta u + \left(\sum F_y \right) \delta v \qquad (11.38)$$

---

[5]Although the Greek letter δ has previously been used to denote displacements themselves, in the remainder of this chapter the symbol δ will be reserved for use as the *virtual operator,* much the same as the letter d is the *differential operator* in dx, dV, and so forth. Capital delta (Δ) will be used for displacement.

Equilibrium, and the Principle of Virtual Displacements. At position $B*$ the pin in Fig. 11.14b is in static equilibrium. From Newton's Second Law,

$$\Sigma F_x = 0, \qquad \Sigma F_y = 0 \qquad (11.39)$$

at the equilibrium position. Therefore, the following statement holds:

> *If a particle is in equilibrium under a system of forces, then for <u>any</u> virtual displacement, the virtual work δW is zero.*

The converse, which is called the *Principle of Virtual Displacements,* is also true.[6]

> *If the virtual work done on a particle is zero for <u>any arbitrary</u> kinematically-admissible virtual displacements, then the particle is in equilibrium.*

Thus, if, for arbitrary virtual displacement from a given configuration,

$$\boxed{\delta W = 0} \qquad \begin{array}{l}\textbf{Principle of Virtual} \\ \textbf{Displacements}\end{array} \qquad (11.40)$$

then the given configuration is an equilibrium configuration. Referring back to Eq. 11.38, if $\delta W = 0$ regardless of the (nonzero) values that we might pick for δu and δv, then it must be true that $\Sigma F_x = \Sigma F_y = 0$.

Virtual Work for Deformable Bodies.

The Principle of Virtual Displacements can be extended to any collection of particles—we just get more virtual displacements and more equilibrium equations, but only one virtual work. For example, if there were two particles, we would get

$$\delta W = \left[\left(\Sigma F_x \right)_1 \delta u_1 + \left(\Sigma F_y \right)_1 \delta v_1 \right] + \left[\left(\Sigma F_x \right)_2 \delta u_2 + \left(\Sigma F_y \right)_2 \delta v_2 \right]$$

But, we are interested in applying virtual work methods to deformable bodies. Let the (deformable) spring AB and (rigid) block C in Fig. 11.15 comprise a system that is in equilibrium under the action of a static load P. Since every particle in the spring AB and block C is in equilibrium, we can say that $\delta W = 0$ applies to the entire system. But the force P is the only external force that can do work, because it is the only external force whose point of application can move. Therefore, the virtual work done by the external force is

$$\delta W_{\text{ext}} = P \delta u$$

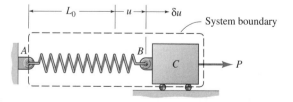

FIGURE 11.15 A system in equilibrium.

---

[6]This is also sometimes called the *Principle of Virtual Work.*

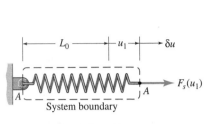

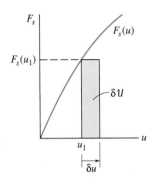

FIGURE 11.16 A nonlinear, elastic spring treated as a deformable system.

(*a*) A deformable system.

(*b*) Force-elongation curve.

In addition, if block C were to undergo a virtual displacement δu, there would be some virtual work done by the internal forces in the deformable spring. Thus, we can state the following *Principle of Virtual Displacements for a deformable system*:

> *If the virtual work done on a deformable system during any kinematically admissible displacements (changes from a given configuration) of the system is zero, then the system is in equilibrium in the given configuration.*

$$\delta \mathcal{W} = \delta \mathcal{W}_{\text{ext}} + \delta \mathcal{W}_{\text{int}} = 0 \qquad (11.41)$$

In Eq. 11.41, $\delta \mathcal{W}_{\text{int}}$ is the virtual work done by internal forces in the system under consideration.

What is $\delta \mathcal{W}_{\text{int}}$ and how is it calculated? Let us isolate just the spring in Fig. 11.15 and call it our "deformable system," as depicted in Fig. 11.16*a*. And let us suppose that this is a nonlinear elastic spring whose force versus elongation curve is shown in Fig. 11.16*b*. The system is in equilibrium in Fig. 11.16*a* because the applied force $F_s(u_1)$ corresponds to the displacement u_1. Therefore, we can say that

$$\delta \mathcal{W}_{\text{ext}} + \delta \mathcal{W}_{\text{int}} = 0$$

We know that $\delta \mathcal{W}_{\text{ext}}|_{u_1} = F_s(u_1)\delta u$, so

$$\delta \mathcal{W}_{\text{int}}|_{u_1} = -F_s(u_1)\delta u$$

That is, the virtual work done by the internal forces in an elastic deformable body when the body undergoes virtual displacements is equal to the negative of the virtual work done by the external forces acting on the deformable body alone. Therefore, referring to Fig. 11.16*b*, we see that the virtual work of internal forces and the strain energy are related by

$$\delta \mathcal{W}_{\text{int}} = -\delta \mathcal{U} \qquad (11.42)$$

where $\delta \mathcal{U}$ is the infinitesimal, imaginary increment in the strain energy that would result from the body undergoing a virtual displacement δu, as illustrated in Fig. 11.16*b*.

We will need to obtain $\delta \mathcal{U}$ when the strain energy is expressed as a function of several displacements, say $\mathcal{U} = \mathcal{U}(q_1, q_2, \ldots, q_n)$, where the q's are independent displacements. Then $\delta \mathcal{U}$ can be determined in the same manner as the differential of $\mathcal{U}$, $d\mathcal{U}$, namely by

$$\delta \mathcal{U} = \frac{\partial \mathcal{U}}{\partial q_1}\delta q_1 + \frac{\partial \mathcal{U}}{\partial q_2}\delta q_2 + \cdots + \frac{\partial \mathcal{U}}{\partial q_n}\delta q_n \qquad (11.43)$$

Principle of Virtual Displacements.
By combining Eqs. 11.41 and 11.42 we arrive at the *Principle of Virtual Displacements applied to elastic bodies,* which can be stated as follows:

> *Among all kinematically admissible configurations of an elastic deformable body, the actual equilibrium configuration is the one that satisfies Eq. 11.44 when the body undergoes arbitrary virtual displacements from that configuration.*

$$\delta W_{\text{ext}} = \delta U$$

Principle of Virtual Displacements (11.44)

A *kinematically admissible configuration* is a configuration that satisfies all relevant kinematic boundary conditions and all relevant equations of deformation compatibility.

In the next example problem we apply the Principle of Virtual Displacements to a problem that could also be solved by the Work-Energy Principle of Section 11.4. Then, in Example Problem 11.7, Eq. 11.44 will be used to solve a deflection problem that cannot be solved by the method of Section 11.4.

■■■■■■■■■■■■■■■■■ EXAMPLE 11.6 ■■■■■■■■■■■■■■■■■■■

For the two-segment rod in Fig. 1, use the Principle of Virtual Displacements for deformable bodies to solve for the following: (a) the displacement u_B of node B; and (b) the internal forces F_1 and F_2 in the two elements of the statically indeterminate rod system.

Fig. 1

The rod properties are A_1, E_1, L_1, and A_2, E_2, L_2, respectively.

Plan the Solution Using the compatibility of displacements at node B and using a strain energy formula from Eq. 11.8, we can determine an expression for $\mathcal{U}(u_B)$, the strain energy of the two-bar system in terms of the displacement of node B. The load P_B does virtual work when node B moves (actually, when it is imagined to move) through a virtual displacement δu_B. Thus, we can apply Eq. 11.44 to get an equation of equilibrium for node B.

Solution
(a) Determine u_B, the displacement of node B.

Strain Energy: It will simplify notation if we use the element stiffness coefficients

$$k_1 = \frac{A_1 E_1}{L_1}, \qquad k_2 = \frac{A_2 E_2}{L_2} \tag{1}$$

Then, using Eq. 11.8, we can write the total strain energy as

$$\mathcal{U} = \mathcal{U}_1 + \mathcal{U}_2 = \frac{1}{2}k_1 e_1^2 + \frac{1}{2}k_2 e_2^2 \tag{2}$$

where e_i is the elongation of rod element i.

Geometry of Deformation: According to the principle of virtual displacements, we are to consider only "kinematically admissible deformations." Therefore, the kinematics of deformation, including the compatibility of the displacements of the two rods at node B, must be incorporated into the strain energy so that all kinematic requirements will be automatically satisfied.

$$e_1 = u_B, \qquad e_2 = -u_B \tag{3}$$

Virtual Work of the External Force: The point of application (node B) of external force P_B is imagined to move through a virtual displacement δu_B (in the same direction as u_B), doing virtual work

$$\delta \mathcal{W}_{\text{ext}} = P_B \delta u_B \tag{4}$$

Principle of Virtual Displacements: From Eq. 11.44,

$$\delta \mathcal{W}_{\text{ext}} = \delta \mathcal{U} \tag{5}$$

Combining Eqs. (2) and (3), we get

$$\mathcal{U} = \frac{1}{2}(k_1 + k_2)u_B^2 \tag{6}$$

Since the strain energy $\mathcal{U}$ is a function of a single displacement u_B, Eq. 11.43 reduces to

$$\delta \mathcal{U} = \frac{d\mathcal{U}}{du_B}\delta u_B = (k_1 + k_2)u_B \delta u_B \tag{7}$$

Substituting Eqs. (4) and (7) into Eq. (5) we get

$$P_B \delta u_B = (k_1 + k_2)u_B \delta u_B$$

or

$$[(k_1 + k_2)u_B - P_B]\delta u_B = 0 \tag{8}$$

The Principle of Virtual Displacements states that the system is in equilibrium if Eq. (8) is satisfied for *any arbitrary virtual displacement,* in this case for arbitrary δu_B. Therefore, the expression in square brackets in Eq. (8) must vanish, giving

$$u_B = \frac{P_B}{k_1 + k_2} \qquad \text{Ans.} \tag{9}$$

where k_1 and k_2 are given in Eqs. 1.

(b) Determine the element forces F_1 and F_2. From Eq. 3.14,

$$F_i = k_i e_i \tag{10}$$

Therefore, combining Eqs. (3), (9), and (10), we get

$$F_1 = \frac{k_1 P_B}{k_1 + k_2}, \qquad F_2 = \frac{-k_2 P_B}{k_1 + k_2} \qquad \text{Ans.} \quad (11)$$

Review the Solution   We could check this solution by using the methods of Chapter 3. Since k_i has the dimensions of F/L, the dimensions in Eq. (9) are correct. The force P_B has to stretch rod AB and also compress rod BC. Therefore, the denominator in Eq. (9) correctly reflects the addition of k_1 and k_2 in resisting P_B.

Castigliano's First Theorem. The Principle of Virtual Displacements, Eq. 11.44, can be used to formulate the equations of equilibrium of any elastic body. When the deformation of the body can be characterized by a finite number of independent displacements, Eq. 11.43 provides a convenient procedure for expressing δU in terms of the virtual displacements, δq_i. The virtual work of external forces can then be expressed in the form

$$\delta W_{\text{ext}} = \sum_{i=1}^{N} P_i \delta q_i \qquad (11.45)$$

where N is the number of independent displacements, q_i. The P_i's are called *generalized forces*. The correct values or expressions for the P_i's are determined by formulating δW_{ext}.

Combining Eqs. 11.43 and 11.45, we get the following form of the Principle of Virtual Displacements:

$$\sum_{i=1}^{N} \left(P_i - \frac{\partial U}{\partial q_i} \right) \delta q_i = 0$$

But, since the above sum is required to be zero for any arbitrary δq_i's, and since the δq_i's are independent, we get

$$\boxed{P_i = \frac{\partial U}{\partial q_i}, \ i = 1, 2, ..., N} \qquad \begin{array}{l} \text{Castigliano's} \\ \text{First Theorem} \end{array} \quad (11.46)$$

This equation is called *Castigliano's First Theorem.*[7] It can be viewed as an alternative form of the Principle of Virtual Displacements when the deformation of the elastic system can be characterized by N discrete displacement coordinates, q_i. Equation 11.46 produces equilibrium equations written in terms of displacements.

The following example problem could be solved by applying the Principle of Virtual Displacements. However, we will use the slightly shorter route of using Castigliano's First Theorem. This example problem demonstrates the power of Castigliano's First Theorem to provide solutions to problems that cannot be solved by the Work-Energy Principle of Section 11.4.

---

[7]The First Theorem and Second Theorem of Italian engineer Alberto Castigliano (1847–1884) appeared in his 1873 dissertation for the engineer's degree at Turin, Italy. [Ref. 11-1]

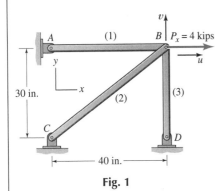

Fig. 1

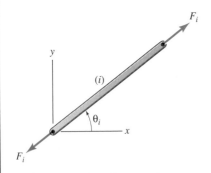

Fig. 2 Notations for truss element.

A three-element truss has the configuration shown in Fig. 1. A horizontal load of 4 kips is applied to the truss through a pin at node B. Each member has a cross-sectional area of 1.2 in², and all of them are made of aluminum ($E = 10 \times 10^3$ ksi). Use Castigliano's First Theorem to determine u and v, the horizontal and vertical displacements, respectively, at node B.

Plan the Solution The strain energy of an element can be written in terms of its elongation by using Eq. 11.8. The elongation of each element can be related to the nodal displacements u and v through Eq. 3.22. Since there are two nodal displacements, we will have two corresponding virtual displacements δu and δv. Using Castigliano's First Theorem, Eq. 11.46, we will get two equilibrium equations to solve simultaneously for u and v.

Solution

Strain Energy: We can borrow part of this solution from Example Problem 3.15 (Section 3.8).

$$k_1 = \left(\frac{AE}{L}\right)_1 = 300 \text{ kips/in.}, \quad \cos\theta_1 = 1.0, \quad \sin\theta_1 = 0.0$$

$$k_2 = \left(\frac{AE}{L}\right)_2 = 240 \text{ kips/in,} \quad \cos\theta_2 = 0.8, \quad \sin\theta_2 = 0.6 \quad (1)$$

$$k_3 = \left(\frac{AE}{L}\right)_3 = 400 \text{ kips/in.}, \quad \cos\theta_3 = 0.0, \quad \sin\theta_3 = 1.0$$

where θ_i is defined in Fig. 2.

Using Eq. 11.8, we can write the total strain energy in the three-bar truss as

$$\mathcal{U} = \mathcal{U}_1 + \mathcal{U}_2 + \mathcal{U}_3 = \frac{1}{2}k_1e_1^2 + \frac{1}{2}k_2e_2^2 + \frac{1}{2}k_3e_3^2 \quad (2)$$

From Eq. 3.22, the elongation of each element can be related to the nodal displacements u and v by

$$e_i = u\cos\theta_i + v\sin\theta_i \quad (3)$$

Then,

$$e_1 = u, \quad e_2 = 0.8u + 0.6v, \quad e_3 = v \quad (4)$$

Combining Eqs. 2 and 4 we get

$$\mathcal{U} = [150u^2 + 120(0.8u + 0.6v)^2 + 200v^2] \text{ kip} \cdot \text{in.} \quad (5)$$

Virtual Work of External Force: The horizontal force P_x acting through a horizontal virtual displacement of δu inches does virtual work given by

$$\delta\mathcal{W}_{\text{ext}} = 4\delta u \text{ kip} \cdot \text{in.}$$

For this case, Eq. 11.45 has the form

$$\delta W_{ext} = P_x \delta u + P_y \delta v \tag{6}$$

where

$$P_x = 4 \text{ kips}, \qquad P_y = 0 \text{ kips} \tag{7}$$

Castigliano's First Theorem: From Eq. 11.46, the two equations of equilibrium for this truss are given by

$$\frac{\partial U}{\partial u} = P_x, \qquad \frac{\partial U}{\partial v} = P_y \tag{8}$$

so, differentiating Eq. (5) with respect to u and v and combining the results with Eqs. (7) and (8), we get

$$453.6u + 115.2v = 4 \text{ kips} \tag{9}$$
$$115.2u + 486.4v = 0 \text{ kips}$$

The left-hand sides of these two equations are exactly the same as those of the *nodal equilibrium equations in terms of nodal displacements* that we got in Eq. (4) of Example Problem 3.15 by employing the *displacement method.* The right-hand sides are just the nodal force components P_x and P_y. The solution of Eqs. (9) is

$$u = 9.38(10^{-3}) \text{ in.} \qquad \text{Ans.} \quad (10)$$
$$v = -2.22(10^{-3}) \text{ in.}$$

Review the Solution   We got essentially the same equations of equilibrium by energy methods as we got by using the displacement method in Example Problem 3.15.

Although only linearly elastic systems have been considered in the example problems used to illustrate the Principle of Virtual Displacements and Castigliano's First Theorem, these methods only require that the system be elastic, not necessarily linear.[8]

Displacement Method. Note that the primary solution in Example Problem 11.6 was a solution for the nodal displacement u_B, and recall from earlier chapters (e.g., Chapters 3, 4, and 7) that this was a characteristic of *displacement-method* solutions. Also, the use of Castigliano's First Theorem in Example Problem 11.7 produced essentially the same two nodal equilibrium equations that were obtained by use of the displacement method in Example Problem 3.15. We can conclude that the Principle of Virtual Displacements and Castigliano's First Theorem provide alternative ways to formulate a *displacement-method solution* for deformable-body mechanics problems.

---

[8]See Ref. 11-2, Section 10.6, for an example that treats a truss with nonlinearly elastic members.

Complementary work, $\mathcal{W}^c$, was defined (for an axially loaded member) in Eq. 11.4. Complementary strain energy was defined, for the same case, in Eq. 11.6 and illustrated in Figs. 11.3 and 11.4. Let us now consider how these concepts can be used to solve deformable-body mechanics problems.

Virtual Forces. Virtual displacements, δu, δq_i, and so forth, were defined in Section 11.5 and used in the formulation of analytical methods based on virtual work. Recall that a virtual displacement was defined as an imaginary, infinitesimal displacement that does not affect the forces in a body and that is consistent with the geometric compatibility conditions of the body. In contrast, a virtual force is defined as follows:

> A **virtual force**, δF, is an infinitesimal, imaginary force that, together with other virtual forces applied to a body in equilibrium, does not affect the equilibrium state of the body.

Figure 11.17 illustrates real forces P_i and virtual forces δP_i (including real couples M_i and virtual couples δM_i) applied to a deformable body. The body in Fig. 11.17 is in equilibrium under the action of the (real) applied loads P_1, P_2, ..., M_1, and so forth. The virtual forces and couples are arbitrary, but they must also satisfy the same equilibrium equations as the real forces. Therefore, when virtual forces are applied as external loads, *virtual stresses* are induced within the body so that equilibrium of every part of the body is maintained. Figure 11.18a shows an axial load P_1 applied to a rod, and Fig. 11.18b depicts the resulting virtual normal stress $\delta\sigma_x$ required at cross-section x to maintain equilibrium of the free body in Fig. 11.18b when P_1 is increased to $P_1 + \delta P$.

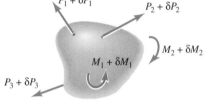

FIGURE 11.17 An illustration of virtual forces.

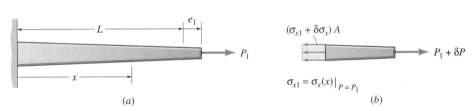

FIGURE 11.18 Virtual force and virtual stress.

Complementary Work and Complementary Strain Energy. *Complementary work* was defined in Eq. 11.4 as

$$\mathcal{W}^c(P_1) = \int_0^{P_1} e(P)\,dP \qquad \begin{matrix}(11.4)\\ \text{repeated}\end{matrix}$$

and *complementary strain energy* was defined as

$$\mathcal{U}^c(P) = \mathcal{W}^c(P)\big|_{\text{done on elastic body}} \qquad \begin{matrix}(11.6)\\ \text{repeated}\end{matrix}$$

That is, when a force is applied to an elastic body, the complementary work that is done on the elastic body is stored in the body as complementary strain energy. Note that complementary virtual work is expressed as a function of force, not as a function of displacement as virtual work is. As indicated in Fig. 11.19b *complementary strain energy density* is the area above the stress-strain curve up to the value of the applied stress.

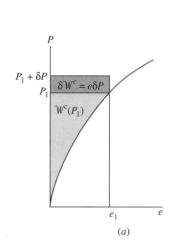
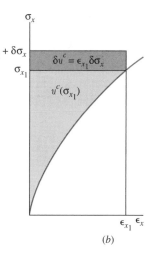

Complementary Energy Methods

FIGURE 11.19
(*a*) Complementary work and complementary virtual work; (*b*) complementary energy density and virtual complementary energy density.

Principle of Virtual Forces. The *Principle of Virtual Forces,* or the *Principle of Complementary Virtual Work,* states that:

> *Among all possible equilibrium states, the actual state of a body is the one that satisfies deformation compatibility.*

In equation form, the *Principle of Virtual Forces* is given by

$$\boxed{\delta \mathcal{W}_{\text{ext}}^c = \delta \mathcal{U}^c}$$

Principle of Virtual Forces (11.47)

In Example Problem 11.8 this principle is applied to a statically determinate problem, illustrating how the Principle of Virtual Forces leads to deformation-compatibility equations. In Example Problem 11.9 it is applied to a statically indeterminate problem.

■■■■■■■■■■■■■■■□ E X A M P L E 1 1 . 8 ■■■■■■■■■■■■■■■■■

Two elastic, but not necessarily linearly elastic, rod elements are connected together at *B* and loaded by axial forces P_B and P_C, as shown in Fig. 1.

The displacements of *B* and *C* under loads P_B and P_C are u_B and u_C respectively. Show that "among all equilibrium states of this system, the state that satisfies Eq. 11.47 is the kinematically admissible state, that is, the state that satisfies deformation compatibility."

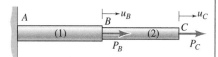

Fig. 1

Plan the Solution The force versus elongation (F_i vs e_i) curve for each (nonlinearly elastic) element can be assumed to be similar to that in Fig. 11.3. The part of the statement that says "among all equilibrium states" can be satisfied by relating internal forces F_1 and F_2 to the external loads P_B and P_C through the use of free-body diagrams and equilibrium equations. The "deformation compatibility" in this problem consists of relating the elongations e_1 and e_2 to the displacements u_B and u_C (e.g., see Chapter 3). The complementary virtual work terms will be $u_B \delta P_B$, $e_1 \delta F_1$, and so forth.

Solution

Equilibrium: Both real forces and virtual forces must satisfy equilibrium. Since this is a statically determinate rod system, we can directly relate

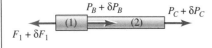

(a) FBD 1.

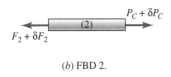

(b) FBD 2.

Fig. 2 Free-body diagrams.

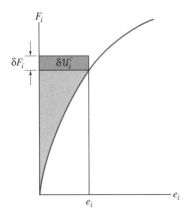

Fig. 3 Force-elongation curve for an element.

the internal forces F_1 and F_2 to the external forces P_B and P_C through equilibrium equations based on the free-body diagrams in Fig. 2. The same is true for virtual forces. Equilibrium must hold for the real forces and also when arbitrary virtual forces are added. Therefore, we must have

$$\left(\sum F \right)_1 = 0: \quad F_1 = P_B + P_C \quad \text{and} \quad \delta F_1 = \delta P_B + \delta P_C$$

$$\left(\sum F \right)_2 = 0: \quad F_2 = P_C \quad \text{and} \quad \delta F_2 = \delta P_C \tag{1}$$

Principle of Virtual Forces: For each element the internal forces are assumed to be related to the elongations by a curve having the form indicated in Fig. 3. Therefore, for element i the virtual force δF_i does an amount of complementary virtual work $e_i \delta F_i$, as indicated in Fig. 3. This is the change in complementary energy in element i.

The Principle of Virtual Forces, Eq. 11.47, states that

$$\delta \mathcal{W}^c = \delta \mathcal{U}^c \tag{2}$$

Writing out the contributions to each side of this equation we get

$$u_B \delta P_B + u_C \delta P_C = e_1 \delta F_1 + e_2 \delta F_2 \tag{3}$$

The virtual forces must satisfy the virtual force equilibrium equations in Eqs. (1). Substituting these into Eq. (3) gives

$$u_B \delta P_B + u_C \delta P_C = e_1 (\delta P_B + \delta P_C) + e_2 (\delta P_C) \tag{4}$$

or

$$(e_1 - u_B) \delta P_B + (e_1 + e_2 - u_C) \delta P_C = 0 \tag{5}$$

Then, since δP_B and δP_C are arbitrary (as long as equilibrium is satisfied; which it is by the step from Eq. (3) to Eq. (4)), Eq. (5) requires that

$$u_B = e_1, \quad u_C = e_1 + e_2 \qquad \textbf{Ans.} \tag{6}$$

These are the equations of deformation compatibility for this problem.

Review the Solution We started by guaranteeing that the virtual forces would always satisfy equilibrium, applied the principle of virtual forces, and arrived at sensible expressions for the kinematic relationships between nodal displacements and elongations. Therefore, we illustrated the statement of the principle of virtual forces, as requested. Although we allowed for possible nonlinear material behavior, this application did not involve material properties directly, since we were not asked to solve for u_B and u_C in terms of the loads.

Next we consider a statically indeterminate problem. This will enable us to illustrate the relationship between the method of virtual forces and the force method, which was discussed in Chapters 3, 4, and 7.

The external force P is applied to the rigid block in Fig. 1 in such a manner that the two rods that support the block undergo equal elongations. Let the rod elements be linearly elastic with flexibility coefficients f_i (see Eq. 3.13). Use the Principle of Virtual Forces, Eq. 11.47: (a) to obtain the compatibility equations for this system, and (b) to obtain the internal forces in the two rods.

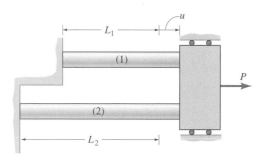

Fig. 1

Plan the Solution   Part (a) should be a repeat, more or less, of Example Problem 11.8. In Part (b) we can employ the linearly-elastic force-elongation relationships, $e_i = f_i F_i$, and see if this gets the desired expressions for F_1 and F_2. This problem is statically indeterminate, however, so we can expect to have to identify one of the F_i's as a redundant force.

Solution

Equilibrium: The block in Fig. 2 must be in equilibrium under the action of the real forces, and also when virtual forces are added to the real forces. Therefore,

$$\sum F = 0: \quad F_1 + F_2 = P \quad \text{and} \quad \delta F_1 + \delta F_2 = \delta P \quad (1)$$

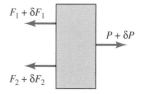

Fig. 2 Free-body diagram.

Since there is only one equilibrium equation for real forces (and, correspondingly only one equilibrium equation for virtual forces), this is a statically-indeterminate problem. Let us select F_1 as the redundant force (and δF_1 as the redundant virtual force). Then, Eqs. (1) can be expressed in the form

$$F_2 = P - F_1 \quad \text{and} \quad \delta F_2 = \delta P - \delta F_1 \quad (2)$$

Principle of Virtual Forces: The principle of virtual forces, Eq. 11.47, states that

$$\delta W_{\text{ext}}^c = \delta U^c \quad (3)$$

Complete Part (a) by determining the compatibility equations: If we write δU^c in terms of e_1 and e_2 directly, as we did in Example Problem 11.8, we will get the requested compatibility equations.

$$u \delta P = e_1 \delta F_1 + e_2 \delta F_2 \quad (4)$$

Substituting Eq. (2b) into Eq. (4), we get

$$u\delta P = e_1 \delta F_1 + e_2(\delta P - \delta F_1) \tag{5}$$

or

$$(u - e_2)\delta P - (e_1 - e_2)\delta F_1 = 0 \tag{6}$$

But, since equilibrium is satisfied for any δP and any δF_1 because of the step from Eq. (4) to Eq. (5), Eq. (6) requires that

$$e_2 = u, \qquad e_1 = e_2 \qquad\qquad \textbf{Ans. (a)} \quad (7)$$

Therefore, the principle of virtual forces has produced a complete set of compatibility equations for this problem.

Complete Part (b) by determining the axial forces F_1 and F_2: If we insert the force-elongation relations for the two rod elements, we will be able to solve for the forces in the rods. In the present problem, the rods are linearly elastic. We need the equations in the form $e = e(F)$. Therefore, let us use Eq. 3.13 written as

$$e_1 = f_1 F_1, \qquad e_2 = f_2 F_2 \tag{8}$$

Substituting Eqs. (2a) and (8) into Eq. (7b) we get

$$f_1 F_1 = f_2(P - F_1)$$

or, solving for the redundant force F_1, we get

$$F_1 = \frac{f_2 P}{f_1 + f_2} \qquad\qquad \textbf{Ans. (b)} \quad (9a)$$

This redundant force can be substituted back into Eq. (2a) to give the other internal force

$$F_2 = \frac{f_1 P}{f_1 + f_2} \qquad\qquad \textbf{Ans. (b)} \quad (9b)$$

Note that the use of the *Principle of Virtual Forces* leads us naturally to a *force-method* solution.

Review the Solution   Using the principle of virtual forces, we arrived at the compatibility equations that we could have written down "by inspection." However, we also arrived at a solution for internal forces that we can verify by noting that the solution for the forces in Eqs. (9) satisfies equilibrium, $F_1 + F_2 = P$ and compatibility $f_1 F_1 = f_2 F_2$.

The Crotti-Engesser Theorem. The Principle of Virtual Displacements, Eq. 11.34, was rephrased as Castigliano's First Theorem by incorporating the expression

$$\delta W_{\text{ext}} = \sum_{i=1}^{N} P_i \delta q_i$$

for the virtual work of external forces. In an analogous manner, when bodies are loaded by discrete forces the complementary virtual work can be written in the form

$$\delta \mathcal{W}^c = \sum_{i=1}^{N} u_i \delta P_i \qquad (11.48)$$

where the P_i's can include couples, and where u_i is the displacement (or rotation) that corresponds to the force (or couple) P_i. Since the complementary strain energy is fundamentally a function of the applied loads, an expression analogous to Eq. 11.43 can be written for $\delta \mathcal{U}^c$, as follows:

$$\delta \mathcal{U}^c = \sum_{i=1}^{N} \frac{\partial \mathcal{U}^c}{\partial P_i} \delta P_i \qquad (11.49)$$

Inserting Eqs. 11.48 and 11.49 into Eq. 11.47, the Principle of Virtual Forces, and noting that the virtual forces are independent, we get

$$\boxed{u_i = \frac{\partial \mathcal{U}^c}{\partial P_i}, \qquad i = 1, 2, \ldots, N}$$
Crotti-Engesser Theorem $\qquad (11.50)$

This is called the *Crotti-Engesser Theorem*.[9] It applies to nonlinearly elastic bodies as well as linearly elastic bodies, as is illustrated by the next example problem.

---

[9]The Crotti-Engesser Theorem is named after the Italian engineer Francesco Crotti (1839–1896) who developed it in 1878, and the German engineer Friedrich Engesser (1848–1931), who independently derived the theorem in 1889. Engesser introduced the notion of *complementary energy*. [Ref. 11-1]

■■■■■■■■■■■■■■■□ E X A M P L E 1 1 . 1 0 □□□□□□□□□□□□□□■■

Axial loads P_B and P_C are applied, as shown in Fig. 1a, to a two-element rod system whose elements both behave according to the nonlinear force-elongation curve in Fig. 1b. Using the Crotti-Engesser Theorem, determine expressions for u_B and u_C in terms of the loads P_B and P_C and the material constant D.

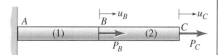

(a) A 2-element rod system.

Plan the Solution The complementary strain energy of an element is the shaded area above the curve in Fig. 1b. We can obtain $\mathcal{U}_i^c(F_i)$ by integration. To get $\mathcal{U}^c(P_B, P_C)$ we must relate the F_i's to the P's by equilibrium equations. Then we can directly apply Eq. 11.50, the Crotti-Engesser Theorem.

Solution
Equilibrium: This is the same configuration as the one in Example Problem 11.8, where we got

$$F_1 = P_B + P_C \quad \text{and} \quad \delta F_1 = \delta P_B + \delta P_C \qquad (1)$$
$$F_2 = P_C \quad \text{and} \quad \delta F_2 = \delta P_C$$

Complementary Strain Energy: From Eqs. 11.4 and 11.6,

$$\mathcal{U}^c(F_i) = \int_0^{F_i} e\, dF = \int_0^{F_i} DF^3 dF = \frac{1}{4} DF_i^4 \qquad (2)$$

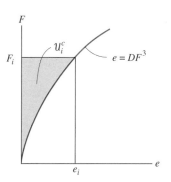

(b) Nonlinear element force-elongation curve.

Fig. 1

For the two-element system, then,

$$\mathcal{U}^c = \frac{1}{4}DF_1^4 + \frac{1}{4}DF_2^4 \qquad (3)$$

which, when combined with Eqs. (1), gives

$$\mathcal{U}^c = \frac{D}{4}[(P_B + P_C)^4 + P_C^4] \qquad (4)$$

Crotti-Engesser Theorem: Since Eq. (4) incorporates both equilibrium and force-elongation information, we apply the Crotti-Engesser Theorem, which takes care of the geometric compatibility aspect of the solution. Then,

$$\left.\begin{aligned} u_B &= \frac{\partial \mathcal{U}^c}{\partial P_B} = D(P_B + P_C)^3 \\[2ex] u_C &= \frac{\partial \mathcal{U}^c}{\partial P_C} = D[(P_B + P_C)^3 + P_C^3] \end{aligned}\right\} \qquad \text{Ans.} \quad (5)$$

Review the Solution   For this simple example, we can see "by inspection" that $u_B = e_1$, and $u_C = e_1 + e_2$. Then, using $e_i = DF_i^3$ from Fig. 1b, and substituting the equilibrium expressions for the F_i's, we get the answers in Eq. (5).

Statically Indeterminate Systems. Example Problem 11.9 treated a statically indeterminate problem from the standpoint of the Principle of Virtual Forces. Let us select a set of N_R redundant forces to use in solving a statically indeterminate problem, and let these be called $R_1, R_2, \ldots, R_{N_R}$. Then the complementary strain energy can be expressed in terms of the applied loads and these redundants. The displacements that correspond to the redundant forces are all zero (i.e., the system is not really cut, but we imagine that it is cut so that the redundants that we have chosen appear as known forces). Therefore, the *Crotti-Engesser theorem applied to statically indeterminate systems* is given by the following two equations. First, the N_R redundant forces are determined by

$$\frac{\partial \mathcal{U}^c(P_1, P_2, \ldots, P_N; R_1, \ldots, R_{N_R})}{\partial R_i} = 0, \quad i = 1, 2, \ldots, N_R \qquad (11.51)$$

Then, the displacements that correspond to the loads P_i can be obtained from

$$u_i = \frac{\partial \mathcal{U}^c(P_1, P_2, \ldots, P_N; R_1, \ldots, R_{N_R})}{\partial P_i}, \quad i = 1, 2, \ldots, N \qquad (11.52)$$

We will now use Eq. 11.51 to re-solve Example Problem 11.9.

For the system in Example Problem 11.9, which consists of two linearly elastic rods connected to a rigid block, use Eq. 11.51 to solve for the redundant force F_1.

Plan the Solution For linearly elastic rods, Eq. 11.8 gives an expression for $\mathcal{U}^c$. If we incorporate the equilibrium equation (see Example Problem 11.9) into this we will get $\mathcal{U}^c$ in terms of P and F_1 ($R_1 \equiv F_1$ = redundant force). Then we can apply Eq. 11.51 to get an equation for F_1.

Solution

Equilibrium: From Eq. (2) of Example Problem 11.9,

$$F_2 = P - F_1 \tag{1}$$

Complementary Strain Energy: From Eq. 11.8,

$$\mathcal{U}^c = \frac{1}{2}f_1 F_1^2 + \frac{1}{2}f_2 F_2^2 \tag{2}$$

Selecting F_1 as the redundant and combining Eqs. (1) and (2), we get

$$\mathcal{U}^c = \frac{1}{2}f_1 F_1^2 + \frac{1}{2}f_2 (P - F_1)^2 \tag{3}$$

Crotti-Engesser Theorem: With F_1 taken as the redundant, Eq. 11.51 becomes

$$\frac{\partial \mathcal{U}^c}{\partial F_1} = 0 \tag{4}$$

Then, the result of combining Eqs. (3) and (4) is

$$f_1 F_1 - f_2 (P - F_1) = 0$$

or

$$F_1 = \frac{f_2 P}{f_1 + f_2} \qquad \textbf{Ans.} \quad (5)$$

which is the same solution (a force-method solution) that we got in Example Problem 11.9.

Review the Solution No further checking is necessary, since we got the same answer that we got in Example Problem 11.9. If we take $\partial \mathcal{U}^c / \partial P$ using Eq. (3), we directly get the equation for the displacement, that is,

$$u = f_2 (P - F_1)$$

into which we can substitute F_1 from Eq. (5).

In this section we examine two important energy methods that may be used for solving statically-indeterminate deflection problems. They are closely related to the complementary-energy methods of Section 11.7, but they are valid only for linearly elastic behavior of deformable bodies. Table 11.2 compares and contrasts the basic characteristics of the strain-energy methods of Section 11.6, the complementary-energy methods of Section 11.7, and the special linear energy methods of this section—Castigliano's Second Theorem and the Unit-Load Method.

As noted in Section 11.2, when the load-deformation behavior of a body is linearly elastic,

$$\mathcal{W}^c = \mathcal{W} \quad \text{and} \quad \mathcal{U}^c = \mathcal{U} \tag{11.53}$$

Therefore, for linear problems, the complementary energy methods of Section 11.7 can be expressed in terms of virtual work and strain energy instead of in terms of complementary virtual work and complementary strain energy. This leads to two very popular methods for calculating displacements of deformable bodies—Castigliano's Second Theorem and the Unit-Load Method.

Castigliano's Second Theorem. We restrict our attention to deformable bodies that behave linearly under the action of independent loads, $P_1, P_2, \ldots, P_N$, plus, possibly, distributed loads and/or additional concentrated loads. Furthermore, let us designate the displacements that correspond to the above N loads as $\Delta_1, \Delta_2, \ldots, \Delta_N$. That is, let Δ_i be the displacement at the point of application of load P_i in the direction of load P_i. As indicated in Fig. 11.4 and Eq. 11.53, when a body behaves linearly, the complementary energy and the strain energy are equal, so the Crotti-Engesser Theorem, Eq. 11.50, may be written in terms of *strain energy expressed as a function of forces*. The result is called *Castigliano's Second Theorem*, which can be stated as follows:

Among all possible equilibrium configurations of a linearly elastic deformable body or system, the actual configuration is the one for which

$$\Delta_i = \frac{\partial \mathcal{U}(P_1, P_2, \ldots, P_N)}{\partial P_i}, \quad i = 1, 2, \ldots, N \qquad \begin{array}{l} \text{Castigliano's} \\ \text{Second} \\ \text{Theorem} \end{array} \tag{11.54}$$

where Δ_i is the displacement corresponding to force P_i, and $\mathcal{U}$ is the strain energy expressed as a function of the loads.

TABLE 11.2   **Comparison of Energy Methods**

Strain Energy Method	Complementary Energy Method	Special Linear Methods
Functional dependencies: $\delta \mathcal{W}$(displ.), $\delta \mathcal{U}$(displ.)	Functional dependencies: $\delta \mathcal{W}^c$(forces), $\delta \mathcal{U}^c$(forces)	Functional dependencies: $\delta \mathcal{W}$(forces), $\delta \mathcal{U}$(forces)
Compatibility implicitly required.	Equilibrium implicitly required.	Equilibrium implicitly required.
Uses virtual displacements.	Uses virtual forces.	Uses virtual forces.
Produces equilibrium equations in terms of displacements.	Produces compatibility equations in terms of forces.	Produces compatibility equations in terms of forces.
Material may be nonlinear.	Material may be nonlinear.	Material must be linear.

Typically, the strain energy is related to the loads through expressions like Eqs. 11.18, 11.23, 11.28, and 11.32, that is, through expressions like

$$\mathcal{U} = \int_0^L \frac{F^2 dx}{2AE} + \int_0^L \frac{T^2 dx}{2GI_p} + \int_0^L \frac{M^2 dx}{2EI} + \int_0^L \frac{f_s V^2 dx}{2GA} \tag{11.55}$$

where $F(x)$, $T(x)$, $M(x)$, and $V(x)$ are related to the P's by equilibrium equations, thus guaranteeing that only equilibrium states are considered. It is convenient to combine Eqs. 11.54 and 11.55 initially, which gives

$$\Delta_i = \int_0^L \frac{F(x)\frac{\partial F(x)}{\partial P_i}dx}{AE} + \int_0^L \frac{T(x)\frac{\partial T(x)}{\partial P_i}dx}{GI_p}$$
$$+ \int_0^L \frac{M(x)\frac{\partial M(x)}{\partial P_i}dx}{EI} + \int_0^L \frac{f_s V(x)\frac{\partial V(x)}{\partial P_i}dx}{GA} \tag{11.56}$$

Of course, if a system consists of several members, Eqs. 11.55 and 11.56 will have to include summations over all members in the system. This will be illustrated in example problems involving planar trusses.

Castigliano's Second Theorem Applied to Statically Determinate Problems.
Figure 11.13 was used to point out a severe limitation of the Work-Energy Principle, namely that it can be used to calculate deflections only when just a single load is applied. We will use the beam in Fig. 11.13 to illustrate the ease with which displacements can be calculated using Castigliano's Second Theorem, even when multiple loads are applied. As noted in the statement of the theorem, we must express the strain energy $\mathcal{U}$ as a function of the loads, incorporating equilibrium in the process.

■■■■■■■■■■■■■■■■□□□ E X A M P L E   1 1 . 1 2 ■□□□□□□□□□□□■■■■■

Use Castigliano's Second Theorem to solve for the slope θ and displacement Δ at end A of the uniform, linearly elastic beam shown in Fig. 1. Neglect shear deformation of the beam.

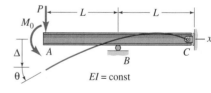

Fig. 1

Plan the Solution We can use Castigliano's Second Theorem, Eq. 11.54, to determine θ and Δ directly. The strain energy for this situation was given in Eq. 11.36 (Section 11.4) in conjunction with Fig. 11.13.

Solution In Eq. 11.36 the strain energy for the beam in Fig. 1 was given as a function of the loads P and M_0 as

$$U = \frac{P^2 L^3}{3EI} + \frac{5PM_0 L^2}{6EI} + \frac{2M_0^2 L}{3EI} \tag{1}$$

Castigliano's Second Theorem, Eq. 11.54, gives the following equations for determining Δ and θ:

$$\Delta = \frac{\partial \mathcal{U}}{\partial P}, \qquad \theta = \frac{\partial \mathcal{U}}{\partial M_0} \tag{2}$$

Therefore, combining Eqs. (1) and (2) we get

$$\left.\begin{array}{l} \Delta = \dfrac{2PL^3}{3EI} + \dfrac{5M_0L^2}{6EI} \\[4mm] \theta = \dfrac{5PL^2}{6EI} + \dfrac{4M_0L}{3EI} \end{array}\right\} \qquad \textbf{Ans.} \quad (3)$$

Review the Solution   The first check we should make is to check the dimensions of each term to be sure that the answers are dimensionally correct. We see that they are. See Example Problem 11.4 for additional magnitude checks that could be performed, for example, using the following data from Table E.1 of Appendix E.

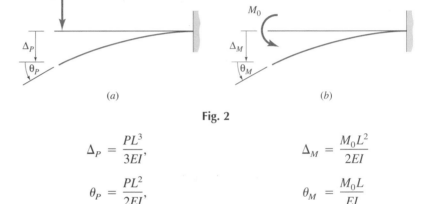

Fig. 2

$$\Delta_P = \frac{PL^3}{3EI}, \qquad\qquad\qquad \Delta_M = \frac{M_0L^2}{2EI}$$

$$\theta_P = \frac{PL^2}{2EI}, \qquad\qquad\qquad \theta_M = \frac{M_0L}{EI}$$

Castigliano's Second Theorem Applied to Statically Indeterminate Problems.

When a structure is statically indeterminate and has independent external loads, $P_1, P_2, \ldots, P_N$, and we write the strain energy in terms of forces using equilibrium relations to give us $F(x)$, $T(x)$, $M(x)$, and so forth, the strain energy will also involve unknown redundant forces $R_1, R_2, \ldots, R_{N_R}$. That is, we will get $\mathcal{U} = \mathcal{U}(P_1, P_2, \ldots, P_N; R_1, R_2, \ldots, R_{N_R})$. As in Eq. 11.51, we get N_R equations that enable us to solve for the redundants. We also get N equations to solve for the displacements under the N external loads, as in Eqs. 11.52. Thus, *Castigliano's Second Theorem* applied to statically indeterminate bodies can be stated as follows:

Among all possible equilibrium configurations of a statically indeterminate, linearly elastic body, the actual equilibrium configuration is the one that satisfies the equations

$$\boxed{\begin{array}{l} 0 = \dfrac{\partial \mathcal{U}}{\partial R_i}, \qquad i = 1, 2, \ldots, N_R \\[5mm] \Delta_i = \dfrac{\partial \mathcal{U}}{\partial P_i}, \qquad i = 1, 2, \ldots, N \end{array}} \qquad (11.57)$$

where $\mathcal{U}$ is the strain energy of the body expressed as a function of the N applied loads P_i and the N_R redundant forces R_i, and where Δ_i is the displacement component of the point of application of load P_i in the direction of that load.[10]

---

[10]Equation 11.57a is sometimes referred to as the *Principle of Least Work.*

To illustrate the use of these equations, we will solve a problem similar to Example Problem 3.15 of Section 3.8. This example problem illustrates the use of a *dummy load* to obtain the displacement when there is no real load that corresponds to the required displacement.

■■■■■■■■■■■■■■■□□ E X A M P L E 1 1 . 1 3 □□□□□■■■■■■■■■■■

A three-element truss has the configuration shown in Fig. 1. Each member has a cross-sectional area A and is made of material whose modulus is E. A horizontal load, P_{xB}, is applied to the truss at node B.

Using Castigliano's Second Theorem in the form of Eqs. 11.57, with F_2 as the redundant force, solve for (a) the member forces F_1, F_2, and F_3; (b) the horizontal displacement u_B at B; and (c) the vertical displacement v_B at B.

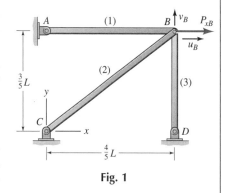

Fig. 1

Plan the Solution To determine F_2 we will need to use the first of Eqs. 11.57, letting the single redundant (R_1) be F_2. We can use the second of Eqs. 11.57 to solve directly for the displacement u_B that corresponds to the horizontal load P_{xB}. Since there is no load at B corresponding to the vertical displacement v_B that is required in Part (c), we will need to introduce a dummy load P_{yB}. After carrying out the differentiation with respect to P_{yB} to get v_B, we can then set $P_{yB} = 0$.

Equation 11.19 gives the strain energy stored in a two-force, linearly elastic rod.

Solution

(a) Use Castigliano's Second Theorem to determine the member forces.

Equilibrium: The strain energy expression that we utilize in Eqs. 11.57 must be expressed in terms of forces in a manner such that equilibrium is automatically satisfied. Therefore, we begin by writing equilibrium equations for the free body in Fig. 2. To determine v_B [in Part (c)] we will need to have a force acting in the y direction at B. Therefore, on the free-body diagram in Fig. 2, a "dummy force" P_{yB} is included. We let F_2 be the redundant force.

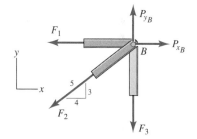

Fig. 2 Free-body diagram.

$$\sum F_x = 0: \qquad F_1 = P_{xB} - \frac{4}{5}F_2 \qquad (1)$$

$$\sum F_y = 0: \qquad F_3 = P_{yB} - \frac{3}{5}F_2$$

Strain Energy: For element i,

$$\mathcal{U}_i = \frac{1}{2}f_i F_i^2 \qquad (2)$$

where $f_i = (L/AE)_i$. Combining Eqs. (1) and (2) and summing for the three-element system, we get

$$\mathcal{U} = \frac{1}{2}f_1\left(P_{xB} - \frac{4}{5}F_2\right)^2 + \frac{1}{2}f_2 F_2^2 + \frac{1}{2}f_3\left(P_{yB} - \frac{3}{5}F_2\right)^2 \qquad (3)$$

Castigliano's Second Theorem: To determine the redundant force F_2, we use the first of Eqs. 11.57. That is,

$$\frac{\partial \mathcal{U}}{\partial F_2} = 0$$

Then,

$$-\frac{4}{5}f_1\left(P_{xB} - \frac{4}{5}F_2\right) + f_2 F_2 - \frac{3}{5}f_3\left(P_{yB} - \frac{3}{5}F_2\right) = 0$$

Setting $P_{yB} = 0$ and solving for F_2, we get

$$F_2 = \frac{20f_1 P_{xB}}{16f_1 + 25f_2 + 9f_3} \tag{4}$$

The flexibility coefficients are

$$f_1 = \frac{L_1}{A_1 E_1} = \frac{4}{5}\left(\frac{L}{AE}\right) = \frac{4}{5}f, \quad f_2 = f, \quad f_3 = \frac{3}{5}f$$

Then,

$$F_2 = \frac{20\left(\frac{4}{5}f\right)P_{xB}}{16\left(\frac{4}{5}f\right) + 25f + 9\left(\frac{3}{5}f\right)} = \frac{10}{27}P_{xB}$$

This answer can be combined with Eqs. (1) to give

$$F_1 = \frac{19}{27}P_{xB}, \quad F_2 = \frac{10}{27}P_{xB}, \quad F_3 = \frac{-2}{9}P_{xB} \quad \textbf{Ans. (a)} \tag{5}$$

(b) Determine the horizontal displacement u_B. We use the second of Eqs. 11.57 in the form

$$u_B = \frac{\partial \mathcal{U}}{\partial P_{xB}} \tag{6}$$

Combining Eqs. (3) and (6) gives

$$u_B = f_1\left(P_{xB} - \frac{4}{5}F_2\right) = \left(\frac{4}{5}f\right)\left(\frac{19}{27}P_{xB}\right)$$

or

$$u_B = \frac{76}{135}\frac{P_{xB}L}{AE} \quad \textbf{Ans. (b)} \tag{7}$$

(c) Determine the vertical displacement v_B. Corresponding to the dummy load P_{yB}, we take

$$v_B = \left.\frac{\partial \mathcal{U}}{\partial P_{yB}}\right|_{P_{yB}=0} \tag{8}$$

Thus,

$$v_B = f_3\left(-\frac{3}{5}F_2\right) = \left(-\frac{3}{5}\right)\left(\frac{3}{5}f\right)\left(\frac{10}{27}P_{xB}\right)$$

or

$$v_B = -\frac{2}{15}\frac{P_{xB}L}{AE} \qquad\qquad \text{Ans. (c)} \quad (9)$$

Review the Solution The element forces in Eqs. (5) all have the dimensions of force, and the displacements in Eqs. (7) and also have the correct dimensions. The signs of F_1 and F_2 indicate tension, which is correct for P_{xB} acting to the right; and with F_2 pulling down on node B, F_3 must be pushing up, as it is. To check further, we could re-solve the problem using the displacement method, as in Example 3.15, or we could use the force method, which should give us the same expression for F_2 that we got in Eq. (4).

The Unit-Load Method. Suppose that we need to calculate the value of the displacement Δ at point C in the beam shown in Fig. 11.20a. To calculate the displacement Δ we could use Eq. 11.56, but instead we will derive and use a variation of Eq. 11.56 called the *unit-load method*.[11] In Fig. 11.20a are shown all of the real loads applied to the beam. (There could be a real load corresponding to the desired displacement Δ, but the load P in Fig. 11.20a could also be a dummy load that is actually zero.)

Since this is a linear problem, the internal forces $F(x)$, $T(x)$, and so forth are linear functions of each of the loads, including P. Therefore, if we evaluate $\partial F(x)/\partial P$, the answer that we get is the same as if we were to evaluate $F(x)$ for single load of $P_u = 1$. This leads to a method for calculating displacements called the *unit-load method*. Figure 11.20b shows the beam with a unit load for the value of P where displacement Δ is to be obtained. Let $F(x)$, $T(x)$, etc. be the internal resultants due to the real loads (Fig. 11.20a); and let $F_u(x)$, $T_u(x)$, etc. be the internal resultants due to the unit load, $P_u = 1$ (Fig. 11.20b). Then Eq. 11.56 can be written,

$$\Delta = \int_0^L \frac{F\,F_u\,dx}{AE} + \int_0^L \frac{T\,T_u\,dx}{GI_p}$$
$$+ \int_0^L \frac{M\,M_u\,dx}{EI} + \int_0^L \frac{f_s V\,V_u\,dx}{GA}$$

Unit-Load Method (11.58)

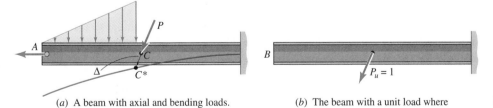

(a) A beam with axial and bending loads.

(b) The beam with a unit load where displacement Δ is required.

FIGURE 11.20 Illustration of the unit-load method.

---

[11]The unit-load method is basically a complementary energy method, and therefore it could be derived directly from the Principle of Virtual Forces, Eq. 11.47. Here we will only apply the method to linearly elastic bodies, so Eq. 11.56 is a more convenient equation on which to base the present derivation of the unit-load method.

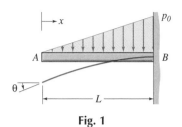

Fig. 1

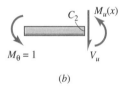

Fig. 2 Unit-load case.

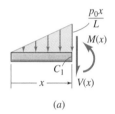

(a)

(b)

Fig. 3 Free-body diagrams.

Use the unit-load method to determine the slope, θ, at the tip of the uniform, linearly elastic cantilever beam with linearly varying distributed load, as shown in Fig. 1. Neglect shear deformation.

Plan the Solution We can use the bending moment term in Eq. 11.58. We will need to put a unit couple at A corresponding to the desired angle θ.

Solution

Unit Load: To calculate the angle θ, we need to place a unit couple at A, as shown in Fig. 2.

Equilibrium: We need $M(x)$ based on the real loads in Fig. 1 and $M_u(x)$ based on the unit couple in Fig. 2. Figure 3 shows the appropriate free-body diagrams.

$$\left(\sum M\right)_{C_1} = 0: \quad M(x) = -\left(\frac{p_0 x}{L}\right)\left(\frac{x}{2}\right)\left(\frac{x}{3}\right) = -\frac{p_0 x^3}{6L} \tag{1a}$$

$$\left(\sum M\right)_{C_2} = 0: \quad M_u(x) = -1 \tag{1b}$$

Unit-Load Calculation: From Eq. 11.58,

$$\theta = \int_0^L \frac{M\,M_u dx}{EI} = \int_0^L \frac{1}{EI}\left(-\frac{p_0 x^3}{6L}\right)(-1)dx \tag{2}$$

$$\theta = \frac{p_0 L^3}{24EI} \qquad \textbf{Ans.} \tag{3}$$

Review the Solution The solution in Eq. 3 is dimensionally correct. For this simple problem we can check the answer in Table E.1 of Appendix E.

As a second example of the unit-load method, and also as an example of a statically indeterminate problem, consider a statically indeterminate truss similar to the one in Example Problem 11.13. (Note that we must employ a unit load corresponding to each redundant force that we choose, and then set the resulting displacements of the redundants to zero.)

For trusses, the unit-load equation can be written as the following sum of member contributions:

$$\Delta = \sum_{i=1}^{N} f_i F_i F_{ui} = \sum_{i=1}^{N} \left(\frac{F F_u L}{AE}\right)_i \qquad \begin{matrix}\textbf{Unit-Load}\\\textbf{Method}\\\textbf{(Trusses)}\end{matrix} \qquad (11.59)$$

A four-element truss has the configuration shown in Fig. 1. Each member has a cross-sectional area A and modulus of elasticity E. Horizontal and vertical loads are applied at node B. Use the unit-load method to determine the member forces F_1 through F_4. Let F_2 and F_4 be the two redundants.

Plan the Solution We will need to write equilibrium equations for the real forces. We can use a free-body diagram for this. Since there are four unknown forces and only two equations of equilibrium, there are two redundant forces, which are to be F_2 and F_4. We can employ the unit-load method of Eq. 11.59, and since there is no displacement at C corresponding to F_2 or at E corresponding to F_4, we need to apply a unit load at each of these places and set the corresponding displacement in the unit-load equation to zero.

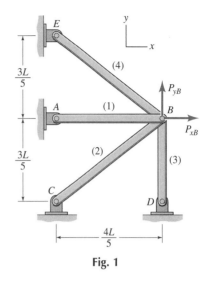

Fig. 1

Solution

Equilibrium: The equilibrium equations relating the real forces can be obtained by use of Fig. 2. Thus,

$$\sum F_x = 0: \qquad F_1 = P_{xB} - \frac{4}{5}F_2 - \frac{4}{5}F_4$$

$$\sum F_y = 0: \qquad F_3 = P_{yB} - \frac{3}{5}F_2 + \frac{3}{5}F_4 \qquad (1)$$

Since we have two redundants, we need two unit-load cases. The free-body diagrams for these are shown in Fig. 3. We will designate the two unit-load cases by $(\cdot)'$ and $(\cdot)''$ superscripts, rather than by subscript u. From Fig. 3a,

$$F_2' = 1, \qquad F_4' = 0$$

$$\sum F_x = 0: \qquad F_1' = -\frac{4}{5} \qquad (2)$$

$$\sum F_y = 0: \qquad F_3' = -\frac{3}{5}$$

From Fig. 3b,

$$F_2'' = 0, \qquad F_4'' = 1$$

$$\sum F_x = 0: \qquad F_1'' = -\frac{4}{5} \qquad (3)$$

$$\sum F_y = 0: \qquad F_3'' = \frac{3}{5}$$

Unit-Load Calculation: Since the members of this truss are uniform, we can use Eq. 11.59, that is

$$\Delta = \sum_{i=1}^{4} f_i F_i F_{ui} \qquad (4)$$

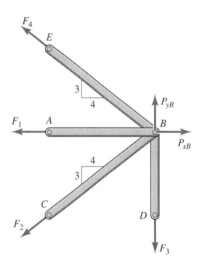

Fig. 2 Free-body diagram with real forces.

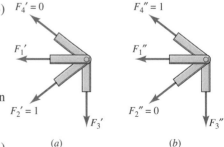

(a)   (b)

Fig. 3 Unit-load free-body diagrams.

But, we have two unit-load cases. Therefore, from Eq. (4) we have

$$0 = \sum_{i=1}^{4} f_i F_i F'_i, \qquad 0 = \sum_{i=1}^{4} f_i F_i F''_i \qquad (5)$$

The flexibility coefficients are given by

$$f_i = \left(\frac{L}{AE}\right)_i$$

so

$$f_1 = \frac{4}{5}\left(\frac{L}{AE}\right) \equiv \frac{4}{5}f, \quad f_2 = f_4 = f, \quad f_3 = \frac{3}{5}f \qquad (6)$$

Substituting Eqs. (1), (2), (3), and (6) into Eqs. (5) we get

$$0 = \left(\frac{4}{5}f\right)\left(P_{xB} - \frac{4}{5}F_2 - \frac{4}{5}F_4\right)\left(-\frac{4}{5}\right) + f\,F_2(1)$$

$$+ \left(\frac{3}{5}f\right)\left(P_{yB} - \frac{3}{5}F_2 + \frac{3}{5}F_4\right)\left(-\frac{3}{5}\right) + 0$$

$$(7)$$

$$0 = \left(\frac{4}{5}f\right)\left(P_{xB} - \frac{4}{5}F_2 - \frac{4}{5}F_4\right)\left(-\frac{4}{5}\right) + 0$$

$$+ \left(\frac{3}{5}f\right)\left(P_{yB} - \frac{3}{5}F_2 + \frac{3}{5}F_4\right)\left(\frac{3}{5}\right) + f\,F_4(1)$$

This gives us two simultaneous algebraic equations to solve for the redundants F_2 and F_4. Once F_2 and F_4 are determined, the remaining member forces F_1 and F_3 can be obtained from Eqs. (1), the equilibrium equations for the real forces. Equations (7) simplify to

$$216F_2 + 37F_4 = 80P_{xB} + 45P_{yB}$$

$$37F_2 + 216F_4 = 80P_{xB} - 45P_{yB} \qquad (8)$$

Solving for F_2 and F_4 we get

$$\left.\begin{array}{l} F_2 = \dfrac{80}{253}P_{xB} + \dfrac{45}{179}P_{yB} \\[3mm] F_4 = \dfrac{80}{253}P_{xB} - \dfrac{45}{179}P_{yB} \end{array}\right\} \qquad \text{Ans.} \quad (9)$$

and, combining Eqs. (1) and (9) we get

$$\left.\begin{array}{l} F_1 = \dfrac{125}{253}P_{xB} \\[3mm] F_3 = \dfrac{125}{179}P_{yB} \end{array}\right\} \qquad \text{Ans.} \quad (10)$$

Review the Solution All answers in Eqs. (9) and (10) obviously have the correct dimension of force. A positive force P_{xB} should put members 1, 2, and 4 in tension; and since the vertical components of the resulting values of F_2 and F_4 balance out, F_3 should not depend on P_{xB}. Equations (9) and (10) exhibit these effects of P_{xB}. A similar analysis shows that Eqs. (9) and (10) properly reflect the effect of P_{yB} also.

To calculate displacements for a truss with many members, it is convenient to tabulate the contributions of the various members to the sum in Eq. 11.59, as illustrated by the following example problem.

■■■■■■■■■■■■■■■■■ E X A M P L E   1 1 . 1 6 ■■■■■■■■■■■■■■■■■

Determine the horizontal displacement u_D at node D that results when load P is applied downward at node C, as shown in Fig. 1. All members have the same value of AE.

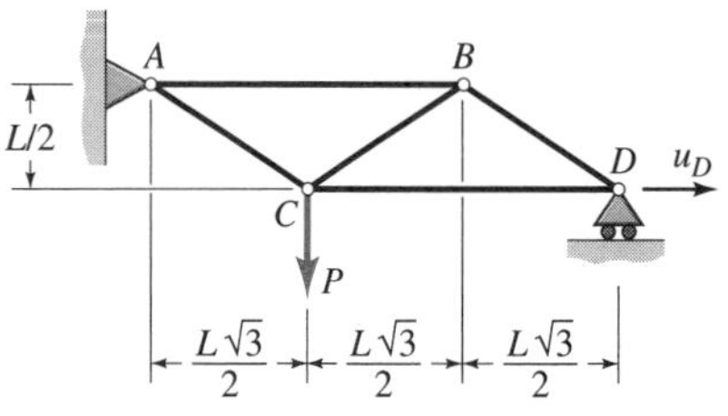

Plan the Solution We will need to use equilibrium of the truss joints to determine the F_i's due to the applied load P, and we will also use equilibrium to compute the F_{ui}'s due to a unit horizontal force at D. Then, u_D will be given by Eq. 11.59.

Fig. 1

Solution

Equilibrium of Real Forces: Either the method of joints or the method of sections can be used to determine the F_i's due to P. The relevant free-body diagrams are not shown, but the results are shown in Fig. 2.

Equilibrium of Virtual Forces due to Unit Load: For a unit load applied at D in the direction of u_D, equilibrium solutions give the F_{ui}'s shown in Fig. 3.

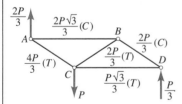

Fig. 2 Real forces F_i.

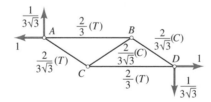

Fig. 3 Forces F_{ui} due to a unit load.

Unit-Load Deflection Calculation: Table 1 gives the values of the terms in the unit-load equation (Eq. 11.59).

$$u_D = \sum_{i=1}^{5} \frac{F_i F_{ui} L_i}{A_i E_i} \qquad (1)$$

$$u_D = -\frac{2}{3}\frac{PL}{AE} \qquad \text{Ans.} \quad (2)$$

597

TABLE 1. Unit-Load Deflection Calculations

(1) Member	(2) L_i	(3) F_i	(4) F_{ui}	(5) $F_i F_{ui} L_i$
AB	$\sqrt{3}L$	$\dfrac{-2P\sqrt{3}}{3}$	$\dfrac{2}{3}$	$-\dfrac{4}{3}PL$
AC	L	$\dfrac{4P}{3}$	$\dfrac{2\sqrt{3}}{9}$	$\dfrac{8\sqrt{3}}{27}PL$
BC	L	$\dfrac{2P}{3}$	$\dfrac{-2\sqrt{3}}{9}$	$\dfrac{-4\sqrt{3}}{27}PL$
BD	L	$\dfrac{-2P}{3}$	$\dfrac{2\sqrt{3}}{9}$	$-\dfrac{4\sqrt{3}}{27}PL$
CD	$\sqrt{3}L$	$\dfrac{P\sqrt{3}}{3}$	$\dfrac{2}{3}$	$\dfrac{2}{3}PL$

$$\sum_{i=1}^{5} = -\frac{2}{3}PL$$

Review the Solution The answer is dimensionally correct. It also makes sense that pulling down on node C would tend to pull node D to the left, so the sign of the answer is reasonable.

■■■■■■■■■■ *11.9 DYNAMIC LOADING; IMPACT

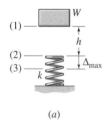

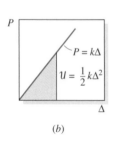

(a)

(b)

FIGURE 11.21 Impact loading—Falling-weight case.

So far we have assumed that all loads are applied slowly until each reaches its maximum value and then remains at this *static-load* value. When forces are applied more rapidly, like wave loading of an offshore oil platform or impact loading of an automobile during a collision, it is necessary to turn to the topic of structural dynamics to determine the time-dependent behavior of the dynamically loaded deformable body.[12] Here we make simplifying assumptions that enable us to see that there is definitely a difference between the response of a deformable body to static loading and its response to dynamic loading.

We begin by employing a "massless" linear spring as the deformable member. Later we will generalize to bars, beams, and other deformable bodies. In the first case we consider a weight W that falls from height h onto the massless, linear spring with spring constant k (i.e., $P = k\Delta$). In the second case we consider a mass that is moving with speed v when it strikes the spring.

Gravitational Potential Energy Converted to Strain Energy of a "Massless" Linear Spring.
Let weight W be dropped from height h onto a massless, linear spring, and assume that no energy is lost during the initial contact of the weight with the spring. Three positions are identified in Fig. 11.21a: (1) the position where the weight is released from rest, (2) the position where the weight initially contacts the spring, and (3) the lowest position reached by the weight, where the spring has its maximum compression, Δ_{max}, and where the weight is (momentarily) stopped.

---
[12]For example, see *Structural Dynamics—An Introduction to Computer Methods,* by Roy R. Craig, Jr., [Ref. 11-3].

If no energy is lost during the impact, we can equate the gravitational potential energy lost by the weight in falling through the distance $(h + \Delta_{max})$ to the increase in strain energy in the spring when it is compressed by an amount Δ_{max}. Then,

$$W(h + \Delta_{max}) = \frac{1}{2}k\Delta_{max}^2 \tag{11.60}$$

Solving for the positive root of this quadratic equation in Δ_{max}, we get

$$\Delta_{max} = \left(\frac{W}{k}\right) + \left[\left(\frac{W}{k}\right)^2 + 2h\left(\frac{W}{k}\right)\right]^{1/2} \tag{11.61}$$

But W/k is the static compression, Δ_{st}, that would occur if the weight W were slowly lowered onto the spring. Therefore, Eq. 11.61 can be recast in the convenient form

$$\Delta_{max} = \Delta_{st}\left[1 + \left(1 + \frac{2h}{\Delta_{st}}\right)^{1/2}\right] \tag{11.62}$$

This equation says that $\Delta_{max} \geq 2\Delta_{st}$, with

$$\Delta_{max} = 2\Delta_{st} \tag{11.63}$$

if the weight is suddenly released when it is just touching the spring, that is, at position (2) in Fig. 11.21a.

At position (3) in Fig. 11.21a, where the spring is at its maximum compression, Δ_{max}, the force exerted on the spring by the weight (and vice versa) has a magnitude

$$P_{max} = k\Delta_{max} \tag{11.64}$$

where Δ_{max} is given by Eq. 11.61 or 11.62.

Kinetic Energy Converted to Strain Energy of a "Massless" Linear Spring.
Figure 11.22 shows a mass that is moving with speed v at the instant when it makes contact with a massless, linear spring. If we assume that no energy is lost in the impact process, then energy is conserved, and the kinetic energy of the mass at position (1) in Fig. 11.22 is converted to strain energy stored in the spring in position (2). Then,

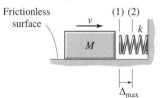

FIGURE 11.22 Impact loading—moving-mass case.

$$\frac{1}{2}Mv^2 = \frac{1}{2}k\Delta_{max}^2 \tag{11.65}$$

or

$$\Delta_{max} = \sqrt{\frac{Mv^2}{k}} \tag{11.66}$$

As in Case A, the maximum force exerted on the spring by the mass is given by Eq. 11.64, with Δ_{max}, in the present case, given by Eq. 11.66.

Impact on Deformable Bodies.
Figure 11.23 depicts a sliding collar that drops from height h, makes contact with a flange on the end of a linearly elastic rod, and stretches the rod by an amount Δ_{max}. The analysis of Case A can be applied to this system and other deformable bodies if we make the following assumptions:

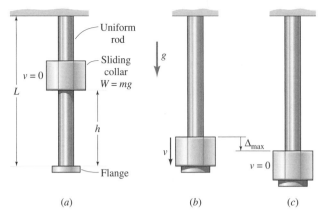

FIGURE 11.23 Impact
loading of a uniform rod.

- The mass of the impacted deformable body is negligible in comparison with the impacting mass.
- The impacting mass is rigid.
- No energy is lost in the impact.

These assumptions lead to conservative answers. That is, the deformation and stresses calculated are greater than the actual values would be if energy losses and other factors hinted at above were completely accounted for.

With the preceding assumptions, we do not have to consider conservation of momentum upon impact or consider stress waves in the impacted body; we can just apply the results obtained previously for a "massless" spring. Equations 11.62 and 11.66 can be used to determine $\Delta_{\max}$ for the respective two types of impact, with an equivalent stiffness, k, determined for the particular elastic body impacted. However, to determine the maximum stress caused by the impact, we must apply the dynamic load $P_{\max}$ of Eq. 11.64 to the particular body impacted (e.g., rod, beam, etc.). The next two example problems illustrate the effect of impact loading on deformable bodies.

■■■■■■■■■■■■■■■■ EXAMPLE 11.17 ■■■■■■■■■■■■■■■■

For the rod and sliding collar of Fig. 11.23, (a) determine an expression for $\Delta_{\max}$ as a function of $W, A, E, h,$ and L. (b) If the weight is dropped from a height of $h = 40\Delta_{st}$, determine the value of the *impact amplification factor* $\Delta_{\max}/\Delta_{st}$. (c) Determine the maximum impact stress $\sigma_{\max}$ in terms of the static stress σ_{st}, the drop height h, and the rod parameters.

Solution
(a) Determine $\Delta_{\max}$, the maximum displacement. We could use either Eq. 11.61 or 11.62, taking Δ to be positive in tension, rather than compression as in the original derivation. Let us use Eq. 11.62.

$$\Delta_{\max} = \Delta_{st}\left[1 + \left(1 + \frac{2h}{\Delta_{st}}\right)^{1/2}\right] \tag{1}$$

If the weight were to be lowered slowly onto the flange of the rod, we would get the static elongation

$$\Delta_{st} = \frac{W}{k} = \frac{WL}{AE} \tag{2}$$

Combining Eqs. (1) and (2), we get the desired expression

$$\Delta_{max} = \frac{WL}{AE}\left[1 + \left(1 + \frac{2AEh}{WL}\right)^{1/2}\right] \qquad \text{Ans. (a)} \quad (3)$$

(b) Determine the impact amplification factor. For $h = 40\Delta_{st}$, Eq. (1) gives

$$\frac{\Delta_{max}}{\Delta_{st}} = 10 \qquad \text{Ans. (b)} \quad (4)$$

(c) Determine the maximum impact stress. Since $\sigma = E\epsilon = E\left(\dfrac{\Delta}{L}\right)$ for axial deformation, Eq. (1) can be converted directly to the following equation for σ_{max}:

$$\sigma_{max} = \sigma_{st}\left[1 + \left(1 + \frac{2Eh}{L\sigma_{st}}\right)^{1/2}\right] \qquad \text{Ans. (c)} \quad (5)$$

It is left as an exercise for the reader (Probs. 11.9-8, 11.9-9) to show that impact loading of a rod with enlarged cross section over a portion of its length produces a higher maximum stress than the stress given by Eq. (5) for a uniform rod. Thus, from the stand-point of elastic energy absorption, a rod with uniform cross section is preferable to a rod with nonuniform cross section.

■■■■■■■■■■■■■■■□ E X A M P L E 1 1 . 1 8 ■■■■■■■■■■■■■■■■

A diver is springing on the diving board shown in Fig. 1. On a particular bounce, the diver reaches a height h above the end of the board. Treat the diver as a rigid mass that falls from a height h, and assume that the diving board is much lighter than the diver (e.g., a fiberglass board might be very light) and is straight when the diver strikes it at the very end. (a) Determine the maximum deflection of the tip of the diving board. Express your answer in terms of W, h, and the parameters of the diving board—E, I, and L. (b) Determine the maximum flexural stress caused by the diver's "impact." Express your answer in terms of the maximum static flexural stress and the parameters of the diving board.

Neglect shear deformation and neglect any energy loss during impact. Also neglect the mass of the diving board, and assume that EI = const.

Plan the Solution We can use the results of Example Problem 11.4 to determine an expression for k to use in Eq. 11.61 for Δ_{max}. In Part (b) we can use Eq. 11.64 to determine the maximum force that the diver exerts on the beam. Then we can use the flexure formula, Eq. 6.13, with the maximum bending moment produced by P_{max}.

Solution
(a) Determine the maximum deflection at the tip of the diving board. This deflection can be determined from Eq. 11.61, which we can write in the form

$$\Delta_{max} = \frac{W}{k}\left[1 + \left(1 + \frac{2kh}{W}\right)^{1/2}\right] \qquad (1)$$

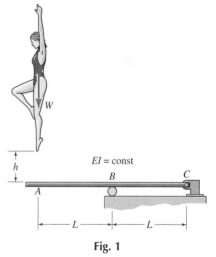

Fig. 1

601

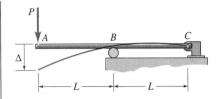

Fig. 2 The deflection caused by a tip load.

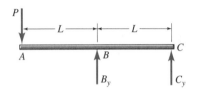

Fig. 3 Free-body diagram.

From Example Problem 11.4, the deflection Δ at the tip of the beam (diving board) in Fig. 2 is

$$\Delta = \frac{2}{3} \frac{PL^3}{EI} \qquad (2)$$

Since k is defined by $P = k\Delta$, for this beam configuration and loading

$$k = \frac{3EI}{2L^3} \qquad (3)$$

Therefore, Eqs. (1) and (3) may be combined, giving the desired answer

$$\Delta_{\max} = \frac{2WL^3}{3EI}\left[1 + \left(1 + \frac{3EIh}{WL^3}\right)^{1/2}\right] \qquad \text{Ans. (a)} \quad (4)$$

(b) Determine the maximum flexural stress. For the free body in Fig. 3,

$$\left(\sum M\right)_B = 0: \qquad C_y = -P$$

Therefore, the beam loading is symmetric about B, and the maximum bending moment along the beam occurs at B, where $M_B = -PL$. We obtain the maximum impact force at the tip, A, by combining Eq. 11.64 and Eq. (3), giving

$$P_{\max} = \left(\frac{3EI}{2L^3}\right)\Delta_{\max} \qquad (5)$$

so the maximum bending moment is

$$(M_B)_{\max} = -\left(\frac{3EI}{2L^2}\right)\Delta_{\max} \qquad (6)$$

Let c be the distance from the neutral axis to the top fibers of the beam. Then, from the flexure formula, Eq. 6.13,

$$\sigma_{\max} = \frac{-(M_B)_{\max}c}{I} = \left(\frac{3Ec}{2L^2}\right)\Delta_{\max} \qquad (7)$$

Let σ_{st} be the maximum flexural stress in the beam under static loading (e.g., with the diver standing on the end of the diving board). Then, $(M_B)_{st} = -WL$, so

$$\sigma_{st} = \frac{WLc}{I} \qquad (8)$$

Finally, we can combine Eqs. (4), (7), and (8) to get the following expression relating $\sigma_{\max}$ and σ_{st}:

$$\sigma_{\max} = \sigma_{st}\left[1 + \left(1 + \frac{3Ehc}{\sigma_{st}L^2}\right)^{1/2}\right] \qquad \text{Ans. (b)} \quad (9)$$

(Note that this expression differs from the answer for axial impact on a rod, whereas the same expression, Eq. 11.62, relates $\Delta_{\max}$ to Δ_{st} for both problems.)

11.10 PROBLEMS ■■■

Problems 11.3-1 through 11.3-15. *These problems deal with the strain energy stored in various members undergoing* **axial deformation***. Assume that the stress distribution is uniform at every cross section, even where there is a sudden change in the cross-sectional area.*

Prob. 11.3-1. A uniform aluminum-alloy rod with cross-sectional area $A = 1$ in² and length $L = 50$ in. is subjected to an axial load P, as shown in Fig. P11.3-1. Let $E_{al} = 10(10^3)$ ksi. (a) Sketch a load versus elongation diagram (i.e., load P versus elongation e) for elongations from 0 in. to 0.250 in. (Hint: Recall Eq. 3.14.) (b) Calculate the strain energy U stored in the rod when $P = 40$ kips, and indicate on your $P - e$ diagram of Part (a) the area that represents this strain energy.

Prob. 11.3-2. A uniform steel rod with a diameter of $d = 30$ mm and a length $L = 1.5$ m is subjected to an axial load P as shown in Fig. P11.3-2. Let $E_{st} = 200$ GPa and $\sigma_Y = 250$ MPa. (a) Determine the minimum value of P at which yielding would occur. (b) Calculate the strain energy U stored in the rod when P reaches the yield load P_Y determined in Part (a).

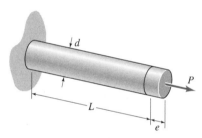

P11.3-1 and P11.3-2

Prob. 11.3-3. The stepped rod shown in Fig. P11.3-3 is subjected to equal axial loads of magnitude P at section B and C. (a) If $A_1 = 2A$, $A_2 = A$, $L_1 = L_2 = L$, and $E_1 = E_2 = E$, determine: (1) the strain energy U_1 in the rod AB, (2) the strain energy U_2 in rod BC, and (3) the total strain energy U. (b) Would <u>less</u> strain energy be stored in rod AC if the rod were uniform, that is, if $A_1 = A_2 = A$? Justify your yes or no answer.

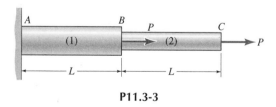

P11.3-3

Prob. 11.3-4. Determine the strain energy U stored in the uniform member shown in Fig. P11.3-4 if the modulus of elasticity is E and the cross-sectional area is A.

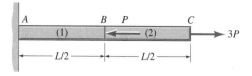

P11.3-4

Prob. 11.3-5. (a) Determine the strain energy U stored in the uniform aluminum rod shown in Fig. P11.3-5(a). Let $E_{al} = 10(10^3)$ ksi and $P_B = 6$ kips. (b) Determine the strain energy stored in the rod if $P_E = 6$ kips and the diameter over half of the length of the rod is increased to $d_2 = 1.25$ in., as illustrated in Fig. P11.3-5b.

Prob. 11.3-6. (a) Determine the value of the force P_E that would cause the stepped rod in Fig. P11.3-6b to have the same elongation as the rod in Fig. P11.3-6a when $P_B = 6$ kips. Let $E_{al} = 10(10^3)$ ksi. (b) Determine the strain energy stored in each of the two rods under the loads specified in Part (a).

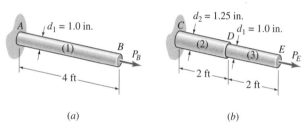

(a) (b)

P11.3-5, P11.3-6, and P11.4-1

Prob. 11.3-7. Let U_a be the strain energy stored in a uniform rod of cross-sectional area A and length L whose modulus of elasticity is E, as illustrated in Fig. P11.3-7a. If the cross-sectional area of the rod is changed to αA ($\alpha < 1$ means that the area is decreased; $\alpha > 1$ means that the area is increased) over a portion of the length equal to λL ($0 < \lambda < 1$), determine an expression for the ratio of strain energy in the stepped bar to the strain energy in the uniform bar, that is, determine an expression for U_b/U_a in terms of the parameters α and λ.

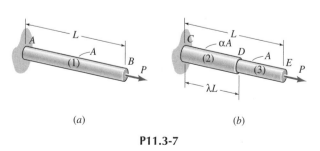

(a) (b)

P11.3-7

Prob. 11.3-8. Two types of steel bolts are being considered for a particular application for which energy-storage capacity is important. Bolt A has threads along its entire length, but bolt B has a larger-diameter shank over a 3-in. portion of its original length. When the nut is tightened, a 3.5-in. section of each bolt will be under a tension of 5 kips. The diameter of the threaded portion to be used in calculating the tensile stress is 0.432 in., and the diameter of the unthreaded shank portion of bolt B is 0.500 in. Letting $E_{st} = 30(10^3)$ ksi for both bolts, determine the strain energy stored in each bolt.

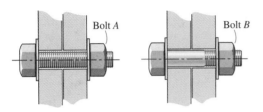

P11.3-8

Prob. 11.3-9. Determine an expression for the strain energy U stored in a solid conical frustum having a maximum diameter of d_A, a minimum diameter of d_B, and a length of L. Let the modulus of elasticity be E and the axial load be P.

P11.3-9 and P11.4-2

Prob. 11.3-10. The width of an 0.5-in.-thick steel plate ($E = 29(10^3)$ ksi) varies linearly from $h_1 = 4$ in. to $h_2 = 2$ in. over a length $L = 12$ in. (a) Determine the strain energy U_a stored in this plate when it is subjected to a load $P = 2$ kips. (b) Determine the strain energy U_b stored in the plate if the load P is increased to the value that causes initial yielding in the plate. Let $\sigma_Y = 36$ ksi.

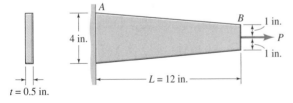

P11.3-10 and P11.4-3

Prob. 11.3-11. The two members of the pin-jointed truss shown in Fig. P11.3-11 have the same modulus of elasticity E and the same cross-sectional area A. Determine the total strain energy U stored in the truss in terms of the load P, the length L, E, and A.

604

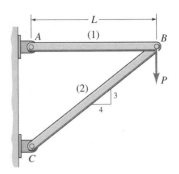

P11.3-11 and P11.4-4

Prob. 11.3-12. The cross-sectional areas of the four members of the pin-jointed truss shown in Fig. P11.3-12 are $A_1 = A_2 = 1$ in², $A_3 = A_4 = 1.5$ in². Determine the total strain energy U stored in the truss when a vertical load $P = 2$ kips is applied at D. All members are made of steel with $E_{st} = 30(10^3)$ ksi.

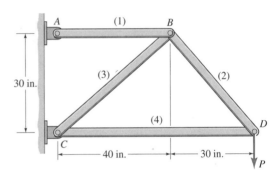

P11.3-12 and P11.4-5

Prob. 11.3-13. The cross-sectional areas of the three members of the pin-jointed truss shown in Fig. P11.3-13 are $A_1 = A_2 = 500$ mm², and $A_3 = 800$ mm². Determine the total strain energy U stored in the truss when a load $P = 5.2$ kN is applied at B, as shown. All members are made of aluminum alloy with $E_{al} = 70$ GPa.

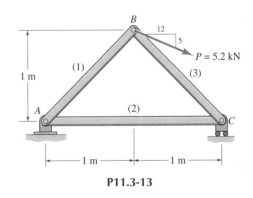

P11.3-13

Prob. 11.3-14. A uniform rod AB of length L and cross-sectional area A hangs under its own weight from a rigid support at A. If the rod is made of linearly elastic material with modulus of elasticity E and specific weight γ (i.e., weight per unit volume), determine the strain energy $\mathcal{U}$ stored in the rod in terms of γ, E, A, and L.

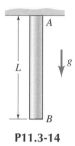

P11.3-14

Prob. 11.3-15. Determine an expression for the strain energy U stored in a vertically hanging solid conical frustum having a maximum diameter d_1, a minimum diameter d_2, and a length L. Let the modulus of elasticity be E and the specific weight (i.e., the weight per unit volume) be γ.

P11.3-15

Problems 11.3-16 through 11.3-25. *These problems deal with the strain energy stored in various members undergoing* **torsional deformation**. *Assume that the stress distribution obeys Eq. 4.11 at every cross section, even where there is a sudden change in the diameter of the cross section.*

Prob. 11.3-16. A uniform aluminum-alloy rod with circular cross section of diameter $d = 1$ in. and length $L = 50$ in. is subjected to an end torque T as shown in Fig. P11.3-16. Let $G = 4(10^3)$ ksi. (a) Sketch a torque versus total angle-of-twist diagram

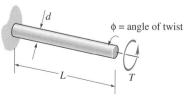

P11.3-16

for torques ranging from 0 kip·in. to 10 kip·in. (Hint: Recall Eq. 4.19.) (b) Calculate the strain energy $\mathcal{U}$ stored in the rod when $T = 4$ kip·in., and indicate on your $T - \phi$ diagram of Part (a) the area that represents this strain energy.

Prob. 11.3-17. The stepped rod in Fig. P11.3-17 is subjected to equal external torques T_0 at section B and at end C. The shear modulus of the rod is G, and the dimensions of the rod are $d_1 = 1.5d$, $d_2 = d$, and $L_1 = L_2 = L$. (a) Determine (1) the strain energy $\mathcal{U}_1$ stored in rod AB, (2) the strain energy $\mathcal{U}_2$ stored in rod BC, and (3) the total strain energy $\mathcal{U}$. (b) Would <u>less</u> strain energy be stored in rod AC if the rod were uniform, that is, if $d_1 = d_2 = d$? Justify your answer.

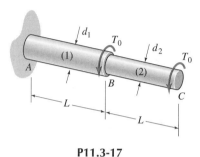

P11.3-17

Prob. 11.3-18. A uniform torsion rod is subjected to an end torque T, as shown in Fig. P11.3-18. The rod has a diameter $d = 30$ mm and a length $L = 1.5$ m. If the maximum shear stress in the rod is 60 MPa, how much strain energy is stored in the rod? Let $G = 70$ GPa.

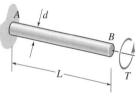

P11.3-18

Prob. 11.3-19. A uniform tubular aluminum-alloy shaft ($E = 3.8(10^3)$ ksi) with outside diameter $d_o = 1.25$ in. and length $L = 2$ ft is subjected to a torque $T = 3$ kip·in. (a) If the maximum shear stress in the shaft is 10 ksi, determine the value of the inside diameter, d_i, of the shaft, and (b) determine the strain energy stored in the shaft.

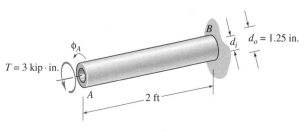

P11.3-19 and P11.4-6

605

Prob. 11.3-20. Determine the strain energy stored in the tapered bar in Fig. P11.3-20. Express your answer in terms of T_0, L, d_A, d_B, and the shear modulus G.

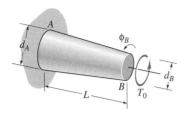

P11.3-20 and P11.4-7

Prob. 11.3-21. A uniform torsion rod is subjected to a distributed torque of constant magnitude t_0 (torque per unit length). Derive an expression for the total strain energy stored in the rod. Express your answer in terms of t_0, d, L, and the shear modulus G.

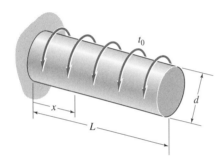

P11.3-21

Prob. 11.3-22. (a) Determine the strain energy $\mathcal{U}$ stored in the uniform aluminum rod AB shown in Fig. P11.3-22a. Let $G = 4(10^3)$ ksi and $T_B = 5$ kip·in. (b) Determine the strain energy stored in the rod CE if $T_E = 5$ kip·in. and the diameter over half the length of the rod is $d_2 = 1.25$ in., as illustrated in Fig. P11.3-22b.

Prob. 11.3-23. (a) Determine the value of the torque T_E that would cause the stepped rod CE in Fig. P11.3-23b to have the same total angle of twist that the uniform rod AB in Fig. P11.3-23a would have when $T_B = 5$ kip·in. Let $G = 4(10^3)$ ksi. (b) Determine the strain energy stored in each of the two rods under the torques specified in Part (a).

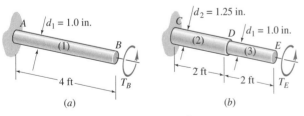

P11.3-22, P11.3-23, and P11.4-8

Prob. 11.3-24. A torque T_B is applied to the uniform rod AC as shown in Fig. P11.3-24. The rod is attached to rigid supports at

ends A and C. Determine the total strain energy $\mathcal{U}(\phi_B)$ stored in the rod. Express your answer in terms of the length L, the polar moment of inertia I_p, the shear modulus G, and the angle of twist ϕ_B. The torque T_B will not appear explicitly in your answer.

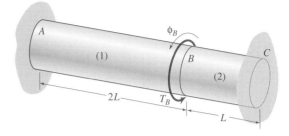

P11.3-24 and P11.4-9

Prob. 11.3-25. Torques T_B and T_C are applied to the uniform rod AD, as shown in Fig. P11.3-25. The rod is attached to rigid supports at A and D. Determine an expression for the total strain energy in the rod, $\mathcal{U}(\phi_B, \phi_C)$, where ϕ_B and ϕ_C are the angles of twist at sections B and C, respectively. Express your answer in terms of ϕ_B, ϕ_C, the length L, the polar moment of inertia I_p, and the shear modulus G. The torques T_B and T_C will not appear explicitly in your answer.

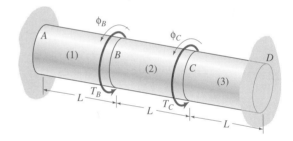

P11.3-25

Prob. 11.3-26. Four gears are attached to a shaft that transmits torques as shown in Fig. P11.3-26. (a) Determine the required diameters d_1 through d_3 if the allowable shear stress for each segment of the shaft is 80 MPa. (b) Determine the strain energy stored in the shaft if the three segments have the diameters determined in Part (a), the shear modulus of the shaft is $G = 80$ GPa, and the shaft lengths are $L_1 = L_2 = L_3 = 1$ m.

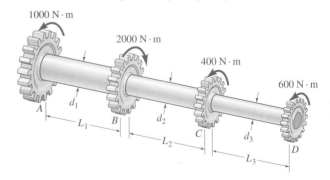

P11.3-26

Prob. 11.3-27. A stepped steel shaft is subjected to the three torques shown in Fig. P11.3-27. (a) Determine the required diameters d_1 through d_3 if the allowable shear stress for each segment of the shaft is 12 ksi. (b) Determine the strain energy stored in the shaft if the three segments have the diameters determined in Part (a) and the shear modulus of the shaft is $G = 11.5(10^3)$ ksi.

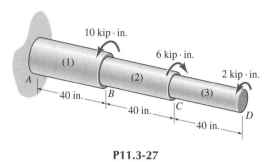

P11.3-27

Prob. 11.3-28. A tubular shaft has the dimensions shown in Fig. P11.3-28 and is made of brass with a shear modulus $G = 40$ GPa. A strain gage mounted at 45° to the axis of the shaft gives a strain reading of $\epsilon = 2.00(10^{-3})$. (a) Determine the value of the applied torque T that produces the given extensional-strain reading. (Recall Section 4.5.) (b) Determine the strain energy stored in the shaft under the stated loading.

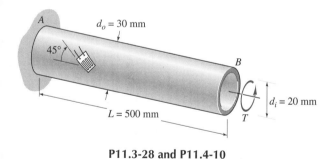

P11.3-28 and P11.4-10

Prob. 11.3-29. The two-segment solid stepped shaft in Fig. P11.3-29 is stress-free when it is welded to rigid structures at ends A and C. The diameters of the segments are $d_1 = 1.250$ in. and $d_2 = 0.875$ in., and the shaft is made of steel with a shear modulus $G = 11(10^3)$ ksi. With torque T_B applied at node B, as shown, a strain gage mounted on segment 1 at 45° to the axis of the shaft gives a strain reading of $\epsilon = 1.20(10^{-3})$. (a) Determine the value of the applied torque T_B that produces the given extensional-strain reading. (Recall Section 4.5.) (b) Determine the strain energy stored in the shaft under the given loading.

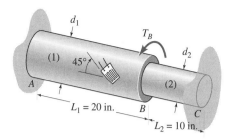

P11.3-29 and P11.4-11

Prob. 11.3-30. A stepped, solid shaft has the dimensions shown in Fig. P11.3-30 and is made of material with shear modulus G. The shaft is stress-free when its ends are fixed to rigid structures at ends A and D. A single torque T_C is applied to the shaft at node C. Determine an expression for the total energy in the rod, $\mathcal{U}(\phi_B, \phi_C)$, where ϕ_B and ϕ_C are the angles of rotation at sections B and C, respectively. Express your answer in terms of ϕ_B, ϕ_C, the length L, the polar moment of inertia I_p, and the shear modulus G. The torque T_C will not appear explicitly in your answer.

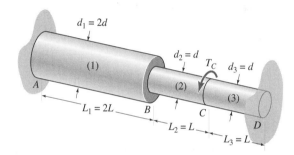

P11.3-30 and P11.4-12

Problems 11.3-31 through 11.3-42 *deal with the strain energy stored in various linearly elastic beams. Except for Probs. 11.3-31 and 11.3-36, all beams are uniform, that is,* EI = *const.*

Prob. 11.3-31. A wood cantilever beam with rectangular cross section has a concentrated load P applied at end A as shown in Fig. P11.3-31(a). (a) Determine an expression for the flexural strain energy, $\mathcal{U}_{\sigma a}$, stored in this uniform beam. (b) To strengthen the beam, planks of the same wood and of width b, thickness $h/4$, and length $L/2$ are bonded to the top and bottom of the right half of the original beam, giving the stepped beam shown in Fig. P11.3-31(b). Determine the flexural strain energy, $\mathcal{U}_{\sigma b}$, for this beam, and discuss why $\mathcal{U}_{\sigma b}$ is less (or more, if that is the case) than $\mathcal{U}_{\sigma a}$.

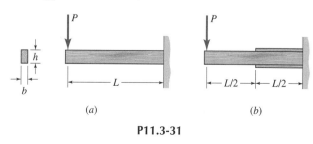

P11.3-31

Prob. 11.3-32. Let the concentrated load P in Example Problem 11.3 be replaced by a uniformly distributed load of intensity w_0 over the entire length L. (a) Determine an expression for the strain-energy ratio, $\mathcal{U}_\tau / \mathcal{U}_\sigma$, similar to the first part of Eq. (7) in this example. (b) Using the dimensions of the wide-flange beam in Example Problem 11.3, simplify your answer in Part (a) to the form given in the second part of Eq. (7) of the example.

For **Problems 11.3-33 through 11.3-35**, *(a) determine the flexural strain energy, $\mathcal{U}_\sigma$, and (b) determine the bending shear strain energy $\mathcal{U}_\tau$.*

Prob. 11.3-33. For the beam in Fig. P11.3-33, $E = 200$ GPa and $G = 80$ GPa.

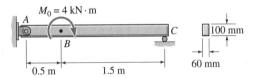

P11.3-33 and P11.4-13

Prob. 11.3-34. For the **W**8 × 21 beam in Fig. P11.3-34, $E = 30(10^3)$ ksi and $G = 11.5(10^3)$ ksi.

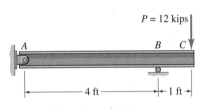

P11.3-34 and P11.4-14

Prob. 11.3-35. For the beam in Fig. P11.3-35, $E = 70$ GPa and $G = 26$ GPa.

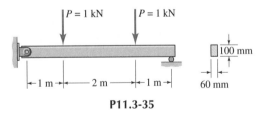

P11.3-35

*Prob. 11.3-36. The cantilever beam AB in Fig. P11.3-36 has a constant depth h and constant modulus of elasticity E, but its width varies linearly from b_0 and $x = 0$ to $3b_0$ at $x = L$. Determine an expression for the flexural strain energy, $\mathcal{U}_\sigma$, stored in this beam when a concentrated transverse load P is applied at $x = 0$.

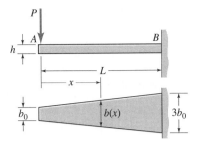

P11.3-36 and P11.4-17

Prob. 11.3-37. A uniform, rectangular beam AC is to be simply supported and is to support a concentrated transverse load P at its midspan, as shown in Fig. P11.3-37(a). The cross-sectional dimensions are b and $2b$. (a) Compare the flexural strain energy $\mathcal{U}_{\sigma b}$ that would be stored in the beam if it is placed in orientation "b" (i.e., with the $2b$ dimension vertical) with the flexural strain energy $\mathcal{U}_{\sigma c}$ that would be stored in the beam if it is placed in orientation "c," with the $2b$ dimension horizontal. (b) Compare the maximum flexural stress values for the two configurations.

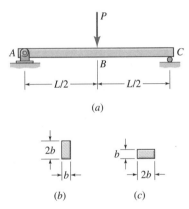

P11.3-37 and P11.3-38

Prob. 11.3-38. (a) Repeat Prob. 11.3-37, but this time compare the shear strain energy $\mathcal{U}_{\tau b}$ that would be stored in the beam if it is placed in orientation "b" (i.e., with the $2b$ dimension vertical) with the shear strain energy $\mathcal{U}_{\tau c}$ that would be stored in the beam if it is placed in orientation "c," with the $2b$ dimension horizontal. (b) Compare the maximum shear stress values for the two configurations.

Prob. 11.3-39. Determine an expression for the flexural strain energy, $\mathcal{U}_\sigma$, stored in the uniform, simply supported beam shown in Fig. P11.3-39. The flexural rigidity, EI, is constant.

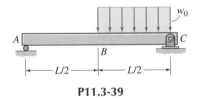

P11.3-39

Prob. 11.3-40. Determine an expression for the flexural strain energy, $\mathcal{U}_\sigma$, stored in the uniform, propped-cantilever beam shown in Fig. P11.3-40. (See Prob. 7.4-1 for this figure.) Express your answer as a function of w_0, the intensity of the applied uniformly distributed load, and R_A, the redundant reaction at A. The flexural rigidity, EI, is constant.

Prob. 11.3-41. Determine an expression for the flexural strain energy, $\mathcal{U}_\sigma$, stored in the fixed-fixed beam AB shown in Fig. P11.3-41. (See Prob. 7.4-5 for this figure.) Express your answer as a function of w_0, the maximum intensity of the linearly varying distributed load, and R_A and M_A, the redundant reactions at $x = 0$. The flexural rigidity, EI, is constant.

*Prob. 11.3-42. A uniform cantilever beam AB has an inclined concentrated load P applied at the centroid of the cross section at end A as shown in Fig. P11.3-42. Determine an expression for the total strain energy, $\mathcal{U}_\sigma$, associated with normal stress σ. Express your answer as a function of the load P, the inclination angle α, the modulus of elasticity E, and the dimensions of the beam — b, h, and L. Start your solution with Eqs. 11.9 and 11.11.

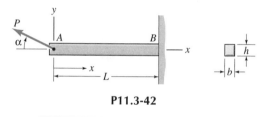

P11.3-42

In **Problems 11.4-1 through 11.4-20** *use the* **Work-Energy Principle** *to calculate the required linear or angular displacement.*

Prob. 11.4-1. Using the data and figures of Prob. 11.3-5, determine the displacement Δ_B due to the load $P_B = 6$ kips acting on the uniform rod in Fig. P11.3-5(a) and the displacement Δ_E due to the load $P_E = 6$ kips acting on the stepped rod in Fig. P11.3-5(b).

Prob. 11.4-2. Derive an expression for the displacement Δ_B due to the axial load P acting at end B of the tapered bar in Fig. P11.3-9. Let $d_A = 2d$ and $d_B = d$.

Prob. 11.4-3. Using the data and figures of Prob. 11.3-10, determine the displacement Δ_B due to a load $P = 2$ kips acting on the tapered bar in Fig. P11.3-10.

Prob. 11.4-4. Determine the vertical displacement Δ_B in the direction of the load P acting on the two-member truss in Fig. P11.3-11.

Prob. 11.4-5. Determine the vertical displacement Δ_B in the direction of the load P acting on the two-member truss in Fig. P11.3-12.

Prob. 11.4-6. Using the data and figure of Prob. 11.3-19, determine the angle of twist, ϕ_A, at end A of the tubular shaft in Fig. P11.3-19.

Prob. 11.4-7. If the end diameters of the tapered shaft in Fig. P11.3-20 are $d_A = 2d$ and $d_B = d$, determine the angle of twist of the shaft, ϕ_B.

Prob. 11.4-8. Use the *Work-Energy Principle* to solve Prob. 11.3-23(a).

Prob. 11.4-9. For the torsion rod in Fig. P11.3-24, determine an expression relating the twist angle ϕ_B and the applied torque T_B.

Prob. 11.4-10. Using the data and figure of Prob. 11.3-28, determine the angle of rotation of end B, ϕ_B, when the strain gage reads $\epsilon = 2.0(10^{-3})$.

Prob. 11.4-11. Using the data and figure of Prob. 11.3-29, determine the angle of rotation of joint B, ϕ_B, when the strain gage reads $\epsilon = 1.20(10^{-3})$.

*Prob. 11.4-12. The stepped, solid shaft in Fig. P11.3-30 is made of a material with shear modulus G and is subjected to a single applied torque T_C. Determine an expression for the resulting angle of rotation, ϕ_C, at joint C.

Prob. 11.4-13. Considering only flexural strain energy and using the data and figure for the uniform rectangular beam in Prob. 11.3-33, determine the angle of rotation θ_B in the direction of the moment $M_0 = 4$ kN·m applied at point B.

Prob. 11.4-14. Considering only flexural strain energy, and using the data and figure for the **W**8 × 21 wide-flange beam in Fig. P11.3-34, determine the vertical displacement Δ_C in the direction of the load $P = 12$ kips acting at end C.

Prob. 11.4-15. The uniform, cantilever beam with rectangular cross section, shown in Fig. P11.4-15, is made of linearly elastic material with modulus of elasticity E and Poisson's ratio ν. Determine expressions for $\Delta_{B\sigma}$ and $\Delta_{B\tau}$ and at end B, the contributions to the displacement at B that are related, respectively, to the flexural strain energy $\mathcal{U}_\sigma$ and the shear strain energy $\mathcal{U}_\tau$.

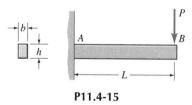

P11.4-15

Prob. 11.4-16. Use the *Work-Energy Principle* to obtain an expression for the deflection Δ_B (flexure only) under the load P for the simply supported beam in Fig. P11.4-16. $EI = $ const.

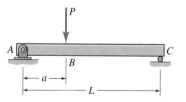

P11.4-16, P11.8-1, and P11.8-26

*Prob. 11.4-17. Determine an expression for the tip deflection Δ_A, for the tapered beam in Prob. 11.3-36.

Prob. 11.4-18. Use the *Work-Energy Principle* to determine the slope θ_A under the couple M_0 applied at end A of the uniform simply supported beam in Fig. P11.4-18. EI = const.

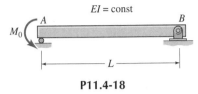

P11.4-18

Prob. 11.4-19. Use the *Work-Energy Principle* to determine the deflection Δ_B under the load P on the **W**150 × 24 wide-flange, simply supported beam in Fig. P11.4-19. E = 200 GPa.

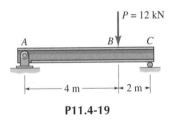

P11.4-19

Prob. 11.4-20. Use the *Work-Energy Principle* to determine the slope θ_A under the couple M_0 = 2100 kips · in. applied at end A of the **W**16 × 100 wide-flange beam in Fig. P11.4-20. Let E = 30 × 10³ ksi.

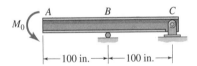

P11.4-20, P11.8-3, and P11.8-27

Problems 11.6-1 through 11.6-10. *Use the* **Principle of Virtual Displacements**, *Eq. 11.44, to solve each of these problems.*

Prob. 11.6-1. (a) For the two segment rod in Fig. P11.6-1, use the *Principle of Virtual Displacements* to determine an expression for the axial displacement u_B at the point of application of the axial force P. (b) Determine the axial forces F_1 and F_2 in segments (1) and (2) respectively.

$$A_1 = 2A, \qquad A_2 = A, \qquad E = \text{const}$$

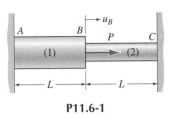

P11.6-1

Prob. 11.6-2. (a) For the two segment rod in Fig. P11.6-2, use the *Principle of Virtual Displacements* to determine an expression for the axial displacement u_B at the point of application of the axial force P. (b) Determine the axial forces F_1 and F_2 in segments (1) and (2) respectively.

$$A_1 = 1.0 \text{ in}^2, \qquad A_2 = 1.2 \text{ in}^2$$
$$E = 10(10^3) \text{ kis}, \qquad P = 20 \text{ kips}$$

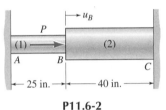

P11.6-2

Prob. 11.6-3. A rigid beam AC is supported by three columns, as shown in Fig. P11.6-3. (a) Use the *Principle of Virtual Displacements* to determine the vertical displacement v (taken positive downward) of the beam when a vertical load P = 540 kN is applied. (b) Determine the axial forces F_1 = F_3 and F_2 carried by the respective columns.

$$A_1 = A_3 = 3500 \text{ mm}^2$$
$$A_2 = 2000 \text{ mm}^2, \qquad E = 200 \text{ GPa}$$

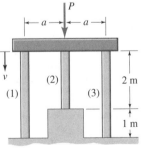

P11.6-3

Prob. 11.6-4. A single axial force P is applied at node C of the three-segment rod shown in Fig. P11.6-4. (a) Using the *Principle of Virtual Displacements*, determine expressions for the axial displacements u_B and u_C of nodes B and C respectively. (b) Determine the forces F_1, F_2, and F_3 in the three segments of the rod.

$$A_1 = A_3 = 2A, \qquad A_2 = A, \qquad E = \text{const}$$

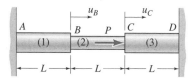

P11.6-4 and P11.6-12

Prob. 11.6-5. Use the *Principle of Virtual Displacements* to solve Prob. 3.5-1.

Problems 11.6-6 through 11.6-10. *For each of the pin-jointed trusses shown, (a) use the* **Principle of Virtual Displacements** *to determine the horizontal and vertical displacements of joint D, u_D and v_D respectively, and (b) determine the axial force in member (1), that is, the force in member AD.*

Prob. 11.6-6. For the truss in Fig. P11.6-6. $A_1 = A_2 = A_3 = 1.0$ in^2, $E = 30(10^3)$ ksi, and $P_x = 10$ kips.

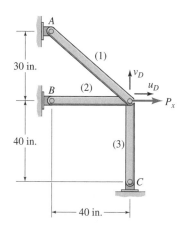

P11.6-6 and P11.6-17

Prob. 11.6-7. For the truss in Fig. P11.6-7, $A_1 = A_2 = 1000$ mm^2, $A_3 = 2000$ mm^2, $E = 70$ GPa, and $P_x = 100$ kN.

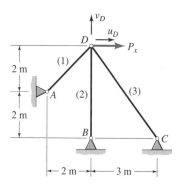

P11.6-7 and P11.6-18

Prob. 11.6-8. For the truss in Fig. P11.6-8, $A_1 = A_2 = A_3 = A$, and $E = $ const.

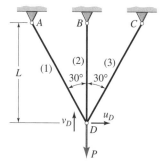

P11.6-8, P11.6-9, and P11.6-19

Prob. 11.6-9. For the truss in Fig. P11.6-9, $A_1 = A_2 = A, A_3 = 2A$, and $E = $ const.

Prob. 11.6-10. For the truss in Fig. P11.6-10, $A_1 = A_2 = A_3 = A$, and $E = $ const.

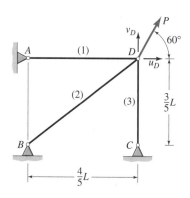

P11.6-10 and P11.6-20

611

Prob. 11.6-11. Use *Castigliano's First Theorem* to solve Prob. 3.5-15. Use u_B and u_C as the two displacement unknowns.

Prob. 11.6-12. Use *Castigliano's First Theorem* to solve Prob. 11.6-4. Use u_B and u_C as the two displacement unknowns.

Prob. 11.6-13. Use *Castigliano's First Theorem* to solve Prob. 3.5-25. Let θ be the displacement variable, and express *Castigliano's First Theorem* in the form:

$$\frac{d\mathcal{U}}{d\theta} = M_0 = P(a + b)$$

Prob. 11.6-14. Use *Castigliano's First Theorem* to solve Prob. 3.5-23. Let θ be the displacement variable, and express *Castigliano's First Theorem* in the form:

$$\frac{d\mathcal{U}}{d\theta} = M_0 = P(a + b)$$

Prob. 11.6-15. The rigid beam AC in Fig. P11.6-15 (see Prob. 3.6-16 for this figure) is supported by three vertical rods that are attached to the beam at points A, B, and C. When the rods are initially attached to the beam, the three rods are stress-free. (a) Use *Castigliano's First Theorem* to solve for the vertical displacements u_A and u_C of points A and C when downward loads $P_A = 8$ kips and $P_C = 2$ kips are applied to the beam at ends A and C, respectively. (b) Calculate the axial stresses σ_1, σ_2, and σ_3 resulting from the given loading.

$$A = 1.0 \text{ in}^2, \quad L = 30 \text{ in.}, \quad E = 30(10^3) \text{ ksi}$$

Prob. 11.6-16. Solve Prob. 11.6-15 with $P_A = 16$ kN and $P_C = 4$ kN and with the following properties of the structure:

$$A = 500 \text{ mm}^2, \quad L = 1 \text{ m}, \quad E = 100 \text{ GPa}$$

Prob. 11.6-17. Use *Castigliano's First Theorem* to solve Prob. 11.6-6.

Prob. 11.6-18. Use *Castigliano's First Theorem* to solve Prob. 11.6-7.

Prob. 11.6-19. Use *Castigliano's First Theorem* to solve Prob. 11.6-9.

Prob. 11.6-20. Use *Castigliano's First Theorem* to solve Prob. 11.6-10.

In **Problems 11.8-1 through 11.8-16** *use* **Castigliano's Second Theorem***, Eq. 11.54 (or Eq. 11.56), to solve each of these statically determinate problems.*

Prob. 11.8-1. Determine an expression for the vertical displacement, Δ_B, under the load at point B on the simply supported uniform beam shown in Fig. P11.8-1. (See Prob. 11.4-16 for this figure.) $EI = $ const.

Prob. 11.8-2. Determine an expression for the rotation angle θ_B produced by the external couple M_0 applied at point B of the simply supported uniform beam shown in Fig. P11.8-2. $EI = $ const.

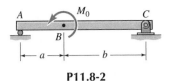

P11.8-2

Prob. 11.8-3. Determine the rotation angle θ_A produced by the external couple M_0 applied at end A of the beam AC shown in Fig. P11.8-3. (See Prob. 11.4-20 for this figure.) $EI = $ const.

Prob. 11.8-4. Determine the vertical deflection Δ_C under the load P applied at end C of the beam in Fig. P11.8-4. Let $E_{wood} = 1,600$ ksi, and see Table D.8 for the cross-sectional dimensions of the 6×8 wood beam.

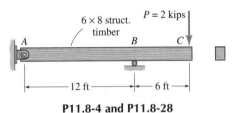

P11.8-4 and P11.8-28

Prob. 11.8-5. Determine the vertical deflection Δ_B under the load P applied at point B to the **W**310 $\times$ 143 structural steel beam in Fig. P11.8-5. Let $E_{steel} = 12$ GPa.

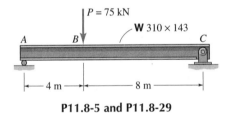

P11.8-5 and P11.8-29

Prob. 11.8-6. Determine an expression for the vertical deflection Δ_A under the load P acting at end A of the uniform beam in Fig. P11.8-6. Let $EI = $ const.

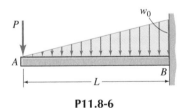

P11.8-6

Prob. 11.8-7. Use *Castigliano's Second Theorem* to solve Prob. 7.6-25.

Prob. 11.8-8. Use *Castigliano's Second Theorem* to solve Prob. 7.6-26.

Prob. 11.8-9. Use *Castigliano's Second Theorem* to solve Prob. 7.6-27.

Prob. 11.8-10. Use *Castigliano's Second Theorem* to solve Prob. 7.6-28.

Prob. 11.8-11. Use *Castigliano's Second Theorem* to solve Prob. 7.6-29.

*Prob. 11.8-12. Two segments, *AB* and *BC*, of 4 × 2 × 0.1875 structural steel tubing are welded together at *B* to form the L-shaped frame *ABC* shown in Fig. P11.8-12. Use *Castigliano's Second Theorem* to determine the vertical displacement Δ_C under the load $P = 500$ lbs at *C*. Let $E = 29(10^3)$ ksi. (Be sure to include the effects of both bending and stretching of segment *AB* on the displacement at *C*.)

$$A = 2.02 \text{ in}^2, \qquad I = 1.29 \text{ in}^4$$

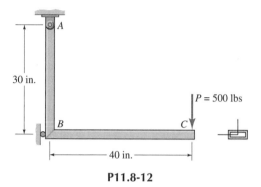

P11.8-12

Prob. 11.8-13. Determine the horizontal component, u_B, of the displacement of point *B* due to flexure of the curved beam shown in Fig. P11.8-13. Assume that the radius of curvature, *R*, of the centerline of the bar is large relative to the radial depth of the cross section so that the straight-beam elastic flexure formula holds. Express your answer in terms of *P*, *R*, *E*, and *I*.

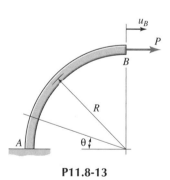

P11.8-13

Prob. 11.8-14. Determine an expression for the vertical displacement Δ_C under the vertical load *P* that acts at end *C* of the equal-leg, right-angle-joined pipe *ABC* shown in Fig. P11.8-14. The points *A*, *B*, and *C* lie in a horizontal plane. Let the modulus of elasticity, *E*, the shear modulus, *G*, and the moment of inertia, *I*, be constant over the lengths *AB* and *BC*, and ignore the joint dimensions at *B*. Neglect shear deformation due to bending of segments *AB* and *BC*. (Note: Segment *AB* undergoes both bending and torsion. The polar moment of inertia is $I_p = 2I$.)

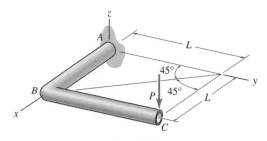

P11.8-14

Prob. 11.8-15. For the truss in Fig. P11.8-15, determine the horizontal displacement u_B due to the horizontal load P_x applied at joint *B*. Let $E = 30(10^3)$ ksi, $P_x = 1.5$ kips, $A_1 = 1$ in², $A_2 = 2$ in², $a = 24$ in., and $\beta = 30°$.

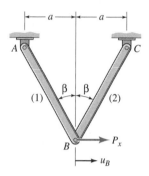

P11.8-15 and P11.8-36

Prob. 11.8-16. For the truss in Fig. P11.8-16, determine the downward vertical displacement v_B due to the weight *W* that is suspended from the pin at *B*. Let $E = 200$ GPa, $A_1 = 800$ mm², $A_2 = 1600$ mm², and $W = 10$ kN.

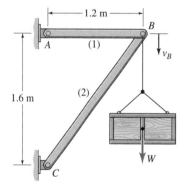

P11.8-16 and P11.8-37

In **Problems 11.8-17 through 11.8-25** *use* **Castigliano's Second Theorem**, *Eq. 11.57, to solve these statically indeterminate problems.*

Prob. 11.8-17. The uniform ($EI = $ const) propped-cantilever beam in Fig. P7.4-3 supports a linearly varying load of maximum intensity w_0 (force per unit length). Use *Castigliano's Second Theorem*: (a) to determine the (redundant) reaction R_A, and (b) to determine the slope, θ_A, of the beam at end *A*.

613

Prob. 11.8-18. For the uniformly loaded **W**8 × 40 propped-cantilever beam in Fig. P7.4-2, (a) determine the (redundant) reaction R_B, and (b) determine the slope of the beam, θ_B, at end B.

Prob. 11.8-19. The uniform (EI = const) propped-cantilever beam in Fig. P7.4-7 supports a concentrated load P at point B. Use *Castigliano's Second Theorem*: (a) to determine the reaction R_A at end A, and (b) to determine the vertical displacement, Δ_B, of the beam at the point of application of load P.

Prob. 11.8-20. The uniform fixed-fixed beam in Fig. P7.4-8 is subjected to a concentrated load P at distance a from end A. Use *Castigliano's Second Theorem*: (a) to determine the reactions R_A and M_A at end A, and (b) to determine the vertical displacement, Δ_B, of the beam at the point of application of load P.

Prob. 11.8-21. The uniform continuous beam in Fig. P7.4-10 supports a uniformly distributed load of intensity w_0 on the span AB. Use *Castigliano's Second Theorem*: (a) to determine the reaction R_B and the central support, and (b) to determine the slope of the beam θ_C at end C.

Prob. 11.8-22. For the non-uniform beam AC in Fig. P7.4-11, use *Castigliano's Second Theorem*: (a) to determine the reactions R_A and M_A at end A, and (b) to determine the slope θ_B of the beam at the central support, B.

Prob. 11.8-23. For the uniform fixed-fixed beam AC in Fig. P7.4-9, use *Castigliano's Second Theorem* to determine the fixed-end reactions R_A and M_A.

Prob. 11.8-24. For the uniform fixed-fixed beam AC in Fig. P7.7-2, use *Castigliano's Second Theorem*: (a) to solve for the fixed-end reactions R_A and M_A, and (b) to solve for the transverse displacement Δ_B at node B.

Prob. 11.8-25. For the nonuniform fixed-fixed beam shown in Fig. P7.7-3, use *Castigliano's Second Theorem*: (a) to solve for the fixed-end reactions R_A and M_A, and (b) to solve for the transverse displacement Δ_B at node B.

Problems 11.8-26 through 11.8-39. *Use the* **Unit Load Method** *to solve each of these statically determinate problems.*

Prob. 11.8-26. Use the *Unit Load Method* to solve Prob. 11.8-1.

Prob. 11.8-27. Use the *Unit Load Method* to solve Prob. 11.8-3.

Prob. 11.8-28. Use the *Unit Load Method* to solve Prob. 11.8-4.

Prob. 11.8-29. Use the *Unit Load Method* to solve Prob. 11.8-5.

Prob. 11.8-30. For the uniform simply supported beam in Fig. P11.8-30, determine the vertical displacement Δ_B at midspan and the slope θ_C at end C. (See Prob. 7.6-24 for Fig. P11.8-30.)

Prob. 11.8-31. For the uniform cantilever beam in Fig. P11.8-31, use the *Unit Load Method* to determine the vertical displacement Δ_A and the slope θ_A at the free end A. The maximum load intensity (at B) is w_0 (force per unit length).

Prob. 11.8-32. Use the *Unit Load Method* to solve Prob. 7.6-25.

Prob. 11.8-33. Use the *Unit Load Method* to solve Prob. 7.6-26.

Prob. 11.8-34. For the beam-rod system described in Prob. 7.6-31 and shown in Fig. P7.6-31, use the *Unit Load Method* to determine the vertical displacement Δ_B and slope θ_B of end B when beam AB is subjected to a uniformly distributed load as shown.

Prob. 11.8-35. For the beam described in Prob. 7.6-32 and shown in Fig. P7.6-32, use the *Unit Load Method* to determine the vertical displacements Δ_B and Δ_C under the load P and at end C, respectively. Consider both bending of the beam AC and stretching of the rod CD.

Prob. 11.8-36. Use the *Unit Load Method* to determine the vertical displacement v_B of joint B of the two-member truss described in Prob. 11.8-15 and shown in Fig. P11.8-15.

Prob. 11.8-37. Use the *Unit Load Method* to determine the horizontal displacement u_B of joint B of the two-member truss described in Prob. 11.8-16 and shown in Fig. P11.8-16.

Prob. 11.8-38. The symmetric four-member pin-jointed truss shown in Fig. P11.8-38 has a height $H = 2$ m and a span $L = 6$ m. A load $P = 60$ kN acts vertically through the joint at B. The cross-sectional area of each tension member is 1200 mm² and of each compression member is 3200 mm². The truss members are made of steel with $E_{steel} = 200$ GPa. Use the *Unit Load Method*: (a) to calculate the vertical displacement, v_B, of joint B, and (b) to calculate the horizontal displacement, u_C, of joint C.

Prob. 11.8-39. Each of the six members of the planar truss shown in Fig. P11.8-39 has an axial rigidity AE. The truss is loaded by equal vertical forces P at joints D and E. Use the *Unit Load Method* to determine the horizontal and vertical components, u_E and v_E, respectively, of the displacement of joint E.

P11.8-38

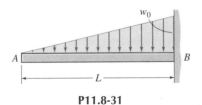

P11.8-31

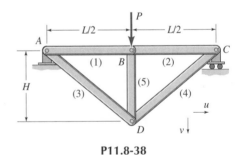

P11.8-39

Problems 11.8-40 through 11.8-55. *Use the* **Unit Load Method** *to solve the following statically indeterminate problems.*

Prob. 11.8-40. By combining Eqs. 11.57 and 11.58, show that the *Unit Load Method* for statically indeterminate structures can be expressed by the following equations:

$$0 = \Delta_{ri} = \int_0^L \frac{F\,F_{ri}\,dx}{AE} + \int_0^L \frac{T\,T_{ri}\,dx}{GI_p}$$
$$+ \int_0^L \frac{M\,M_{ri}\,dx}{EI} + \int_0^L \frac{f_s\,V V_{ri}\,dx}{GA}, \; i = 1, 2, \ldots, N_R$$

$$\Delta_{ui} = \int_0^L \frac{F\,F_{ui}\,dx}{AE} + \int_0^L \frac{T\,T_{ui}\,dx}{GI_p}$$
$$+ \int_0^L \frac{M\,M_{ui}\,dx}{EI} + \int_0^L \frac{f_s\,V V_{ui}\,dx}{GA}, \; i = 1, 2, \ldots, N$$

where F_{ri}, T_{ri}, and so forth, are the distributions of force, torque, and so forth, due to a unit value of the redundant R_i; and F_{ui}, T_{ui}, and so forth, are the distributions of force, torque, and so forth, due to a unit value of force (or couple) P_i where the linear displacement (or angular displacement) is to be determined. (The latter unit ''forces'' may be ''dummy forces'' if no actual load is applied where displacement is to be determined.)

Prob. 11.8-41. Use the *Unit Load Method* to solve Prob. 11.8-17. Let R_A be the redundant force.

Prob. 11.8-42. Use the *Unit Load Method* to solve Prob. 11.8-18. Let R_B be the redundant force.

Prob. 11.8-43. Use the *Unit Load Method* to solve Prob. 11.8-21. Let R_B be the redundant force.

Prob. 11.8-44. Use the *Unit Load Method* to solve Prob. 11.8-23. Let R_A and M_A be the redundant force.

Prob. 11.8-45. Use the *Unit Load Method* to solve Prob. 11.8-25. Let R_A and M_A be the redundant force.

Prob. 11.8-46. Use the *Unit Load Method* to determine the (redundant) reactions R_A and M_A for the uniform fixed-fixed beam with linearly varying load, as shown in Fig. P7.4-5. Let $EI =$ const.

Prob. 11.8-47. For the uniformly loaded beam in Fig. P7.4-6, use the *Unit Load Method*: (a) to determine the (redundant) force F_2 in the hanger rod, and (b) to determine the vertical displacement, v_B, of the rod-supported end B.

Prob. 11.8-48. For the two-span beam AC in Fig. P7.7-10, use the *Unit Load Method*: (a) to determine the two (redundant) reactions R_A and R_B, and (b) to determine the slope, θ_A, of the beam at end A.

Prob. 11.8-49. For the nonuniform two-span continuous beam in Fig. P7.7-12, use the *Unit Load Method*: (a) to determine the (redundant) reaction R_A, and (b) to determine the slope angle, θ_A, at end A.

Prob. 11.8-50. For the nonuniform two-span continuous beam in Fig. P7.7-13, use the *Unit Load Method*: (a) to determine the (redundant) reaction R_A, and (b) to determine the slope angle, θ_A, at end A.

Prob. 11.8-51. For the nonuniform two-span continuous beam in Fig. P7.7-15, use the *Unit Load Method*: (a) to determine the

(redundant) reaction R_A, and (b) to determine the slope angle, θ_A, at end A.

*Prob. 11.8-52. For the two-span nonuniform beam AC in Fig. P7.7-4, use the *Unit Load Method*: (a) to determine three redundant reactions, R_A, M_A, and R_B; and (b) to determine the slope angle, θ_B, at the central support B.

Prob. 11.8-53. The truss members in Fig. P3.8-13 have the following axial rigidities: $(AE)_2 = (AE)_3 = AE$, and $(AE)_1 = 2AE$. A horizontal load P is applied to the truss at joint A. (a) Use the *Unit Load Method* to determine the (redundant) force F_1 in member (1); then, determine the other two members forces. (b) Use the *Unit Load Method* to determine u_A and v_A, the horizontal displacement and the vertical displacement, respectively, of joint A.

Prob. 11.8-54. Each of the three truss members in Fig. P3.8-16 has a length L and modulus of elasticity E. The cross-sectional areas of the members are $A_1 = A_2 = A$, and $A_3 = 2A$. A horizontal load P acts on the truss at joint A. (a) Use the *Unit Load Method* to determine the (redundant) force F_3 in member (3); then, determine the other two member forces. (b) Use the *Unit Load Method* to determine u_A and v_A, the horizontal displacement and the vertical displacement, respectively, of joint A.

Prob. 11.8-55. For the truss in Fig. P3.8-19: $A_1 = A_2 = A_3 = 1.0$ in^2, $E_1 = E_2 = E_3 = 30(10^3)$ ksi, and $P = 15$ kips. (a) Use the *Unit Load Method* to determine the (redundant) force F_1 in member (1); then, determine the other two members forces. (b) Use the *Unit Load Method* to determine u_A and v_A, the horizontal displacement and the vertical displacement, respectively, of joint A.

Prob. 11.9-1. If, instead of being released from rest at position (1) in Fig. 11.21a, the weight W has a downward speed v at position (1), determine an expression (similar to Eq. 11.62) for the maximum deflection of the spring.

Prob. 11.9-2. A uniform rod like the one shown in Fig. 11.23 has a diameter d and length L, and it is to be used to absorb the impact loading from a falling mass. Determine the total amount of elastic strain energy that can be absorbed by the rod if it is made of the following metals: (a) 2014-T6 aluminum alloy, (b) 6061-T6 aluminum alloy, (c) ASTM-A36 structural steel, and (d) Ti-6AL-4V titanium alloy.

Let $d = 1$ in. and $L = 40$ in., and express your answers in units of kip $\cdot$ in.

Prob. 11.9-3. Repeat Prob. 11.9-2 for a rod whose dimensions are $d = 30$ mm and $L = 1$ m. Express your answers in units of kN $\cdot$ m.

In **Problems 11.9-4 through 11.9-11** *assume that the material remains linearly elastic and that the three assumptions regarding* Impact on Deformable Bodies *are satisfied.*

Prob. 11.9-4. As shown in Fig. P11.9-4, a weight $W = 200$ lb is dropped from a height $h = 20$ in. and lands on the top of a 4-in.-square aluminum alloy post ($E = 10 \times 10^3$ ksi and $\sigma_Y = 60$ ksi) whose length is $L = 2$ ft. The aluminum post rests on a rigid base. Determine the maximum compressive stress in the post, the maximum shortening of the post, and the impact magnification factor, Δ_{max}/Δ_{st}.

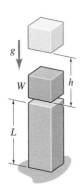

P11.9-4 and P11.9-5

Prob. 11.9-5. From what height h_1 must the weight $W = 200$ lb in Prob. 11.9-4 be dropped if the impact causes a maximum stress in the post of $\sigma_Y/4$? From what height h_2 must it be dropped if the impact causes a maximum stress in the post of $\sigma_Y/2$?

Prob. 11.9-6. The collar mass in Fig. 11.23 slides down a uniform 6061-T6 aluminum-alloy rod of diameter d, impacting against the flange at the bottom of the rod. Determine the maximum stress in the rod if: (a) the collar is released from rest at distance $h = h_a$ above the collar; (b) the collar is released from rest at $h = 0$, that is, when it is just in contact with the collar; and (c) the collar is lowered slowly onto the flange. Let $m = 50$ kg, $d = 25$ mm, $L = 1.25$ m, and $h_a = 0.5$ m.

Prob. 11.9-7. Repeat Prob. 11.9-6 Parts (a) through (c) using the following data for the rod and the mass: $W = 500$ lb, $d = 2$ in., $L = 50$ in., and $h_a = 20$ in. (d) From what height would the 500-lb weight have to be dropped to cause the rod to yield? (e) What weight, if dropped from the height $h_a = 20$ in., would cause the rod to yield?

*Prob. 11.9-8. The stepped rod in Fig. P11.9-8 has cross-sectional areas $2A$ and A. (a) Determine an expression for the maximum elongation of the rod as a function of W, E, A, L, and h. Neglect the energy stored in the short transition segment, and neglect

stress-concentration effects due to the change of cross section. (Stress Concentration is discussed in Section 12.2.) Hint: Use the methods of Section 3.5 to determine an equivalent stiffness factor, k, for the stepped bar. (b) Determine an expression for the maximum impact axial stress, σ_{max}, in terms of the maximum static stress, σ_{st}, the drop height h, and the rod dimensions and modulus of elasticity. (c) Determine an expression for the ratio of the maximum impact stress determined in Part (b) to the maximum stress that a uniform bar of cross-sectional area A would experience under impact loading by a weight W dropped from height h.

Prob. 11.9-9. The stepped ASTM-A36 structural steel rod in Fig. P11.9-9 has equal-length segments of length L, with cross-sectional areas A (diameter $= d$) and $2A$. Determine the maximum axial impact stress in the rod if a collar of weight W is dropped from height h above the flange at the bottom of the rod. Neglect the energy stored in the short transition segment, and neglect stress-concentration effects due to the change of cross section. (The topic of Stress Concentration is discussed in Section 12.2.) Hint: Use the methods of Section 3.5 to determine an equivalent stiffness factor, k, for the stepped rod. Let $d = 0.5$ in., $L = 20$ in., $W = 20$ lb, and $h = 2$ in.

Prob. 11.9-10. A drop test is used to test the impact performance of automobile bumpers. Assume that the bumper is a uniform simply-supported beam and that mass m is dropped from height h, impacting the beam at distance aL from support A ($0 < a \leq 1/2$), as shown in Fig. P11.9-10. (a) Derive an expression that relates the maximum flexural stress due to impact, σ_{max}, to the drop height and other parameters: W, E, I, c, and L. (b) Specialize your answer to Part (a) for the case $a = 1/2$.

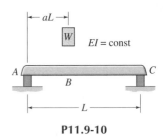

P11.9-10

Prob. 11.9-11. A mass m is dropped from height h, impacting the uniform cantilever beam AC at point B, at a distance aL from the cantilevered end, A. Determine an expression that relates the maximum tip deflection, $(\Delta_C)_{max}$, to the drop height h and location ($0 < a \leq 1$) and to other parameters: m, E, I, c, and L.

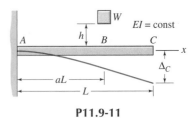

P11.9-11

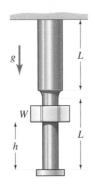

P11.9-8 and P11.9-9

SPECIAL TOPICS RELATED TO DESIGN

12

12.1 INTRODUCTION

In this chapter you will be introduced to several important topics that you will encounter again in courses on design of structures and/or machines. Section 12.2, *Stress Concentrations,* discusses the effect of holes, notches, or changes of cross section, on the stresses due to axial loading, torsion, or bending. Section 12.3, *Failure Theories,* indicates how mechanical properties obtained from simple uniaxial tension or compression testing may be used to predict yielding or brittle fracture of members subjected to more complex loading conditions. Finally, in Section 12.4, *Fatigue and Fracture,* you will learn how repeated loading and unloading of a member causes small cracks to grow in length, eventually leading to a fatigue failure.

12.2 STRESS CONCENTRATIONS

The key formulas

$$\sigma = \frac{F}{A}, \qquad \tau = \frac{T\rho}{I_p}, \qquad \sigma = -\frac{My}{I}$$

enable one to calculate the normal stress due to axial loading, the shear stress due to torsion, and the flexural stress due to bending, respectively. These formulas may be used only so long as the cross section of the member is relatively uniform, that is, there are no abrupt changes in cross section. In Section 3.3 it was noted that these formulas do not hold in the immediate vicinity of points of application of load, but that away from that vicinity, *St. Venant's Principle* ensures the validity of the above formulas for bodies of uniform cross section. The increase of stress due to localized application of a load or due to nonuniformity of the cross section is called a *stress concentration.* In the design of load-bearing members it is important to avoid stress concentrations whenever possible and to properly account for them when they do exist, particularly if the member is made of brittle material or is subjected to a fluctuating load (see Section 12.4). We will briefly examine the effect of stress concentrations in members subjected to axial loading, torsional loading, and bending.

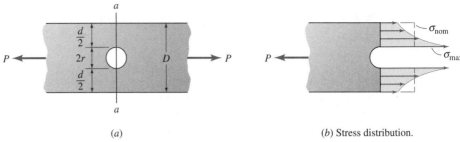

(a)   (b) Stress distribution.

FIGURE 12.1 Stress concentration in an axially loaded flat bar with circular hole.

Stress Concentration—Axial Loading.

The last picture in the color insert depicts the stress-concentration effects due to a centrally-located circular hole in an axially-loaded flat bar, as computed by use of the finite element method. Holes, notches, or abrupt changes in cross section produce stress concentrations in axially-loaded members.

Figure 12.1b shows the stress concentration due to a *centrally-located circular hole* in a flat bar in tension. The maximum normal stress occurs at the edge of the hole on the cross section $a - a$, which passes through the center of the hole. In Figure 12.2 a *shoulder fillet* of radius r is used to smooth the transition between the wider portion of the bar and the narrower portion. The maximum normal stress for this case occurs on section $b - b$, where the fillet joins the narrower part of the bar.

The stress distribution due to a stress concentration like the ones in Figs. 12.1 and 12.2 may be determined analytically through the use of the theory of elasticity [Ref. 12-1], or numerically by using finite element analysis [Refs. 12-2, 12-3] (See the last color-insert photo.), or experimentally through the use of photoelasticity [Ref. 12-4]. The exact distribution of stress is not of great importance, but the maximum value of stress is very important. This maximum stress, σ_{max}, may be related to the average stress on the <u>net</u> cross section, the *nominal stress* σ_{nom}, by defining the *stress-concentration factor* as

$$K = \frac{\sigma_{max}}{\sigma_{nom}} \tag{12.1}$$

For linearly elastic behavior, the stress-concentration factor is a function of geometry of the member and type of load applied (i.e., axial, torsion, bending). Formulas for the stress-concentration factor for many types of loading and geometry are given in Ref. 12-5; and many graphs may be found in Ref. 12-6, from which the graphs in this section were

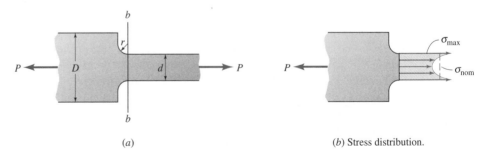

(a)   (b) Stress distribution.

FIGURE 12.2 Stress concentration in an axially loaded flat bar with abrupt change in cross section.

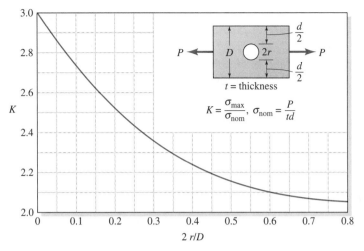

FIGURE 12.3 Stress-concentration factor K for a flat bar with centrally located circular hole (tension). (Adapted from Ref. 12-6. Reprinted by permission of J. Wiley & Sons, Inc.)

obtained. Values of the stress-concentration factor given in these references and in the figures presented in this section are based on linearly elastic behavior and are valid only as long as the computed value σ_{max} does not exceed the proportional limit of the material.

Figure 12.3 shows the effect of hole-size on the stress-concentration factor for a flat bar with centrally-located circular hole. Note that as the radius of the hole approaches zero, the stress-concentration factor approaches the value three. So, a very small hole can have a very damaging effect on a member. Fatigue cracks, which are discussed in Section 12.4, frequently are initiated at just such small holes.

From Fig. 12.4 it may be observed that as $r \rightarrow 0$ the stress concentration factor increases rapidly. With no fillet (i.e., for $r = 0$) the theoretical stress concentration factor would be infinitely large because of the sharp 90° reentrant corner. Therefore, good design practice requires that such sharp corners be avoided and that generous fillets be provided.

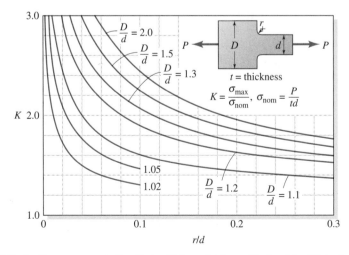

FIGURE 12.4 Stress concentration factor K for a flat bar with shoulder fillets (tension). (Adapted from Ref. 12-6. Reprinted by permission of J. Wiley & Sons, Inc.)

■■■■■■■■■■■■■■■□□ EXAMPLE 12.1 ■■■■■■■■■■■■■■■■

An aluminum bar has the dimensions shown in Fig. 1. If the allowable stress is $\sigma_{allow} = 200$ MPa, determine the maximum axial force P_{allow} that can be carried by the bar.

Solution   Figure 12.4 may be used to estimate the stress concentration factor for this bar with shoulder fillets. For the bar in Fig. 1,

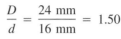

$$\frac{D}{d} = \frac{24 \text{ mm}}{16 \text{ mm}} = 1.50$$

$$\frac{r}{d} = \frac{4 \text{ mm}}{16 \text{mm}} = 0.25$$

Since there is a curve for $D/d = 1.50$, we can directly estimate that $K = 1.74$.

The allowable load is based on the average stress in the smaller cross section, so

$$P_{allow} = \sigma_{nom}A_{min} = \sigma_{nom}(80 \text{ mm}^2)$$

and when the stress reaches σ_{allow} at the points of stress concentration,

$$K\sigma_{nom} = \sigma_{allow} = 200 \text{ MPa}$$

Therefore,

$$P_{allow} = \left(\frac{200 \text{ MPa}}{1.74}\right)(80 \text{ mm}^2) = 9.20 \text{ kN}$$

$$P_{allow} = 9.20 \text{ kN} \qquad \textbf{Ans.}$$

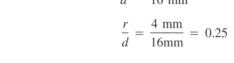

t = thickness = 5 mm

Fig. 1

(a) Flanged connection.

(b) Gear on shaft.

(c) Stepped shaft.

FIGURE 12.5   Three situations that produce stress concentration in circular shafts subjected to torsion.

Stress Concentration—Torsion. The torsion formula $\tau = \dfrac{T\rho}{I_p}$, which was derived in Chapter 4, may be used to determine the shear-stress distribution on the cross section of a homogeneous linearly elastic rod with uniform circular cross section. In order for this formula to be valid, the cross section in question must not be near a point where torque is applied to the rod by a gear, pulley, or flange, or to a point where there is a sudden change in the diameter of the cross section. Figure 12.5 shows three such situations where stress concentration occurs in circular rods loaded in torsion.

Applications like the two shown in Fig. 12.5a and 12.5b are considered in courses on machine design. Here we will only consider the stress concentration in the vicinity of a change in cross section, as illustrated in Fig. 12.5c and Fig. 12.6. Based on St. Venant's Principle, the shear-stress distribution at sections $a - a$ and $c - c$, which are more than 1 diameter away from the region of diameter change, may be determined by the torsion formula

$$\tau = \frac{T\rho}{I_p} \qquad (12.2)$$

These linear shear-stress distributions are illustrated in Fig. 12.6b and 12.6d, respectively.

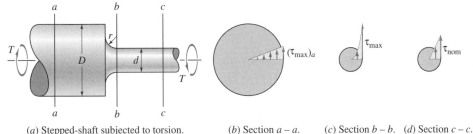

(a) Stepped-shaft subjected to torsion.   (b) Section a – a.   (c) Section b – b.   (d) Section c – c.

FIGURE 12.6   A stepped torsion rod.

But, as indicated in Fig. 12.6c, the maximum shear stress in a stepped torsion rod with shoulder fillet occurs at section $b - b$, where the fillet joins the smaller-diameter portion of the rod. The *stress-concentration factor for torsion* is defined as

$$K = \frac{\tau_{max}}{\tau_{nom}} = \tau_{max}\left(\frac{\pi d^3}{16T}\right) \tag{12.3}$$

since the nominal maximum shear stress would occur at the outer fibers of the smaller cross section, as indicated in Fig. 12.6d.

Figure 12.7 gives the value of the stress-concentration factor K for various geometries of stepped shafts. As was discussed above for axial loading of a stepped bar, the stress-concentration factor for torsion of a stepped shaft approaches infinity as the fillet radius r approaches zero. Therefore, in order to minimize the effect of the torsional shear-stress concentration characterized by Fig. 12.7, one should provide a fillet with the largest radius that is practical for the particular design. (The application of the stress-concentration factor to torsion problems is virtually identical to the application to axial deformation problems, as illustrated in Example Problem 12.1. Therefore, no torsion example is provided, although there are homework exercises on this topic.)

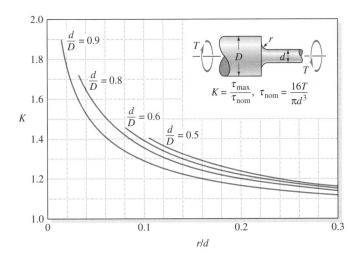

FIGURE 12.7   Torsional stress-concentration factor K for a stepped shaft with shoulder fillet. (Adapted from Ref. 12-6. Reprinted by permission of J. Wiley & Sons, Inc.)

Stress Concentration—Bending.   In Chapter 6 the flexure formula, $\sigma = -My/I$, was derived for bending of a uniform beam having a plane of symmetry. This formula is valid only so long as there is no abrupt change in cross section and no concentrated load or reaction at or very near the cross section of interest. Here we will consider only two cases where stress concentrations occur in beams—the case of a rectangular beam with

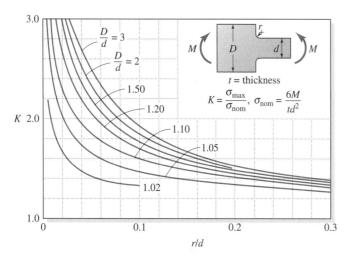

FIGURE 12.8 Stress-concentration factors for pure bending of flat bars with fillets. (Adapted from Ref. 12-6. Reprinted by permission of J. Wiley & Sons, Inc.)

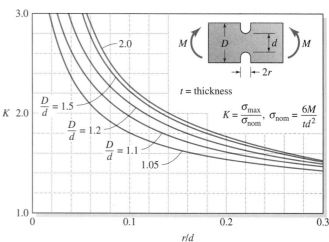

FIGURE 12.9 Stress-concentration factors for pure bending of flat bars with U-shaped notches. (Adapted from Ref. 12-6. Reprinted by permission of J. Wiley & Sons, Inc.)

abrupt change of cross section (Fig. 12.8), and that of a rectangular beam with symmetric, U-shaped notches (Fig. 12.9). Other cases are considered in Refs. [12-5, 12-6].

The *stress-concentration factor for bending* is defined as

$$K = \frac{\sigma_{max}}{\sigma_{nom}} = \sigma_{max}\left(\frac{td^2}{6M}\right) \qquad (12.4)$$

with σ_{nom} being the nominal maximum stress at the reduced cross section, based on the flexure formula. Figures 12.8 and 12.9 present values of the stress-concentration factor K based on the data from Ref. [12-6]. These figures clearly indicate the desirability of avoiding fillets and notches with small radius r.

The stress-concentration factors presented in Figs. 12.8 and 12.9 are based on linearly elastic behavior, so values of σ_{max} obtained by using Eq. 12.4 are valid only if the proportional limit of the material is not exceeded. Otherwise, if the material is ductile, plastic deformation will occur in the vicinity of the stress concentration, resulting in stresses that are smaller than the value of σ_{max} given by Eq. 12.4.

■■■■■■■■■■■■■■■■□ E X A M P L E 1 2 . 2 □■■■■■■■■■■■■■■■■

The flat bars in Fig. 1 are subjected to pure bending. By what percentage can the maximum moment in the bar be increased by removing material in order to convert the deep grooves of Fig. 1*a* to the semicircular grooves as illustrated in Fig. 1*b*? Assume that both bars have the same thickness t and are made of the same material (i.e., they have the same allowable stress).

Solution The stress concentration factors for the two beams in Fig. 1 can be estimated from the curves in Fig. 12.9. The required parameters are

$$\frac{D_1}{d_1} = \frac{D_2}{d_2} = 1.5, \qquad \frac{r_1}{d_1} = \frac{1}{8}, \qquad \frac{r_2}{d_2} = \frac{1}{4}$$

622

From Fig. 12.9 we estimate that

$$K_1 = 2.06, \qquad K_2 = 1.61$$

From Eq. 12.4,

$$\sigma_{max} = K\sigma_{nom} = K\left(\frac{6M}{td^2}\right) \qquad (1)$$

But, since $\sigma_{max} = \sigma_{allow}$ is the same for both bars, and since they have the same dimensions t and d, Eq. (1) gives

$$\frac{M_2}{M_1} = \frac{K_1}{K_2}$$

so, the increase in moment capacity is given by

$$\frac{M_2 - M_1}{M_1} = \frac{K_1}{K_2} - 1 = \frac{K_1 - K_2}{K_2} = \frac{2.06 - 1.61}{1.61} = 0.28 \qquad \textbf{Ans.}$$

Therefore, by decreasing the sharpness of the notches we can increase the applied moment by about 28% without increasing the value of the maximum stress in the beam.

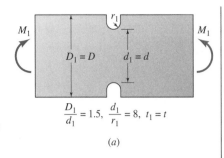

$$\frac{D_1}{d_1} = 1.5, \ \frac{d_1}{r_1} = 8, \ t_1 = t$$

(a)

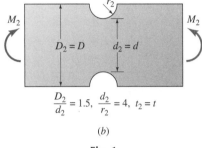

$$\frac{D_2}{d_2} = 1.5, \ \frac{d_2}{r_2} = 4, \ t_2 = t$$

(b)

Fig. 1

*12.3 FAILURE THEORIES

■■■■■■■■■■■■

The design engineer is faced with two distinct tasks. The first task is to analyze the behavior of proposed designs subjected to specified loadings. For simple structural members and machine components the formulas in this book may be used to calculate stress and deformation. If there are stress concentrations, the procedures of Section 12.2 must be applied. For more complex members the *finite element method* [Refs. 12-2, 12-3] is generally used to obtain the stress distribution and deformation. In some cases solutions may be obtained by using the theory of elasticity or the theory of plates and shells. The other important task of the design engineer is to determine what values of stress and/or deformation would constitute *failure* of the object being designed.[1] That is the subject we address in this section.

If a tension test is performed on a specimen of *ductile material,* the specimen may be said to fail when the axial stress reaches the yield stress σ_Y, that is, the *criterion of failure* is *yielding.* If the specimen is made of *brittle material,* the usual failure criterion is *brittle fracture* at the ultimate tensile stress σ_U. But a machine component or structural element is invariably subjected to a multiaxial state of stress, for which it is more difficult to designate what value of stress produces failure.

A tension test is relatively easy to perform using the procedures described in Section 2.4, and test results are published for many materials. But in order to apply the results of a tension test (or a compression test, or a torsion test) to a member that is subjected to

---

[1]In the remainder of this section we will only consider failures due to excessive static stress.

multiaxial loading, it is necessary to consider the actual mechanism of failure. That is, was failure caused by the maximum normal stress reaching a critical value? Or was it due to maximum shear stress that reached a critical value, or to strain energy or some other quantity having reached its critical value? In the tension test, the criterion for failure can be easily stated in terms of the principal (tensile) stress σ_1, but for multiaxial stress we must consider the actual cause of the failure and say what combinations of stress would constitute failure.

In this section we consider four theories of failure.[2] Two of these apply to materials that behave in a ductile manner, that is, to materials that yield before they fracture. The other two theories apply to brittle materials. For plane stress, the failure theories are expressed in terms of the principal stresses σ_1 and σ_2. For triaxial states of stress, σ_1, σ_2, and σ_3 are used.

Ductile Materials. Two theories of failure for ductile materials will be discussed, the maximum-shear-stress theory and the maximum-distortion-energy theory.

Maximum-Shear-Stress Theory:[3] When a flat bar of ductile material, like mild steel, is tested in tension, it is observed that the mechanism that is actually responsible for yielding is *slip*, that is, shearing along planes of maximum shear stress at 45° to the axis of the member. Initial yielding is associated with the appearance of the first slip line on the surface of the specimen, and as the strain increases, more slip lines appear until the entire specimen has yielded. If this slip is assumed to be the actual mechanism of failure, then the stress that best characterizes this failure is the shear stress on the slip planes. Figure 12.10 shows a Mohr's circle of stress for this uniaxial stress state, indicating that the shear stress on the slip planes has a magnitude of $\sigma_Y/2$. Therefore, if it is postulated that in a ductile material under any state of stress (uniaxial, biaxial, or triaxial), failure occurs when the shear stress on any plane reaches the value $\sigma_Y/2$, then the *failure criterion* for the *maximum-shear-stress theory* may be stated as

$$\tau_{\substack{\text{abs} \\ \text{max}}} = \frac{\sigma_Y}{2} \tag{12.5}$$

where σ_Y is the yield stress determined by a simple tension test. Using Eq. 8.39 we can express Eq. 12.5 in terms of principal stresses as

$$\boxed{\sigma_{\text{max}} - \sigma_{\text{min}} = \sigma_Y} \tag{12.6}$$

where σ_{max} is the maximum principal stress and σ_{min} is the minimum principal stress.[4]

For the case of plane stress, the *maximum-shear-stress failure criterion* may be stated in terms of the in-plane principal stresses σ_1 and σ_2 as follows:[5]

$$\left. \begin{array}{l} |\sigma_1| = \sigma_Y \text{ if } |\sigma_1| \geq |\sigma_2| \\ |\sigma_2| = \sigma_Y \text{ if } |\sigma_2| \geq |\sigma_1| \end{array} \right\} \text{ and } \sigma_1 \text{ and } \sigma_2 \text{ have same sign} \tag{12.7}$$

$$|\sigma_1 - \sigma_2| = \sigma_Y \qquad \text{if } \sigma_1 \text{ and } \sigma_2 \text{ have opposite signs}$$

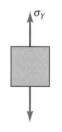

(a) Mohr's circle for $\sigma_1 = \sigma_Y$.

$\left(\dfrac{\sigma_Y}{2}, -\dfrac{\sigma_Y}{2}\right)$

$\sigma_2 = \sigma_3 = 0$

$\sigma_1 = \sigma_Y$

$\left(\dfrac{\sigma_Y}{2}, \dfrac{\sigma_Y}{2}\right)$

(b) Principal stress element.

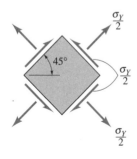

(c) Maximum shear-stress element.

FIGURE 12.10 Principal stresses and maximum shear stresses for a uniaxial stress test.

---

[2]The theories presented here apply only to homogeneous, isotropic materials.

[3]The names of C. A. Coulomb, H. Tresca, and J. J. Guest are associated with this theory of failure [Ref. 12-7].

[4]Note that this theory of failure ignores the normal stresses acting on the planes of maximum shear stress.

[5]For the case of plane stress, the out-of-plane principal stress is called σ_3 ($\sigma_3 \equiv 0$), even though it is not necessarily the minimum principal stress.

Equations 12.7 may be represented in the convenient graphical forms shown in Fig. 12.11. For a member undergoing plane stress, the state of stress at every point in the body can be represented by a *stress point* (σ_1, σ_2) in the $\sigma_1 - \sigma_2$ plane, as indicated in Fig. 12.11.[6] If the state of stress for any point in the body corresponds to a stress point that lies outside the hexagon of Fig. 12.11 or on its boundary, failure is said to have occurred according to the maximum-shear-stress theory.

Maximum-Distortion-Energy Theory:[7] Although the maximum-shear-stress theory provides a reasonable hypothesis for yielding in ductile materials, the maximum-distortion-energy theory correlates better with test data and is therefore generally preferred. In this theory, yielding is assumed to occur when the energy associated with change of shape of a body undergoing multiaxial loading is equal to the energy of distortion in a tensile specimen when yielding occurs at the uniaxial yield stress σ_Y.

Consider the strain energy stored in an element of volume, like the one shown in Fig. 12.12a. The strain energy density due to multiaxial loading is given by Eq. 11.14, which can be written, using the three principal axes, in the form

$$ u = \frac{1}{2}(\sigma_1\epsilon_1 + \sigma_2\epsilon_2 + \sigma_3\epsilon_3) \tag{12.8} $$

Combining Eq. 12.8 with Hooke's Law (Eq. 2.34 with $\Delta T = 0$) we get

$$ u = \frac{1}{2E}[\sigma_1^2 + \sigma_2^2 + \sigma_3^2 - 2\nu(\sigma_1\sigma_2 + \sigma_2\sigma_3 + \sigma_1\sigma_3)] \tag{12.9} $$

A portion of this strain energy can be associated with the *change of volume* of the element, and the remainder of the strain energy is associated with *change of shape,* that is, with *distortion.* The change of volume is produced by the average stress $\sigma_{avg} = \frac{1}{3}(\sigma_1 + \sigma_2 + \sigma_3)$, as illustrated in Fig. 12.12b. The net stresses shown in Fig. 12.12c produce distortion without any change of volume.

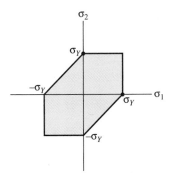

•Experimental data from tension test.

FIGURE 12.11 Failure hexagon for the maximum-shear-stress theory (plane stress).

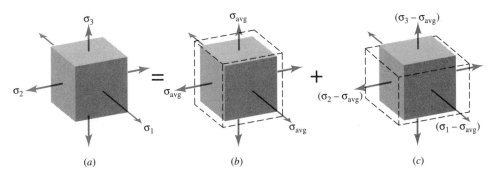

FIGURE 12.12 (*a*) Triaxial stress state. (*b*) Stresses producing volume change. (*c*) Stresses producing distortion.

---

[6]Although σ_1 and σ_2 are principal stresses, Eq. 12.7 and Fig. 12.11 do not require that the principal axes be labeled such that $\sigma_1 \geq \sigma_2$, as was done in Chapter 8.

[7]This theory is also called the *maximum-octahedral-shear-stress theory.* Credit for this theory is generally given to M. T. Huber, R. von Mises, and H. Hencky, although it was earlier conjectured by J. Clerk Maxwell [Ref. 12-7].

625

Experiments have shown that materials do not yield when they are exposed to *hydro-static stresses*[8] of extremely large magnitude. Therefore, it has been postulated that the stresses that actually cause yielding are the stresses that produce distortion. This hypothesis constitutes the *maximum-distortion-energy yield (failure) criterion,* which states:

Yielding of a ductile material occurs when the distortion energy per unit volume equals or exceeds the distortion energy per unit volume when the same material yields in a simple tension test.

When the distortion-producing stresses of Fig. 12.12c are substituted into Eq. 12.9 we get the following expression for the *distortion-energy density,*

$$u_d = \frac{1}{12G}[(\sigma_1 - \sigma_2)^2 + (\sigma_2 - \sigma_3)^2 + (\sigma_1 - \sigma_3)^2] \tag{12.10}$$

The distortion energy density in a tensile test specimen at the yield stress σ_Y is

$$(u_d)_Y = \frac{1}{6G}\sigma_Y^2 \tag{12.11}$$

since $\sigma_1 = \sigma_Y$ and $\sigma_2 = \sigma_3 = 0$. Therefore, yielding occurs when the distortion energy for general loading, given by Eq. 12.10, equals or exceeds the value of $(u_d)_Y$ in Eq. 12.11. Therefore, the *maximum-distortion-energy failure criterion* can be stated in terms of the three principal stresses as

$$\frac{1}{2}[(\sigma_1 - \sigma_2)^2 + (\sigma_2 - \sigma_3)^2 + (\sigma_1 - \sigma_3)^2] = \sigma_Y^2 \tag{12.12a}$$

In terms of the normal stresses and shear stresses on three arbitrary mutually orthogonal planes, the maximum-distortion-energy failure criterion can be shown to have the form[9]

$$\frac{1}{2}[(\sigma_x - \sigma_y)^2 + (\sigma_y - \sigma_z)^2 + (\sigma_x - \sigma_z)^2 + 6(\tau_{xy}^2 + \tau_{yz}^2 + \tau_{xz}^2)] = \sigma_Y^2 \tag{12.12b}$$

For the case of plane stress, the corresponding expressions for the *maximum-distortion-energy yield criterion* can easily be obtained from Eqs. 12.12 by setting $\sigma_3 = \sigma_z = \tau_{xz} = \tau_{yz} = 0$. In terms of the principal stresses, then,

$$\sigma_1^2 - \sigma_1\sigma_2 + \sigma_2^2 = \sigma_Y^2 \tag{12.13}$$

This is the equation of an ellipse in the $\sigma_1 - \sigma_2$ plane, as depicted in Fig. 12.13. For comparison purposes, the failure hexagon for the maximum-shear-stress yield theory is also shown in dashed lines in Fig. 12.13. At the six vertices of the hexagon the two failure theories coincide; that is, both theories predict that yielding will occur if the state of (plane) stress at a point corresponds to any one of these six stress states. Otherwise, the maximum-shear-stress theory gives the more conservative (i.e., smaller-valued) estimate of the stresses required to produce yielding, since the hexagon falls either on or inside the ellipse.

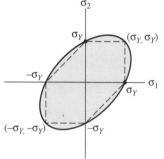

● Experimental data from tension test.
− − Maximum-shear-stress criterion.

FIGURE 12.13 Failure ellipse for the maximum-distortion-energy theory (plane stress).

---

[8]Figure 12.12b represents a hydrostatic state of stress, that is, equal stresses in all three principal directions.
[9]See Sections 78 and 90 of Ref. 12-1.

A convenient way to apply the maximum-distortion-energy theory is to take the square root of the left-hand side of Eq. 12.12a (or Eq. 12.12b) to form an equivalent stress quantity that is called the *Mises equivalent stress*. Either of the following two equations can be used to compute the Mises equivalent stress, σ_M:

$$\sigma_M = \frac{\sqrt{2}}{2}[(\sigma_1 - \sigma_2)^2 + (\sigma_2 - \sigma_3)^2 + (\sigma_1 - \sigma_3)^2]^{1/2} \tag{12.14a}$$

or

$$\sigma_M = \frac{\sqrt{2}}{2}[(\sigma_x - \sigma_y)^2 + (\sigma_y - \sigma_z)^2 + (\sigma_x - \sigma_z)^2 + 6(\tau_{xy}^2 + \tau_{yz}^2 + \tau_{xz}^2)]^{1/2}$$

$$\tag{12.14b}$$

For the case of plane stress, the corresponding expressions for the Mises equivalent stress can easily be obtained from Eqs. 12.14 by setting $\sigma_3 = \sigma_z = \tau_{xz} = \tau_{yz} = 0$.

By comparing the value of the Mises equivalent stress at any point with the value of the tensile yield stress, σ_Y, it can be determined whether yielding is predicted to occur according to the maximum-distortion-energy theory of failure. Therefore, the Mises equivalent stress is widely used when calculated stresses are presented in tabular form or in the form of color stress plots, as has been done for the finite element analysis results shown in the last picture in the color-photo insert.

■■■■■■■■■■■■■■■■ E X A M P L E   1 2 . 3 ■■■■■■■■■■■■■■■■■

A force P_0 kips applied by a lever arm to the shaft in Fig. 1 produces stresses at the critical point A having the values shown on the element in Fig. 1. Determine the load $P_S \equiv c_S P_0$ that would cause the shaft to fail according to the maximum-shear-stress theory, and determine the load $P_D \equiv c_D P_0$ that would cause failure according to the maximum-distortion-energy theory. The shaft is made of steel with $\sigma_Y = 36$ ksi.

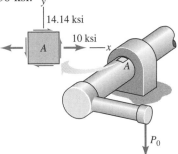

Fig. 1

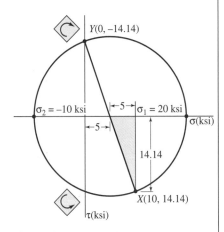

Fig. 2 Mohr's circle for stress state at point A.

Solution   It will be helpful if we construct a Mohr's circle for the plane stress state in Fig. 1 and also sketch the failure envelopes for the two failure theories.

Since σ_1 is positive and σ_2 is negative, we only need to sketch the fourth quadrant of the failure envelope. This is shown in Fig. 3. Since the stresses at point A are proportional to load, the stresses due to any load cP_0 will lie along the radial line identified in Fig. 3 as the *load line*. This line passes through the origin of the $\sigma_1 - \sigma_2$ plane and through the stress point ($\sigma_{1P} = 20$ ksi,

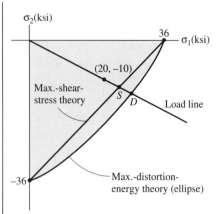

σ_2(ksi)

36 σ_1(ksi)

(20, −10)

Max.-shear-stress theory

S D Load line

−36

Max.-distortion-energy theory (ellipse)

Fig. 3 Fourth quadrant of failure envelopes.

$\sigma_{2P} = -10$ ksi) that corresponds to the principal stresses due to load P_0. Failure according to the maximum-shear-stress theory occurs at the stress state marked S in Fig. 3, and failure according to the maximum-distortion-energy theory occurs at point D.

Maximum-Shear-Stress Theory. Point S is the intersection of the load line given by

$$\sigma_1 = -2\sigma_2 \tag{1}$$

and the maximum-shear-stress boundary line

$$\sigma_1 - \sigma_2 = \sigma_Y = 36 \text{ ksi} \tag{2}$$

Solving Eqs. (1) and (2) for σ_1 and σ_2 we get

$$\sigma_{1S} = 24 \text{ ksi}, \qquad \sigma_{2S} = -12 \text{ ksi}$$

Combining these values with σ_1 and σ_2 of Fig. 2 we get

$$c_S = \frac{P_S}{P_0} = \frac{\sigma_{1S}}{\sigma_{1P}} = \frac{24 \text{ ksi}}{20 \text{ ksi}} = 1.2$$

$$c_S = 1.2 \qquad\qquad \text{Ans.}$$

Maximum-Distortion-Energy Theory: Point D in Fig. 3 is the intersection of the load line, given by Eq. (1), and the parabola given by Eq. 12.13. Thus,

$$\sigma_{1D}^2 - \sigma_{1D}\sigma_{2D} + \sigma_{2D}^2 = \sigma_Y^2 = (36 \text{ ksi})^2 \tag{3}$$

Combining Eqs. (1) and (3) gives

$$7\sigma_{2D}^2 = (36 \text{ ksi})^2$$

Then, since $\sigma_1 > 0$ and $\sigma_2 < 0$, we get

$$\sigma_{1D} = 27.21 \text{ ksi}, \qquad \sigma_{2D} = -13.61 \text{ ksi}$$

Comparing these stresses with the principal stresses produced by P_0 gives

$$c_D = \frac{P_D}{P_0} = \frac{\sigma_{1D}}{\sigma_{1P}} = \frac{27.21 \text{ ksi}}{20 \text{ ksi}} = 1.36$$

$$c_D = 1.36 \qquad\qquad \text{Ans.}$$

We could get this same result by noting that the factor of safety against yielding, according to the maximum-distortion-energy yield criterion is given by

$$FS_d = \sigma_Y/\sigma_M$$

where σ_M is the Mises equivalent stress. For the original load P_0,

$$\sigma_M = \sigma_1^2 - \sigma_1\sigma_2 + \sigma_2^2 = [(20)^2 - (20)(-10) + (-10)^2] = 26.46 \text{ ksi}$$

Then,

$$FS_d = \frac{\sigma_Y}{\sigma_M} = \frac{36 \text{ ksi}}{26.46 \text{ ksi}} = 1.36$$

In summary, a 20% increase in the load would cause failure according to the maximum-shear-stress theory, but a 36% increase would be required to cause failure according to the maximum-distortion-energy theory. That is, under load P_0 the member would have a factor of safety $FS_s = 1.2$ with respect to failure according to the maximum-shear-stress theory and a factor of safety of $FS_d = 1.36$ with respect to failure according to the maximum-distortion-energy theory of failure.

Brittle Materials. Two theories of failure for brittle materials are presented, the maximum-normal-stress theory and Mohr's failure theory.

Maximum-Normal-Stress Theory:[10] It was stated in Section 2.4 and illustrated in Fig. 2.15b, that in a tension test, a brittle material fails suddenly by *fracture*, without prior yielding. And it was stated in Section 4.5 and illustrated in Fig. 4.19b, that in a torsion test, a bar made of brittle material also fails by fracture on planes of maximum tensile stress. Experiments have shown that the value of the normal stress on the failure plane for this biaxial state of stress is not significantly different from the fracture stress σ_U in a uniaxial tensile test. Therefore, the hypothesis of the *maximum-normal-stress theory* is that an object made of brittle material will fail when the maximum principal stress in the material reaches the ultimate normal stress that the material can sustain in a uniaxial tension test. This theory also assumes that compression failures occur at the same ultimate stress value as do tension failures.

For the case of *plane stress,* the *maximum-normal-stress failure criterion* is given by the equations

$$|\sigma_1| = \sigma_U \quad \text{or} \quad |\sigma_2| = \sigma_U \qquad (12.15)$$

These equations may be plotted on the $\sigma_1 - \sigma_2$ plane, as shown in Fig. 12.14.[11]

Mohr's Failure Criterion:[12] If the ultimate compressive strength of a brittle material is not equal to its ultimate strength in tension, the maximum-normal-stress theory should not be used. An alternative failure theory was proposed by Otto Mohr and is called *Mohr's failure criterion.* Figure 12.15a shows Mohr's circles for a uniaxial tensile test and for a uniaxial compression test for a brittle material having a tensile ultimate strength σ_{TU} and an ultimate strength in compression of σ_{CU}. By Mohr's theory, when σ_1 and σ_2 have the

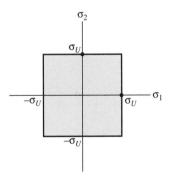

• Experimental data from tension test.

FIGURE 12.14 Failure diagram for the maximum-normal-stress theory (plane stress).

---

[10] Also called *Rankine's Theory* after W. J. M. Rankine (1820–1872), an eminent professor of engineering at Glasgow University in Scotland.

[11] As in Figs. 12.11 and 12.13, σ_1 is not necessarily the greater principal stress.

[12] This theory is named for the German engineer Otto Mohr (1835–1918), the developer of Mohr's circle.

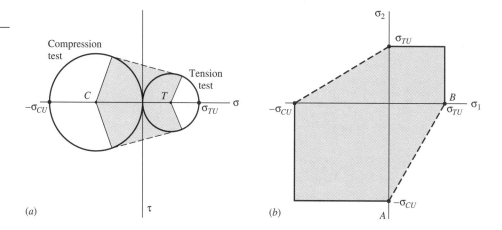

FIGURE 12.15 Mohr's failure criterion (plane stress). *(a)*   *(b)*

same sign, failure occurs if either of the following stress limits is reached:

$$\sigma_{max} = \sigma_{TU} \quad \text{or} \quad \sigma_{min} = -\sigma_{CU} \quad\quad (12.16)$$

These equations are plotted as solid-line boundaries in Fig. 12.15b. For cases where σ_1 and σ_2 have opposite signs, Mohr proposed that the failure boundary be determined by drawing tangents to the tension and compression circles, as illustrated by the dashed lines in Fig. 12.15a. It can be shown (see Example 12.4) that the principal stresses for all circles that have centers on the σ-axis and are tangent to the dashed tangent lines in Fig. 12.15a plot as points on the dashed-line boundaries in Fig. 12.15b. If torsion test data or other plane-stress failure data are available, the failure boundaries in the second and fourth quadrants of Fig. 12.6b can be modified to incorporate these experimental data (e.g., see *Modified Mohr Theory* in Ref. [12-8]).

■■■■■■■■■■■■■■■■ E X A M P L E 1 2 . 4 ■■■■■■■■■■■■■■■■

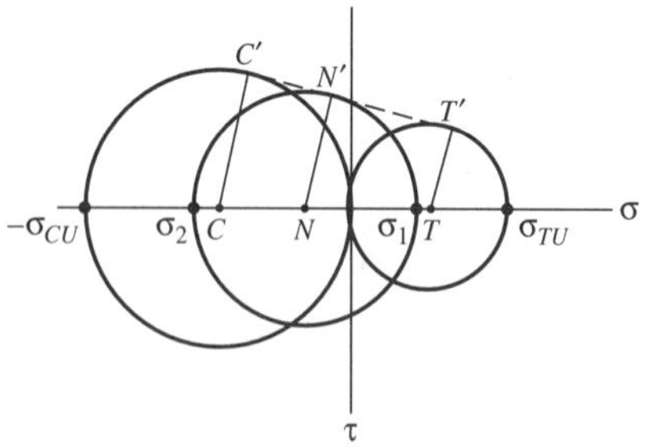

Fig. 1

Show that, for $\sigma_1 > 0$ and $\sigma_2 < 0$, the stress points that lie on the dashed line AB in Fig. 12.15b correspond to the principal stresses on Mohr's circles that have centers lying between points C and T in Fig. 12.15a and are tangent to the dashed lines in this figure.

Solution Let us redraw Fig. 12.15a and add an intermediate circle as specified in the problem statement (Fig. 1). The equation of the dashed-line AB in Fig. 12.15b is

$$\sigma_2 = \sigma_{CU}\left(\frac{\sigma_1}{\sigma_{TU}} - 1\right) \quad\quad (1)$$

We are to prove that this equation is the equation that relates the principal stresses σ_1 and σ_2 for the circle with center at N in Fig. 1. From Fig. 1,

$$\sigma_1 = \sigma_n + \overline{NN'}, \quad \sigma_2 = \sigma_n - \overline{NN'} \quad\quad (2)$$

where σ_n is the normal stress corresponding to point N. Also, from Fig. 1, the centers of the tension-test Mohr's circle and the compression-test Mohr's circle are

$$\sigma_t = \frac{\sigma_{TU}}{2}, \qquad \sigma_c = -\frac{\sigma_{CU}}{2} \qquad\qquad (3)$$

and their radii are

$$\overline{TT'} = \frac{\sigma_{TU}}{2}, \qquad \overline{CC'} = \frac{\sigma_{CU}}{2} \qquad\qquad (4)$$

Let

$$\sigma_n = \sigma_c + k(\sigma_t - \sigma_c)$$

That is,

$$\sigma_n = -\frac{\sigma_{CU}}{2} + k\left(\frac{\sigma_{TU}}{2} + \frac{\sigma_{CU}}{2}\right) \qquad\qquad (5)$$

Then, the radius $\overline{NN'}$ will be linearly related to the radii $\overline{CC'}$ and $\overline{TT'}$ by the equation

$$\overline{NN'} = \overline{CC'} - k(\overline{CC'} - \overline{TT'})$$

or

$$\overline{NN'} = \frac{\sigma_{CU}}{2} - k\left(\frac{\sigma_{CU}}{2} - \frac{\sigma_{TU}}{2}\right) \qquad\qquad (6)$$

Combining Eqs. (2), (5), and (6), we get

$$\sigma_1 = k\sigma_{TU}, \qquad \sigma_2 = \sigma_{CU}(k - 1) \qquad\qquad (7)$$

The elimination of k from Eqs. (7) produces Eq. 1, the desired equation of the dashed line AB in Fig. 12.15b. QED.

*12.4 FATIGUE AND FRACTURE

At some time or other you have probably held a paper clip in your hands and bent the paper-clip wire back and forth several times until the wire finally broke in two. The failure did not occur when the paper clip was first bent, even though the wire experienced very large plastic deformation. Instead, failure occurred after a few reversals of flexural stress in the wire. This type of failure is called *fatigue failure*.[13] If the failure occurs after a few cycles of loading, perhaps up to a thousand cycles, it is called *low-cycle fatigue*. However, many metal components experience fatigue failure only after millions of stress cycles. In any case, when failure occurs at a stress level that is less than the level that would produce fracture under a single static application of load, the failure is called a fatigue failure.

---

[13]Fatigue failure of axles of railway cars was a source of great concern and study in the 1800s. The term *fatigue* was used by Poncelet to denote this type of failure due to cyclic stresses. A. Wöler (1819–1914), a German railway engineer, is credited with developing the first fatigue testing machine. [Ref. 12-7]

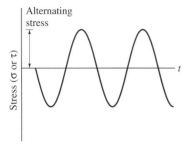

(a) Fully reversed stress.

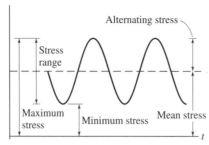
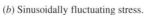

(b) Sinusoidally fluctuating stress.

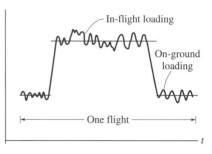

(c) Ground-air-ground stress history for one airplane flight.

FIGURE 12.16 Typical cyclic stresses that can produce fatigue failure.

The types of fluctuating stresses that lead to fatigue failure in metals are illustrated in Fig. 12.16. Figure 12.16*a* is the type of fully reversed stress that would be experienced by a railway-car axle as the train moves at constant speed along the track. Figure 12.16*b* shows a superposition of a (constant) mean stress and a sinusoidally varying stress; Fig. 12.16*c* illustrates the more complicated type of fluctuating stress experienced, for example, by an airplane wing component during a single flight. The fracture that occurs as a result of such fluctuating stresses usually begins at a point of stress concentration (Section 12.2) like the edge of the fastener hole in the splice plate shown in Fig. 12.17*a* and the airplane wing leading-edge nose cap in Fig. 12.17*b*. The crack is initiated in a region of high stress intensity, usually at some microscopic flaw or imperfection. Stress cycles cause the fatigue crack to grow slowly in size until the crack reaches a *critical crack length,* at which point the crack propagates at an explosive rate leaving the component unable to sustain load. Figure 12.17*c* shows the typical "beach-sand markings" of a fatigue fracture surface. Along the bottom edge of this photo, two points of initiation of the fatigue crack can be identified; the "ridges" on the fracture surface indicate the progress of the crack as it propagated with successive cycles of stress.

To characterize the behavior of a material under repeated cycles of loading, *fatigue tests* at various levels of stress are performed, and the results are plotted as an *S-N diagram,*

(a) Fatigue crack in a splice plate.

(b) Fatigue crack in a wing nose cap.

(c) The surface of a fatigue crack.

FIGURE 12.17 Examples of fatigue cracks. (Used with permission, Lockheed Fort Worth Company, 1993.)

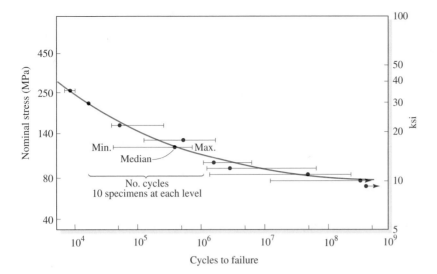

FIGURE 12.18 An S-N diagram for 7075-T6 aluminum-alloy notched tensile specimens. (From Ref. 12-10.)

or *endurance curve*.[14] Figure 12.18 is an S-N diagram based on a series of tests on nominally identical test specimens. (Note that the diagram is a log-log plot.) Several tests (e.g., ten tests at each stress level for Fig. 12.18) are conducted with cyclic stress whose maximum amplitude is slightly less than the ultimate static strength of the material. The number of cycles of stress required to cause fracture at that stress level is recorded for each test. Similar test series are conducted at progressively lower levels of stress, and the results plotted as the number of cycles to cause fatigue failure in each test. The scatter band for the tests at various stress levels are indicated on Fig. 12.18.

The endurance curves for some materials have the form illustrated in Fig. 12.19, where there is a stress level, called the *fatigue level*, or *endurance limit*, of the material, below which a virtually infinite number of stress cycles can be sustained without resulting in a fatigue failure. Many steel alloys exhibit this type of behavior. Fatigue limits for steel alloys, based on 10^7 stress cycles, are typically 35%–60% of the ultimate tensile strength of the steel.[15] As indicated in Fig. 12.19, notches or other imperfections obviously degrade

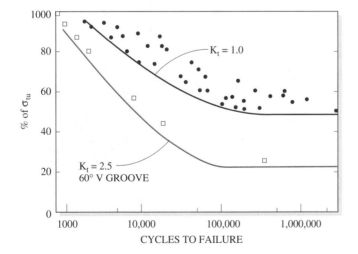

FIGURE 12.19 An S-N diagram for steel round-bar rotating-beam specimens. (From Ref. 12-11, maintained and published by CINDAS/Purdue University under Cooperative Research and Development Agreement with the U.S. Air Force. Used with permission.)

---

[14]Test procedures to produce S-N diagrams are described in ASTM Standards E466-E468. [Ref. 12-9]

[15]*Structural Alloys Handbook*. [Ref. 12-11]

fatigue strength. Some materials, including many aluminum alloys, do not exhibit a clearly defined endurance limit. In such cases, if the material continues to undergo cyclic stressing, it will eventually fail, regardless of how small the stress is. Tabulated ''endurance limits'' for aluminum alloys indicate that they can sustain up to 5×10^8 stress cycles if the stress does not exceed about 25% of their ultimate tensile strength.[16]

From Figs. 12.18 and 12.19 and the preceding discussion, it is clear that *fatigue strength* must be taken into consideration in the design of a component that is to be subjected to cycles of stress during its service lifetime. One approach to fatigue design is called the *safe-life design philosophy.* Using endurance limit data and data from fatigue tests of the actual structural component under realistic stress cycles (see Fig. 12.16*c*), the design engineer establishes a safe lifetime for the component, that is, the number of stress cycles that the component will be allowed to experience before it must be retired from service.

As the desire has arisen to extend the useful lifetime of machines and structures (e.g., commercial and military aircraft), and as the discipline of *fracture mechanics* has matured, a new design philosophy has gained in importance—the *fail-safe design philosophy,* or *damage-tolerant design philosophy.* This philosophy involves the use of extensive fatigue testing, including careful observation of the initiation and propagation of cracks, to determine the complex relationship between cyclic loading and crack propagation. Then, with proper inspection, fatigue cracks can be prevented from reaching critical length and damaged parts can be replaced before leading to catastrophic failure of the entire structure or machine.[17] Figure 12.20 shows an airplane horizontal tail section undergoing full-scale fatigue testing to identify the fatigue-critical parts of the structure and to determine how many stress cycles could be applied before some component of the tail would fail by fatigue. This information enables the airplane's designers to ensure that no fatigue failure will occur while the aircraft is in service.

In your future courses on machine-design or design of structures, you will have the opportunity to study and apply *safe-life design* and *fail-safe design* in much greater depth.

FIGURE 12.20 An airplane horizontal tail section undergoing full-scale fatigue testing. (Photo used with permission of the Boeing Company, 1995.)

---

[16]*Aluminum Standards and Data.* [Ref. 12-12] (See the source of these data for restrictions on their use in design.)

[17]Reference 12-13 discusses fracture mechanics and describes fatigue-fracture control strategies. Reference 12-14 discusses fatigue design methodologies in general, and Reference 12-15 applies the fail-safe and safe-life fatigue-design methods to aircraft structural design.

12.5 PROBLEMS ■■

Problems 12.2-1 through 12.2-8. *In solving these stress-concentration problems, assume that the axial load is centrally applied at the ends of the flat bar and that the material remains linearly elastic.*

Prob. 12.2-1. For the flat tension bar with central hole shown in Fig. P12.2-1, let $D = 3$ in., $t = 0.5$ in., and $P = 8$ kips. Determine the maximum normal stress in the bar for the following hole diameters: 0.5 in., 1.0 in., and 2.0 in.

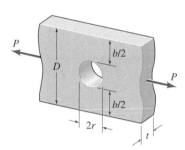

P12.2-1 through P12.2-4

Prob. 12.2-2. For the flat tension bar with central hole shown in Fig. P12.2-2, let $r = 20$ mm, $t = 12$ mm, and $P = 40$ kN. Determine the maximum normal stress in the bar for the following bar widths: $D = 100$ mm, $D = 125$ mm, and $D = 150$ mm.

Prob. 12.2-3. For the flat tension bar with central hole shown in Fig. P12.2-3, let $r = 15$ mm, $t = 20$ mm, and $\sigma_{max} = 75$ MPa. Determine the maximum axial load that can be applied to the bar for the following bar widths: $D = 130$ mm, $D = 150$ mm, and $D = 170$ mm.

Prob. 12.2-4. For the flat tension bar with central hole shown in Fig. P12.2-4, let $D = 2$ in., $t = 0.5$ in., and $\sigma_{max} = 12$ ksi. Determine the maximum axial load that can be applied to the bar for the following hole diameters: 0.75 in., 1.0 in., and 1.50 in.

Prob. 12.2-5. For the flat tension bar with stepped cross section shown in Fig. P12.2-5, the fillets are quarter circles having the largest radius that is consistent with the two widths D and d (i.e., $r = (D-d)/2$). Let $D = 4.0$ in. and $t = 1.0$ in. (a) If the smaller width is $d = 3$ in. and the axial load is $P = 15$ kips, what is the maximum normal stress in the bar? (b) If the bar must be able to

carry an axial load $P = 15$ kips without the maximum normal stress exceeding $\sigma_{max} = 8$ ksi, what is the minimum width, d, that can be used for the reduced-width cross section?

Prob. 12.2-6. For the flat tension bar with stepped cross section shown in Fig. P12.2-6, the fillets are quarter circles of radius $r = 10$ mm. Let $D = 80$ mm and $t = 20$ mm for this bar. (a) If the smaller width is $d = 50$ mm and the axial load is $P = 25$ kN, what is the maximum normal stress in the bar? (b) If the bar must be able to carry an axial load $P = 35$ kN without the maximum normal stress exceeding $\sigma_{max} = 60$ MPa, what is the minimum width, d, of the reduced-width cross section?

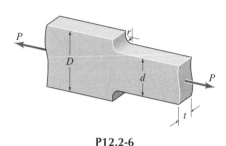

P12.2-6

Prob. 12.2-7. For the flat tension bar with stepped cross section shown in Fig. P12.2-7, the hole is centered in the wide portion of the bar and the fillets are quarter circles. Let $D = 4$ in., $d = 2.5$ in., and $t = 0.5$ in. for this bar, and let the maximum axial load be $P = 10$ kips. What is the maximum hole radius and what is the minimum fillet radius that may be accommodated if the maximum normal stress in the bar is not to exceed $\sigma_{max} = 18$ ksi? Select radii that are multiples of 0.05 in.

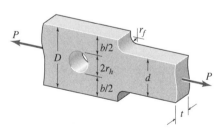

P12.2-7 and P12.2-8

Prob. 12.2-8. For the flat tension bar with stepped cross section shown in Fig. P12.2-8, the hole is centered in the wide portion of the bar and the fillets are quarter circles. Let $D = 100$ mm, $t = 20$ mm, and $d = 65$ mm for this bar, and let the maximum axial load be $P = 70$ kN. What is the maximum hole radius and what is the minimum fillet radius that may be accommodated if the maximum normal stress in the bar is not to exceed $\sigma_{max} = 125$ MPa? Select radii that are multiples of 1.0 mm.

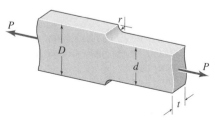

P12.2-5

Problems 12.2-9 through 12.2-14. *In solving these torsional stress-concentration problems, consider stepped, circular shafts as shown here, and assume that the material remains linearly elastic.*

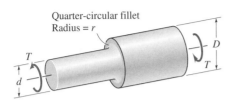

Quarter-circular fillet
Radius = r

P12.2-9 through P12.2-14

Prob. 12.2-9. A stepped torsion bar with major diameter $D = 65$ mm and minor diameter $d = 40$ mm is subjected to a torque $T = 100$ N·m. (a) Determine the maximum shear stress in the shaft for the following size fillets: $r_1 = 5$ mm and $r_2 = 10$ mm. (b) Compare the shear stress results you obtained in Part (a) with the maximum shear stress the shaft would experience if it were of uniform diameter $D = d = 40$ mm.

Prob. 12.2-10. A stepped torsion bar has a major diameter $D = 1.5$ in. and a full quarter-circular fillet (i.e., $r = (D-d)/2$). It is subjected to a torque $T = 1200$ lb·in. (a) Determine the maximum shear stress in the shaft for the following minor diameters: $d_1 = 1$ in., and $d_2 = 1.25$ in. (b) Compare the shear stress results you obtained in Part (a) with the maximum shear stresses the shafts would experience if they were of uniform diameters $D_1 = d_1 = 1$ in. and $D_2 = d_2 = 1.25$ in., respectively.

Prob. 12.2-11. A stepped torsion bar with major diameter $D = 2$ in. and minor diameter $d = 1.25$ in. is subjected to a torque $T = 3750$ lb·in. If the allowable maximum shear stress in the shaft is $\tau_{max} = 12$ ksi, what is the minimum radius that may be used for the quarter-circular fillet at the junction of the two segments of the shaft? The fillet radius must be chosen as some multiple of 0.05 in. Show your calculations for at least three different radii.

Prob. 12.2-12. A stepped torsion bar with major diameter $D = 50$ mm and (quarter-circular) fillet radius 7 mm is subjected to a torque $T = 400$ N·m. If the allowable maximum shear stress in the shaft is $\tau_{max} = 70$ MPa, what is the minimum diameter that may be used for the smaller-diameter segment of the shaft? The diameter must be chosen as some multiple of 1 mm. Show your calculations for at least three different diameters.

Prob. 12.2-13. A stepped torsion bar with major diameter $D = 2$ in. and minor diameter $d = 1.5$ in. has a full quarter-circular fillet at the junction of the two segments (i.e., $r = (D - d)/2$). If the allowable maximum shear stress in the shaft is $\tau_{max} = 16$ ksi, and the shaft rotates at a constant angular speed of $\omega = 500$ rpm, what is the maximum power (hp) that may be delivered by the shaft?

Prob. 12.2-14. In Example Prob. 4.8 it was determined that a $\frac{5}{8}$-in.-diameter shaft would be required to transmit 10 hp at 875 rpm if the allowable maximum shear stress in the shaft is 20 ksi.

Can a stepped shaft with major diameter $D = 1$ in. and minor diameter $d = 0.625$ in. (i.e., $\frac{5}{8}$ in.) be used instead of a uniform $\frac{5}{8}$-in.-diameter shaft? If so, what is the minimum radius that may be used for the quarter-circular fillet at the junction of the two segments of the shaft? The fillet radius must be chosen as some multiple of $\frac{1}{16}$ in. Show your calculations for at least three different fillet radii.

Problems 12.2-15 through 12.2-20. *In solving these stress-concentration problems, assume that the material remains linearly elastic.*

Prob. 12.2-15. The stepped-cross-section beam in Fig. P12.2-15 has a constant thickness $t = 15$ mm, a major depth $D = 60$ mm, and a minor depth $d = 40$ mm; and it is subjected to a bending moment $M = 100$ N·m. (a) Determine the maximum bending (normal) stress in the beam for the following sizes of quarter-circular fillets: $r_1 = 5$ mm, and $r_2 = 10$ mm. (b) Compare the bending stress results you obtained in Part (a) with the maximum flexural stress the beam would experience if it were of uniform diameter $D = d = 40$ mm.

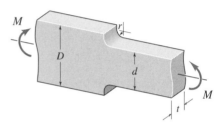

P12.2-15, P12.2-16, and P12.2-17

Prob. 12.2-16. The stepped-cross-section beam in Fig. P12.2-16 has a major depth $D = 1.5$ in., and it has full quarter-circular fillets joining the major depth and the minor depth (i.e., $d = D - 2r$). The beam has a constant thickness $t = 0.5$ in., and is subjected to a bending moment $M = 1200$ lb·in. (a) Determine the maximum bending (normal) stress in the beam for the following minor depths: $d_1 = 1$ in., and $d_2 = 1.25$ in. (b) Compare the bending stress results you obtained in Part (a) with the maximum flexural stresses the beams would experience if they were of uniform depths $D_1 = d_1 = 1$ in. and $D_2 = d_2 = 1.25$ in., respectively.

Prob. 12.2-17. The stepped-cross-section beam in Fig. P12.2-17 has a major depth $D = 2.5$ in., and it has full quarter-circular fillets joining the major depth and the minor depth (i.e., $d = D - 2r$). The beam has a constant thickness $t = 1.0$ in., and is subjected to a bending moment $M = 5$ kip·in. If the allowable maximum normal stress is $\sigma_{max} = 12$ ksi, determine the smallest minor depth d that may be used. The depth d must be chosen as some multiple of 0.10 in. Show your calculations for at least three different values of d.

Prob. 12.2-18. The beam in Fig. P12.2-18 has a constant thickness $t = 0.75$ in., a minor depth $d = 2$ in., and symmetrical notches with semicircular roots of radius $r = 0.125$ in. The beam is subjected to a bending moment $M = 2500$ lb·in. (a) Determine

636

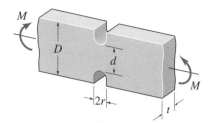

P12.2-18, P12.2-19, and P12.2-20

the maximum bending (normal) stress in the beam for the following major depths: $D_1 = 2.25$ in., $D_2 = 2.5$ in., and $D_3 = 2.75$ in. (b) Compare the bending stress results you obtained in Part (a) with the maximum flexural stress that a uniform beam of depth $D = d = 2$ in. (without notches) would experience if subjected to the same bending moment, $M = 2500$ lb·in.

Prob. 12.2-19. The beam in Fig. P12.2-19 has a constant thickness $t = 12$ mm, a major depth $D = 50$ mm, and symmetrical notches with semicircular roots of radius $r = 5$ mm. It is subjected to a bending moment $M = 100$ N·m. (a) Determine the maximum bending (normal) stress in the beam for the following minor depths: $d_1 = 30$ mm, $d_2 = 35$ mm, and $d_3 = 40$ mm. (b) Compare the bending stress results you obtained in Part (a) with the maximum flexural stress the beams would experience if they were uniform beams (i.e., no notches), with $D_1 = d_1 = 30$ mm, $D_2 = d_2 = 35$ mm, and $D_3 = d_3 = 40$ mm, respectively.

Prob. 12.2-20. The beam in Fig. P12.2-20 has a constant thickness $t = 0.5$ in., a major depth $D = 2$ in., a major depth $d = 1.5$ in., and symmetrical notches with semicircular roots of radius r. The beam is subjected to a bending moment $M = 1000$ lb·in. If the allowable maximum normal stress is $\sigma_{max} = 12$ ksi, determine the smallest notch radius r that may be used. The radius r must be chosen as some multiple of $\frac{1}{32}$ in. Show your calculations for at least three different values of r.

Problems 12.3-1 through 12.3-13. *In solving these problems, assume that the members are made of materials that behave in a* **ductile** *manner.*

Prob. 12.3-1. At a point in a thin plate the stresses σ_x, σ_y, and τ_{xy} are known, and the corresponding principal stresses, σ_1 and σ_2, are of opposite sign. Express the failure criterion of the maximum-shear-stress theory in terms of σ_x, σ_y, and τ_{xy}.

Prob. 12.3-2. At a point in a thin plate the stresses σ_x, σ_y, and τ_{xy} are known, and the corresponding principal stresses, σ_1 and σ_2, are the opposite sign. Express the failure criterion of the maximum-distortion-energy theory in terms of σ_x, σ_y, and τ_{xy}.

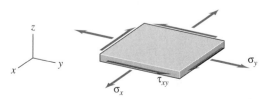

P12.3-1 and P12.3-2

Prob. 12.3-3. Would the maximum-shear-stress theory of failure predict failure (yielding) if the components of plane stress at a point in a steel structural member were to reach the values shown in Fig. P12.3-3? What is the value of the Mises equivalent stress for the given state of plane stress? Would failure be predicted by the maximum-distortion-energy theory?

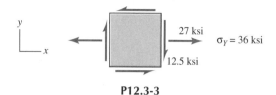

P12.3-3

Prob. 12.3-4. Solve Prob. 12.3-3 for the stresses indicated in Fig. 12.3-4.

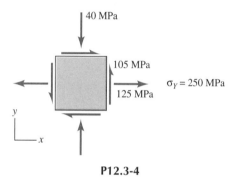

P12.3-4

Prob. 12.3-5. Solve the problem stated in Example Problem 12.3 if the force P_0 produces the stresses depicted in Fig. P12.3-5.

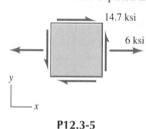

P12.3-5

Prob. 12.3-6. Would the maximum-shear-stress theory of failure predict failure (yielding) if the components of plane stress at a point in a member with yield stress σ_Y were to reach the values shown in Fig. P12.3-6? Would failure be predicted by the maximum-distortion-energy theory?

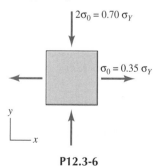

P12.3-6

Prob. 12.3-7. The components of plane stress at a point on the surface of a member made of soft bronze (σ_Y = 175 MPa) are shown in Fig. P12.3-7. (a) For this state of stress, what is the factor of safety, FS_s, as predicted by the failure criterion of the maximum-shear-stress theory of failure? (b) What is the value of the Mises equivalent stress for the given state of plane stress, and what factor of safety, FS_d, is predicted by the failure criterion of the maximum-distortion-energy theory of failure?

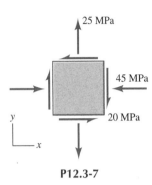

25 MPa

45 MPa

20 MPa

y

x

P12.3-7

Prob. 12.3-8. The components of stress at a highly stressed point on the brass propeller shaft of a power boat (σ_Y = 60 ksi) are shown in Fig. P12.3-8. (a) For this state of stress, what factor of safety, FS_s, is predicted by the failure criterion of the maximum-shear-stress theory of failure? (b) What is the value of the Mises equivalent stress for the given state of plane stress, and what is the factor of safety, FS_d, predicted by the failure criterion of the maximum-distortion-energy theory of failure?

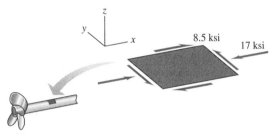

z

y

x

8.5 ksi

17 ksi

P12.3-8

Prob. 12.3-9. A section of double-extra-strong pipe made of steel with yield strength σ_Y = 50 ksi is subjected to a bending moment M = 35 kip·in. and torque T = 175 kip·in., as shown in Fig. P12.3-9. The outer diameter of the pipe is d_o = 3.5 in., and the inner diameter is d_i = 2.3 in. (a) For this loading of the pipe, what factor of safety, FS_s, is predicted by the failure criterion of the maximum-shear-stress theory of failure? (b) What is the value of the Mises equivalent stress for the given state of plane stress, and what is the factor of safety, FS_d, predicted by the failure criterion of the maximum-distortion-energy theory of failure?

*Prob. 12.3-10. A section of steel pipe (σ_Y = 340 MPa) has an inner diameter d_i = 60 mm and is subjected to a bending moment M = 7 kN·m and torque T = 7.8 kN·m, as shown in Fig. P12.3-10. This member is to be designed in accordance with the max-

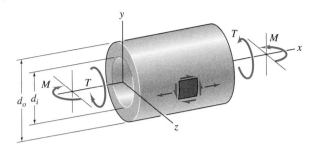

y

T

M

x

d_o d_i

z

P12.3-9 and P12.3-10

imum-distortion-energy criterion of failure, with a factor of safety FS_d = 2.0. To the nearest millimeter, what outer diameter, d_o, is required?

*Prob. 12.3-11. A square-cross-section aluminum-alloy bar (σ_Y = 40 ksi) is subjected to a compressive axial force of magnitude P = 48 kips and a torque T = 13 kip·in., as shown in Fig. P12.3-11. This member is to be designed in accordance with the maximum-shear-stress criterion of failure, with a factor of safety FS_s = 2.0. To the nearest 0.1 in., what is the minimum allowable cross-sectional dimension, b? (Remember to use the theory of torsion of noncircular prismatic bars, which is discussed in Section 4.8.)

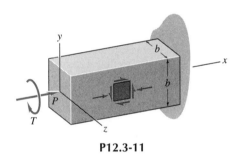

y

b

x

b

P

T

z

P12.3-11

Prob. 12.3-12. A rod with circular cross section and yield stress σ_Y is subjected to a bending moment M and a torque T, as shown in Fig. P12.3-12. Bending occurs in the xz plane. Express the maximum-shear-stress criterion of failure for this bar in terms of M, T, the rod diameter d, and the yield stress σ_Y.

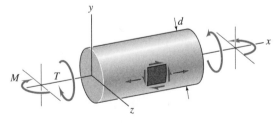

y

d

x

M

T

z

P12.3-12

*Prob. 12.3-13. The handlebar of a bench-press machine has the "longhorn" shape shown in Fig. P12.3-13. You are to analyze the configuration depicted in Fig. P12.3-13b, where the (curved)

(a)

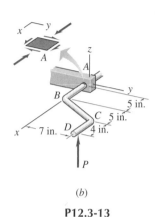

(b)

P12.3-13

*Prob. 12.3-15.** After several failures occurred in cast-iron bearing housings in the vicinity of the point marked A on the housing shown in Fig. P12.3-15a, a decision was made to use strain-gage rosettes (Section 8.12) to determine the operating stresses, and then to perform a failure analysis using Mohr's failure criterion. During a prolonged period of operation, the most critical combination of stresses was determined to be the stresses shown in Fig. P12.3-15b, and the tensile and compressive ultimate strengths of the cast iron were determined to be $\sigma_{TU} = 170$ MPa and $\sigma_{CU} = 655$ MPa, respectively. (a) Determine the principal stresses σ_1 and σ_2 corresponding to the stress state depicted in Fig. P12.3-15b. (b) Construct a Mohr failure diagram, like the one in Fig. 12.15b, for the cast iron. (c) Using the results you obtained in Parts (a) and (b), can you explain why failures have been occurring in the cast-iron bearing housings? Show your calculations.

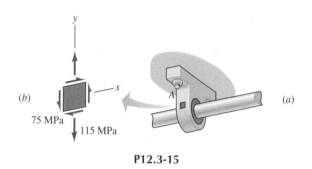

P12.3-15

axis of the handlebar lies in the xy plane and the force P exerted by the athlete's hand acts in the z direction. The handlebar is formed from a solid circular-cross-section steel bar having a diameter $d = 1$ and yield strength $\sigma_Y = 50$ ksi. If the handlebar is to be designed according to the maximum-shear-stress failure criterion with a factor of safety of $FS_s = 2.5$, what is the allowable total force (i.e., $2P$) that should be advertised in the specifications for this bench-press machine?

Problems 12.3-14 through 12.3-16. *In solving these problems, assume that the members are made of materials that behave in a* **brittle** *manner.*

Prob. 12.3-14. A rod with circular cross section and ultimate tensile strength σ_{TU} is subjected to a tensile force P and a torque T, as shown in Fig. P12.3-14. Express the maximum-normal-stress criterion of failure for this bar in terms of P, T, the rod diameter d, and the ultimate tensile strength, σ_{TU}.

*Prob. 12.3-16.** A ceramic rod with diameter d $= 1.5$ in. is subjected to a compressive axial force of magnitude $P = 53$ kips and a torque of magnitude $T = 13.3$ kip·in., acting as shown in Fig. P12.3-16. The tensile and compressive ultimate strengths of the ceramic material from which the rod is made are $\sigma_{TU} = 25$ ksi and $\sigma_{CU} = 95$ ksi, respectively. (a) Determine the principal stresses σ_1 and σ_2 corresponding to the state depicted in Fig. P12.3-16. (b) Construct a Mohr failure diagram, like the one in Fig. 12.15b, for the ceramic material. (c) Construct a load line, like the one in Fig. 3 of Example Problem 12.3. (d) Finally, with respect to failure according to Mohr's failure criterion, what is the factor of safety, FS_m, of the ceramic rod under the given loading? (Hint: You will need to determine the distance along the load line fom the origin of the $\sigma_1 - \sigma_2$ Mohr failure plane to the appropriate segment of the Mohr failure boundary.)

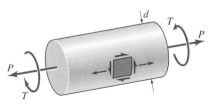

P12.3-14

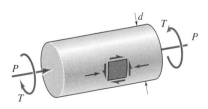

P12.3-16

APPENDICES

NUMERICAL ACCURACY; APPROXIMATIONS

A.1 NUMERICAL ACCURACY; SIGNIFICANT DIGITS

The engineering quantities that enter into deformable-body mechanics problems (e.g., force, length, strain) can usually be measured to an *accuracy* of about 1 part in 100 (1%) or, in some cases, perhaps to 1 part in 1000 (0.1%). In engineering calculations the accuracy of a number is indicated by the number of significant digits used in stating the number. A *significant digit* is any digit from 1 to 9, or any zero that is not used to show the position of the decimal point. For example, the numbers 27, 4.5, 0.30, and 0.0091 each has two significant digits. These could also be written in powers-of-ten form as 27, 45×10^{-1}, 30×10^{-2}, and 91×10^{-4}, again indicating that each number has two significant digits. Zeros immediately to the left of the decimal point can lead to some ambiguity regarding the number of digits that are significant. For example, to indicate that the number 30,000 has two significant digits, rather than just one, it would be preferable to write it as 30×10^3.

It is tempting, when one uses a calculator or computer to make engineering computations, to record all of the digits that are displayed in the computed result, but this could give an unwarranted impression of the true accuracy of the number. For example, it would be reasonable to see the value of a force stated as $F = 426$ lb, but it would be quite unreasonable to see the same force recorded as $F = 426.379$ lb.

In engineering problems five types of numbers are encountered:

- Exact numbers (e.g., the 32 in the formula $I_p = \pi d^4/32$).
- "Formally exact" numbers (e.g., the value of π, or the value of $\sin \theta$).[1]
- Given data.
- Intermediate results of calculations.
- Final results of calculations.

In this book the final results of calculations are reported according to the following rules that are standard practice in engineering:

---

[1]The values of π and of trigonometric functions are calculated to many significant digits (ten or more) within the calculator or computer.

- Numbers that begin with the digits 2 through 9 are recorded to three significant digits.
- Numbers that begin with the digit 1 are recorded to four significant digits.

Given data are also assumed to be accurate to the number of significant digits indicated above, even though fewer digits may actually be stated. Intermediate results, if recorded for use in further calculations, are recorded to several additional digits in order to preserve numerical accuracy.

A.2 APPROXIMATIONS

Power-series approximation formulas of two types are useful (e.g., in the discussion of extensional strain in Section 2.3)—a *trigonometric approximation* and a *binomial-expansion approximation*. Series expansions of the trigonometric functions $\sin \theta$, $\cos \theta$, and $\tan \theta$ are:

$$\sin \theta = \theta - \frac{\theta^3}{3!} + \frac{\theta^5}{5!} - \cdots$$

$$\cos \theta = 1 - \frac{\theta^2}{2!} + \frac{\theta^4}{4!} - \cdots$$

$$\tan \theta = \theta + \frac{\theta^3}{3} + \frac{2\theta^5}{15} + \cdots$$

Therefore, if $\theta \ll 1$, we can replace the trigonometric functions by the approximations:

$$\sin \theta \doteq \theta, \quad \cos \theta \doteq 1, \quad \tan \theta \doteq \theta \qquad \text{(A-1)}$$

A small quantity β may appear in an expression of the form $(1 + \beta)^n$. The series expansion for this expression is

$$(1 + \beta)^n = 1 + n\beta + \frac{n(n - 1)}{2!}\beta^2 + \cdots$$

Therefore, if $\beta \ll 1$, the approximation

$$(1 + \beta)^n \doteq 1 + n\beta \qquad \text{(A-2)}$$

may be used. Equations (A-1) and (A-2) will be useful in many strain-displacement analyses to reduce complex, nonlinear strain-displacement equations to simpler linear, small-displacement forms (e.g., see Example Prob. 2.3).

APPENDIX

SYSTEMS OF UNITS

B

B.1 INTRODUCTION ■■■■■■■■■■

The physical quantities encountered in science and engineering (e.g., force, mass, length, time, and temperature) must be expressed in some system of units. The *International System of Units*[2] (SI) was established by international agreement to provide a uniform system of units for measurement throughout the world. In the United States, the *U.S. Customary System of Units* (USCS) is the most commonly used system. Other important systems of units are non-SI forms of the metric system and the British Imperial System of Units. For the foreseeable future, engineers in the United States should be able to use either USCS or SI units, so both are used throughout this book.

There are three classes of units: base units, derived units, and supplementary units. The *base units* are defined in terms of specific physical standards. For example, the standard unit of *mass*, the kilogram (kg), is defined by a bar of platinum-iridium alloy that is kept at the International Bureau of Weights and Measures in Sèvres, France. The *second*(s) is defined as the duration of 9 192 631 770 periods of the radiation corresponding to the transition between the two hyperfine levels of the ground state of the cesium 133 atom. The *meter* (m) is defined to be the distance traveled by light in a vacuum during a time interval 1/299 792 458 of a second. Plane angles and solid angles comprise a second group of units, called *supplementary units*. Most units are *derived units*, which are related by an algebraic formula to base units and, in some instances, to supplementary units.

B.2 SI UNITS ■■■■■■■■■■

The International System of Units has four base units that are of importance in mechanics: The unit of mass is the *kilogram* (kg), the unit of length is the *meter* (m), the unit of time is the *second* (s), and the unit of temperature is the *kelvin* (K).[3] The unit of force is the newton (N). This derived unit is based on Newton's Second Law, $F = Ma$. Thus, a *newton*

---

[2]*Guide for the Use of the International System of Units*, National Institute of Standards and Technology (NIST) Special Publication 811, September 1991.

[3]Neither the word *degrees* nor the symbol ° is part of the name of the SI unit of temperature, as they are in other systems of units (e.g., degrees Fahrenheit, °F).

is defined as the force required to give one kilogram of mass an acceleration of one meter per second squared:

$$1 \text{ N} = 1 \text{ kg} \cdot \text{m/s}^2$$

The SI units that are pertinent to the mechanics of deformable bodies are listed, according to class, in Table B-1.

TABLE B-1. SI Units

Class	Quantity	Name of Unit	SI Symbol	Unit Formula
Base	Length	meter	m	—
	Mass	kilogram	kg	—
	Time	second	s	—
	Temperature	kelvin	K	—
Derived	Area	square meter	m^2	—
	Volume	cubic meter	m^3	—
	Force	newton	N	$kg \cdot m/s^2$
	Stress, pressure	pascal	Pa	N/m^2
	Work, energy	joule	J	$N \cdot m$
	Power	watt	W	$N \cdot m/s$
Supplementary	Plane angle	radian	rad	—

In the SI system (and in other metric systems), prefixes denoting powers of ten are used to modify basic units (e.g., millimeter, kilonewton). Table B-2 lists the prefixes that are preferred in SI usage. Prefixes that do not signify powers of 10^3 or 10^{-3} (e.g., centi = 10^{-2}) should be avoided. In units represented by fractions, use a prefix in the numerator, rather than in the denominator (except that the symbol for kilogram (kg) may appear in the denominator). For example, if the calculation of stress using $\sigma = \dfrac{F}{A}$ were to give the result $\sigma = 12.6 \text{ N/mm}^2$, this result should be recorded as $12.6 \text{ MN/m}^2 = 12.6 \text{ MPa}$ to avoid having a prefix in the denominator.

TABLE B-2. SI Prefixes

Prefix	Symbol	Multiplication Factor	Exponential Form
giga	G	1 000 000 000	10^9
mega	M	1 000 000	10^6
kilo	k	1 000	10^3
milli	m	0.001	10^{-3}
micro	μ	0.000 001	10^{-6}
nano	n	0.000 000 001	10^{-9}

In SI usage, as shown in the third column of Table B-2, a space, rather than a comma, is used to separate the digits of a numeric value into groups of three, both before and after the decimal point. When there are four digits, the space is optional (e.g., 5943 m, or 5 943 m).

In the U.S. Customary System of Units (sometimes called the British gravitational system) the base units are the following: the foot (ft) for length, the pound (lb) for force, the second (s) for time, and the degree Fahrenheit (°F) for temperature. The foot is defined by U.S. statute as exactly 0.3048 m. The pound is defined as the weight at sea level and at a latitude of 45° of a platinum standard that is kept at the Bureau of Standards in Washington, DC. This platinum standard has a mass of 0.453 592 43 kg.

In the U.S. Customary System of Units, the unit of mass is a derived unit, called the *slug*. One slug is the mass that would be accelerated at 1 ft/sec squared by a force of 1 lb:

$$1 \text{ slug } = 1 \text{ lb} \cdot \text{s}^2/\text{ft}$$

Conversion factors between SI units and USCS units are given in the table inside the front cover of this book.

GEOMETRIC PROPERTIES OF PLANE AREAS

C.1 FIRST MOMENTS OF AREA; CENTROID

Definitions. The solutions of most problems in this book involve one or more geometric properties of plane areas[4]—area, centroid, second moment, etc. The total *area* of a plane surface enclosed by bounding curve B is defined by the integral

$$A = \int_A dA \qquad (C\text{-}1)$$

which is understood to mean a summation of differential areas dA over two spatial variables, such as y and z in Fig. C-1.

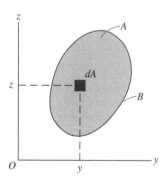

FIGURE C-1. A plane area.

The *first moments* of the area A about the y and z axes, respectively, are defined as

$$Q_y = \int_A z\,dA, \quad Q_z = \int_A y\,dA \qquad (C\text{-}2)$$

---

[4]The word *area* is used in two senses: In one sense, the word refers to the portion of a plane surface that lies within a prescribed bounding curve, like the area bounded by the closed curve B in Fig. C-1; in the second sense, the word refers to the quantity of surface within the bounding curve [Eq. (C-1)].

Q_y and Q_z are called first moments because the distances z and y appear to the first power in the defining integrals.

The *centroid* of an area is its "geometric center." The coordinates $(\bar{y}, \bar{z})$ of the centroid C (Fig. C-2) are defined by the first-moment equations

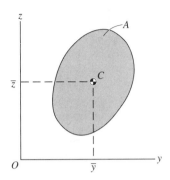

FIGURE C-2. Location of the centroid of an area.

$$\bar{y}A = \int_A y\,dA, \quad \bar{z}A = \int_A z\,dA \qquad \text{(C-3)}$$

For simple geometric shapes (e.g., rectangles, triangles, circles) there are closed-form formulas for the geometric properties of plane areas. A number of these are given in a table inside the back cover of this book.[5] An area may possess one of the three symmetry properties illustrated in Fig. C-3. If an area has *one axis of symmetry*, like the z axis for the C-section in Fig. C-3a, the centroid of the area lies on that axis. If the area has *two axes of symmetry*, like the wide-flange shape in Fig. C-3b, then the centroid lies at the intersection of those axes. Finally, if the area is *symmetric about a point*, like the Z-section in Fig. C-3c, the center of symmetry is the centroid of the area.

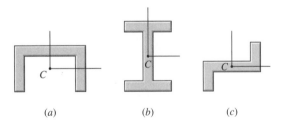

FIGURE C-3. Three types of area symmetry.

Composite Areas. Many structural shapes are composed of several parts, each of which is a simple geometric shape. For example, each of the areas in Fig. C-3 can be treated as a composite area made up of three rectangular areas. Since the integrals in Eqs. (C-1) through (C-3) represent summations over the total area A, they can be evaluated by summing the contributions of the constituent areas A_i, giving

---

[5]The reader may consult textbooks on integral calculus or statics for exercises in evaluating the integrals in Eqs. C-1 through C-3 for specific shapes.

$$A = \sum_i A_i, \quad Q_z = \bar{y}A = \sum_i \bar{y}_i A_i, \quad Q_y = \bar{z}A = \sum_i \bar{z}_i A_i \qquad \text{(C-4)}$$

Note that $\bar{y}$ in Fig. C-4 can be determined directly from the symmetry of the figure about the z' axis.

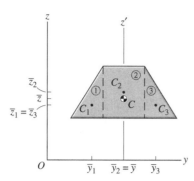

FIGURE C-4. A composite area.

COMPOSITE-AREA PROCEDURE FOR LOCATING THE CENTROID

1. Divide the composite area into simpler areas for which there exist formulas for area and for the coordinates of the centroid. (See the table inside the back cover.)
2. Establish a convenient set of reference axes (y, z).
3. Determine the area, A, using Eq. (C-4a).
4. Calculate the coordinates of the composite centroid, $(\bar{y}, \bar{z})$, using Eqs. (C-4b, c).

■■■■■■■■■■■■■■□□ EXAMPLE C-1 □□■■■■■■■■■■■■■■

Locate the centroid of the L-shaped area in Fig. 1.

Solution A—Addition Method Following the procedure outlined above, we divide the L-shaped area into two rectangles, as shown in Fig. 2. The y and z axes are located along the outer edges of the area, with the origin at the lower-left corner. Since the composite area consists of only two areas, the composite centroid, C, lies between C_1 and C_2 on the line joining the two centroids, as illustrated in Fig. 2.

Area: From Eq. (C-4a),

$$A = A_1 + A_2 = (6t)(t) + (8t)(t) = 14t^2 \qquad (1)$$

Centroid: From Eqs. (C-4b) and (C-4c),

$$\bar{y}A = \bar{y}_1 A_1 + \bar{y}_2 A_2 = (t/2)(6t^2) + 5t(8t^2) = 43t^3$$

$$\bar{y} = \frac{43t^3}{14t^2} = \frac{43}{14}t = 3.07t \qquad \text{Ans.} \quad (2)$$

$$\bar{z}A = \bar{z}_1 A_1 + \bar{z}_2 A_2 = (3t)(6t^2) + (t/2)(8t^2) = 22t^3$$

$$\bar{z} = \frac{22t^3}{14t^2} = \frac{22}{14}t = 1.57t \qquad \text{Ans.} \quad (3)$$

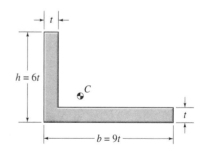

Fig. 1

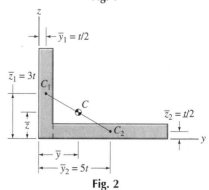

Fig. 2

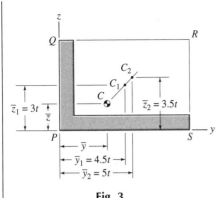

Fig. 3

Solution B—Subtraction Method Sometimes (although not in this particular example) it is easier to solve composite-area problems by treating the area as the net area obtained by subtracting one or more areas from a larger area. Then, in Eqs. (C-4), the A_i's of the removed areas are simply taken as negative areas. This method will now be applied to the L-shaped area in Fig. 1 by treating it as a larger rectangle from which a smaller rectangle is to be subtracted (Fig. 3). Area A_1 is the large rectangle $PQRS$; area A_2 is the smaller unshaded rectangle. The composite centroid, C, lies along the line joining the two centroids, C_1 and C_2, but it does not fall between them.

Area: From Eq. (C-4a),

$$A = A_1 + A_2 = (9t)(6t) + [-(8t)(5t)] = 14t^2 \qquad (4)$$

Centroid: From Eqs. (C-4b) and (C-4c),

$$\bar{y}A = \bar{y}_1 A_1 + \bar{y}_2 A_2 = (4.5t)(54t^2) + [(5t)(-40t^2)] = 43t^3$$

$$\bar{y} = 43t^3/14t^2 = 3.07t \qquad \text{Ans.} \quad (5)$$

$$\bar{z}A = \bar{z}_1 A_1 + \bar{z}_2 A_2 = (3t)(54t^2) + [(3.5t)(-40t^2)] = 22t^3$$

$$\bar{z} = 22t^3/14t^2 = 1.57t \qquad \text{Ans.} \quad (6)$$

C.2 MOMENTS OF INERTIA OF AN AREA ■■■■■■■■■■■■

Definitions of Moments of Inertia.
The *moments of inertia* of a plane area (Fig. C-5) about axes y and z in the plane are defined by the integrals

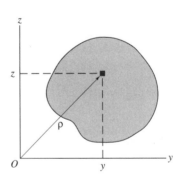

FIGURE C-5. A plane area.

$$I_y = \int_A z^2 dA, \quad I_z = \int_A y^2 dA \qquad (C-5)$$

These are called the *moment of inertia with respect to the y axis* and the *moment of inertia with respect to the z axis*, respectively. Since each integral involves the square of the distance of the elemental area dA from the axis involved, these quantities are called *second moments of area*. These moments of inertia appear primarily in formulas for bending of beams (see Chapter 6).

The moments of inertia defined in Eqs. (C-5) are with respect to axes that lie in the plane of the area under consideration. The second moment of area about the x axis, that is, with respect to the origin O, is called the *polar moment of inertia* of the area. It is defined by

$$I_p = \int_A \rho^2 dA \qquad (C\text{-}6)$$

Since, by the Pythagorean theorem, $\rho^2 = y^2 + z^2$, I_p is related to I_y and I_z by

$$I_p = I_y + I_z \qquad (C\text{-}7)$$

Since Eqs. (C-5) and (C-6) involve squares of distances, I_y, I_z, and I_p are always positive. All have the dimension of (length)^4 — in^4, mm^4, etc.

A table listing formulas for coordinates of the centroid and for moments of inertia of a variety of shapes may be found inside the back cover of this book. The most useful formulas for moments of inertia and for polar moment of inertia are derived here.

Moments of Inertia of a Rectangle: For the rectangle in Fig. C-6a, Eq. (C-5a) gives

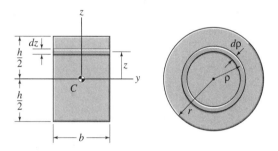

(*a*) A rectangular area.   (*b*) A circular area.

FIGURE C-6. Notation for calculating moments of inertia and polar moment of inertia.

$$I_y = \int_A z^2 dA = \int_{-h/2}^{h/2} z^2(b\,dz) = b\frac{z^3}{3}\bigg|_{-h/2}^{h/2} = \frac{bh^3}{12}$$

where the y axis passes through the centroid and is parallel to the two sides of length b. I_z may be derived in an analogous manner, so the moments of inertia of a rectangle for the two centroidal axes parallel to the sides of the rectangle are:

$$I_y = \frac{bh^3}{12}, \quad I_z = \frac{hb^3}{12} \qquad (C\text{-}8)$$

Polar Moment of Inertia of a Circle about its Center: Letting $dA = 2\pi\rho d\rho$, the area of the dark-shaded ring in Fig. C-6b, and using Eq. (C-6), we can determine the polar moment of inertia of a circle about its center:

$$I_p = \int_A \rho^2 dA = \int_0^r \rho^2 (2\pi\rho d\rho) = \frac{\pi r^4}{2}$$

$$\boxed{I_p = \frac{\pi r^4}{2} = \frac{\pi d^4}{32}} \tag{C-9}$$

Radii of Gyration. A length called the *radius of gyration* is defined for each moment of inertia by the formulas

$$\boxed{r_y = \sqrt{\frac{I_y}{A}}, \quad r_z = \sqrt{\frac{I_z}{A}}} \tag{C-10}$$

These lengths are used to simplify several formulas in Chapters 6 and 10. If these formulas are written in the form

$$I = Ar^2$$

then it is clear that the radius of gyration is the distance at which the entire area could be concentrated and still give the same value, I, of moment of inertia about a given axis.

Parallel-Axis Theorems for Moments of Inertia. Let the (y, z) pair of axes be parallel to centroidal axes (y_c, z_c), as shown in Fig. C-7. The centroid, C, is located with respect to the (y, z) axes by the centroidal coordinates $(\bar{y}, \bar{z})$. Since, from Fig. C-7, $z = \bar{z} + z_c$, the moment of inertia I_y is given by

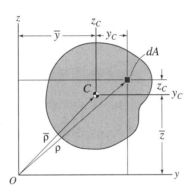

FIGURE C-7. Notation for deriving parallel-axis theorems.

$$I_y = \int_A z^2 dA = \int_A (\bar{z} + z_c)^2 dA$$

$$= \bar{z}^2 A + 2\bar{z}\int_A z_c dA + \int_A z_c^2 dA$$

$$\boxed{I_y = \bar{z}^2 A + I_{yc}} \tag{C-11a}$$

The term $\int_A z_c dA$ vanishes since the y_c axis passes through the centroid; the term I_{yc} is the *centroidal moment of inertia* about the y_c axis. The $\bar{z}^2 A$ term is the moment of inertia that area A would have about the y axis if all of the area were to be concentrated at the centroid.

Since this term is always zero or positive, the centroidal moment of inertia is the minimum moment of inertia with respect to all parallel axes.

By the same procedure that was used to obtain Eq. (C-11a), we get

$$I_z = \bar{y}^2 A + I_{z_c}$$ (C-11b)

Equations C-11 are called *parallel-axis theorem for moments of inertia.*

As a simple example of calculations based on the parallel-axis theorem, let us determine the moment of inertia of the rectangle in Fig. C-8 about the y' axis along an edge of length b. From Eq. (C-11a),

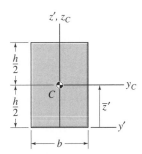

FIGURE C-8. A rectangular area with two sets of axes.

$$I_y = (\bar{z}')^2 A + I_{y_c} = \left(\frac{h}{2}\right)^2 (bh) + \frac{bh^3}{12} = \frac{bh^3}{3}$$ (C-12)

In a similar manner, a *parallel-axis theorem for the polar moment of inertia* may be derived. From Eq. (C-6) and Fig. C-7, the polar moment of inertia about point O is

$$I_{pO} = \int_A \rho^2 dA = \int_A (y^2 + z^2) dA$$

$$= \int_A [(\bar{y} + y_c)^2 + (\bar{z} + z_c)^2] dA$$

$$= \int_A (\bar{y}^2 + 2\bar{y}y_c + y_c^2 + \bar{z}^2 + 2\bar{z}z_c + z_c^2) dA$$

$$I_{pO} = \bar{\rho}^2 A + I_{pC}$$ (C-13)

since $\bar{y}^2 + \bar{z}^2 = \bar{\rho}^2$, $\int_A y_c dA = \int_A z_c dA = 0$, and $\int_A (y_c^2 + z_c^2) dA = I_{pC}$. Note that Eq. (C-13) follows easily from Eq. (C-7) and Eqs. (C-11).

Moments of Inertia of Composite Areas.

The moments of inertia of a composite area, like the one in Fig. C-4, may be computed by summing the contributions of the individual areas:

$$I_y = \sum_i (I_y)_i, \quad I_z = \sum_i (I_z)_i$$ (C-14)

As an efficient procedure for calculating moments of inertia of composite areas, the following is suggested.

1. Divide the composite area into simpler areas for which there exist formulas for centroidal coordinates and moments of inertia. (See the table inside the back cover.)

2. Locate the centroid of each constituent area and establish centroidal reference axes (y_c, z_c) parallel to the given (y, z) axes.

3. Employ Eqs. (C-11) to compute the moments of inertia of the constituent areas with respect to the (y, z) axes and Eq. (C-14) to sum them.

The next example problem illustrates this procedure.

EXAMPLE C-2

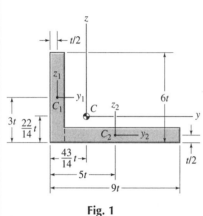

Fig. 1

Determine the centroidal moment of inertia I_y for the L-shaped section in Example C-1. (Here, in Fig. 1, the origin of the (y, z) reference frame is at the centroid of the composite area. The centroidal reference axes for the rectangular "legs" of the L-shaped area are (y_1, z_1) and (y_2, z_2), respectively.)

Solution We can combine Eqs. (C-14) with the parallel axis theorems, Eqs. (C-11), to compute the required moments of inertia.

$$I_y = (I_y)_1 + (I_y)_2 = [(I_{yc})_1 + A_1 \bar{z}_1^2] + [(I_{yc})_2 + A_2 \bar{z}_2^2] \qquad (1)$$

where $(I_{yc})_i$ is the moment of inertia of area A_i about its own centroidal y axis, and $\bar{z}_i$ is the z-coordinate of the centroid C_i measured in the (y, z) reference frame with origin at the composite centroid, C. Referring to Fig. 1, we get

$$
\begin{aligned}
I_y &= (I_y)_1 + (I_y)_2 \\
&= \left[\frac{1}{12}(t)(6t)^3 + (t)(6t)\left(3t - \frac{22}{14}t \right)^2 \right] \\
&\quad + \left[\frac{1}{12}(8t)(t)^3 + (t)(8t)\left(\frac{1}{2}t - \frac{22}{14}t \right)^2 \right] \\
&= 18t^4 + \frac{600}{49}t^4 + \frac{2}{3}t^4 + \frac{450}{49}t^4 \\
&= \frac{842}{21}t^4 = 40.1t^4 \qquad \text{Ans.} \quad (2)
\end{aligned}
$$

C.3 PRODUCT OF INERTIA OF AN AREA

Definition of Product of Inertia. Another geometric property of plane areas is called the *product of inertia*, which is defined by (refer to Fig. C-1)

$$I_{yz} = \int_A yz \, dA \qquad (C-15)$$

The product of inertia is required in the study of bending of unsymmetric beams (Section 6.6).

As an example, let us determine the product of inertia of a rectangular area with respect to two sets of axes (Fig. C-9).

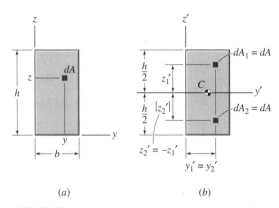

(a)   (b)

FIGURE C-9. A rectangular area with two sets of axes.

From Eq. (C-15) and Fig. C-9(a),

$$I_{yz} = \int_A yz \, dA = \int_0^h \int_0^b yz \, dy \, dz = \frac{b^2 h^2}{4} \tag{C-16}$$

Now consider the product of inertia with respect to the (y', z') axes in Fig. C-9b. The y' axis is an axis of symmetry, and it passes through the centroid C. As is clear from Fig. C-9b, when either reference axis is an axis of symmetry of the area, like the y' axis in this figure, the product of inertia is zero, since

$$I_{y'z'} = \int_A y'z' \, dA$$

and, because of symmetry (since $y'_2 = y'_1$, but $z'_2 = -z'_1$), the contributions of dA_1 and dA_2 to the integral cancel each other. Therefore,

$$I_{yz} = 0 \tag{C-17}$$

if <u>either</u> the y axis or the z axis is an axis of symmetry of the area.

Parallel-Axis Theorem for Product of Inertia of an Area. The procedure used to derive parallel-axis theorems for moments of inertia, leading to Eqs. (C-11) and (C-13), may be applied to derive a parallel-axis theorem for products of inertia. From Eq. (C-15) and Fig. C-7,

$$
\begin{aligned}
I_{yz} &= \int_A yz \, dA = \int_A (\bar{y} + y_c)(\bar{z} + z_c) \, dA \\
&= \bar{y}\bar{z}A + \bar{y} \int_A z_c \, dA + \bar{z} \int_A y_c \, dA + I_{y_c z_c}
\end{aligned}
$$

Therefore, since y_c and z_c are coordinates in a centroidal reference frame, the *parallel-axis theorem for products of inertia of an area* is

$$\boxed{I_{yz} = \bar{y}\bar{z}A + I_{y_cz_c}} \tag{C-18}$$

Just as for the moments of inertia, I_{yz} has one term that represents the product of inertia of an area A concentrated at the centroid, plus a *centroidal product of inertia* $I_{y_cz_c}$.

Product of Inertia for Composite Areas. The summations for moments of inertia in Eqs. (C-14) are readily extended to the product of inertia of an area composed of several constituent areas:

$$\boxed{I_{yz} = \sum_i (I_{yz})_i} \tag{C-19}$$

■■■■■■■■■■■■■□□□ EXAMPLE C-3 □□□□□□□□□□■■■■■■■■

For the L-shaped area in Example C-2, use the composite-area procedure to determine the centroidal product of inertia, I_{yz}. (Note: Here the (y, z) reference frame is a centroidal reference frame for the whole area; (y_1, z_1) and (y_2, z_2) are centroidal reference frames for the constituent areas A_1 and A_2, respectively.) The centroidal product of inertia relative to the (y, z) axes is given by

$$I_{yz} = (I_{y_1z_1} + A_1\bar{y}_1\bar{z}_1) + (I_{y_2z_2} + A_2\bar{y}_2\bar{z}_2)$$

It is very important to note that $\bar{y}_1$, $\bar{z}_1$, etc., are signed values, that is, some of them could be negative. By Eq. (C-17), $I_{y_1z_1} = I_{y_2z_2} = 0$. Therefore,

$$I_{yz} = A_1\bar{y}_1\bar{z}_1 + A_2\bar{y}_2\bar{z}_2$$

$$= (6t^2)\left(-\frac{43}{14}t + \frac{1}{2}t\right)\left(3t - \frac{22}{14}t\right)$$

$$+ (8t^2)\left(5t - \frac{43}{14}t\right)\left(-\frac{22}{14}t + \frac{1}{2}t\right)$$

or

$$I_{yx} = -\frac{270}{7}t^4 \qquad \textbf{Ans.}$$

Note that, since the centroid C_1 lies in the second quadrant and C_2 lies in the fourth quadrant, both A_1 and A_2 make negative contributions to I_{yz}.

C.4 AREA MOMENTS OF INERTIA ABOUT INCLINED AXES; PRINCIPAL MOMENTS OF INERTIA ■■■■■■■■■□□

In some applications, especially in unsymmetric bending of beams (Section 6.6), it is necessary to determine the moments and products of inertia relative to inclined axes (y', z') when I_y, I_z, and I_{yz} are known. The *coordinate transformation* relating coordinates (y', z') to coordinates (y, z) can be deduced from Fig. C-10.

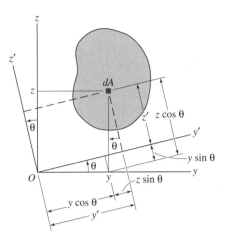

FIGURE C-10. Transformation of coordinates in a plane.

The angle θ is measured **positive counterclockwise from y to y'** (and z to z').

$$y' = y \cos \theta + z \sin \theta \qquad (\text{C-20})$$
$$z' = -y \sin \theta + z \cos \theta$$

From Eqs. (C-5), (C-15), and (C-20),

$$I_{y'} = \int_A (z')^2 \, dA = \int_A (-y \sin \theta + z \cos \theta)^2 \, dA \qquad (\text{C-21})$$
$$I_{y'z'} = \int_A y'z' \, da = \int_a (y \cos \theta + z \sin \theta)(-y \sin \theta + z \cos \theta) \, dA$$

Expanding each of the above integrands and recognizing that $\int_A y^2 \, dA = I_z$, and so forth, we get

$$I_{y'} = I_y \cos^2 \theta + I_z \sin^2 \theta - 2I_{yz} \sin \theta \cos \theta$$
$$I_{y'z'} = (I_y - I_z) \sin \theta \cos \theta + I_{yz}(\cos^2 \theta - \sin^2 \theta)$$

These equations may be simplified by using the trigonometric identities $\sin 2\theta = 2 \sin \theta \cos \theta$ and $\cos 2\theta = \cos^2 \theta - \sin^2\theta$. Thus,

$$I_{y'} = \frac{I_y + I_z}{2} + \frac{I_y - I_z}{2}\cos 2\theta - I_{yz} \sin 2\theta \qquad (\text{C-22})$$
$$I_{y'z'} = \frac{I_y - I_z}{2}\sin 2\theta + I_{yz} \cos 2\theta$$

Note the similarity between these equations and the stress-transformation equations, Eqs. 8-5.[6]

---

[6]There is a sign difference between the τ_{nt}-type terms and the I_{yz}-type terms, however.

Principal Moments of Inertia. From Eqs. (C-22) it may be seen that $I_{y'}$ and $I_{y'z'}$ depend on the angle θ. We will now determine the orientations of the y' axis for which $I_{y'}$ takes on its maximum and minimum values. The axes having these orientations are called the *principal axes of inertia* of the area, and the corresponding moments of inertia are called the *principal moments of inertia*. To each point O in an area, there is a specific set of principal axes passing through that point. The principal axes that pass through the centroid of the area, called the *centroidal principal axes*, are the most important. The orientations of the centroidal principal axes for several unequal-leg angles are given in Appendix D.6.

The moment of inertia $I_{y'}$ will have a maximum, or minimum, value if the y' axis is oriented at an angle $\theta = \theta_p$ that satisfies the equation

$$\frac{dI_{y'}}{d\theta} = -2\left(\frac{I_y - I_z}{2}\right)\sin 2\theta - 2I_{yz}\cos 2\theta = 0$$

Therefore,

$$\tan 2\theta_p = \frac{-I_{yz}}{\left(\dfrac{I_y - I_z}{2}\right)} \tag{C-23}$$

Figure C-11 illustrates how to use the tangent value given by Eq. (C-23) to determine the angles θ_p. There are two distinct angles that satisfy Eq. (C-23). As illustrated by Fig. C-11, these two values of $2\theta_p$, labeled $2\theta_{p1}$ and $2\theta_{p2}$, differ by 180°, so the principal axes are oriented at 90° to each other (as they must be).

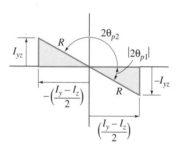

FIGURE C-11. Orientation of the principal axes of inertia.

From Fig. C-11, the hypotenuse of either of the shaded triangles is given by

$$R = \sqrt{\left(\frac{I_y - I_z}{2}\right)^2 + I_{yz}^2} \tag{C-24}$$

Also, from Fig. C-11, the angles $2\theta_{p1}$ and $2\theta_{p2}$ satisfy

$$\sin 2\theta_{p1} = \frac{-I_{yz}}{R}, \qquad \cos 2\theta_{p1} = \frac{\left(\dfrac{I_y - I_z}{2}\right)}{R} \tag{C-25a}$$

$$\sin 2\theta_{p2} = \frac{I_{yz}}{R}, \qquad \cos 2\theta_{p2} = \frac{-\left(\dfrac{I_y - I_{yz}}{2}\right)}{R} \tag{C-25b}$$

Substituting these sines and cosines into the equation for $I_{y'}$, Eq. (C-22a), we get the following expressions for the two principal moments of inertia:

$$I_{max} \equiv I_{p1} = \frac{I_y + I_z}{2} + \sqrt{\left(\frac{I_y - I_z}{2}\right)^2 + I_{yz}^2}$$

$$I_{min} \equiv I_{p2} = \frac{I_y + I_z}{2} - \sqrt{\left(\frac{I_y - I_z}{2}\right)^2 + I_{yz}^2}$$

(C-26)

If Eqs. (C-25a) or Eqs. (C-25b) are substituted into Eq. (C-22b), it is found that

$$I_{p1 p2} = 0 \qquad \text{(C-27)}$$

That is, the product of inertia with respect to the principal axes of inertia is equal to zero.

By adding Eqs. (C-26a) and (C-26b) we get

$$I_{p1} + I_{p2} = I_y + I_z \qquad \text{(C-28)}$$

Thus, the sum of the moments of inertia about any pair of mutually perpendicular axes passing through a given point in a given plane is a constant.

■■■■■■■■■■■■■■■□ E X A M P L E C - 4 ■□□■□■□■■□■□■■■■

For the L-shaped area in Fig. 1 of Example C-2, (a) Determine the orientation of the centroidal principal axes and show the orientation on a sketch. (b) Determine the principal moments of inertia.

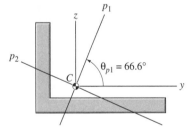

$$I_y = \frac{5894}{147}t^4 = 40.10t^4, \quad I_z = \frac{33,103}{294}t^4 = 112.60t^4$$

$$I_{yz} = \frac{-270}{7}t^4 = -38.57t^4$$

Fig. 1. Principal axes of inertia.

Solution
(a) From Eq. (C-23),

$$\tan 2\theta_p = \frac{-I_{yz}}{\left(\dfrac{I_y - I_z}{2}\right)} = \frac{-\left(\dfrac{-270}{7}\right)}{\dfrac{11,788 - 33,103}{2(294)}} = -1.064$$

$$2\theta_{p1} = 133.22°, \quad 2\theta_{p2} = -46.78°$$

Then, as illustrated in Fig. 1,

$$\theta_{p1} = 66.6°, \quad \theta_{p2} = -23.4° \qquad \text{Ans.}$$

(b) From Eq. (C-26a),

$$I_{p_1} = \frac{I_y + I_z}{2} + \sqrt{\left(\frac{I_y - I_z}{2}\right)^2 + I_{yz}^2}$$

$$= \frac{40.10t^4 + 112.60t^4}{2} + \sqrt{\left(\frac{40.10t^4 - 112.60t^4}{2}\right)^2 + (-38.57t^4)^2}$$

$$= 129.28t^4$$

or

$$I_{p_1} = 129.3t^4 \qquad \text{Ans.}$$

Similarly, from Eq. (C-26b),

$$I_{p_2} = 23.4t^4 \qquad \text{Ans.}$$

Mohr's Circle for Moments and Products of Inertia. Equations (C-22) have the same basic form as Eqs. 8.5, which were used to develop Mohr's circle for stress.[7] Therefore, by a procedure that is virtually identical to that in Section 8.5, it can be shown that a Mohr's circle plotted as in Fig. C-12 can be used to compute $I_{y'}$ and $I_{y'z'}$ for any (y', z') axes located at angle θ counterclockwise from the given (y, z) axes. And the Mohr's circle provides a convenient way to calculate the orientation of the principal axes of inertia and the principal moments of inertia, I_{p_1} and I_{p_2}, given moments of inertia I_y and I_z and the corresponding product of inertia I_{yz}. To an angle θ measured counterclockwise (or clockwise) on the planar area A, there corresponds an angle 2θ measured counterclockwise (or clockwise) on Mohr's circle.

The following procedure will facilitate your calculation of moments and products of inertia with respect to rotated axes.

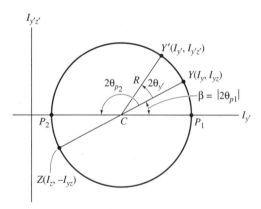

FIGURE C-12. Mohr's circle for moments and products of inertia.

---

[7]There is a difference between the signs preceding the τ_{xy}-type terms in Eqs. 8.5 and the signs preceding the corresponding I_{yz}-type terms in Eqs. (C-22). Thus, for Mohr's circle for moments and products of inertia, the $I_{y'z'}$ axis is positive upward, not positive downward, as it was for Mohr's circle for stress.

1. Establish a set of Mohr's-circle axes ($I_{y'}$, $I_{y'z'}$), as shown in Fig. C-12. (Note that the positive $I_{y'z'}$ axis is counterclockwise 90° from the $I_{y'}$ axis, unlike the τ_{nt} axis for Mohr's circle of stress in Chapter 8.)
2. Plot points $Y:(I_y, +I_{yz})$ and $Z:(I_z, -I_{yz})$, respectively.
3. Draw a straight line joining points Y and Z. The intersection of the YZ line with the $I_{y'}$ axis is the center of the Mohr's circle passing through points Y and Z.

4. Point Y', located at angle $2\theta_{y'}$ counterclockwise from the line CY, as shown in Fig. C-12, locates the point whose coordinates are ($I_{y'}$, $+I_{y'z'}$).
5. Points P_1 and P_2 locate the two principal axes at $2\theta_{p_1}$ and $2\theta_{p_2}$, respectively, as shown in Fig. C-12. The principal moments of inertia are I_{p_1} and I_{p_2}, which are also given by Eqs. (C-26).

■■■■■■■■■■■■■■□ EXAMPLE C-5 □□□□□□□□□□■■■■■

(a) Draw the Mohr's circle for the centroidal moments and products of inertia for the L-shaped area in Fig. 1 of Example C-2, given that:

$$I_y = \frac{5894}{147}t^4 = 40.10t^4, \quad I_z = \frac{33,103}{294}t^4 = 112.60t^4$$

$$I_{yz} = \frac{-270}{7}t^4 = -38.57t^4$$

(b) Use the Mohr's circle constructed in Part (a) to compute the principal moments of inertia I_{p_1} and I_{p_2} and to locate the principal axes. Show the orientation of the principal axes on a sketch.

Solution

(a) Sketch Mohr's circle and calculate the principal moments of inertia. Points Y and Z are plotted and Mohr's circle is then drawn (Fig. 1). From the circle,

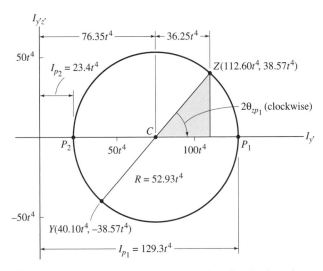

Fig. 1. Mohr's circle for centroidal inertias of an L-shaped area.

$$I_{\text{avg.}} = \frac{40.10t^4 + 112.60t^4}{2} = 76.35t^4$$

$$R = \sqrt{\left(\frac{112.60t^4 - 40.10t^4}{2}\right)^2 + (38.57t^4)^2} = 52.93t^4$$

$$I_{p_1} = I_{\text{avg.}} + R = 129.3t^4$$

$$I_{p_2} = I_{\text{avg.}} - R = 23.4t^4$$

Ans. (a)

(b) Determine the orientation of the principal axes and show them on a sketch.

$$\tan |2\theta_{zp_1}| = \frac{38.57t^4}{\left(\dfrac{112.60t^4 - 40.10t^4}{2}\right)} = 1.064$$

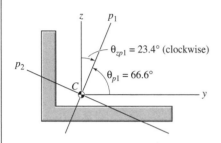

$\theta_{zp1} = 23.4°$ (clockwise)

$\theta_{p1} = 66.6°$

Therefore, $2\theta_{zp_1} = 46.78°$ (clockwise), so

$$\theta_{zp_1} = \theta_{yp_2} = 23.4° \text{ clockwise} \qquad \text{Ans. (b)}$$

Note that the orientations of the principal axes in Fig. 2 are such that the contributions to $I_{p_1p_2}$ of the areas in the four quadrants cancel out, giving $I_{p_1p_2} = 0$.

Fig. 2. Principal axes of inertia.

The results obtained from Mohr's circle are the same as those obtained by the use of formulas in Example C-4. However, mistakes are less likely to be made if Mohr's circle is carefully drawn and it is recalled that an angle 2θ on Mohr's circle corresponds to an angle θ on the planar area A, and that <u>angles are taken in the same sense on Mohr's circle as on the planar area</u>.

APPENDIX

D

SECTION PROPERTIES OF SELECTED STRUCTURAL SHAPES

Tables D.1 through D.10 give the cross-sectional properties of structural shapes made of aluminum, steel and wood. The tables for steel shapes and for aluminum shapes were compiled from more extensive tables given in the following references; they are used with permission:

- Table D.1 and Tables D.3-D.7: *Manual of Steel Construction—Load and Resistance Factor Design*, American Institute of Steel Construction, Inc., 400 N. Michigan Ave., Chicago, IL, 60611-4185.
- Table D.2: *Metric Properties of Structural Shapes*, American Institute of Steel Construction, Inc., 400 N. Michigan Ave., Chicago, IL, 60611-4185.
- Tables D.9 and D.10: *The Aluminum Design Manual*, The Aluminum Association Inc., 900 19th Street, NW, Washington, DC, 20006.

Nomenclature

I = moment of inertia

J = polar moment of inertia (pipe sections)

S = elastic section modulus

Z = plastic section modulus

$r = \sqrt{I/A}$ = radius of gyration

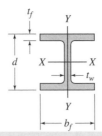

D.1. Properties of Steel Wide-Flange (W) Shapes (U.S. Customary Units)

| | | | Flange | | Web | Elastic Properties | | | | | | Plastic Modulus |
| | | | | | | Axis X − X | | | Axis Y − Y | | | |
Designation*	Area A in²	Depth d in.	Width b_f in.	Thickness t_f in.	Thickness t_w in.	I_x in⁴	S_x in³	r_x in.	I_y in⁴	S_y in³	r_y in.	Z_x in³
W36 × 230	67.6	35.90	16.470	1.260	0.760	15000	837	14.9	940	114	3.73	943
× 150	44.2	35.85	11.975	0.940	0.625	9040	504	14.3	270	45.1	2.47	581
W33 × 201	59.1	33.68	15.745	1.150	0.715	11500	684	14.0	749	95.2	3.56	772
× 130	38.3	33.09	11.510	0.855	0.580	6710	406	13.2	218	37.9	2.39	467
W30 × 173	50.8	30.44	14.985	1.065	0.655	8200	539	12.7	598	79.8	3.43	605
× 90	26.4	29.53	10.400	0.610	0.470	3620	245	11.7	115	22.1	2.09	283
W27 × 146	42.9	27.38	13.965	0.975	0.605	5630	411	11.4	443	63.5	3.21	461
× 84	24.8	26.71	9.960	0.640	0.460	2850	213	10.7	106	21.2	2.07	244
W24 × 94	27.7	24.31	9.065	0.875	0.515	2370	196	9.79	94.4	20.9	1.95	224
× 62	18.2	23.74	7.040	0.590	0.430	1550	131	9.23	34.5	9.80	1.38	153
W21 × 101	29.8	21.36	12.290	0.800	0.500	2420	227	9.02	248	40.3	2.89	253
× 73	21.5	21.24	8.295	0.740	0.455	1600	151	8.64	70.6	17.0	1.81	172
× 50	14.7	20.83	6.530	0.535	0.380	984	94.5	8.18	24.9	7.64	1.30	110
W18 × 130	38.2	19.25	11.160	1.200	0.670	2460	256	8.03	278	49.9	2.70	291
× 76	22.3	18.21	11.035	0.680	0.425	1330	146	7.73	152	27.6	2.61	163
W18 × 60	17.6	18.24	7.555	0.695	0.415	984	108	7.47	50.1	13.3	1.69	123
× 50	14.7	17.99	7.495	0.570	0.355	800	88.9	7.38	40.1	10.7	1.65	101
W16 × 100	29.4	16.97	10.425	0.985	0.585	1490	175	7.10	186	35.7	2.51	198
× 67	19.7	16.33	10.235	0.665	0.395	954	117	6.96	119	23.2	2.46	130
× 50	14.7	16.26	7.070	0.630	0.380	659	81.0	6.68	37.2	10.5	1.59	92.0
× 40	11.8	16.01	6.995	0.505	0.305	518	64.7	6.63	28.9	8.25	1.57	72.9
W14 × 176	51.8	15.22	15.650	1.310	0.830	2140	281	6.43	838	107	4.02	320
× 120	35.3	14.48	14.670	0.940	0.590	1380	190	6.24	495	65.7	3.74	212
× 82	24.1	14.31	10.130	0.855	0.510	882	123	6.05	148	29.3	2.48	139
× 53	15.6	13.92	8.060	0.660	0.370	541	77.8	5.89	57.7	14.3	1.92	87.1
× 26	7.69	13.91	5.025	0.420	0.255	245	35.3	5.65	8.91	3.54	1.08	40.2
W12 × 152	44.7	13.71	12.480	1.400	0.870	1430	209	5.66	454	72.8	3.19	243
× 96	28.2	12.71	12.160	0.900	0.550	833	131	5.44	270	44.4	3.09	147
× 65	19.1	12.12	12.000	0.605	0.390	533	87.9	5.28	174	29.1	3.02	96.8
× 50	14.7	12.19	8.080	0.640	0.370	394	64.7	5.18	56.3	13.9	1.96	72.4
× 35	10.3	12.50	6.560	0.520	0.300	285	45.6	5.25	24.5	7.47	1.54	51.2
× 22	6.48	12.31	4.030	0.425	0.260	156	25.4	4.91	4.66	2.31	0.847	29.3
W10 × 60	17.6	10.22	10.080	0.680	0.420	341	66.7	4.39	116	23.0	2.57	74.6
× 45	13.3	10.10	8.020	0.620	0.350	248	49.1	4.32	53.4	13.3	2.01	54.9
× 30	8.84	10.47	5.810	0.510	0.300	170	32.4	4.38	16.7	5.75	1.37	36.6
× 12	3.54	9.87	3.960	0.210	0.190	53.8	10.9	3.90	2.18	1.10	0.785	12.6
W 8 × 48	14.1	8.50	8.110	0.685	0.400	184	43.3	3.61	60.9	15.0	2.08	49.0
× 40	11.7	8.25	8.070	0.560	0.360	146	35.5	3.53	49.1	12.2	2.04	39.8
× 35	10.3	8.12	8.020	0.495	0.310	127	31.2	3.51	42.6	10.6	2.03	34.7
× 21	6.16	8.28	5.270	0.400	0.250	75.3	18.2	3.49	9.77	3.71	1.26	20.4
× 15	4.44	8.11	4.015	0.315	0.245	48.0	11.8	3.29	3.41	1.70	0.876	13.6
W 6 × 25	7.34	6.38	6.080	0.455	0.320	53.4	16.7	2.70	17.1	5.61	1.52	18.9
× 20	5.87	6.20	6.020	0.365	0.260	41.1	13.4	2.66	13.3	4.41	1.50	14.9
× 16	4.74	6.28	4.030	0.405	0.260	32.1	10.2	2.60	4.43	2.20	0.966	11.7
× 15	4.43	5.99	5.990	0.260	0.230	29.1	9.72	2.56	9.32	3.11	1.46	10.8

*W(nominal depth in inches) × (weight in pounds per foot)

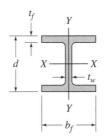

D.2. Properties of Steel Wide-Flange (W) Shapes (SI Units)

Designation*	Area A mm²	Depth d mm	Flange Width b_f mm	Flange Thickness t_f mm	Web Thickness t_w mm	Axis $X-X$ I_x 10⁶mm⁴	Axis $X-X$ S_x 10³mm³	Axis $X-X$ r_x mm	Axis $Y-Y$ I_y 10⁶mm⁴	Axis $Y-Y$ S_y 10³mm³	Axis $Y-Y$ r_y mm	Plastic Modulus Z_x 10³mm³
W920×342	43 600	912	418.0	32.0	19.30	6250	13 700	379	390	1870	94.6	15 400
×223	28 500	911	304.0	23.9	15.90	3770	8280	364	112	737	62.7	9540
W840×299	38 100	855	400.0	29.2	18.20	4790	11 200	355	312	1560	90.5	12 600
×193	24 700	840	292.0	21.7	14.70	2780	6620	335	90.3	618	60.5	7620
W760×257	32 800	773	381.0	27.1	16.60	3420	8850	323	250	1310	87.3	9930
×134	17 000	750	264.0	15.5	11.90	1500	4000	297	47.7	361	53.0	4630
W690×217	27 700	695	355.0	24.8	15.40	2340	6730	291	185	1040	81.7	7570
×125	16 000	678	253.0	16.3	11.70	1190	3510	273	44.1	349	52.5	4010
W610×140	17 900	617	230.0	22.2	13.10	1120	3630	250	45.1	392	50.2	4150
× 92	11 800	603	179.0	15.0	10.90	646	2140	234	14.4	161	34.9	2510
W530×150	19 200	543	312.0	20.3	12.70	1010	3720	229	103	660	73.2	4150
×109	13 900	539	211.0	18.8	11.60	667	2470	219	29.5	280	46.1	2830
× 74	9490	529	166.0	13.6	9.65	410	1550	208	10.4	125	33.1	1810
W460×193	24 700	489	283.0	30.5	17.00	1020	4170	203	115	813	68.2	4760
×113	14 400	463	280.0	17.3	10.80	556	2400	196	63.3	452	66.3	2670
× 74	9460	457	190.0	14.5	9.02	333	1460	188	16.6	175	41.9	1650
W410×149	19 000	431	265.0	25.0	14.90	619	2870	180	77.7	586	63.9	3250
×100	12 700	415	260.0	16.9	10.00	398	1920	177	49.5	381	62.4	2130
× 74	9510	413	180.0	16.0	9.65	275	1330	170	15.6	173	40.5	1510
× 60	7600	407	178.0	12.8	7.75	216	1060	169	12.0	135	39.7	1200
W360×262	33 400	387	398.0	33.3	21.10	894	4620	164	350	1760	102	5260
×179	22 800	368	373.0	23.9	15.00	575	3130	159	207	1110	95.3	3480
×122	15 500	363	257.0	21.7	13.00	365	2010	153	61.5	479	63.0	2270
× 79	10 100	354	205.0	16.8	9.40	227	1280	150	24.2	236	48.9	1430
× 39	4960	353	128.0	10.7	6.48	102	578	143	3.75	58.6	27.5	661
W310×226	28 900	348	317.0	35.6	22.10	596	3430	144	189	1190	80.9	3980
×143	18 200	323	309.0	22.9	14.00	348	2150	138	113	731	78.8	2420
× 97	12 300	308	305.0	15.4	9.91	248	1590	135	81.2	531	77.3	1770
× 74	9480	310	205.0	16.3	9.40	165	1060	132	23.4	228	49.7	1190
× 52	6670	318	167.0	13.2	7.62	119	748	134	10.3	123	39.3	841
× 33	4180	313	102.0	10.8	6.60	65.0	415	125	1.92	37.6	21.4	480
W250× 89	11 400	260	256.0	17.3	10.70	143	1100	112	48.4	378	65.2	1230
× 67	8560	257	204.0	15.7	8.89	104	809	110	22.2	218	50.9	901
× 45	5700	266	148.0	13.0	7.62	71.1	535	112	7.03	95	35.1	602
× 18	2280	251	101.0	5.3	4.83	22.5	179	99.3	0.919	18.2	20.1	208
W200× 71	9100	216	206.0	17.4	10.20	76.6	709	91.7	25.4	247	52.8	803
× 59	7580	210	205.0	14.2	9.14	61.2	583	89.9	20.4	199	51.9	653
× 52	6640	206	204.0	12.6	7.87	52.7	512	89.1	17.8	175	51.8	569
× 31	3980	210	134.0	10.2	6.35	31.3	298	88.7	4.10	61.2	32.1	335
× 22	2860	206	102.0	8.0	6.22	20.0	194	83.6	1.42	27.8	22.3	222
W150× 37	4730	162	154.0	11.6	8.13	22.2	274	68.5	7.07	91.8	38.7	310
× 24	3060	160	102.0	10.3	6.60	13.4	168	66.2	1.83	35.9	24.5	192
× 22	2860	152	152.0	6.6	5.84	12.1	159	65.0	3.87	50.9	36.8	176

*W(nominal depth in mm) × (mass in kg/m)

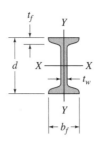

D.3. Properties of American Standard (S) Beams (U.S. Customary Units)

Designation*	Area A in²	Depth d in.	Flange Width b_f in.	Flange Thickness t_f in.	Web Thickness t_w in.	Axis $X-X$ I_x in⁴	Axis $X-X$ S_x in³	Axis $X-X$ r_x in.	Axis $Y-Y$ I_y in⁴	Axis $Y-Y$ S_y in³	Axis $Y-Y$ r_y in.	Plastic Modulus Z_x in³
S24 × 100	29.3	24.00	7.245	0.870	0.745	2390	199	9.02	47.7	13.2	1.27	240
× 90	26.5	24.00	7.125	0.870	0.625	2250	187	9.21	44.9	12.6	1.30	222
× 80	23.5	24.00	7.000	0.870	0.500	2100	175	9.47	44.2	12.1	1.34	204
S20 × 96	28.2	20.30	7.200	0.920	0.800	1670	165	7.71	50.2	13.9	1.33	198
× 75	22.0	20.00	6.385	0.795	0.635	1280	128	7.62	29.8	9.32	1.16	153
S18 × 70	20.6	18.00	6.251	0.691	0.711	926	103	6.71	24.1	7.72	1.08	125
× 54.7	16.1	18.00	6.001	0.691	0.461	804	89.4	7.07	20.8	6.94	1.14	105
S15 × 50	14.7	15.00	5.640	0.622	0.550	486	64.8	5.75	15.7	5.57	1.03	77.1
× 42.9	12.6	15.00	5.501	0.622	0.411	447	59.6	5.95	14.4	5.23	1.07	69.3
S12 × 50	14.7	12.00	5.477	0.659	0.687	305	50.8	4.55	15.7	5.74	1.03	61.2
× 35	10.3	12.00	5.078	0.544	0.428	229	38.2	4.72	9.87	3.89	0.980	44.8
S10 × 35	10.3	10.00	4.944	0.491	0.594	147	29.4	3.78	8.36	3.38	0.901	35.4
× 25.4	7.46	10.00	4.661	0.491	0.311	124	24.7	4.07	6.79	2.91	0.954	28.4
S 8 × 23	6.77	8.00	4.171	0.426	0.441	64.9	16.2	3.10	4.31	2.07	0.798	19.3
× 18.4	5.41	8.00	4.001	0.426	0.271	57.6	14.4	3.26	3.73	1.86	0.831	16.5
S 6 × 17.25	5.07	6.00	3.565	0.359	0.465	26.3	8.77	2.28	2.31	1.30	0.675	10.6
× 12.5	3.67	6.00	3.332	0.359	0.232	22.1	7.37	2.45	1.82	1.09	0.705	8.47
S 4 × 9.5	2.79	4.00	2.796	0.293	0.326	6.79	3.39	1.56	0.903	0.646	0.569	4.04
× 7.7	2.26	4.00	2.663	0.293	0.193	6.08	3.04	1.64	0.764	0.574	0.581	3.51

*S(nominal depth in inches) × (weight in pounds per foot)

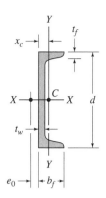

D.4. Properties of American Standard (C) Channels (U.S. Customary Units)

Designation*	Area A in^2	Depth d in.	Flange Width b_f in.	Flange Thickness t_f in.	Web Thickness t_w in.	Centroid Loc'n. x_c in.	Shear Center Loc'n. e_0 in.	Axis $X-X$ I_x in^4	Axis $X-X$ S_x in^3	Axis $X-X$ r_x in.	Axis $Y-Y$ I_y in^4	Axis $Y-Y$ S_y in^3	Axis $Y-Y$ r_y in.
C15 × 50	14.7	15.00	3.716	0.650	0.716	0.798	0.583	404	53.8	5.24	11.0	3.78	0.867
× 40	11.8	15.00	3.520	0.650	0.520	0.777	0.767	349	46.5	5.44	9.23	3.37	0.886
× 33.9	9.96	15.00	3.400	0.650	0.400	0.787	0.896	315	42.0	5.62	8.13	3.11	0.904
C12 × 30	8.82	12.00	3.170	0.501	0.510	0.674	0.618	162	27.0	4.29	5.14	2.06	0.763
× 25	7.35	12.00	3.047	0.501	0.387	0.674	0.746	144	24.1	4.43	4.47	1.88	0.780
× 20.7	6.09	12.00	2.942	0.501	0.282	0.698	0.870	129	21.5	4.61	3.88	1.73	0.799
C10 × 30	8.82	10.00	3.033	0.436	0.673	0.649	0.369	103	20.7	3.42	3.94	1.65	0.669
× 25	7.35	10.00	2.886	0.436	0.526	0.617	0.494	91.2	18.2	3.52	3.36	1.48	0.676
× 20	5.88	10.00	2.739	0.436	0.379	0.606	0.637	78.9	15.8	3.66	2.81	1.32	0.692
× 15.3	4.49	10.00	2.600	0.436	0.240	0.634	0.796	67.4	13.5	3.87	2.28	1.16	0.713
C 9 × 20	5.88	9.00	2.648	0.413	0.448	0.583	0.515	60.9	13.5	3.22	2.42	1.17	0.642
× 15	4.41	9.00	2.485	0.413	0.285	0.586	0.682	51.0	11.3	3.40	1.93	1.01	0.661
× 13.4	3.94	9.00	2.433	0.413	0.233	0.601	0.743	47.9	10.6	3.48	1.76	0.962	0.669
C 8 × 18.75	5.51	8.00	2.527	0.390	0.487	0.565	0.431	44.0	11.0	2.82	1.98	1.01	0.599
× 13.75	4.04	8.00	2.343	0.390	0.303	0.553	0.604	36.1	9.03	2.99	1.53	0.854	0.615
× 11.5	3.38	8.00	2.260	0.390	0.220	0.571	0.697	32.6	8.14	3.11	1.32	0.781	0.625
C 7 × 14.75	4.33	7.00	2.299	0.366	0.419	0.532	0.441	27.2	7.78	2.51	1.38	0.779	0.564
× 12.25	3.60	7.00	2.194	0.366	0.314	0.525	0.538	24.2	6.93	2.60	1.17	0.703	0.571
× 9.8	2.87	7.00	2.090	0.366	0.210	0.540	0.647	21.3	6.08	2.72	0.968	0.625	0.581
C 6 × 13	3.83	6.00	2.157	0.343	0.437	0.514	0.380	17.4	5.80	2.13	1.05	0.642	0.525
× 10.5	3.09	6.00	2.034	0.343	0.314	0.499	0.486	15.2	5.06	2.22	0.866	0.564	0.529
× 8.2	2.40	6.00	1.920	0.343	0.200	0.511	0.599	13.1	4.38	2.34	0.693	0.492	0.537
C 5 × 9	2.64	5.00	1.885	0.320	0.325	0.478	0.427	8.96	3.56	1.83	0.632	0.450	0.489
× 6.7	1.97	5.00	1.750	0.320	0.190	0.484	0.552	7.49	3.00	1.95	0.479	0.378	0.493
C 4 × 7.25	2.13	4.00	1.721	0.296	0.321	0.459	0.386	4.59	2.29	1.47	0.433	0.343	0.450
× 5.4	1.59	4.00	1.584	0.296	0.184	0.457	0.502	3.85	1.93	1.56	0.319	0.283	0.449
C 4 × 6	1.76	3.00	1.596	0.273	0.356	0.455	0.322	2.07	1.38	1.08	0.305	0.268	0.416
× 5	1.47	3.00	1.498	0.273	0.258	0.438	0.392	1.85	1.24	1.12	0.247	0.233	0.410
× 4.1	1.21	3.00	1.410	0.273	0.170	0.436	0.461	1.66	1.10	1.17	0.197	0.202	0.404

*C(nominal depth in inches) × (weight in pounds per foot)

A-27

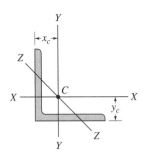

D.5. Properties of Steel Angle Sections—Equal Legs (U.S. Customary Units)

Size and Thickness in.	Weight per Ft. lb	Area A in²	Axis $X - X$				Axis $Y - Y$				Axis $Z - Z$	
			I_x in⁴	S_x in³	r_x in.	y_c in.	I_y in⁴	S_y in³	r_y in.	x_c in.	r_z in.	tan α
L8×8×1	51.0	15.0	89.0	15.8	2.44	2.37	89.0	15.8	2.44	2.37	1.56	1.000
$\frac{3}{4}$	38.9	11.4	69.7	12.2	2.47	2.28	69.7	12.2	2.47	2.28	1.58	1.000
$\frac{1}{2}$	26.4	7.75	48.6	8.36	2.50	2.19	48.6	8.36	2.50	2.19	1.59	1.000
L6×6×1	37.4	11.0	35.5	8.57	1.80	1.86	35.5	8.57	1.80	1.86	1.17	1.000
$\frac{3}{4}$	28.7	8.44	28.2	6.66	1.83	1.78	28.2	6.66	1.83	1.78	1.17	1.000
$\frac{1}{2}$	19.6	5.75	19.9	4.61	1.86	1.68	19.9	4.61	1.86	1.68	1.18	1.000
$\frac{3}{8}$	14.9	4.36	15.4	3.53	1.88	1.64	15.4	3.53	1.88	1.64	1.19	1.000
L5×5×$\frac{7}{8}$	27.2	7.98	17.8	5.17	1.49	1.57	17.8	5.17	1.49	1.57	0.973	1.000
$\frac{3}{4}$	23.6	6.94	15.7	4.53	1.51	1.52	15.7	4.53	1.51	1.52	0.975	1.000
$\frac{1}{2}$	16.2	4.75	11.3	3.16	1.54	1.43	11.3	3.16	1.54	1.43	0.983	1.000
$\frac{3}{8}$	12.3	3.61	8.74	2.42	1.56	1.39	8.74	2.42	1.56	1.39	0.990	1.000
L4×4×$\frac{3}{4}$	18.5	5.44	7.67	2.81	1.19	1.27	7.67	2.81	1.19	1.27	0.778	1.000
$\frac{1}{2}$	12.8	3.75	5.56	1.97	1.22	1.18	5.56	1.97	1.22	1.18	0.782	1.000
$\frac{3}{8}$	9.8	2.86	4.36	1.52	1.23	1.14	4.36	1.52	1.23	1.14	0.788	1.000
$\frac{1}{4}$	6.6	1.94	3.04	1.05	1.25	1.09	3.04	1.05	1.25	1.09	0.795	1.000
L3$\frac{1}{2}$×3$\frac{1}{2}$×$\frac{1}{2}$	11.1	3.25	3.64	1.49	1.06	1.06	3.64	1.49	1.06	1.06	0.683	1.000
$\frac{3}{8}$	8.5	2.48	2.87	1.15	1.07	1.01	2.87	1.15	1.07	1.01	0.687	1.000
$\frac{1}{4}$	5.8	1.69	2.01	0.794	1.09	0.968	2.01	0.794	1.09	0.968	0.694	1.000
L3×3×$\frac{1}{2}$	9.4	2.75	2.22	1.07	0.898	0.932	2.22	1.07	0.898	0.932	0.584	1.000
$\frac{3}{8}$	7.2	2.11	1.76	0.833	0.913	0.888	1.76	0.833	0.913	0.888	0.587	1.000
$\frac{1}{4}$	4.9	1.44	1.24	0.577	0.930	0.842	1.24	0.577	0.930	0.842	0.592	1.000
L2$\frac{1}{2}$×2$\frac{1}{2}$×$\frac{1}{2}$	7.7	2.25	1.23	0.724	0.739	0.806	1.23	0.724	0.739	0.806	0.487	1.000
$\frac{3}{8}$	5.9	1.73	0.984	0.566	0.753	0.762	0.984	0.566	0.753	0.762	0.487	1.000
$\frac{1}{4}$	4.1	1.19	0.703	0.394	0.769	0.717	0.703	0.394	0.769	0.717	0.491	1.000
L2×2×$\frac{3}{8}$	4.7	1.36	0.479	0.351	0.594	0.636	0.479	0.351	0.594	0.636	0.389	1.000
$\frac{1}{4}$	3.19	0.938	0.348	0.247	0.609	0.592	0.348	0.247	0.609	0.592	0.391	1.000
$\frac{1}{8}$	1.65	0.484	0.190	0.131	0.626	0.546	0.190	0.131	0.626	0.546	0.398	1.000

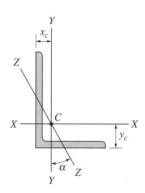

D.6. Properties of Steel Angle Sections—Unequal Legs (U.S. Customary Units)

Size and Thickness in.	Weight per Ft. lb	Area A in²	Axis $X-X$				Axis $Y-Y$				Axis $Z-Z$	
			I_x in⁴	S_x in³	r_x in.	y_c in.	I_y in⁴	S_y in³	r_y in.	x_c in.	r_z in.	tan α
L8×6×1	44.2	13.0	80.8	15.1	2.49	2.65	38.8	8.92	1.73	1.65	1.28	0.543
¾	33.8	9.94	63.4	11.7	2.53	2.56	30.7	6.92	1.76	1.56	1.29	0.551
½	23.0	6.75	44.3	8.02	2.56	2.47	21.7	4.79	1.79	1.47	1.30	0.558
L8×4×1	37.4	11.0	69.6	14.1	2.52	3.05	11.6	3.94	1.03	1.05	0.846	0.247
¾	28.7	8.44	54.9	10.9	2.55	2.95	9.36	3.07	1.05	0.953	0.852	0.258
½	19.6	5.75	38.5	7.49	2.59	2.86	6.74	2.15	1.08	0.859	0.865	0.267
L6×4×¾	23.6	6.94	24.5	6.25	1.88	2.08	8.68	2.97	1.12	1.08	0.860	0.428
½	16.2	4.75	17.4	4.33	1.91	1.99	6.27	2.08	1.15	0.987	0.870	0.440
⅜	12.3	3.61	13.5	3.32	1.93	1.94	4.90	1.60	1.17	0.941	0.877	0.446
L5×3×½	12.8	3.75	9.45	2.91	1.59	1.75	2.58	1.15	0.829	0.750	0.648	0.357
⅜	9.8	2.86	7.37	2.24	1.61	1.70	2.04	0.888	0.845	0.704	0.654	0.364
¼	6.6	1.94	5.11	1.53	1.62	1.66	1.44	0.614	0.861	0.657	0.663	0.371
L4×3×½	11.1	3.25	5.05	1.89	1.25	1.33	2.42	1.12	0.864	0.827	0.639	0.543
⅜	8.5	2.48	3.96	1.46	1.26	1.28	1.92	0.866	0.879	0.782	0.644	0.551
¼	5.8	1.69	2.77	1.00	1.28	1.24	1.36	0.599	0.896	0.736	0.651	0.558
L3½×2½×½	9.4	2.75	3.24	1.41	1.09	1.20	1.36	0.760	0.704	0.705	0.534	0.486
⅜	7.2	2.11	2.56	1.09	1.10	1.16	1.09	0.592	0.719	0.660	0.537	0.496
¼	4.9	1.44	1.80	0.755	1.12	1.11	0.777	0.412	0.735	0.614	0.544	0.506
L3×2×½	7.7	2.25	1.92	1.00	0.924	1.08	0.672	0.474	0.546	0.583	0.428	0.414
⅜	5.9	1.73	1.53	0.781	0.940	1.04	0.543	0.371	0.559	0.539	0.430	0.428
¼	4.1	1.19	1.09	0.542	0.957	0.993	0.392	0.260	0.574	0.493	0.435	0.440

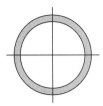

D.7. Properties of Standard-Weight Steel Pipe* (U.S. Customary Units)

Dimensions					Properties					
Nominal Diameter in.	Outside Diameter in.	Inside Diameter in.	Wall Thickness in.	Weight per ft. (plain ends) lbs.	Area in^2	I in^4	S in^3	r in.	J in^4	Z in^3
$\frac{1}{2}$	0.840	0.622	0.109	0.85	0.250	0.017	0.041	0.261	0.034	0.059
$\frac{3}{4}$	1.050	0.824	0.113	1.13	0.333	0.037	0.071	0.334	0.074	0.100
1	1.315	1.049	0.133	1.68	0.494	0.087	0.133	0.421	0.175	0.187
$1\frac{1}{4}$	1.660	1.380	0.140	2.27	0.669	0.195	0.235	0.540	0.389	0.324
$1\frac{1}{2}$	1.900	1.610	0.145	2.72	0.799	0.310	0.326	0.623	0.620	0.448
2	2.375	2.067	0.154	3.65	1.07	0.666	1.561	0.787	1.33	0.761
$2\frac{1}{2}$	2.875	2.469	0.203	5.79	1.70	1.530	1.06	0.947	3.06	1.45
3	3.500	3.068	0.216	7.58	2.23	3.02	1.72	1.16	6.03	2.33
$3\frac{1}{2}$	4.000	3.548	0.226	9.11	2.68	4.79	2.39	1.34	9.58	3.22
4	4.500	4.026	0.237	10.79	3.17	7.23	3.21	1.51	14.5	4.31
5	5.563	5.047	0.258	14.62	4.30	15.2	5.45	1.88	30.3	7.27
6	6.625	6.065	0.280	18.97	5.58	28.1	8.50	2.25	56.3	11.2
8	8.625	7.981	0.322	28.55	8.40	72.5	16.8	2.94	145.	22.2
10	10.750	10.020	0.365	40.48	11.9	161.	29.9	3.67	321.	39.4
12	12.750	12.000	0.375	49.56	14.6	279.	43.8	4.38	559.	57.4

*Steel pipe is also available in extra-strong and double-extra-strong weights. Based on a table compiled by the American Institute of Steel Construction and published in the *Manual of Steel Construction.*

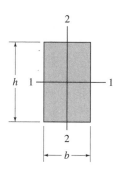

D.8. Properties of Structural Lumber* (U.S. Customary Units)
Sectional properties of American Standard Lumber (S4S)*

Nominal Dimensions $b \times h$	Net Dimensions $b \times h$	Area $A = bh$	Axis 1 − 1		Axis 2 − 2		Weight per Lineal Foot
			Moment of Inertia $I_1 = \dfrac{bh^3}{12}$	Section Modulus $S_1 = \dfrac{bh^2}{6}$	Moment of Inertia $I_2 = \dfrac{hb^3}{12}$	Section Modulus $S_2 = \dfrac{hb^2}{6}$	
in.	in.	in^2	in^4	in^3	in^4	in^3	lb
2×4	1.5×3.5	5.25	5.36	3.06	0.98	1.31	1.3
2×6	1.5×5.5	8.25	20.80	7.56	1.55	2.06	2.0
2×8	1.5×7.25	10.88	47.63	13.14	2.04	2.72	2.6
2×10	1.5×9.25	13.88	98.93	21.39	2.60	3.47	3.4
2×12	1.5×11.25	16.88	177.98	31.64	3.16	4.22	4.1
3×4	2.5×3.5	8.75	8.93	5.10	4.56	3.65	2.1
3×6	2.5×5.5	13.75	34.66	12.60	7.16	5.73	3.3
3×8	2.5×7.25	18.13	79.39	21.90	9.44	7.55	4.4
3×10	2.5×9.25	23.13	164.89	35.65	12.04	9.64	5.6
3×12	2.5×11.25	28.13	296.63	52.73	14.65	11.72	6.8
4×4	3.5×3.5	12.25	12.51	7.15	12.51	7.15	3.0
4×6	3.5×5.5	19.25	48.53	17.65	19.65	11.23	4.7
4×8	3.5×7.25	25.38	111.15	30.66	25.90	14.80	6.2
4×10	3.5×9.25	32.38	230.84	49.91	33.05	18.89	7.9
4×12	3.5×11.25	39.38	415.28	73.83	40.20	22.97	9.6
6×6	5.5×5.5	30.25	76.3	27.7	76.3	27.7	7.4
6×8	5.5×7.5	41.25	193.4	51.6	104.0	37.8	10.0
6×10	5.5×9.5	52.25	393.0	82.7	131.7	47.9	12.7
6×12	5.5×11.5	63.25	697.1	121.2	159.4	58.0	15.4
8×8	7.5×7.5	56.25	263.7	70.3	263.7	70.3	13.7
8×10	7.5×9.5	71.25	535.9	112.8	334.0	89.1	17.3
8×12	7.5×11.5	86.25	950.5	165.3	404.3	107.8	21.0

*S4S = surfaced four sides.
All properties and weights are for dressed lumber sizes.
Specific weight = 35 lb/ft^3.
Axes 1 − 1 and 2 − 2 are principal centroidal axes.
Based on a table compiled by the National Forest Products Association.

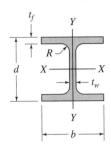

D.9. Properties of Aluminum Association Standard I-Beams (U.S. Customary Units)

Size							Section Properties					
				Flange Thickness	Web Thickness	Fillet Radius	Axis $X - X$			Axis $Y - Y$		
Depth d in.	Width b in.	Area* in^2	Weight† lb/ft	t_f in.	t_w in.	R in.	I in^4	S in^3	r in.	I in^4	S in^3	r in.
3.00	2.50	1.392	1.637	0.20	0.13	0.25	2.24	1.49	1.27	0.52	0.42	0.61
3.00	2.50	1.726	2.030	0.26	0.15	0.25	2.71	1.81	1.25	0.68	0.54	0.63
4.00	3.00	1.965	2.311	0.23	0.15	0.25	5.62	2.81	1.69	1.04	0.69	0.73
4.00	3.00	2.375	2.793	0.29	0.17	0.25	6.71	3.36	1.68	1.31	0.87	0.74
5.00	3.50	3.146	3.700	0.32	0.19	0.30	13.94	5.58	2.11	2.29	1.31	0.85
6.00	4.00	3.427	4.030	0.29	0.19	0.30	21.99	7.33	2.53	3.10	1.55	0.95
6.00	4.00	3.990	4.692	0.35	0.21	0.30	25.50	8.50	2.53	3.74	1.87	0.97
7.00	4.50	4.932	5.800	0.38	0.23	0.30	42.89	12.25	2.95	5.78	2.57	1.08
8.00	5.00	5.256	6.181	0.35	0.23	0.30	59.69	14.92	3.37	7.30	2.92	1.18
8.00	5.00	5.972	7.023	0.41	0.25	0.30	67.78	16.94	3.37	8.55	3.42	1.20
9.00	5.50	7.110	8.361	0.44	0.27	0.30	102.02	22.67	3.79	12.22	4.44	1.31
10.00	6.00	7.352	8.646	0.41	0.25	0.40	132.09	26.42	4.24	14.78	4.93	1.42
10.00	6.00	8.747	10.286	0.50	0.29	0.40	155.79	31.16	4.22	18.03	6.01	1.44
12.00	7.00	9.925	11.672	0.47	0.29	0.40	255.57	42.60	5.07	26.90	7.69	1.65
12.00	7.00	12.153	14.292	0.62	0.31	0.40	317.33	52.89	5.11	35.48	10.14	1.71

*Areas and section properties listed are based on nominal dimensions.

†Weight per foot is based on nominal dimensions and a density of 0.098 pound per cubic inch, which is the density of alloy 6061.

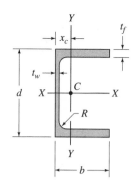

D.10. Properties of Aluminum Association Standard Channels (U.S. Customary Units)

Size				Flange Thickness	Web Thickness	Fillet Radius	Section Properties						
							Axis X − X			Axis Y − Y			
Depth d in.	Width b in.	Area* in^2	Weight† lb/ft	t_f in.	t_w in.	R in.	I in^4	S in^3	r in.	I in^4	S in^3	r in.	x_c in.
2.00	1.00	0.491	0.577	0.13	0.13	0.10	0.288	0.288	0.766	0.045	0.064	0.303	0.298
2.00	1.25	0.911	1.071	0.26	0.17	0.15	0.546	0.546	0.774	0.139	0.178	0.397	0.471
3.00	1.50	0.965	1.135	0.20	0.13	0.25	1.41	0.94	1.21	0.22	0.22	0.47	0.49
3.00	1.75	1.358	1.597	0.26	0.17	0.25	1.97	1.31	1.20	0.42	0.37	0.55	0.62
4.00	2.00	1.478	1.738	0.23	0.15	0.25	3.91	1.95	1.63	0.60	0.45	0.64	0.65
4.00	2.25	1.982	2.331	0.29	0.19	0.25	5.21	2.60	1.62	1.02	0.69	0.72	0.78
5.00	2.25	1.881	2.212	0.26	0.15	0.30	7.88	3.15	2.05	0.98	0.64	0.72	0.73
5.00	2.75	2.627	3.089	0.32	0.19	0.30	11.14	4.45	2.06	2.05	1.14	0.88	0.95
6.00	2.50	2.410	2.834	0.29	0.17	0.30	14.35	4.78	2.44	1.53	0.90	0.80	0.79
6.00	3.25	3.427	4.030	0.35	0.21	0.30	21.04	7.01	2.48	3.76	1.76	1.05	1.12
7.00	2.75	2.725	3.205	0.29	0.17	0.30	22.09	6.31	2.85	2.10	1.10	0.88	0.84
7.00	3.50	4.009	4.715	0.38	0.21	0.30	33.79	9.65	2.90	5.13	2.23	1.13	1.20
8.00	3.00	3.526	4.147	0.35	0.19	0.30	37.40	9.35	3.26	3.25	1.57	0.96	0.93
8.00	3.75	4.923	5.789	0.41	0.25	0.35	52.69	13.17	3.27	7.13	2.82	1.20	1.22
9.00	3.25	4.237	4.983	0.35	0.23	0.35	54.41	12.09	3.58	4.40	1.89	1.02	0.93
9.00	4.00	5.927	6.970	0.44	0.29	0.35	78.31	17.40	3.63	9.61	3.49	1.27	1.25
10.00	3.50	5.218	6.136	0.41	0.25	0.35	83.22	16.64	3.99	6.33	2.56	1.10	1.02
10.00	4.25	7.109	8.360	0.50	0.31	0.40	116.15	23.23	4.04	13.02	4.47	1.35	1.34
12.00	4.00	7.036	8.274	0.47	0.29	0.40	159.76	26.63	4.77	11.03	3.86	1.25	1.14
12.00	5.00	10.053	11.822	0.62	0.35	0.45	239.69	39.95	4.88	25.74	7.60	1.60	1.61

*Areas and section properties listed are based on nominal dimensions.

†Weight per foot is based on nominal dimensions and a density of 0.098 pound per cubic inch, which is the density of alloy 6061.

APPENDIX

DEFLECTIONS AND SLOPES OF BEAMS; FIXED-END ACTIONS

E

E.1. Deflections and Slopes of Uniform Cantilever Beams*

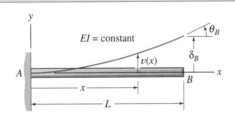

Notation

$v(x)$ = deflection in the y direction

$v'(x)$ = slope of the deflection curve

$\delta_B \equiv v(L)$ = deflection at end B

$\theta_B \equiv v'(L)$ = slope at end B

1

$$v = \frac{M_0 x^2}{2EI} \qquad v' = \frac{M_0 x}{EI}$$

$$\delta_B = \frac{M_0 L^2}{2EI} \qquad \theta_B = \frac{M_0 L}{EI}$$

2

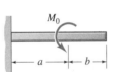

$$v = \frac{M_0 x^2}{2EI} \qquad v' = \frac{M_0 x}{EI} \qquad 0 \le x \le a$$

$$v = \frac{M_0 a}{2EI}(2x - a) \qquad v' = \frac{M_0 a}{EI} \qquad a \le x \le L$$

$$\delta_B = \frac{M_0 a}{2EI}(2L - a) \qquad \theta_B = \frac{M_0 a}{EI}$$

3

$$v = \frac{P x^2}{6EI}(3L - x) \qquad v' = \frac{P x}{2EI}(2L - x)$$

$$\delta_B = \frac{P L^3}{3EI} \qquad \theta_B = \frac{P L^2}{2EI}$$

4

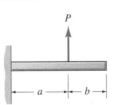

$$v = \frac{P x^2}{6EI}(3a - x) \qquad v' = \frac{P x}{2EI}(2a - x) \qquad 0 \le x \le a$$

$$v = \frac{P a^2}{6EI}(3x - a) \qquad v' = \frac{P a^2}{2EI} \qquad a \le x \le L$$

$$\delta_B = \frac{P a^2}{6EI}(3L - a) \qquad \theta_B = \frac{P a^2}{2EI}$$

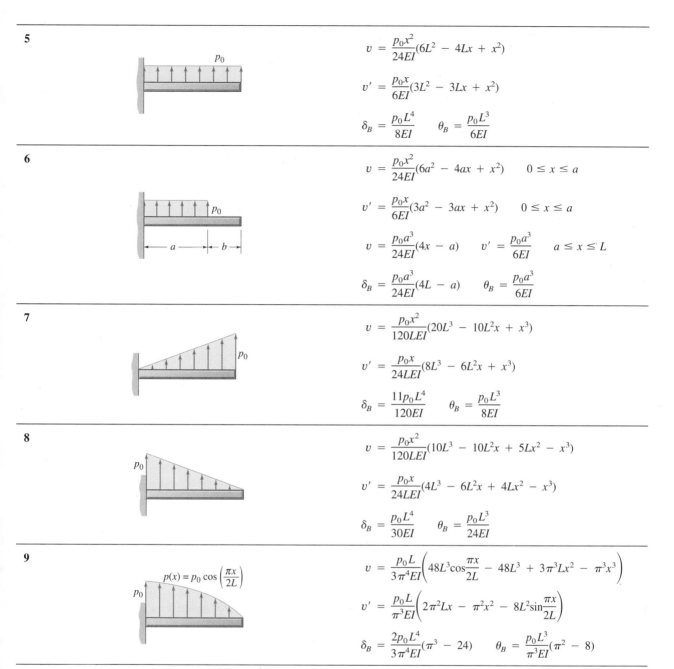

5

$$v = \frac{p_0 x^2}{24EI}(6L^2 - 4Lx + x^2)$$

$$v' = \frac{p_0 x}{6EI}(3L^2 - 3Lx + x^2)$$

$$\delta_B = \frac{p_0 L^4}{8EI} \qquad \theta_B = \frac{p_0 L^3}{6EI}$$

6

$$v = \frac{p_0 x^2}{24EI}(6a^2 - 4ax + x^2) \qquad 0 \le x \le a$$

$$v' = \frac{p_0 x}{6EI}(3a^2 - 3ax + x^2) \qquad 0 \le x \le a$$

$$v = \frac{p_0 a^3}{24EI}(4x - a) \qquad v' = \frac{p_0 a^3}{6EI} \qquad a \le x \le L$$

$$\delta_B = \frac{p_0 a^3}{24EI}(4L - a) \qquad \theta_B = \frac{p_0 a^3}{6EI}$$

7

$$v = \frac{p_0 x^2}{120LEI}(20L^3 - 10L^2 x + x^3)$$

$$v' = \frac{p_0 x}{24LEI}(8L^3 - 6L^2 x + x^3)$$

$$\delta_B = \frac{11 p_0 L^4}{120EI} \qquad \theta_B = \frac{p_0 L^3}{8EI}$$

8

$$v = \frac{p_0 x^2}{120LEI}(10L^3 - 10L^2 x + 5Lx^2 - x^3)$$

$$v' = \frac{p_0 x}{24LEI}(4L^3 - 6L^2 x + 4Lx^2 - x^3)$$

$$\delta_B = \frac{p_0 L^4}{30EI} \qquad \theta_B = \frac{p_0 L^3}{24EI}$$

9

$$v = \frac{p_0 L}{3\pi^4 EI}\left(48L^3 \cos\frac{\pi x}{2L} - 48L^3 + 3\pi^3 Lx^2 - \pi^3 x^3\right)$$

$$v' = \frac{p_0 L}{\pi^3 EI}\left(2\pi^2 Lx - \pi^2 x^2 - 8L^2 \sin\frac{\pi x}{2L}\right)$$

$$\delta_B = \frac{2p_0 L^4}{3\pi^4 EI}(\pi^3 - 24) \qquad \theta_B = \frac{p_0 L^3}{\pi^3 EI}(\pi^2 - 8)$$

*Beam-deflection theory is covered in Chapter 7. The sign convention used here is the same as in Chapter 7.

Notation

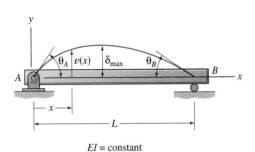

$v(x)$ = deflection in the y direction

$v'(x)$ = slope of the deflection curve

$\theta_A \equiv v'(0)$ = slope (angle) at end A

$\theta_B \equiv -v'(L)$ = angle of rotation at end B

x_m = distance from end A to the point of maximum deflection

$\delta_C \equiv |v(L/2)|$ = deflection at the center of the beam

$\delta_{max} \equiv \max |v(x)|$ = maximum deflection

1

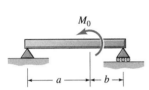

$$v = \frac{M_0 x}{6LEI}(2L^2 - 3Lx + x^2)$$

$$v' = \frac{M_0}{6LEI}(2L^2 - 6Lx + 3x^2)$$

$$\theta_A = \frac{M_0 L}{3EI} \qquad \theta_B = \frac{M_0 L}{6EI}$$

$$x_m = L\left(1 - \frac{\sqrt{3}}{3}\right) \text{ and } \delta_{max} = \frac{M_0 L^2}{9\sqrt{3}EI}$$

2

$$v = \frac{-M_0 x}{6LEI}(6aL - 3a^2 - 2L^2 - x^2) \qquad 0 \le x \le a$$

$$v' = \frac{-M_0}{6LEI}(6aL - 3a^2 - 2L^2 - 3x^2) \qquad 0 \le x \le a$$

$$\theta_A = \frac{-M_0}{6LEI}(6aL - 3a^2 - 2L^2) \qquad \theta_B = \frac{-M_0}{6LEI}(3a^2 - L^2)$$

3

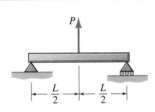

$$v = \frac{Px}{48EI}(3L^2 - 4x^2) \qquad 0 \le x \le \frac{L}{2}$$

$$v' = \frac{P}{16EI}(L^2 - 4x^2) \qquad 0 \le x \le \frac{L}{2}$$

$$\delta_C = \delta_{max} = \frac{PL^3}{48EI} \qquad \theta_A = \theta_B = \frac{PL^2}{16EI}$$

4

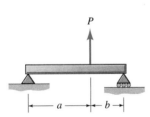

$$v = \frac{Pbx}{6LEI}(L^2 - b^2 - x^2) \qquad 0 \le x \le a$$

$$v' = \frac{Pb}{6LEI}(L^2 - b^2 - 3x^2) \qquad 0 \le x \le a$$

$$\theta_A = \frac{Pab(L + b)}{6LEI}$$

$$\theta_B = \frac{Pab(L + a)}{6LEI}$$

If $a \ge b$, $x_m = \sqrt{\frac{L^2 - b^2}{3}}$ and $\delta_{max} = \frac{Pb(L^2 - b^2)^{3/2}}{9\sqrt{3}LEI}$

5

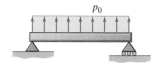

$$v = \frac{p_0 x}{24EI}(L^3 - 2Lx^2 + x^3)$$

$$v' = \frac{p_0}{24EI}(L^3 - 6Lx^2 + 4x^3)$$

$$\delta_C = \delta_{max} = \frac{5p_0 L^4}{384EI} \qquad \theta_A = \theta_B = \frac{p_0 L^3}{24EI}$$

6

$$v = \frac{p_0 x}{24LEI}(a^4 - 4a^3 L + 4a^2 L^2 + 2a^2 x^2 - 4aLx^2 + Lx^3) \qquad 0 \le x \le a$$

$$v' = \frac{p_0}{24LEI}(a^4 - 4a^3 L + 4a^2 L^2 + 6a^2 x^2 - 12aLx^2 + 4Lx^3) \qquad 0 \le x \le a$$

$$v = \frac{p_0 a^2}{24LEI}(-a^2 L + 4L^2 x + a^2 x - 6Lx^2 + 2x^3) \qquad a \le x \le L$$

$$v' = \frac{p_0 a^2}{24LEI}(4L^2 + a^2 - 12Lx + 6x^2) \qquad a \le x \le L$$

$$\theta_A = \frac{p_0 a^2}{24LEI}(2L - a)^2 \qquad \theta_B = \frac{p_0 a^2}{24LEI}(2L^2 - a^2)$$

7

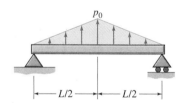

$$v = \frac{p_0 x}{360LEI}(7L^4 - 10L^2 x^2 + 3x^4)$$

$$v' = \frac{p_0}{360LEI}(7L^4 - 30L^2 x^2 + 15x^4)$$

$$\theta_A = \frac{7p_0 L^3}{360EI} \qquad \theta_B = \frac{p_0 L^3}{45EI}$$

$$x_m = 0.5193\,L \qquad \delta_{max} = 0.00652\frac{p_0 L^4}{EI}$$

8

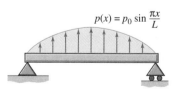

$$v = \frac{p_0 x}{960LEI}(5L^2 - 4x^2)^2 \qquad 0 \le x \le \frac{L}{2}$$

$$v' = \frac{p_0}{192LEI}(5L^2 - 4x^2)(L^2 - 4x^2) \qquad 0 \le x \le \frac{L}{2}$$

$$\delta_C = \delta_{max} = \frac{p_0 L^4}{120LEI} \qquad \theta_A = \theta_B = \frac{5p_0 L^3}{192EI}$$

9

$p(x) = p_0 \sin \dfrac{\pi x}{L}$

$$v = \frac{p_0 L^4}{\pi^4 EI} \sin\left(\frac{\pi x}{L}\right)$$

$$v' = \frac{p_0 L^3}{\pi^3 EI} \cos\left(\frac{\pi x}{L}\right)$$

$$\delta_C = \delta_{max} = \frac{p_0 L^4}{\pi^4 EI} \qquad \theta_A = \theta_B = \frac{p_0 L^3}{\pi^3 EI}$$

*Beam-deflection theory is covered in Chapter 7. The sign convention used here is the same as in Chapter 7.

E.3. Fixed-End Actions for Uniform Beams*

Loading	Shear and Moment Reactions†	
	End A	End B
1	$R_A = \dfrac{6M_0 a}{L^2}\left(1 - \dfrac{a}{L}\right)$	$R_B = -\dfrac{6M_0 a}{L^2}\left(1 - \dfrac{a}{L}\right)$
	$M_A = M_0\left(-1 + 4\dfrac{a}{L} - 3\dfrac{a^2}{L^2}\right)$	$M_B = \dfrac{M_0 a}{L}\left(2 - 3\dfrac{a}{L}\right)$
2 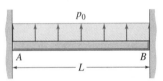	$R_A = -\dfrac{Pb^2}{L^3}(3a + b)$	$R_B = -\dfrac{Pa^2}{L^3}(a + 3b)$
	$M_A = -\dfrac{Pab^2}{L^2}$	$M_B = \dfrac{Pba^2}{L^2}$
3 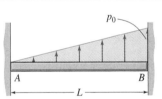	$R_A = -\dfrac{p_0 L}{2}$	$R_B = -\dfrac{p_0 L}{2}$
	$M_A = -\dfrac{p_0 L^2}{12}$	$M_B = \dfrac{p_0 L^2}{12}$
4 	$R_A = -\dfrac{3p_0 L}{20}$	$R_B = -\dfrac{7p_0 L}{20}$
	$M_A = -\dfrac{p_0 L^2}{30}$	$M_B = \dfrac{p_0 L^2}{20}$
5 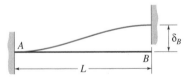	$R_A = \dfrac{6EI}{L^2}\theta_B$	$R_B = -\dfrac{6EI}{L^2}\theta_B$
	$M_A = \dfrac{2EI}{L}\theta_B$	$M_B = \dfrac{4EI}{L}\theta_B$
6	$R_A = -\dfrac{12EI}{L^3}\delta_B$	$R_B = \dfrac{12EI}{L^3}\delta_B$
	$M_A = -\dfrac{6EI}{L^2}\delta_B$	$M_B = -\dfrac{6EI}{L^2}\delta_B$

*This table provides fixed-end actions for use with Section 7.7 and Appendix G6.

†For all fixed-end actions tabulated, the positive sense for shear and moment reactions are the same as those shown in the first table entry.

MECHANICAL PROPERTIES OF SELECTED ENGINEERING MATERIALS

Mechanical properties of engineering materials vary significantly as a result of heat treatment, mechanical working, moisture content, and various other factors. The properties listed in Tables F.1 through F.3 are representative values that are intended for educational purposes only, not for commercial design of members. Additional information is available from several other sources, including the following: *Annual Book of ASTM*, American Society for Testing Materials, Philadelphia, PA; *Metals Handbook* ASM International, Materials Park, OH; *Aluminum and Aluminun Alloys* and *Aluminum Standards and Data*, The Aluminum Association, Washington, DC; *Wood Handbook*, U.S. Department of Agriculture, Washington, DC; *Marks' Standard Handbook for Mechanical Engineers*, McGraw-Hill, Inc., New York, NY; and *CRC Materials Science and Engineering Handbook*, CRC Press, Inc., Boca Raton, FL.

TABLE F.1. Specific Weight and Mass Density

Material	Specific Weight (γ)			Mass Density (ρ)	
	lb/in^3	lb/ft^3	kN/m^3	slugs/ft^3	kg/m^3
Aluminum Alloys	0.096–0.103	165–180	26–28	5.2–5.5	2600–2800
Alloy 2014-T6	0.101	175	27	5.4	2800
Alloy 6061-T6	0.098	170	27	5.3	2700
Brass	0.301–0.313	520–540	82–85	16–17	8300–8600
Red Brass (85% Cu, 35% Zn)	0.313	540	85	17	8600
Cast Iron	0.252–0.266	435–460	68–72	13–14	7000–7400
Gray, ASTM-A48	0.260	450	71	14	7200
Malleable, ASTM-A47	0.264	456	72	14	7300
Steel	0.284	490	77	15.2	7850
Titanium	0.162	280	44	8.7	4500
Concrete					
Plain	0.081–0.087	140–150	22–24	4.4–4.7	2200–2400
Lightweight	0.052–0.067	90–115	14–18	2.8–3.6	1400–1800
Reinforced	0.087	150	24	4.7	2400
Glass	0.087–0.104	150–180	24–28	4.7–5.6	2400–2900
Plastics					
Nylon, type 6/6 (molding cpd.)	0.041	70	11	2.2	1100
Polycarbonate	0.043	75	12	2.3	1200
Vinyl, rigid PVC	0.048–0.052	82–90	12.9–14.1	2.6–2.8	1320–1440
Wood	0.014–0.026	25–45	3.9–7.1	0.78–1.4	400–720
Douglas Fir	0.019	32	5.0	1.0	510
Southern Pine	0.022	38	6.0	1.2	610

TABLE F.2. Modulus of Elasticity, Poisson's Ratio, and Coefficient of Thermal Expansion

Material	Modulus of Elasticity (E)		Poisson's Ratio (ν)	Coefficient of Thermal Expansion (α)	
	10^3 ksi	GPa		10^{-6}/°F	10^{-6}/°C
Aluminum Alloys	10.0–11.4	70–79	0.33	11.7–13.3	21–24
Alloy 2014-T6	10.6	73	0.33	12.8	23.0
Alloy 6061-T6	10.0	70	0.33	13.1	23.6
Brass					
Red Brass (85% Cu, 15% Zn)					
Cold-rolled	15	100	0.34	10.4	19
Annealed	15	100	0.34	10.4	19
Cast Iron					
Gray, ASTM-A48	10	70	0.22	6.7	12
Malleable, ASTM-47	24	165	0.27	6.7	12
Steel					
Structural, ASTM-A36	29	200	0.29	6.5	12
Stainless, AISI 302					
Cold-rolled	28	195	0.30	9.6	17
Annealed	28	195	0.30	9.6	17
High-strength, low alloy, ASTM-A242	29	200	0.29	6.5	12
Quenched & tempered, ASTM-A514	29	200	0.29	6.5	12
Titanium					
Alloy (6% Al, 4% V)	16.5	115	0.33	5.3	9.5
Concrete[1]			0.1–0.2		
Medium Strength	3.6	25	—	5.5	10
High Strength	4.5	31	—	5.5	10
Glass	8.7	60	0.2–0.3	3–6	5–11
Plastics					
Nylon, type 6/6 (molding cpd.)	0.4	2.8	0.4	17	30
Polycarbonate	0.35	2.4	—	3.8	6.8
Vinyl, rigid PVC	0.4	2.8	—	28–33	50–60
Wood[2]					
Douglas Fir	1.75	12	—	—	—
Southern Pine	1.75	12	—	—	—

[1]Concrete properties are for compression.
[2]Timber properties are for loading parallel to the grain.

TABLE F.3. Yield Strength, Ultimate Strength, and Percent Elongation in 2 Inches

Material	Yield Strength $(\sigma_Y)^{1,2}$		Ultimate Strength $(\sigma_U)^1$		Percent Elongation over 2 in. Gage Length
	ksi	MPa	ksi	MPa	
Aluminum Alloys					
Alloy 2014-T6	60	410	70	480	13
Alloy 6061-T6	40	275	45	310	17
Brass					
Red Brass (85% Cu, 15% Zn)					
Cold-rolled	60	410	75	520	4
Annealed	15	100	40	275	50
Cast Iron					
Gray, ASTM-A48	—	—	25	170	0.5
Malleable, ASTM-A47	33	230	50	345	10
Steel					
Structural, ASTM-A36	36	250	58	400	20
Stainless, AISI 302					
Cold-rolled	75	520	125	860	12
Annealed	38	260	95	655	50
High-strength, low alloy, ASTM-A242	50	345	70	480	22
Quenched & tempered, ASTM-A514	100	690	110	760	18
Titanium					
Alloy (6% Al, 4% V)	120	830	130	900	10
Concrete[3]					
Medium Strength	—	—	4	28	—
High Strength	—	—	6	40	—
Glass[4]	—	—	(4)	(4)	—
Plastics[5]					
Nylon, type 6/6 (molding cpd.)	8.0	55	11	75	50
Polycarbonate	8.5	60	9.5	65	110
Vinyl (rigid PVC)	6	40	7	50	1–10
Wood[5]					
Douglas Fir	—	—	7.5	50	—
Southern Pine	—	—	8.5	60	—

[1]For ductile metals, the strengh in compression is generally assumed to be equal to the tensile strength.
[2]For most metals, this is the 0.2% offset value.
[3]Concrete properties are for loading in compression.
[4]Glass properties vary widely. For example, glass fibers may have tensile strengths to 1,000 ksi (7000 MPa) or more.
[5]Timber properties are for loading in compression parallel to the grain.

APPENDIX

COMPUTATIONAL MECHANICS; THE *MechSOLID* COMPUTER SOFTWARE

G

Throughout this textbook it has been noted that the digital computer is an essential tool for solving complex problems in solid mechanics, and that systematic problem-solving procedures are required to facilitate the solution of such problems. Although there are special courses (e.g., *Matrix Structural Analysis* and *Finite Element Analysis*) that are devoted to the application of the computer in solving various solid-mechanics problems, in this appendix you will be introduced to the use of the computer in solving problems in axial deformation, torsion, bending, and combined-loading of the type treated in Chapters 3 through 9 of this book. The *displacement method* forms the basis for such systematic computer solutions. Since a key step in these solutions involves solving a set of simultaneous algebraic equations, Appendix G.2 indicates how a set of simultaneous algebraic equations can be expressed in matrix form for solution on a pocket calculator or a digital computer.

For each of the topics covered in Appendices G.1 through G.7 there is an associated computer program in the *MechSOLID* collection of programs.[1] Both Macintosh and Windows versions of *MechSOLID* are available.[2]

The seven computer programs that comprise the *MechSOLID* suite of programs are:

AREAPROP—Computation of Geometric Properties of Plane Areas (Appendix C and Chapter 6)

LINSOLVE—Solution of Simultaneous Linear Algebraic Equations (General)

AXIALDEF—Solution of Axial Deformation Problems (Chapter 3)

TORSDEF—Solution of Torsional Deformation Problems (Chapter 4)

V/M-DIAG—Plotting of Shear and Moment Diagrams (Chapter 5)

BEAMDEF—Solution of Beam Deflection Problems (Chapter 7)

MOHR—Plotting of Mohr's Circles for Stress and Strain (Chapters 8 and 9)

---

[1]The *MechSOLID* computer program may be obtained by consulting the following World Wide Web address: http://www.ae.utexas.edu/~craig.

[2]Windows is a registered trademark of the Microsoft Corporation, and Macintosh is a registered trademark of Apple Computer, Inc.

For axial deformation, a complete "hand" solution and the corresponding AXIALDEF solution are provided here in Appendix G; *MechSOLID* results are given for examples solved by the use of other programs. Homework assignments that require use of *Mech-SOLID* computer programs follow regular homework problems for the following sections: 3.5, 3.6, 3.7, 4.4, 5.5, 7.7, 8.5, 8.10, 8,12, and 9.4. A number of these *MechSOLID* computer exercises are stated as *design-type* problems, where the student is asked to change one or more of the system parameters (e.g., the cross-sectional area of a member) and determine how this modification affects the system's response to given loading.

G.1 AREAPROP—COMPUTATION OF GEOMETRIC PROPERTIES OF PLANE AREAS

Appendices C.1 through C.4 review the mathematical formulas for locating centroids of plane areas, determining the orientation of their principal axes, and calculating various moments and products of inertia. Most structural members have a cross section that is a special case of the general area shown in Fig. G.1-1, for example **W** shapes, **C** shapes, equal- and unequal-leg angles, **Z** shapes, and so forth.

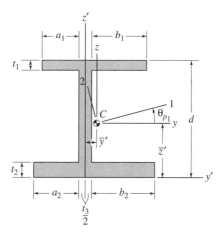

FIGURE G.1-1. Notation for a general cross-sectional area.

Based on equations presented in Appendices C.1 through C.4, the *MechSOLID* computer program option AREAPROP computes the following geometric properties:

- Cross-sectional area: A
- Location of the centroid C: $(\bar{y}', \bar{z}')$
- Moments and product of inertia relative to the (y', z') reference frame: $I_{y'}$, $I_{z'}$, and $I_{y'z'}$
- Moments and product of inertia relative to the centroidal (y, z) reference frame: I_y, I_z, and I_{yz}
- Orientation of the centroidal principal axes: θ_{p1}
- Centroidal principal moments of inertia: I_1 and I_2

Use the AREAPROP computer program to compute the geometric properties of the L-shaped cross-sectional area shown in Example C-1. Let $t = 10$ mm, giving the area shown in Fig. 1.

    The AREAPROP input window is shown. The cross section is then displayed, and an opportunity is given for correction of the input data. Once the cross section input data is accepted, AREAPROP computes the area properties and produces a plot of the cross section, showing the location of the centroid and the orientation of the centroidal principal axes.

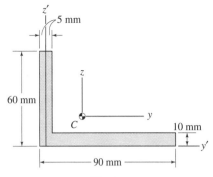

Fig. 1

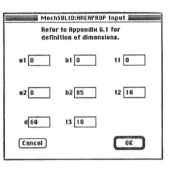

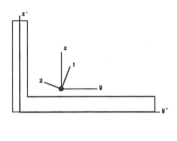

AREAPROP: Results

A = 1.450e+03
Centroid location (2.784e+01 , 1.534e+01)

Moments and Product of Inertia relative to the (y',z') reference frame:
Iy' = 7.483e+05
Iz' = 2.435e+06
Iy'z' = 2.019e+05

Moments and Product of Inertia relative to the (y,z) reference frame:
Iy = 4.069e+05
Iz = 1.310e+06
Iyz = –4.177e+05

Orientation of the principal axes:
θ1 = 6.862e+01, θ2 = 1.586e+02
Imax = 1.474e+06, Imin = 2.434e+05

G.2 LINSOLVE—SOLUTION OF SIMULTANEOUS ALGEBRAIC EQUATIONS EXPRESSED IN MATRIX NOTATION

■■■■■■■■■■

Many of the problems in this textbook require the solution of a set of simultaneous linear algebraic equations—equilibrium equations, force-temperature-deformation equations, and deformation-compatibility equations. In many cases, these can be systematically reduced to the solution of one equation for one unknown displacement or force. However, in many cases, several simultaneous equations have to be solved for an equal number of unknowns. For instance, in Example 3.4 the two simultaneous equations

$$(k_1 + k_2)u_B - k_2u_C = P_B \qquad \text{(G.2-1)}$$
$$-k_2u_B + (k_2 + k_3)u_C = P_C$$

must be solved for the two unknown displacements u_B and u_C.

It is convenient to write such sets of simultaneous linear algebraic equations in matrix format.[3] For example, three equations in three unknowns can be written in expanded form as

$$K_{11}u_1 + K_{12}u_2 + K_{13}u_3 = P_1$$
$$K_{21}u_1 + K_{22}u_2 + K_{23}u_3 = P_2 \qquad \text{(G.2-2)}$$
$$K_{31}u_1 + K_{32}u_2 + K_{33}u_3 = P_3$$

where the K_{ij}'s are the *coefficients*; P_1, P_2, and P_3 are *known values*; and u_1, u_2, and u_3 are the three *unknowns* to be determined by solving the three simultaneous linear algebraic equations, Eqs. G.2-2.[4] Such equations may be written in the matrix form

$$[K]\{u\} = \{P\} \qquad \text{(G.2-3)}$$

where

$$[K] = \begin{bmatrix} K_{11} & K_{12} & K_{13} \\ K_{21} & K_{22} & K_{23} \\ K_{31} & K_{32} & K_{33} \end{bmatrix}, \quad \{u\} \equiv \begin{Bmatrix} u_1 \\ u_2 \\ u_3 \end{Bmatrix}, \quad \{P\} \equiv \begin{Bmatrix} P_1 \\ P_2 \\ P_3 \end{Bmatrix} \qquad \text{(G.2-4)}$$

Given as input the values of the elements K_{ij} of the $N \times N$ matrix $[K]$ and the N elements of the vector $\{P\}$, the **MechSOLID** computer program option LINSOLVE obtains the solution vector $\{u\}$.

■■■■■■■■■■■■■■■■□ E X A M P L E G - 2 □■■■■■■■■■■■■■■■■■■

In Example 3.15 the following equations are solved for the unknowns u_1 and u_2.

$$453.6u_1 + 115.2u_2 = -19.20$$
$$115.2u_1 + 486.4u_2 = -14.40$$

with the units of u_1 and u_2 being inches. Then,

$$[K] = \begin{bmatrix} 453.6 & 115.2 \\ 115.2 & 486.4 \end{bmatrix}, \quad \{P\} = \begin{Bmatrix} -19.20 \\ -14.40 \end{Bmatrix}$$

Using LINSOLVE we obtain

$$\{u\} \equiv \begin{Bmatrix} u_1 \\ u_2 \end{Bmatrix} = \begin{Bmatrix} -3.704(10^{-2}) \text{ in.} \\ -2.083(10^{-2}) \text{ in.} \end{Bmatrix} \qquad \textbf{Ans.}$$

---

[3]See, for example, a textbook on Matrix Algebra or Linear Algebra such as *Elementary Linear Algebra*, Seventh edition, by H. Anton, John Wiley & Sons, Inc., New York, NY, 1994.

[4]The reader may consult a textbook on matrix algebra or linear algebra for conditions under which either no solution, or no unique solution, of Eqs. G.2-2 exists.

As an illustration of how the computer can be programmed in a systematic way to solve solid-mechanics problems, let us formulate a *displacement-method solution* of an axial-deformation problem for the multielement member shown in Fig. G.3-1.[5]

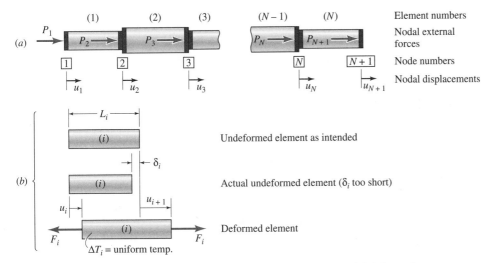

FIGURE G.3-1. Notation for a member undergoing axial deformation.

Based on Sections 3.5, 3.6, and 3.7, we will combine the treatment of axial-deformation problems involving applied external loads P_i (Section 3.5), thermal strains due to temperature changes ΔT_i (Section 3.6), and geometric "misfits" δ_i (Section 3.7). We consider only a uniaxial arrangement of linearly elastic, uniform elements (i), rigidly connected together at nodes $\boxed{i}$, as shown in Fig. G.3-1. Either node 1 or node ($N+1$) <u>must</u> be restrained; both may be restrained (giving a statically indeterminate problem).

Nodal Equilibrium: For each of the ($N+1$) nodes we can draw a free-body diagram and write an equilibrium equation.

$$\sum F_x = 0: \qquad F_{i-1} - F_i = P_i \quad i = 1, 2, \ldots, (N+1) \qquad (G.3\text{-}1)$$

Element Force-Temperature-Deformation Behavior: From Eq. 3.21, we can write, for each of the N elements,

$$F_i = k_i(e_i - \alpha_i L_i \Delta T_i) \quad i = 1, 2, \ldots, N \qquad (G.3\text{-}2)$$

where $k_i = \left(\dfrac{AE}{L}\right)_i$ is the *axial stiffness coefficient*.

---

[5]In earlier chapters of this book, nodes (joints) were labeled with capital letters and elements were labeled with numbers. For computer solutions, which usually involve solving sets of simultaneous linear algebraic equations, it will be necessary to label nodes with numbers rather than with letters.

Geometry of Deformation: As indicated in Fig. G.3-1*b*, let us assume that element (*i*) was intended to be L_i in length, but that it was actually fabricated δ_i <u>too short</u>. Also, as is indicated in Fig. G.3-1, let the displacements u_i be measured from the <u>intended</u> original nodal positions $\boxed{i}$. Using Fig. G.3-1*b* we can relate the total elongation, e_i, of element (*i*) to the nodal displacements u_i and u_{i+1} at its two ends and to the original length "error" δ_i (which is positive if the element is initially shorter than the intended length L_i; negative if it is initially too long). Then,

$$e_i = (u_{i+1} - u_i) + \delta_i \quad i = 1, 2, \ldots, N \tag{G.3-3}$$

Nodal Equilibrium Equations in Terms of Nodal Displacements:

Using the *displacement method* of solution, as discussed in Sections 3.5, 3.6, and 3.7, we can combine Eqs. G.3-1 through G.3-3 and write, for each node

$$-k_{i-1}u_{i-1} + (k_{i-1} + k_i)u_i - k_i u_{i+1} = (P_i)_{\text{eff}} \tag{G.3-4a}$$

$$(P_i)_{\text{eff}} \equiv P_i - k_{i-1}(\delta_{i-1} - \alpha_{i-1}L_{i-1}\Delta T_{i-1}) + k_i(\delta_i - \alpha_i L_i \Delta T_i) \tag{G.3-4b}$$

Given applied nodal loads P_i, element temperature increments ΔT_i, element length "errors" δ_i, and end conditions on u_1 and/or u_{N+1}, we can solve the set of simultaneous equations, Eqs. G.3-4, for the unknown nodal displacements. Note that P_1 is an unknown *reaction* if u_1 is known, and similarly for P_{N+1}. **Hence, Eqs. G.3-4 are written only for those nodes whose displacement u_i is unknown.**

Once the unknown nodal displacements have been obtained by solving Eqs. G.3-4, the (internal) element forces F_i may be obtained by combining Eqs. G.3-2 and G.3-3, giving

$$F_i = k_i(u_{i+1} - u_i + \delta_i - \alpha_i L_i \Delta T_i) \quad i = 1, 2, \ldots, N \tag{G.3-5}$$

Before giving an example of computer solution of an axial-deformation problem, let us set up and solve the following example problem "by hand."

▪▪▪▪▪▪▪▪▪▪▪▪▪▪ EXAMPLE G-3 ▫▫▫▫▫▫▫▫▫▫▫▫▫▫

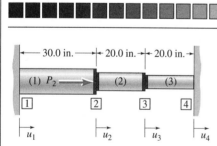

Fig. 1 A statically indeterminate assemblage.

Three uniform elements make up the statically indeterminate axial-deformation assemblage in Fig. 1.

The physical properties of the elements are:

Element	Area A_i	(Intended) Length L_i	Modulus of Elasticity E_i	Coefficient of Thermal Expansion α_i	Gap δ_i
1	2.00 in^2	30.00 in.	$15(10^3)$ ksi	—	—
2	1.50 in^2	20.00 in.	$30(10^3)$ ksi	—	0.01 in.
3	1.00 in^2	20.00 in.	$30(10^3)$ ksi	$7(10^{-6})/°F$	—

Member (2) is actually fabricated 0.01 in. too short (i.e., $\delta_2 = 0.01$ in.), so it must be stretched in order to weld it to adjacent elements (1) and (3).

(a) After the three members are welded to the rigid surfaces at nodes ① and ④ and welded together at nodes ② and ③, the forces used to assemble the system are released. What are the resulting locations of nodes ② and ③? What are the ''initial stresses'' induced in the three elements? (b) If an external force $P_2 = 10$ kips is subsequently applied at node ②, and element (3) is <u>cooled</u> by $100°F$, what are the final positions of nodes ② and ③? What are the final axial stresses in the three elements?

Plan the Solution For Parts (a) and (b) we can write the *nodal equilibrium equations in terms of nodal displacements* (Eqs. G.3-4) for the non-restrained nodes ② and ③. Then, we can use Eq. G.3-5 to solve for each internal axial force F_i, which can then be divided by the corresponding area A_i to give the stress σ_i.

Solution
(a) Solve for the initial displacements u_{2a} and u_{3a}. Then solve for the initial axial stresses σ_{1a}, σ_{2a}, and σ_{3a}. The element stiffness coefficients are:

$$k_1 = \left(\frac{AE}{L}\right)_1 = \frac{(2.00 \text{ in}^2)(15 \times 10^3 \text{ ksi})}{30.00 \text{ in.}} = 1000 \text{ kips/in.}$$

$$k_2 = \left(\frac{AE}{L}\right)_2 = \frac{(1.50 \text{ in}^2)(30 \times 10^3 \text{ ksi})}{20.00 \text{ in.}} = 2250 \text{ kips/in.}$$

$$k_3 = \left(\frac{AE}{L}\right)_3 = \frac{(1.00 \text{ in}^2)(30 \times 10^3 \text{ ksi})}{20.00 \text{ in.}} = 1500 \text{ kips/in.}$$

Note that the initial length error $\delta_2 = 0.01$ in. is ignored in computing k_2 because of the small effect that it would have on k_2. From Eqs. G.3-4 we get

Node ②: $\qquad (k_1 + k_2)u_{2a} - k_2 u_{3a} = k_2 \delta_2$ $\qquad$ (1a)

Node ③: $\qquad -k_2 u_{2a} + (k_2 + k_3)u_{3a} = -k_2 \delta_2$ $\qquad$ (1b)

When the values of the k's and δ_2 are inserted into Eqs. (1), the resulting *equilibrium equations in terms of nodal displacements* are:

$$(1000 + 2250)u_{2a} - 2250u_{3a} = (2250 \text{ kips/in.})(0.01 \text{ in.})$$

$$-2250u_{2a} + (2250 + 1500)u_{3a} = -(2250 \text{ kips/in.})(0.01 \text{ in.})$$

or

$$3250u_{2a} - 2250u_{3a} = 22.50 \text{ kips} \qquad (2a)$$

$$-2250u_{2a} + 3750u_{3a} = -22.50 \text{ kips} \qquad (2b)$$

Using LINSOLVE to solve Eqs. (2), we get the following *initial displacements*:

$$\left. \begin{array}{l} u_{2a} = 0.00474 \text{ in.} \\ u_{3a} = -0.00316 \text{ in.} \end{array} \right\} \qquad \text{Ans. (a)} \quad (3)$$

We can use Eq. G.3-5 to solve for the *initial axial forces*. The result is

$$F_{1a} = F_{2a} = F_{3a} = 4.74 \text{ kips} \tag{4}$$

(Note that the forces induced in the elements are all the same, since there are no external forces applied at the nodes.) The *initial stresses* are

$$\left. \begin{aligned} \sigma_{1a} &= \frac{F_{1a}}{A_1} = \frac{4.74 \text{ kips}}{2.00 \text{ in}^2} = 2.37 \text{ ksi} \\[2ex] \sigma_{2a} &= \frac{F_{2a}}{A_2} = \frac{4.74 \text{ kips}}{1.50 \text{ in}^2} = 3.16 \text{ ksi} \\[2ex] \sigma_{3a} &= \frac{F_{3a}}{A_3} = \frac{4.74 \text{ kips}}{1.00 \text{ in}^2} = 4.74 \text{ ksi} \end{aligned} \right\} \quad \textbf{Ans. (a)} \quad (5)$$

(b) Solve for the final displacements u_{2b} and u_{3b} and for the final axial stresses σ_{1b}, σ_{2b}, and σ_{3b} that result when the external force $P_2 = 100$ kips and element temperature change $\Delta T_3 = -100°F$ are included. Only the right-hand sides of Eqs. (1) change. Then, from Eqs. G.3-4,

Node $\boxed{2}$: $\qquad (k_1 + k_2)u_{2a} - k_2 u_{3a} = k_2 \delta_2 + P_2 \qquad$ (6a)

Node $\boxed{3}$: $\quad -k_2 u_{2a} + (k_2 + k_3)u_{3a} = -k_2 \delta_2 - k_3 \alpha_3 L_3 \Delta T_3 \qquad$ (6b)

When the values of k_1, etc. are inserted into Eqs. (6), the resulting *equilibrium equations in terms of nodal displacements* are

$$3250u_{2b} - 2250u_{3b} = 22.50 \text{ kips} + 10.00 \text{ kips} = 32.50 \text{ kips} \tag{7a}$$

$$-2250u_{2b} + 3750u_{3b} = -22.50 \text{ kips} + 21.00 \text{ kips} = -1.50 \text{ kips} \tag{7b}$$

The *final displacements* are obtained by solving Eqs. (7). Thus,

$$\left. \begin{aligned} u_{2b} &= 0.0166 \text{ in.} \\ u_{3b} &= 0.0096 \text{ in.} \end{aligned} \right\} \quad \textbf{Ans. (b)} \quad (8)$$

We can use Eq. G.3-5 to solve for the *final axial forces*. The result is

$$F_{1b} = 16.6 \text{ kips}, \quad F_{2b} = 6.6 \text{ kips}, \quad F_{3b} = 6.6 \text{ kips}$$

The *final stresses*, then, are

$$\left. \begin{aligned} \sigma_{1b} &= \frac{F_{1b}}{A_1} = \frac{16.6 \text{ kips}}{2.00 \text{ in}^2} = 8.3 \text{ ksi} \\[2ex] \sigma_{2b} &= \frac{F_{2b}}{A_2} = \frac{6.6 \text{ kips}}{1.50 \text{ in}^2} = 4.4 \text{ ksi} \\[2ex] \sigma_{3b} &= \frac{F_{3b}}{A_3} = \frac{6.6 \text{ kips}}{1.00 \text{ in}^2} = 6.6 \text{ ksi} \end{aligned} \right\} \quad \textbf{Ans. (b)} \quad (9)$$

The *MechSOLID* computer program option AXIALDEF employs the displacement method to solve linear axial-deformation problems for uniaxial arrangements of up to five elements (i.e., $N_{max} = 5$).

EXAMPLE G-4

Use AXIALDEF to solve Example G-3.

Solution   Shown here are most of the "screens" that were generated in the process of solving Example G-3 on a Macintosh computer. Similar screens would be generated by the MS-Windows version of *MechSOLID*.

1. *Select the program from the Main Menu; then initialize the solution.* With the computer's mouse the user selects the desired program option. The user next indicates the number of axial-deformation elements, and may give a title for the AXIALDEF solution.

```
╔════════ MechSOLID ════════╗
║  ○ AREAPROP                ║
║  ○ LINSOLVE                ║
║  ◉ AXIALDEF        ╔═══════════ AXIALDEF ═══════════╗
║  ○ TORSDEF         ║                                ║
║  ○ V/M-DIAG        ║  Enter the number of           ║
║  ○ BEAMDEF         ║  Bar elements.      [3      ]  ║
║  ○ MOHR            ║                                ║
║                    ║  Model Name:        [Example G-4]║
║      ( Begin )     ║                                ║
║                    ║  ( Cancel )         (  OK  )   ║
╚════════════════════╚════════════════════════════════╝
```

2. *Define the elements.* The user enters data into the pre-defined element data boxes and tabs to the next box. An opportunity is provided later to change the element data. Two sample input screens are shown here. Note that the value of δ_2 is input as the "gap."

```
╔═══ Bar Element Definition ═══╗    ╔═══ Bar Element Definition ═══╗
║      Element Number: 2       ║    ║      Element Number: 3       ║
║                              ║    ║                              ║
║  Length              [20  ]  ║    ║  Length              [20  ]  ║
║  Area                [1.5 ]  ║    ║  Area                [1   ]  ║
║  Modulus of                  ║    ║  Modulus of                  ║
║  Elasticity          [30e3]  ║    ║  Elasticity          [30e3]  ║
║  Coefficient of              ║    ║  Coefficient of              ║
║  Thermal Expansion   [0.0 ]  ║    ║  Thermal Expansion   [7e-6]  ║
║  Delta Temperature   [0.0 ]  ║    ║  Delta Temperature   [0.0 ]  ║
║  Gap                 [0.01]  ║    ║  Gap                 [0.0 ]  ║
║                              ║    ║                              ║
║  ( Cancel )     (  OK  )     ║    ║  ( Cancel )     (  OK  )     ║
╚══════════════════════════════╝    ╚══════════════════════════════╝
```

3. *Review the model and correct the data if necessary.* The user is presented with a listing of all element data that has been entered and is given an opportunity to correct the data. This summary of Model data may be

printed later. (See Step 6.) The solution proceeds once the input data has been accepted as correct.

```
═══════════════════════════ Model ═══════════════════════════
                              Example G-4
                          AXIALDEF: Model
        Element                    Modulus of    Delta
        Number    Length    Area   Elasticity    Temp     Alpha     Gap
          1      3.00e+01  2.000e+00  1.500e+04  0.00e+00  0.0e+00  0.0e+00
          2      2.00e+01  1.500e+00  3.000e+04  0.00e+00  0.0e+00  1.0e-02
          3      2.00e+01  1.000e+00  3.000e+04  0.00e+00  7.0e-06  0.0e+00




        ● Accept Values
        ○ Change Element Number  [    ▼ ]
                                                        [ OK ]
```

4. *Define the applied loads and specified displacement boundary conditions.* The user indicates the number of each, and appropriate data-entry screens are presented to the user. Changes may be made to the input data later, provided that corresponding force/displacement data has been input initially. (For example, the value $P_2 = 0$ is entered so that the value can be changed to $P_2 = 10$ kips for Part (b).)

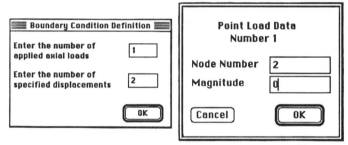

```
═══ Boundary Condition Definition ═══        ┌─────────────────────────────┐
                                             │        Point Load Data      │
Enter the number of          [ 1 ]          │         Number 1            │
applied axial loads                          │                             │
                                             │ Node Number    [ 2 ]        │
Enter the number of          [ 2 ]          │                             │
specified displacements                      │ Magnitude      [ 0| ]       │
                                             │                             │
                     [ OK ]                  │ ( Cancel )        [ OK ]     │
                                             └─────────────────────────────┘
```

5. *Review the input data and correct it if necessary.* The user is presented with a listing of all load and boundary-condition data that has been entered and is given an opportunity to change the data. This Boundary Condition input data may be printed later. (See Step 6.) The solution proceeds once the input data has been accepted as correct.

```
═══════════════════════ Boundary Conditions ═══════════════════════
                              Example G-4
                    AXIALDEF: Boundary Conditions
                      Concentrated                      Specified
        Number   Node    Load      Number    Node    Displacement
          1       2    0.000e+00      1        1       0.00e+00
                                      2        4       0.00e+00




        ● Accept Values
        ○ Change Force Number        [    ▼ ]
        ○ Change Displacement Number
                                                        [ OK ]
```

6. *Print the input and the output.* The Model, Boundary Conditions, Results, Displacement Diagram, and Force Diagram windows may be selected from the Windows pull-down menu and printed.

Example G-4
AXIALDEF: Model

Element Number	Length	Area	Modulus of Elasticity	Delta Temp	Alpha	Gap
1	3.00e+01	2.000e+00	1.500e+04	0.00e+00	0.0e+00	0.0e+00
2	2.00e+01	1.500e+00	3.000e+04	0.00e+00	0.0e+00	1.0e-02
3	2.00e+01	1.000e+00	3.000e+04	0.00e+00	7.0e-06	0.0e+00

Example G-4
AXIALDEF: Boundary Conditions

Number	Node	Concentrated Load	Number	Node	Specified Displacement
1	2	0.000e+00	1	1	0.00e+00
			2	4	0.00e+00

Example G-4
AXIALDEF: Results

Node Number	Displacement
1	0.000e+00
2	4.737e-03
3	-3.158e-03
4	0.000e+00

Element Number	Internal Force	Normal Stress
1	4.737e+00	2.368e+00
2	4.737e+00	3.158e+00
3	4.737e+00	4.737e+00

EXAMPLE G-4
AXIALDEF: Displacement Diagram

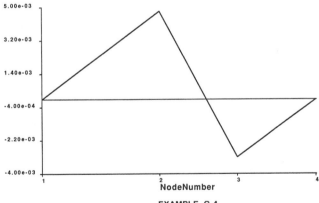

EXAMPLE G-4
AXIALDEF: Force Diagram

A-53

7. *Change input values and re-run the program.* In the present case, Part (b) will be run by changing the values of the element temperature ΔT_3 and the applied load P_2 and re-running the program. From the File pull-down menu, choose Select Previous Model. Data input windows for the change of the temperature of element (3) and the modification to the axial force P_2 are shown. Finally, the input and output "windows" for Part (b) are shown.

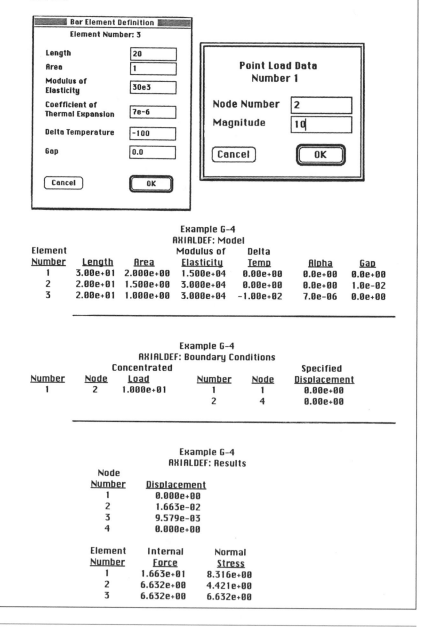

```
                             Example G-4
                           AXIALDEF: Model
Element                   Modulus of      Delta
Number    Length    Area  Elasticity      Temp       Alpha      Gap
   1     3.00e+01  2.000e+00  1.500e+04   0.00e+00  0.0e+00   0.0e+00
   2     2.00e+01  1.500e+00  3.000e+04   0.00e+00  0.0e+00   1.0e-02
   3     2.00e+01  1.000e+00  3.000e+04  -1.00e+02   7.0e-06   0.0e+00
```

```
                             Example G-4
                     AXIALDEF: Boundary Conditions
                     Concentrated                          Specified
Number    Node         Load          Number    Node      Displacement
   1       2         1.000e+01          1        1        0.00e+00
                                         2        4        0.00e+00
```

```
                             Example G-4
                          AXIALDEF: Results
        Node
        Number      Displacement
          1         0.000e+00
          2         1.663e-02
          3         9.579e-03
          4         0.000e+00

        Element     Internal      Normal
        Number       Force        Stress
          1         1.663e+01    8.316e+00
          2         6.632e+00    4.421e+00
          3         6.632e+00    6.632e+00
```

G.4 TORSDEF—DISPLACEMENT-METHOD SOLUTION OF TORSIONAL-DEFORMATION PROBLEMS

Because of the close similarity of torsion problems to axial-deformation problems (e.g., see Table 4.1) the procedures described in this appendix are very similar to the procedures described in Appendix G.3. However, no thermal strains or geometric "misfits" are considered in the present discussion of torsion.

Let us formulate a displacement-method procedure for solving torsion problems for the multi-element torsion member shown in Fig. G.4-1.[6] The development here is based on Sections 4.3 and 4.4, but is restricted to a straight-line assemblage of N uniform, linearly elastic torsion elements (i) rigidly connected to one another at nodes $\boxed{i}$. Either node 1 or node $N + 1$ <u>must</u> be restrained; both may be restrained (giving a statically indeterminate problem).

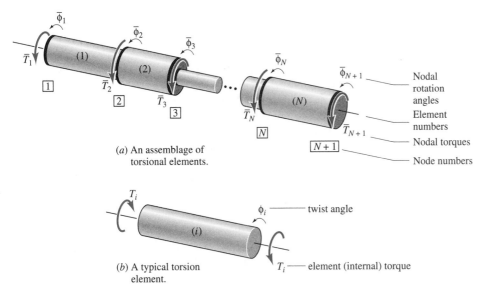

(a) An assemblage of torsional elements.

(b) A typical torsion element.

FIGURE G.4-1. Notation for an assemblage of torsion elements and for a typical torsion element.

Nodal Equilibrium: Except for elements 1 and N at the ends of the assemblage, elements (i) and ($i + 1$) are connected together by node $\boxed{i}$. For each of the nodes, we can draw a free-body diagram and write an equilibrium equation:

$$\sum T = 0: \qquad T_{i-1} - T_i = \bar{T}_i \quad i = 1, 2, \ldots, (N + 1) \qquad \text{(G.4-1)}$$

Element Torque-Twist Behavior: From Eq. 4.19, p. 185, we can write the following *torque-twist equation* for each of the N elements:

$$T_i = k_{ti}\phi_i \quad i = 1, 2, \ldots, N \qquad \text{(G.4-2)}$$

where ϕ_i is the angle of twist of element i and k_{ti} is the *torsional stiffness coefficient*, given by $k_{ti} = \left(\dfrac{GJ}{L}\right)_i$.

---

[6]In Appendix G.3, the symbol P_i was used for an external force applied at node $\boxed{i}$ and the symbol F_i was used for the internal axial force in element (i). Similarly, u_i was used to signify the axial displacement of node $\boxed{i}$ and e_i denoted the elongation of element (i). Here we use $\bar{T}_i$ and $\bar{\phi}_i$ for the applied (external) nodal torque and rotation angle, respectively, and we use T_i and ϕ_i for the internal torque in element (i) and the twist angle of element (i).

Geometry of Deformation: As indicated in Fig. G.4-1, we can observe that the twist angle of element i is given in terms of the rotation angles at its ends by

$$\phi_i = \overline{\phi}_{i+1} - \overline{\phi}_i \quad i = 1, 2, \ldots, N \tag{G.4-3}$$

Nodal Equilibrium Equations in Terms of Nodal Displacements: Using the *displacement method* of solution, as discussed in Section 4.4, we can combine Eqs. G.4-1 through G.4-3 and write for each node

$$-k_{t(i-1)}\overline{\phi}_{i-1} + (k_{t(i-1)} + k_{ti})\overline{\phi}_i - k_{ti}\overline{\phi}_{i+1} = \overline{T}_i \tag{G.4-4}$$

Given applied nodal torques $\overline{T}_i$ and end conditions on $\overline{\phi}_1$ and/or $\overline{\phi}_{N+1}$, we can solve the set of simultaneous equations, Eqs. G.4-4, for the unknown nodal rotations. Note that $\overline{T}_1$ is an unknown *reaction* if $\overline{\phi}_1$ is known, and similarly for $\overline{T}_{N+1}$ and $\overline{\phi}_{N+1}$. **Hence, Eqs. G.4-4 are written only for those nodes whose displacement $\overline{\phi}_i$ is unknown.**

Once the nodal rotations have been obtained by solving Eqs. G.4-4, the (internal) element torques T_i may be obtained by combining Eqs. G.4-2 and G.4-3, giving

$$T_i = k_{ti}(\overline{\phi}_{i+1} - \overline{\phi}_i) \quad i = 1, 2, \ldots, N \tag{G.4-5}$$

Example Problem 4.5 on page 188 is a three-element, statically indeterminate torsion problem that is solved by the above displacement-method procedure.

The *MechSOLID* computer program option TORSDEF employs the displacement method outlined above to solve torsional-deformation problems for uniaxial arrangements of up to five elements ($N_{\max} = 5$). Input for TORSDEF is very similar to input for AXIALDEF, so no TORSDEF example is given.

■■■■■■■■■■■ G.5 V/M DIAG—SHEAR AND BENDING-MOMENT DIAGRAMS

This appendix discusses application of the *discontinuity-function method* of Section 5.5 for constructing shear and moment diagrams. Applications here are limited to statically determinate beams; shear and moment diagrams for statically indeterminate beams are considered in Appendix G.6. To be statically determinate, a beam must either be supported on a pair of simple supports, as in Fig. G.5-1, or be cantilevered at one end, as in Fig. G.5-2a or Fig. G.5-2b. Loading is restricted to concentrated forces, concentrated couples, and linearly varying (or constant) distributed loading. Expressions for the "load" $p(x)$, the shear $V(x)$, and the bending moment $M(x)$ for these beams are illustrated below and are incorporated in the V/M-DIAG option of the *MechSOLID* collection of computer programs.

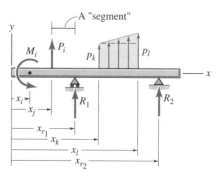

FIGURE G.5-1. A statically determinate beam with simple supports.

Beams with Simple Supports. A typical statically determinate beam with simple supports is shown in Fig. G.5-1. Using "Load" entries from Table 5.2 we form the following expression for $p(x)$, taking the "loads" in order from left to right on the beam:

$$
\begin{aligned}
p(x) = {} & -M_i \langle x - x_i \rangle^{-2} + P_j \langle x - x_j \rangle^{-1} \\
& + R_1 \langle x - x_{r1} \rangle^{-1} + p_k \langle x - x_k \rangle^{0} - p_l \langle x - x_l \rangle^{0} \\
& + \frac{(p_l - p_k)}{(x_l - x_k)} \langle x - x_k \rangle^{1} - \frac{(p_l - p_k)}{(x_l - x_k)} \langle x - x_l \rangle^{1} \\
& + R_2 \langle x - x_{r2} \rangle^{-1}
\end{aligned}
\tag{G.5-1}
$$

Equation G.5-1 can be integrated with the aid of Eqs. 5.15 or the entries from the "Shear" and "Moment" columns of Table 5.2 to give

$$
\begin{aligned}
V(x) = {} & \int_{0-}^{x} p(\xi) d\xi \\
= {} & -M_i \langle x - x_i \rangle^{-1} + P_j \langle x - x_j \rangle^{0} \\
& + R_1 \langle x - x_{r1} \rangle^{0} + p_k \langle x - x_k \rangle^{1} - p_l \langle x - x_l \rangle^{1} \\
& + \frac{1}{2} \frac{(p_l - p_k)}{(x_l - x_k)} \langle x - x_k \rangle^{2} - \frac{1}{2} \frac{(p_l - p_k)}{(x_l - x_k)} \langle x - x_l \rangle^{2} \\
& + R_2 \langle x - x_{r2} \rangle^{0}
\end{aligned}
\tag{G.5-2}
$$

$$
\begin{aligned}
M(x) = {} & \int_{0-}^{x} V(\xi) d\xi \\
= {} & -M_i \langle x - x_i \rangle^{0} + P_j \langle x - x_j \rangle^{1} \\
& + R_1 \langle x - x_{r1} \rangle^{1} + \frac{p_k}{2} \langle x - x_k \rangle^{2} - \frac{p_l}{2} \langle x - x_l \rangle^{2} \\
& + \frac{1}{6} \frac{(p_l - p_k)}{(x_l - x_k)} \langle x - x_k \rangle^{3} - \frac{1}{6} \frac{(p_l - p_k)}{(x_l - x_k)} \langle x - x_l \rangle^{3} \\
& + R_2 \langle x - x_{r2} \rangle^{1}
\end{aligned}
\tag{G.5-3}
$$

To evaluate the reactions R_1 and R_2 we can write equilibrium equations for the entire beam. These can be easily formulated by simply writing the following *closure conditions* on shear and moment:[7]

$$
\begin{aligned}
V(L^+) = {} & + P_j + R_1 + R_2 + p_k (L - x_k) - p_l (L - x_l) \\
& + \frac{1}{2} \frac{(p_l - p_k)}{(x_l - x_k)} (L - x_k)^2 - \frac{1}{2} \frac{(p_l - p_k)}{(x_l - x_k)} (L - x_l)^2 \\
= {} & 0
\end{aligned}
\tag{G.5-4}
$$

$$
\begin{aligned}
M(L^+) = {} & -M_i + P_j(L - x_j) + R_1(L - x_{r1}) + R_2(L - x_{r2}) \\
& + \frac{p_k}{2} (L - x_k)^2 - \frac{p_l}{2} (L - x_l)^2 \\
& + \frac{1}{6} \frac{(p_l - p_k)}{(x_l - x_k)} (L - x_k)^3 - \frac{1}{6} \frac{(p_l - p_k)}{(x_l - x_k)} (L - x_l)^3 \\
= {} & 0
\end{aligned}
\tag{G.5-5}
$$

---

[7] See Section 7.5.

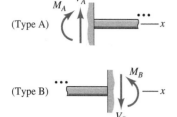

FIGURE G.5-2. Statically determinate cantilever beams.

Equation G.5-4 corresponds to a summation of forces in the y direction, and Eq. G.5-5 corresponds to a summation of moments about the right end of the beam ($x = L$). These two equations can be solved simultaneously for the two reactions, R_1 and R_2.

Statically Determinate Cantilever Beams. Typical statically determinate beams with cantilever supports are shown in Figs. G.5-2. A beam with cantilever support at $x = 0$ will be designated as a "Type A" beam; a beam with cantilever support at $x = L$ will be designated as "Type B." Only the support reactions are given explicitly; concentrated loads and distributed loads would be treated just as they were for the beam with simple supports.

Type A—Cantilever Support at $x = 0$: From Table 5.2 expressions for the load, shear, and moment may be obtained as follows:

$$p(x) = M_A\langle x \rangle^{-2} + V_A\langle x \rangle^{-1} + \ldots \tag{G.5-6}$$

$$V(x) = M_A\langle x \rangle^{-1} + V_A\langle x \rangle^{0} + \ldots \tag{G.5-7}$$

$$M(x) = M_A\langle x \rangle^{0} + V_A\langle x \rangle^{1} + \ldots \tag{G.5-8}$$

The reactions V_A and M_A may be obtained by solving the two *closure equations*:

$$V(L^+) = V_A + \ldots = 0 \tag{G.5-9}$$

$$M(L^+) = M_A + V_A L + \ldots = 0 \tag{G.5-10}$$

which are the equilibrium equation for summation of forces in the y direction and the equilibrium equation for summation of moments about $x = L$, respectively.

Type B—Cantilever Support at $x = L$: From Table 5.2, expressions for the load, shear, and moment may be obtained as follows:

$$p(x) = \ldots - M_B\langle x - L \rangle^{-2} - V_B\langle x - L \rangle^{-1} \tag{G.5-11}$$

$$V(x) = \ldots - M_B\langle x - L \rangle^{-1} - V_B\langle x - L \rangle^{0} \tag{G.5-12}$$

$$M(x) = \ldots - M_B\langle x - L \rangle^{0} - V_B\langle x - L \rangle^{1} \tag{G.5-13}$$

The reactions V_B and M_B may be obtained by solving the two *closure equations*:

$$V(L^+) = \ldots - V_B = 0 \tag{G.5-14}$$

$$M(L^+) = \ldots - M_B = 0 \tag{G.5-15}$$

which are the equilibrium equation for summation of forces in the y direction and the equilibrium equation for summation of moments about $x = L$, respectively.

As noted at the end of Section 5.5, Macauley-bracket expressions like those in Eqs. G.5-1 through G.5-3 are very easy to incorporate into a computer program, since each bracket corresponds to an *If-Then* type statement. The **MechSOLID** computer program option V/M-DIAG employs the discontinuity-function method outlined previously to obtain shear and moment diagrams for beams with up to five "segments."

Using the *MechSOLID* program V/M-DIAG, the reader should get the following shear- and moment-diagram results for the beam in Example Problem 5.7.

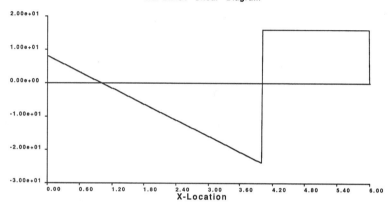

V/M-DIAG: Results
Simply Supported Beam

Left Reaction: 8.000e+00
Right Reaction: 4.000e+01

Maximum Shear	1.600e+01	at X=4.01e+00
Minimum Shear	-2.397e+01	at X=4.00e+00
Maximum Moment	4.000e+00	at X=9.96e-01
Minimum Moment	-3.190e+01	at X=4.00e+00

V/M-DIAG: Shear Diagram

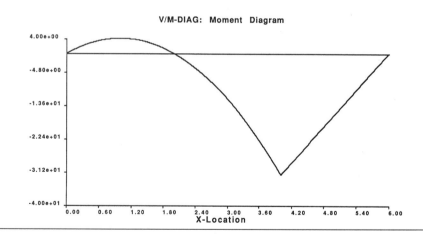

V/M-DIAG: Moment Diagram

G.6 BEAMDEF—DISPLACEMENT-METHOD SOLUTION OF BEAM-DEFLECTION PROBLEMS

This appendix introduces you to the *MechSOLID* computer program option BEAMDEF, which solves beam-deflection problems by using the displacement-method procedure discussed in Section 7.7. We will consider only straight beams consisting of one or more (up to five) elements, labeled (n), connected to each other by nodes that are labeled ⬛, as shown in Fig. G.6-1.

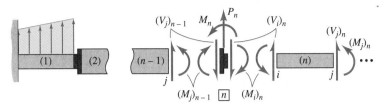

FIGURE G.6-1. Typical beam elements and nodes; a free-body diagram of node ⬛

Nodal Equilibrium: Summing forces and moments on the free-body diagram of node ⬛ in Fig. G.6-1, we get the following *nodal equilibrium equations*:

$$\sum F = 0: \qquad (V_j)_{n-1} + (V_i)_n = P_n \qquad \text{Equilibrium} \qquad \text{(G.6-1a)}$$

$$\sum M = 0: \qquad (M_j)_{n-1} + (M_i)_n = M_n \qquad \text{(G.6-1b)}$$

Element Force-Deformation Behavior: Shown in Fig. G.6-2a (see also Fig. 7.14a), is a typical Bernoulli-Euler beam element with linearly varying load, and in Fig. G.6-2b (also Fig. 7.14b) is shown the deflection curve of this element.

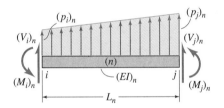

(a) The distributed loading and the stress resultants V and M acting on a typical beam element.

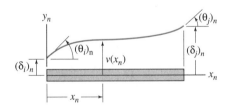

(b) The deflection curve, $v(x_n)$, of a typical beam element (n).

FIGURE G.6-2. A typical uniform beam element with transverse loading.

The *force-deformation relations* for the uniform Bernoulli-Euler beam element are given by Eqs. 7.19, repeated here as Eqs. G.6-2.

$$V_i = \frac{EI}{L^3}(12\delta_i + 6L\theta_i - 12\delta_j + 6L\theta_j) - \frac{7p_iL}{20} - \frac{3p_jL}{20}$$

$$M_i = \frac{EI}{L^2}(6\delta_i + 4L\theta_i - 6\delta_j + 2L\theta_j) - \frac{p_iL^2}{20} - \frac{p_jL^2}{30}$$

$$V_j = \frac{EI}{L^3}(-12\delta_i - 6L\theta_i + 12\delta_j - 6L\theta_j) - \frac{3p_iL}{20} - \frac{7p_jL}{20}$$

$$M_j = \frac{EI}{L^2}(6\delta_i + 2L\theta_i - 6\delta_j + 4L\theta_j) + \frac{p_iL^2}{30} + \frac{p_jL^2}{20}$$

Force-deformation equations with fixed-end forces (G.6-2)

All values in Eqs. G.6-2 are for a particular element, say element (n), so it is implied that $V_i \equiv (V_i)_n$, $EI \equiv (EI)_n$, etc.

Geometry of Deformation; Compatibility: It will be assumed that there are no "misfits" to be considered so that, where elements $(n-1)$ and (n) are connected together by node 𝐧, there is continuity of displacement and continuity of slope. That is,

Displ: $$(\delta_j)_{n-1} = (\delta_i)_n \equiv \delta_n \qquad \text{(G.6-3a)}$$

Compatibility

Slope: $$(\theta_j)_{n-1} = (\theta_i)_n \equiv \theta_n \qquad \text{(G.6-3b)}$$

Nodal Equilibrium Equations in Terms of Nodal Displacements: For each node 𝐧, we substitute, in *displacement-method* fashion, Eqs. G.6-3 into Eqs. G.6-2, and the results into Eqs. G.6-1 to obtain a set of *nodal equilibrium equations expressed in terms of the nodal displacements δ_n and nodal slopes θ_n.* (See, for instance, Examples 7.18 and 7.19 of Section 7.7.)

Deflection Curve: Once the nodal displacements have been determined by solving the nodal equilibrium equations for the unknown nodal displacements and slopes, the deflection curve can be formulated on an element-by-element basis using Eq. 7.20, repeated here as Eq. G.6-4.

$$\begin{aligned}v(x) = &\;\delta_i\left[1 - 3\left(\frac{x}{L}\right)^2 + 2\left(\frac{x}{L}\right)^3\right] + L\theta_i\left[\left(\frac{x}{L}\right) - 2\left(\frac{x}{L}\right)^2 + \left(\frac{x}{L}\right)^3\right] \\ &+ \delta_j\left[3\left(\frac{x}{L}\right)^2 - 2\left(\frac{x}{L}\right)^3\right] + L\theta_j\left[-\left(\frac{x}{L}\right)^2 + \left(\frac{x}{L}\right)^3\right] \\ &+ \frac{p_iL^4}{120EI}\left[2\left(1 - \frac{x}{L}\right)^2 - 3\left(1 - \frac{x}{L}\right)^3 + \left(1 - \frac{x}{L}\right)^5\right] \\ &+ \frac{p_jL^4}{120EI}\left[2\left(\frac{x}{L}\right)^2 - 3\left(\frac{x}{L}\right)^3 + \left(\frac{x}{L}\right)^5\right]\end{aligned} \qquad \text{(G.6-4)}$$

where, again, all values in Eqs. G.6-4 are for a particular element, say element (n), so it is implied that $x \equiv x_n$, $\delta_i \equiv (\delta_i)_n$, $EI \equiv (EI)_n$, and so forth.

Shear and Moment Diagrams: Once the nodal displacement and slope at each node have been obtained, Eqs. G.6-2 may be used to determine the shear force $(V_i)_n$ and the bending moment $(M_i)_n$ for each element (n). Then, expressions for the shear $V(x_n)$ and moment $M(x_n)$ may be employed to plot shear and moment diagrams on an element-by-element basis.

Writing equilibrium equations for the free-body diagram in Fig. G.6-3, we get the following expressions for the transverse shear force $V(x_n)$ and bending moment $M(x_n)$:

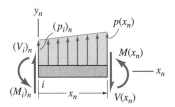

FIGURE G.6-3. A free-body diagram of a portion of beam element (n).

$$\sum F_y = 0: \quad V(x_n) = (V_i)_n + (p_iL)_n \left[\left(\frac{x_n}{L}\right) - \left(\frac{1}{2}\right)\left(\frac{x_n}{L}\right)^2 \right] \quad \text{(G.6-5a)}$$

$$+ (p_jL)_n \left(\frac{1}{2}\right)\left(\frac{x_n}{L}\right)^2$$

$$\sum M = 0: \quad M(x_n) = (V_i)_n x_n - (M_i)_n \quad \text{(G.6-5b)}$$

$$+ (p_iL^2)_n \left[\left(\frac{1}{2}\right)\left(\frac{x_n}{L}\right)^2 - \left(\frac{1}{6}\right)\left(\frac{x_n}{L}\right)^3 \right]$$

$$+ (p_jL^2)_n \left(\frac{1}{6}\right)\left(\frac{x_n}{L}\right)^3$$

The *MechSOLID* computer program option BEAMDEF employs the displacement method outlined above to solve linear beam deflection problems for straight-line arrangements of up to five beam elements.

■■■■■■■■■■■■■■□ E X A M P L E   G - 6 ■■■■■■■■■■■■■■

Use BEAMDEF to solve Example Problem 7.19. Let $L = 180$ in., $E = 30(10^3)$ ksi, $I = 75$ in⁴, and $w_0 = 80$ lb/in. (Neglect the weight of the beam.) The computed results and the diagrams for transverse deflection and shear are shown in the following BEAMDEF output.

Example G-6
BEAMDEF: Results
*** Displacement and Slope Values ***

Node	X	Deflection	Slope
1	0.00e+00	0.000e+00	0.000e+00
2	1.80e+02	0.000e+00	1.571e-03
3	3.60e+02	0.000e+00	-7.855e-04

*** Shear and Moment Values ***

Element	VI	MI	VJ	MJ
1	8.509e+03	2.945e+05	5.891e+03	-5.891e+04
2	3.273e+02	5.891e+04	-3.273e+02	-3.765e-15

	Max Value	X-Location	Min Value	X-Location
Deflection	5.442e-02	2.556e+02	-8.623e-02	1.008e+02
Shear	8.509e+03	0.000e+00	-5.747e+03	1.782e+02
Moment	1.580e+05	1.062e+02	-2.945e+05	0.000e+00

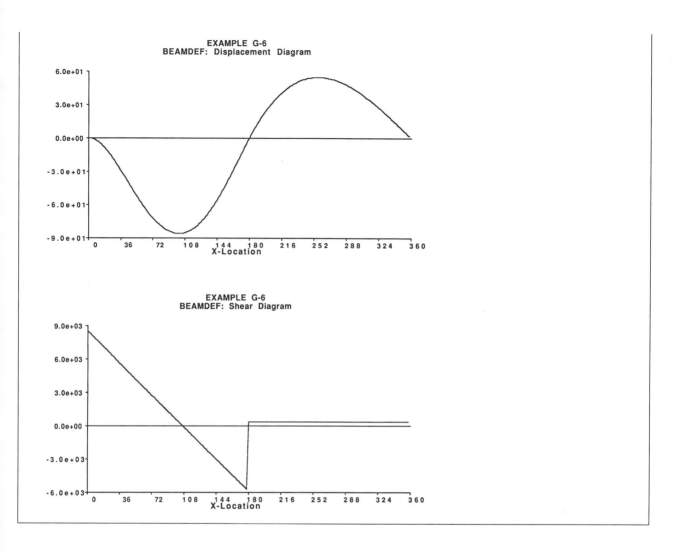

EXAMPLE G-6
BEAMDEF: Displacement Diagram

EXAMPLE G-6
BEAMDEF: Shear Diagram

G.7 MOHR—MOHR'S CIRCLE FOR STRESS; MOHR'S CIRCLE FOR STRAIN; STRAIN ROSETTES

This appendix describes the use of the computer program MOHR to solve stress-transformation and strain-transformation problems. The Mohr's-circle theory on which this appendix is based is covered in Sections 8.5, 8.10, and 8.12.

Mohr's Circle for Plane Stress. The notation for Mohr's circle for plane stress is given in Fig. G.7-1 (Fig. 8.16 repeated). The *stress-transformation equations* that relate the stresses σ_n and τ_{nt} on an arbitrarily oriented face to the stresses on the (x, y) faces — σ_x, σ_y, τ_{xy} — are (Eqs. 8.5 repeated):

$$\sigma = \left(\frac{\sigma_x + \sigma_y}{2}\right) + \left(\frac{\sigma_x - \sigma_y}{2}\right) \cos 2\theta + \tau_{xy} \sin 2\theta \qquad \text{(G.7-1a)}$$

$$\tau_{nt} = -\left(\frac{\sigma_x - \sigma_y}{2}\right) \sin 2\theta + \tau_{xy} \cos 2\theta \qquad \text{(G.7-1b)}$$

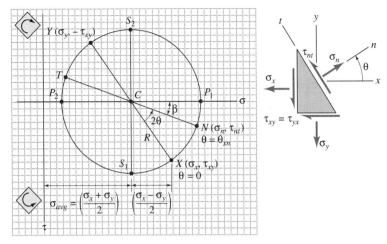

FIGURE G.7-1. The notation for Mohr's circle for plane stress.

Mohr's circle for stress is a circle of radius R_σ with its center at $(\sigma_{avg}, 0)$ in a (σ, τ) plane, as shown in Fig. G.7-1.[8]

$$\sigma_{avg} = \frac{\sigma_x + \sigma_y}{2} \tag{G.7-2a}$$

$$R_\sigma = \sqrt{\left(\frac{\sigma_x - \sigma_y}{2}\right)^2 + \tau_{xy}^2} \tag{G.7-2b}$$

The two *principal planes* correspond to points P_1 and P_2 on the circle, and the corresponding *principal stresses* are given by (Eqs. 8.13 repeated)

$$\sigma_1 = \sigma_{avg} + R_\sigma \tag{G.7-3a}$$

$$\sigma_2 = \sigma_{avg} - R_\sigma \tag{G.7-3b}$$

The *maximum (minimum) in-plane shear stress* acts on the plane that corresponds to point S_1 (S_2) on the circle. The magnitude of the maximum (minimum) shear stress is

$$\tau_{max} = R_\sigma \tag{G.7-4}$$

An example of the use of the MOHR option of **MechSOLID** to construct a Mohr's stress circle is given at the end of Appendix G.7.

Mohr's Circle for Strains in a Plane.

The notation for Mohr's circle for strains in a plane is given in Fig. G.7-2 (Fig. 8.30 repeated). The *strain-transformation equations* that relate the strains ϵ_n, ϵ_t, and γ_{nt} for arbitrarily oriented (n, t) axes to the strains ϵ_x, ϵ_y, and γ_{xy} for the (x, y) reference frame are given by (Eqs. 8.54 repeated):

---

[8]Note that the positive τ axis points <u>downward</u>; that the t-axis is 90° counterclockwise from the n axis; that the angle θ is measured counterclockwise from the x axis to the n axis; and that τ_{nt} points in the $+t$ direction on the $+n$ face.

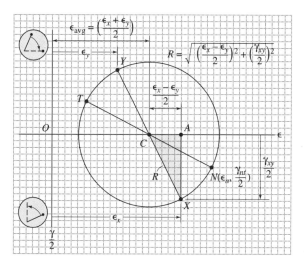

FIGURE G.7-2. The notation for Mohr's circle for strain.

$$\epsilon_n = \left(\frac{\epsilon_x + \epsilon_y}{2}\right) + \left(\frac{\epsilon_x - \epsilon_y}{2}\right)\cos 2\theta + \frac{\gamma_{xy}}{2}\sin 2\theta \qquad \text{(G.7-5a)}$$

$$\frac{\gamma_{nt}}{2} = -\left(\frac{\epsilon_x - \epsilon_y}{2}\right)\sin 2\theta + \frac{\gamma_{xy}}{2}\cos 2\theta \qquad \text{(G.7-5b)}$$

As shown in Section 8.10, *Mohr's circle for strain* is a circle of radius R_ϵ with its center at $(\epsilon_{\text{avg}}, 0)$, as shown in Fig. G.7-2, where

$$\epsilon_{\text{avg}} = \frac{\epsilon_x + \epsilon_y}{2} \qquad \text{(G.7-6a)}$$

$$R_\epsilon = \sqrt{\left(\frac{\epsilon_x - \epsilon_y}{2}\right)^2 + \left(\frac{\gamma_{xy}}{2}\right)^2} \qquad \text{(G.7-6b)}$$

The *in-plane principal strains* are given by

$$\epsilon_1 = \epsilon_{\text{avg}} + R_\epsilon \qquad \text{(G.7-7a)}$$

$$\epsilon_2 = \epsilon_{\text{avg}} - R_\epsilon \qquad \text{(G.7-7b)}$$

and the magnitude of the maximum (minimum) in-plane shear strain is

$$\gamma_{\text{max}} = 2R_\epsilon \qquad \text{(G.7-8)}$$

An example of the use of the MOHR option of *MechSOLID* to construct a Mohr's strain circle is given below.

Strain Rosette Analysis.
Figure G.7-3 (Fig. 8.36 repeated) gives the notation for a strain rosette with arbitrary angles.[9] Equations G.7-9 (Eqs. 8.60 repeated) may be used to

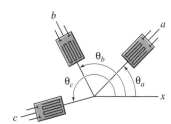

FIGURE G.7-3. The notation for a strain rosette with arbitrary angles.

[9]See Fig. 8.35*b* for photos of an *equiangular rosette* and a *rectangular rosette*.

take the three extensional strains $- \epsilon_a$, ϵ_b, and $\epsilon_c -$ measured by using a strain rosette; compute the corresponding strains $- \epsilon_x$, ϵ_y, and γ_{xy}; and plot a Mohr's circle representing the state of strain at the gage location.

$$\epsilon_a = \epsilon_x \cos^2 \theta_a + \epsilon_y \sin^2 \theta_a + \gamma_{xy} \sin \theta_a \cos \theta_a$$

$$\epsilon_b = \epsilon_x \cos^2 \theta_b + \epsilon_y \sin^2 \theta_b + \gamma_{xy} \sin \theta_b \cos \theta_b \qquad \text{(G.7-9)}$$

$$\epsilon_c = \epsilon_x \cos^2 \theta_c + \epsilon_y \sin^2 \theta_c + \gamma_{xy} \sin \theta_c \cos \theta_c$$

Example G-7 illustrates the use of the MOHR option of *MechSOLID* to process Eqs. G.7-9 and construct a Mohr's strain circle.

■■■■■■■■■■■■■■■□ E X A M P L E G - 7 □□□□□□□□□□□□□□■■■

Use the MOHR computer program option to construct a Mohr's strain circle for the rosette data given in Example Problem 8.7. The Mohr's circle below may be compared with Fig. 2 of Example Problem 8.7. The 45°-rosette strains are: $\epsilon_a = -700\mu$, $\epsilon_b = 0$, and $\epsilon_c = -100\mu$. Let the orientation of the "a" gage be along the x-axis.

MOHR: Results
Strain Rosette Results

The Principal Plane is θ = 63.43
The Principal Strains are ε1 = 1.000e+02 and ε2 = -9.000e+02
The max In-Plane Shear Strain is γmax = 1.000e+03
it occurs on the θ = 18.43 degree plane.

At the strain gage location:
εx = -7.0000e+02 εy = -1.0004e+02 γxy = 8.0007e+02

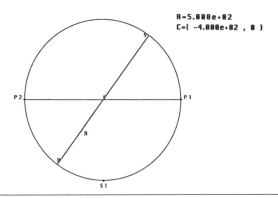

R=5.000e+02
C=(-4.000e+02 , 0)

ANSWERS TO SELECTED ODD-NUMBERED PROBLEMS

CHAPTER 1

1.4-1. (a) $C_x = 4$ kN, $C_y = 1$ kN, $E_x = -4$ kN, $E_y = 3$ kN;
  (b) $F_3 = \frac{8}{3}$ kN, $F_4 = \frac{5}{3}$ kN, $F_5 = -5$ kN

1.4-5. (a) $A_x = -P \sin 20°$, $A_y = \frac{1}{2}P \cos 20°$, $C_y = \frac{1}{2}P \cos 20°$;
  (b) $F_1 = 0.613P$, $F_2 = 0.543P$, $F_3 = -0.543P$

1.4-7. (a) $A_x = 0$, $A_y = \frac{4}{9}w_0L$, $D_y = \frac{2}{9}w_0L$;
  (b) $F_B = 0$, $V_B = \frac{1}{9}w_0L$, $M_B = \frac{5}{54}w_0L^2$

1.4-9. (a) $B_y = \frac{32}{3}$ kips, $C_x = 0$, $C_y = \frac{7}{3}$ kips;
  (b) $F_D = 0$, $V_D = \frac{35}{12}$ kips, $M_D = -\frac{5}{4}$ kip·ft

1.4-13. (a) $V_1(x) = -\frac{5}{3}x^2$ lb, $M_1(x) = -\frac{5}{9}x^3$ lb·in.;
  (b) $V_D = 25$ lb, $M_D = 75$ lb·in.

CHAPTER 2

2.2-3. (a) $F_R = 180$ kips, (b) $y_R = 5.6$ in.

2.2-5. (a) $F(x) = \frac{3}{4}\sigma_0 bh$, (b) $y_R = \frac{7}{18}h$

2.2-7. $\sigma_1 = 4.79$ ksi, $\sigma_2 = 4.89$ ksi, $\sigma_3 = 3.33$ ksi

2.2-13. $\sigma_{CD} = 140$ MPa

2.2-15. $\sigma_1 = 820$ psi, $\sigma_2 = -492$ psi

2.3-1. $\epsilon = 625\ \mu$

2.3-3. $\epsilon(\theta) = (3 + \sqrt{2} - (2 + \sqrt{2}) \cos\theta - \sqrt{2}\sin\theta)^{1/2} - 1$

2.3-7. $\epsilon_1 = 250\ \mu$, $\epsilon_2 = -1500\ \mu$

2.3-9. $\epsilon = 0.122$ in./in.

2.3-11. $\epsilon_1 = 0.0173$ in./in., $\epsilon_2 = -0.0176$ in./in.

2.3-15. (a) $d_i^* = 4.065$ in., (b) $\Delta L = 0.0454$ ft

2.3-19. $\theta = \tan^{-1}\left[\dfrac{L}{a}(\alpha_2 \Delta T_2 - \alpha_1 \Delta T_1)\right]$

2.4-1. $\epsilon = 2.00(10^{-3})$ in./in.

2.4-3. (b) $E = 30.0(10^3)$ ksi, (c) $\sigma_{YS} = 54$ ksi

2.6-1. (a) $E = 10.2(10^3)$ ksi, (b) $L^* = 2.01$ in.

2.6-5. $E = 3.40(10^3)$ ksi, $\nu = 0.20$

2.6-9. (a) $b = 0.915$ m, (b) $E = 200$ GPa, (c) $\nu = 0.33$

2.7-1. $d = 0.564$ in.

2.7-7. $\tau_{avg} = 7.55$ ksi

2.7-9. $\tau_{avg} = 112.5$ psi

2.7-11. (a) $\gamma_{xy} = 0.0125$ in./in.,
  (b) $\gamma_{x'y'} = -0.0125$ in./in.

2.7-13. (a) $\tau = 31.3$ kPa, (b) $\gamma = 0.0521$ m/m, (c) $\delta = 1.30$ mm

2.8-1. (a) $\theta_{na} = 63.4°$ (or $-63.4°$), (b) $\theta_{nb} = 26.6°$ (or $-26.6°$)

2.8-3. $\sigma_n = 115.4$ MPa, $\tau_{nt} = -66.6$ MPa; $\sigma_t = 38.5$ MPa, $\tau_{tn} = 66.6$ MPa

2.8-5. (a) $\sigma_n = 11.09$ ksi, (b) $\sigma_x = 3.99$ ksi, $\tau_{xy} = 5.32$ ksi

2.8-7. (a) $P = 19.81$ kips, (b) $|\tau_{nt}| = 3.90$ kips, (c) $\tau_{max} = 4.95$ ksi

2.8-9. $(b_{min})_\sigma = 20$ mm

2.9-1. (a) $\sigma_{xD} = 16$ ksi, (b) $F_x = 384$ kips,
(c) $M_z = -\frac{128}{3}$ kip·in.

2.9-3. (a) $F_x = 3\,b^2\sigma_0$, (b) $y_P = \frac{5}{6}b$

2.9-5. $\tau_{\max} = \dfrac{2T}{\pi r^3}$

2.9-9. $\epsilon_x = -0.01\,(y/a)^2$

2.9-11. $\epsilon_x = 0.10\,(y/a)$

2.9-13. $\epsilon_y = -0.01\,(x/a)^3$

2.9-15. $\epsilon_x = \dfrac{\partial u}{\partial x}$

2.9-17. $\epsilon_n = \dfrac{1}{L}\,(\delta_x \sin\theta + \delta_y \cos\theta)$

2.10-1. $\sigma_x = \dfrac{E}{1 - \nu^2}\,(\epsilon_x + \nu\epsilon_y)$,

$\sigma_y = \dfrac{E}{1 - \nu^2}\,(\epsilon_y + \nu\epsilon_x)$

2.10-3. $\epsilon_x = 214.3\mu$, $\epsilon_y = -33.3\mu$, $\epsilon_z = 28.6\mu$;
$\gamma_{xy} = 247.6\mu$, $\gamma_{xz} = 185.7\mu$, $\gamma_{yz} = 123.8\mu$

2.10-7. (a) $\sigma_y = -\nu\sigma_0$, (b) $\Delta a = \dfrac{a\sigma_0}{E}\,(\nu^2 - 1)$,

(c) $\Delta t = \dfrac{\nu t\sigma_0}{E}\,(1 + \nu)$

2.10-9. (a) $\epsilon_z = -171.4\,\mu$,
(b) $\sigma_x = 149.5$ MPa, $\sigma_y = -35.2$ MPa,
$\tau_{xy} = 15.4$ MPa,
(c) $e = 228(10^{-6})\dfrac{m^3}{m^3}$

2.12-1. (a) $t_{\min} = 0.447$ in., (b) $d_{\min} = 0.345$ in.

2.12-3. $P_{\text{allow}} = 2.22$ kips (shear in pin)

2.12-7. $d_w = \frac{1}{8}$ in., $d_p = \frac{1}{8}$ in.

2.12-9. $d_r = 0.638$ in., $d_p = 0.660$ in.

2.12-11. $w_{\text{allow}} = 2.72$ kips

2.12-15. $A_1 = 0.384$ in^2, $A_2 = 0.360$ in^2,
$A_3 = 0.800$ in^2

CHAPTER 3

3.2-1. (a) $\sigma_{x1} = 1.5$ ksi, $\sigma_{x2} = 4.5$ ksi,
(b) $y_P = 0.75$ in., (c) $e = 0.0075$ in.

3.2-3. (a) $\sigma_{xm} = 3.28$ ksi, $\sigma_{xb} = 7.56$ ksi,
(b) $e = 1.009(10^{-2})$ in.

3.2-5. (a) $\sigma_{xs} = -2.84$ ksi, $\sigma_{xc} = -0.341$ ksi,
(b) $e = -1.364(10^{-2})$ in.

3.2-9. (a) $e = \dfrac{\gamma L^2}{2E}$, (b) $P = \dfrac{\gamma AL}{2}$

3.2-15. $e = \dfrac{3PL}{bh(E_1 + 2E_2)}$

3.2-17. $u(x) = \dfrac{\gamma}{E}\left(Lx - \dfrac{x^2}{2}\right)$

3.5-1. $u_B = 2.44(10^{-2})$ in., $u_C = 1.171(10^{-2})$ in.

3.5-3. $u_A = 0.1312$ in., $u_B = 0.1056$ in.,
$u_C = 0.0576$ in.

3.5-7. (a) $d_2 = 20.4$ mm, (b) $u_C = 1.552$ mm

3.5-11. $L_1 = 3L_2$

3.5-13. (a) $u_B = \dfrac{2PL}{7EA}$, $u_C = \dfrac{-PL}{7EA}$;

(b) $\sigma_1 = \dfrac{P}{7A}$, $\sigma_2 = \dfrac{-3P}{7A}$, $\sigma_3 = \dfrac{P}{7A}$

3.5-19. (a) $d_b = 36.1$ mm, (b) $e = 0.377$ mm

3.5-23. (a) $\theta = 8.60(10^{-5})$ rad,
(b) $\sigma_1 = 6.45$ ksi, $\sigma_2 = 2.58$ ksi

3.5-25. (a) $\theta = 2.94(10^{-3})$ rad,
(b) $\sigma_1 = 102.9$ MPa, $\sigma_2 = 147.1$ MPa

3.5-27. (a) $u_A = 1.158(10^{-2})$ in.,
$u_D = 1.053(10^{-3})$ in.;
(b) $\sigma_1 = 5.79$ ksi, $\sigma_2 = 3.68$ ksi,
$\sigma_3 = 0.526$ ksi

3.6-1. $\sigma_{\max} = 6.50$ ksi, $\tau_{\max} = 3.25$ ksi

3.6-3. (a) $\sigma_1 = -13.40$ ksi, $\sigma_2 = -7.44$ ksi,
(b) $u = -2.13(10^{-3})$ in.

3.6-5. $\sigma_1 = \dfrac{-A_2 E_1 E_2\,\Delta T\,(\alpha_1 L_1 + \alpha_2 L_2)}{A_1 E_1 L_2 + A_2 E_2 L_1}$

$\sigma_2 = \dfrac{-A_1 E_1 E_2\,\Delta T\,(\alpha_1 L_1 + \alpha_2 L_2)}{A_1 E_1 L_2 + A_2 E_2 L_1}$

3.6-7. $\sigma_1 = -\left(\dfrac{2}{7}\right)E\alpha\Delta T$, $\sigma_2 = -\left(\dfrac{1}{7}\right)E\alpha\Delta T$,

$\sigma_3 = -\left(\dfrac{2}{7}\right)E\alpha\Delta T$

3.6-9. $\sigma_1 = 0.753$ ksi, $\sigma_2 = \sigma_3 = 2.34$ ksi

3.6-11. (a) $\theta = -8.85(10^{-5})$ rad,
    (b) $\sigma_1 = 6.63$ ksi, $\sigma_2 = 2.66$ ksi,
       $\sigma_3 = 10.29$ ksi

3.6-13. $\Delta T = 76.1°F$

3.6-15. (a) $\theta = 5.76\,(10^{-4})$ rad,
    (b) $\sigma_1 = 20.2$ MPa, $\sigma_2 = 28.8$ MPa,
       $\sigma_3 = 19.59$ MPa

3.6-19. (a) $F_1 = 5.49$ kips, $F_2 = -5.49$ kips,
    (b) $e = 3.37\,(10^{-2})$ in.

3.7-1. (a) $u_B = 2.43\,(10^{-2})$ in.,
    (b) $\sigma_1 = 14.59$ ksi, $\sigma_2 = 45.4$ ksi

3.7-3. (a) $u_B = 0.4\delta$,
    (b) $\sigma_1 = -\dfrac{3E\delta}{5L}$, $\sigma_2 = -\dfrac{2E\delta}{5L}$

3.7-5. $\sigma_1 = 38.2$ MPa, $\sigma_2 = 19.09$ MPa

3.7-7. $\sigma_1 = 38.0$ MPa, $\sigma_2 = -12.00$ MPa

3.7-11. $\sigma_1 = 92.6$ MPa, $\sigma_2 = 157.4$ MPa

3.7-15. $\sigma_1 = -2.53$ ksi, $\sigma_2 = 6.95$ ksi

3.7-19. $\sigma_1 = 9.85$ ksi, $\sigma_2 = 30.6$ ksi

3.8-3. $u_C = 5.22$ mm, $v_C = 4.57$ mm

3.8-5. $u_C = 0.1587$ mm, $v_C = 9.05$ mm

3.8-7. (a) $\sigma_1 = 9.04$ ksi, $\sigma_2 = -2.86$ ksi;
    (b) $u_C = 2.38\,(10^{-2})$ in.,
       $v_C = -4.13\,(10^{-2})$ in.

3.8-11. $\theta = 75.0°$

3.8-15. (a) $u_A = \dfrac{375}{887}\alpha_2 L\Delta T$, $v_A = \dfrac{500}{2661}\alpha_2 L\Delta T$;
    (b) $F_1 = F_3 = \dfrac{320}{887}AE\alpha_2\Delta T$,
       $F_2 = -\dfrac{512}{887}AE\alpha_2\Delta T$

3.8-17. (a) $u_A = 1.875$ mm, $v_A = -0.217$ mm;
    (b) $F_1 = 37.5$ kN, $F_2 = 22.5$ kN,
       $F_3 = -22.5$ kN

3.8-19. (a) $u_A = -2.05\,(10^{-3})$ in.,
       $v_A = -4.90\,(10^{-2})$ in.;
    (b) $F_1 = 9.78$ kips, $F_2 = 10.11$ kips,
       $F_3 = 0.853$ kips

3.8-21. (a) $u_A = 3.20\,(10^{-2})$ in.,
       $v_A = 2.84\,(10^{-3})$ in.;
    (b) $\sigma_1 = 6.07$ ksi, $\sigma_2 = -5.36$ ksi,
       $\sigma_3 = 7.51$ ksi

3.9-1. (a) $P_{Y1} = 15$ kips, (b) $u_C = 0.1333$ in.

3.9-3. (a) $P_Y = 43.2$ kips, $u_Y = 0.024$ in.;
    (b) $P_U = 57.6$ kips, $u_U = 0.048$ in.

3.9-5. (a) $P_Y = 145.7$ kips, $u_Y = 5.76\,(10^{-2})$ in.;
    (b) $P_U = 187.2$ kips, $u_U = 9.00\,(10^{-2})$ in.

CHAPTER 4

4.2-1. (a) $\tau_{max} = 8$ ksi, $\tau_{min} = 4.8$ ksi;
    (b) $T = 10.94$ kip·in.
    (c) $T = 10.94$ kip·in.,
    (d) $\%T_{add} = 14.89\%$ $\%W_{add} = 56.3\%$

4.2-3. (a) $\tau_{max} = 9.55$ ksi, (b) $\%T_{core} = 6.25\%$,
    (c) $\%W_{core} = 25.0\%$

4.2-5. $\%W_{red} = 77.6\%$

4.2-7. $T_{allow} = 7.64$ kip·in. (ϕ controls)

4.2-13. (a) $\tau_{max}(x) = \dfrac{8t_0}{\pi d^3}\left(L - 2x + \dfrac{x^2}{L}\right)$,
    (b) $\phi_B = \dfrac{16t_0 L^2}{3\pi G d^4}$

4.2-15. (a) $\tau_{max}(x) = \dfrac{32t_0 L}{\pi^2 d^3}\left[1 - \sin\left(\dfrac{\pi x}{2L}\right)\right]$,
    (b) $\phi_B = \dfrac{64t_0 L^2}{\pi^2 G d^4}\left(1 - \dfrac{2}{\pi}\right)$

4.4-1. (a) $\tau_{max} = (\tau_{max})_1 = 2.29$ ksi,
       $\phi_C = 4.24\,(10^{-4})$ rad

4.4-3. (a) $T_D = 1.386$ kip·in.,
    (b) $\tau_{max} = (\tau_{max})_2 = 1.443$ ksi

4.4-5. (a) $\tau_{max} = (\tau_{max})_1 = 4.51$ ksi,
       $\phi_A = -3.47\,(10^{-2})$ rad,
    (c) $k_t = 34.6\ \dfrac{\text{kip·in.}}{\text{rad}}$

4.4-7. (a) $\tau_{max} = (\tau_{max})_1 = 4.07$ ksi,
    (b) $\phi_{C/A} = -6.79\,(10^{-4})$ rad

4.4-9. $k_t = 104.7\ \dfrac{\text{kip·in.}}{\text{rad}}$

4.4-11. $k_t = 149.1 \dfrac{\text{kip}\cdot\text{in.}}{\text{rad}}$

4.4-17. (a) $\tau_{\max} = \dfrac{8t_0 L}{\pi d^3}$, (b) $\phi_A = \dfrac{12 t_0 L}{\pi G d^4}$

4.4-21. (a) $T_1 = -444$ lb·in., $T_2 = 3560$ lb·in.,
(b) $(\tau_{\max})_1 = (\tau_{\max})_2 = 2.26$ ksi,
(c) $\phi_B = 3.77\,(10^{-3})$ rad

4.4-23. (a) $T_1 = 6.52$ kN·m, $T_2 = -3.48$ kN·m;
(b) $(\tau_{\max})_1 = 164.0$ MPa, $(\tau_{\max})_2 = 82.0$ MPa;
(c) $\phi_B = 8.41\,(10^{-2})$ rad

4.4-25. (a) $(\tau_{\max})_1 = 6.27$ ksi, $(\tau_{\max})_2 = 3.92$ ksi;
(b) $(\tau_{\max})_1 = 8.58$ ksi, $(\tau_{\max})_2 = 1.610$ ksi

4.4-27. $L_2 = 1.2 L_1$

4.4-29. (a) $\phi_A = 0.1205$ rad,
(b) $(\tau_{\max})_1 = 11.29$ ksi, $(\tau_{\max})_2 = 6.69$ ksi

4.4-31. $t = 0.1362$ in.

4.4-33. (a) $T_A = T_D = 1.386$ kip·in.,
(b) $\tau_{\max} = (\tau_{\max})_2 = 1.443$ ksi,
(c) $\phi_C = -2.85\,(10^{-3})$ rad

4.5-1. (a) $T_{\max} = 314$ lb·in.
(b) $\tau_{\theta x}$ is shear stress on longitudinal planes parallel to the grain, along which the wood is weak in shear. But $\tau_{x\theta} = \tau_{\theta x}$. The wood would split along longitudinal surfaces.

4.5-3. $(\sigma_{\max})_T = 37.7$ MPa @ $\theta = -45°$,
$(\sigma_{\max})_C = -37.7$ MPa @ $\theta = 45°$,
$\tau_{\max} = 37.7$ MPa @ $0°$

4.5-7. (a) $T = 125.7$ N·m,
(b) $(\sigma_{\max})_T = 80.0$ MPa

4.5-9. (a) $T = 722$ N·m, (b) $(\sigma_{\max})_T = 84.0$ MPa

4.5-11. (a) $\gamma_{\max} = 1.479\,(10^{-3})$,
(b) $(\epsilon_{\max})_T = 7.39\,(10^{-4})$ rad

4.5-13. (a) $\tau_{\max} = 97.0$ MPa, (b) $G = 40.4$ GPa

4.6-1. $P_{\max} = 18.94$ hp

4.6-3. (a) $\tau_{\max} = 3.87$ ksi,
(b) $\phi = 2.58\,(10^{-3})$ rad

4.6-5. $d = 65$ mm

4.6-7. Select shaft C: $(\tau_{\max})_C = 18.57$ ksi

4.7-1. (a) $\tau_{\max} = 6.25$ ksi, (b) $J = 42.7$ in^4

4.7-3. (a) $\tau_{\max} = 2.94$ ksi, (b) $J = 2220$ in^4

4.7-7. (a) $\tau_{\max} = 17.64$ MPa,
(b) $J = 2.92\,(10^{-5})$ m^4

4.7-11. (a) $\tau_{\max}(\alpha) = \dfrac{2T\,(\alpha + 1)^2}{t L_m^2\, \alpha}$, (b) $\alpha = 1$

4.7-13. (a) $\dfrac{\tau_c}{\tau_s} = \dfrac{\pi}{4}$, (b) $\dfrac{J_c}{J_s} = \dfrac{16}{\pi^2}$

4.8-1. (a) $\tau_{\max} = 2.40$ ksi,
(b) $\phi = 6.45\,(10^{-3})$ rad, (c) $A = 0.816$ in^2

4.8-3. (a) $\tau_{\max} = 7.64$ ksi, (b) $\phi = 0.0208$ rad,
(c) $(\%\tau_{\max})_{\text{decr}} = 29.3\%$

4.8-5. (a) $\dfrac{(\tau_{\max})_{\text{rect}}}{(\tau_{\max})_{\text{ell}}} = 2.29$, (b) $\dfrac{(\phi/L)_{\text{rect}}}{(\phi/L)_{\text{ell}}} = 1.112$

4.8-7. (a) $\tau_{\max} = 19.02$ MPa, (b) $\phi = 0.1188$ rad

4.9-1. (b) $T = 126.1$ kip·in., (c) $\phi = 0.1536$ rad

4.9-3. (a) $T_Y = 3.04$ kN·m < 4 kN·m $< T_P = 4.06$ kN·m,
(b) $r_Y = 9.62$ mm, (c) $\phi = 0.332$ rad

4.9-5. (a) $T_Y = 0.684$ kip·in., $T_P = 0.821$ kip·in.,
(b) % Area yielded $= 34.7\%$,
(c) $\phi_P = 6.67\,(10^{-2})$ rad

4.9-7. $T = \dfrac{\pi \tau_Y}{6}\left(4 r_o{}^3 - \dfrac{3 r_i^4}{r_Y} - r_Y{}^3\right)$

CHAPTER 5

5.2-1. (a) $V_{B^-} = -P$, $M_{B^-} = -Pa$;
(b) $V_C = \dfrac{-P}{2}$, $M_C = \dfrac{-3Pa}{2}$

5.2-3. (a) $A_y = -84$ lb, $C_y = 44$ lb,
(b) $V_{B^+} = 16$ lb, $M_{B^+} = -672$ lb·in.,
(c) $V_{C^-} = 16$ lb, $M_{C^-} = -480$ lb·in.

5.2-5. $V_C = \dfrac{-M_0}{4a}$, $M_C = \dfrac{5M_0}{4}$

5.2-7. $V_A = 4$ kips, $M_A = -8$ kip·ft

5.2-9. $V_B = \dfrac{p_0 L}{12}$, $M_B = \dfrac{p_0 L^2}{36}$

5.2-11. (a) $a = \dfrac{\sqrt{2} - 1}{2} = 0.207$

(b) $|M_B| = M_C = |M_D| =$
$$\left(\frac{3}{8} - \frac{\sqrt{2}}{4}\right) wL^2 = 0.0214\ wL^2$$

5.2-13. $V_E = 31\ \text{lb},\ M_E = 123\ \text{lb} \cdot \text{in.}$

5.2-15. **(a)** $V_{B-} = 2\ \text{kips},\ M_{B-} = 8\ \text{kip} \cdot \text{ft},$
(b) $V_{B+} = -3\ \text{kips},\ M_{B+} = 18\ \text{kip} \cdot \text{ft}$

5.2-17. $V(x) = \dfrac{p_0 L}{6}\left[1 - 3\left(\dfrac{x}{L}\right)^2\right],$

$M(x) = \dfrac{p_0 L x}{6}\left[1 - \left(\dfrac{x}{L}\right)^2\right]$

5.2-19. **(a)** $V_1(x) = (14.33 - x)\ \text{kips},$
$M_1(x) = \left(14.33x - \dfrac{x^2}{2}\right)\ \text{kip} \cdot \text{ft},$
(b) $V_2(x) = (6.33 - x)\ \text{kips},$
$M_2(x) = \left(48 + 6.33x - \dfrac{x^2}{2}\right)\ \text{kip} \cdot \text{ft}$

5.2-21. **(a)** $V_1(x) = \left(46 - \dfrac{5x^2}{3}\right)\ \text{lb},$

$M_1(x) = \left(46x - \dfrac{5x^3}{9}\right)\ \text{lb} \cdot \text{in.},$

(b) $V_2(x) = -14\ \text{lb},$
$M_2(x) = (240 - 14x)\ \text{lb} \cdot \text{in.}$

5.2-23. **(a)** $V_1(x) = -1.5\ \text{kips},$
$M_1(x) = -0.75\ x^2\ \text{kip} \cdot \text{ft},$
(b) $V_2(x) = 8\ \text{kips},$
$M_2(x) = (8x - 112)\ \text{kip} \cdot \text{ft},$
(c) $V_C = 8\ \text{kips},\ M_C = 48\ \text{kip} \cdot \text{ft}$

5.2-25. **(a)** $W_g = 120\ \text{lb},$ **(b)** $p_0 = 20\ \text{lb/ft},$
(c) $V(x) = (20x - 40)\ \text{lb},$
$M(x) = (40 - 40x + 10x^2)\ \text{lb} \cdot \text{ft}$

5.2-27. **(a)** $A_x = 500\ \text{lb},\ A_y = -223\ \text{lb},\ D_y = 223\ \text{lb},$
(b) $F_2(x) = 200\ \text{lb},\ V_2(x) = 177\ \text{lb},$
$M_2(x) = (177x - 9600)\ \text{lb} \cdot \text{in.}$

5.2-29. **(a)** $A_x = -2.14\ \text{kN},\ A_y = -3.62\ \text{kN},$
$B = -4.8\ \text{kN}$
(b) $F_1(x) = 2.14\ \text{kN},\ V_1(x) = -3.62\ \text{kN},$
$M_1(x) = -3.62x\ \text{kN} \cdot \text{m}$

5.2-33. For $0 < x < \dfrac{2L}{3}$: $V_1(x) = A_y - \dfrac{p_0 x^2}{2L},\ M_1(x) =$

$A_y x - \dfrac{p_0 x^3}{6L},$ For $\dfrac{2L}{3} < x < L$: $V_2(x) =$

$$\dfrac{p_0 L}{2}\left[1 - \left(\dfrac{x}{L}\right)^2\right] - 2A_y,\ M_2(x) =$$
$$\dfrac{p_0 L^2}{6}\left[3\left(\dfrac{x}{L}\right) - \left(\dfrac{x}{L}\right)^3 - 2\right] +$$
$$2A_y(L - x)$$

5.4-1. $V_{max} = 1\ \text{kip},\ V_{min} = -1\ \text{kip},\ M_{max} = 0,$
$M_{min} = -3\ \text{kip} \cdot \text{ft}$

5.4-3. $V_{max} = 2\ \text{kN},\ V_{min} = 0,\ M_{max} = 1\ \text{kN} \cdot \text{m},$
$M_{min} = -3\ \text{kN} \cdot \text{m}$

5.4-5. $V_{max} = 0,\ V_{min} = -\dfrac{3p_0 L}{4},\ M_{max} = \dfrac{7p_0 L^2}{24},$

$M_{min} = 0$

5.4-7. $V_{max} = \dfrac{2p_0 L}{3},\ V_{min} = -\dfrac{5p_0 L}{6},$

$M_{max} = 0.1881\ p_0 L^2,\ M_{min} = 0$

5.4-9. $V_{max} = 60\ \text{lb},\ V_{min} = -84\ \text{lb},\ M_{max} = 0,$
$M_{min} = -672\ \text{lb} \cdot \text{in.}$

5.4-11. $V_{max} = 0,\ V_{min} = -\dfrac{M_0}{4a},\ M_{max} = \dfrac{3M_0}{2},$

$M_{min} = -\dfrac{M_0}{2}$

5.4-13. $V_{max} = 0,\ V_{min} = -5P,\ M_{max} = 0,$
$M_{min} = -9Pa$

5.4-15. $V_{max} = \dfrac{43}{3}\ \text{kips},\ V_{min} = -\dfrac{35}{3}\ \text{kips},$
$M_{max} = \dfrac{1225}{18}\ \text{kip} \cdot \text{ft},\ M_{min} = 0$

5.4-17. $V_{max} = 46\ \text{lb},\ V_{min} = -64\ \text{lb},$
$M_{max} = 161.1\ \text{lb} \cdot \text{in.},\ M_{min} = 0$

5.4-19. $V_{max} = \dfrac{p_0 L}{6},\ V_{min} = -\dfrac{p_0 L}{3},$

$M_{max} = \dfrac{p_0 L^2}{9\sqrt{3}},\ M_{min} = 0$

5.4-21. $V_{max} = 8\ \text{kips},\ V_{min} = -12\ \text{kips},$
$M_{max} = 48\ \text{kip} \cdot \text{ft},\ M_{min} = -48\ \text{kip} \cdot \text{ft}$

Answers are provided only for Part (b) of the following problems from Sect. 5.5.

5.5-1. $P(x) = -P\langle x \rangle^{-1} + \dfrac{P}{2}\langle x - a \rangle^{-1} + \dfrac{5P}{2}\langle x - 3a \rangle^{-1} - 2P\langle x - 4a \rangle^{-1}$

5.5-3. $p(x) = 28\ \text{kN}\langle x \rangle^{-1} - 10\ \text{kN}\ \langle x - 1\text{m} \rangle^{-1} - 20\ \text{kN}\ \langle x - 2\text{m} \rangle^{-1} - 20\ \text{kN}\ \langle x - 3\text{m} \rangle^{-1} + 22\ \text{kN}\ \langle x - 5\text{m} \rangle^{-1}$

5.5-5. $p(x) = \dfrac{p_0 L}{4}\langle x \rangle^{-1} - p_0 [\langle x \rangle^0 - \langle x - L \rangle^0] + \dfrac{3p_0 L}{4}\left\langle x - \dfrac{2L}{3}\right\rangle^{-1}$

5.5-7. $p(x) = 0.5\,\text{kN}\,\langle x \rangle^{-1} - 4\,\text{kN}\cdot\text{m}\,\langle x - 2\text{m}\rangle^{-2} + 1.5\,\text{kN}\,\langle x - 4\text{m}\rangle^{-1} - 1\,\text{kN/m}\,[\langle x - 4\text{m}\rangle^0 - \langle x - 6\text{m}\rangle^0]$

5.5-9. $p(x) = 46\,\text{lb}\,\langle x \rangle^{-1} - \frac{10}{3}\,\text{psi}\,[\langle x^1 \rangle - \langle x - 6\,\text{in.}\rangle^1] + 20\,\text{lb/in.}\,\langle x - 6\,\text{in.}\rangle^0 - 50\,\text{lb}\,\langle x - 8\,\text{in.}\rangle^{-1} + 64\,\text{lb}\,\langle x - 10\,\text{in.}\rangle^{-1}$

5.5-11. $p(x) = \dfrac{7p_0 L}{24}\langle x \rangle^{-2} - \dfrac{3p_0 L}{4}\langle x \rangle^{-1} + p_0\langle x \rangle^0 - \dfrac{2p_0}{L}\left[\left\langle x - \dfrac{L}{2}\right\rangle^1 - \langle x - L \rangle^1\right]$

CHAPTER 6

6.2-1. (a) $\epsilon_{max} = 1.041(10^{-3})\dfrac{\text{in.}}{\text{in.}}$,

(b) $\epsilon_{max} = 5.21(10^{-4})\dfrac{\text{in.}}{\text{in.}}$

6.2-3. (a) $\rho = 32.07$ in.,

(b) $\epsilon_{max} = 1.949(10^{-3})\dfrac{\text{in.}}{\text{in.}}$

6.2-5. $\kappa' = \nu\kappa$

6.3-1. $\sigma_{max} = \dfrac{Eh}{2r + h}$

6.3-3. $h_{max} = 0.0497$ in.

6.3-5. (a) $(\sigma_{max})_T = 180$ MPa, (b) $\rho = 58.3$ m, (c) $\delta_{max} = 137.1$ mm

6.3-7. (a) $(\sigma_{max})_T = 1.208$ ksi, (b) $F_{top} = -7.56$ kips $= 7.56$ kips (C)

6.3-9. (a) $(M_z)_{max} = 1.868(10^3)$ kip·in, (b) $(M_y)_{max} = 644$ kip·in.

6.3-11. $\sigma_A = -2.16$ ksi, $\sigma_B = 4.20$ ksi

6.3-13. (a) $\sigma_A = -1700$ psi, (b) $\sigma_C = 1400$ psi, (c) $F_{top} = -11.55$ kips

6.3-17. $(\sigma_{max})_C = 1.864$ ksi

6.3-19. (a) $(\sigma_{maxT})_{B^-} = 10.50$ ksi, (b) $(\sigma_{maxC})_C = -9.96$ ksi

6.3-21. $(\sigma_{maxT})_A = 787$ psi, $(\sigma_{maxC})_A = -2910$ psi

6.3-27. $\sigma_{max} = 24.7$ MPa

6.3-29. $\sigma_{max} = 28.3$ MPa

6.3-31. $\sigma_{max} = 17.91$ MPa

6.4-1. $S_{min} = 38.4$ in^3, select a **W**12×35 beam

6.4-3. $S_{min} = 48.0$ in^3, select a **W**10×45 beam

6.4-5. $S_{min} = 66.5$ in^3, select a 6×10 beam

6.4-7. $S_{min} = 28.8$ in^3, select a **W**14×26 beam

6.4-9. $S_{min} = 28.8$ in^3, select an **L**$8\times8\times1$ angle

6.4-11. $b = \left(\dfrac{\sqrt{3}}{6}\right)d$, $h = \left(\dfrac{\sqrt{6}}{3}\right)d$

6.5-1. $(\sigma_s)_{max} = 10.62$ ksi, $(\sigma_w)_{max} = 0.531$ ksi

6.5-3. $(\sigma_s)_{max} = 8.52$ ksi, $(\sigma_w)_{max} = 0.395$ ksi

6.5-5. $P_{allow} = 28.9$ kN

6.5-7. (a) $(\sigma_c)_{max} = 65.6$ MPa, $(\sigma_n)_{max} = -87.3$ MPa, (b) $\rho = 4.15$ m

6.6-1. $\sigma_A = 955$ psi, $\sigma_B = 256$ psi, $\sigma_D = -955$ psi, $\sigma_E = -256$ psi

6.6-3. (a) $\beta = 2.52°$, (b) $\sigma_{max} = \sigma_A = 9.64$ ksi

6.6-5. $\sigma_{max} = 2.69$ ksi (**W**12×50), $\sigma_{max} = 3.86$ ksi (**S**12×50)

6.6-7. (a) $\beta = 7.64°$, (b) $\sigma_{max} = 5.86$ ksi (2.5° misalignment)

6.6-9. $M_{max} = 29.2$ kip·in.

6.6-11. (a) $\beta = -75.5°$, (b) $\sigma_{max} = 6.18$ ksi

6.6-15. See problem statement.

6.7-1. (a) $M_Y = 24$ kip·in., $M_P = 36$ kip·in., $f = 1.5$

6.7-3. $M_Y = 2250$ kip·in., $M_P = 2750$ kip·in., $f = 1.218$

6.7-5. $M_Y = 2740$ kip·in., $M_P = 3070$ kip·in., $f = 1.120$

6.7-11. $M_Y = \dfrac{\sigma_Y \pi d^3}{32}$, $M_P = \dfrac{\sigma_Y d^3}{6}$, $f = 1.697$

6.7-13. $M_F = 2890$ kip·in.

6.7-17. $\sigma_x(y = 0^-) = \sigma_Y$, $\sigma_x(y = d/2) =$
$$\left(\frac{16 - 3\pi}{3\pi}\right)\sigma_Y$$

6.8-1. $V_{max} = 13.9$ kips

6.8-3. $L_{max} = \dfrac{3\pi d^2 \tau_{allow}}{8w}$

6.8-5. $\sigma_{max} = 10.67$ MPa, $\tau_{max} = 400$ kPa

6.8-7. $\sigma(0, 30\text{ mm}) = 43.2$ MPa,
$\tau(0, 30\text{ mm}) = 2.22$ MPa

6.8-9. $P_{max} = P_\sigma = 360$ N

6.10-1. $\tau_A = 1.073$ ksi, $\tau_B = 2.07$ ksi

6.10-3. $\tau_a/\tau_b = 1.34$

6.10-5. $\tau_w = \dfrac{V(25t^2 - 4y^2)}{68t^4}$

6.10-7. $\tau_A = 7.23$ MPa, $\tau_B = 12.39$ MPa

6.10-9. $(\tau_{max})_a = 2.81(\tau_{max})_b$

6.10-11. (a) $\sigma_{max} = 4.21$ ksi, (b) $\tau_{max} = 2.57$ ksi

6.10-13. (a) $\sigma_{max} = 17.35$ ksi, (b) $\tau_{max} = 1.048$ ksi

6.10-15. $(w_0)_{max} = 128.0$ lb/ft

6.10-17. $P_{max} = 8.28$ kN

6.11-1. $(\tau_{glue})_{max} = 213$ psi

6.11-3. $V_{allow} = 360$ kips

6.11-5. (a) $I = 2200$ in^4, $S = 180.0$ in^3,
(b) $V_{max} = 325$ kips

6.11-9. $(\Delta x)_{max} = 105.1$ mm

6.11-11. $(\Delta x)_{max} = 70.9$ mm

6.11-15. $V_s = 839$ lb

(Note: As in Fig. 4 of Example Prob. 6.19, the solutions to the homework problems for Section 6.12 assume that flange shear continues to the midpoint of the web-flange junction.)

6.12-1. $e = 0.953$ in.

6.12-3. $e = 0.869$ in. (Table D.4 gives a value for eccentricity of the **C**12×30 channel section as

$e = 0.929$ in. (from web centerline). This value takes into account the nonrectangular flange shape.)

6.12-7. $e = \dfrac{1544}{549}a$

CHAPTER 7

7.3-1. (a) $v(x) = \dfrac{M_0 L^2}{6EI}\left[\left(\dfrac{x}{L}\right) - \left(\dfrac{x}{L}\right)^3\right]$,

(b) $\theta_A = \dfrac{M_0 L}{6EI}$,

(c) $\delta = v\left(\dfrac{L\sqrt{3}}{3}\right) = \dfrac{\sqrt{3}}{27}\left(\dfrac{M_0 L^2}{EI}\right)$

7.3-3. (a) $M_A = 1237$ kip·in.,
(b) $\sigma_{max} = 19.13$ ksi

7.3-5. (a) $v(x) = \dfrac{w_0 L^4}{360EI}\left[-7\left(\dfrac{x}{L}\right) + 10\left(\dfrac{x}{L}\right)^3 - 3\left(\dfrac{x}{L}\right)^5\right]$,

(b) $\theta_A = -\dfrac{7w_0 L^3}{360EI}$,

(c) $\delta = 0.00652\left(\dfrac{w_0 L^4}{EI}\right)$

7.3-7. (a) $v_a(x) = \dfrac{w_0 L^4}{5760EI}\left[-37\left(\dfrac{x}{L}\right) + 40\left(\dfrac{x}{L}\right)^3\right]$,

$v_b(x) = \dfrac{w_0 L^4}{5760EI}\left[-37\left(\dfrac{x}{L}\right) + 40\left(\dfrac{x}{L}\right)^3 - 48\left(\dfrac{2x - L}{L}\right)^5\right]$,

(b) $\delta_B = \dfrac{3w_0 L^4}{1280EI}$

7.3-11. (a) $v_a(x) = \dfrac{PL^3}{3EI}\left[\left(\dfrac{x}{L}\right) - \left(\dfrac{x}{L}\right)^3\right]$, $v_b(x) = $
$\dfrac{PL^3}{6EI}\left[-3 + 11\left(\dfrac{x}{L}\right) - 9\left(\dfrac{x}{L}\right)^2 + \left(\dfrac{x}{L}\right)^3\right]$,

(b) $\delta_C = \dfrac{4PL^3}{EI}$, (c) $\delta = \dfrac{2\sqrt{3}}{27}\left(\dfrac{PL^3}{EI}\right)$

7.3-13. (a) $v(x) = \dfrac{w_0 L^4}{120 EI}\left[1 - 5\left(\dfrac{x}{L}\right) - \left(\dfrac{L-x}{L}\right)^5\right]$,

(b) $\theta_B = \dfrac{w_0 L^3}{24 EI}$, $\delta_B = \dfrac{w_0 L^4}{30 EI}$

7.3-19. $\delta = 0.307$ in.

7.3-21. (a) $\delta = 0.1420$ in., (b) $\sigma_{max} = 10.17$ ksi

7.3-23. (a) $\delta = 0.1537$ in., (b) $\sigma_{max} = 16.02$ ksi

7.3-25. (a) $v(x) = \dfrac{M_0 L^2}{2EI}\left[\left(\dfrac{x}{L}\right)^2 - \left(\dfrac{x}{L}\right)^3\right]$,

(b) $\theta_A = 0$, (c) $\delta = \dfrac{2 M_0 L^2}{27 EI}$

7.3-27. (a) $v(x) = \dfrac{-w_0}{EI}\left(\dfrac{L}{\pi}\right)^4 \sin\left(\dfrac{\pi x}{L}\right)$,

(b) $\theta_A = \dfrac{-w_0}{EI}\left(\dfrac{L}{\pi}\right)^3$

7.3-29. (a) $v_a(x) = \dfrac{w_0 L^4}{2880 EI}\left[-41\left(\dfrac{x}{L}\right) + 80\left(\dfrac{x}{L}\right)^3 - 48\left(\dfrac{x}{L}\right)^5\right]$,

$v_b(x) = \dfrac{w_0 L^4}{1440 EI}\left[3 - 43\left(\dfrac{x}{L}\right) + 60\left(\dfrac{x}{L}\right)^2 - 20\left(\dfrac{x}{L}\right)^3\right]$,

(b) $\delta_B = |v_a(L/2)| = \dfrac{w_0 L^4}{240 EI}$

7.3-33. See Prob. 7.3-13.

7.4-1. (a) $R_A = \dfrac{3 w_0 L}{8}$, $R_B = \dfrac{5 w_0 L}{8}$, $M_B = \dfrac{-w_0 L}{8}$,

(b) $v(x) = \dfrac{w_0 L^4}{48 EI}\left[-\left(\dfrac{x}{L}\right) + 3\left(\dfrac{x}{L}\right)^3 - 2\left(\dfrac{x}{L}\right)^4\right]$

7.4-3. (a) $R_A = \dfrac{w_0 L}{10}$, $R_B = \dfrac{2 w_0 L}{5}$, $M_B = -\dfrac{w_0 L^2}{15}$,

(b) $v(x) = \dfrac{w_0 L^4}{120 EI}\left[-\left(\dfrac{x}{L}\right) + \right.$

$2\left(\dfrac{x}{L}\right)^3 - \left(\dfrac{x}{L}\right)^5$,

(c) $\delta = \dfrac{2\sqrt{5}\, w_0 L^4}{1875 EI}$

7.4-5. (a) $R_A = \dfrac{3 w_0 L}{20}$, $M_A = -\dfrac{w_0 L^2}{30}$

$R_B = \dfrac{7 w_0 L}{20}$, $M_B = \dfrac{-w_0 L^2}{20}$,

$v(x) = \dfrac{w_0 L^4}{120 EI}\left[-2\left(\dfrac{x}{L}\right)^2 + 3\left(\dfrac{x}{L}\right)^3 - \left(\dfrac{x}{L}\right)^5\right]$,

(b) $\delta = 1.309\,(10^{-3})\dfrac{w_0 L^4}{EI}$

7.4-7. (a) $R_A = \dfrac{P}{2}\left[2 - 3\left(\dfrac{a}{L}\right) + \left(\dfrac{a}{L}\right)^3\right]$,

$R_C = \dfrac{P}{2}\left[3\left(\dfrac{a}{L}\right) - \left(\dfrac{a}{L}\right)^3\right]$,

$M_C = \dfrac{PL}{2}\left[-\left(\dfrac{a}{L}\right) + \left(\dfrac{a}{L}\right)^3\right]$,

(b) $v_a(x) = \dfrac{P x^3}{12 EI}\left[2 - 3\left(\dfrac{a}{L}\right) + \left(\dfrac{a}{L}\right)^3\right] + \dfrac{PL^2 x}{4 EI}\left[-\left(\dfrac{a}{L}\right) + 2\left(\dfrac{a}{L}\right)^2 - \left(\dfrac{a}{L}\right)^3\right]$,

$v_b(x) = \dfrac{P x^3}{12 EI}\left[-3\left(\dfrac{a}{L}\right) + \left(\dfrac{a}{L}\right)^3\right] + \dfrac{P a x^2}{2 EI} - \dfrac{PL^2 x}{4 EI}\left[\left(\dfrac{a}{L}\right) + \left(\dfrac{a}{L}\right)^3\right] + \dfrac{P a^3}{6 EI}$

7.4-11. (a) $R_A = \dfrac{-3 w_0 L}{80}$, $R_B = \dfrac{5 w_0 L}{16}$,

$R_C = \dfrac{9 w_0 L}{40}$, $M_A = \dfrac{w_0 L^2}{160}$,

(b) $v_a(x) = \dfrac{w_0 L^4}{320 EI}\left[\left(\dfrac{x}{L}\right)^2 - 2\left(\dfrac{x}{L}\right)^3\right]$,

$v_b(x) = \dfrac{w_0 L^4}{480 EI}\left[-2 + 14\left(\dfrac{x}{L}\right) - 33\left(\dfrac{x}{L}\right)^2 + 31\left(\dfrac{x}{L}\right)^3 - 10\left(\dfrac{x}{L}\right)^4\right]$

7.4-13. See Prob. 7.4-3.

7.4-17. $R_A = \dfrac{w_0 L}{24}$, $R_B = \dfrac{7w_0 L}{24}$, $M_B = \dfrac{-w_0 L^2}{24}$,

$v(x) = \dfrac{w_0 L^4}{720EI}\left[3\left(\dfrac{x}{L}\right) + 5\left(\dfrac{x}{L}\right)^3 - 2\left(\dfrac{x}{L}\right)^6\right]$

7.4-19. $R_A = \dfrac{48 p_0 L}{\pi^4}(\pi - 4)$,

$M_A = \dfrac{4 p_0 L^2}{\pi^4}[24 - 4\pi - \pi^2]$,

$R_B = \dfrac{2 p_0 L}{\pi^4}(96 - 24\pi - \pi^3)$,

$M_B = \dfrac{32 p_0 L^2}{\pi^4}(\pi - 3)$,

$v(x) = \dfrac{8 p_0 L^4}{\pi^4 EI}\left[-2 + 2\cos\left(\dfrac{\pi x}{2L}\right) + \right.$

$\left. (6 - \pi)\left(\dfrac{x}{L}\right)^2 + (\pi - 4)\left(\dfrac{x}{L}\right)^3\right]$

7.5-1. (a) $\theta(x) = \dfrac{p_0}{EI}\left[\dfrac{3L^2}{8}\langle x\rangle^1 - \dfrac{L}{4}\langle x\rangle^2 + \right.$

$\left. \dfrac{1}{6}\left(\left\langle x - \dfrac{L}{2}\right\rangle^3 - \langle x - L\rangle^3\right)\right]$, $v(x) = $

$\dfrac{p_0}{EI}\left[\dfrac{3L^2}{16}\langle x\rangle^2 - \dfrac{L}{12}\langle x\rangle^3 + \dfrac{1}{24}\left(\left\langle x - \right.\right.\right.$

$\left.\left.\left. \dfrac{L}{2}\right\rangle^4 - \langle x - L\rangle^4\right)\right]$,

(b) $\delta_B = \dfrac{7 p_0 L^4}{192EI}$, $\delta_C = \dfrac{41 p_0 L^4}{384EI}$

7.5-3. (a) $\theta(x) = \dfrac{1}{EI}\left[-70 \text{ kip} \cdot \text{ft } \langle x\rangle^1 + \right.$

$11 \text{ kips } \langle x\rangle^2 - \dfrac{2}{3}\dfrac{\text{kips}}{\text{ft}}\left(\langle x\rangle^3 - \langle x - \right.$

$\left.\left. 5 \text{ ft}\rangle^3\right) - 1 \text{ kip } \langle x - 10 \text{ ft}\rangle^2\right]$

$v(x) = \dfrac{1}{EI}\left[-35 \text{ kip} \cdot \text{ft } \langle x\rangle^2 + \right.$

$\dfrac{11}{3}\text{ kips } \langle x\rangle^3 - \dfrac{1}{6}\dfrac{\text{kips}}{\text{ft}}\left(\langle x\rangle^4 - \langle x - \right.$

$\left.\left. 5 \text{ ft}\rangle^4\right) - \dfrac{1}{3}\text{ kip } \langle x - 10 \text{ ft}\rangle^3\right]$

(b) $\delta_B = 0.0225$ in., $\delta_C = 0.0603$ in.

7.5-5. (a) $\theta(x) = \dfrac{w_0}{EI}\left[\dfrac{L}{12}\langle x\rangle^2 - \dfrac{1}{12L}\left(\langle x\rangle^4 - \right.\right.$

$\left\langle x - \dfrac{L}{2}\right\rangle^4\right) + \dfrac{1}{6}\left\langle x - \dfrac{L}{2}\right\rangle^3 + \dfrac{L}{24}\langle x - $

$\left. L\rangle^2 - \dfrac{41L^3}{2880}\langle x\rangle^0\right]$,

$v(x) = \dfrac{w_0}{EI}\left[\dfrac{L}{36}\langle x\rangle^3 - \dfrac{1}{60L}\left(\langle x\rangle^5 - \right.\right.$

$\left\langle x - \dfrac{L}{2}\right\rangle^5\right) + \dfrac{1}{24}\left\langle x - \dfrac{L}{2}\right\rangle^3 + $

$\left. \dfrac{L}{72}\langle x - L\rangle^3 - \dfrac{41L^3}{2880}\langle x\rangle^1\right]$,

(b) $v(L/4) = \dfrac{-193 w_0 L^4}{61440EI}$

7.5-11. (a) $R_A = \dfrac{w_0 L}{10}$, $R_B = \dfrac{2w_0 L}{5}$, $M_B = -\dfrac{w_0 L^2}{15}$,

(b) $\theta(x) = \dfrac{w_0}{EI}\left[\dfrac{L}{20}\langle x\rangle^2 - \right.$

$\left. \dfrac{1}{24L}\langle x\rangle^4 - \dfrac{L^3}{120}\langle x\rangle^0\right]$,

$v(x) = \dfrac{w_0}{EI}\left[\dfrac{L}{60}\langle x\rangle^3 - \right.$

$\left. \dfrac{1}{120L}\langle x\rangle^5 - \dfrac{L^3}{120}\langle x\rangle^1\right]$

7.5-13. (a) $R_A = \dfrac{w_0 L}{24}$, $R_B = \dfrac{7w_0 L}{24}$, $M_B = -\dfrac{w_0 L^2}{24}$,

(b) $\theta(x) = \dfrac{w_0}{EI}\left[\dfrac{L}{48}\langle x\rangle^2 - \dfrac{1}{60L^2}\langle x\rangle^5 + \right.$

$\left. \dfrac{7L}{48}\langle x - L\rangle^2 + \dfrac{L^2}{24}\langle x - L\rangle^1 - \dfrac{L^3}{240}\langle x\rangle^0\right]$,

$v(x) = \dfrac{w_0}{EI}\left[\dfrac{L}{144}\langle x\rangle^3 - \dfrac{1}{360L^2}\langle x\rangle^6 + \right.$

$\left. \dfrac{7L}{144}\langle x - L\rangle^3 + \dfrac{L^2}{48}\langle x - L\rangle^2 - \dfrac{L^3}{240}\langle x\rangle^1\right]$

7.5-15. (a) $R_A = -\dfrac{p_0 L}{28}$, $M_A = \dfrac{p_0 L^2}{105}$,

$R_B = -\dfrac{3 p_0 L}{14}$, $M_B = \dfrac{p_0 L^2}{42}$,

(b) $\theta(x) = \dfrac{p_0}{EI}\left[\dfrac{L^2}{105}\langle x\rangle^1 - \dfrac{L}{56}\langle x\rangle^2 + \right.$

$$\frac{1}{120L^3} \langle x \rangle^6 - \frac{3L}{28} \langle x - L \rangle^2 -$$

$$\frac{L^2}{42} \langle x - L \rangle^1 \Bigg],$$

$$v(x) = \frac{p_0}{EI} \Bigg[\frac{L^2}{210} \langle x \rangle^2 - \frac{L}{168} \langle x \rangle^3 +$$

$$\frac{1}{840L^3} \langle x \rangle^7 - \frac{L}{28} \langle x - L \rangle^3 -$$

$$\frac{L^2}{84} \langle x - L \rangle^2 \Bigg]$$

7.6-1. $\theta_A = \dfrac{3PL^2}{8EI}, \ \delta_A = \dfrac{7PL^3}{48EI}$

7.6-3. $\theta_C = \dfrac{P_0L^2}{4EI}, \ \delta_C = \dfrac{P_0L^3}{8EI}$

7.6-5. $\theta_C = \dfrac{7p_0L^3}{48EI}, \ \delta_C = \dfrac{41p_0L^4}{384EI}$

7.6-7. $\theta_A \equiv |v_A{}'(0)| = \dfrac{3PL^2}{32EI}, \ \delta_C \equiv |v(L/2)| = \dfrac{11PL^3}{384EI}$

7.6-9. $\theta_A = 0, \ \delta_B \equiv |v(L/2)| = \dfrac{PL^3}{192EI}$

7.6-15. $\theta_C \equiv |v'(2L)| = \dfrac{5PL^2}{6EI}, \ \delta_C \equiv |v(2L)| = \dfrac{2PL^2}{3EI}$

7.6-17. $\theta_B \equiv v'(L) = \dfrac{p_0}{EI} \left(\dfrac{L}{\pi} \right)^3 (\pi^2 - 8),$

$\qquad v_B \equiv v(L) = \dfrac{2p_0}{3EI} \left(\dfrac{L}{\pi} \right)^4 (\pi^3 - 24)$

7.6-19. $\theta_B \equiv |v'(L)| = \dfrac{13w_0L^3}{192EI},$

$\qquad \delta_B \equiv |v(L)| = \dfrac{7w_0L^4}{128EI}$

7.6-21. $\theta_A \equiv |v'(0)| = \dfrac{3w_0L^3}{128EI}, \ \delta_B \equiv v(L/2) = \dfrac{5w_0L^4}{768EI}$

7.6-25. $\theta_C \equiv |v'(10 \text{ ft})| = 6.60 \ (10^{-4}) \text{ rad},$
$\qquad \delta_C \equiv |v(10 \text{ ft})| = 0.0603 \text{ in.}$

7.6-27. $\delta_B \equiv |v(4 \text{ ft})| = 0.430 \text{ in.},$
$\qquad \delta_C \equiv |v(8 \text{ ft})| = 1.289 \text{ in.}$

7.6-29. $\theta_A \equiv |v'(0)| = 4.19 \ (10^{-3}) \text{ rad},$
$\qquad \delta_B \equiv |v(5 \text{ m})| = 14.97 \text{ mm}$

7.6-33. See Prob. 7.4-1.

7.6-35. (a) $R_A = 25$ kips, $M_A = -40$ kip·ft,
$\qquad R_B = 15$ kips

7.6-37. See Prob. 7.4-3.

7.6-39. (a) $R_A = -\dfrac{p_0L}{2}, \ M_A = -\dfrac{p_0L^2}{12},$

$\qquad v(x) = \dfrac{p_0L^4}{24EI} \left[\left(\dfrac{x}{L}\right)^2 - 2\left(\dfrac{x}{L}\right)^3 + \left(\dfrac{x}{L}\right)^4 \right]$

7.6-41. See Prob. 7.4-5.

7.6-49. (a) $R_A = \dfrac{7w_0L}{32}, \ R_B = \dfrac{5w_0L}{16}, \ R_C = \dfrac{-w_0L}{32},$

$\qquad$ **(b)** $v_a(x) = \dfrac{w_0L^4}{768EI} \left[-3\left(\dfrac{x}{L}\right) + \right.$

$\qquad \left. 28\left(\dfrac{x}{L}\right)^3 - 32\left(\dfrac{x}{L}\right)^4 \right],$

$\qquad v_b(x) = \dfrac{w_0L^4}{768EI} \left[-3 + 11\left(\dfrac{x}{L}\right) - \right.$

$\qquad \left. 12\left(\dfrac{x}{L}\right)^2 + 4\left(\dfrac{x}{L}\right)^3 \right]$

7.6-51. See Prob. 7.4-19.

7.6-53. $T = \dfrac{7w_0AL^4}{128(AL^3 + 3HI)}$

7.7-1. $R_A = \dfrac{3M_0}{2L}$

7.7-3. (a) $\theta_B = \dfrac{P_0L^2}{37EI}, \ \delta_B = \dfrac{8P_0L^3}{111EI},$

$\qquad$ **(b)** $R_A = -\dfrac{26P_0}{37}, \ M_A = -\dfrac{14P_0L}{37}$

7.7-5. (a) $\theta_B = \dfrac{w_0L^3}{96EI},$

$\qquad$ **(b)** $R_A = \dfrac{9w_0L}{16}, \ M_A = \dfrac{5w_0L^2}{48}, \ R_B = \dfrac{w_0L}{2}$

7.7-7. (a) $\theta_B = -\dfrac{w_0L^3}{24EI},$

$\qquad$ **(b)** $R_A = \dfrac{w_0L}{4}, \ M_A = 0, \ R_B = \dfrac{27w_0L}{16}$

7.7-9. (a) $\theta_B = -\dfrac{p_0L^3}{240EI}, \ \delta_B = \dfrac{17p_0L^4}{480EI},$

$\qquad$ **(b)** $R_A = -\dfrac{19p_0L}{20}, \ M_A = -\dfrac{73p_0L^2}{240}$

7.7-15. (a) $\theta_A = \dfrac{w_0 L^3}{96EI}$, $\theta_B = -\dfrac{w_0 L^3}{16EI}$, $\theta_C = \dfrac{11 w_0 L^3}{96EI}$,

(b) $R_A = \dfrac{3 w_0 L}{16}$

7.7-17. $P_A = \dfrac{12 EI \delta_A}{L^3}$, $M_A = \dfrac{6 EI \delta_A}{L^2}$, $P_B = -P_A$,

$M_B = M_A$, $v(x) = \delta_A \left[1 - 3\left(\dfrac{x}{L}\right)^2 + 2\left(\dfrac{x}{L}\right)^3 \right]$

CHAPTER 8

8.3-1. $\sigma_n = 35.4$ ksi, $\tau_{nt} = 10.42$ ksi

8.3-3. $\sigma_n = -1150$ psi, $\tau_{nt} = -6400$ psi

8.3-5. $\sigma_{x'} = 35.4$ ksi, $\sigma_{y'} = 0.240$ ksi, $\tau_{x'y'} = 10.42$ ksi

8.3-7. $\sigma_{x'} = 41.9$ MPa, $\sigma_{y'} = -25.9$ ksi, $\tau_{x'y'} = -26.6$ ksi

8.3-9. $\sigma_{x'} = -1150$ psi, $\sigma_{y'} = -850$ psi, $\tau_{x'y'} = -6400$ psi

8.3-11. $\sigma_x = 4000$ psi, $\tau_y = -1000$ psi, $\tau_{xy} = 0$

8.3-13. $\theta = 45°$, $\sigma_{y'} = 0$

8.3-15. $P = 12$ kN

8.3-19. $\sigma_0 = 150.1$ psi

8.4-1. (a) $\sigma_1 = 5400$ psi, $\sigma_2 = -7400$ psi, $\theta_{p1} = -25.7°$,
(b) $\tau_{max} = 6400$ psi, $\sigma_{s1} = -1000$ psi, $\theta_{s1} = -70.7°$

8.4-3. (a) $\sigma_1 = 27.0$ ksi, $\sigma_2 = -3.0$ ksi, $\theta_{p1} = -63.4°$,
(b) $\tau_{max} = 15.0$ ksi, $\sigma_{s1} = 12.0$ ksi, $\theta_{s1} = 71.6°$

8.4-5. (a) $\sigma_1 = 14.0$ ksi, $\sigma_2 = -6.0$ ksi, $\theta_{p1} = 71.6°$,
(b) $\tau_{max} = 10.0$ ksi, $\sigma_{s1} = 4.0$ ksi, $\theta_{s1} = 26.6°$

8.4-7. (a) $\sigma_1 = 38.2$ ksi, $\sigma_2 = -2.6$ ksi, $\theta_{p1} = 75.3°$,
(b) $\tau_{max} = 20.4$ ksi, $\sigma_{s1} = 17.8$ ksi, $\theta_{s1} = 30.3°$

8.4-9. (a) $\sigma_1 = 14.9$ ksi, $\sigma_2 = -2.9$ ksi, $\theta_{p1} = 31.7°$,

(b) $\tau_{max} = 8.9$ ksi, $\tau_{s1} = 6.0$ ksi, $\theta_{s1} = -13.3°$

8.4-13. $\sigma_{p1} = 35$ MPa, $\sigma_{p2} = -65$ MPa, $\tau = 43.3$ MPa

8.4-15. $\sigma = 130$ MPa, $\tau = 72$ MPa, $\sigma_{p1} = 178$ MPa

8.5-1. See Prob. 8.3-1.

8.5-3. See Prob. 8.3-3.

8.5-5. See Prob. 8.3-5.

8.5-7. See Prob. 8.3-7.

8.5-9. See Prob. 8.3-9.

8.5-13. See Prob. 8.4-1.

8.5-15. See Prob. 8.4-3.

8.5-17. (a) $\sigma_1 = 18$ MPa, $\sigma_2 = -8$ MPa, $\theta_{p1} = 11.3°$,
(b) $\tau_{max} = 13$ MPa, $\theta_{s1} = -33.7°$

8.5-19. (a) $\sigma_1 = 12.36$ MPa, $\sigma_2 = -32.4$ MPa, $\theta_{p1} = 76.7°$,
(b) $\tau_{max} = 22.4$ MPa, $\theta_{s1} = 31.7°$

8.5-23. See Prob. 8.3-13

8.5-29. See Prob. 8.4-13

8.5-31. See Prob. 8.4-15

8.6-5. $\sigma_1 = 16.06$ MPa, $\sigma_2 = -56.1$ MPa, $\sigma_3 = \sigma_z = 20$ MPa

8.6-7. $\sigma_1 = 96.6$ MPa, $\sigma_2 = -16.57$ MPa, $\sigma_3 = \sigma_z = -40$MPa

8.7-1. (b) $\tau_{abs\,max} = 6.5$ ksi

8.7-3. (b) $\tau_{abs\,max} = 6.83$ ksi

8.7-5. (b) $\tau_{abs\,max} = 50$ MPa

8.7-7. (b) $\sigma_1 = 17.66$ ksi, $\sigma_2 = 6.34$ ksi, $\sigma_3 = -2$ ksi,
(c) $\tau_{abs\,max} = 9.83$ ksi

8.9-7. (a) $\epsilon_x = -125\mu$, $\epsilon_y = 37.5\mu$,
(b) $\epsilon_n = \epsilon_t = -43.75\mu$, $\gamma_{nt} = 162.5\mu$

8.9-9. (a) $\epsilon_x = 100\mu$, $\epsilon_y = 500\mu$, (b) $\gamma_{nt} = 346\mu$,
(c) $\nu = 1/3$

8.10-1. (a) $\epsilon_x = 160\mu$, $\epsilon_y = -48\mu$,
(b) $\epsilon_n = \epsilon_t = 56\mu$, (c) $\gamma_{nt} = 208\mu$

8.10-3. $E = 10(10^3)$ ksi, $\nu = 1/3$

8.10-7. (a) $\epsilon_n = 607\mu$, $\epsilon_t = 443\mu$, $\gamma_{nt} = 615\mu$

8.10-9. (a) $\epsilon_n = 85\mu$, $\epsilon_t = -35\mu$, $\gamma_{nt} = -50\mu$

8.10-13. (a) $\epsilon_1 = 50\mu$, $\epsilon_2 = -450\mu$, $\theta_{p1} = 18.43°$,
(b) $\gamma_{max} = 500\mu$

8.10-15. (a) $\epsilon_1\ 230\mu$, $\epsilon_2 = -30\mu$, $\theta_{p1} = 78.7°$,
(b) $\gamma_{max} = 260\mu$

8.10-19. $\epsilon_x = 4000\mu$, $\epsilon_y = 0$, $\gamma_{xy} = 0$. Principal axes are the same for stress and strain.

8.12-3. (a) $\epsilon_x = 175\mu$, $\epsilon_y = -105\mu$, $\gamma_{xy} = 210\mu$
(b) $\epsilon_1 = 210\mu$, $\epsilon_2 = -140\mu$,
(c) $\theta_{xp1} = 18.43°$

8.12-5. (a) $\epsilon_x = -270\mu$, $\epsilon_y = 670\mu$, $\gamma_{xy} = 342\mu$,
(b) $\epsilon_1 = 700\mu$, $\epsilon_2 = -300\mu$, $\gamma_{max} = 1000\mu$

8.12-9. (a) $\epsilon_x = \epsilon_a$, $\epsilon_y = \epsilon_b + \epsilon_c - \epsilon_a$, $\gamma_{xy} = \epsilon_b - \epsilon_c$

8.12-13. (a) $\epsilon_x = 592\mu$, $\epsilon_y = 0$, $\gamma_{xy} = 1367\ \mu$,
(b) $\sigma_x = 46.5$ MPa, $\sigma_y = 15.35$ MPa, $\tau_{xy} = 36.0$ MPa

8.12-15. $T = \left[\dfrac{\pi E(r_0^4 - r_i^4)}{4r_0(1 + \nu)} \right]\epsilon_t$

CHAPTER 9

9.2-1. (a) $\sigma_a = 8000$ psi, $\sigma_h = 16,000$ psi,
(b) $F_w/L_w = 4000$ lb/in.

9.2-3. (a) $\sigma_a = 28.1$ MPa, $\sigma_n = 56.2$ MPa,
(b) $F_w/L_w = 112.5$ kN/m,
(c) $\tau_{abs\ max} = 28.1$ MPa

9.2-7. (a) $\sigma_a = 25$ MPa, $\sigma_h = 50$ MPa,
(b) $\sigma_n = 29.5$ MPa, $\sigma_t = 45.5$ MPa,
$\tau_{nt} = -9.6$ MPa

9.2-9. $t_{min} = 1$mm

9.2-13. (a) $(t_{cyl})_{min} = 3.52$ mm,
(b) $(t_{sph})_{min} = 1.76$ mm,
(c) $(t_{weld})_{min} = 2.81$ mm

9.4-1. $\sigma_{1A} = 0.0$ ksi, $\sigma_{2A} = -24.6$ ksi
$\sigma_{1B} = 0.0$ ksi, $\sigma_{2B} = -41.3$ ksi

9.4-3. $\sigma_{1B} = 54.2$ psi, $\sigma_{2B} = -40.6$ psi

9.4-9. $\sigma_{1A} = -2.94$ MPa, $\sigma_{2A} = 1.76$ MPa,
$\sigma_{1B} = 0$, $\sigma_{2B} = -85.5$ MPa

9.4-13. $\sigma_{1A} = 1880$ psi, $\sigma_{2A} = -1940$ psi,
$(\tau_{max})_A = 1910$ psi

9.4-15. $T = 8.16$ kip · in.

CHAPTER 10

10.1-1. $P_{cr} = \dfrac{kL}{4}$

10.1-3. $P_{cr} = \dfrac{kL}{6}$

10.2-1. $P_{cr} = 43.9$ kips

10.2-5. $W10 \times 60$

10.2-7. $P_{cr} = \dfrac{\pi^3}{4}\left(\dfrac{Er^4}{L^3}\right)$, $P_{cr2} = \dfrac{\pi^4}{12}\left(\dfrac{Er^4}{L^3}\right)$,
$P_{cr3} = \dfrac{\sqrt{3}\pi^4}{18}\left(\dfrac{Er^4}{L^3}\right)$; $P_{cr3} > P_{cr2} > P_{cr1}$

10.2-9. $I_{min} \geq 1.26$ in^4 → $L4 \times 4 \times \frac{3}{8}$ has $I_{min} \equiv I_z = Ar_z^2 = 1.76$ in^4

10.2-11. $\Delta T_{cr} = \dfrac{\pi^2 I}{\alpha A L^2}$

10.2-13. (a) $W_a = \dfrac{5}{6}\dfrac{\pi^2 EI}{L^2}$, (b) $P_b = \dfrac{\pi^2 EI}{L^2}$

10.2-15. $T_{max} = 21.1$ kips

10.2-17. (a) $W_{cr} = 18.20$ kips, (b) $d_{min} = 1.04$ in.

10.3-1. $(P_{cr})_a = 54.7$ kips, $(P_{cr})_b = 111.6$ kips,
$(P_{cr})_c = 219$ kips

10.3-3. $(P_{cr})_a = 35.9$ kN, $(P_{cr})_b = 73.3$ kN,
$(P_{cr})_c = 143.7$ kN

10.3-5. $FS = 25.2$

10.3-7. $P_{cr} = 503$ kN

10.3-9. (a) $\Delta T_{cr} = \dfrac{4\pi^2}{\alpha(L/r)^2}$, (b) $\Delta T_{cr} = 111.6$ °F

10.3-11. $w_{cr} = \dfrac{\pi^2 E_w b^4}{6a(KL)^2}$

10.4-1. (a) $v_{max} = e(\sec \lambda L - 1)$,
(b) $M_{max} = Pe \sec \lambda L$

10.4-3. (a) $v_{max} = 9.29$ mm,
(b) $M_{max} = 3.93$ kN · m

10.4-5. $L_{max} = 6.37$ m

10.4-7. (a) $\sigma_{max} = 18.69$ kips,
(b) $P_{allow} = 1.029$ kips

10.4-9. (a) $d_{min} = 1.4$ in., (b) $e_{max} = 0.9$ in.

10.4-11. (a) $\sigma_{max} = 31.7$ ksi, (b) $P_{allow} = 15.21$ kips,
(c) $d_0 = 5.0$ in.

10.5-1. (a) $\delta_{max} = 0.353$ in., (b) $\sigma_{max} = 5.31$ ksi

10.5-3. (a) $\delta_{max} = \dfrac{b}{3}$, (b) $\sigma_{max} = \dfrac{\pi^2 Eb^2}{16L^2}$

10.5-5. (a) $\sigma_{max} = 110.2$ MPa,
(b) $P_{allow} = 189.2$ kN, (c) $d_0 = 120$ mm

10.6-1. (a) $P_t = 195.3$ kips, (b) $L_{ec} = 18.95$ ft

10.7-1. $(P_{allow})_{12'} = 191.9$ kips,
$(P_{allow})_{16'} = 161.0$ kips, $(P_{allow})_{24'} = 88.4$ kips

10.7-3. $(P_{allow})_{18'} = 177.2$ kips,
$(P_{allow})_{22'} = 123.6$ kips, $(P_{allow})_{26'} = 88.5$ kips

10.7-5. $(P_{allow})_{4m} = 439$ kN, $(P_{allow})_{5m} = 290$ kN,
$(P_{allow})_{6m} = 201$ kN

10.7-7. (a) $L_{max} = 28.3$ ft, (b) $L_{max} = 13.68$ ft

10.7-9. (a) $L_{max} = 25.4$ ft, (b) $L_{max} = 20.4$ ft

10.7-11. $(P_{allow})_{K = 0.6} = 137.1$ kips,
$(P_{allow})_{K = 0.7} = 106.2$ kips,
$(P_{allow})_{K = 0.8} = 81.3$ kips

10.7-13. $(P_{allow})_{3'} = 120.2$ kips, $(P_{allow})_{4'} = 76.3$ kips,
$(P_{allow})_{5'} = 48.8$ kips

10.7-15. (a) $b_{min} = 3$ in., (b) $b_{min} = 3\frac{1}{8}$ in.

10.7-17. $b_{min} = 1\frac{5}{8}$ in.

10.7-19. (a) $P_{allow} = 123.5$ kips, (b) $t_{min} = \frac{9}{16}$ in.

10.7-21. (a) $L_{max} = 9.68$ ft, (b) $L_{max} = 7.91$ ft,
(c) $L_{max} = 6.53$ ft

10.7-23. $(P_{allow})_{6'} = 20.1$ kips, $(P_{allow})_{8'} = 13.82$ kips,
$(P_{allow})_{10'} = 8.84$ kips

CHAPTER 11

11.3-1. (b) $\mathcal{U} = 4$ kip·in.

11.3-3. (a) $\mathcal{U}_1 = \dfrac{P^2L}{AE}$, (b) $\mathcal{U}_2 = \dfrac{P^2L}{2AE}$,
(c) $\mathcal{U}_T = \dfrac{3P^2L}{2AE}$

11.3-7. $\dfrac{\mathcal{U}_b}{\mathcal{U}_a} = \dfrac{\alpha + \lambda - \alpha\lambda}{\alpha}$

11.3-9. $\mathcal{U} = \dfrac{2P^2L}{\pi E d_A d_B}$

11.3-11. $\mathcal{U} = \dfrac{21P^2L}{8AE}$

11.3-13. $\mathcal{U} = 0.701$ N·m

11.3-17. (a) $\mathcal{U}_1 = \dfrac{1024T_0^2L}{81\pi d^4 G}$, $\mathcal{U}_2 = \dfrac{16T_0L^2}{\pi d^4 G}$,
$\mathcal{U} = \dfrac{2320T_0L^2}{81\pi d^4 G}$,
(b) No, more would be stored.

11.3-25. $\mathcal{U} = \dfrac{GI_p}{L}(\phi_B{}^2 - \phi_B\phi_C + \phi_C{}^2)$

11.3-27. (a) $d_1 = 1.366$ in., $d_2 = 1.193$ in.,
$d_3 = 0.947$ in.,
(b) $\mathcal{U} = 0.412$ kip·in.

11.3-31. (a) $\mathcal{U}_{\sigma a} = \dfrac{2P^2L^3}{Ebh^3}$, $\mathcal{U}_{\sigma b} = \dfrac{88P^2L^3}{108Ebh^3}$

11.3-39. $\mathcal{U}_\sigma = \dfrac{17w_0L^5}{15360EI}$

11.4-1. $\Delta_B = 0.0367$ in., $\Delta_E = 0.0301$ in.

11.4-3. $\Delta_B = 5.74 \, (10^{-4})$ in.

11.4-7. $\phi = \dfrac{28}{3\pi}\dfrac{T_0L}{Gd}$

11.4-11. $\phi = 0.0768$ rad

11.4-13. $\theta_B = 1.167 \, (10^{-3})$ rad

11.4-15. $\Delta_{B_\sigma} = \dfrac{4PL^3}{Ebh^3}$, $\Delta_{B_\tau} = \dfrac{12PL(1 + \nu)}{5Ebh}$

11.6-1. (a) $u_B = \dfrac{PL}{3AE}$, (b) $F_1 = \dfrac{2P}{3}$, $F_2 = \dfrac{P}{3}$

11.6-3. $v = 0.810$ mm, $F_1 = -189$ kN,
$F_2 = -162$ kN

11.6-7. (a) $u_D = 4.46$ mm, $v_D = 0.099$ mm,
$F_1 = 79.8$ kN

11.6-13. (a) $\theta = 2.94 \, (10^{-3})$ rad,
(b) $\sigma_1 = 102.9$ MPa, $\sigma_2 = 147.1$ MPa

11.6-17. (a) $u_D = 9.54 \, (10^{-3})$ in.,
$v_D = 2.84 \, (10^{-3})$ in.,
(b) $F_1 = 3.56$ kips

11.8-1. $\Delta_B = \dfrac{Pa^2L}{3EI}\left(1 - \dfrac{a}{L}\right)^2$

11.8-3. $\theta_A = 6.26\,(10^{-3})$ rad

11.8-7. $\Delta_B = 511$ mm

11.8-13. $\Delta_B = \dfrac{PR^3}{EI}\left(\dfrac{3\pi}{4} - 2\right)$

11.8-15. $\Delta_B = 3.60\,(10^{-3})$ in.

11.8-17. (a) $R_A = \dfrac{w_0 L}{10}$, (b) $\theta_A = -\dfrac{w_0 L^3}{120EI}$

11.8-21. (a) $R_B = \dfrac{5}{16} w_0 L$, (b) $\theta_C = -\dfrac{w_0 L^3}{768EI}$

11.8-31. $\Delta_A = \dfrac{w_0 L^4}{30EI}$, $\theta_A = \dfrac{w_0 L^3}{24EI}$

11.8-37. $u_B = 0.0562$ mm

11.8-41. (a) $R_A = \dfrac{w_0 L}{10}$, (b) $\theta_A = \dfrac{-w_0 L^3}{120EI}$

11.8-43. (a) $R_B = \dfrac{5}{16} w_0 L$, (b) $\theta_C = -\dfrac{w_0 L^3}{768EI}$

11.8-51. (a) $R_A = \dfrac{3}{16} w_0 L$, (b) $\theta_A = \dfrac{w_0 L^3}{96EI}$

11.9-1. $\Delta_{\max} = \Delta_{st} + \Delta_{st}\left[1 + \dfrac{1}{\Delta_{st}}\left(\dfrac{v^2}{g} + 2h\right)\right]^{1/2}$

11.9-3. (a) $\mathcal{U} = 0.814$ kN·m,
(b) $\mathcal{U} = 0.382$ kN·m,
(c) $\mathcal{U} = 0.110$ kN·m,

(d) $\mathcal{U} = 0.212$ kN·m

11.9-5. $h_1 = 21.6$ in., $h_2 = 86.3$ in.

11.9-7. (a) $\sigma_{\max} = 35.8$ ksi, (e) $W = 622$ lb

11.9-9. $\sigma_{\max} = 19.95$ ksi

CHAPTER 12

12.2-1. $(\sigma_{\max})_a = 16.4$ ksi, $(\sigma_{\max})_b = 18.4$ ksi,
$(\sigma_{\max})_c = 33.12$ ksi

12.2-3. $(P_{\max})_a = 61.0$ kN, $(P_{\max})_b = 71.4$ kN,
$(P_{\max})_c = 82.0$ kN

12.2-5. (a) $\sigma_{\max} = 9.15$ ksi, (b) $d_{\min} = 3.43$ in.

12.2-9. (a) $(\tau_{\max})_1 = 10.3$ MPa, $(\tau_{\max})_2 = 9.5$ MPa,
(b) $\tau_{\max} = 8.0$ MPa, $(\tau_{\max})_1$ is 29% greater

12.2-11. $r_{\min} = 0.25$ in.

12.2-13. $P_{\max} = 67$ hp

12.2-15. (a) $(\sigma_{\max})_1 = 41.5$ MPa, $(\sigma_{\max})_2 = 35.2$ MPa,
(b) $\sigma_{\max} = 25$ MPa, $(\sigma_{\max})_1$ is 66% greater

12.3-1. $(\sigma_x - \sigma_y)^2 + 4\tau_{xy}^2 = \sigma_Y^2$

12.3-3. Fails, according to maximum-shear-stress theory, but not according to maximum-distortion-energy theory. $\sigma_M = 34.6$ ksi

12.3-5. $FS_s = 1.20$, $FS_d = 1.38$

12.3-9. (a) $FS_s = 0.96$, (b) $FS_d = 1.10$

12.3-15. $FS_m = 1.05$ (a very small margin of safety)

REFERENCES

1-1 Meriam, J. L., and Kraige, L. G., *Statics*, John Wiley & Sons, Inc., New York, NY, 1992.

1-2 Crandall, S. H., et. al., *An Introduction to the Mechanics of Solids*, McGraw-Hill, Inc., New York, 1978.

2-1 Timoshenko, S. P., *History of Strength of Materials*, McGraw-Hill, Inc., New York, 1953.

2-2 Drucker, D. C., *Introduction to Mechanics of Deformable Solids*, McGraw-Hill, Inc., New York, 1967.

2-3 Ramaley, D., and McHenry, D., *Stress-Strain Curves for Concrete Strained Beyond Ultimate Load*, Lab. Rept. No. Sp-12, U.S. Bureau of Reclamation, Denver, CO, 1947. (Reprinted in: Ferguson, P. M., Breen, J. E., and Jirsa, J. O., *Reinforced Concrete Fundamentals*, 5th edition, John Wiley & Sons, Inc., New York, 1988.)

2-4 Lee, S. M., ed., *International Encyclopedia of Composites*, VCH Publishers, Inc., New York, 1989.

2-5 Carswell, T. S., and Nason, H. K., ''Effect of Environmental Conditions on the Mechanical Properties of Organic Plastics,'' *Symposium on Plastics*, American Society for Testing Materials, Philadelphia, 1944. (Reprinted in: Callister, W. D. Jr., *Materials Science and Engineering*, 3rd edition, John Wiley & Sons, Inc., New York, 1994.)

2-6 Gordon, J. E., *Structures, or Why Things Don't Fall Down*, Da Capo Press, New York, 1978.

2-7 Shinozuka, M., and Yao, J. T. P., eds., *Probabilistic Methods in Structural Engineering*, American Society of Civil Engineers, New York, 1981.

2-8 Madsen, H. O., Krenk, S., and Lind, N. C., *Methods of Structural Safety*, Prentice-Hall, Inc., Englewood Cliffs, NJ, 1986.

2-9 Lewis, E. E., *Introduction to Reliability Engineering*, John Wiley & Sons, New York, 1987.

2-10 Rao, S. S., *Reliability-Based Design*, McGraw-Hill, Inc., New York, 1992.

2-11 *Manual of Steel Construction—Load & Resistance Factor Design*, American Institute of Steel Construction, One East Wacker Drive, Suite 3100, Chicago, 60601-2001.

3-1 Timoshenko, S. P., and Goodier, J. N., *Theory of Elasticity*, 3rd edition, McGraw-Hill, Inc., New York, 1970.

3-2 Timoshenko, S. P., *History of Strength of Materials*, McGraw-Hill, Inc., New York, 1953.

4-1 Oden, J. T., and Ripperger, E. A., *Mechanics of Elastic Structures*, 2nd edition, McGraw-Hill, Inc., New York, 1981.

4-2 Timoshenko, S. P., and Goodier, J. N., *Theory of Elasticity*, 3rd edition, McGraw-Hill, Inc., New York, 1970.

6-1 Timoshenko, S. P., *History of Strength of Materials*, McGraw-Hill, Inc., New York, 1953.

6-2 *Manual of Steel Construction—Allowable Stress Design*, and *Manual of Steel Construction—Load & Resistance Factor Design*, American Institute of Steel Construction, One East Wacker Drive, Suite 3100, Chicago, IL, 60601-2001.

6-3 *The Aluminum Design Manual*, The Aluminum Association, 900 19th Street, NW, Suite 300, Washington, DC, 20006.

6-4 Oden, J. T., and Ripperger, E. A., *Mechanics of Elastic Structures*, 2nd edition, McGraw-Hill, Inc., New York, 1981.

6-5 Allen, D. H., and Haisler, W. E., *Introduction to Aerospace Structural Analysis*, John Wiley & Sons, Inc., New York, 1985.

6-6 Timoshenko, S. P., and Goodier, J. N., *Theory of Elasticity*, 3rd edition, McGraw-Hill, Inc., New York, 1970.

6-7 Popov, E. P., *Engineering Mechanics of Solids*, Prentice Hall, Englewood Cliffs, NJ, 1990.

7-1 Pilkey, W. D., ''Clebsch's Method for Beam Deflection,'' *J. of Engineering Education*, Vol. 54, No. 5, January 1964, pp. 170–174.

8-1 Timoshenko, S. P., and Goodier, J. N., *Theory of Elasticity*, 3rd edition, McGraw-Hill, Inc., New York, 1970.

8-2 Timoshenko, S. P., *History of Strength of Materials*, McGraw-Hill, Inc., New York, 1953.

8-3 Anton, H., *Elementary Linear Algebra*, 7th edition, John Wiley & Sons, Inc., New York, 1994.

8-4 Beckwith, T. G., Marangoni, R. D., and Lienhard, J. H., *Mechanical Measurements*, 5th edition, Addison-Wesley, Reading, MA, 1993.

9-1 Kraus, H., *Thin Elastic Shells*, John Wiley & Sons, Inc., New York, 1967.

9-2 Timoshenko, S. P., and Woinowsky-Krieger, S., *Theory of Plates and Shells*, McGraw-Hill, Inc., New York, 1959.

9-3 Rabia, H., *Oilwell Drilling Engineering—Principles and Practice*, Graham & Trotman, Ltd., London, 1985.

10-1 Timoshenko, S. P., and Gere, J. M., *Theory of Elastic Stability*, 2nd edition, McGraw-Hill, Inc., New York, 1961.

10-2 Brush, D. O., and Almroth, B. O., *Buckling of Bars, Plates, and Shells*, McGraw-Hill, Inc., New York, 1975.

10-3 Chen, W. F., and Lui, E. M., *Structural Stability—Theory and Implementation*, Elsevier, New York, 1987.

10-4 Ziegler, H., *Principles of Structural Stability*, Blaisdell Publishing Co., Waltham, MA, 1968.

10-5 Galambos, T. V., ed., *Guide to Stability Design Criteria for Metal Structures*, 4th edition, John Wiley & Sons, Inc., New York, 1988.

10-6 Duberg, J. E., ''Inelastic Buckling,'' Chapter 52 in *Handbook of Engineering Mechanics*, edited by W. Flügge, McGraw-Hill, Inc., New York, 1962.

10-7 *Manual of Steel Construction—Allowable Stress Design*, and *Manual of Steel Construction—Load & Resistance Factor Design*, American Institute of Steel Construction, One East Wacker Drive, Suite 3100, Chicago, IL, 60601-2001.

10-8 *The Aluminum Design Manual*, The Aluminum Association, 900 19th Street, NW, Washington, DC, 20006.

10-9 *National Design Specifications for Wood Construction*, American Forest and Paper Association, American Wood Council, Engineering Division, 1111 19th Street, NW, Washington, DC, 20036.

11-1 Timoshenko, S. P., *History of Strength of Materials*, McGraw-Hill, Inc., New York, 1953.

11-2 Gere, J. M., and Timoshenko, S. P., *Mechanics of Materials*, 3rd edition, PWS-KENT Publishing Company, Boston, 1984.

11-3 Craig, R. R. Jr., *Structural Dynamics—An Introduction to Computer Methods*, John Wiley & Sons, Inc., New York, 1981.

12-1 Timoshenko, S. P., and Goodier, J. N., *Theory of Elasticity*, 3rd edition, McGraw-Hill, Inc., New York, 1970.

12-2 Zienkiewicz, O. C., and Taylor, R. L., *The Finite Element Method, Vol. 1—Basic Formulation and Linear Problems*, 4th edition, McGraw-Hill Book Company, London, England, 1989.

12-3 Cook, R. D., *Finite Element Modeling for Stress Analysis*, John Wiley & Sons, Inc., New York, 1995.

12-4 Kuske, A., and Robertson, G., *Photoelastic Stress Analysis*, John Wiley & Sons, Inc., New York, 1974.

12-5 Roark, R. J., and Young, W. C., *Roark's Formulas for Stress and Strain*, 6th edition, McGraw-Hill, Inc., New York, 1988.

12-6 Peterson, R. E., *Stress Concentration Factors*, John Wiley & Sons, Inc., New York, 1974.

12-7 Timoshenko, S. P., *History of Strength of Materials*, McGraw-Hill, Inc., New York, 1953.

12-8 Juvinall, R. C., *Fundamentals of Machine Component Design*, John Wiley & Sons, Inc., New York, 1983.

12-9 *ASTM Standards*, American Society for Testing and Materials, 1916 Race Street, Philadelphia, PA 19103-1187.

12-10 Hardrath, H. F., Utley, E. C., and Guthrie, D. E., *Rotating-Beam Fatigue Tests of Notched and Unnotched 7075-T6 Aluminum-Alloy Specimens Under Stress of Constant and Varying Amplitudes*, NASA TN D-210, December 1959. (Reprinted in: Hertzberg, R. W., *Deformation and Fracture Mechanics of Engineering Materials*, 3rd edition, John Wiley & Sons, Inc., New York, 1989.)

12-11 *Structural Alloys Handbook*, CINDAS/USAF CRDA, Purdue University, 1293 Potter Engineering Center, West Lafayette, IN 47907-1293.

12-12 *Aluminum Standards and Data*, The Aluminum Association, 900 19 Street, NW, Washington, DC, 1993.

12-13 Rolfe, S. T., and Barsom, J. M., *Fracture and Fatigue Control in Structures, Applications of Fracture Mechanics*, Prentice-Hall, Inc., Englewood Cliffs, NJ, 1977.

12-14 Osgood, C. C., *Fatigue Design*, Pergamon Press, Inc., Elmsford, NY, 1982.

12-15 Niu, M. C. Y., *Airframe Structural Design*, Technical Book Company, Los Angeles, 1988.

INDEX

A

Absolute maximum shear stress, 453–458
  plane stress case, 454–456
Accuracy, numerical, A-2–A-3
Allowable-stress design, 65
  beams, 276
Aluminum-alloy columns, buckling of, 538–539
Angle of rotation, 175, 189
Angle of twist, 180, 185, 203, 206, 209
Approximations, A-3
AREAPROP computer program, A-44–A-45
Areas, plane, geometric properties of, A-7–A-22
  AREAPROP computer program, A-44–A-45
  table of, (See inside back cover)
AXIALDEF computer program, 114, 129, 134,
    A-47–A-54
Axial deformation, 94–172
  analogy between axial deformation and torsion, 174
  basic theory of, 95–100
  defined, 94
  displacement method and, 105–108, 109, 134,
    A-47–A-54
  elastic behavior of uniform members, 102–103
  elastic-plastic analysis of statically indeterminate
    structures, 142–147
  elastic strain energy and, 561–562
  equilibrium and: *See* Equilibrium
  force method and, 109–113
  geometric misfits, 129–133
  geometry of deformation, 95–96, 102
    elastic-plastic analysis of statically indeterminate
      structures, 143
    geometric misfits and, 129–133
    inelastic deformation, 141
    kinematic assumptions, 95
    thermal stresses and, 118–119

  of homogeneous linearly elastic member, 99–100
  of uniform linearly elastic member, 102–103
  inelastic, 140–151
    elastic-plastic analysis of statically indeterminate
      structures, 142–147
    equilibrium equations, 141
    fundamental equations and, 140–142
    geometry of deformation, 141
    material behavior and, 141–142
    uniform end-loaded element, 142
  material behavior and, 96, 102–103
    elastic-plastic analysis of statically indeterminate
      structures, 143–144
    inelastic deformation, 141–142
  residual stress, 148
  Saint-Venant's principle and, 101
  stress resultants and, 22–23, 96–97
  thermal stress and, 118–129
  torsion, analogy between axial deformation and, 174
  uniform, in structures, 103–118
Axial loading
  normal stress due to, 19–20
  stress concentrations and, 618–620
Axial (normal) stress: *See* Normal stress

B

Ball joints, 10
Beam-column behavior, 526–528
BEAMDEF computer program, 409, 412, A-59 A-63
Beams: *See also* Deflection of beams
  bending of, elastic strain energy and, 563–565
  Bernoulli-Euler assumptions, 260, 266
  built-up, shear in, 323–326
  defined, 229
  deflection of: *See* Deflection of beams
  deformation terminology, 258–259

Beams (*continued*)

    discontinuity functions and

        for deflection, 385–390, A-56–A-59

        representing equilibrium, 243–250

    equilibrium of, 229–257

        bending moment and, 235–250

            discontinuity functions, 246–250, A-56–A-59

        loads and, 235–238

            discontinuity functions, 246–250, A-56–A-59

        Macaulay functions, 244–246

        shear force and, 235–243

            discontinuity functions, 246–250, A-56–A-59

        singularity functions, 246

        using finite free-body diagrams (FBD), 231–235

        using infinitesimal free-body diagrams (FBD), 235–238

    multi-interval, 362

    nonhomogeneous, flexural stress in, 280–286

    sign conventions

        for displacements, 263, 359

        for external loads, 235

        for stress resultants, 230–231, 359

    statically determinate

        slope and deflection by integration, 364–375

        slope and deflection by superposition, 392–396, 406–412

    statically indeterminate

        slope and deflection by integration, 375–384

        slope and deflection by superposition, 397–405, 406–412

    strain-displacement analysis, 259–265

    strain-displacement equation, 262, 280, 295

    strain in

        extensional strain (longitudinal), 260–265

        shear strain distribution, 306

        transverse strain, 262

    strength, design of beams for, 274–279

    stresses in, 258–357, 493–496

        allowable-stress design, 276

        Bernoulli-Euler beam theory and, 260

        closed, thin-wall beams, 320–322

        from combined loading, 493–498

        discrete shear connectors, 324–326

        doubly symmetric beams with inclined loads, 287–288

        elastic-plastic bending, 296–304

        flexural stress

            in linearly elastic beams, 266–274

            in nonhomogeneous beams, 280–286

                direct method, 280–283

                transformed-section method, 283–286

        flexure formula, 268

        inelastic bending and, 295–304

        linear shear connectors, 323

        moment-curvature equation, 268

        moment-curvature formulas for inelastic bending, 301–302

        neutral axis

            for elastic-plastic bending, 297

            for symmetric bending, 260

            for unsymmetric bending, 290–291

        neutral surface, 259–260

        principal axes of inertia, 288–289, A-18–A-20

        pure bending, 259–260

        residual stresses, 302–304

        shear center, 326–332

        shear flow, 306–308

            in the flanges, 314–315

            in the web, 315–316

        shear in built-up beams, 323–326

        shear stress

            distribution of, 305–306, 310

            formula, 308–309

                limitations on, 310–312

            and shear flow, 304–310

            in web-flange beams 313–320

        thin-wall beams, 312–322

        transformed-section method, flexural stress in nonhomogeneous beams, 283–286

        unsymmetric bending, 287–295

            maximum flexural stresses, 291–295

            principal-axis method, 289–295

    thin-wall, stresses in, 312–322

Bending

    pure, in beams, 259–260

    stress concentrations and, 621–623

Bending moments, 11, 230–231

    and beams, equilibrium of, 235–243

        discontinuity functions representing, 243–250, A-56–A-59

Bernoulli-Euler beam theory, 260

Biaxial stress, 431

Body forces, 8–9

Boundary conditions (BC), deflection of beams, 362–364

Brittle materials, 34–35

    failure theories for, 629–631

    in compression, 560

Brooklyn Bridge, 2–3

Buckling of columns: *See* Columns, buckling of

Built-up beams, shear in, 323–326

C

Cable supports, 10

Cantilever supports, 10, 230

Cartesian components of stress, 57–60

Castigliano's theorems
first, strain energy methods, 577–579
second, 588–593

Centrally loaded columns, design of, 536–541

Circular bars/members, torsion of: *See* Torsion

Clebsch's Method, 385n

Closed, thin-wall beams, stresses in, 320–322

Codes, for design of deformable bodies, 65

Columns
buckling of, 511–553
aluminum-alloy columns, 538–539
beam-column behavior, 526–528
buckling load
ideal fixed-pinned column, 520–522
ideal pin-ended column, 517
centrally loaded columns, design of, 536–541
defined, 513–514
design procedure for, 540
eccentric loading, 526–531
effective length of columns, 522–525
end condition effects, 519–526
equilibrium and, 515, 520
differential equation of, 515–518
stability of, 512–514
Euler buckling load, 517
ideal fixed-pinned column, 520–522
ideal pin-ended column, 514–519
imperfections in columns, 532–533
inelastic buckling of ideal columns, 533–536
secant formula, 528–531
structural steel columns, 537–538
timber columns, 539–540
design procedure for, 540

Combined loading, stresses due to, 487–510

Complementary energy methods: *See* Energy methods

Composite materials, 35–36

Compressive stress, 13, 19, 21
brittle material and, 500
columns, buckling under, 515, 517, 524, 529, 532, 536–540
in concrete, 29–30, 35
maximum, in beams, 291–295
tests, 29–30, 35

Computer-Aided-Design (CAD), 4, 510

Computer exercises, 160–161, 163–164, 166–167, 221–222, 257, 427–428, 481, 484–485, 486, 507, 510

Computer programs, 3–4, 114, 118, 129, 134, 191, 250, 412, 450, 469, A-43–A-66

Concentrated loads, 8–9

Connections, types of, 10

Constant-strain experiment, 38

Constant-stress experiment, 37–38

Constitutive behavior of materials, 5: *See* Material behavior

Constraints, 9

Continuity conditions (CC), deflection of beams, 362–364

Conversion factors, between SI units and U.S. Customary units, (See inside front cover)

Core, of cross sections, 500

Creep, 37–38

Crotti-Engesser theorem, 584–586

Cup and cone fracture, 32

Cylindrical pressure vessels, stresses in, 488–490

D

Deflection of beams, 358–428
deflections and slopes of uniform cantilever beams, table of, A-34–A-35
deflections and slopes of uniform simply-supported beams, table of, A-36–A-37
deflection curve, 358
boundary conditions (BC), 362–364
continuity conditions (CC), 362–364
differential equations of, 359–364
load-deflection equation, 361–362
moment-curvature equation, 359–361
discontinuity functions determining, 385–390
fixed-end actions for uniform beams, table of, A-38
fixed-end solutions, 382
slope and deflection by integration
statically determinate beams, 364–375
statically indeterminate beams, 375–384
slope and deflection by superposition
differential-load superposition, 396–497
displacement method, 406–412
procedure, 407
force-deformation equations for uniform beam element, 406-407
force method, 391–405
procedure, 391–392
linearly varying distributed load, 406–412
statically determinate beams, 392–396
statically indeterminate beams, 397–405
uniform Bernoulli-Euler beam element, 406–412
work-energy principle for calculating, 567–571

Deformable bodies/Deformation, 1–2
axial: *See* Axial deformation
bending: *See* Deflection of beams
design and: *See* Design
geometry of: *See* Geometry of deformation

Deformable bodies *(continued)*
    impact loading on, 599–602
    mechanics of, 1–5
    strain: *See* Strain
    superplastic, 34–35
    terminology, beams, 258–259
    torsion: *See* Torsion
    virtual work for, 573–574
Design
    computer-aided, 4, 510
    damage-tolerant, 634
    of deformable bodies, 64–69
        codes for, 65
        factor of safety and allowable stress, 64–67
        philosophies for, 67–69, 634
    fail-safe, 634
    failure theories, 623–631
        brittle materials, 629–631
        ductile materials, 624–629
    fatigue and fracture, 631–634
    load and resistance factor design (LRFD), 69
    probability-based, 68–69
    properties of materials, 33–35
    safe-life, 634
    special topics on, 617–623
    stress concentrations and: *See* Stress concentrations
Differential-load superposition, slope and deflection by, 396–397
Digits, significant, A-2–A-3
Direct method, flexural stress in nonhomogeneous beams, 280–283
Discontinuity effects, in pressure vessels, thin-wall, 491
Discontinuity functions, and beams
    deflection of, 385–390
    equilibrium of, 243–250, A-56–A59
Discrete shear connectors, in beams, 324–326
Displacement, 95
    strain and, analysis of, 25–27, 95–96
Displacement method
    axial deformation, 109, A-47–A-48
    planar trusses, 137
    slope and deflection of beams, 406–412, A-59–A-63
    strain energy and, 579
    torsion of circular shafts, 188–191, A-54–A-56
Distortion-energy theory, maximum, 625–629
Distributed loads on beams, linearly varying, 406–407, A-60–A-61
Doubly symmetric beams with inclined loads, stresses in, 287–288
Ductile materials, 34–35
    failure theories for, 624–629
Dynamic loading, 598–602

E
Eccentric loading, columns, 526–531
Effective length of columns, 522–526
Effective slenderness ratio, 524, 540
Elasticity
    elastic behavior of materials, 36–37
    elastic-plastic analysis
        bending of beams, 296–304
        of statically indeterminate structures, 142–144
        torque-twist analysis, 210–211
    linear, 39–40
        Hooke's Law for isotropic materials, 58–60
        strain energy and, 560–561
    limits on, 36–37
    modulus of: *See* Young's modulus
Elementary beam theory, 493–494
Element force-deformation behavior
    axial-deformation, 102–103, A-47
    Bernoulli-Euler beam, 406–407, A-60–A-61
    torsion, 184–185, A-55
    truss, 134–135
End closure effects, of pressure vessels, thin-wall, 491
End condition effects, columns, buckling of, 519–526
Endurance curve, 633
Endurance limit, 633–634
Energy methods, 554–616
    axial deformation, elastic strain and, 561–562
    beams, bending of
        elastic strain and , 563–565
        flexural strain energy, 563–564
        shear-strain energy, 564–565
    Castigliano's first theorem, 577–579
    Castigliano's second theorem, 588–593
    complementary, 580–587
        complementary work and complementary strain energy, 580–581
        Crotti-Engesser theorem, 584–587
        Principle of Virtual Forces, 581–584, 586–587
        virtual forces and, 580–584
    displacement method related to, 579
    dynamic loading, 598–602
    elastic strain energy, 560–566
    flexural strain energy, 563–564
    force method related to, 584
    impact loading, 598–602
        on deformable bodies, 599–600
    linearly elastic behavior, 557–559
        strain energy density and, 560–561
    shear-strain energy, 565–566
    strain energy
        and complementary energy, 554–559

elastic, for various types of loading, 561–566
  density of, for linearly elastic bodies, 560–561
  methods, 575–579
torsion of circular members, 562–563
unit-load method, 593–598
virtual displacements, 572
virtual forces, 580
virtual work, 571–574
work and complementary work, 555–557
work-energy principle for calculating deflections,
  567–571
Equilibrium, 4, 8, 134
  axial deformation and
    elastic-plastic analysis of statically indeterminate
      structures, 142–143
    geometric misfits, 129–134
    inelastic deformation and, 141
    planar trusses, 135
    stress resultants and, 96–98
    structures with external loads, 102–118
    thermal stress and, 121
  of beams: See Beams
  of buckled columns, 512–514, 515–516, 520
  of circular bars in torsion, 178–179, 202, 209
  differential equation of, and columns, buckling of,
    515–518
  inelastic bending of beams, 296
  requirements, shear and, 45–46, 57–58
  stability of, and columns, buckling of, 512–514
  static, 7–15
    external loads, 8–9
    free-body diagram (FBD), 11–12
    internal resultants, 10–11
    member connections, support reactions and, 9–10
  torsion of circular bars, 178–179
    inelastic, 209
    uniform torsion members, 185–194
  and transformation of stress, 429, 433
  and virtual displacements, 573
Euler buckling load, 517
Extensional strain, 25–27, 95–96, 261
  defined, 53–55

F

Factor of safety, 64–67
Fail-safe design, 634
Failure, 511
  buckling, 511
  fatigue and fracture, 631–634
  theories, 623–631
    brittle materials, 629–631
    ductile materials, 624–629

Fatigue
  and fracture, 631–634
  level, 633–634
  strength, 634
Finite-element method, (See color-photo insert), 3–4,
  496, 554, 623, A-43
Fixed-end solutions, deflection of beams, 382, A-38
Flexibility coefficient
  axial deformation, 102–103
  torsion, 185
Flexibility method, 109n
Flexural stress
  in linearly elastic beams, 266–274
  in nonhomogeneous beams, 280–286
Flexure formula, 268
Fluids, mechanics of, 1
Force method
  axial deformation, 109–113, 116–117, 126–128
  planar trusses, 135–137
  slope and deflection by superposition, 391–405
Forces: See also Element force-deformation behavior
  normal, 11, 20, 51
  resultant: See Stress
  shear, 11, 42, 51
Fourth-order method for beam deflection, 364–370,
  373–375, 377–379
Fracture
  brittle, 35, 196
  cup and cone, 32
  ductile, 196
  fatigue and, 631–634
  stress, 32, 33
  torsion testing and, 196
Free-body diagrams (FBD), 8, 11–12
  finite, and beams, equilibrium of, 231–235
  infinitesimal, and beams, equilibrium of,
    235–237
  fourth-order solutions and, 375
Fully plastic moment, 297–298
Fundamental equations
  axial deformation, elastic-plastic analysis,
    142–144
  axial deformation, inelastic, 140–142
  bending of beams, inelastic, 295–296
  circular torsion members
    inelastic, 209
    linearly elastic, 179–180

G

General state of stress
  absolute maximum shear stress, 453–458
  principal stresses, 452–453

Geometric compatibility/incompatibility, 129: *See* Geometry of Deformation
Geometric misfits, axial deformation and, 129–134
Geometric properties of plane areas, (See inside back cover)
Geometry of deformation, 4, 25
    axial deformation, 102–103, 134
        geometric "misfits" and, 129–134
        strain-displacement analysis, 95–96
        thermal stresses and, 118–119, 134
    of circular bars, in torsion, 174–176
        inelastic, 209
    elastic-plastic analysis of statically indeterminate structures, 143
    inelastic axial deformation, 141
    inelastic bending of beams, 295–296
    planar trusses, 135
    in relationship between E and G, 63
    torsion of circular bars, 174–176
        inelastic, 209
    uniform torsion members, 185–187
Gravitational potential energy, 598–599

H

Hardening, strain, 32
Homogeneous linearly elastic member, 99
Hooke's Law, 39–41, 58–61, 195
    for isotropic materials, 58–61
    for shear, 47, 195
Hoop stress, 489–490

I

Ideal columns
    fixed-pinned, buckling load of, 520–522
    inelastic buckling of, 533–536
    pin-ended, buckling load of, 514–519
Impact loading, 598–602
    on deformable bodies, 599–602
Inclined loads, doubly symmetric beams with, 287–288
Inclined planes, stresses on, 47–49
Inelasticity
    in axial deformation: *See* Axial deformation
    and beams, stresses in, 295–304
    of ideal columns, 533–536
    in torsion of circular bars, 208–214
Inertia, principal axes of, beams, 288–289, A-18–A-20
Initial stresses, 129n
Integration, slope and deflection by
    statically determinate beams, 364–375
    statically indeterminate beams, 375–384
Internal resultants, 10–11, 50–53
Isotropic materials, Hooke's Law for, 58–61

K

Kern, of a cross section, 500
Kinetic energy, converted to strain energy, 599

L

Length, effective, of columns, 523
Linear elasticity, 39–41
    axial deformation and, 99–100
    in beams, flexural stress and, 266–274
    energy methods and, 557–558
    and Hooke's Law for isotropic materials, 58–61
    and strain energy density, 560–561
    of torsion members, 176–177, 184–185
Linearly varying distributed load: *See* Distributed loads on beams
Line loads, 8–9
LINSOLVE computer program, A-45–A-46
Load and resistance factor design (LRFD), 69, 145–147
Load-deflection equation, for beams, 361–362
Loads
    allowable, and factor of safety, 64
    allowable, for columns, 530, 537–541
    in beams
        distribution of, effect on shear stress, 310–312
        equilibrium of, 235–238
            discontinuity functions representing, 243–250
        external, 8–9
Longitudinal plane of symmetry (LPS), 258–259, 287
Longitudinal strain, in beams, 260–262
Longitudinal stress, in cylindrical pressure vessels, 488–489
Lüders' bands, 48

M

Macaulay functions, 244–246
"Massless" linear springs, and strain energy, 598–599
Material behavior, 5: *See also* Mechanical properties of materials
    axial deformation and, 96, 102–103, 121, 141–142
    circular bars in torsion
        linearly elastic, 176
        inelastic, 209
    elastic: *See* Elasticity
    force-temperature-deformation behavior, 118–119, 121, 134
    inelastic
        axial deformation and, 141–142
        bending of beams, 296
        torsion of circular bars, 209
        in relationship between E and G, 63
Maximum-distortion-energy theory, ductile materials, 625–629

Maximum elastic moment, 297
Maximum-normal-stress theory of failure, brittle
       materials, 629
Maximum shear stress
  absolute, 453–458
  in-plane, 440–443
Maximum-shear-stress theory of failure, ductile
       materials, 624–625
Maximum stress, in beams
  allowable-stress design and, 276
  under combined flexure and shear, 493–496
  unsymmetric bending, 291–295
Mechanical properties of materials, 31–33: *See also*
       Material behavior
  design properties, 33–35
  elasticity and plasticity, 36–39
  Hooke's Law for linearly elastic materials, 39–40,
       58–60
  in shear, 47
  tables of, A-39–A-42
  temperature effects on, 38–39
  time dependent, 37–39
Mechanics
  defined, 1
  deformable body: *See* Deformable bodies,
       mechanics of
MechSOLID, 114, 129, 134, 160, 163, 166, 191, 221,
       257, 427, 250, 450, 469, 485, 486, A-43–A66
Member connections, support reactions and, 9–10
Misfits, geometric, axial deformation and, 129–134
Modulus of elasticity: *See* Young's modulus
MOHR computer program, 450, 469, 484, 486,
       A-63–A-66
Mohr's circle
  for moments and products of inertia, A-20–A-22
  for plane stress
    derivation of, 444–446
    properties of, 446–450
  for two-dimentional strain, 463–469
Mohr's failure criterion, 629–630
Moment diagram, bending, 238–243
  V/M DIAG computer program, A-56–A-59
Moments
  bending: *See* Bending moments
  fully plastic, 247
  maximum elastic, 297
  moment-curvature equation, 268
    beams
      deflection of, 359–361
      inelastic bending of, 301–302
  resultant, 11
Multi-interval beams, 362

N
Necking, 32
Neutral axis, 260, 262, 290–291, 297
Noncircular prismatic bars, torsion of, 206–208
Nonhomogeneous beams, bending of, 280–286
  direct method, 280–283
  transformed-section method, 283–286
Normal, stress, 19–24
  average, 23
  axial, 23–24, 96
  defined, 20–21, 49–50
  flexural, in linearly elastic beams, 266–274
  flexural, inelastic, in beams, 296
  on inclined plane, 47–49
  maximum, theory of failure for brittle
      materials, 629
  positive in tension, 13
  uniform, 23–24
Notation convention
  transformation of stress and strain, 432–433
  uniform torsion members, 185–186, A-55

O
Offset yield stress, 33
Orthogonal faces, plane stress on, 435–438

P
Percent elongation, 34
Percent reduction in area, 34
Perfectly plastic zone, 32, 141, 210, 296
Permanent set, 36–37, 302–303
Philosophies, design, for deformable bodies, 67–69
Pin supports, 10, 363
Piobert's bands, 48
Planar trusses: *See* Trusses, planar
Plane of symmetry, longitudinal (LPS), 258–259, 287
Plane strain
  Mohr's circle for two dimensional strain, 463–469
  three-dimensional analysis, 470
  transformation of stress and strain, 459–463
Plane stress, 430–431
  absolute maximum shear stress and, 454–456
  and principal stresses, 438–444, 451
  Mohr's circle for, 444–450
  three-dimensional analysis, 470–471
  transformation of stress and strain, 432–438
Plasticity, 36–37
Plastic moment (fully), 297–300
Plastics and composites, 35–36
Poisson's-ratio effect, 40–41, 262
Polar moment of inertia, 178, A-11–A-12
Power transmission shafts, 198–200

Pressure vessels, thin-wall, 488–493
  absolute maximum stress in, 490–491
  cylindrical, 488–490
  discontinuity effects, 491
  end closure effects, 491
  spherical, 490
Prestrain effects, 129n
Principal axes of inertia, beams, 288–289,
      A-18–A-20
Principal stresses, 438–440
  three-dimensional stress states, 450–453
Principle of Virtual Displacements, 575–577
Principle of Virtual Forces, 581–584, 586–587
Prismatic bars, torsion of noncircular, 206–208
Probability-based design, 68–69
Problem-solving procedures, 5–7
Properties, geometric, of plane areas, A-7–A-22
  AREAPROP, computer program for, A-44–A-45
  table of, (See inside back cover)
Properties, mechanical, of selected engineering materials,
      A-39–A-42
Properties, section, of selected structural shapes,
      A-23–A-33
Pure bending, beams, 259–260
Pure shear, 45–46, 431

R
Redundant reactions, 375
Redundant supports, 8
Residual strain, 36
Residual stress
  axial deformation, 148–151
  in beams, 302–303
  torsion of circular bars, 212–214
Resultants: See Stress
Right-hand rule, 51, 175
Rigidity
  of connections, 10
  torsional, 179
Rod supports, 10
Roebling, Johann, 3
Roebling, Washington, 3
Roller supports, 10, 230, 363

S
Safe-life design, 634
Saint-Venant's principle, 101, 493, 617
Secant formula, columns, 528–531
Second-order method for beam deflection, 364–366,
      370–372, 375–377
Shear, 42–47
  in built-up beams, 323–326

direct, 42–45
  equilibrium requirements, 45–46
  material properties in, 47
  pure, 45–46, 431
Shear center, of beams, 326–332
Shear diagrams, 238–243
  V/M DIAG computer program, A-56–A-59
Shear flow
  shear stress and, in beams, 304–310
  in thin-wall beams, 313–322
  in thin-wall torsion members, 201–203
Shear force, 11, 51
  direct, 42
  and beams, equilibrium of, 235–243
    discontinuity functions representing, 243–250
Shear modulus, 47, 176
  torsion testing and, 196
  Young's modulus and, 62–63
Shear strain, 46–47
  defined, 55–56
Shear-strain distribution, in beams, 306
Shear-strain energy, elastic strain and, 564–565
Shear stress, 42–46
  absolute maximum, 453–458
  defined, 49–50
  distribution of, 177, 305–306
    in bending, 310, 311, 313
    in torsion, 177, 201, 206–207, 210
  equilibrium requirements, 57–58
  formula, 308–309, 310–312
    limitations on, 310–312
      cross-sectional shape, effect of, 310–311
      length of beam, effect of, 312
      load distribution on beam, effect of, 310–312
  maximum
    absolute, 453–458
    ductile materials, 624–625
    in-plane, 440–443
  and shear flow, 304–310
  in web-flange beams, 313–320
Sign conventions
  for axial deformation, 95–96, A-47
  for beams, 230–231, A-60
  for extensional strain, 25
  for normal stress and shear stress, 50, 57, 432–433
  for shear strain, 46
  for torsion, 175, 179, 185–186, A-55
Singularity functions, 246
Slenderness ratio, effective, 523–524, 536
Slip, 37
Slip bands, 48
Slope and deflection: See Deflection of beams

S-N diagram, 632–633
Solids, mechanics of, 1
Spherical pressure vessels, stress and, 490
Stability of equilibrium, 512–514
Statically determinate conditions, 8
  beams
    slope and deflection by integration, 364–375
    slope and deflection by superposition, 392–396
    Castigliano's second theorem applied to, 589–590
    residual stress and, 148
Statically indeterminate conditions, 8, 106
  beams
    slope and deflection by integration, 375–384
    slope and deflection by superposition, 397–405
    Castigliano's second theorem applied to, 590–593
    Principle of Virtual Displacements applied to, 586–587
      residual stress and, 148
Static equilibrium: See Equilibrium
Stiffness, 2, 33–34
Stiffness coefficient
  for uniform axial-deformation members, 103
  for torsion of uniform circular members, 185
Stiffness method, 108, 109n
Strain: See also Stress-strain behavior
  constant, experiment, 38
  displacement analysis: See Strain-displacement analysis
    distribution of, 2
    circular bars in torsion, 195–196
    extensional, in axial deformation, 96
    extensional, in bending of beams, 264, 295, 297
    shear, in bending of beams, 306
    shear, in torsion of circular bars, 176, 195–196
  extensional, 25–27, 95–96
    defined, 53–55
  longitudinal, in beams, 260–262
  measurement of, 471–475
  Mohr's circle, for two-dimensional, 463–469
  plane
    defined, 459
    three-dimensional analysis, 470
    transformation of stress and strain, 459–463
  residual, 36
  rosettes, 471–475
  shear, 46–47
    defined, 55–56
  thermal, 27–29, 119–121
  three-dimensional analysis, 469–471
  transformation of: See Transformation of stress
    and strain
  transverse, in beams, 262
  two-dimensional, Mohr's circle for, 463–469,
    A-64–A-66

Strain-displacement analysis
  for axial deformation, 25–27, 95–96
  for beams, 259–265, 295–296
  for circular bars in torsion, 175–176
Strain energy
  and complementary energy, 557–566
  flexural, elastic strain and, 563–564
  of ''massless'' linear spring
    gravitational potential energy converted to, 598–599
    kinetic energy converted to, 599
  methods, 575–579
Strain hardening, 32
Strength, 2, 33
  design of beams for, 274–279
Stress: See also Stress-strain behavior
  allowable, 64–67
  axial: See Normal stress
  in beams: See Beams
  biaxial, 431
  buckling
    allowable, 537–541
    effective slenderness ratio and, 523–524, 536
    of eccentrically loaded columns, 529
    of ideal pin-ended columns, 517
    of imperfect columns, 532
  cartesian components of, 57–58
  from combined loading, 487–510
    beams, stress distribution in, 493–498
    pressure vessels, thin-wall, 488–498
  compressive: See Compressive stress
  constant, experiment, 37–38
  distribution of, 2
    circular bars in torsion, 194–197
    flexural, in bending of beams, 266–268, 297
    normal, in axial deformation, 96
    shear, in bending of beams, 310–311, 313
    shear, in torsion, 177–178, 201, 206–207, 210–212
  fracture, 32, 33
  general state of
    absolute maximum shear stress, 453–458
    principal stresses, 452–453
  hoop, in cylindrical shells, 489–490
  on inclined plane, 47–49
  longitudinal, in cylindrical shells, 488–489
  Mohr's Circle for, 444–450, A-63–A-64
  normal: See Normal stress
  offset yield, 33
  plane, 430–431
    three-dimensional analysis, 470–471
    transformation of: See Transformation of stress and
      strain
  pressure vessels, thin-wall, 488–493

Stress (*continued*)

principal

Mohr's circle and, 447

plane stress case, 438–440, 451

three-dimensional stress states, 452–453

residual

axial deformation, 148–151

beams, 302–304

torsion of circular bars, 212–214

resultants, 11, 21–23, 50–53

axial deformation and, 23, 96–98

of circular bars in torsion, 178–179

of thin-wall torsion members, 202–203

shear: *See* Shear stress

tensile: *See* Tensile stress

thermal, in axial deformation, 118–129, 134

transformation of : *See* Transformation of stress and strain

true, 32

ultimate (strength), 32, 33

uniaxial, 431

Stress concentrations, 617–623

axial loading, 618–620

bending, 621–623

torsion, 620–621

Stress contours, in beams, 496

Stress relaxation, 38

Stress-strain behavior, 2: *See also* Material behavior

of circular bars in torsion, 176–177, 209

diagrams of, 30–36

linear, 39–41, 58–60

stress-strain-temperature equation, for axial deformation, 118, 121

time-dependent, 37–38

Stress trajectories, in beams, 495–496

Structural steel columns, buckling of, 537–538

Superplastic deformation, 34–35

Superposition, slope and deflection by: *See* Deflection of beams

Supports

reactions, and member connections, 9–10

types of, 10, 363

Surface loads, 8–9

T

Temperature effects

on properties of materials, 38–39

stress-strain-temperature equation, for axial deformation, 118, 121

thermal strain, 27–29

Tensile stress, 21

in beams, maximum, 291–295

tests, 29–30

Thermal effects: *See* Temperature effects

Thermal strain, 27–29

Thermal stress, in axial deformation, 118–129, 134

equilibrium equation and, 121

Thin-wall members

beams, stress in, 312–322

torsion members, stress in, 201–205

Three-dimensional analysis

plane strain, 470

plane stress, 470–471

strain, 469–471

stress states, 430, 450–453

Timber columns, buckling of, 539–540

Time-dependent stress-strain behavior, 37–38

Torque, 11, 51, 178, 202–203: *See also* Torsion

internal (resisting), 175

notation convention, 185–186, A-55

sign conventions, 175, 179, 185–186, A-55

Torque-twist equation, 178, 180, 185, 210–212, A-55

TORSDEF computer program, 191, 221–222, A-54–A-56

Torsion, 173–228

and axial deformation, analogy between, 174

of circular bars/members

angle of rotation, 175, 180, 186, A-55

angle of twist, 180, 185–186, 203, 207, A-55

basic theory for circular members, 174–184

elastic-plastic torque-twist analysis, 210–214

elastic strain energy and, 562–563

equilibrium, stress resultants and, 178–179, 202–203, 209

geometry of deformation and, 174–176, 209

inelastic, 208–214

power transmission shafts, 198–200

residual stress, 212–214

shear flow, 201–203

sign convention for, 175, 179, A-55

strain-displacement analysis, 175–176

strain distribution, 176, 195–196, 209

stress distribution, 177–178, 194–197, 210–211

stress resultants and equilibrium, 178–179, 202–203, 209

stress-strain behavior, 176–177, 194–197, 209–212

torsion testing and, 194–197

unloading; residual stress, 212–214

formula, 179

linearly elastic behavior of a uniform torsion member, 184–185

of noncircular prismatic bars, 206–208

sign convention for, 175, 179

stress concentrations and, 620–621
of thin-wall members, 201–205
uniform torsion members
   assemblages of, 185–194
   element force-deformation behavior, 185–186
   equilibrium, 186–187, 188, 191–192
   fundamental equations, 185
   linearly elastic behavior of, 184–185
   notation convention, 185–186
   sign convention, 185–186
Torsional rigidity, 179
Transformation of stress and strain, 429–486
  plane strain, 459
  plane stress, 430–438
   Mohr's circle for, 444–450
    derivation of, 444–446
    MOHR computer program, A-63–A-66
    properties of, 446–450
   notation and sign convention, 432–433
   on orthogonal faces, 435–438
   state of, at a point, 432
  principal stresses, 438–440
  shear stress
   absolute maximum, 453–458
   maximum in-plane, 440–443
  strain measurement, 471–475
  strain rosettes, 471–475
  three-dimensional strain analysis, 469–471
  three-dimensional stress states–principal stresses,
   450–453
Transformed-section method, flexural stress and,
  283–286
Transverse strain, in beams, 262
True stress, 32
Trusses, planar, 134–140
  displacement method for, 137–140
  element force-temperature-deformation behavior,
   135
  equilibrium of, 135
  force method for, 135–137
  geometry of deformation, 135
  unit-load method for, 594–593
Twist angle: *See* Angle of twist
Twisting moment: *See* Torque
Two-dimensional state of stress, 430
Two-dimensional strain, Mohr's circle for, 463–469

U
Ultimate strength, 32, 33
Uniaxial stress, 431
Uniform axially loaded element
  inelastic, 142
  linearly elastic, 102–102, A-47
  planar truss, 134–135
  with temperature change, 121
Uniform Bernoulli-Euler beam element, 406–412,
  A-60–A-61
Uniform end-loaded torsion element, 184–185, A-55
Uniform normal (axial) stress, 22–24, 102–118
Unit doublet function, 245
Unit impulse function, 245
Unit-load method, 593–598
Unit Macaulay function, 245
Unit ramp function, 245
Unit step function, 245
Units, systems of, A-4–A-6
  conversion of, (See inside front cover), A-6
  SI, A-4–A-5
  U.S. Customary, A-6
Unloading
  axial deformation, 148–151
  bending of beams, 302–304
  torsion of circular bars, 212–214
Unsymmetric bending, beams, 287–295

V
Virtual displacements, 572–573
  Principle of Virtual Displacements, 575–577
Virtual forces, 580–584
  Principle of Virtual Forces, 581–584
Virtual work, 571–579
V/M DIAG computer program, A-56–A-59

W
Web-flange beams, shear stress in, 313–319
Work and complementary work, 554–561
Work-energy principle for calculating deflections,
  567–571

Y
Yield point, 32
Yield strength, 32–33
Young's modulus, 32, 98–99
  and shear modulus, 62–63

GEOMETRIC PROPERTIES OF PLANE AREAS*

Notation†

A = area

$\bar{y}', \bar{z}'$ = coordinates of centroid C

I_y, I_z = moments of inertia with respect to the y and z axes, respectively

$I_{y'}$ = moment of inertia with respect to the y' axis

$I_p = I_y + I_z$ = polar moment of inertia about the centroid

I_{yz} = product of inertia with respect to the y and z axes

---

*Basic formulas for area properties may be found in Appendix C.

†Note: The y and z axes are centroidal axes.

Rectangle

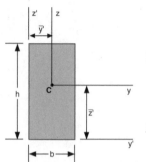

$$A = bh \qquad \bar{y}' = \frac{b}{2}$$

$$\bar{z}' = \frac{h}{2} \qquad I_y = \frac{bh^3}{12}$$

$$I_z = \frac{hb^3}{12} \qquad I_{yz} = 0$$

$$I_{y'} = \frac{bh^3}{3} \qquad I_p = \frac{bh}{12}(b^2 + h^2)$$

Circle

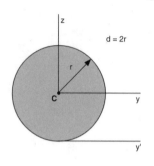

$$A = \pi r^2 = \frac{\pi d^2}{4}$$

$$I_y = I_z = \frac{\pi r^4}{4} = \frac{\pi d^4}{64}$$

$$I_{yz} = 0$$

$$I_{y'} = \frac{5\pi r^4}{4} = \frac{5\pi d^4}{64}$$

$$I_p = \frac{\pi r^4}{2} = \frac{\pi d^4}{32}$$

Right Triangle

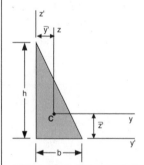

$$A = \frac{bh}{2} \qquad \bar{y}' = \frac{b}{3}$$

$$\bar{z}' = \frac{h}{3} \qquad I_y = \frac{bh^3}{36}$$

$$I_z = \frac{hb^3}{36} \qquad I_{yz} = -\frac{b^2 h^2}{72}$$

$$I_{y'} = \frac{bh^3}{12} \qquad I_p = \frac{bh}{36}(b^2 + h^2)$$

Circular Ring (t/r small)

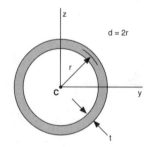

$$A = 2\pi r t = \pi d t$$

$$I_y = I_z = \pi r^3 t = \frac{\pi d^3 t}{8}$$

$$I_{yz} = 0$$

$$I_p = 2\pi r^3 t = \frac{\pi d^3 t}{4}$$

Triangle

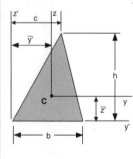

$$A = \frac{bh}{2} \qquad \bar{y}' = \frac{b + c}{3}$$

$$\bar{z}' = \frac{h}{3} \qquad I_y = \frac{bh^3}{36}$$

$$I_z = \frac{bh}{36}(b^2 - bc + c^2)$$

$$I_{yz} = -\frac{bh^2}{72}(b - 2c)$$

$$I_{y'} = \frac{bh^3}{12}$$

$$I_p = \frac{bh}{36}(h^2 + b^2 - bc + c^2)$$

Semicircle

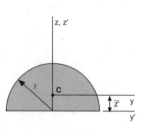

$$A = \frac{\pi r^2}{2} \qquad \bar{z}' = \frac{4r}{3\pi}$$

$$I_y = \frac{(9\pi^2 - 64)r^4}{72\pi}$$

$$I_z = \frac{\pi r^4}{8} \qquad I_{yz} = 0$$

$$I_{y'} = \frac{\pi r^4}{8}$$